APPLIED OPTICS GROUP.

Optical Phase Conjugation

QUANTUM ELECTRONICS — PRINCIPLES AND APPLICATIONS

A Series of Monographs

EDITED BY

PAUL F. LIAO
Bell Telephone Laboratories
Murray Hill, New Jersey

PAUL KELLEY
Lincoln Laboratory
Massachusetts Institute of Technology
Lexington, Massachusetts

N. S. Kapany and J. J. Burke. OPTICAL WAVEGUIDES, 1972

Dietrich Marcuse. THEORY OF DIELECTRIC OPTICAL WAVEGUIDES, 1974

Benjamin Chu. LASER LIGHT SCATTERING, 1974

Bruno Crosignani, Paolo Di Porto, and Mario Bertolotti. STATISTICAL PROPERTIES OF SCATTERED LIGHT, 1975

John D. Anderson, Jr. GASDYNAMIC LASERS: AN INTRODUCTION, 1976

W. W. Duley. CO_2 LASERS: EFFECTS AND APPLICATIONS, 1976

Henry Kressel and J. K. Butler. SEMICONDUCTOR LASERS AND HETEROJUNCTION LEDs, 1977

H. C. Casey and M. B. Panish. HETEROSTRUCTURE LASERS: PART A, FUNDAMENTAL PRINCIPLES; PART B, MATERIALS AND OPERATING CHARACTERISTICS, 1978

Robert K. Erf (Ed.). SPECKLE METROLOGY, 1979

Marc D. Levenson. INTRODUCTION TO NONLINEAR LASER SPECTROSCOPY, 1982

David S. Kliger (Ed.). ULTRASENSITIVE LASER SPECTROSCOPY, 1983

Robert A. Fisher (Ed.). OPTICAL PHASE CONJUGATION, 1983

John F. Reintjes. NONLINEAR OPTICAL PARAMETRIC PROCESSES IN LIQUIDS AND GASES

YOH-HAN PAO
Case Western Reserve University
Cleveland, Ohio

Founding Editor
1972–1979

Optical Phase Conjugation

EDITED BY

Robert A. Fisher

Los Alamos National Laboratory
Los Alamos, New Mexico

ACADEMIC PRESS, INC.
Harcourt Brace Jovanovich, Publishers
Orlando San Diego New York
Austin Boston London Sydney
Tokyo Toronto

ACADEMIC PRESS, INC.
Orlando, Florida 32887

United Kingdom Edition published by
ACADEMIC PRESS, INC. (LONDON) LTD.
24/28 Oval Road, London NW1 7DX

Library of Congress Cataloging in Publication Data

Main entry under title:

Optical phase conjugation.

(Quantum electronics--principles and applications)
Bibliography: p.
Includes index.
1. Nonlinear optics. 2. Backscattering. 3. Photon echoes. 4. Optical wave guides. I. Fisher, Robert A. II. Series.
QC446.2.067 537.5 82-8871
ISBN 0-12-257740-X AACR2

PRINTED IN THE UNITED STATES OF AMERICA

87 88 89 9 8 7 6 5 4 3 2

For Andrea, Andy, and Derek

Contents

Contributors

Numbers in parentheses indicate the pages on which the authors' contributions begin.

R. L. ABRAMS (211), Hughes Research Laboratories, Malibu, California 90265

JOHN C. AuYEUNG (285),* Jet Propulsion Laboratory, California Institute of Technology, Pasadena, California 91109

PIERRE A. BELANGER (465), Département de Physique LROL, Faculté des Sciences et de Génie, Laval University, Quebec, Canada G1K 7P4

JACK FEINBERG (417), Department of Physics, University of Southern California, Los Angeles, California 90007

ROBERT A. FISHER (1, 79, 597), Los Alamos National Laboratory, Los Alamos, New Mexico 87545

AMOS HARDY (465), Weizmann Institute of Science, Rehovot, Israel

R. W. HELLWARTH (127, 169), Departments of Electrical Engineering and Physics, University of Southern California, Los Angeles, California 90007

R. K. JAIN (307), Hughes Research Laboratories, Malibu, California 90265

M. B. KLEIN (307), Hughes Research Laboratories, Malibu, California 90265

J. F. LAM (211), Hughes Research Laboratories, Malibu, California 90265

P. F. LIAO (211), Bell Telephone Laboratories, Holmdel, New Jersey 07733

R. C. LIND (211), Hughes Research Laboratories, Malibu, California 90265

JOHN H. MARBURGER (99), Office of the President, State University of New York at Stony Brook, Stony Brook, New York 11794

THOMAS R. O'MEARA (537), Hughes Research Laboratories, Malibu, California 90265

DAVID M. PEPPER (23, 537), Hughes Research Laboratories, Malibu, California 90265

N. F. PILIPETSKII (135, 445), Institute for Problems in Mechanics, USSR Academy of Sciences, Moscow 117526, USSR

V. V. SHKUNOV (135, 445), Institute for Problems in Mechanics, USSR Academy of Sciences, Moscow 117526, USSR

A. E. SIEGMAN (465), Department of Electrical Engineering, and Edward L. Ginzton Laboratory, Stanford University, Stanford, California 94305

D. G. STEEL (211), Hughes Research Laboratories, Malibu, California 90265

*Present address: Newport Corporation, Fountain Valley, California 92708.

A. N. SUDARKIN (445), Institute for Problems in Mechanics, USSR Academy of Sciences, Moscow 117526, USSR

B. R. SUYDAM (79), Los Alamos National Laboratory, Los Alamos, New Mexico 87545

JEFFREY O. WHITE (537), Department of Applied Physics, California Institute of Technology, Pasadena, California 91125

AMNON YARIV (1, 23), Division of Engineering and Applied Science, California Institute of Technology, Pasadena, California 91125

B. YA. ZEL'DOVICH (135, 445), Institute for Problems in Mechanics, USSR Academy of Sciences, Moscow 117526, USSR

Preface

My introduction to phase conjugation took place at Lawrence Livermore Laboratory on May 22, 1972. Two visitors from Moscow's P. N. Lebedev Physical Institute mentioned that a team headed by Dr. B. Ya. Zel'dovich had observed that the stimulated Brillouin backscattering process in a CS_2-filled waveguide displayed an extremely curious property. If a ground-glass distorting element was placed in the beam path in front of the CS_2 cell, the backscattered light still displayed excellent spatial quality after returning through the distorter. Somehow, the distortions introduced on the first pass had been "undone." At the time, no explanation was offered for this strange new phenomenon, and I did not appreciate its startling consequences.

Five years later I had the good fortune to visit with Dr. Zel'dovich. When I asked him about the experiments mentioned so long ago in Livermore, his interest and excitement convinced me that these effects were more than mere laboratory curiosities; a high-quality optical beam could apparently be transported through a low-quality optical system without any consequent degradation in beam quality, and light emanating from a small spot could be programmed to return to that very same spot.

By the end of 1977 interest in this topic had grown remarkably. It was found that this strange "conjugate wave" could be generated by using nonlinear effects as diverse as degenerate four-wave mixing, stimulated Brillouin scattering, stimulated Raman scattering, and photon echoes. We now know that almost *any* nonlinear optical effect can convert an incoming light beam into one with these remarkable image-restoring properties.

This book appears at a time of intense activity in optical phase conjugation. We chose not to await the maturation of the field, but instead to provide this material in time to be useful in its development. We have tried very hard to elucidate and interrelate the various nonlinear phenomena which can be used for optical phase conjugation. Much attention has been given to overall continuity, yet any deficiencies must be considered mine. Both active researchers and graduate students should find this book helpful, and the extensive bibliography will provide guidance for anyone wishing to examine material beyond the scope of the book.

At the book's inception we planned two chapters on stimulated backscattering: an experimental chapter by N. F. Pilipetskii, V. V. Shkunov, and B. Ya. Zel'dovich, and a theoretically oriented one by R. W. Hellwarth. Shortly thereafter, deteriorating Soviet–American relations made it uncertain whether a Soviet contribution would be received. I suggested, therefore, that Professor Hellwarth expand his chapter to include descriptions of some of the more important experiments. When the book was nearly complete, the Soviet contributions arrived unexpectedly. We decided, nevertheless, to leave Professor Hellwarth's contribution unchanged; Chapters 6 and 7 therefore contain some overlapping information. In retrospect this is fortunate, because the book now presents the first clear comparison of the two somewhat distinct philosophies that have evolved in the field.

The lion's share of credit for this volume must go to the authors, who are specialists in their fields and who sacrificed valuable research time to prepare this review material. Special thanks are due to Dr. R. L. Abrams, Dr. D. M. Pepper, and Professor A. E. Siegman, who contributed to the book as a whole.

It has been a pleasure to watch this book develop.

1 Introduction

Amnon Yariv†
California Institute of Technology
Pasadena, California

Robert A. Fisher
Los Alamos National Laboratory
Los Alamos, New Mexico

Optical phase conjugation involves the use of any of a large variety of nonlinear optical phenomena to exactly reverse the direction of propagation of a light beam. In this first chapter we present an introduction to the

† Work supported by the Army Research Office and the Air Force Office of Scientific Research.

Optical Phase Conjugation

ISBN 0-12-257740-X

pertinent nonlinear optical phenomena and we describe the properties of phase-conjugate waves.

The chapter opens with a discussion of the properties of phase-conjugate waves, in which the reflection of a phase conjugator is compared with that of a conventional mirror. Various details relevant to all subsequent chapters are then presented, including (1) notational and dimensional conventions, (2) the nonlinear wave equation and its reduction to the slowly varying envelope approximation (SVEA), (3) the ways in which various nonlinear responses can couple light waves that intersect in a medium, (4) and the nature of the various nonlinear optical phenomena. A list of pertinent nonlinear optics texts is provided, followed by a proof of the distortion-undoing properties of phase-conjugate waves. The chapter ends with a short historical overview.

I. What Is Phase Conjugation and What Does It Do?

Optical phase conjugation is a technique that incorporates nonlinear optical effects to precisely reverse both the direction of propagation and the overall phase factor for each plane wave in an arbitrary beam of light. The process can be regarded as a unique kind of "mirror" with very unusual image-transformation properties. A beam reflected by a phase conjugator retraces its original path. This remarkable autoretracing property suggests widespread application to problems associated with passing high-quality optical beams through nonuniform distorting media. This chapter will only identify general properties of the process; the remainder of this book is devoted to describing specific details and applications.

Figure 1 illustrates the remarkable difference between phase-conjugate reflection and conventional mirror reflection. A conventional plane mirror (Fig. 1a) changes the sign of the **k**-vector component normal to the mirror surface while leaving the tangential component unchanged. An incoming light ray can thus be redirected arbitrarily by suitably tilting the mirror. On the other hand, the phase conjugator (Fig. 1b) causes an inversion of the vector quantity **k**, so that the incident ray exactly returns upon itself, independent of the orientation of the conjugator. A simple extension of Fig. 1 indicates that an incident diverging beam would be conjugated to become a converging beam and that an incident converging beam would be conjugated to become a diverging beam.

Compare the two arrangements depicted in Fig. 2. In Fig. 2a, a light ray is redirected by a phase-conjugate reflector. A thin wedge is about to be

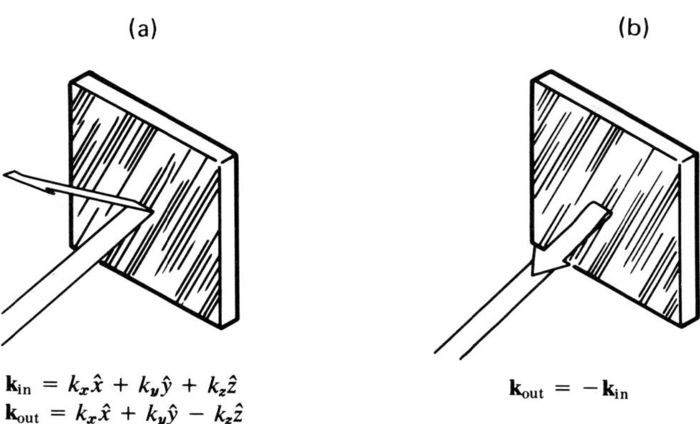

(a)
(b)
$\mathbf{k}_{\text{in}} = k_x\hat{x} + k_y\hat{y} + k_z\hat{z}$
$\mathbf{k}_{\text{out}} = k_x\hat{x} + k_y\hat{y} - k_z\hat{z}$
$\mathbf{k}_{\text{out}} = -\mathbf{k}_{\text{in}}$

Fig. 1. Comparison of (a) a mirror reflection and (b) a phase-conjugate reflection. The mirror reflection reverses the **k**-vector component normal to its surface, whereas the phase conjugator reverses the vector quantity **k**. Tipping the conventional mirror will change the reflected direction, whereas tipping the phase-conjugating mirror will not. (Reproduced with permission from I. J. Bigio *et al.*, *Proc. Int. Conf. Lasers 1978,* STS Press, McLean, Virginia, p. 532.)

put into the beam. In Fig. 2b, the wedge has been placed into the beam, thereby slightly redirecting the beam on its way to the conjugator. Note that the conjugate beam returns to the same spot on the wedge and that the return beam is subsequently deflected by the wedge to continue retracing the path of the incoming beam. Comparison of these figures shows the essence of the phase-conjugation process; the properties of the conjugate beam are *not at all* impaired by the interposition of a distorting medium (the wedge) as long as all the distorted light strikes the conjugator. This argument clearly extends to an array of randomly oriented wedges and thus to any phase distorter. This remarkable property indicates that, through optical phase conjugation, a high-quality optical beam can be double passed through a poor-quality optical system with no loss in beam quality.

In addition to the aberration-correcting properties, phase conjugators can be used for pointing and tracking. To visualize pointing applications, consider what would happen if one were looking into a phase-conjugating mirror: an observer would see his or her face in a *conventional* mirror but not in a phase conjugator. This is because any light emanating from a particular point on the face would be returned by the conjugator to that same point, thereby not entering the viewer's eye. The only light seen by the observer would be that which had struck the conjugator after emanation as a diffuse reflection of room light scattered from the cornea covering the

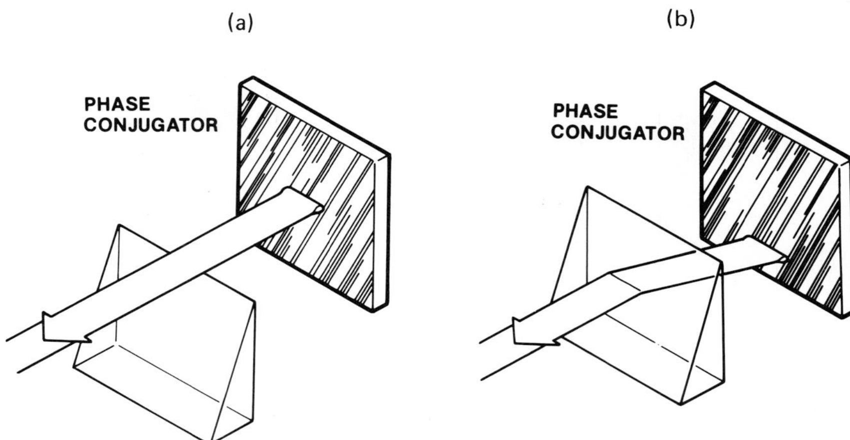

Fig. 2. Pictorial demonstration of simplest correction. In (a), a wedge is about to be placed between the conjugator and the light source. In (b), the wedge has been moved into the beam, thereby deflecting the incoming beam downward. The conjugate beam is redirected to hit the same spot on the wedge, and is then deflected by the wedge to continue on its route directly back to the light source. After the returning beam has passed through the prism, it is identical to the return beam that would be there if the prism had been absent. Thus, the properties of the returning beam are not at all affected by the presence of the wedge as long as the conjugator is large enough to intercept the light that the wedge has deflected.

pupil of either eye. If the observer increased the illumination of one eye (perhaps by using a flashlight), the entire conjugator would appear, to that eye only, to become relatively brighter. Obviously, the viewer's observations would be unaltered if an aberrating medium were placed between the viewer and the conjugator. Thus we have the essential feature of the pointing application. As Fig. 3 shows, a small glint from a diffusely illuminated target can propagate through an inhomogeneous intervening medium (such as a turbulent atmosphere). The light that enters the optical aperture of the device can be amplified in a possibly distorting laser amplifier. If the amplified beam were then to impinge upon a conjugator, a "reflected," or conjugate, beam would be generated to pass in the reverse direction through the amplifier and then through the intervening distorting medium and finally to strike the target. This use of phase conjugation is an alternative to the conventional adaptive optics[1] techniques for aiming a

[1] Adaptive optics involves sensing wave-front distortions and using a computer to feed correcting instructions to a mirror. These corrections can take the form of electromechanical distortions in the mirror or electro-optical deformation of the phase front. Such systems are costly, cumbersome, and slow. [See, for example, Hardy (1978).]

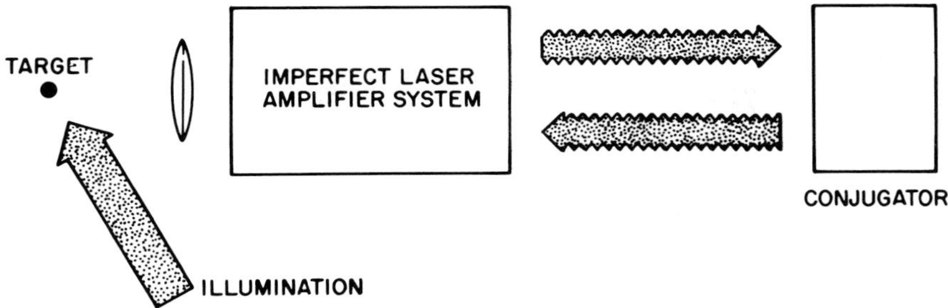

Fig. 3. Pictorial description of a pointing application. Instead of trying to aim a laser at a small target, one could set up the arrangement depicted here. A small auxiliary laser illuminates a broad area, and some of the scattered light is gathered in the optical system and amplified in the gain medium. The phase conjugator then returns the beam to pass through the gain medium and onto the target. This scheme can greatly reduce the required aiming precision, and can compensate for any static phase distortions in the amplifier system, optical system, or intervening atmosphere. (Reproduced with permission from I. J. Bigio *et al.*, *Proc. Int. Conf. Lasers 1978,* STS Press, McLean, Virginia, p. 532.)

powerful laser at a small target. Such a problem is encountered in laser fusion research and in other applications. Phase conjugation can also correct for random variations in birefringence, a correction that cannot be performed by adaptive optics methods.

Although useful for an introductory description, the simple ray pictures in Figs. 1 and 2 do not completely specify the conjugation process. In addition to reversing each **k** vector, the device complex-conjugates the overall multiplicative electric field amplitude associated with each plane wave in the beam. Phase-conjugate distortion correction requires *both* of these properties. A mathematical proof of phase conjugation's distortion-undoing properties appears in Section V, after our discussion of the wave equation.

II. Notation and Systems of Units

Because of the large variety of notational and dimensional conventions, mistakes are often made. Problems include whether frequencies are negative or positive, what multipliers are used in front of field amplitudes, etc. This section identifies the most common conventions and points out some of the potential difficulties.

A. Notation

In classical optics it has been reasonable to write harmonically varying fields as $A \cos(\omega t - \mathbf{k} \cdot \mathbf{r})$. Because classical optics is a linear discipline, there is no harm in replacing $\cos(\omega t - \mathbf{k} \cdot \mathbf{r})$ with $\exp[-i(\omega t - \mathbf{k} \cdot \mathbf{r})]$ or in changing $-i$ to $+i$. Unfortunately, in topics such as ours we must be much more cautious, because nonlinear optical effects (which cause optical phase conjugation) depend upon the strength of the optical fields and because negative and positive-frequency terms "mix" with each other.

In this chapter, we shall describe a plane wave traveling in the positive z direction by

$$E(z, t) = \tfrac{1}{2}\mathscr{E}(z, t) \exp[\pm i(\omega t - kz)] + \text{c.c.}, \tag{1}$$

where ω is the (radian) frequency of the carrier and $k = n_0\omega/c$. Here n_0 is the linear index of refraction (the index in the *absence* of nonlinearities). The abbreviation c.c. denotes the complex conjugate, and the $\pm$ symbol indicates that some authors use one sign and some the other. In this section and in Section III, every affected equation will carry either the $\pm$ or the $\mp$ sign; the reader should elect to read all the upper symbols or all the lower symbols. The corresponding expression for a plane wave traveling in the negative z direction is obtained by changing $-kz$ to $+kz$ in Eq. (1).

The symbol $\mathscr{E}$ denotes the slowly varying complex-amplitude envelope of the pulse. A phase-modulated (or frequency-modulated) pulse would be described by the relationship

$$\mathscr{E}(z, t) = a(z, t) \exp[-i\, \delta\phi(z,t)], \tag{2}$$

where $a(z, t)$ is real and $\delta\phi$ is the phase-modulation function.

B. Systems of Units

The research community is presently divided on the question of units. Many laser scientists prefer to use Gaussian units, but international pressure has been mounting to teach courses using mks units (now called SI units). The literature is now fairly evenly divided in the use of Gaussian or SI units; The electrical engineering community seems especially committed to the SI system. With waves as described by Eq. (1), intensity and electric field strength are related by

$$I = (cn/8\pi)\mathscr{E}^2 \qquad \text{(Gaussian)},$$

$$I = \tfrac{1}{2}\sqrt{\epsilon/\mu}\,\mathscr{E}^2 \qquad \text{(SI)}.$$

C. Variations in the Literature and Variations from Chapter to Chapter

Deviations from this chapter's notational conventions concern modifications of Eq. (1) through elimination of the factor $\frac{1}{2}$, elimination of the abbreviation c.c., inconsistent uses of the $\pm$ signs, and the replacement of $-i$ with j. Because these variations permeate the literature and because most chapter authors feel much more comfortable using their own notational conventions, some variations will appear in this book.

D. Additional Reading on Conversion of Units

The interested reader can learn more about the SI system and its relationship to other systems of units in:

D. Halliday and R. Resnick, "Physics." Wiley, 1962 (see, in particular, Appendixes F and G).

J. D. Jackson, "Classical Electrodynamics," 2nd ed. Wiley, 1975 (see, in particular, the Appendix).

E. A. Mechtly, "The International System of Units." NASA SP-7012, Washington, D. C., 1969.

C. H. Page, The international system of units (SI), *Phys. Teach.* **9,** 379, 1971.

J. M. Stone, "Radiation and Optics." McGraw-Hill, New York, 1963 (see especially Appendix A).

III. The Wave Equation and Its Reduction to the Slowly Varying Envelope Approximation (SVEA) Form

In this section we obtain the wave equation from Maxwell's equations; we then reduce the wave equation to a first-order equation. Note that Maxwell's equations describe only *half* of the nonlinear optics problem; they only show how a nonlinear polarization generates another wave through an equation in which only derivatives of the electric field appear on the left-hand side and only the induced nonlinear polarization appears as a driving term on the right-hand side. Maxwell's equations do *not* address how a nonlinear polarization is generated by electric fields; this matter is discussed in Section IV.

A. Derivation of the Wave Equation

The four Maxwell equations are

$$
\begin{array}{ll}
\text{(Gaussian)} & \text{(SI)} \\
\nabla \cdot \mathbf{D} = 4\pi\rho, & \nabla \cdot \mathbf{D} = \rho, \\
\nabla \cdot \mathbf{B} = 0, & \nabla \cdot \mathbf{B} = 0, \\
\nabla \times \mathbf{H} = \dfrac{4\pi}{c}\mathbf{J} + \dfrac{1}{c}\dfrac{\partial \mathbf{D}}{\partial t}, & \nabla \times \mathbf{H} = \mathbf{J} + \dfrac{\partial \mathbf{D}}{\partial t}, \\
\nabla \times \mathbf{E} = -\dfrac{1}{c}\dfrac{\partial \mathbf{B}}{\partial t}, & \nabla \times \mathbf{E} = -\dfrac{\partial \mathbf{B}}{\partial t}.
\end{array}
\tag{3}
$$

We assume that the material is homogeneous, nonmagnetic, and nonconducting and that there are no free charges.

We write the displacement vector **D** as

$$
\begin{array}{ll}
\mathbf{D} = \mathbf{E} + 4\pi\mathbf{P} & \text{(Gaussian)}, \\
\mathbf{D} = \epsilon_0\mathbf{E} + \mathbf{P} & \text{(SI)}
\end{array}
\tag{4}
$$

where **P** is the polarization, which we divide into its linear (L) and nonlinear (NL) components

$$
\mathbf{P} = \mathbf{P}_{\mathrm{L}} + \mathbf{P}_{\mathrm{NL}}. \tag{5}
$$

We first take the curl of the last of Eqs. (3). We then use the operator identity $\nabla \times \nabla \mathbf{x} = -\nabla^2 + \nabla\nabla\cdot$. Throughout most of this book, we assume that all electric fields are perpendicular to their corresponding **k** vectors (transversality), giving $\nabla \cdot \nabla = 0$. We then obtain for plane waves

$$
\begin{array}{ll}
\nabla^2\mathbf{E} - \left(\dfrac{n_0}{c}\right)^2 \dfrac{\partial^2\mathbf{E}}{\partial t^2} = \dfrac{4\pi}{c^2}\dfrac{\partial^2\mathbf{P}_{\mathrm{NL}}}{\partial t^2} & \text{(Gaussian)}, \\
\nabla^2\mathbf{E} - \mu_0\epsilon\dfrac{\partial^2\mathbf{E}}{\partial t^2} = \mu_0\dfrac{\partial^2\mathbf{P}_{\mathrm{NL}}}{\partial t^2} & \text{(SI)}.
\end{array}
\tag{6}
$$

Note that n_0 and ϵ pertain, respectively, to the *linear* index of refraction and the *linear* medium permittivity. Although in the Gaussian system the subscript 0 identifies the linear portion of a constitutive quantity, this subscript cannot be similarly used in the SI system because it has been traditionally used to indicate free-space values.

We now have the desired wave equation. It shows how a (possibly new) electric field evolves in the presence of a nonlinear polarization. In the absence of a nonlinear polarization, the right-hand sides are zero, and this wave equation reduces to the standard *linear* wave equation in which many waves simultaneously present in an unbounded material pass right through each other without any mutual influence and without generating any new waves. Coupling between waves can *only* come about through an interaction that creates a *nonlinear* polarization.

Throughout this book, electromagnetic waves will be treated as classical fields. The interaction is known as semiclassical when the nonlinear material is described in quantum-mechanical terms and when the waves are treated as classical quantities.

In the remainder of this chapter, we shall, for notational simplicity, assume that all electric fields and polarizations are plane polarized in the same direction. This allows us to drop the vector notation and to treat nonlinear susceptibility tensors as scalar qualities.

B. Reduction to SVEA Form

It is often both cumbersome and unnecessary to manipulate the second-order nonlinear wave equation; consequently, one usually reduces it to a simpler first-order equation. This reduction is valid only when the features of the pulse (the envelope and instantaneous phase) vary little during an optical period. Under this "slowly varying" approximation, one uses the symbols $\mathscr{E}$ and $\delta\phi$ to denote quantities that are averaged over an optical period.

To develop a first-order expression, we assume plane waves and write the nonlinear polarization in a special way. We shall *single out* that particular portion of the nonlinear polarization (called the phase-matched portion) whose modulation function propagates with the *same* frequency and wave vector (same ω and k) as the modulation function of the electric field

$$P_{\mathrm{NL}} = \tfrac{1}{2}\mathscr{P}(z, t) \exp[\pm i(\omega t - kz)] + \text{c.c.} + (\text{other terms}). \tag{7}$$

The other terms in Eq. (7) can be neglected because they correspond to other modulation frequencies or other wave vectors and do not couple to the particular electric field. The power transferred to the field is the volume integral of $E \cdot P$, which, when both E and P vary harmonically in space and time, leads to an increase (or decrease) in $\mathscr{E}$ *only* if E and P have the same modulation frequencies and wave vectors. This simplification arises from the orthogonality of complex exponential functions.

Equation (7) factors out of the polarization the fast variations in both space and time. Some authors factor out only the fast temporal variation, as will be seen in Chapter 2.

We are now ready to make the slowly varying envelope approximation

(SVEA).[2] We insert Eqs. (1) and (7) into Eq. (6) and apply the slowly varying conditions

$$|k^2\mathscr{E}| \gg \left|k\frac{\partial\mathscr{E}}{\partial z}\right| \gg \left|\frac{\partial^2\mathscr{E}}{\partial z^2}\right|. \tag{8}$$

These conditions could, equivalently, be rewritten by substituting t for z and ω for k.

After applying the slowly varying conditions, multiply $\exp[\mp i(\omega t - kz)]$ times the resulting equation and integrate over one optical period to obtain our desired result

$$\begin{aligned} \left(\frac{\partial}{\partial z} + \frac{n_0}{c}\frac{\partial}{\partial t}\right)\mathscr{E} &= \mp i\frac{2\pi k}{n_0^2}\mathscr{P} \qquad \text{(Gaussian)}, \\ \left(\frac{\partial}{\partial z} + \sqrt{\mu_0\epsilon}\frac{\partial}{\partial t}\right)\mathscr{E} &= \mp i\frac{\omega}{2}\sqrt{\frac{\mu_0}{\epsilon}}\mathscr{P} \qquad \text{(SI)}. \end{aligned} \tag{9}$$

This is the SVEA equation for a plane wave traveling in the $+z$ direction. For a plane wave traveling in the $-z$ direction, we have

$$\begin{aligned} \left(\frac{\partial}{\partial z} - \frac{n_0}{c}\frac{\partial}{\partial t}\right)\mathscr{E} &= \pm i\frac{2\pi k}{n_0^2}\mathscr{P} \qquad \text{(Gaussian)}, \\ \left(\frac{\partial}{\partial z} - \sqrt{\mu_0\epsilon}\frac{\partial}{\partial t}\right)\mathscr{E} &= \pm i\frac{\omega}{2}\sqrt{\frac{\mu_0}{\epsilon}}\mathscr{P} \qquad \text{(SI)}. \end{aligned} \tag{10}$$

Equations (9) and (10) now show how a given nonlinear polarization drives a desired electric field as long as both have the same modulation frequencies and wave vectors.

We must now generalize this discussion to include the case in which many waves are present in the medium. We decompose the electric field function into a hopefully finite number of plane waves as

$$E(\mathbf{r}, t) = \tfrac{1}{2}\sum_i \mathscr{E}_i(\mathbf{r}, t)\exp[\pm i(\omega_i t - \mathbf{k}_i\cdot\mathbf{r})] + \text{c.c.} \tag{11}$$

Here each component of the wave is written as in Eq. (1), but note that the propagation directions are now arbitrary (not necessarily along the $+z$ direction). Equation (11) indicates that the set $\{\mathscr{E}_i\}$ denotes the slowly varying envelope functions of the compound wave. Similarly, we write the nonlinear polarization as

$$P_{\text{NL}}(\mathbf{r}, t) = \tfrac{1}{2}\sum_i \mathscr{P}_i(\mathbf{r}, t)\exp[\pm i(\omega_i t - \mathbf{k}_i\cdot\mathbf{r})] + \text{c.c.} + (\text{other terms}). \tag{12}$$

[2] The SVEA is sometimes called the adiabatic or Born approximation.

In generalizing Eq. (7), we have separated out *only* the terms in the nonlinear polarization that couple to the waves in Eq. (11). Again, the set $\{\mathscr{P}_i\}$ denotes the slowly varying envelope functions of the nonlinear polarization. The results of the SVEA [Eqs. (9) and (10)], when generalized along *each* propagation direction, must *independently apply* for *each* subscript i.

The meaning of the SVEA should now be especially clear; the growth of a wave modulated at a particular frequency and **k** vector is determined *only* by the nonlinear polarization that has a modulation *at that same frequency and* **k** *vector*. The task of developing coupled equations for any particular situation often involves examining the general expression for the nonlinear polarization and then picking out *only* the portion oscillating at the appropriate frequency and **k** vector.[3] The associated slowly varying envelope function is then inserted into the right-hand side of Eq. (9) or Eq. (10).

IV. Material Nonlinearities and Their Coupling of Optical Waves

In this section we shall introduce some of the nonlinear optical effects that are useful for optical phase conjugation. Many of these topics are of sufficient interest to warrant treatment in a separate chapter later in this book. Only an elementary guideline is presented here; the interested reader should consult the list of additional reading material at the end of this section.

A. Generalized Expansion of Polarization

We write the field-dependent susceptibility χ of a material in the power expansion

$$\chi(E) = \chi^{(1)} + \chi^{(2)}E + \chi^{(3)}E^2 + \cdots . \tag{13}$$

Here the susceptibility is divided into different terms. Similarly, the polarization of the material is found by multiplying E times $\chi(E)$:

$$P(E) = E\chi(E) = \chi^{(1)}E + \chi^{(2)}E^2 + \chi^{(3)}E^3 + \cdots . \tag{14}$$

In Eqs. (13) and (14) the symbol E denotes the total field, which is often

[3] As described just before Eq. (9), each $\mathscr{P}_i$ is obtained after integration over a wavelength. Although $\mathscr{P}_i$ can often be obtained by inspection, the problems addressed in Chapter 8 will require carrying out the integration to obtain the proper polarization.

made up of a number of different waves at different frequencies, polarizations, **k** vectors, etc. Thus there are usually many terms in the expression for E (and more for E^2, etc.). For this reason, each χ in the previous expansion has several components. To be specific, each nonlinear susceptibility is identified by the superscript n (to denote the order), and each is written as a function of several frequencies, the first frequency being that of the induced nonlinear polarization and the remaining n frequencies being those of the n input light waves. We will spell out all frequency dependencies in this section; in practice, however, they are often neglected, because they are cumbersome and because their nature is usually clearly implied. In this chapter, we disregard complications that arise when all waves are not plane polarized in the same direction.

1. Linear Terms

The $\chi^{(1)}$ terms in Eqs. (13) and (14) correspond to *linear* optical properties, including index of refraction, absorption, gain, and birefringence. These properties constitute the subject of classical optics, and these coefficients cannot be responsible for the production of a conjugate wave because a nonlinear optical process is needed to couple waves in a material. In a linear medium, waves pass through each other without influencing each other. Of course, the saturation of these "linear" effects is a nonlinear process described by higher-order terms.

2. Second-Order Terms

The $\chi^{(2)}$ terms in Eqs. (13) and (14) correspond to second-order effects, which, in general, can be called three-wave mixing. These effects are:

second-harmonic generation	$\chi^{(2)}(2\omega;\ \omega,\ \omega)$
optical rectification	$\chi^{(2)}(0;\ \omega,\ -\omega)$
parametric mixing	$\chi^{(2)}(\omega_1 \pm \omega_2;\ \omega_1,\ \pm\omega_2)$
Pockels effect	$\chi^{(2)}(\omega;\ \omega,\ 0)$

These effects occur only in materials that lack inversion symmetry. In second-harmonic generation and parametric mixing, the conversion efficiency (whether or not the newly created polarization radiates efficiently) is determined by phase-matching conditions; as indicated by Eq. (9), the process will be efficient only when the newly created nonlinear polarization is modulated at the appropriate frequency and **k** vector. These conditions can be adjusted by changing propagation angles, light frequencies, and material temperatures. The role of three-wave mixing in optical phase conjugation will be discussed in Chapter 2.

3. *Third-Order Terms*

The $\chi^{(3)}$ effects occur independently of whether or not a material possesses inversion symmetry. Included among these effects are those especially popular for production of a phase-conjugate wave: stimulated Brillouin scattering, stimulated Raman scattering, and degenerate four-wave mixing in Kerr-like media. Some important effects are:

third-harmonic generation	$\chi^{(3)}(3\omega;\ \omega,\ \omega,\ \omega)$
nondegenerate four-wave mixing	$\chi^{(3)}(\omega_1 + \omega_2 \pm \omega_3;\ \omega_1,\ \omega_2,\ \pm\omega_3)$
Raman scattering	$\chi^{(3)}(\omega \pm \Omega;\ \omega,\ -\omega,\ \omega \pm \Omega)$
instantaneous ac Kerr effect (degenerate four-wave mixing)	$\mathrm{Re}\ \chi^{(3)}(\omega;\ \omega,\ \omega,\ -\omega)$
Brillouin scattering	$\chi^{(3)}(\omega \pm \Omega;\ \omega,\ -\omega,\ \omega \pm \Omega)$
dc Kerr effect	$\mathrm{Re}\ \chi^{(3)}(\omega;\ \omega,\ 0,\ 0)$
two-photon absorption	$\mathrm{Im}\ \chi^{(3)}(\omega;\ \omega,\ -\omega,\ \omega)$
dc-induced harmonic generation	$\chi^{(3)}(2\omega;\ \omega,\ \omega,\ 0)$

4. *Specific Third-Order Effects*

Here we describe some of the four-wave (third-order) nonlinear phenomena listed in Subsection A.3.

a. Stimulated Brillouin scattering. In a material in which the polarizability is a function of pressure (the electrostrictive effect), light can be scattered by a sound wave. Through this mechanism, pressure waves can couple to light waves. Stimulated Brillouin scattering can be thought of as the coupling of two processes: (1) the input and scattered light waves interfere to produce a sound wave electrostrictively, and (2) the input wave Bragg scatters from the sound wave to produce the scattered light wave. These two couplings reinforce one another, and above a particular threshold condition, a large fraction of an input light beam can be converted into the scattered light wave.

For a sound wave of frequency Ω and wave vector $\mathbf{q}$ and an input light of frequency ω and wave vector $\mathbf{k}$, the scattered light frequency is $\omega - \Omega$ and the wave vector $\mathbf{k} - \mathbf{q}$. Because of these two constraints, the frequency by which the scattered light is shifted depends on the angle between the incident and scattered optical beams. The frequency shift can be thought of as a Doppler shift from the Bragg mirror moving at the speed of sound, and the change in wave vector can be thought of as first-order Bragg scattering from the periodic sound wave. For backward scattering, the frequency shift is maximized and is twice the refractive index times the ratio of the material's speed of sound to the vacuum speed of light. In most materials, the fractional optical frequency shift is

$\sim 10^{-4}-10^{-5}$. In Chapters 6 and 7, phase conjugation via stimulated Brillouin scattering will be discussed.

b. Stimulated Raman scattering. This is the scattering of light waves from molecular vibrations or from longitudinal-optical phonons in a solid. Generally, the frequencies of molecular vibrations are between 500 and several thousand inverse centimeters; consequently, the Stokes-scattered light is downshifted by this amount. Whereas stimulated Brillouin scattering couples light waves through the electrostrictive effect, stimulated Raman scattering couples light waves because the polarizability changes with a molecular coordinate. In contrast to Brillouin scattering, the Raman frequency shift is nearly independent of scattering angle. Phase conjugation via stimulated Raman scattering will be discussed in Chapters 6 and 7.

c. n_2 (or "Kerr-like") effects. This is a $\chi^{(3)}$ (or third-order) effect in which the speed of light in a transparent material depends linearly upon the light intensity. In a liquid, such an effect can arise from a forced orientation of elongated molecules along an applied electric field; if the electric field vector is not along a principal axis of the individual molecule's polarizability tensor, then the induced dipole is not parallel to the field, and the torque is nonzero, forcing the partial alignment of the molecule's high polarizability axis along the direction of the applied electric field. This increases the refractive index that a linearly polarized light wave experiences and thereby decreases the speed of the light wave. Other mechanisms that can contribute to the nonlinear refractive index of a material include off-resonant excitation of narrow-band absorbers and distortion of the electronic distribution in a material. These two effects are closely related.

A Kerr-like effect is characterized by an intensity-dependent index of the form

$$n = n_0 + n_2 \langle E^2 \rangle, \tag{15}$$

where the brackets denote a temporal average. For a single pulse $\mathscr{E}(z, t)$ traveling through a Kerr-like medium, $n = n_0 + \frac{1}{2} n_2 \mathscr{E}^2(z, t)$. The nonlinear index coefficient n_2 is related to $\chi^{(3)}(\omega, \omega, \omega, -\omega)$ by

$$n_2 = 12(\pi/n_0)\chi^{(3)}(\omega, \omega, \omega, -\omega). \tag{16}$$

The factor 12 in Eq. (16) includes a degeneracy factor of 3, which arises because there are three identical terms in which $-\omega$ occupies one of the last three arguments. Note that n_2 and $\chi^{(3)}$ have the same sign.[4]

[4] In the Gaussian system, E and P are given in electrostatic units, and B and M in electromagnetic units. Since n_2 multiplies an electrostatic unit quantity, authors using the Gaussian system usually denote n_2 as an electrostatic unit quantity; n_2 would have the same numerical value if it were treated as a Gaussian quantity.

This intensity-dependent index is responsible for self-focusing, in which a transverse (x, y) intensity variation in a beam propagating in the z direction causes a corresponding transverse variation in refractive index, which can be thought of as a lens formed in the material. For a spatially smooth beam whose intensity peaks on axis, a positive n_2 induces a positive lens, which further concentrates the light, causing the formation of a stronger lens, etc. Eventually the beam collapses upon itself, until other nonlinear mechanisms (breakdown or stimulated scattering) come into play. For a material with a negative n_2, self-defocusing occurs.

In addition to the spatial collapse (or growth) of a beam, the intensity-dependent index will vary in time if the optical field envelope does. Thus a pulse of light will experience a time-varying refractive index, and this pulse will acquire a form of frequency modulation known as self-phase modulation. When multiple beams pass through such a material, interference sets up (through Eq. (5) phase gratings that can scatter beams into new directions or into each other. This nonlinearity provides the basis for degenerate four-wave mixing, which will be discussed in Chapters 2–5.

B. Resonances: Where a Generalized Expansion Fails

When a material exhibits an absorption (or gain) feature, the situation is so complex that the susceptibility expansion [Eq. (13)] may no longer be meaningful, because terms of arbitrarily high order may be significant. The proper treatment of such systems involves solving the material's dynamic equations of motion, leading to effects such as adiabatic rapid passage, optical free-induction decay, self-induced transparency, and photon echoes.

The simplest resonance is associated with an isolated two-level system in which the linear absorption coefficient is a symmetric function of frequency that peaks at ω_0. The width of the function may come from many broadening mechanisms, including those associated with a dipole relaxation time T_1, a homogeneous lifetime T_2, and an inhomogeneous lifetime T_2^*. The parameter T_1 is the time of energy loss in the system, and T_2 the dephasing (not necessarily energy loss) time for an individual ion, atom, or molecule. The distribution function of absorbers $g(\Delta)$ (where $\Delta = \omega - \omega_0$), is determined by crystal strains, velocity distributions, etc.; T_2^* is the reciprocal of the width of $g(\Delta)$. A system is considered homogeneously broadened if T_2 is less than both T_1 and T_2^* and is inhomogeneously broadened if T_2^* is less than both T_1 and T_2. Even this complicated picture is unrealistically simplified. Typical resonances involve more than two levels and more complicated coupling (e.g., optical pumping and phase-changing or velocity-changing collisions) between the various levels.

Chapter 8 treats optical phase conjugation in resonant systems from the standpoint of a steady-state analysis, and Chapter 9 deals with conjugation in the transient regime.

C. Additional Reading on Nonlinear Optics

In the field of lasers and nonlinear optics, a number of excellent textbooks and reference books are available; however, optical phase conjugation is so new that it has yet to receive any attention. Nevertheless, the beginning reader should consult the following books to study the fundamental nonlinear optical effects without which phase conjugation would not be possible.

L. Allen and J. H. Eberly, "Optical Resonance and Two-Level Atoms." Wiley, New York, 1975.

G. Birnbaum, "Optical Masers." Academic Press, New York, 1964.

N. Bloembergen, "Nonlinear Optics." Benjamin, New York, 1965.

P. N. Butcher, "Nonlinear Optical Phenomena." Ohio State Univ. Engineering Bull. No. 200, 1965.

W. S. C. Chang, "Principles of Quantum Electronics." Addison-Wesley, Reading, Massachusetts, 1969.

I. L. Fabelinskii, "Molecular Scattering of Light." Plenum, New York, 1968.

V. M. Fain and Ya. I. Khanin, "Quantum Electronics." MIT Press, Cambridge, Massachusetts, 1969.

D. C. Hanna, M. A. Yuratich, and D. Cotter, "Nonlinear Optics of Free Atoms and Molecules." Springer-Verlag, Berlin and New York, 1980.

S. F. Jacobs, M. Sargent III, and M. O. Scully, "Adaptive Optics and Short Wavelength Sources." Addison-Wesley, Reading, Massachusetts, 1978.

P. L. Kelley, B. Lax, and P. E. Tannenwald (eds.), "Physics of Quantum Electronics." McGraw-Hill, New York, 1966.

V. S. Letokhov and V. P. Chebotayev, "Nonlinear Laser Spectroscopy." Springer-Verlag, Berlin and New York, 1977.

D. Marcuse, "Principles of Quantum Electronics." Academic Press, New York, 1980.

R. H. Pantell and H. E. Puthoff, "Fundamentals of Quantum Electronics." Wiley, New York, 1969.

G. Placzek, "The Rayleigh and Raman Scattering," UCRL Translation 526 (L). Available from Federal Scientific Clearinghouse, 1934.

M. Sargent III, M. O. Scully, and W. E. Lamb, Jr., "Laser Physics." Addison-Wesley, Reading, Massachusetts, 1974.

Y. R. Shen (ed.), "Nonlinear Infrared Generation." Springer-Verlag, Berlin and New York, 1977.
A. E. Siegman, "An Introduction to Lasers and Masers." McGraw-Hill, New York, 1971.
A. Yariv, "Quantum Electronics." Wiley, New York, 1967.
A. Yariv, "Introduction to Optical Electronics." Holt, New York, 1971.

V. The Distortion-Undoing Properties of a Phase-Conjugate Wave

We now present a mathematical proof of the precise distortion-undoing nature of the conjugation process. This proof closely follows that of Yariv (1977). Consider a monochromatic wave propagating through a composite medium whose electric permittivity (dielectric constant) is given by $\epsilon(\mathbf{r})$. The real quantity $\epsilon(\mathbf{r})$ represents the presence in the beam of passive linear components, such as lenses, wedges, or the presence of distorting media, such as a turbulent atmosphere. The scalar near-forward $(+z)$ propagating beam is taken as

$$E_1(\mathbf{r}, t) = \tfrac{1}{2}\mathscr{E}_1(\mathbf{r}) \exp[-i(\omega t - kz)] + \text{c.c.} \tag{17}$$

In the limit of slow spatial and temporal variations, the scalar wave equation obeyed by Eq. (17) is taken as

$$\nabla^2 \mathscr{E}_1 + [\omega^2 \mu \epsilon(\mathbf{r}) - k^2]\mathscr{E}_1 + 2ik\, \partial \mathscr{E}_1/\partial z = 0. \tag{18}$$

Let us next, as a purely mathematical operation, consider the complex conjugate of Eq. (18)

$$\nabla^2 \mathscr{E}_1^* + [\omega^2 \mu \epsilon(\mathbf{r}) - k^2]\mathscr{E}_1^* - 2ik\, \partial \mathscr{E}_1^*/\partial z = 0. \tag{19}$$

Equation (19) is also the *same* wave equation applied to a wave propagating in the $-z$ direction of the form

$$E_2(\mathbf{r}, t) = \tfrac{1}{2}\mathscr{E}_2 \exp[-i(\omega t + kz)] + \text{c.c.}, \tag{20}$$

provided we put

$$\mathscr{E}_2(\mathbf{r}) = a\mathscr{E}_1^*(\mathbf{r}), \tag{21}$$

where a is any constant. Thus a backward-going wave $\mathscr{E}_2$ whose complex amplitude is everywhere the complex conjugate of $\mathscr{E}_1$ (within an arbitrary multiplicative constant) satisfies the same wave equation obeyed by $\mathscr{E}_1$. This means that if, after traversing a distorting medium with beam E_1, we can generate a beam E_2 that is the phase conjugate of E_1 (within a multiplicative constant), then E_2 will propagate backwards and its amplitude $\mathscr{E}_2$ will *remain everywhere the complex conjugate of* $\mathscr{E}_1$. *Therefore, its wave fronts everywhere coincide with those of* E_1.

The incident wave E_1 and the reflected conjugate wave E_2 have the same time dependence, namely, $\exp(-i\omega t)$. It is only the spatial part $\mathscr{E}_1(\mathbf{r})\exp(ikz)$ that is complex conjugated. However, the previous proof given holds only when $\epsilon(\mathbf{r})$ is real. When the propagation medium is lossy or amplifying, $\epsilon(\mathbf{r})$ is complex and the proof does not apply. The exception is the case where the loss or gain is independent of $\mathbf{r}$. In this case the loss or gain behavior can be harmlessly factored out of the propagation equation.

In Chapter 14, generalizations of the distortion-correction process when the distorter itself exhibits a nonlinearity will be discussed.

VI. Review Material on Optical Phase Conjugation

Although this book is designed to provide fundamental review material concerning optical phase conjugation, the interested reader should note that seven reviewlike publications have appeared recently.

A. Yariv, Phase-conjugate optics and real-time holography, *IEEE J. Quantum Electron.* **QE-14,** 650 (1978).

A. Yariv, "Comments on 'Phase-conjugate optics and real-time holography'; author's reply," *IEEE J. Quantum Electron.* **QE-15,** 524 (1979).

D. M. Pepper, "Phase-Conjugate Optics," Ph.D. thesis, California Institute of Technology, 1980.

R. A. Fisher and B. J. Feldman, Optical phase conjugation. *In* "1980 McGraw-Hill Yearbook of Science and Technology." McGraw-Hill, New York.

C. R. Giuliano, Applications of optical phase conjugation, *Phys. Today* April, 1981.

D. M. Pepper, Nonlinear optical phase conjugation, *Laser Focus,* January 1982, p. 71.

D. M. Pepper, Nonlinear optical phase conjugation, *Opt. Eng.* **21,** 156 (1982).

In addition to these publications, a Russian-language collection of papers entitled "Wavefront Reversal of Optical Radiation in Nonlinear Media" was edited in 1979 by the Soviet Academy of Applied Physics, 603600 Gorkii, GSP-120, Uljanov Str. 46. We are not aware of any plans to produce an English version of this collection.

VII. Historical Overview

Optical phase conjugation is a fascinating subject, with great promise for application to image transmission in fibers, four-wave optical parametric oscillation, coherent image amplification, optical filtering, photon echoes, dispersion compensation, distortion compensation, novel laser resonator design, pointing and tracking, image processing, saturation spectroscopy, and high-resolution microscopy, to name a few possibilities.

Although the study of optical phase conjugation unifies one's understanding of various nonlinear optical effects, the subject does not have a simple evolutionary history. Searching for a progenitor of optical phase conjugation appears to be as futile as trying to find the father of a new litter of alley cats.

One of the earliest related developments occurred in the field of holography, when Kogelnik (1965) pointed out that conventional holographic techniques could be used for imaging through static inhomogeneous media. The development of transient holography (also know as real-time or dynamic holography) was based on the recognition that one could first "write" a grating (hologram) by interfering (in an appropriate material) two beams derived from the same laser. The hologram could subsequently "read," using a third wave at a possibly different wavelength. The pioneering work in this field includes that of Gerritsen (1967), Woerdman (1970), Amodei (1971), and Stepanov *et al.* (1971). The writing of the grating was through modification of some parameter (such as absorption or index of refraction) in the medium by the two interfering beams, and much of the early interest in this field concerned the use of this technique to probe the dynamics of the various physical processes, such as electron excitation, diffusion, and bleaching. The mathematical models describing the nonlinear optical interactions treated the reading and writing of the holograms as simultaneous but physically distinct, i.e., as noninteracting.

With apparently no ties to the work in transient holography, a major increase in effort occurred in the Soviet Union after the discovery that stimulated backward scattering (Brillouin, Raman, or Rayleigh) could produce phase-conjugate waves. Pioneering efforts include those of Zel'dovich and Shkunov (1977) and Zel'dovich *et al.* (1972, 1977), Nosach *et al.* (1972), Bel'dyugin *et al.* (1976), Kochemasov and Nikolaev (1977), and Wang and Giuliano (1978). These stimulated inelastic processes and their role in optical phase conjugation will be addressed in Chapters 6 and 7.

The study of elastic parametric scattering processes in nonlinear media produced a third exceedingly productive discipline, which led directly to the present surge of activity in the field. Here the concepts and mathematical techniques of nonlinear optics [see, e.g., Chiao *et al.* (1966)] were applied to the field of transient holography. Major contributions include Yariv (1976a, 1977b), and Hellwarth (1977). A first-order coupled-wave approach was applied to degenerate four-wave mixing by Yariv and Pepper (1977) and, independently, by Bloom and Bjorklund (1977), who presented the first direct demonstration. The coupled-wave approach led to the prediction and demonstration of amplification and oscillation in transmission and reflection modes and to the substitution of phase-conjugate mirrors for one of the conventional mirrors in laser resonators (AuYeung *et al.*, 1979) and wide field-of-view optical filtering (Pepper and Abrams, 1978). All of these developments that use Kerr-like nonlinearities will be discussed in Chapters 2–5. The extension of these four-wave mixing geometries to nonlinearities other than Kerr-like was both rapid and successful. As will be discussed in Chapter 8, nonlinear saturation of resonant (or near-resonant) media was explored by Bloom *et al.* (1978), Liao *et al.* (1978), Liao and Bloom (1978), Abrams and Lind (1978), and Fu and Sargent (1979). As will be discussed in Chapter 9, Shtyrkov *et al.* (1978) and Shiren (1978) further extended these concepts to show the phase-conjugate nature of the photon echo effects.

It is now understood that virtually any nonlinear optical effect can be used to generate the conjugate of a given wave, as indicated by the remarkable variety of phenomena described in the this book. We hope that even the reader not intending to use optical phase conjugation will come away with a richer understanding of the various curious effects that constitute the field of nonlinear optics.

Although most readers will turn first to the chapters closest to their work, we urge that the other chapters be examined as well, so that the common thread underlying all of these effects will be apparent.

References

Abrams, R. L., and Lind, R. C. (1978). Degenerate four-wave mixing in absorbing media, *Opt. Lett.* **2,** 94; (1978). **3,** 205.

Amodei, J. M. (1971). *RCA Rev.* **32,** 185.

AuYeung, J., Fekete, D., Pepper, D. M., and Yariv, A. (1979). A theoretical and experimental investigation of the modes of optical resonators with phase-conjugate mirrors, *IEEE J. Quantum Electron.* **QE-15,** 1180.

Bel'dyugin, I. M., Galushkin, M. G., Zemskov, E. M., and Mandrosov, V. I. (1976). Complex conjugation of fields in stimulated Brillouin scattering, *Kvantovaya Elektron.* **3,** 2467 [*English transl.: Sov. J. Quantum Electron.* **6,** 1349 (1976)].

Bloom, D. M., and Bjorklund, G. C. (1977). Conjugate wave-front generation and image reconstruction by four-wave mixing, *Appl. Phys. Lett.* **31,** 592.

Bloom, D. M., Liao, P. F., and Economou, N. P. (1978). Observation of amplified reflection by degenerate four-wave mixing in atomic sodium vapor. *Opt. Lett.* **2,** 58.

Chiao, R. Y., Kelley, P. L., and Garmire, E. (1966). Stimulated four-photon intraction and its influence on stimulated Rayleigh-Wing scattering, *Phys. Rev. Lett.* **17,** 1158.

Fu, T. Y., and Sargent, M., III (1979). Effects of signal detuning on phase conjugation, *Opt. Lett.* **4,** 366.

Gerritsen, H. J. (1967). Nonlinear effects in image formation, *Appl. Phys. Lett.* **10,** 237.

Hardy, J. W. (1978). *Proc. IEEE* **66,** 651.

Hellwarth, R. W. (1977). Generation of time-reversed wave fronts by nonlinear refraction, *J. Opt. Soc. Am.* **67,** 1.

Kochemasov, G. G., and Nikolaev, V. D. (1977). Reproduction of the spatial amplitude and phase distributions of a pump beam in stimulated Brillouin scattering, *Kvantovaya Elektron.* **4,** 115 [*English transl.: Sov. J. Quantum Electron.* **7,** 60 (1977)].

Kogelnik, H. (1965). Holographic image projection through inhomogeneous media, *Bell Syst. Tech. J.* **44,** 2451 (1965).

Liao, P. F., and Bloom, D. M. (1978). Continuous-wave backward-wave generation by degenerate four-wave mixing in ruby, *Opt. Lett.* **3,** 4.

Liao, P. F., Bloom, D. M., and Economou, N. P. (1978). cw optical wave-front conjugation by saturated absorption in atomic sodium vapor, *Appl. Phys. Lett.* **32,** 813.

Nosach, O. Yu., Popovichev, V. I., Ragul'skiy, V. V., and Faizullov, F. S. (1972). Cancellation of phase distortions in an amplifying medium with a Brillouin mirror, *Zh. Eksp. Teor. Fiz. Pis'ma Red.* **16,** 617 [*English transl.: Sov. Phys. JETP* **16,** 435 (1972)].

Pepper, D. M., and Abrams, R. L. (1978). Narrow optical bandpass filter via nearly degenerate four-wave mixing, *Opt. Lett.* **3,** 212.

Shiren, N. S. (1978). Generation of time-reversed optical wave fronts by backward-wave photon echoes, *Appl. Phys. Lett.* **33,** 299.

Shtyrkov, E. I., and Samartsev, V. V. (1978). Dynamic holograms on the superposition states of atoms, *Phys. Status Solidi* (*A*) **45,** 647.

Stepanov, B. I., Ivakin, E. V., and Rubanov, A. S. (1971). Recording two-dimensional and three-dimensional dynamic holograms in bleachable substances, *Dokl. Akad. Nauk SSSR* **196,** 567 [*English transl.: Sov. Phys.-Dokl.-Tech. Phys.* **16,** 46 (1971)].

Wang, V., and Giuliano, C. R. (1978). Correction of phase aberrations via stimulated Brillouin scattering, *Opt. Lett.* **2,** 4.

Woerdman, J. P. (1971). Formation of a transient free-carrier hologram in Si, *Opt. Commun.* **2,** 212.

Yariv, A. (1976a). Three-dimensional pictorial transmission in optical fibers, *Appl. Phys. Lett.* **28,** 88.

Yariv, A. (1976b). On transmission and recovery of three-dimensional image information in optical waveguides, *J. Opt. Soc. Am.* **66,** 301.

Yariv, A. (1977). Compensation for atmospheric degradation of optical beam transmission by nonlinear optical mixing, *Opt. Commun.* **21,** 49.

Yariv, A., and Pepper, D. M. (1977). Amplified reflection, phase conjugation, and oscillation in degenerate four-wave mixing, *Opt. Lett.* **1,** 16.

Zel'dovich, B. Ya., Popovichev, V. I., Ragul'skiy, V. V., and Faizullov, F. S. (1972). Connection between the wave fronts of the reflected and exciting light in stimulated Mandel'shtam-Brillouin scattering, *Zh. Eksp. Teor. Fiz. Pis'ma Red.* **15,** 160 [*English Transl.: Sov. Phys. JETP* **15,** 109 (1972)].

Zel'dovich, B. Ya., Mel'nikov, N. A., Pilipetskii, N. F., and Ragul'skiy, V. V. (1977). Observation of wave-front inversion in stimulated Raman scattering of light, *Zh. Eksp. Teor. Fiz. Pis'ma Red.* **25,** 41 [*English transl.: JETP Lett.* **25,** 36].

Zel'dovich, B. Ya., and Shkunov, V. V. (1977). Wavefront reproduction in stimulated Raman scattering, *Kvantovaya Elektron.* **4,** 1090 [*English transl.: Sov. J. Quantum Electron.* **7,** 610 (1977)].

2 *Optical Phase Conjugation Using Three-Wave and Four-Wave Mixing via Elastic Photon Scattering in Transparent Media*

David M. Pepper†
Hughes Research Laboratories
Malibu, California

Amnon Yariv
California Institute of Technology
Pasadena, California

† Work supported by Hughes Research Laboratories.

Optical Phase Conjugation

ISBN 0-12-257740-X

I. Introduction

In this chapter we undertake two tasks: we develop the basic mathematical foundations needed to treat the various phenomena and applications categorized under the name of optical phase conjugation (OPC), and we briefly review selected basic experiments performed to date that demonstrate the underlying phenomena.

We will limit our treatment to two classes of (nonlinear) *elastic* photon scattering processes that can generate phase-conjugate replicas: three-wave and four-wave mixing. By definition, these processes leave the nonlinear medium in the *same* quantum state before and after the optical interaction. This is in contrast to (nonlinear) *inelastic* photon scattering processes (e.g., stimulated Brillouin and Raman scattering), which leave the atomic (or molecular) medium in a different final state as a result of the photon coupling. This latter class of interactions, which can also give rise to wave-front reversal, is the topic of Chapters 6 and 7.

In this chapter, we restrict ourselves to considering nondispersive, unsaturable media as constituting the nonlinear medium. That is, the interacting fields are at optical frequencies far removed from any material resonances. (Resonant media are considered in Chapters 8 and 9.) In four-wave mixing, the optical nonlinearity (which typically requires high optical intensities) is manifested by a material index of refraction (the nonlinear index) or, equivalently, a phase velocity dependent upon the local field intensity. The term ''Kerr-like media'' is generally used to describe this class of materials; the nonlinear index is given by a numerical constant or tensor, which may be polarization dependent. The coupling of the interacting optical fields is characterized by a third-order nonlinear optical susceptibility. In three-wave mixing, the field coupling is described by a numerical constant or tensor, i.e., the second-order nonlinear optical susceptibility.

The common denominator of phenomena and applications in OPC is that a number of optical waves—usually two or three, but possibly more—are incident on a nonlinear medium, which then radiates a new wave whose frequency is equal to some particular combination (i.e., sum and difference) of the input frequencies. The spatial amplitude of the radiated field is proportional to the complex conjugate of one of the input amplitudes—hence the name "optical phase conjugation."

It is a source of wonder to the present authors that this basic optical mixing process can give rise to so many fascinating phenomena and applications. These include image transmission in fibers, four-wave optical parametric oscillation, coherent image amplification, optical filtering, conjugate photon echoes, dispersion compensation, correction for propagation distortion in passive media and in laser resonators, image processing, temporal signal processing, and projection. These applications will be taken up separately in Chapter 14. The starting point for treating most of the phenomena listed is a simple nonlinear optical formulation of the wave interaction, whose derivation and implications are the concern of Section II of this chapter.

The first experiments in what we now call optical phase conjugation were performed by Gerritsen (1967) and Staebler and Amodei (1972). These researchers first introduced the concept of a grating produced in a medium owing to the interference of two light beams and the subsequent diffraction of the beams from their own grating. Amodei also realized the possibility of power exchange between the two beams. The next step was taken independently by Woerdman (1971) and Stepanov *et al.* (1971). Their experiments, which demonstrated compensation of propagation distortion, relied on a qualitative analogy to holography.

The present developments, insight, and intense activity in optical phase conjugation directly followed the application of the powerful formalism of nonlinear optics to the discipline by Yariv (1976a,b, 1977), Hellwarth (1977a), Yariv and Pepper (1977), Bloom and Bjorklund (1977), and Abrams and Lind (1978). With this new understanding, many new phenomena were predicted, including coherent image transmission, image amplification, a new type of parametric oscillation, and high-resolution optical filtering. The resonant nature of the interaction, especially in gaseous media or atomic vapors, when the applied frequency is near that of an atomic (or molecular) transition, also led to a strong connection between optical phase conjugation and high-resolution nonlinear spectroscopy.

The exciting field of optical phase conjugation using stimulated emission processes, pioneered by Zel'dovich and his collaborators (1972), has followed a separate course. Its realm of application includes experiments involving pulsed, high-intensity lasers, to which it is especially well suited. Conjugation via stimulated scattering will be considered separately in Chapters 6 and 7.

II. Theoretical Formalism of Optical Phase Conjugation in Transparent Media

In this section we will treat the coupling of the interacting optical fields via a given set of material nonlinearities. This will lead us to the basic coupled-mode equations describing the spatial (and/or temporal) evolution of the desired conjugated wave front. The analysis will emphasize four-wave mixing; three-wave mixing will be treated only briefly, for historical and comparative purposes. The fundamental drawbacks of three-wave mixing (in terms of its spatial bandwidth), which will surely affect its potential usefulness, will be shown to be circumvented by the four-wave scheme. In our discussion of four-wave mixing, we will examine the effects of nonlinear phase shifts, linear losses, and phase matching (leading to nearly degenerate four-wave mixing). This section will conclude with a discussion of the operational equivalence of real-time holography and degenerate four-wave mixing. It is this property—the multiplicative spatial-field nature of the interaction—that opens the door to myriad image processing and optical computing applications of Optical Phase Conjugation (see Chapter 14).

A. Coupling of the Wave Equation to the Nonlinearities of the Medium

A material medium subject to an optical field $\mathbf{E}(t)$ will develop a polarization $\mathbf{P}(t)$. If the response is immediate, $\mathbf{P}(t)$ is a single-valued function of $\mathbf{E}(t)$, or $\mathbf{P} = f(\mathbf{E})$. In this case we can derive the polarization from an energy function

$$U(\mathbf{E}) = -\tfrac{1}{2}\epsilon_0\chi_{ij}\,E_iE_j - \tfrac{2}{3}d_{ijk}\,E_iE_jE_k - \chi_{ijk\ell}E_iE_jE_kE_\ell + \cdots . \qquad (1)$$

The ith component of the medium polarization is thus given by

$$P_i = -\frac{\partial U}{\partial E_i} = \epsilon_0\chi_{ij}E_j + 2d_{ijk}E_jE_k + 4\chi_{ijk\ell}E_jE_kE_\ell + \cdots , \qquad (2)$$

where a summation over repeated indices is implied. In Eq. (2) we used the fact that all the $\chi_{ijk\ell}$ tensor elements that result from a mere reordering of the subscripts are equal to each other, because a shuffling of the subscripts in Eq. (1) has no physical significance.[1]

In what follows, we will concentrate on steady-state four-wave mixing

[1] In general, the induced polarization may not be an instantaneous function of the inducing electric field. In this case we may still use a relation such as (1), but the nonlinear third-order susceptibility $\chi_{ijk\ell}$ will be a function of the frequencies as well as of their ordering.

processes (using the term proportional to $\chi_{ijk\ell}$ in Eq. (2)) that can generate phase-conjugate replicas of given incident waves. Steady-state three-wave mixing (which makes use of the term proportional to d_{ijk} in Eq. (2)) and its application to OPC will be treated in Section II.C.

Consider the case of a field $\mathbf{E}(t)$ made up of four waves, as in the case of phase conjugation via four-wave mixing:

$$\mathbf{E}(t) = \sum_{\alpha=1,2,\mathrm{p},\mathrm{c}} \tfrac{1}{2}\, \vec{\mathscr{E}}_\alpha(\omega_\alpha, \mathbf{r}) \exp[i(\omega_\alpha t - \mathbf{k}_\alpha \cdot \mathbf{r})] + \text{c.c.} \tag{3}$$

By $\mathscr{E}_{\alpha,j}(\omega_\alpha, \mathbf{r})$ we mean the slowly varying (complex) amplitude of the jth (Cartesian) vector component of the αth field, oscillating at ω_α at point $\mathbf{r}$. Being a complex quantity implies that it contains the phase information of the wave.

For reasons that will soon become apparent, we define three of the four waves given by Eq. (3) to be input fields: $\vec{\mathscr{E}}_1(\omega_1, \mathbf{r})$ and $\vec{\mathscr{E}}_2(\omega_2, \mathbf{r})$ are the pump waves, and $\vec{\mathscr{E}}_\mathrm{p}(\omega_\mathrm{p}, \mathbf{r})$ is the probe wave. The resultant nonlinearly generated output wave (as we subsequently show) is given by $\vec{\mathscr{E}}_\mathrm{c}(\omega_\mathrm{c}, \mathbf{r})$ and is referred to as the conjugate wave, or the time-reversed replica of the probe wave.

The term $\chi_{ijk\ell}E_jE_kE_\ell$ in Eq. (2) will, in general, give rise to nonlinear polarization terms oscillating at $\omega_j \pm \omega_k \pm \omega_\ell$. In four-wave phase-conjugate optical processes, we are interested only in terms which have a frequency resulting from the addition of two frequencies (the pump wave) and a subtraction of the third (the probe wave): $\omega_\mathrm{c} = \omega_1 + \omega_2 - \omega_\mathrm{p}$. Using the form (3) in the last term of (2) and isolating the terms at $\omega_1 + \omega_2 - \omega_\mathrm{p}$ results in

$$\begin{aligned}\mathscr{P}_i(\omega_\mathrm{c} = \omega_1 + \omega_2 - \omega_\mathrm{p}, \mathbf{r}) = 6\chi_{ijk\ell}\mathscr{E}_{1,j}(\omega_1, \mathbf{r})\mathscr{E}_{2,k}(\omega_2, \mathbf{r})\\ \times \mathscr{E}^*_{\mathrm{p},\ell}(\omega_\mathrm{p}, \mathbf{r}) \exp[-i(\mathbf{k}_1 + \mathbf{k}_2 - \mathbf{k}_\mathrm{p}) \cdot \mathbf{r}],\end{aligned} \tag{4}$$

where $\mathscr{P}_i$ is the ith component of the *complex* polarization amplitude at point $\mathbf{r}$.[2] The *real* polarization at point $\mathbf{r}$ is given by

$$\begin{aligned}P_i(\mathbf{r}, t) &= \tfrac{1}{2}\, \mathscr{P}_i(\omega_\mathrm{c} = \omega_1 + \omega_2 - \omega_\mathrm{p}, \mathbf{r}) \exp[i(\omega_1 + \omega_2 - \omega_\mathrm{p})t + \text{c.c.}\\ &= \tfrac{1}{2}\, 6\chi_{ijk\ell}\mathscr{E}_{1,j}(\omega_1, \mathbf{r})\mathscr{E}_{2,k}(\omega_2, \mathbf{r})\\ &\quad \times \mathscr{E}^*_{\mathrm{p},\ell}(\omega_\mathrm{p}, \mathbf{r}) \exp\{i[(\omega_1 + \omega_2 - \omega_\mathrm{p})t - (\mathbf{k}_1 + \mathbf{k}_2 - \mathbf{k}_\mathrm{p}) \cdot \mathbf{r}]\}\\ &\quad + \text{c.c.}\end{aligned} \tag{5}$$

There is a factor of 6 in Eqs. (4) and (5) because there are six different

[2] In this chapter, we will use $\vec{\mathscr{P}}$ to denote the polarization with only the *frequency* term factored out. Our $\vec{\mathscr{P}}$ is *not* slowly varying in space, as opposed to that in Chapter 1. This will enable us to show the phase-matching conditions explicitly.

combinations of the form $\mathscr{E}_{1,j}(\omega_1, \mathbf{r})\mathscr{E}_{2,k}(\omega_2, \mathbf{r})\mathscr{E}^*_{\mathrm{p},\ell}(\omega_\mathrm{p}, \mathbf{r})$ in the product $\chi_{ijk\ell}E_j(t)E_k(t)E_\ell(t)$, where $\mathbf{E}(t)$ is given by Eq. (3).

We thus write, dropping the $\mathbf{r}$ labeling, which will be understood,

$$\mathscr{P}_i(\omega_\mathrm{c} = \omega_1 + \omega_2 - \omega_\mathrm{p}) = 6\chi_{ijk\ell}(-\omega_\mathrm{c}, \omega_1, \omega_2, -\omega_\mathrm{p})\mathscr{E}_{1,j}(\omega_1)\mathscr{E}_{2,k}(\omega_2) \times \mathscr{E}^*_{\mathrm{p},\ell}(\omega_\mathrm{p}) \exp[-i(\mathbf{k}_1 + \mathbf{k}_2 - \mathbf{k}_\mathrm{p}) \cdot \mathbf{r}]. \quad (6)$$

Note that a complex-conjugate field amplitude such as $\mathscr{E}^*_\mathrm{p}(\omega_\mathrm{p})$ is always accompanied by a negative-frequency term $\exp(-i\omega_\mathrm{p}t)$.

The form of the fourth-rank susceptibility tensor $\chi_{ijk\ell}$ is determined by the symmetry properties of the medium. Unlike the third-rank optical susceptibility tensor d_{ijk}, which can lead to nonlinear effects such as second harmonic generation, and which is nonvanishing primarily in asymmetric crystals, $\chi_{ijk\ell}$ can be nonvanishing in all material media. Note that although the tensor is fourth rank it involves, according to Eq. (4), a third power of the electric field, and is thus referred to as a third-order nonlinearity. For a comprehensive reference on the symmetry and physics of $\chi_{ijk\ell}$, see Butcher (see also Chapter 1, Section IV.C).

Numerous physical phenomena give rise to third-order optical nonlinearities characterized by $\chi_{ijk\ell}$. Some of these effects involve inelastic photon scattering processes, such as Raman and Brillouin scattering, and elastic photon scattering processes, such as gain saturation and the ac Kerr effect. In what follows, we assume a Kerr-like effect, with $\chi_{ijk\ell}$ being a constitutive parameter of the medium, and enquire about the electromagnetic consequences of the nonlinear polarizations so generated. Because we assume that χ is a constant, it follows that the nonlinear medium responds instantaneously to the presence of the interacting fields.

To derive the interaction equation between four electromagnetic waves in a material medium that can lead to the generation of phase-conjugated wave fronts, we follow the procedure of Yariv and Pepper (1977). The geometry is illustrated in Fig. 1. The nonlinear (Kerr-like) medium is assumed to be transparent and nondispersive. The pump waves $\vec{\mathscr{E}}_1$ and $\vec{\mathscr{E}}_2$ are assumed to be intense and counterpropagating. A weak probe wave $\vec{\mathscr{E}}_\mathrm{p}$ is incident at an arbitrary angle; we desire to obtain a phase-conjugate replica of this probe wave. The output conjugate wave $\vec{\mathscr{E}}_\mathrm{c}$ which is generated by the nonlinear interaction, is also considered to be weak. We shall assume that all three input frequencies are equal: $\omega_1 = \omega_2 = \omega_\mathrm{p} \equiv \omega$. This situation is referred to as degenerate four-wave mixing. We will limit our attention to the practical case where the depletion of pump-wave photons due to the nonlinear interaction is negligible, so that $|\vec{\mathscr{E}}_1| = |\vec{\mathscr{E}}_2| =$ const. (Nonsaturable linear losses of the fields will be considered in Section II.D.) We will, however, allow for the effect of the nonlinear interaction on the *phases* of $\vec{\mathscr{E}}_1$ and $\vec{\mathscr{E}}_2$. The four waves in the medium are of the

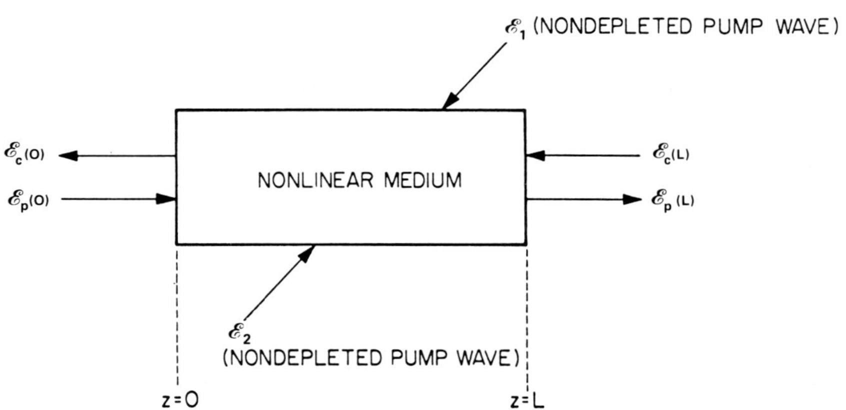

Fig. 1. The basic geometry of phase conjugation by four-wave mixing. The counterpropagating pump-wave amplitudes $\mathscr{E}_1$ and $\mathscr{E}_2$ are assumed to be nondepleted. The probe wave $\mathscr{E}_p$ is, in general, incident at an arbitrary angle to the nonlinear medium. As a result of the nonlinear optical interaction, the conjugate wave $\mathscr{E}_c$ is generated. [After Yariv and Pepper (1977).]

form

$$\mathbf{E}_\alpha(\mathbf{r}, t) = \tfrac{1}{2}\vec{\mathscr{E}}_\alpha(\mathbf{r}, t) \exp[i(\omega t - \mathbf{k}_\alpha \cdot \mathbf{r})] + \text{c.c.}, \qquad \alpha = 1, 2, \text{p}, \text{c}, \tag{7}$$

where the subscript α is a field label. We will limit ourselves to the steady state, so that all the temporal derivatives vanish; the transient analysis of four-wave mixing is treated in Chapter 3.

Let us first consider the wave equation for the ith polarization (or Cartesian) component of the conjugate wave $\mathscr{E}_{c,i}(z)$. From the SVEA of Chapter 1 (Section III.B), we have

$$\frac{d\mathscr{E}_{c,i}}{dz} = i\frac{\omega}{2}\sqrt{\frac{\mu}{\epsilon}}\,\mathscr{P}_i \exp[i\mathbf{k}_c \cdot \mathbf{r}], \tag{8}$$

where $\mathscr{P}_i$ is given by Eq. (6) and has the same state of polarization as $\mathscr{E}_{c,i}$. As discussed in Chapter 1, for a cumulative nondestructive buildup of $\mathscr{E}_{c,i}$, the (rapidly varying) phase factor of the product $\mathscr{P}_i \exp(i\mathbf{k}_c \cdot \mathbf{r})$ must be near zero; otherwise, successive contributions to $\mathscr{E}_{c,i}$ along z will not add up in phase. This can happen if in Eq. (8) we use the field combination

$$\begin{aligned}\mathscr{P}_i(\omega = \omega + \omega - \omega) = {} & 6\chi_{ijk\ell}(-\omega, \omega, \omega, -\omega) \\ & \times \mathscr{E}_{1,j}(\omega)\mathscr{E}_{2,k}(\omega)\mathscr{E}^*_{p,\ell}(\omega) \\ & \times \exp[-i(\mathbf{k}_1 + \mathbf{k}_2 - \mathbf{k}_p) \cdot \mathbf{r}].\end{aligned} \tag{9}$$

Thus, the total (rapidly varying) phase factor in Eq. (8) becomes

$$\mathscr{P}_i \exp[+i\mathbf{k}_c \cdot \mathbf{r}] \exp[-i(\mathbf{k}_1 + \mathbf{k}_2 - \mathbf{k}_p) \cdot \mathbf{r}]. \tag{10}$$

Because waves 1 and 2 are counterpropagating, $\mathbf{k}_1 + \mathbf{k}_2 \simeq 0$ (the reason why $\mathbf{k}_1 + \mathbf{k}_2$ may not be exactly zero will be discussed). The total phase factor in Eq. (10) is identically equal to zero if $\mathbf{k}_c = -\mathbf{k}_p$; thus, the nonlinearly generated field, $\mathscr{E}_{c,i}$ propagates in opposition to the incident probe field $\mathscr{E}_{p,\ell}$.

Before continuing, we will simplify the field labeling and polarization designations. We will henceforth dispense with the Cartesian subscripts. With regard to polarizations, in practice we choose a given set of polarizations (j, k, ℓ) for the input waves $\mathscr{E}_{1,j}$, $\mathscr{E}_{2,k}$, and $\mathscr{E}_{p,\ell}$, respectively; the nonlinear interaction generates an output wave $\mathscr{E}_{c,i}$ with a unique polarization i, determined by the existence of the appropriate nonvanishing tensor element(s) of $\chi_{ijk\ell}$ and by the experimental conditions. We will therefore assume that the wave $\vec{\mathscr{E}}_2$ for example is polarized along some Cartesian axis $\hat{\mathscr{E}}_2$. The same holds for $\vec{\mathscr{E}}_c$, $\vec{\mathscr{E}}_p$, and $\vec{\mathscr{E}}_1$, so that from now on we will use the subscripts both as wave and as field polarization labels. Hence, we will drop the vector notation. Thus, the αth field $\mathbf{E}_\alpha$ becomes

$$\mathbf{E}_\alpha \rightarrow E_\alpha = \tfrac{1}{2}\mathscr{E}_\alpha \exp[i(\omega_\alpha t - \mathbf{k}_\alpha \cdot \mathbf{r})] + \text{c.c.}$$

We now wish to analyze a situation where, in Eq. (10), the sum of $\mathbf{k}_1$ and $\mathbf{k}_2$ is not equal to zero, even if the fields of $\mathscr{E}_1$ and $\mathscr{E}_2$ are counterpropagating. As we will show, this can arise from asymmetric, nonlinear phase shifts due to unequal pump-wave intensities. As a starting point for this discussion, we examine the various third-order field products synchronous in space and time. As an example, consider the field $\mathscr{E}_c$. Using Eq. (3) in the third term of Eq. (2), one can show that the phase-matched terms involving the field $\mathscr{E}_c$ are

$$(\mathscr{E}_1\mathscr{E}_1^* + \mathscr{E}_2\mathscr{E}_2^* + \mathscr{E}_p\mathscr{E}_p^* + \mathscr{E}_c\mathscr{E}_c^*)\mathscr{E}_c.$$

Sets of products similar to those contained in the parenthesis apply to the three other fields.

Using these terms in the expression for the nonlinear polarization and inserting them into the slowly varying wave equations [Eqs. (9) and (10) of Chapter 1] yields

$$\begin{aligned}\frac{d\mathscr{E}_p}{dz} = &-i\frac{\omega}{2}\sqrt{\frac{\mu}{\epsilon}}\,\chi^{(3)}(|\mathscr{E}_1|^2 + |\mathscr{E}_2|^2 + |\mathscr{E}_p|^2 + |\mathscr{E}_c|^2)\mathscr{E}_p \\ &-i\frac{\omega}{2}\sqrt{\frac{\mu}{\epsilon}}\,\chi^{(3)}\mathscr{E}_1\mathscr{E}_2\mathscr{E}_c^* \exp[-i(\mathbf{k}_1 + \mathbf{k}_2)\cdot\mathbf{r}],\end{aligned} \tag{11}$$

$$\frac{d\mathscr{E}_c}{dz} = i\frac{\omega}{2}\sqrt{\frac{\mu}{\epsilon}}\,\chi^{(3)}(|\mathscr{E}_1|^2 + |\mathscr{E}_2|^2 + |\mathscr{E}_p|^2 + |\mathscr{E}_c|^2)\mathscr{E}_c$$
$$+ i\frac{\omega}{2}\sqrt{\frac{\mu}{\epsilon}}\,\chi^{(3)}\mathscr{E}_1\mathscr{E}_2\mathscr{E}_p^* \exp[-i(\mathbf{k}_1 + \mathbf{k}_2)\cdot\mathbf{r}], \qquad (12)$$

where we have defined $\chi^{(3)} \equiv 6\chi_{ijk\ell}$.

Because we have limited ourselves at the outset to the case where the pump energy depletion is negligible, it would seem at first sight that we may proceed to solve Eqs. (11) and (12) by taking $\mathscr{E}_1$ and $\mathscr{E}_2$ as constants. A problem arises, however, owing to the presence of the term $\mathbf{k}_1 + \mathbf{k}_2$ in the exponent. At low intensities the magnitudes of $\mathbf{k}_1$ and $\mathbf{k}_2$ are equal, $k_1 = k_2 = k = (\omega/c)n$; because the directions are opposite, we have $\mathbf{k}_1 + \mathbf{k}_2 = 0$. At high intensities, however, the magnitudes of $\mathbf{k}_1$ and $\mathbf{k}_2$ are modified, allowing the sum to become finite. To understand this modification, consider the wave equation that describes the evolution of the pump waves $\mathscr{E}_1$ and $\mathscr{E}_2$. Let us take the ζ axis as the direction of propagation of $\mathscr{E}_1$. Equation (9) of Chapter 1 becomes

$$\frac{d\mathscr{E}_1}{d\zeta} = -i\frac{\omega}{2}\sqrt{\frac{\mu}{\epsilon}}\,\mathscr{P}e^{ik\zeta}, \qquad k \equiv \omega\sqrt{\mu\epsilon}, \qquad (13)$$

and the equation for $\mathscr{E}_2$ (which propagates in the $-\zeta$ direction) becomes

$$\frac{d\mathscr{E}_2}{d\zeta} = i\frac{\omega}{2}\sqrt{\frac{\mu}{\epsilon}}\,\mathscr{P}e^{-ik\zeta}. \qquad (14)$$

The driving polarization $\mathscr{P}$ on the right-hand sides of Eqs. (13) and (14) has, in general, many synchronous terms. To simplify matters (without loss of physical significance), we will ignore all those terms involving the weak fields of 3 and 4, and retain only terms involving the strong fields 1 and 2.

We thus take the total field as

$$E = \tfrac{1}{2}\mathscr{E}_1 \exp[i(\omega t - k\zeta)] + \tfrac{1}{2}\mathscr{E}_2 \exp[i(\omega t + k\zeta)] + \text{c.c.} \qquad (15)$$

We note that, from Eq. (3), the product $E_jE_kE_\ell$ contains 156 terms. Of these, 27 are synchronous with $\mathscr{E}_1 \exp[i(\omega t - k\zeta)]$, and thus should be included in $\mathscr{P}$ and inserted in the right-hand side of Eq. (13). The dominant terms can be grouped together to form the third-order nonlinear polarization

$$P = \tfrac{1}{2}\chi^{(3)}(9\mathscr{E}_1|\mathscr{E}_1|^2 + 18\mathscr{E}_1|\mathscr{E}_2|^2) \exp[i(\omega t - k\zeta)] + \text{c.c.}, \qquad (16)$$

Because $|\mathscr{E}_p|$, $|\mathscr{E}_c| \ll |\mathscr{E}_1|$, $|\mathscr{E}_2|$, we have neglected other synchronous terms such as $|\mathscr{E}_c|^2\mathscr{E}_1$. Recalling that $P = \frac{1}{2}\mathscr{P}e^{i\omega t} + \text{c.c.}$, we obtain

$$\mathscr{P} = \chi^{(3)}(9|\mathscr{E}_1|^2 + 18|\mathscr{E}_2|^2)\mathscr{E}_1 \exp[-ik\zeta],$$

which, when substituted in Eq. (13), gives

$$\frac{d\mathscr{E}_1}{d\zeta} = -i\frac{\omega}{2}\sqrt{\frac{\mu}{\epsilon}}\,\chi^{(3)}(9|\mathscr{E}_1|^2 + 18|\mathscr{E}_2|^2)\mathscr{E}_1,$$

so that

$$\mathscr{E}_1(\zeta) = \mathscr{E}_1(0)\exp\left[-i\frac{\omega}{2}\sqrt{\frac{\mu}{\epsilon}}\,\chi^{(3)}(9|\mathscr{E}_1|^2 + 18|\mathscr{E}_2|^2)\zeta\right].$$

The magnitude of the propagation vector $\mathbf{k}_1$ is thus modified according to

$$k_1 = \frac{\omega}{c}n + \frac{\omega}{2}\sqrt{\frac{\mu}{\epsilon}}\,\chi^{(3)}(9|\mathscr{E}_1|^2 + 18|\mathscr{E}_2|^2). \tag{17}$$

The same reasoning leads to

$$\frac{d\mathscr{E}_2}{d\zeta} = i\frac{\omega}{2}\sqrt{\frac{\mu}{\epsilon}}\,\chi^{(3)}(9|\mathscr{E}_2|^2 + 18|\mathscr{E}_1|^2)\mathscr{E}_2, \tag{18}$$

so that

$$\mathscr{E}_2(\zeta) = \mathscr{E}_2(0)\exp\left[i\frac{\omega}{2}\sqrt{\frac{\mu}{\epsilon}}\,\chi^{(3)}(9|\mathscr{E}_2|^2 + 18|\mathscr{E}_1|^2)\zeta\right]$$

and

$$k_2 = \frac{\omega}{c}n + \frac{\omega}{2}\sqrt{\frac{\mu}{\epsilon}}\,\chi^{(3)}(9|\mathscr{E}_2|^2 + 18|\mathscr{E}_1|^2). \tag{19}$$

The sum $\mathbf{k}_1 + \mathbf{k}_2$ in Eqs. (9), (11), and (12) is thus

$$\mathbf{k}_1 + \mathbf{k}_2 = \hat{e}_\zeta\,\tfrac{9}{2}\omega\sqrt{\mu/\epsilon}\,\chi^{(3)}(|\mathscr{E}_2|^2 - |\mathscr{E}_1|^2). \tag{20}$$

Note that if $|\mathscr{E}_2| \neq |\mathscr{E}_1|$, i.e., if the pump waves have unequal intensities, the sum in Eq. (20) is finite. Thus, according to Eqs. (11) and (12) the interaction terms involving $\mathscr{E}_1\mathscr{E}_2\mathscr{E}_p^*$ and $\mathscr{E}_1^*\mathscr{E}_2^*\mathscr{E}_c$ are *not* phase matched, because $\mathbf{k}_1 + \mathbf{k}_2 \neq 0$. This phase mismatch is almost always deleterious. In what follows we shall assume that $|\mathscr{E}_1| = |\mathscr{E}_2|$, as is the case in most experiments. The very special case where all four waves are collinear has been solved exactly for $|\mathscr{E}_1| \neq |\mathscr{E}_2|$ and will be considered in Chapter 4.

Putting $\mathbf{k}_1 + \mathbf{k}_2 = 0$ in Eqs. (11) and (12) leads to

$$\frac{d\mathscr{E}_c}{dz} = i\frac{\omega}{2}\sqrt{\frac{\mu}{\epsilon}}\,\chi^{(3)}(|\mathscr{E}_1|^2 + |\mathscr{E}_2|^2)\mathscr{E}_c + i\frac{\omega}{2}\sqrt{\frac{\mu}{\epsilon}}\,\chi^{(3)}\mathscr{E}_1\mathscr{E}_2\mathscr{E}_p^*, \tag{21}$$

where we assumed that $|\mathscr{E}_1|^2 + |\mathscr{E}_2|^2 \gg |\mathscr{E}_p|^2 + |\mathscr{E}_c|^2$. The first term on the right-hand side of Eq. (21) merely modifies the phase of $\mathscr{E}_c$ and does not contribute to the energy exchange. We can simplify Eq. (21) by factoring out this phase dependence. We define

$$\mathscr{E}_p(z) \equiv \mathscr{E}_p'(z) \exp\left[-i\frac{\omega}{2}\sqrt{\frac{\mu}{\epsilon}}\,\chi^{(3)}(|\mathscr{E}_1|^2 + |\mathscr{E}_2|^2)z\right],$$
$$\mathscr{E}_c(z) \equiv \mathscr{E}_c'(z) \exp\left[i\frac{\omega}{2}\sqrt{\frac{\mu}{\epsilon}}\,\chi^{(3)}(|\mathscr{E}_1|^2 + |\mathscr{E}_2|^2)z\right]. \tag{22}$$

Equation (21) becomes

$$\frac{d\mathscr{E}_c'}{dz} = i\frac{\omega}{2}\sqrt{\frac{\mu}{\epsilon}}\,\chi^{(3)}\mathscr{E}_1\mathscr{E}_2(\mathscr{E}_p')^*, \tag{23}$$

and similarly

$$\frac{d\mathscr{E}_p'}{dz} = -i\frac{\omega}{2}\sqrt{\frac{\mu}{\epsilon}}\,\chi^{(3)}\mathscr{E}_1\mathscr{E}_2(\mathscr{E}_c')^*. \tag{24}$$

These equations, first derived by Yariv and Pepper (1977), and independently by Bloom and Bjorklund (1977), are the starting point for much of the formalism of optical phase conjugation. Their solutions in different physical situations will be discussed in the remainder of this chapter. We will see that the process of degenerate four-wave mixing, using the geometry of Fig. 1, allows us to generate phase-conjugate replicas for *any* arbitrary spatial input probe wave (at frequency ω). It is this property—the ability of the conjugator to "time-reverse" a general wave front—that makes this interaction potentially useful for a host of interesting applications.

Before pursuing this line of thought, we will digress somewhat, and treat the case of phase conjugation via three-wave mixing. The resultant limitations on the acceptance angle (and hence the ultimate spatial bandwidth) of the probe wave, as imposed by phase-matching constraints, will enable us to appreciate the elegance and simplicity, and hence the utility, of the four-wave process.

B. Phase Conjugation by Three-Wave Mixing: A Digression

Although most the discussion in this book concerns optical phase conjugation by four-wave mixing, we could benefit from a brief discussion of phase conjugation by three-wave mixing. It will serve (a) to demonstrate that second-order optical nonlinearities, i.e., three-wave mixing can be used to perform phase conjugation, and (b) to point out the crucial role played by phase matching. Also, historically, the first application of the formalism of nonlinear optics to phase-conjugate optics was based on three-wave mixing.

Three-wave mixing can be used to generate phase-conjugate replicas of optical beams, as was first pointed out by Yariv (1976a,b, 1977). The scheme exploits the second-order optical nonlinearity [the second term on the right-hand side of Eq. (2)]. A nonzero second-order nonlinearity can occur, for example, in a crystal lacking inversion symmetry. In such crystals, the presence of input fields

$$\begin{aligned} \mathbf{E}_1 &= \tfrac{1}{2}\,\vec{\mathscr{E}}(\omega_1)\exp[i(\omega_1 t - \mathbf{k}_1\cdot\mathbf{r})] + \text{c.c.},\\ \mathbf{E}_\mathrm{p} &= \tfrac{1}{2}\,\vec{\mathscr{E}}(\omega_\mathrm{p})\exp[i(\omega_\mathrm{p} t - \mathbf{k}_\mathrm{p}\cdot\mathbf{r})] + \text{c.c.}, \end{aligned} \tag{25}$$

where $\mathbf{E}_1$ and $\mathbf{E}_\mathrm{p}$ are, respectively, the pump and probe waves, induces in the medium a nonlinear optical polarization, the ith component of which is given by

$$P_i(\mathbf{r}, t) = \tfrac{1}{2}\mathscr{P}_i \exp[i(\omega_1 - \omega_\mathrm{p})t] + \text{c.c.}, \tag{26}$$

Their polarization oscillates at the difference frequency $\omega_1 - \omega_\mathrm{p}$, where

$$\mathscr{P}_i = d_{ijk}\mathscr{E}_j(\omega_1)\mathscr{E}_k^*(\omega_\mathrm{p})\exp[-i(\mathbf{k}_1 - \mathbf{k}_\mathrm{p})\cdot\mathbf{r}]. \tag{27}$$

Such a polarization acting as a source in the wave equation will radiate a new wave $\mathscr{E}_i(\omega_\mathrm{c})$ at frequency $\omega_\mathrm{c} = \omega_1 - \omega_\mathrm{p}$, with an amplitude proportional to $\mathscr{E}_k^*(\omega_\mathrm{p})$, i.e., to the complex conjugate of the spatial amplitude of the low-frequency probe wave at ω_p.

A practical problem, which has prevented wide-scale application of this technique, is that of phase matching. A necessary condition for a phase-coherent cumulative buildup of the conjugate-field radiation at $\omega_1 - \omega_\mathrm{p}$ is that the wave vector at this new frequency be equal to $\mathbf{k}_1 - \mathbf{k}_\mathrm{p}$, i.e.,

$$\mathbf{k}(\omega_1 - \omega_\mathrm{p}) = \mathbf{k}_1(\omega_1) - \mathbf{k}_\mathrm{p}(\omega_\mathrm{p}). \tag{28}$$

This condition can be satisfied along only *one* direction in a crystal by using, for example, the optical anisotropy to compensate for the (linear) refractive index dispersion. An example of such matching is shown in Fig. 2.

For the sake of illustration, we choose the case where $\omega_1 = 2\omega_\mathrm{p}$, so that $\omega_1 - \omega_\mathrm{p} = \omega_\mathrm{p}$. The crystal orientation and polarizations of the waves can be chosen so that, given a plane wave $\vec{\mathscr{E}}(\omega_\mathrm{p})$ propagating parallel to $\vec{\mathscr{E}}(\omega_1)$, perfect phase matching occurs. The new wave at $\omega_1 - \omega_\mathrm{p}$ thus propagates parallel to $\vec{\mathscr{E}}(\omega_\mathrm{p})$. In nearly all the applications envisaged so far for optical phase conjugation, the input wave $\vec{\mathscr{E}}(\omega_\mathrm{p})$ is *not* a plane wave, but pos-

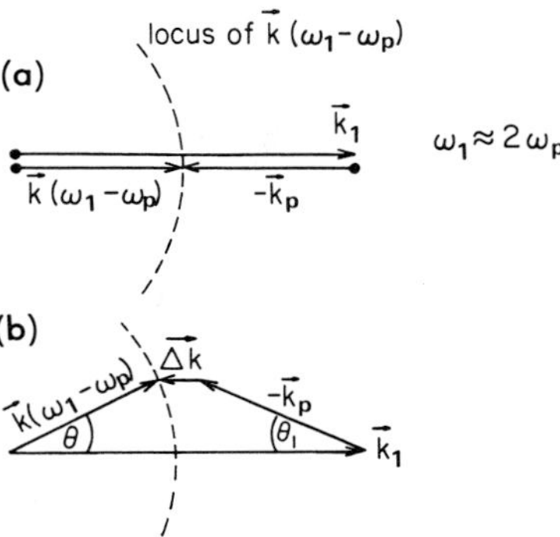

Fig. 2. Phase conjugation by three-wave mixing. (a) Plane waves $\vec{\mathscr{E}}_1$ and $\vec{\mathscr{E}}_p$ with propagation vectors $\mathbf{k}_1$ and $\mathbf{k}_p$, respectively, mix in a crystal to give a phase-matched wave at $\omega_1 - \omega_p$. (b) A different plane-wave component of the wave at ω_p cannot give rise to a phase-matched output.

sesses a complex wave front. Such a wave can be considered as made up of a continuous superposition of plane waves, all at ω_p, each with a $\mathbf{k}_p$ vector. The $\mathbf{k}_p$ vectors occupy some finite solid angle in space, the so-called angular field of view of the input wave. To obtain a complex-conjugate replica of the total input wave $\vec{\mathscr{E}}(\omega_p)$ it is necessary and sufficient that each plane-wave component be complex conjugated separately and that the generated components bear the same ratio to each other in the new beam (at $\omega_1 - \omega_p$) as they did in the input beam $\vec{\mathscr{E}}(\omega_p)$. Consider, for example, the fate of some other plane-wave component at ω_p whose direction of propagation is not parallel to $\mathbf{k}_1$, as shown in Fig. 2b. The phase-matching relationship cannot be satisfied, and the minimum mismatch $\mathbf{\Delta k}$ that can be tolerated is that shown, which occurs in this simple case when $\theta = \theta_1$. If $\mathbf{\Delta k} \cdot \mathbf{L} \simeq \pi$, where $|\mathbf{L}|$ is the interaction length in the crystal, then the efficiency with which this new wave component is generated is reduced to zero.

One major practical consequence of this limitation is that the angular field of view of the input field $\vec{\mathscr{E}}(\omega_p)$, whose phase-conjugate replica is sought, must be limited to such small values as to be of no interest in most applications. An impressively difficult experiment performed by Avizonis *et al.* (1977) succeeded in demonstrating the basic principles, as well as the inherent limitations, of this technique.

Because of these basic difficulties,[3] phase conjugation via three-wave mixing is not seriously pursued. In the remainder of Section II we will consider only the case of phase conjugation via four-wave mixing.

[3] We also note that phase conjugation via degenerate, *forward-going four-wave* mixing, where ω_1(3-wave) $\rightarrow$ ω_1(4-wave) + ω_2(4-wave), suffers from similar phase-matching constraints (see Section III.E.3 for experimental details).

C. Phase Conjugation by Four-Wave Mixing: The Basic Solution

As discussed in Section II. A (recall Fig. 1), the pump waves $\mathscr{E}_1$ and $\mathscr{E}_2$ propagate in opposition to each other, along some arbitrary direction. The input probe wave $\mathscr{E}_p$ is usually considerably less intense than $\mathscr{E}_1$ or $\mathscr{E}_2$. The result of the nonlinear interaction is the generated wave $\mathscr{E}_c$, called the conjugate wave. Our notation is that used in Chapter 1. In the limit where the depletion of waves $\mathscr{E}_1$ and $\mathscr{E}_2$ can be neglected, and taking $|\mathscr{E}_1| = |\mathscr{E}_2|$ as well as $\omega_1 = \omega_2 = \omega_p = \omega_c = \omega$ (degenerate four-wave mixing), the interaction between $\mathscr{E}_p$ and $\mathscr{E}_c$, using the results of Section II.A, is given by

$$\frac{d\mathscr{E}_p}{dz} = -i\kappa^*\mathscr{E}_c^*, \qquad \frac{d\mathscr{E}_c}{dz} = i\kappa^*\mathscr{E}_p^*, \tag{29}$$

where

$$\begin{aligned} \kappa^* &= (\omega/2)\sqrt{\mu/\epsilon}\,\chi^{(3)}\mathscr{E}_1\mathscr{E}_2\,(\mathrm{SI}), \\ \kappa^* &= (2\pi\omega/cn)\,\chi^{(3)}\mathscr{E}_1\mathscr{E}_2 \quad (\mathrm{Gaussian}). \end{aligned} \tag{30}$$

The primed notation of Section II.A has been dropped, because the difference between, say, $\mathscr{E}_p$ and $\mathscr{E}_p'$ at some plane z involves an inconsequential phase shift.

If we specify the complex amplitudes $\mathscr{E}_c(L)$ and $\mathscr{E}_p(0)$ of the two weak waves at their respective input planes $z = L$ and $z = 0$, the solution of Eq. (29) given by Yariv and Pepper (1977) and Bloom and Bjorklund (1977) is

$$\begin{aligned} \mathscr{E}_p(z) &= -i\frac{|\kappa|\sin(|\kappa|z)}{\kappa\cos(|\kappa|L)}\mathscr{E}_c^*(L) + \frac{\cos[|\kappa|(z-L)]}{\cos(|\kappa|L)}\mathscr{E}_p(0), \\ \mathscr{E}_c(z) &= \frac{\cos(|\kappa|z)}{\cos(|\kappa|L)}\mathscr{E}_c(L) + i\frac{\kappa^*\sin[|\kappa|(z-L)]}{|\kappa|\cos(|\kappa|L)}\mathscr{E}_p^*(0). \end{aligned} \tag{31}$$

In the practical case of phase conjugation, $\mathscr{E}_p(0)$ is finite, whereas $\mathscr{E}_c(L) = 0$. In this case, the nonlinearly reflected wave at the input plane ($z = 0$) is

$$\mathscr{E}_c(0) = -i[(\kappa^*/|\kappa|)\tan(|\kappa|L)]\mathscr{E}_p^*(0), \tag{32}$$

whereas at the output plane ($z = L$) the probe wave is

$$\mathscr{E}_p(L) = \mathscr{E}_p(0)/\cos(|\kappa|L). \tag{33}$$

Equations (32) and (33) constitute our main result. They show that at $z = 0$ the nonlinearly generated reflected field $\mathscr{E}_c(0)$ is proportional to the complex conjugate of the incident field $\mathscr{E}_p(0)$ multiplied by a factor $-i[(\kappa^*/|\kappa|)\tan(|\kappa|L)]$.

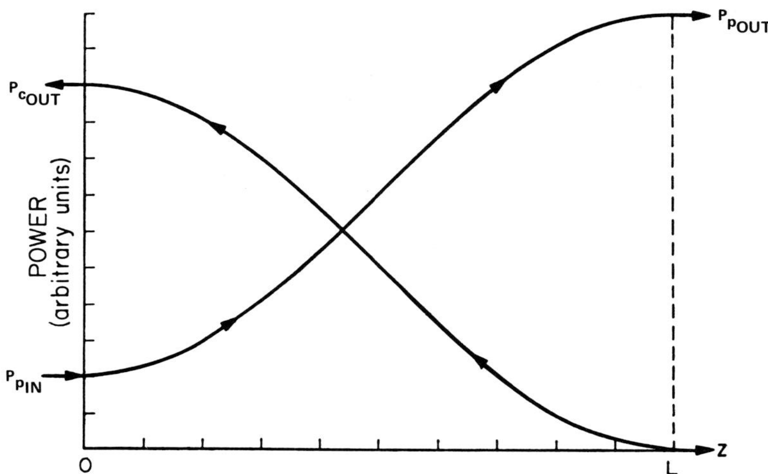

Fig. 3. Amplification by four-wave mixing. The transmission gain T is $|\mathscr{E}_p(L)/\mathscr{E}_p(0)|^2$, the conjugate-wave (power) gain R is $|\mathscr{E}_c(0)/\mathscr{E}_p(0)|^2$. For this calculation $T = 5$ and $R = 4$. [After Yariv (1978).]

The field distribution inside the interaction region for a value of $|\kappa|L$ satisfying

$$\pi/4 < |\kappa|L < 3\pi/4 \tag{34}$$

is shown in Fig. 3. In this regime, the intensity of the reflected wave exceeds that of the input wave, and the device functions as both an *amplifying phase-conjugate reflector* and a *coherent transmission amplifier.*

When

$$|\kappa|L = \pi/2, \tag{35}$$

$$\mathscr{E}_c(0)/\mathscr{E}_p(0) = \infty, \qquad \mathscr{E}_p(L)/\mathscr{E}_p(0) = \infty, \tag{36}$$

which corresponds to oscillation; that is, finite outputs exist (at $z = 0$ and $z = L$) for zero input intensity of the probe wave. Thus, the four-wave mixing process is capable of oscillation without mirror feedback. The field distribution at the oscillation condition is illustrated in Fig. 4.

Oscillation without mirrors due to backward-wave propagation is well known in the microwave art, where it is used in backward-traveling wave oscillators. It has been proposed in the case of three-wave parametric interactions by Harris (1966) and in the case of Brillouin scattering by Sealer and Hsu (1965).

The main advantage of four-wave mixing over the three-wave method

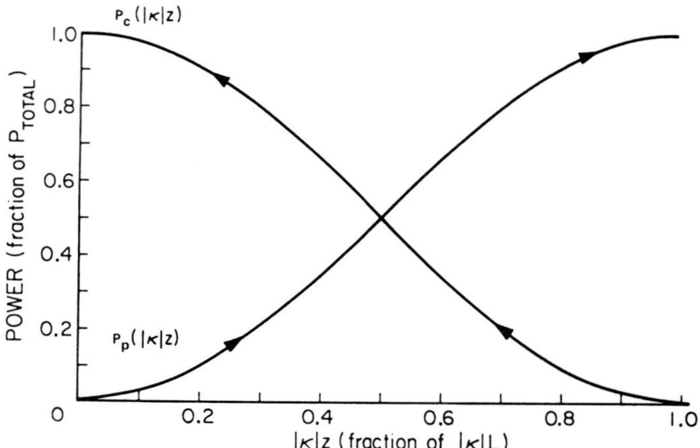

Fig. 4. Field distribution inside the interaction region when the oscillation condition $|\kappa|L \simeq \pi/2$ is satisfied. [After Yariv and Pepper (1977).]

described in Section II.B involves phase matching. For the present case, the condition for a cumulative buildup of $\mathscr{E}_c$ is that

$$\mathbf{k}_1 + \mathbf{k}_2 = \mathbf{k}_p + \mathbf{k}_c. \tag{37}$$

In the canonical geometry of Fig. 1, $\mathbf{k}_1 + \mathbf{k}_2 = 0$. Thus, the output wave $\mathscr{E}_c$ will be generated with its propagation vector $\mathbf{k}_c = -\mathbf{k}_p$; i.e., it will propagate exactly opposite to $\mathscr{E}_p$. In addition, the solution (32) shows that, at the input plane, $\mathscr{E}_c(0)$ is proportional to the complex conjugate of $\mathscr{E}_p(0)$.

These two key properties of the interaction—the reversal of the **k** vector and the complex conjugation of the spatial amplitude—are *independent* of the direction of the input wave, that is, of $\mathbf{k}_p$. It follows directly from the linearity of Eqs. (29) that if the input wave is not planar but possesses a complex wave front (but is still of a single frequency ω), each of the plane-wave components making up the complex wave front is reversed and conjugated. The result of the nonlinear interaction is a backward wave at ω whose wave fronts everywhere *coincide* with those of the input wave.[4] Stated formally: if the (weak) input probe wave at $z < 0$ is given by

$$E_p(\mathbf{r}, t) = \mathrm{Re}[\mathscr{E}_p(x, y, z) \exp i(\omega t - \mathbf{k} \cdot \mathbf{r})], \tag{38}$$

[4] This statement assumes that the input plane of the conjugator is infinite in extent; hence, diffractive effects are not considered.

the generated wave at $z < 0$ is

$$E_c(\mathbf{r}, t) = \mathrm{Re}[-i(\kappa^*/|\kappa|) \tan(|\kappa|L)\, \mathscr{E}_p^*(x, y, z) \exp i(\omega t + \mathbf{k} \cdot \mathbf{r})], \quad (39)$$

which is a *coherent, amplified* (if $|\kappa|L > \pi/4$), *conjugate* replica of the input field.

Numerical Estimate of κ

It follows from Eq. (39) that the parameter that characterizes the nonlinear optical coupling in a given experimental setup is $|\kappa|L$, because the ratio of the phase-conjugated power to that of the incident beam is $\tan^2(|\kappa|L)$. To appreciate the orders of magnitude involved, consider the case of CS_2. Here $\chi_{yxxy}(-\omega, \omega, \omega, -\omega) \simeq 2.73 \times 10^{-13}$ esu, or 3.37×10^{-32} in SI units.[5] The use of χ_{yxxy} makes it possible to have the pump beams ($\mathscr{E}_1$, $\mathscr{E}_2$) polarized at right angles to $\mathscr{E}_p$ and $\mathscr{E}_c$, as will be discussed in Section III.B.

Using Eq. (30) and $\mathscr{E}_1 = \mathscr{E}_2 = \sqrt{2I_1/\epsilon_0 nc}$, we obtain

$$|\kappa| = \frac{2\pi}{n\epsilon_0\lambda}\sqrt{\frac{\mu}{\epsilon}}\,\chi^{(3)}I_1 = 2.68 \times 10^{14}\frac{\chi^{(3)}}{n^2\lambda}\,I_1,$$

where I_1 is expressed in watts per square centimeter. Using $\chi^{(3)} = 6\chi_{yxxy} = 2.025 \times 10^{-31}$ in SI units, $\lambda = 10^{-6}$ m, and $n = 1.62$, we obtain

$$|\kappa| = 0.21 I_1 \text{ MW cm}^{-2}\text{ m}^{-1}.$$

To obtain values of $|\kappa|L \simeq 1$ with, say, $L \simeq 10$ cm requires intensities in the range of tens of megawatts per square centimeter, presently available only from pulsed lasers. The use of resonantly enhanced $\chi^{(3)}$ reduces the power levels to those available from cw (gas) lasers (see Chapter 8).

From a heuristic viewpoint, the geometry of Fig. 1 can be viewed as a "magic box" that retroreflects any ray incident upon it from *any* direction and reverses its phase: the so-called phase-conjugate mirror (PCM). More fundamentally, it retroreflects any plane wave with a reflection coefficient that can assume any value between zero and infinity. (In the case of very large amplification we may no longer assume that $\mathscr{E}_1$ and $\mathscr{E}_2$ are not depleted, and a more exact solution is required; see Chapter 4.)

One of the more astounding consequences of our analytical results concerns the use of a phase-conjugate mirror (PCM) as a laser reflector. This possibility, analyzed and demonstrated experimentally first by AuYeung *et al.* (1979b), is interesting both for its unique new properties and for the fact that many basic rigorous results can be deduced by inspection. Con-

[5] We use $\chi^{(3)}_{\mathrm{SI}} = \chi^{(3)}_{\mathrm{esu}}/3^4 \times 10^{17}$.

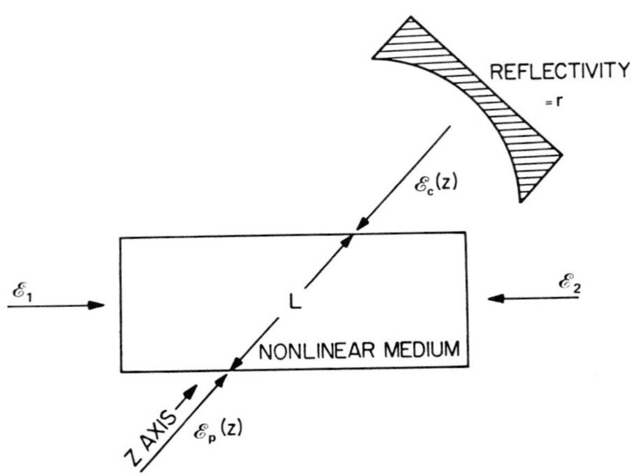

Fig. 5. A laser oscillator whose resonator is bounded by one conventional reflector and one phase-conjugate mirror. [After Yariv and Pepper (1977).]

sider, for example, the laser geometry sketched in Fig. 5. An observer situated within the resonator sees to his right a phase-conjugate mirror with an amplitude reflection coefficient of $\tan(|\kappa|L)$ and to his left a conventional mirror with an amplitude reflectance r. Obviously, a Gaussian beam whose radius of curvatue at the mirror is equal to that of the mirror is a *self-consistent beam solution,* because it is retroreflected and hence unchanged by the PCM. Because the phase shift accrued in passage between the two mirrors reverses sign upon reflection from the PCM, the resonator oscillation frequency (at ω) is *independent* of the separation between the two reflectors and for *all orders of transverse* Gaussian beam modes.

The self-consistent round-trip unity gain condition for oscillation threshold becomes

$$|r| \tan(|\kappa|L) = 1. \tag{40}$$

In terms of the pump intensities $I_{1,2} = (nc/2)\epsilon_0|\mathscr{E}_{1,2}|^2$, the threshold condition is given by

$$(I_{1,2})\text{th} = (\epsilon_0^2 n^2 c^2/\omega\chi^{(3)}L) \tan^{-1}(1/|r|). \tag{41}$$

Note that in the limit of $r = 0$, i.e., no regular reflector or output coupler, the oscillation condition is $|\kappa|L = \pi/2$ [recall Eq. (36)], whereas for a perfectly reflecting mirror (i.e., $r = 1$) $|\kappa|L = \pi/4$.

A detailed account of the theory of lasers employing phase-conjugate mirrors is given in Chapter 13.

D. Phase Conjugation in Kerr-like Media Having Nonsaturable Background Losses

In many situations of interest, the linear (nonsaturable) absorption of the interacting fields must be considered in addition to the Kerr-like nonlinearity. This situation can occur if the nonlinear interaction takes place in an elongated, relatively transparent medium, e.g., an optical waveguide possessing both linear (Rayleigh) scattering losses and a nonlinear index. In what follows, we will consider the effects of linear losses upon all four interacting fields. For simplicity, we shall assume that all fields are collinear, with an interaction length L.[6] Accounting for linear losses, we modify Eqs. (29) to read

$$\frac{d\mathscr{E}_{\mathrm{p}}}{dz} = -i\kappa^* e^{-\alpha L/2}\mathscr{E}_{\mathrm{c}}^* - \frac{\alpha}{2}\mathscr{E}_{\mathrm{p}}, \qquad \frac{d\mathscr{E}_{\mathrm{c}}}{dz} = i\kappa^* e^{-\alpha L/2}\mathscr{E}_{\mathrm{p}}^* + \frac{\alpha}{2}\mathscr{E}_{\mathrm{c}}, \tag{42}$$

where α is the (linear) nonsaturating background (intensity) absorption coefficient or scattering loss that characterizes the nonlinear medium. The new terms on the far right are needed to obtain the simple exponential attenuation in the free propagation case ($\kappa = 0$). The factor $\exp(-\alpha L/2)$ in the coupling terms is due to the fact that now

$$\mathscr{E}_1\mathscr{E}_2 \rightarrow \mathscr{E}_1(0)e^{-\alpha z/2}\mathscr{E}_2(L)e^{\alpha(z-L)/2} = \mathscr{E}_1(0)\mathscr{E}_2(L)e^{-\alpha L/2},$$

so that κ in Eq. (42) is given by

$$\kappa = (\omega/2)\sqrt{\mu/\epsilon}\,\chi^{(3)}\mathscr{E}_1^*(0)\mathscr{E}_2^*(L). \tag{43}$$

The solution of Eq. (42) is facilitated by a change of variables

$$\mathscr{E}_{\mathrm{p}} \rightarrow A(z)e^{-\alpha z/2}, \qquad \mathscr{E}_{\mathrm{c}} \rightarrow B(z)e^{\alpha(z-L)/2}. \tag{44}$$

Equations (42) become

$$\frac{dA^*}{dz} = i\kappa B e^{\alpha(z-L)}, \tag{45}$$

$$\frac{dB}{dz} = i\kappa^* A^* e^{-\alpha z}. \tag{46}$$

The solution of these equations subject to boundary conditions $B(L) = 0$ and a given $A(0)$ is

$$A(z) = \frac{2e^{\alpha z/2}A(0)\{-(\alpha/2)\sin[\kappa_{\mathrm{eff}}(z-L)] + \kappa_{\mathrm{eff}}\cos[\kappa_{\mathrm{eff}}(z-L)]\}}{\alpha\sin(\kappa_{\mathrm{eff}}L) + 2\kappa_{\mathrm{eff}}\cos(\kappa_{\mathrm{eff}}L)}, \tag{47}$$

[6] An extension of this analysis in which each pair of counterpropagating waves travels a different distance in the interaction volume is treated by Pepper (1980).

$$B(z) = \frac{e^{-\alpha z/2} 2i\kappa^* A^*(0) \sin[\kappa_{\text{eff}}(z - L)]}{\alpha \sin(\kappa_{\text{eff}} L) + 2\kappa_{\text{eff}} \cos(\kappa_{\text{eff}} L)}, \tag{48}$$

where

$$\kappa_{\text{eff}} = \sqrt{|\kappa|^2 e^{-\alpha L} - (\alpha/2)^2}. \tag{49}$$

The complex reflection coefficient of the phase conjugator is thus

$$r = \frac{\mathscr{E}_c(0)}{\mathscr{E}_p^*(0)} = \frac{B(0)e^{-\alpha L/2}}{A^*(0)} = \frac{-2i\,\kappa^*\, e^{-\alpha L/2} \tan(\kappa_{\text{eff}} L)}{\alpha \tan(\kappa_{\text{eff}} L) + 2\kappa_{\text{eff}}}. \tag{50}$$

It follows that the (mirrorless) oscillation condition, $r = \infty$, now becomes

$$\tan[\sqrt{|\kappa|^2 e^{-\alpha L} - (\alpha/2)^2}\, L] = -2\frac{\sqrt{|\kappa|^2 e^{-\alpha L} - (\alpha/2)^2}}{\alpha}. \tag{51}$$

By increasing $|\kappa|$ ($\propto \mathscr{E}_1 \mathscr{E}_2$), i.e., by increasing the pump intensity, oscillation can be achieved at any value of the loss constant α. A necessary condition for oscillation is

$$|\kappa| e^{-\alpha L/2} > \alpha/2, \tag{52}$$

so that, beyond a certain value of loss, the increase in length (for a given pump intensity) results in a decreased reflectivity.

For small losses, $\alpha \ll |\kappa|$, the oscillation condition (51) becomes

$$(|\kappa| L)_{\text{osc}} = \sqrt{\pi^2/4 + (\alpha L/2)^2}\; e^{\alpha L/2}. \tag{53}$$

We therefore see that the value of the nonlinear gain $|\kappa| L$ required for a given nonlinear reflection coefficient (or for the onset of mirrorless oscillation) increases rapidly from that of the lossless case when *linear* losses are also present. This is an important consideration in the design of practical phase conjugators, especially when long optical fibers or strongly absorbing nonlinear materials are used. *Nonlinear* losses (due to the conversion of pump photons to the conjugate and probe fields) will be discussed in Chapter 4.

E. Optical Phase Conjugation via Nearly Degenerate Four-Wave Mixing

So far, we have considered the case where the three input frequencies are equal, $\omega_1 = \omega_2 = \omega_p = \omega$, so that the output frequency $\omega_c = \omega_1 + \omega_2 - \omega_p = \omega$, the so-called degenerate frequency interaction. In this section, we consider a nondegenerate case, where $\omega_1 = \omega_2 = \omega$, $\omega_p \neq \omega$. The analysis follows closely that of Pepper and Abrams (1978). The basic geometry and notation are similar to those in Fig. 1, except that the weak input probe field $\mathscr{E}_p$ is at a frequency $\omega + \delta$, so that the output frequency $\omega_c = \omega + \omega - (\omega + \delta) = \omega - \delta$. We also assume that $|\delta/\omega| \ll 1$: the so-called nearly degenerate four-wave mixing interaction.

The basic geometry is shown in Fig. 6. The fields are taken as

$$E_i(\mathbf{r}, t) = \tfrac{1}{2}\mathscr{E}_i(r_i)\exp[i(\omega_i t - \mathbf{k}_i \cdot \mathbf{r})] + \text{c.c.}, \tag{54}$$

where r_i is the distance along $\mathbf{k}_i$. Without loss of generality, we assume that $\chi^{(3)}$ is polarization, frequency, and spatially (i.e., homogeneous and isotropic) invariant. The nonlinear polarization that couples fields $\mathscr{E}_p$ and $\mathscr{E}_c$ is given by

$$\begin{aligned} P_{\mathrm{NL}}(\omega_c = \omega - \delta) = \tfrac{1}{2}\chi^{(3)}\mathscr{E}_1\mathscr{E}_2\mathscr{E}_p^* &\times \exp(i\{[\omega + \omega - (\omega + \delta)]t \\ &- [\mathbf{k}_1 + \mathbf{k}_2 - \mathbf{k}_p]\cdot \mathbf{r}\}) + \text{c.c.} \end{aligned} \tag{55}$$

Because $\mathbf{k}_1 + \mathbf{k}_2 = 0$, the resultant nonlinearly generated conjugate output field that minimizes the phase mismatch will propagate along a direction opposite that of $\mathscr{E}_p$, as shown in Fig. 6. Forming a similar nonlinear polarization at frequency ω_p, and following the procedure of Yariv and Pepper (1977), results in the set of coupled-mode equations

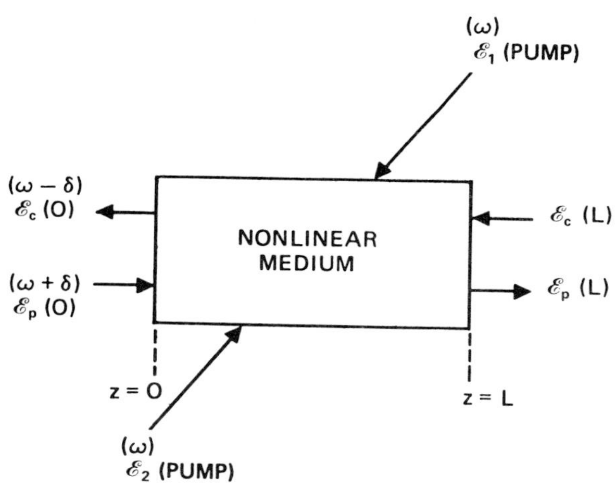

Fig. 6. The basic geometry of phase conjugation via nearly degenerate four-wave mixing. In this case, an incident probe wave, whose frequency $\omega + \delta$ is slightly detuned from that of the pump waves (both at frequency ω), will result in a conjugate wave with an "inverted" frequency $\omega_c = \omega - \delta$. [After Pepper and Abrams (1978).]

$$\frac{d\mathscr{E}_p}{dz} = -i\kappa_p^* \mathscr{E}_c^* \exp(i\,\Delta k\, z),$$
$$\frac{d\mathscr{E}_c}{dz} = i\kappa_c^* \mathscr{E}_p^* \exp(i\,\Delta k\, z), \tag{56}$$

where $\kappa_\ell^* = (\omega_\ell/2)\sqrt{\mu/\epsilon}\;\chi^{(3)}\mathscr{E}_1\mathscr{E}_2$ gives the complex coupling coefficient. Owing to the frequency difference of $\mathscr{E}_p$ and $\mathscr{E}_c$, there is now a nonzero phase mismatch Δk, the magnitude of which is given by

$$|\Delta k| = 2n\pi(|\Delta\lambda|/\lambda^2) = 2n|\delta|/c. \tag{57}$$

As we will show, this phase mismatch results in a decreased phase-conjugate reflectivity if the frequency detuning δ differs appreciably from zero. Thus, the conjugator will behave as a narrow-bandpass mirror or optical filter. Note that $|\Delta\lambda|$ is the wavelength difference between the fields $\mathscr{E}_p$ and $\mathscr{E}_c$. In arriving at Eq. (56), we assumed that the pump fields $\mathscr{E}_{1,2}$ were nondepleted, and we used the SVEA technique, $|d^2\mathscr{E}_i/dz^2| \ll |k_i d\mathscr{E}_i/dz|$.

The solutions to Eqs. (56) using the boundary conditions $\mathscr{E}_p(z = 0) \equiv \mathscr{E}_p(0)$ and $\mathscr{E}_c(z = L) \equiv \mathscr{E}_c(L)$ are

$$\begin{aligned}\mathscr{E}_p(z) = {}& [\exp(i\,\Delta kz/2)/D]\{-i\kappa_p^* \exp(i\,\Delta kL/2)\sin(\beta z)\,\mathscr{E}_c^*(L)\\ &+ (\beta\cos[\beta(z-L)] + (-i\,\Delta k/2)\sin[\beta(z-L)])\,\mathscr{E}_p(0)\},\\ \mathscr{E}_c(z) = {}& [\exp(i\,\Delta kz/2)/D]\{\exp(-i\,\Delta kL/2)[\beta\cos(\beta z)\\ &- (i\,\Delta k/2)\sin(\beta z)]\,\mathscr{E}_c(L) + i\kappa_c^*\sin[\beta(z-L)]\,\mathscr{E}_p^*(0)\},\end{aligned} \tag{58}$$

where $D \equiv \beta\cos(\beta L) - (i\,\Delta k/2)\sin(\beta L)$ and $\beta \equiv \sqrt{\kappa_p\kappa_c^* + (\Delta k/2)^2}$.

For the filter application, we assume that $\mathscr{E}_c(L) = 0$, with only a single weak input at $z = 0$, the probe wave. In this case, the nonlinearly reflected wave at the input plane ($z = 0$) becomes

$$\mathscr{E}_c(0) = \frac{-i\kappa_c^*\tan(\beta L)\,\mathscr{E}_p^*(0)}{\beta - (i\,\Delta k/2)\tan(\beta L)}. \tag{59}$$

We can now appreciate several filter characteristics of the four-wave interaction. First $\mathscr{E}_c(0) \propto \mathscr{E}_p^*(0)$, implying the near time-reversed nature of the filter output.[7] Hence, through spatial filtering, the signal-to-noise ratio of the filter can be improved. For example, passing the input signal (to be

[7] Because $\Delta k \neq 0$, the net phase acquired by a probe wave will not be exactly canceled by the conjugate wave.

In most cases, the residual phase shift acquired over a given path is inconsequential. The presence of higher-order modes in a phase conjugate resonator is, however, a situation where these phase shifts come into play (see Chapter 13 and 14).

filtered) through various optical elements (spatial filters, lenses, etc.) will result in a time-reversed, filtered field; in contrast, undesirable noise terms (e.g., Rayleigh-scattered fields) will be minimized on passage through the given optical train. Second, the conjugate wave $\mathscr{E}_c(0)$ can be greater in amplitude than the input field (i.e., amplification) for the proper range of κ_ℓ and Δk. Also, from Eq. (55), we note that the output frequency is downshifted by the same amount the input frequency is upshifted with respect to the pump wavelength, and vice versa.

A useful limit is that of weak nonlinear coupling, i.e., $|\kappa_\ell/\Delta k| \ll 1$. In this case, the power reflection coefficient, defined as

$$R \equiv |\mathscr{E}_c(0)/\mathscr{E}_p(0)|^2, \tag{60}$$

becomes

$$R \to |\kappa L|^2[\sin(\Delta k\ L/2)]^2/(\Delta k\ L/2)^2. \tag{61}$$

In Eq. (61) and in what follows, we set $\kappa_p \simeq \kappa_c \equiv \kappa$, because $|\delta/\omega| \ll 1$. We note that the $\mathrm{sinc}^2(x)$ dependence of Eq. (61) is a typical result of coherent, phase-mismatched, nondepleted interactions (see, for example, Chapter 10 of Yariv, 1975). This result can be obtained directly by assuming in Eqs. (56) that $d\mathscr{E}_p/dz = 0$ (i.e., weak perturbation of the input fields) and simply integrating the remaining differential equation for $\mathscr{E}_c(z)$.

The filter-bandpass characteristic is obtained by using Eq. (59) in the defining expression for the power-reflection coefficient, Eq. (60), yielding

$$R = \frac{|\kappa L|^2 \tan^2(\beta L)}{|\kappa L|^2 + (\Delta k\ L/2)^2 \sec^2(\beta L)}. \tag{62}$$

Using Eq. (62), we plot in Fig. 7 the power-reflection coefficient R versus a normalized wavelength-detuning parameter Ψ for several values of the nonlinear gain $|\kappa|L$. By definition, $\Psi = (\Delta\lambda/2)(2nL/\lambda^2)$, which is also equal to the phase mismatch $\Delta k\ L$ divided by 2π. The wavelength-detuning parameter $\Delta\lambda/2$ corresponds to the difference in wavelength of the probe field $\mathscr{E}_p$ relative to the pump fields $\mathscr{E}_{1,2}$. For an interaction length L of 1 cm, a wavelength λ of 0.5 μm, and a linear index of refraction n of 1.62 (that of CS_2), a value of unity in Fig. 7 corresponds to a wavelength detuning $\Delta\lambda/2$ of 0.0772 Å, or $\Delta\nu = 9.26$ GHz. Note that the $\mathrm{sinc}^2(x)$ nature of the response holds only in the limit of weak nonlinear coupling, $|\kappa/\Delta k| \ll 1$. As $|\kappa|L$ increases, the bandpass becomes more sharply peaked, with the zeros of the response occurring at decreasing values of the frequency offset. This is because the amplitude of the output wave increases with $|\kappa|L$, and because the zeros of the tangent in Eq. (62) occur at smaller values of $\Delta kL/2$ as $|\kappa|L$ increases. As $|\kappa|L > \pi/4$, the filter is seen

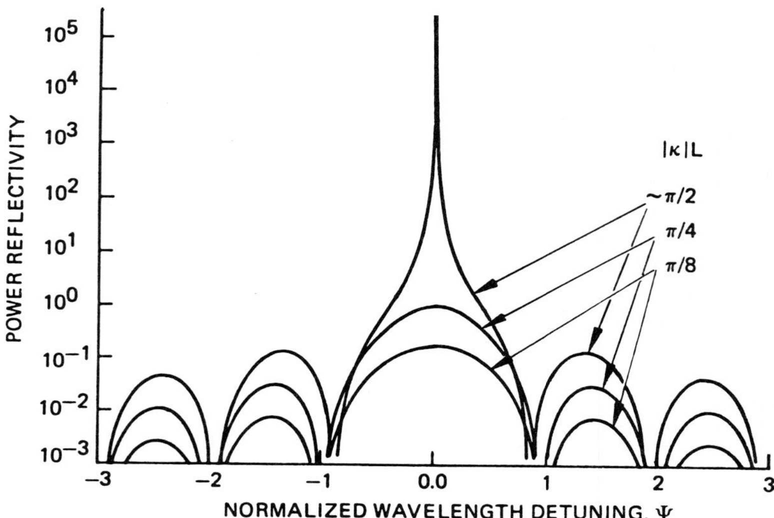

Fig. 7. Power reflectivity R versus normalized wavelength detuning Ψ for several valuse of nonlinear gain $|\kappa|L$. For the example given in the text, unity along the abscissa corresponds to $\Delta\lambda/2 = 0.0772$ Å, or $\Delta\nu = 9.26$ GHz. $\Psi = (\Delta\lambda/2)(2nL/\lambda^2)$. [After Pepper and Abrams (1978).]

to exhibit a power-reflection coefficient exceeding unity over regions of the bandpass.

The wavelength response of the filter becomes apparent if we recast the family of curves in Fig. 7 normalized to unity power reflectivity, as shown in Fig. 8. Several prominent features are to be noted. First, as $|\kappa|L$ increases, the bandwidth dramatically decreases. Second, the side-lobe structure of the filter also decreases with increasing nonlinear gain, yielding a sharper, better-defined, and amplified bandpass. Physically, these features follow if we recognize that the filter is analogous to a real-time distributed Bragg-reflecting resonator with an internal gain medium. Because the finesse of a resonator increases as we add gain to the cavity, a sharper response (or Q) results. We see that, as the oscillation condition is satisfied, the bandpass approaches zero, being ultimately limited by the linewidth and/or the coherence length of the pump sources.

Because the net phase mismatch is a function of the total interaction length L, one can realize an extremely narrow filter bandpass by using an elongated nonlinear medium, for example, a long optical fiber. Moreover, with an optical fiber one can realize reasonable values of the nonlinear coupling coefficient κ with modest pump powers (owing to the small cross sections of the fiber). Optical phase conjugation using optical fibers will be discussed in more detail in Chapter 5.

In addition to the ability of phase-matching constraints to yield an optical filter, the use of *resonantly enhanced* nonlinear media can exhibit even sharper filter characteristics owing to the narrow linewidth of the transition(s) involved. Furthermore, resonantly enhanced nonlinear media typically require much less pump power to achieve reasonable nonlinear gains. (See Chapter 8 for more details.)

The above formalism is also applicable to the analysis of multiwavelength conjugation processes. As an example, consider a situation where the pump waves $\mathscr{E}_1$, $\mathscr{E}_2$ and/or the probe field $\mathscr{E}_p$ consist of a group of closely spaced optical frequency components, as in the case of a multiline molecular or chemical laser. Now, owing to the nonlinear interaction

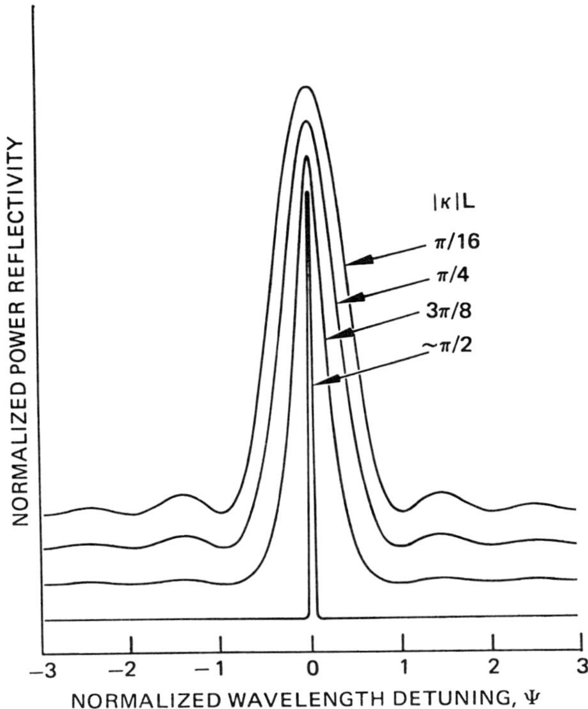

Fig. 8. Power reflectivity normalized wavelength detuning Ψ for several values of nonlinear gain $|\kappa|L$. All curves are normalized to unity power reflectivity to emphasize the frequency bandpass of the interaction. [After Pepper and Abrams (1978).]

coupling these three (multifrequency) incident fields, there exists a multiplicity of sum and difference frequencies, yielding a set of degenerate as well as nearly degenerate phase-conjugate wave combinations. In general, the nondegenerate modes are undesirable because of their imperfect phase-reversal properties and/or undesirable conjugate-wave frequency components. Using the above theoretical developments, we can predict and hence calculate the power ratio of the desirable (i.e., degenerate, time-reversed propagation) to the undesirable output from the phase conjugator as a function of the experimental parameters $\mathscr{E}_{1,2}$, L, etc.

In conclusion, the process of nearly degenerate four-wave mixing can yield an active, narrow optical bandpass filter. The *interaction* has a large ($\sim 4\pi$) field of view and a frequency response that depends on both the interaction length and the pump intensity for a given medium. The actual field of view for a given *device* will, however, be limited by the specific optical detection scheme used and the degree of spatial filtering. The tradeoff is that spatial filtering of the time-reversed, phase-conjugate output wave can be used to discriminate against unwanted noise sources at the expense of the field of view of the system. In either case, the filter can yield an amplified bandpass (for increased nonlinear gains).

The complex filter reflectivity, Eq. (59), can be useful in analyzing many other OPC applications. Specifically, it provides insight into the transient properties of a phase conjugator (see Chapter 3), and can serve in evaluating several temporal domain applications of OPC (see Chapter 14).

F. Comparisons of Holography with Four-Wave Mixing

There is a fundamental relationship between holography and the four-wave phase conjugation process based upon the geometry of Fig. 1. To demonstrate this analogy, consider the conventional holographic procedure illustrated in Fig. 9. The first step, the conventional recording of a thin hologram using the interference between a reference beam $\mathscr{E}_1$ and a signal beam $\mathscr{E}_p$, is shown in Fig. 9a. The resulting transmission function is

$$T \propto (\mathscr{E}_p + \mathscr{E}_1)(\mathscr{E}_p^* + \mathscr{E}_1^*) = |\mathscr{E}_p|^2 + |\mathscr{E}_1|^2 + \mathscr{E}_p\mathscr{E}_1^* + \mathscr{E}_1\mathscr{E}_p^*, \qquad (63)$$

where $\mathscr{E}_1(x, y)$ and $\mathscr{E}_p(x, y)$ denote the complex amplitudes of the reference and object fields, respectively, in the hologram plane $z = 0$.

In the reconstruction step, the hologram is illuminated by a single reference wave $\mathscr{E}_2$ impinging from the right in a direction opposite to that of $\mathscr{E}_1$, as shown in Fig. 9b. We thus have $\mathscr{E}_2 = \mathscr{E}_1^*$, so that the diffracted field to

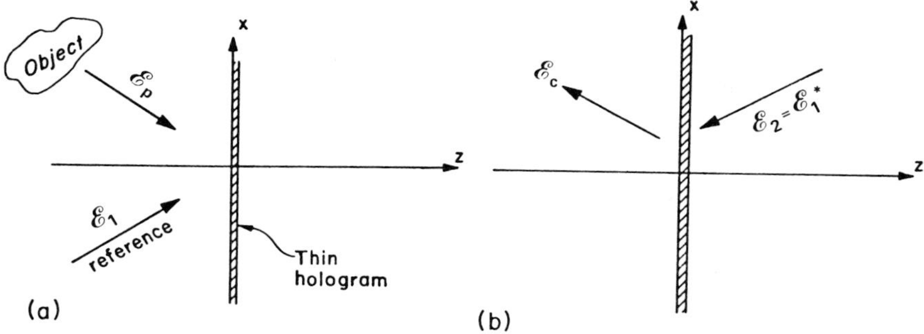

Fig. 9. A holographic (a) exposure and (b) reconstruction. [After Yariv (1978).]

the left of the hologram is

$$\begin{aligned}\mathscr{E}_c &= T\mathscr{E}_2 \propto (|\mathscr{E}_p|^2 + |\mathscr{E}_1|^2 + \mathscr{E}_p\mathscr{E}_1^* + \mathscr{E}_p^*\mathscr{E}_1)\mathscr{E}_1^* \\ &= (|\mathscr{E}_p|^2 + |\mathscr{E}_1|^2)\mathscr{E}_1^* + (\mathscr{E}_1^*)^2\,\mathscr{E}_p + |\mathscr{E}_1|^2\,\mathscr{E}_p^*.\end{aligned} \tag{64}$$

The first term on the right-hand side of Eq. (64) is proportional to the incident field $\mathscr{E}_2$ $(=\mathscr{E}_1^*)$ and is of no interest in this discussion. The term $(\mathscr{E}_1^*)^2\mathscr{E}_p$ will, in a thick hologram, have a phase factor $\exp[i(2\mathbf{k}_1 - \mathbf{k}_p)\cdot\mathbf{r}]$ and is thus phase mismatched; i.e., it will not radiate. The term of interest is

$$\mathscr{E}_c \propto |\mathscr{E}_1|^2\,\mathscr{E}_p^* = |\mathscr{E}_1\mathscr{E}_2|\mathscr{E}_p^*, \tag{65}$$

which at $z < 0$ corresponds to a time-reversed phase-conjugate replica of the original object field $\mathscr{E}_p$. In holographic terms, this output wave is the pseudoscopic image of $\mathscr{E}_p$. Relation (65), derived for conventional holography, is to be compared with that for four-wave mixing. For $|\kappa|L \ll 1$ (weak hologram) we may rewrite Eq. (32) as (using $\tan\theta \sim \theta$ for small θ)

$$\mathscr{E}_c(0) = -i(\kappa^*L)\,\mathscr{E}_p^*(0) \propto \mathscr{E}_1\mathscr{E}_2\mathscr{E}_p^*(0),$$

which is identical in form to Eq. (65). As a matter of fact, if we superimpose Figs. 9a and 9b, the resulting figure will be identical to Fig. 1, which shows the basic geometry for four-wave mixing. There is, however, a practical difference of major importance. In four-wave mixing, it is not necessary to interrupt the process to develop the hologram, then to place it back in position, and finally to illuminate it in the reconstruction process. All of these steps occur in real time. Using the grating point of view in holography, we can represent the process of phase conjugation, shown in Fig. 1, as a simultaneous recording and reading of two sets of gratings, as illustrated in Fig. 10. The first grating is shown in Fig. 10a,

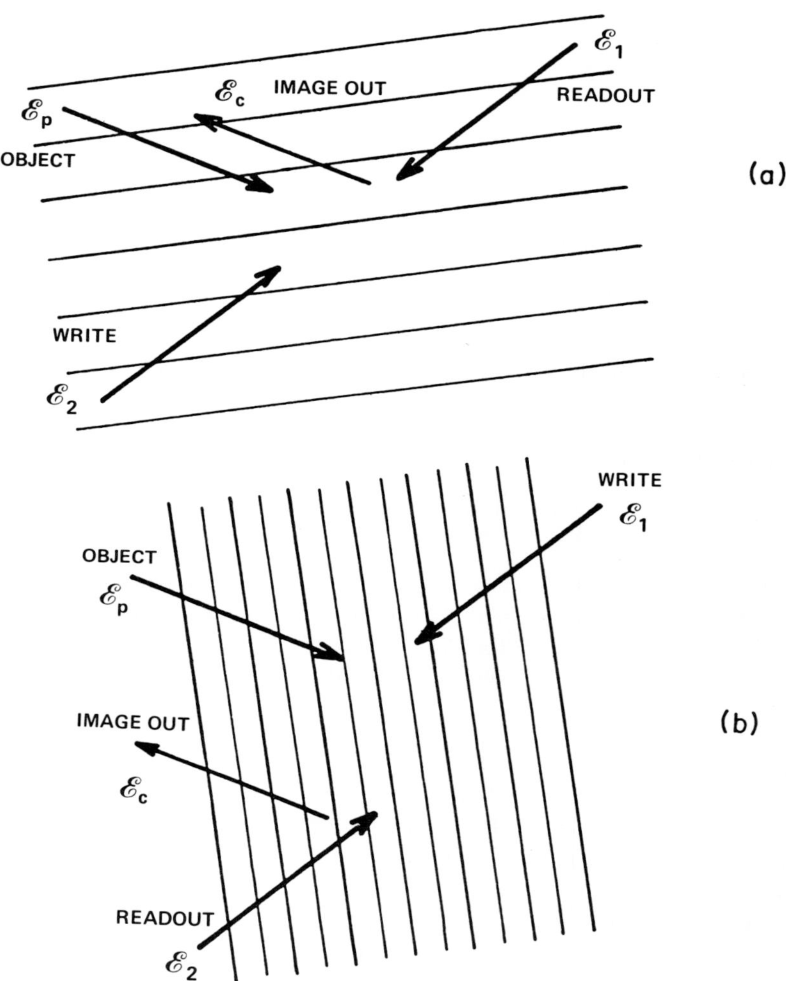

Fig. 10. Schematic diagrams of the two simultaneous spatial gratings formed by the probe wave and each of the pump waves, leading to the pseudoscopic image, or conjugate wave $\mathscr{E}_c$. (a) The waves $\mathscr{E}_p$ and $\mathscr{E}_2$ interfere to form a grating, which is read out by pump wave $\mathscr{E}_1$. (b) The grating is formed by $\mathscr{E}_p$ and $\mathscr{E}_1$, and read out by pump wve $\mathscr{E}_2$. [After Pepper (1982).]

and is produced by the interference between the signal (or probe) wave $\mathscr{E}_p$ and the pump wave $\mathscr{E}_2$ (note the large grating period). The phase-conjugate return wave $\mathscr{E}_c \propto \mathscr{E}_p^*$ is produced by Bragg diffraction of the pump wave $\mathscr{E}_1$ from this grating. At the same time, as shown in Fig. 10b, another component of $\mathscr{E}_c$ is generated by Bragg diffraction of the other pump wave $\mathscr{E}_2$ from a small-period grating produced by the interference of $\mathscr{E}_p$ with the pump wave $\mathscr{E}_1$. The total radiated conjugate field $\mathscr{E}_c$ is due to the coherent superposition of both processes. Those processes are both accounted for by the formalism leading to the coupled-mode equations (29).

The above discussion reveals the close analogy between phase-conjugate four-wave mixing and holography. This analogy cannot be carried too far, because there are basic differences. In conventional holography, one modifies in a semipermanent fashion some gross feature of the recording medium by exposing the medium to the periodic interference pattern of the object and reference beams. The modified feature may involve the transparency of the medium (by causing the development of silver grains on changing the atomic populations) or its index of refraction (phase hologram). The object and reference beams must therefore be of the same frequency; otherwise their interference leads to a moving grating, and thus to a uniform exposure of the recording medium, which washes out the hologram.

We saw in Section II.E, that the pump waves $\mathscr{E}_{1,2}$ and signal wave $\mathscr{E}_p$ need not be of the same frequency. This causes the gratings of Fig. 10 to move with a uniform velocity. The frequency shift between $\mathscr{E}_p$ and $\mathscr{E}_c$ can then be viewed as a Doppler shift from a moving grating.

Another difference between holography and four-wave mixing concerns the tensor properties of the nonlinear susceptibility $\chi_{ijk\ell}$. In an isotropic medium, for example, the *vector* form of the nonlinear polarization becomes

$$\mathbf{P}_{\mathrm{NL}} = \tfrac{1}{2}[A(\mathbf{E}_1 \cdot \mathbf{E}_p^*)\mathbf{E}_2 + B(\mathbf{E}_2 \cdot \mathbf{E}_p^*)\mathbf{E}_1 + C(\mathbf{E}_1 \cdot \mathbf{E}_2)\mathbf{E}_p^*] + \text{c.c.} \tag{66}$$

The coefficients A, B, and C depend on the properties of the specific nonlinear medium (such as the Doppler-broadened linewidth and magnetic coherence effects) and the relative angles of the interacting fields. In Chapter 8 the physics of these constants is discussed in more detail. These constants are manifested in the tensor elements of $\chi_{ijk\ell}$. (The form of $\mathbf{P}_{\mathrm{NL}}$ is not quite so simple in nonisotropic media.)

The first two terms in Eq. (66) have analogs in the above-mentioned grating picture. The first term, for example, leads to a spatial grating (stationary in time) formed by the interference of the fields $\mathbf{E}_1$ and $\mathbf{E}_p^*$; this grating is read out by the field $\mathbf{E}_2$. The second term is similar to the first. Both terms require that the field polarization of the waves responsible for the formation of the grating have a nonzero overlap (by virtue of the dot product). This requirement is also necessary in conventional holography to generate a (spatial) interference pattern in the recording medium.

The third term in Eq. (66) has no holographic analog. This term involves the dot product of the *pump waves*. The grating formed by the two pump waves is *not* equivalent to a spatial interference pattern; what is formed is a *temporally* modulated grating stationary in space. It is the probe wave that scatters off this "breathing" grating (at 2ω), which leads to the conjugate wave. Hence, one can realize a conjugate wave via this mechanism even if the probe wave is *orthogonally* polarized with respect to *both* pump waves. The necessary conditions are that the last term exist in Eq. (66) (i.e., $C \neq 0$) and that the pumps have a nonzero overlapping polarization component (i.e., $\mathbf{E}_1 \cdot \mathbf{E}_2 \neq 0$). This effect can be useful in isolating the probe–conjugate pair from the pump waves using polarization discrimination techniques (see Section III.B for experimental details). In addition, because this grating is stationary in space, the system is free from the deleterious atomic-motionally-induced (or frequency-induced) washout effects.

Another fundamental difference exists between conventional holography and four-wave mixing. It can be revealed by a quantum-mechanical consideration of the four-wave mixing process. The basic relationship (dropping the c.c. and assuming copolarized fields)

$$P_{\mathrm{NL}} = \chi^{(3)} E_1 E_2 E_{\mathrm{p}}^*$$

is consistent with an interaction Hamiltonian between the EM fields and the atomic medium of the form $\mathbf{P}_{\mathrm{NL}} \cdot \mathbf{E}$, or

$$\mathcal{H}' = \chi^{(3)} E_1 E_2 E_{\mathrm{p}}^* E_{\mathrm{c}}^*. \tag{67}$$

We consider a system comprised of an atomic medium characterized by the nonlinear coefficient $\chi^{(3)}$ and four electromagnetic wave modes $\mathscr{E}_1$, $\mathscr{E}_2$, $\mathscr{E}_{\mathrm{p}}$, and $\mathscr{E}_{\mathrm{c}}$. Each field is characterized by a harmonic-oscillator wave function $|n\rangle$, where n is the number of quanta in the mode. The total EM field is thus given by $|n_1, n_2, n_{\mathrm{p}}, n_{\mathrm{c}}\rangle$. The electric field of, say, mode $\mathscr{E}_1$ can be written in the quantum-mechanical operator form

$$\mathbf{E}_1(\mathbf{r}, t) = -i\hat{e}_1 \sqrt{\frac{\hbar\omega_1}{2V\epsilon}}\,(a_1{}^\dagger e^{-i\mathbf{k}\cdot\mathbf{r}} - a_1 e^{i\mathbf{k}\cdot\mathbf{r}}), \tag{68}$$

where V is the quantization volume, $a_1{}^\dagger$ the creation operator for field 1, and a_1 the corresponding annihilation operator. We can associate the field amplitude $\mathscr{E}_1$ with $a_1{}^\dagger$ and $\mathscr{E}_1^*$ with a_1, because of the Heisenberg representation of $a_j{}^\dagger$ and a_j, which shows them to be of the form $a_j{}^\dagger(t) = a_j{}^\dagger(0)e^{i\omega_j t}$ and $a_j(t) = a_j(0)e^{-i\omega_j t}$. The Hamiltonian (67) is thus of the form

$$\mathcal{H}' \propto a_1{}^\dagger a_2{}^\dagger a_{\mathrm{p}} a_{\mathrm{c}}. \tag{69}$$

According to Fermi's golden rule, $\mathcal{H}'$ can cause transitions of the type

$$\langle n_1 - 1, n_2 - 1, n_{\mathrm{p}} + 1, n_{\mathrm{c}} + 1 | a_1{}^\dagger a_2{}^\dagger a_{\mathrm{p}} a_{\mathrm{c}} | n_1, n_2, n_{\mathrm{p}}, n_{\mathrm{c}}\rangle \tag{70}$$

between an initial state with n_1, n_2, n_p, and n_c quanta (photons) in modes $\mathscr{E}_1$, $\mathscr{E}_2$, $\mathscr{E}_p$, and $\mathscr{E}_c$, respectively, and a final state in which the number of quanta is $n_1 - 1$, $n_2 - 1$, $n_p + 1$, and $n_c + 1$, respectively. The basic process therefore involves an annihilation of one photon from mode 1 (at ω_1) and one from mode 2 (at ω_2) and a simultaneous creation of one photon at ω_p and one at ω_c. This is consistent with the classical result, Eqs. (32) and (33), which shows that the input wave $\mathscr{E}_p$ and output wave $\mathscr{E}_c$ are more intense at their output planes than at their input planes. This increase in power is, of course, at the expense of the two pump modes.

The conservation of photon energy takes the form

$$\omega_1 + \omega_2 = \omega_p + \omega_c \tag{71}$$

and the conservation of photon momentum the form

$$\mathbf{k}_1 + \mathbf{k}_2 = \mathbf{k}_p + \mathbf{k}_c; \tag{72}$$

both conditions were derived earlier by our classical approach.

In holography, on the other hand, waves $\mathscr{E}_p$ and $\mathscr{E}_1$ are not present in the medium at the same time as waves $\mathscr{E}_2$ and $\mathscr{E}_c$, so that the above picture does not apply. This leads to profound differences in the quantum optical fluctuations associated with these processes. Initial theoretical investigations have been made in this area by Yuen and Shapiro (1979). Shapiro (1980) has since proposed the use of these quantum optical processes, via the so-called two-photon coherent states (TCS) [see, e.g., Yuen (1976)], in such diverse applications as low-noise quantum optical detectors and waveguide taps with minimal insertion losses. In addition, Caves (1980) has proposed the use of a TCS scheme as a part of a low-noise, high-sensitivity laser interferometer–gravitational detector. In Chapter 14 these applications are discussed in greater detail.

III. Experimental Studies of Optical Phase Conjugation via Elastic Photon Scattering

In Section II, we discussed the basic concepts and theory of phase conjugation using nonlinear optical mixing in nonresonant, transparent media. The nonlinear coupling constant κ characterizing the interaction between the probe and conjugate waves is a function of the nonlinear optical susceptibility $\chi^{(3)}$, and a property of the phase-conjugator material. The nonlinear optical susceptibility is assumed to be a nondispersive (i.e., independent of frequency), unsaturable numerical constant or tensor. In four-wave mixing, this process is mediated by the nonlinear refractive index (the optical Kerr effect). In this section we will focus on the experimental aspects and observation of (forward and backward) optical phase conjugation via three- and four-wave mixing in such transparent, nonlinear media. The experimental results will be compared with the preceding theoretical models when appropriate.

In what follows, we will review the relevant details of experimental configurations necessary to realize conjugate wavefront generation. Techniques to be discussed include the angular separation of the pump and probe waves, as well as polarization discrimination (where the tensor aspects of $\chi^{(3)}$ are exploited). The observation of optical parametric oscillation via degenerate four-wave mixing will then be reviewed. This will be followed by a discussion of novel experimental geometries such as laser intracavity phase conjugation and four-wave mixing in optical waveguides. We will then review such potentially important demonstrations as wave-front reversal of arbitrarily polarized probe waves and multiwavelength optical phase conjugation; the former may find use in the dynamic compensation for laser systems incorporating undesirable poor-quality stress-induced birefringent optical elements, and the latter may be of significance in high-power, multiline (e.g., chemical or exciplex) laser schemes.

Classes of experiments involving resonant media (such as atomic vapors and semiconductors) will be introduced primarily for the sake of comparison, completeness, and historical perspective. Detailed theoretical and experimental discussions of these specific media are relegated to subsequent chapters. Owing to the abundance of recent experimental investigations (see the comprehensive bibliography at the end of this book for a listing of current works), the results discussed in this section should be considered as representative; space limitations make it impractical to review all the published material.

Before discussing various experimental details and configurations, we mention some of the material species (and associated physical phenomena) useful as nonlinear media for nonresonant phase conjugation via four-wave mixing. As mentioned in Section II.F, the nonlinear interaction can be viewed as a real-time holographic process involving the spatial and/or temporal interference (or beating) of two input fields within a medium. This results in a corresponding modulation of the refractive index of the material which is read out, or diffracted, by the third input (pump) field.

There are basically two classes of effects that can yield the desired refractive index modulation. The first involves the response of a material to the *local field strength* (or intensity). This phenomenon is called the Kerr effect. Examples of physical processes that can be assigned to this category include *orientation effects,* such as occur in glasses, liquids, or liquid crystals [see Hellwarth (1977) for details]; *saturation effects,* where atoms and/or molecules can have their refractive index modified owing to the local field intensity (we assume operation far from material resonances, so that the medium is essentially lossless and broadband); and *band-structure distortion effects,* occurring in semiconductors (see Chapter 12), where the incident optical field modifies the local energy-gap configuration, which can, in turn, affect the refractive index of the semiconductor (we again assume operation well within the band gap, far removed from any resonances). These effects can be phenomenologically cast in terms of a nonlinear refractive index n_2. The total local index is therefore given by

$$n = n_0 + n_2 \langle |\mathbf{E}|^2 \rangle, \tag{73}$$

where n_0 is the linear refractive index of the material and $\langle |\mathbf{E}|^2 \rangle$ the time-averaged local intensity (which can be due to any number of optical fields present in the medium). The nonlinear susceptibility for these materials is in the range of $\sim 10^{-3}$–10^{-12} esu, whereas the response times are in the range of $\sim 10^{-9}$–10^{-12} sec. Chang (1981) reviews classes of such fast-responding nonlinear refractive index species.

The second class of effects responsible for nonresonant four-wave nonlinear optical interactions involves the response of a material to *nonlocal fields*. Physical processes relating to this phenomenon include *thermal effects,* where even the mild absorption of light can lead to a temperature gradient, which, through the temperature-dependent refractive index, can result in a spatial modulation of the index (i.e., a phase grating) [see, e.g., Martin *et al.* (1979)]; *photorefractive effects,* where the presence of an optical interference pattern in a material can modify the refractive index via the electro-optic effect [see, e.g., Feinberg *et al.* (1980), and Chapter 11]; and *electrostrictive effects,* where electric field gradients can modify the density of nonabsorbing scattering centers, and hence spatially modulate the refractive index. Crystals such as $Bi_{12}SiO_{20}$, BGO, $LiNbO_3$, and $BaTiO_3$ are examples of photorefractive materials, whereas vapors, aerosols, and suspensions [see, e.g., Palmer (1979, 1980) and Smith *et al.* (1981)] are examples of electrostrictive media. Although one cannot actually represent these processes directly in terms of a nonlinear susceptibility (owing to the nonlocality of the interaction), their "effective" susceptibilities can be rather large ($\sim 10^{-2}$–10^{-5} esu), with typically slow response times (10–10^{-4} sec).

Optical phase conjugation has been observed in spectral regions ranging from the ultraviolet, throughout the visible, and into the infrared; has been reported in numerous states of matter, including gases, liquids, solids and semiconductors, plasmas, liquid crystals, and aerosols; and has used cw, pulsed, and both monochromatic and multiwavelength sources. [See Pepper (1982) for a tabulation of these parameters; see also the bibliography at the end of this volume.]

A. Phase Conjugation via Angular Separation of Pump and Probe–Conjugate Fields

In this subsection, we will discuss experimental geometries that employ an angular separation of the pump waves from the probe–conjugate fields to observe phase-conjugate replicas. We limit the present discussion to degenerate four-wave mixing processes. We describe a typical experimental demonstration showing the properties of the phase-conjugate wave, along with parametric dependencies of the interaction (e.g., the pump intensity and interaction length), which will be compared with theory.

Two typical experimental geometries that have been used to demonstrate the aberration-correction properties of the phase-conjugate wave are shown in Fig. 11. The output from a (pulsed or cw) laser is divided into three beams $\mathscr{E}_1$, $\mathscr{E}_2$, and $\mathscr{E}_p$, which are directed onto a medium possessing a Kerr-like third-order nonlinear susceptibility $\chi^{(3)}$. For simplicity, we consider the case where all three fields have parallel polarizations. Two of the three beams form the (intense) counterpropagating pump fields ($\mathscr{E}_{1,2}$). The third (weak) field is used as the probe wave ($\mathscr{E}_p$), whose conjugate replica is sought. The two geometries differ in that the counterpropagating pump wave $\mathscr{E}_2$ is generated either separately (Fig. 11a) or by retroreflecting the first pump wave $\mathscr{E}_1$ after passage through the nonlinear medium (Fig. 11b). The choice of these two geometries depends on experimental parameters such as the laser coherence length and the linear losses within the nonlinear medium.

The phase-conjugate wave $\mathscr{E}_c$, which will retrace the path of the probe wave $\mathscr{E}_p$, is detected by observing its transmission through beam splitter BS_1. Note that the probe wave interacts with the nonlinear medium at some finite angle with respect to the axis of the pump waves. This leads to an angular offset of the conjugate wave from the direction of propagation of the pump waves, thus enabling the spatial discrimination of the various fields. The cost of this convenience is a reduction of the overlap region of the intracting fields, thus limiting the interaction length and hence the nonlinear reflection coefficient attainable. This constraint can be relaxed via polarization discrimination of the various fields; this technique is discussed in the next subsection.

A striking demonstration of the conjugate nature of $\mathscr{E}_c$ can be realized by placing a phase aberrator ϕ in the path of $\mathscr{E}_p$. This aberrator is typically a piece of etched (in, e.g., HF) or frosted glass. Typical results of such a

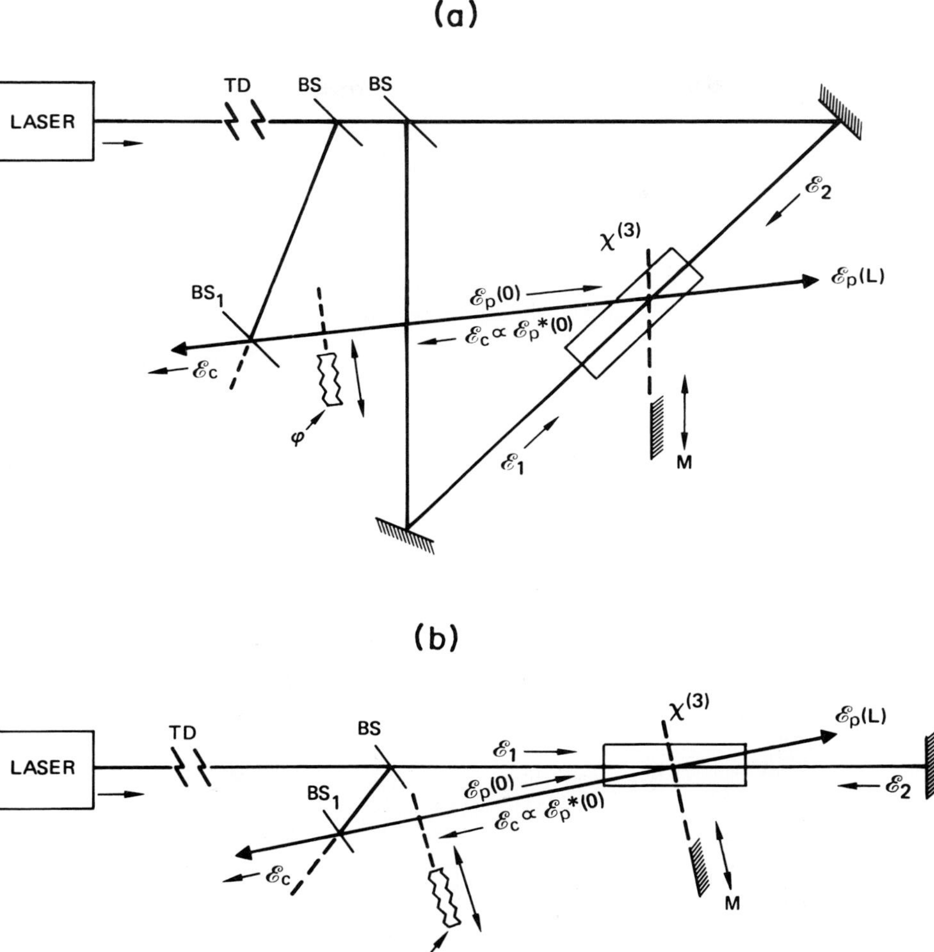

Fig. 11. Typical experimental geometries for the observation of optical phase conjugation via four-wave mixing. M_i = mirror, BS_i = beam splitter, ϕ = phase aberrator, TD = time delay (for pulsed operation) or optical isolator (for cw operation). [After Pepper (1982).]

demonstration, using the geometry of Fig. 11a, are shown in Fig. 12. For this representative experiment [Jain and Lind (1983)], a semiconductor-doped glass sheet was used as the nonlinear medium, and a Q-switched ruby laser provided the source for the interacting fields. Figure 12a is a photograph (taken in the far field) of the probe beam before it impinges

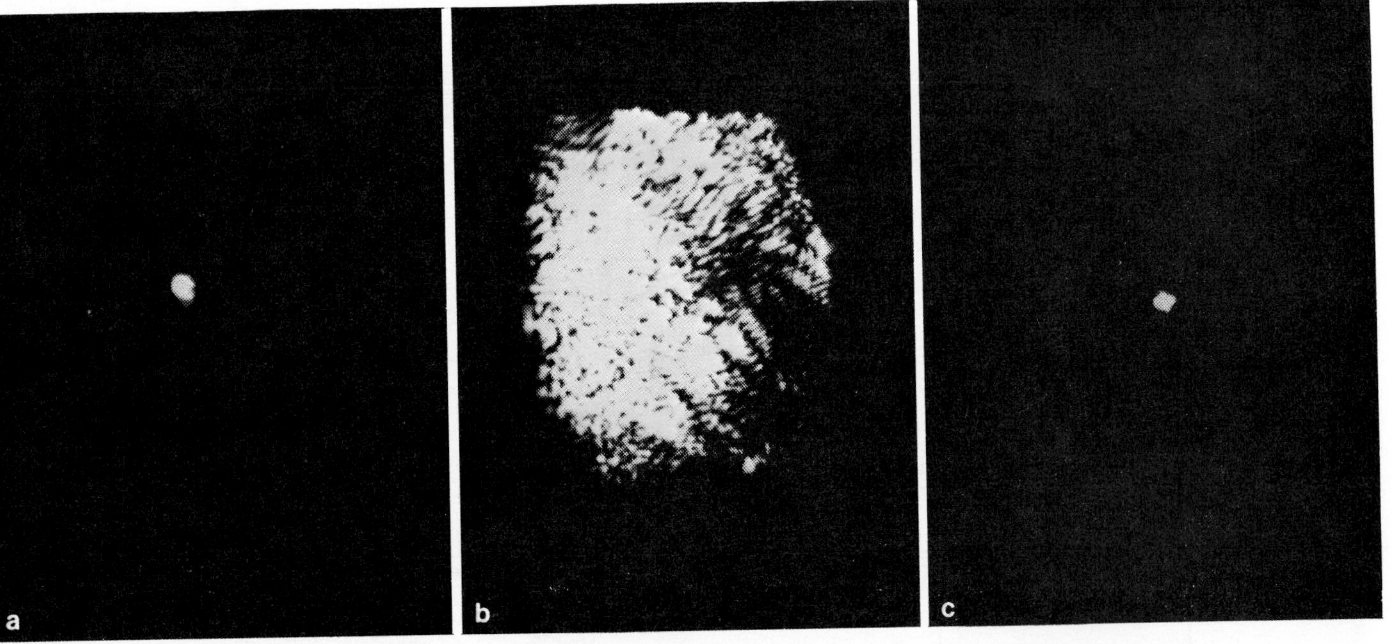

Fig. 12. The ability of the conjugate beam to correct for distortions: (a) input laser beam, (b) distorted beam after single passage through the aberrator, (c) conjugate beam after reverse passage back through the aberrator. [After Jain and Lind as appeared in Giuliano (1981).]

upon the aberrator. To appreciate the degree of aberration that the etched plate imposes on the probe field, a photograph of the probe wave after passage through the aberrator is shown in Fig. 12b. Figure 12c is a photograph of the conjugate beam, taken after interaction of the aberrated probe beam with the conjugator, subsequent passage of the nonlinearly generated conjugate wave back through the aberrator, and propagation through the beam splitter BS_1. The restoration of the originally aberrated beam back to a *near perfect* TEM_{00} *spot is most convincing*. Note that if the conjugator is replaced with an ordinary plane mirror (M), the incident (initially aberrated) probe beam is reflected and, after passage back through the distorter, becomes doubly aberrated.

There are other, less dramatic tests that further experimentally verify the nature of the nonlinear interaction responsible for the generation of the conjugate wave. One test is to examine the dependence of the conjugate-wave reflection coefficient R and probe-wave transmission coefficient T as a function of the pump-wave intensity I. From Eqs. (30), (32), and (33), we see that these observables scale as $R \sim \tan^2(|\kappa|L)$ and $T \sim \sec^2(|\kappa|L)$, where $|\kappa| \propto I$. A typical experimental verification of this behavior is shown in Fig. 13, where R and T are plotted as a function of I. For this representative experiment [Bloom *et al.* (1978)], a cell containing sodium vapor was used as the nonlinear medium; a pulsed dye laser (tuned near but not at the sodium D_1 resonance at 5895 Å) served as the source. The solid and dashed lines correspond to the above functional dependence of T and R, respectively, when, in the theory, a value of the nonlinear index[8] $n_2 = 1.1 \times 10^{-8}$ esu is used. The deviation of theory from experiment (at $I > 30$ kW/cm^2) is due to the fact that the sodium resonance, being essentially a two-level system, is a saturable medium. In this regime, terms higher than $\chi^{(3)}$ in the nonlinear polarization [Eq. (2)] come into play. (Such atomic and molecular resonances are treated in detail in Chapter 8.)

The experimental data in Fig. 13 also demonstrate that the four-wave mixing process is capable of yielding an *amplified* phase-conjugate replica of the input probe wave. This is in accordance with the result derived in Section II C for values of the nonlinear gain $|\kappa|L$ greater than $\pi/4$. In addition, the experimental results demonstrate the ability of the nonlinear interaction to yield (coherent) image amplification of the forward-going probe wave, also in agreement with the theory presented above.

Another verification of the nonlinear nature of the interaction involves the temporal and spatial nature of the conjugate wave. Because the non-

[8] The nonlinear index is related to the third-order susceptibility as $\chi^{(3)} = (n_0 n_2)/(2\pi)$ (in esu), where n_0 is the linear index of refraction of the material.

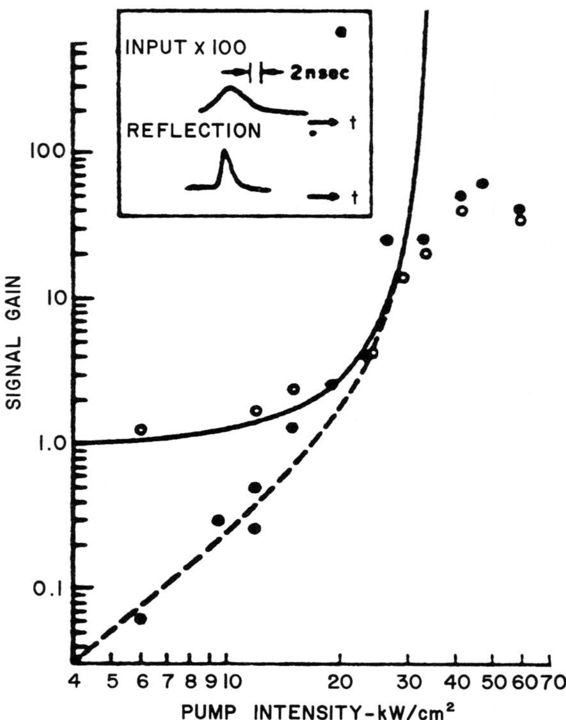

Fig. 13. Signal gain versus pump intensity: —●—, the measured gain of the backward wave; —○—, the measured gain of the transmitted probe wave. The curve is the theoretical fit to the transmitted (backward) wave data. Inset: oscilloscope traces of the input probe pulse and backward-generated wave. [After Bloom *et al.* (1978).]

linear polarization that gives rise to the time-reversed wave is a product of three fields [recall Eq. (4)], the conjugate-wave intensity should be proportional to the (integrated) temporal product of these input waves. Assuming an instantaneous response of the nonlinearity, the conjugate wave should be temporally shorter than the input pulses. This behavior is clearly shown in the inset of Fig. 13, where the temporal dependence of the probe and conjugate waves is displayed.

The spatial nature of the conjugate wave can also reveal information about the nonlinear interaction. If one assumes Gaussian spatial profiles for the three input waves, the near-field spatial extent (i.e., the spot size) of the conjugate wave should be reduced. This conclusion follows if one models the phase-conjugate mirror as being Gaussian tapered in a spatial sense (in terms of the nonlinear reflection coefficient) by virtue of being formed by a pair of spatially Gaussian pump waves [see, for example, Pepper (1980, Appendix 7B)].

We conclude this section by briefly mentioning various experimental constraints necessary for the realization of efficient, time-reversed wave-front generation. First, the source laser should operate in a single longitudinal and transverse mode. This maximizes the laser coherence length, thus enabling one to use longer interaction lengths as well as relaxing path-length difference constraints. Second, in nonguided geometries, it is necessary to have nearly planar (e.g., spatially filtered) pump waves for efficient and precise conjugation. Finally, one must take precautions to isolate the pump laser from the experimental apparatus; retroreflection of the various beams back into the laser may affect its mode properties and/or coherence characteristics.

Several other factors concerning the experimental geometry must be considered. First, the propagation directions of the two pump waves must be antiparallel to optimize the conjugate refection coefficient R. This follows from an examination of Fig. 8: if $\mathbf{k}_1 + \mathbf{k}_2 \neq 0$ (even if $\omega_1 = \omega_2$), a phase mismatch results, thereby reducing the efficiency of the PCM. More important, angular offsets of the pump waves can result in a conjugate wave that does not exactly retrace the path of the incident probe wave. This can lead to imperfect compensation (see Chapter 14). We also note that, for mirrorless oscillation to occur (i.e., $R \rightarrow \infty$), the phase mismatch must be identically equal to zero. Next, in most pulsed experiments, the maximum temporal overlap of the three waves in the nonlinear medium is necessary for efficient conjugate-wave generation. In addition, there are system tradeoffs with respect to the experimental geometry. The optimal choice of the two geometries shown in Fig. 11 depends primarily on the nonsaturable linear losses α in the nonlinear medium. For example, if $\exp(\alpha L) > \frac{1}{4}$, where L is the interaction length, then the geometry in Fig. 11b is preferred. With respect to the aperture of the conjugator, diffraction effects become important if the minimum desired resolution element is smaller than $\mathrm{L}\lambda/a$, where L, λ, and a, are the propagation distance (between the conjugator and aberrator), wavelength, and conjugator aperture, respectively. Care must also be taken with respect to the ratio of the amplitudes of the pump waves, $\mathscr{E}_1/\mathscr{E}_2$, and to the maximum pump wave intensity. The former parameter affects the value of R (especially in resonantly enhanced systems), whereas the latter parameter can result in undesirable nonlinear effects (nonlinear phase shifts, self-focusing, stimulated effects, etc.).

Also of significance is the probe-to-pump-wave amplitude ratio: nonlinear pump depletion effects, which occur at large reflection coefficients, can lead to degradation of the contrast level of the PCM. The optimum ratios of the various fields are dependent on both the physical nature of the nonlinear susceptibility and the nonlinear gain, given a fixed laser output power. As an example, consider the small nonlinear gain limit, where $\tan(|\kappa|L) \sim |\kappa|L$. The conjugate intensity $|\mathscr{E}_c|^2$ therefore scales as $|\mathscr{E}_1\mathscr{E}_2\mathscr{E}_p|^2$. In this limit, one requires equal intensities for the three input waves to optimize $|\mathscr{E}_c|^2$. Finally, one must be cognizant of the tensor aspects of the susceptibility (and therefore the polarization states of the input fields) and of its temporal response time.

The above experimental requirements can be satisfied in most experimental situations. The many successful experimental demonstrations of phase conjugation to date in a host of nonlinear media and using a variety of pulsed and cw laser sources attest to this fact.

B. Phase Conjugation via Polarization Discrimination of Pump and Probe–Conjugate Fields

As discussed in Section III.A, the requirement of angular separation between the pump and probe–conjugate field pairs necessarily limits their mutual overlap within the nonlinear medium. This geometrical constraint reduces the maximum attainable nonlinear reflection coefficient for a given pump-wave intensity. This situation can be circumvented by using a collinear pump–probe geometry and *polarization discrimination* to distinguish between the pump and conjugate waves. This technique allows one to realize long interaction lengths, limited now by the generally less restrictive laser coherence length and/or the linear loss within the nonlinear medium. The use of orthogonally polarized beams is possible if the nonlinear medium possesses a third-order susceptibility tensor with appreciable off-diagonal tensor elements. Below, we discuss a typical experimental geometry and the underlying phenomenological theory, and conclude with experimental details and results. The experiment that will serve as an example was performed by Pepper *et al.* (1978). In their experiment, the nonlinear medium was carbon disulfide, an isotropic transparent (Kerr-type) liquid; a passively Q-switched ruby laser, operating in a single longitudinal and transverse mode, served as the source. This experiment represents the first observation of phase-conjugate reflection coefficients in excess of unity using a nonresonant, Kerr-like medium as the conjugator.

The experimental arrangement is shown in Fig. 14. It consists of a CS_2 cell, 40 cm long, sandwiched between two laser Glan prisms P_1 and P_3. This nonlinear medium is pumped simultaneously by laser beams $\mathscr{E}_1$ and $\mathscr{E}_2$, both of the same frequency ω and linearly polarized in the x direction

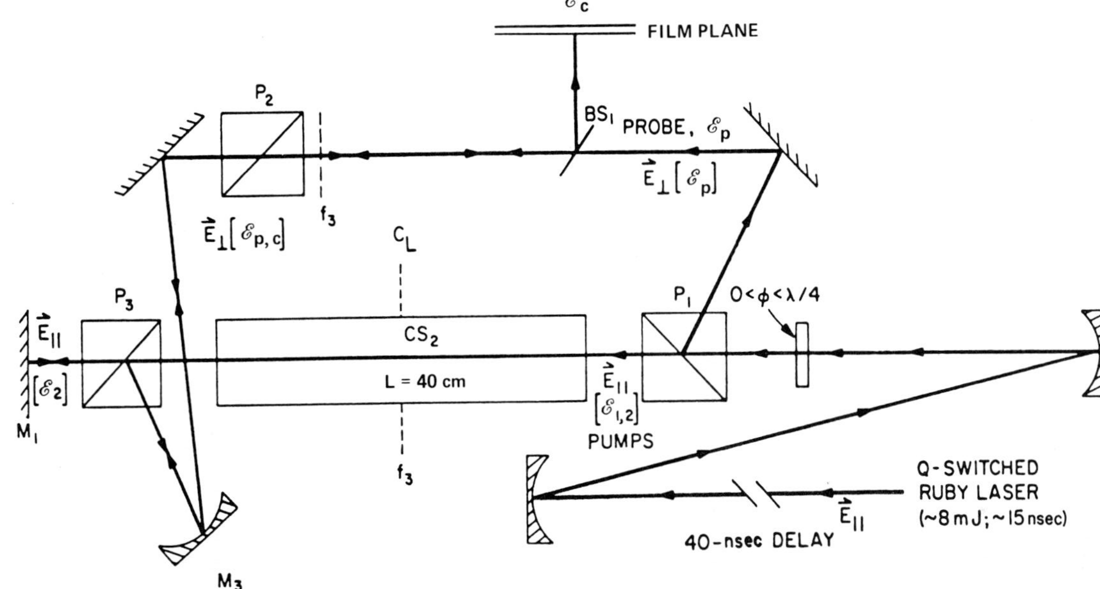

Fig. 14. Experimental apparatus for the observation of phase-conjugate replicas using polarization discrimination to separate the pumps from the probe–conjugate waves. Laser Glan prisms P_i are used to separate these pairs of orthogonally polarized fields. A variable wave plate ϕ is used to generate the (weak) probe wave having a desired amplitude. [After Pepper *et al.* (1978).]

(parallel, π polarization). These two beams form the pump waves, which propagate in opposition to each other. Simultaneously, a probe beam $\mathscr{E}_p$, which is orthogonally polarized in the y direction (s polarization), and also of frequency ω, is introduced from the left through the laser Glan prism P_3 and propagates in a direction collinear with and antiparallel to $\mathscr{E}_1$. The wave plate ϕ determines the pump-to-probe-wave amplitude ratio. The prisms P_1 and P_3 constrain the probe beam to interact only within the CS_2 cell. The mirror M_3 (with focus at plane f_3) ensures that the probe wave is contained within the pump-beam volume.

The experiment consists of (a) measuring the intensity of the nonlinearly reflected beam $\mathscr{E}_c$ as a function of the pumping intensity ($\propto \mathscr{E}_1\mathscr{E}_2$) and (b) establishing that the output field $\mathscr{E}_c$ is the phase conjugate of $\mathscr{E}_p$.

To understand the coupling among the triplet $\mathscr{E}_1$, $\mathscr{E}_2$, and $\mathscr{E}_p$ in the CS_2 cell, we must consider the tensor aspects of $\chi^{(3)}$. It can be shown [recall Eq. (9)] that the nonlinear coupling between the various waves for the above-mentioned experimental conditions is given by

$$\mathscr{P}_y(\omega_c = \omega = \omega_1 + \omega_2 - \omega_p) \propto \chi^{(3)}_{yxxy}\mathscr{E}_{1,x}(\omega_1 = \omega) \times \mathscr{E}_{2,x}(\omega_2 = \omega)\mathscr{E}^*_{p,y}(-\omega_p = -\omega). \quad (74)$$

The conjugate wave $\mathscr{E}_c$, which is radiated by $\mathscr{P}_y$, is therefore linearly polarized in the y direction. This output field is thus polarized parallel to that of the probe field and orthogonal to that of the pump waves. Hence, it can easily be coupled out of the system (and separable from the pump waves) through the Glan prism P_1 and beam splitter BS_1.

The use of a collinear geometry for both the probe–conjugate fields and the pump waves $\mathscr{E}_{1,2}$ enables one to use very long interaction path lengths and thus realize high gains (and possibly oscillation). This geometry entails the use of the off-diagonal tensor element $\chi^{(3)}_{yxxy}$, which, in the case of CS_2, is related to the diagonal susceptibility element[9] by $\chi^{(3)}_{yxxy}/\chi^{(3)}_{xxxx} \sim 0.706$. This somewhat reduced value of the coupling is more than compensated for by the increased interaction length afforded by the collinear geometry.

The phase-conjugate reflection coefficient was measured photographically, and the results were fitted to a theoretical curve having the form given by Eq. (32). A least-squares fit to the data, with the susceptibility as a parameter, yielded a value for $\chi^{(3)}_{yxxy}$ consistent with the generally accepted one.

The phase-conjugate nature of the reflected wave, i.e., $\mathscr{E}_c(0) \propto \mathscr{E}^*_p(0)$, was established through the use of the mirror M_3 (see Fig. 14). This mirror

[9] Hellwarth (1977b) discusses the detailed nature of third-order susceptibilities in transparent media.

focuses the collimated input probe beam $\mathscr{E}_p$ on the midplane f_3. A phase-conjugate reflection that is a time-reversed replica of the input wave should emerge from the CS_2 cell with virtual emanation from the focal plane at f_3, and should thus be collimated. This was verified from beam spot photographs taken by reflecting $\mathscr{E}_c$ from BS_1 in the film plane.

The phase-conjugate nature of the reflected wave was also established through its temporal (pulse shortening), spatial (well-defined spot size), and frequency (via Fabry–Perot spectra) characteristics. These tests, along with other diagnostics, revealed the absence of competing, undesirable nonlinear processes, such as stimulated effects or self-focusing.

C. One-Mirror-Assisted Optical Parametric Oscillation

Here, we describe an experiment that demonstrated a one-mirror-assisted optical parametric oscillation mode in a degenerate four-wave mixing geometry. From the coupled-mode analysis described in Section II C, we concluded that the oscillation condition for the geometry considered (i.e., an unloaded phase-conjugate resonator) is given by

$$|\kappa|L = \tan^{-1}(1/|r|), \tag{75}$$

where r is the amplitude reflectivity of the *added mirror* (see Fig. 5).

Equation (75) states that the oscillation condition is reduced by a factor of 2 (compared with the case without an external mirror) to a value of $|\kappa|L = \pi/4$ when the external mirror is totally reflecting ($|r| = 1.0$). This value of the nonlinear gain implies that the onset of oscillation will occur when the conjugate-mirror reflection coefficient is also totally reflecting (or $|\mathscr{E}_c/\mathscr{E}_p| = 1.0$). Physically, this geometry forms a passive resonator, bounded by a conventional mirror on one end and a conjugate mirror on the other.

Self-oscillation was observed by Pepper *et al.* (1978) using the apparatus shown in Fig. 15. The difference between this and the previous experimental geometry (Fig. 14) is twofold. First, prism P_1 is used to couple out the desired (s-polarized) oscillation signal, while this same prism passes the pump beams. Second, the absence of prism P_3 now allows the totally reflecting flat mirror M_1 to serve a dual function: it (1) retroreflects pump beam $\mathscr{E}_1$, thus providing for its counterpropagating component $\mathscr{E}_2$, and (2) serves as a reflector for the orthogonally polarized oscillation field.

At pumping intensities exceeding 8.8 MW/cm^2 and with no input probe field [$\mathscr{E}_p(0) = 0$], an intense (s-polarized) *oscillation* pulse emerged from

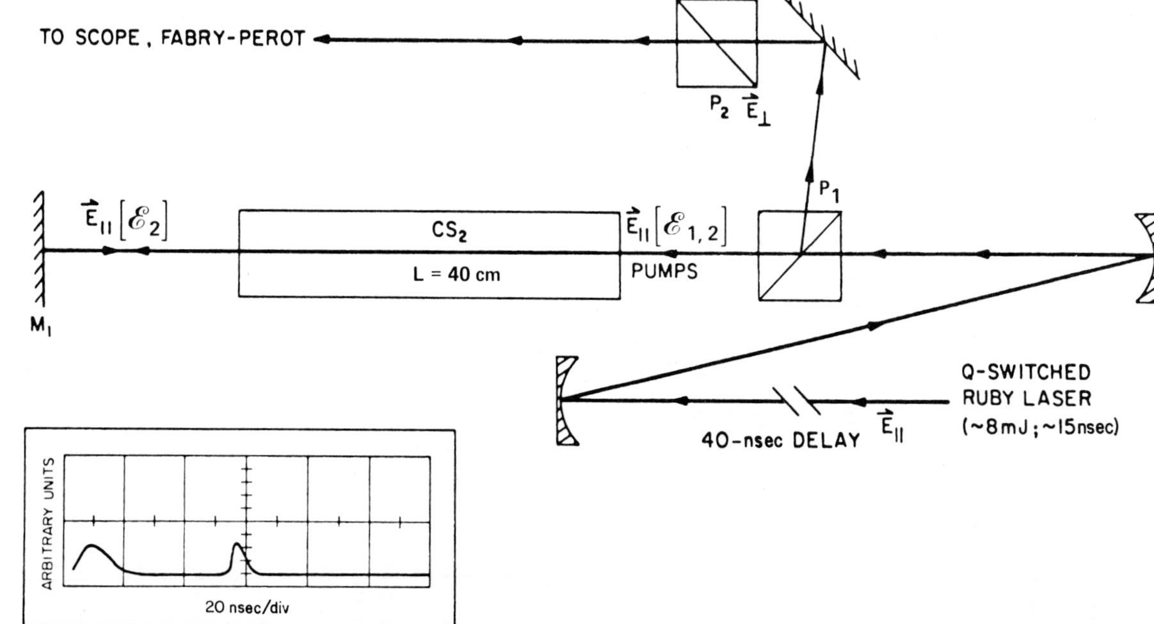

Fig. 15. Experimental apparatus used for the observation of one-mirror-assisted optical parametric oscillation. A passive phase-conjugate resonator is formed by mirror M_1 and a phase-conjugate mirror (the CS_2 cell). The oscillator output is polarized orthogonally to the pump waves and coupled out by the Glan prism P_1. Inset: Time evolution of the pump wave (left) and oscillator output (right). [After Pepper *et al.* (1978).]

prism P_1. The oscillation pulse energy was approximately 1% of the pump energy. The spot size of the oscillation beam was smaller than that of the input pump beam, because the nonlinear coupling constant κ is proportional to the product of two beams $\mathscr{E}_1$ and $\mathscr{E}_2$, each of which has a Gaussian spatial profile.

A typical set of temporal pulse shapes is shown in Fig. 15. The first pulse is the source laser output; the second pulse (properly delayed and of arbitrary amplitude) is the output due to the one-mirror-assisted parametric oscillation mode. The nonlinearity of the interaction is shown by the pulse shortening of the oscillator output. Because the nonlinear gain involves the temporal overlap of the pump beams, the evolution of the gain in the time domain is essentially the temporal convolution of the two Gaussian pulses. This results in a nonlinear gain with a sharper and shorter Gaussian temporal characteristic. Fabry–Perot spectra of these signals verified the degenerate frequency nature of the process. Finally, the threshold behavior of the oscillator output was confirmed by the nonlinear increase of its output as a function of input pump energy. This and the previous experiment clearly demonstrate the utility of polarization discrimination techniques in such nonlinear optical interactions.

D. Observation of Phase-Conjugate Replicas Employing Novel Geometries

In this section, we review several novel experimental geometries in which optical phase conjugation has been demonstrated. We first discuss a class of configurations where optical phase conjugation was realized *within* a laser resonator. In this case, the intense intracavity circulating fields provide the required counterpropagating pump waves. We next discuss a demonstration of phase conjugation in a multimode optical waveguide filled with a nonlinear medium. The properties of optical waveguides—confinement of the waveguide eigenmodes over long distances, and the small transverse dimensions—enable one to obtain efficient phase-conjugate fields with optical (pump-wave) powers orders of magnitude lower than those required in similar bulk media.

1. Laser Intracavity Phase Conjugation

Here we discuss an experiment that demonstrated the possibility of realizing efficient phase conjugation using nonlinear media within a laser cavity. The existence of the counterpropagating waves within an oscillating (standing wave) optical resonator provides a natural set of pump waves necessary to realize a nonlinear optical phase conjugator via four-wave

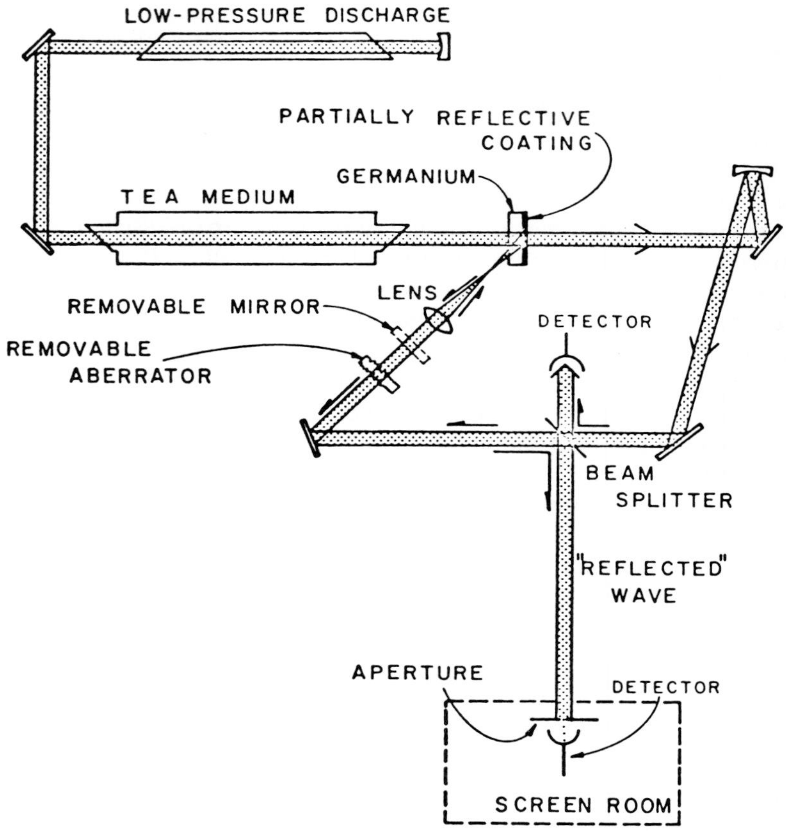

Fig. 16. Experimental apparatus used to observe laser intracavity optical phase conjugation. A low-pressure–high-pressure hybrid CO_2 laser is used in a conventional cavity configuration, except that the germanium output coupler is turned around so that the germanium is *within* the optical cavity. The output from the laser is redirected back through the inner surface of the germanium, overlapping the active region in the sample. The internal angle of overlap is $\sim 10°$. The phase-conjugate reflection occurs within this region of overlap in the germanium. [After Bergmann *et al.* (1978).]

mixing. These fields are in fact autoaligned, in that the laser itself will not oscillate unless the two fields propagate in opposition to each other. Moreover, in this geometry the intracavity waves are typically more intense than the actual laser output. Hence, by placing a nonlinear medium within the laser, greater nonlinear gains can be realized in general.

To illustrate these notions, we describe an experiment performed by Bergmann *et al.* (1978), which used a pulsed CO_2 laser at a wavelength of

10.6 μm. The geometry is sketched in Fig. 16. Germanium served as the intracavity nonlinear medium, with the laser output serving as the probe wave. The source was a pulsed, hybrid low-pressure–high-pressure TEA CO_2 laser operating on the P(20) line. As depicted in Fig. 16, a slightly wedged, intrinsic, polycrystalline germanium flat output mirror (R = 95%) was *reversed* so that both the antireflection-coated surface and the 0.5-cm-thick germanium substrate were *internal* to the optical cavity. Thus, the substrate of the germanium output coupler *itself* served as the nonlinear medium. Germanium was chosen because it is readily available and has an exceptionally large third-order susceptibility ($\chi^{(3)} \simeq 1.5 \times 10^{-10}$ esu). The laser output, which served as the probe wave, was routed by mirrors through a ZnSe beam splitter to impinge upon the active region of the germanium output coupler. The phase-conjugated beam was directed by the beam splitter through an adjustable aperture into a high-speed HgCdTe detector for diagnostics. A removable, flat, 100% reflective mirror could be placed in front of the lens to provide a reference reflection, against which the backscattered signals were compared.

The peak power of the phase-conjugated beam corresponded to a reflectivity of 2%, in accord with theoretical predictions. The canonical test for phase conjugation—the placement of a severely aberrating element in the beam line—had little effect on the nonlinearly reflected signal after repassing the aberrating element, whereas the same aberrator in front of the reference mirror reduced the reflected signal by more than two orders of magnitude.

2. *Phase Conjugation via Four-Wave Mixing in Optical Waveguides*

We now briefly consider phase conjugation via degenerate four-wave mixing in optical waveguides, a matter discussed in greater detail in Chapter 5. Thus far, we have considered nonlinear optical interactions in bulk media. Because of the magnitude of $\chi^{(3)}$ in the media discussed, high optical intensities are required to realize practical nonlinear gains; hence, in bulk media, high-peak-power, Q-switched laser sources and focusing optics must be used. If, on the other hand, we assume the interaction to occur in an optical waveguide (or optical fiber) possessing the same Kerr-like nonlinear susceptibility, we can then realize similar nonlinear reflection coefficients relative to the bulk with a $\sim 10^6$ reduction in rerequired pump powers. This results from the fact that an ideal optical waveguide is capable of maintaining a set of confined eigenmodes over great lengths (~meters), with small transverse dimensions (~micrometers) and with low loss. Therefore, efficient image compensation, filtering, informa-

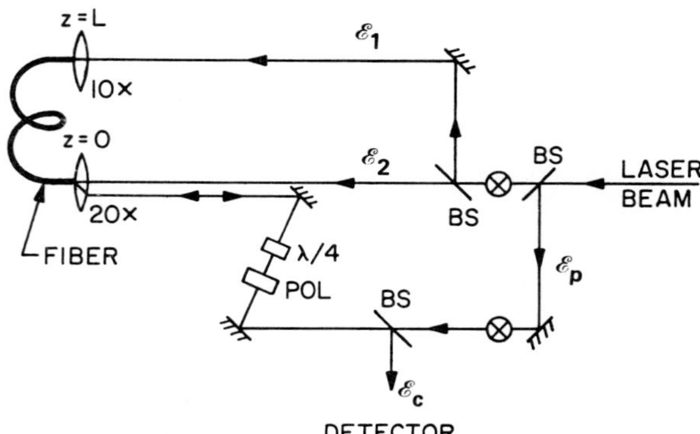

Fig. 17. Experimental apparatus used to observe backward-wave generation via four-wave mixing in a CS_2-filled optical fiber. The optical source is a cw argon ion laser. Microscope objectives are used to couple the various waves to the fiber. A quarter-wave plate (λ/4), combined with a linear polarizer (POL), is used to prevent undesirable Fresnel-reflected beams from reaching the detector; the desired time-reversed wave, however, passes through the isolator. Modulation techniques are also used to detect the conjugate wave (⊗ = chopper). [After AuYeung *et al.* (1979a).]

tion processing, and other optical phase-conjugation applications can be achieved in optical fibers using modest pump powers (~watts), even on a cw basis. Indeed, prior to their use in optical phase conjugation, optical fibers provided an excellent configuration for the realization of many other nonlinear optical processes (e.g., stimulated effects, pulse compression, and parametric interactions).

In this geometry, several experimental parameters must be considered to optimize the overall phase-conjugate efficiency. One desires to (a) match the number of probe-beam spatial resolution elements to the number of guided eigenmodes of the fiber, (b) ensure that the coherence length of the laser(s) employed is greater than any path-length differences throughout the system, and (c) limit the overall fiber length to less than α^{-1}, where α is the linear absorption coefficient (recall Section II D).

To illustrate a typical experimental geometry, we cite the work of AuYeung *et al.* (1979a). As sketched in Fig. 17, a 6-m-long, CS_2-filled optical fiber (4 μm i.d.) was used as the nonlinear medium, and a cw argon ion laser (operating in a single transverse and longitudinal mode) was used as the source. Polarization and temporal discrimination, as well as spatial mode selection, were used to separate the probe and conjugate waves from the pump beams. The measured nonlinear reflectivity ($R \sim 0.45\%$) was consistent with theoretical predictions, which included linear loss effects.

In Chapters 5, 6, and 7, theoretical and experimental aspects of phase conjugation in optical waveguide geometries are considered in detail.

E. Novel Demonstrations of Phase Conjugation via Four-Wave Mixing

Here, we review three novel experimental demonstrations of optical phase conjugation: wave-front inversion of nonuniformly polarized optical beams, multiwavelength phase conjugation, and forward-going phase conjugation. In Chapter 14, applications of OPC that exploit the notions borne out by these successful demonstrations are considered.

1. Phase Conjugation of Nonuniformly Polarized Fields

Thus far, we have discussed wave-front conjugation of aberrated fields having a uniform polarization state. In many situations, there may also exist undesirable time-varying, *anisotropic*, nonuniform phase shifts, i.e., polarization distortions. These effects are likely to occur during propagation through thermally or mechanically generated stress-induced birefringent media (e.g., lenses, wave plates, or solid-state laser amplifier materials) and other imperfect optical elements. This state of affairs imposes an additional requirement upon the phase conjugator: it must be capable of inverting not only the wave fronts of a given field but also its polarization state. For example, the phase-conjugate mirror must reflect a right-handed circularly polarized (RHCP) wave into a RHCP wave, and vice versa. Recall that a conventional mirror reflects an incoming RHCP field into a left-handed circularly polarized wave.

The realization of inhomogeneous polarization conjugation places constraints upon the polarization states and propagation directions of the pump waves, as well as upon the symmetry properties and orientation of the nonlinear medium. Although several pump-wave and material combinations can yield the desired result, we consider here two practical geometries that are simple to evaluate. One scheme, proposed by Zel'dovich and Shkunov (1979), involves the use of counterpropagating pump waves both *linearly* polarized in the same sense; a probe wave of arbitrary polarization propagates along the direction *parallel* to the *electric field vector* of the pump waves. In this configuration (assuming that the nonlinear medium is characterized by an optical susceptibility with equal off-diagonal tensor elements), the nonlinear interaction appears isotropic with respect to any probe-wave polarization. Although this scheme is rigorously satisfied for only one probe-beam propagation direction, it is nonetheless important for certain applications.

A second scheme involves the use of counterpropagating pump waves *circularly* polarized in the same sense (e.g., both RHCP) interacting with a nonuniformly polarized probe wave in an isotropic nonlinear medium. The ability of this scheme to yield a time-reversed probe polarization state can be qualitatively understood by considering the conservation of photon linear and angular momentum, and the fact that there is no momentum transfer to an ideal (lossless) nonlinear medium.

An experiment that demonstrated the ability of a phase conjugator to compensate for both nonuniform polarization and inhomogeneous phase distortions was performed by Martin *et al.* (1980). Using the second of the above schemes, Martin and co-workers employed the canonical four-wave mixing geometry of Fig. 11b. The nonlinear medium was CS_2, and a pulsed, frequency-doubled Nd:YAG laser ($\lambda = 532$ nm) was the source. A polarization distorter consisting of an array of randomly oriented quarter-wave plates was placed in the path of the probe wave. The ability of the scheme to faithfully reproduce various image-bearing probe waves (in both phase and polarization) after interacting with the CS_2 cell and double-passing the aberrator was experimentally verified, thereby supporting the utility of the scheme.

2. *Multiwavelength Optical Phase Conjugation*

In experiments performed by Depatie and Haueisen (1980) and Steel and Lam (1980), the generation of phase-conjugated wave fronts was observed by using a multiline laser as the source. The realization of such a scheme is important for such applications as phase-conjugate resonators employing multiline gain media and multiline laser amplifier schemes. In most situations, given a set of input probe wavelengths, one desires a conjugator that could conjugate each input wavelength component *independently*. Owing to the sum and difference frequency processes involved in the four-wave mixing process, a multitude of potential near-conjugated waves occur at frequencies that can differ from those of the input probe field. Hence the evaluation of the efficiency of both the degenerate and nearly degenerate conjugate field sets is necessary for given system applications (recall Section II.E).

The experiments demonstrating this process used the basic four-wave mixing geometry shown in Fig. 11b. In one experiment [Depatie and Haueisen (1980)], the pump and probe waves were provided by a multiline pulsed DF chemical laser at 4 μm. The nonlinear element was a 5-mm-long crystal of optical-grade-quality germanium, which can be viewed as a broadband, *nonresonant* medium for the wavelengths used in the experiment. The conjugate wave was spectrally analyzed by using a 200-

line/mm grating in conjunction with an infrared vidicon. The relatively wide spacing of the most prominent DF laser lines, coupled with the interaction length in the Ge crystal, ensured a low efficiency of the undesirable cross-coupled conjugate modes (by virtue of phase-matching constraints). This was experimentally verified upon inspection of the spectral (frequency and amplitude) properties of the conjugate wave. The second experiment [Steel and Lam (1980)] demonstrated multiline conjugation in a *resonant* nonlinear medium (SF_6), using a multiline CO_2 laser. Similar techniques and diagnostics were employed, yielding results consistent with theoretical predictions.

3. *Forward-Going Optical Phase Conjugation*

In an experiment performed by Heer and Griffen (1979), phase-conjugated waves were generated in the *forward* direction via degenerate four-wave mixing. The basic geometry, shown in Fig. 18b, resembles that of phase conjugation via three-wave mixing (recall from Section II B, that $\omega_c = \omega_1 - \omega_p = 2\omega - \omega$) shown in Fig. 18a, with the pump wave at 2ω replaced with two pump photons at ω, i.e., 2ω(3-wave) $\rightarrow$ ω(4-wave) + ω(4-wave).[10] Because of the (near) copropagating pump and probe waves, one can use angular or polarization separation techniques to distinguish the desired conjugate wave from the various input fields. This geometry results in a conjugate wave that radiates in the near-forward direction, as shown in Fig. 18. We note, however, that this process suffers from both phase-matching constraints (thus limiting the spatial bandwidth of the conjugate return) and finite-interaction-length-induced distortion effects.[11] Therefore, such forward-going conjugation processes are not expected to find much practical use in spatial domain applications of optical phase conjugation.

The experiment performed by Heer and Griffen (1979) used a thin (2 mm thick) cell containing atomic sodium vapor as the nonlinear medium and a dye laser, tuned in the vicinity of the Na D line, as the source. The conjugation property of the interaction was verified by inserting lenses and/or wires in the path of the input probe wave and observing their effects on the forward-going output conjugate wave. In addition, the maximum input acceptance angle of the probe wave, as imposed by phase-matching constraints, was experimentally confirmed.

[10] A similar geometry employing nitrobenzene and a pulsed ruby laser was used by Carman *et al.* (1966) in the early days of nonlinear optics experiments, but the phase-conjugate nature of the interaction was not recognized or fully appreciated at that time.

[11] These drawbacks apply to all forward-going conjugation processes involving elastic optical parametric interactions, which include forward-going four-wave mixing, and three-wave mixing (to be discussed next). See Hopf *et al.* (1979).

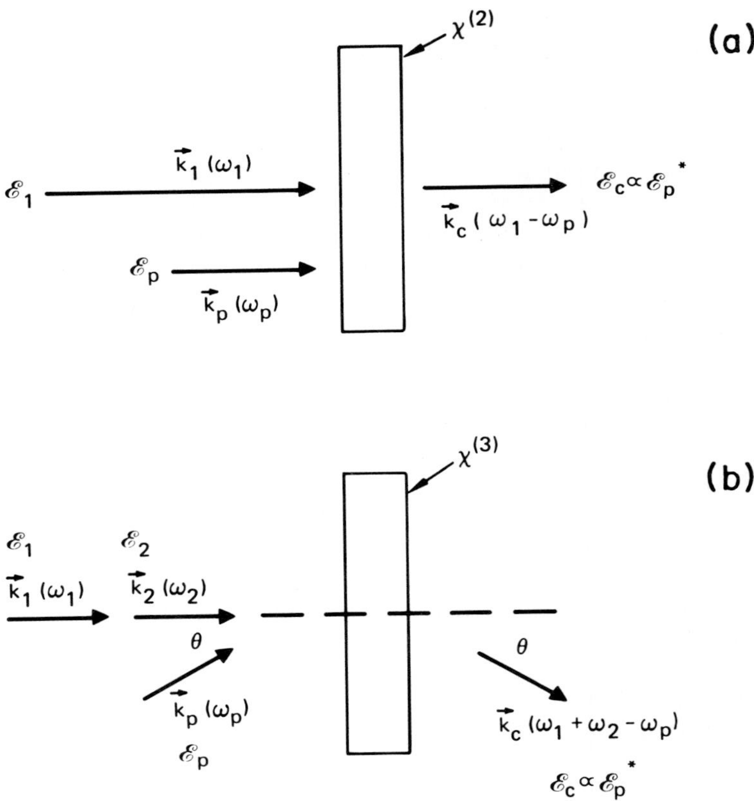

Fig. 18. Phase conjugation by (a) forward-going three-wave mixing and (b) degenerate four-wave mixing. The utility of this geometry is limited by phase-matching constraints. The angular spread of the probe wave $\mathscr{E}_p$ is limited to $\theta \lesssim (\lambda/\ell)^{1/2}$. In Heer's experiment (b), a thin cell ($\ell \sim 2$ mm) containing atomic sodium vapor was used as the nonlinear medium; a dye laser tuned in the vicinity of the sodium D line was used as the source for the various interacting fields. [After Heer and Griffen as depicted in Pepper, 1982.]

F. Observation of Optical Phase Conjugation Using Three-Wave Mixing

As discussed in Section II.B, optical phase conjugation can be realized by using a three-wave mixing process. The experiment discussed here involved the work of Avizonis *et al.* (1977). Because the process in question involves a nonlinear mixing interaction of the form $\omega_{\text{out}} = 2\omega - \omega$, two different input frequencies are required. In the experiment, as shown in

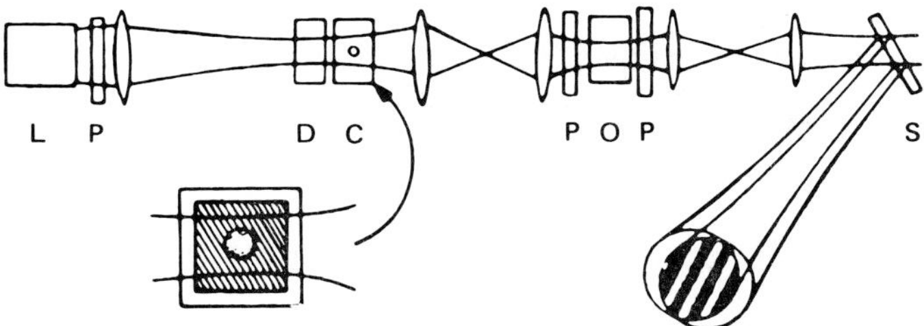

Fig. 19. Phase conjugation by three-wave mixing. The output of a Nd:YAG laser L and its second harmonic D (the doubler, $LiIO_3$) provide the probe and pump beams, respectively. The probe beam is distorted upon passage through a cell C, which contains an index-matching fluid (at 0.53 μm). The conjugator O consists of a crystal of lithium formate, and is sandwiched between two Glan prisms P. The conjugate wave is analyzed by using a wedged shearing interferometer S. [After Avizonis *et al.* (1977).]

Fig. 19, these fields were obtained with a doubled Nd:YAG laser (at 0.53 μm) and its fundamental (at 1.06 μm). The doubling element was a lithium iodate crystal, and the conjugating nonlinear medium was a lithium formate crystal. The interaction can be viewed as a special case of an optical parametric amplifier (OPA), using a planar pump wave at frequency 2ω, with both the signal and the idler waves at frequency ω. The output wave (i.e., the idler) is the desired phase-conjugate replica of the input probe field (i.e., the signal); all fields propagate in the same direction.

The interaction involved the use of type-II phase matching (signal and pump waves polarized in the same direction, idler polarized orthogonally), which resulted in the generation of a conjugate wave whose polarization is othogonal to, and thus separable from, the probe wave. Type-I phase matching will not work here, because both the signal and the idler would have the same frequency and polarization, and could therefore not easily be distinguishable. As shown in Fig. 19, the conjugator was sandwiched between a pair of Glan prisms, which allowed the proper pump and signal polarizations to interact in the lithium formate crystal. A phase distortion (of 0.75 λ) was imposed on the 1.06-μm signal wave upon passage through a cell containing a glass rod and in index-matching fluid (matched at 0.53 μm). These components were chosen to produce a negligible phase distortion at 0.53 μm (i.e., the pump wave), by virtue of the matching fluid. The conjugate nature of the output (idler) wave was established by imaging it onto a wedged shearing interferometer, with subsequent viewing from an IR vidicon–TV monitor. The resultant straight-fringe lines indicated the compensation effect of the conjugator. Finally, the phase-matching restrictions (analogous to those in the forward-going four-wave experiment above) were evident upon inspection of the fringe pattern as a function of the angular extent of the input signal beam.

IV. Conclusion

In this chapter, we have discussed the theory and selected experimental demonstrations of optical phase conjugation using elastic photon scattering processes in transparent nonlinear media. The material presented provides a general background to these important processes, and enables one to appreciate and hence analyze the myriad potential applications of OPC; it also provides a framework for the evaluation of phase conjugation in more complex atomic (and molecular) systems. These topics constitute the major portion of the chapters that follow.

Acknowledgment

The authors wish to thank Mrs. Cheryl Krieger for her meticulous typing of the manuscript and Dr. Janet Lent for her support.

References

Abrams, R. L., and Lind, R. C. (1978). Degenerate four-wave mixing in absorbing media, *Opt. Lett.* **2,** 94; **3,** 205.

AuYeung, J., Fekete, D., Pepper, D. M., Yariv, A., and Jain, R. K. (1979a). Continuous backward-wave generation by degenerate four-wave mixing in optical fibers, *Opt. Lett.* **4,** 42.

AuYeung, J., Fekete, D., Pepper, D. M., and Yariv, A. (1979b). A theoretical and experimental investigation of the modes of optical resonators with phase-conjugate mirrors, *IEEE J. Quantum Electron.* **QE-15,** 1180.

Avizonis, P., Hopf, F. A., Bomberger, W. D., Jacobs, S. F., Tomita, A., and Womack, K. H. (1977). Optical phase conjugation in a lithium formate crystal, *Appl. Phys. Lett.* **31,** 435.

Bergmann, E. E., Bigio, I. J., Feldman, B. J., and Fisher, R. A. (1978). High-efficiency pulsed 10.6 μm phase-conjugate reflection via degenerate four-wave mixing, *Opt. Lett.* **3,** 82.

Bloom, D. M., and Bjorklund, G. C. (1977). Conjugate wave-front generation and image reconstruction by four-wave mixing, *Appl. Phys. Lett.* **31,** 592.

Bloom, D. M., Liao, P. F., and Economoo, M. P. (1978). Observation of amplified reflection by degenerate four-wave mixing in atomic sodium vapor, *Opt. Lett.* **2,** 58.

Carman, R. L., Chiao, R. Y., and Kelley, P. L. (1966). Observation of degenerate stimulated four-photon interaction and four-wave parametric amplification, *Phys. Rev. Lett.* **17,** 1281.

Caves, C. M. (1981). Quantum-mechanical noise in an interferometer, *Phys. Rev. D* **23,** 1693.

Chang, T. Y. (1981). Fast self-induced refractive index changes in optical media: A survey, *Opt. Eng.* **20,** 220.

Depatie, D., and Haueisen, D. (1980). Multiline phase conjugation at 4 μm in germanium, *Opt. Lett.* **5,** 252.

Feinberg, J., Heiman, D., Tanguay, A. R., Jr., and Hellwarth, R. W. (1980). Photorefractive effects and light-induced charge migration in barium titanate, *J. Appl. Phys.* **51,** 1297; **52,** 537 (1981).

Gerritsen, J. J. (1967). Nonlinear effects in image formation, *Appl. Phys. Lett.* **10,** 237.

Harris, S. E. (1966). Proposed backward wave oscillation in the infrared, *Appl. Phys. Lett.* **9,** 114.

Heer, C. V., and Griffen, N. C. (1979). Generation of a phase-conjugate wave in the forward direction with thin Na-vapor cells, *Opt. Lett.* **4,** 239.

Hellwarth, R. W. (1977a). Generation of time-reversed wave fronts by nonlinear refraction, *J. Opt. Soc. Am.* **67,** 1.

Hellwarth, R. W. (1977b). Third order optical susceptibilities of liquids and solids, *Prog. Quantum Electron.* **5,** 1.

Hopf, F. A., Tomita, A., Womack, K. H., and Jewell, J. L. (1979). Optical distortion in nonlinear phase conjugation by three-wave mixing, *J. Opt. Soc. Am.* **69,** 968.

Jain, R. K. and Lind, R. C. (1983). Degenerate four-wave mixing in semiconductor-doped glass. *J. Opt. Soc. Am.*, to be published in Special issue on phase conjugation.

Martin, G., and Hellwarth, R. W. (1979). Infrared-to-optical image conversion by Bragg reflection from thermally induced index gratings, *Appl. Phys. Lett.* **34,** 371.

Martin, G., Lam, L. K., and Hellwarth, R. W., (1980). Generation of time-reversed replica of a nonuniformly polarized image-bearing optical beam, *Opt. Lett.* **5,** 185.

Palmer, A. J. (1979). Nonlinear optics in radiatively cooled vapors, *Opt. Commun.* **30,** 104.

Palmer, A. J. (1980). Nonlinear optics in aerosols, *Opt. Lett.* **5,** 54.

Pepper, D. M. (1980). Phase conjugate optics: On the theory, observation, and utilization of temporally-reversed wavefronts as generated via nonlinear optical parametric interactions, Ph.D. Thesis, California Institute of Technology, Michigan Univ. Microfilm #8014305.

Pepper, D. M. (1982). (Guest Ed.), Special issue on nonlinear optical phase conjugation, *Opt. Eng.* **21,** Number (2).

Pepper, D. M., and Abrams, R. L. (1978). Narrow optical bandpass filter via nearly degenerate four-wave mixing, *Opt. Lett.* **3,** 212.

Pepper, D. M., Fekete, D., and Yariv, A. (1978). Observation of amplified phase-conjugate reflection and optical parametric oscillation by degenerate four-wave mixing in a transparent medium, *Appl. Phys. Lett.* **33,** 41.

Sealer, D. A., and Hsu, H. (1965). Stimulated Brillouin scattering as a parametric interaction, *IEEE J. Quantum Electron.* **QE-1,** 116.

Shapiro, J. H. (1980). Optical waveguide tap with infinitesimal insertion loss, *Opt. Lett.* **5,** 351.

Smith, P. W., Ashkin, A., and Tomlinson, W. J. (1981). Four-wave mixing in an artificial Kerr medium, *Opt. lett.* **6,** 284.

Staebler, D. L., and Amodei, A. J. (1972). Coupled-wave analysis of holographic storage in $LiNbO_3$, *J. Appl. Phys.* **43,** 1042.

Steel, D. G., and Lam, J. F. (1980). Multiline phase conjugation in resonant materials, *Opt. Lett.* **5,** 297.

Stepanov, B. I., Ivakin, E. V., and Rubanov, A. S. (1971). Recording two-dimensional and three-dimensional dynamic holograms in bleachable substances, *Dokl. Akad. Nauk SSSR* **196,** 567 [*English transl.: Sov. Phys.-Dokl. Tech. Phys.* **15,** 46 (1971)].

Woerdman, J. P. (1971). Formation of a transient free carrier hologram in Si, *Opt. Commun.* **2,** 212.

Yariv, A. (1975). "Quantum Electronics," 2nd ed. Wiley, New York.

Yariv, A. (1976a). Three-dimensional pictorial transmission in optical fibers, *Appl. Phys. Lett.* **28,** 88.

Yariv, A. (1976b). On transmission and recovery of three-dimensional image information in optical waveguides, *J. Opt. Soc. Am.* **66,** 301.

Yariv, A. (1977). Compensation for atmospheric degradation of optical beam transmission by nonlinear optical mixing, *Opt. Commun.* **21,** 49.

Yariv, A., and Pepper, D. M. (1977). Amplified reflection, phase conjugation, and oscillation in degenerate four-wave mixing, *Opt. Lett.* **1,** 16.

Yuen, H. P. (1976). Two-photon coherent states of the radiation field, *Phys. Rev. A* **13,** 2226.

Yuen, H. P., and Shapiro, J. H. (1979). Generation and detection of two-photon coherent states in degenerate four-wave mixing, *Opt. Lett.* **4,** 334.

Zel'dovich, B. Ya., and Shkunov, V. V. (1979). Spatial-polarization wavefront reversal in four-photon interaction, *Kvantovaya Elektron.* **6,** 629 [*English transl.: Sov. J. Quantum Electron.* **9,** 379 (1979)].

Zel'dovich, B. Ya., Popovichev, V. I., Ragul'skii, V. V., and Faizullov, F. S. (1972). Connection between the wave fronts of the reflected and exciting light in stimulated Mandel'shtam-Brillouin scattering, *Zh. Eksp. Teor. Fiz. Pis. Red.* **15,** 160 [*English transl.: Sov. Phys. JETP* **15,** 109 (1972)].

3 Transient Response of Kerr-like Phase Conjugators†

B. R. Suydam
Robert A. Fisher

Los Alamos National Laboratory
Los Alamos, New Mexico

I. Introduction

In Chapter 2, the phase-conjugating properties of an optical Kerr-like medium[1] pumped by a pair of counterpropagating waves were discussed. The reduced (SVEA) wave equations were used to derive the conjugate reflectivity, but temporal derivatives were missing from the left-hand sides of the starting equations. What has been described thus far is therefore a *steady-state* theory, in which the pumps, probe, and conjugate envelope functions may not vary in time. Thus the steady-state expression $\tan(\kappa L)$ cannot automatically be applied to situations involving time-varying pulses.

† Work performed under the auspices of the U. S. Department of Energy.

[1] By a "Kerr-like" medium we mean an almost transparent material in which the *only* nonlinear property is that the speed of light depends linearly upon the optical intensity.

Optical Phase Conjugation

ISBN 0-12-257740-X

In this chapter, we will restore those missing temporal derivatives and we will redo the work of the previous chapter, with special interest in considering how conjugate pulses might differ from the probe pulses that generated them.

Throughout this chapter, we will assume that the slowly varying envelope approximation is valid, that the exactly-equal-frequency pump waves are not depleted, that in the medium they remain exactly counterpropagating equal-amplitude plane waves, and that they are significantly stronger than the probe and conjugate waves. We also assume that, because of their low intensity, neither the conjugate wave nor the probe wave experiences self-focusing or self-phase-modulation. We further assume that the optical Kerr effect responds instantaneously, and that all waves are plane polarized in the same direction.

In this chapter, κL is generally not restricted to small values. Therefore, the forward-going (probe) and backward-going (conjugate) wave equations must be solved in a consistent manner. This is in contrast to the temporal processing aspects of Chapter 14, which only considers the $\kappa L \ll 1$ case.

II. The Transient Formalism for cw-Pumped Conjugators

In this section, we develop the interrelationships between the steady-state and transient behavior of cw-pumped Kerr-like conjugators.[2] We use Laplace transform techniques to obtain an expression for the conjugate response to input pulses of arbitrary form. For stable conjugator operation (in which the *continuous-wave* conjugate reflectivity is finite for all probe input frequencies), the expression reduces to an antilinear Fourier transform relationship that is readily adaptable to computer simulation.

We will consider the case in which an antireflection-coated conjugator of length L (extending in the z direction) is pumped by two equal-intensity counterpropagating pump waves of frequency ω. Because it will be needed in Section IV, we make the assumption that, in addition to the optical Kerr effect, the material has some background nonsaturable (linear) attenuation with an absorption coefficient α.

A. General Formalism

A probe wave E_p propagates toward the right and impinges on the dielectric at $z = 0$. Nonlinear interaction with the pump waves generates a

[2] This section and Section III are adapted, with permission, from Fisher *et al.* (1981).

conjugate wave E_c propagating toward the left. We represent the fields as

$$\begin{aligned} E_p &= \tfrac{1}{2}\mathscr{E}_p \exp[-i\omega(t - z/v)] + \text{c.c.}, \\ E_c &= \tfrac{1}{2}\mathscr{E}_c \exp[-i\omega(t + z/v)] + \text{c.c.} \end{aligned} \tag{1}$$

Here $\mathscr{E}_p$ and $\mathscr{E}_c$ are slowly varying functions of z and t, and v is the linear phase velocity of light in the dielectric. We do not require that the probe wave be exactly at the frequency of the pump. If E_p is centered on $\omega + \Delta$, then E_c will be centered on $\omega - \Delta$, and the envelopes $\mathscr{E}_p$ and $\mathscr{E}_c$ will carry factors $\exp[-i\Delta(t - z/v)]$ and $\exp[+i\Delta(t + z/v)]$, respectively, in addition to other z and t dependence. As long as $\Delta \ll \omega$, inclusion of these exponential factors will in no way affect the validity of the slowly varying envelope approximation, from which we obtain

$$\begin{aligned} \frac{\partial \mathscr{E}_p^*}{\partial z} + \frac{1}{v}\frac{\partial \mathscr{E}_p^*}{\partial t} + \tfrac{1}{2}\alpha \mathscr{E}_p^* &= i\kappa^* \mathscr{E}_c, \\ \frac{\partial \mathscr{E}_c}{\partial z} - \frac{1}{v}\frac{\partial \mathscr{E}_c}{\partial t} - \tfrac{1}{2}\alpha \mathscr{E}_c &= i\kappa \mathscr{E}_p^*, \end{aligned} \tag{2}$$

where the asterisk denotes the complex conjugate and κ is the complex coupling coefficient at $\omega = \omega_0$. Note that for no absorption ($\alpha = 0$) the steady state ($\partial/\partial t = 0$) equations reduce to the four-wave mixing equations of Chapter 2. For the assumptions of this chapter, κ may without loss of generality be taken as a real constant depending not only upon the pump amplitudes but also, because of the absorption, upon the pump-wave path length in the conjugator. These equations have been discussed by Bobroff (1965), Bobroff and Haus (1967), Marburger (1978), Rigrod *et al.* (1980), Zel'dovich *et al.* (1980), Fisher *et al.* (1981), Shockley (1981), and Shakir (1981).

A known pulse of radiation is introduced as a probe and enters at $z = 0$, so we have $\mathscr{E}_p(0, t) = F(t)$ and $\mathscr{E}_c(L, t) = 0$. The function $F(t)$ is known. Equations (2) together with the boundary conditions define the problem.

We solve our equations by (two-sided) Laplace transforming in t. We denote transformed functions by lower-case letters; thus,

$$e_p^*(z, s) = \frac{1}{\sqrt{2\pi}} \int_{-\infty}^{-\infty} \mathscr{E}_p^*(z, t) e^{-st}\, dt, \tag{3}$$

and so on. Note that $e_p^*(z, s)$ is the Laplace transform of $\mathscr{E}_p^*(z, t)$, *not* the complex conjugate of $e_p(z, s)$. We restrict our attention to probe pulses of finite energy; thus $F(t)$ belongs[3] to L_2. We shall further require that $F(t) \to 0$ sufficiently rapidly as $t \to -\infty$ that $\int F(t)e^{-\nu t}\, dt$ converges for all

[3] A function is said to belong to L_2 if the square of its modulus is integrable over the range in question (here the infinite interval).

positive values of ν. These restrictions on $F(t)$ guarantee that its transform $f(s)$ has no singularity in the half plane $\mathrm{Re}(s) \geq 0$.

Laplace-transforming Eqs. (2) gives

$$\begin{aligned} \frac{de_{\mathrm{p}}^{*}}{dt} + \frac{\sigma}{v} e_{\mathrm{p}}^{*} - i\kappa e_{\mathrm{c}} &= 0, \\ \frac{de_{\mathrm{c}}}{dt} - \frac{\sigma}{v} e_{\mathrm{c}} - i\kappa e_{\mathrm{p}}^{*} &= 0, \end{aligned} \tag{4}$$

where

$$\sigma \equiv s + \tfrac{1}{2}\alpha v.$$

Equations (4) are linear equations with constant coefficients, so their general solution is a linear combination of the two functions $\exp(\pm i\beta z)$, where

$$\beta \equiv [\kappa^2 - (\sigma/v)^2]^{1/2}. \tag{5}$$

Forming such a solution and applying the boundary conditions $e_{\mathrm{p}}^{*}(0, s) = f^{*}(s)$ and $e_{\mathrm{c}}(L, s) = 0$, we readily find for the conjugate wave

$$\begin{aligned} e_{\mathrm{c}}(z, s) &= h(z, \sigma) f^{*}(s), \\ h(z, \sigma) &= \frac{-i\kappa \sin[\beta(L - z)]}{\beta \cos(\beta L) + (\sigma/v)\sin(\beta L)}. \end{aligned} \tag{6}$$

Finally, inverse transforming back to the time domain, we have

$$\mathscr{E}_{\mathrm{c}}(z, t) = \frac{-i}{\sqrt{2\pi}} \int_{\gamma - i\infty}^{\gamma + i\infty} h(z, \sigma) f^{*}(s) e^{st}\, ds, \tag{7}$$

where γ is subject solely to the restriction that all singularities of the integrand lie to the left of the path of integration. For $z = 0$ and $\alpha = 0$, $h(z, \sigma)$ is formally identical to the steady-state four-wave mixing filter function described in Section II of Chapter 2. This is not surprising if we remember that we started with a pair of linear equations with constant coefficients.

It is always very useful to know the δ-function (or impulse) response $H_{\alpha}(t)$ of a linear system such as ours. In our case, this is calculated by setting $z = 0$ and $f(t) = \delta(t)$ (which yields $f^{*}(s) = [2\pi]^{-1/2}$); thus

$$H_{\alpha}(t) = -\frac{i}{2\pi} \int_{\gamma - i\infty}^{\gamma + i\infty} h(0, \sigma) e^{st}\, ds, \tag{8}$$

where the subscript on the H denotes that there is linear nonsaturating attenuation in the conjugator. By the Faltung theorem, we could also recast Eq. (7) as a convolution, but for computational purposes Eq. (7) is preferable. Note that Eq. (8) can be rewritten as

$$H_\alpha(t) = \exp(-\tfrac{1}{2}\alpha v t)H_0(t). \tag{9}$$

$H_0(t)$ is the δ-function response in the absence of background attenuation, and will be discussed in Section III.

Note that in Eqs. (7)–(9) there is no reason why v, κ, and α cannot depend on the complex frequency s. Thus these transient techniques can be applied to dispersive or to resonant media.

Because the line of integration in Eq. (7) must lie to the right of all singularities, we must know something about their location. Our restrictions on $F(t)$ guarantee that $f(s)$ has no singularities for $\mathrm{Re}(s) > 0$, so only the singularities of $h(z, \sigma)$ need concern us. In Appendix C of Fisher *et al.* (1981), these singularities are discussed in detail, with the following results:

1. The singularities are all poles.
2. Let $X_n(\alpha L)$ denote the nth root[4] of the equation $\tan X = 2X/(\alpha L)$. Then if $\kappa L < [X_1^2 + (\frac{1}{2}\alpha L)^2]^{1/2}$, all the poles are to the left of the imaginary s axis. Note that if $\alpha = 0$, then $X_1 = \pi/2$, and this condition reduces to $\kappa L < \pi/2$. Note also that if $\alpha L \geq 2\kappa L$, then all the poles lie to the left of the imaginary s axis, whatever the values of κ and L.
3. If $[X_n^2 + (\frac{1}{2}\alpha L)^2]^{1/2} < \kappa L < [(X_{n+1})^2 + (\frac{1}{2}\alpha L)^2]^{1/2}$, then there are exactly n poles in the right half plane. These poles are all simple and lie on the real s axis.

Let there be exactly n poles in the right half plane and denote their location by $\sigma = \sigma_k$, $k = 1, 2, \ldots, n$. Let h_k denote the residue of $h(z, \sigma)$ at $\sigma = \sigma_k$. We now form the function $g(z, \sigma)$, defined by

$$g(z, \sigma) = h(z, \sigma) - \sum_{k=1}^{n} \frac{h_k}{\sigma - \sigma_k}. \tag{10}$$

Because the poles $\sigma = \sigma_k$ are all simple, $g(z, \sigma)$ has no poles in the right half plane, and we may therefore set $\gamma = 0$ in the integral over g. Thus, if we solve Eq. (10) for $h(z, \sigma)$ and set the result into Eq. (7), we obtain our main result

$$\mathscr{E}_c(z, t) = \frac{1}{\sqrt{2\pi}} \int_{-\infty}^{\infty} g(z, -i\Omega) f^*(\tfrac{1}{2}\alpha v - i\Omega) e^{-i\Omega t}\, d\Omega + \sum_{k=1}^{n} h_k f^*(s_k) e^{s_k t}. \tag{11}$$

Here $\Omega = -\mathrm{Im}(s)$.

[4] These roots are tabulated, for example, in Abramowitz and Stegun (1972).

B. Specialization to Stable Operation

The integral in Eq. (11) connects the input and conjugate waves through an antilinear Fourier transform relationship. In discussing the integral, we shall restrict our attention to the case of stable operation, in which $\kappa L < [X_1^2 + (\frac{1}{2}\alpha L)^2]^{1/2}$. Treatment of the unstable case would be of limited validity, because the exponential growth of the conjugate wave will eventually violate the condition that the conjugate wave be weaker than the pump wave, and because the effect of spontaneous optical parametric emission is not taken into account. Such a quantum-mechanical effect can be thought of as the conjugate reflection of zero-point photons, a matter not at all considered in our classical starting equations.

We therefore confine our attention to the stable region. Clearly, $n = 0$; the sum in Eq. (11) disappears, and $g(z, \sigma) = h(z, \sigma)$.

We shall denote the Fourier transform of a function by a tilde, for example,

$$\tilde{\mathscr{E}}_{\mathrm{p}}(\Omega) = (2\pi)^{-1/2} \int_{-\infty}^{\infty} \mathscr{E}_{\mathrm{p}}(t') \exp(i\Omega t')\, dt',$$

where $\Omega = \omega - \omega_0$. Accordingly, $f^*(-i\Omega) = \tilde{\mathscr{E}}_{\mathrm{p}}^*(\Omega) = [\tilde{\mathscr{E}}_{\mathrm{p}}(-\Omega)]^*$. We then write Eq. (11) in the form most amenable to numerical analysis

$$\mathscr{E}_{\mathrm{c}}(z, t) = \frac{1}{\sqrt{2\pi}} \int_{-\infty}^{\infty} h(z, \tfrac{1}{2}\alpha v - i\Omega)[\tilde{\mathscr{E}}_{\mathrm{p}}(-\Omega)]^* e^{-i\Omega t}\, d\Omega. \tag{12}$$

This equation is our desired prescription for numerically computing the conjugate waveforms. *First* one Fourier-transforms the input pulse, *then* one multiplies the antilinear transfer function h by the complex conjugate of the Fourier component at frequency $-\Omega$, and *finally* one inverse-Fourier-transforms the product to obtain the output waveform. Note that with this equation one can, for an arbitrary input pulse, readily apply standard fast-Fourier numerical techniques first to obtain the function $[\tilde{\mathscr{E}}_{\mathrm{p}}(-\Omega)]^*$ and then to evaluate the integral in Eq. (12). The relationship derived here for stable operation is precisely that anticipated by Yariv *et al.* (1979).

The Fourier transform of Eq. (12), $\tilde{\mathscr{E}}_{\mathrm{c}}(z, \Omega) = h(z, \frac{1}{2}\alpha v - i\Omega)[\tilde{\mathscr{E}}_{\mathrm{p}}(-\Omega)]^*$ shows that the spectrum of the conjugate wave at upshifted frequency Ω is equal to the antilinear transfer function times the complex conjugate of the input spectral component evaluated at the downshifted frequency $-\Omega$. This is precisely the cw relationship described by Eq. (6) of Pepper and Abrams (1978) and by Eq. (59) of Chapter 2.

Note the difference between the above development and the more conventional *linear* optical system described by a transfer function $T(z, \Omega)$.

The linear counterpart of Eq. (12) contains, instead, the product $T(z, \Omega)\tilde{\mathscr{E}}_p(\Omega)$, which means that input at one frequency corresponds to output at the *same* frequency. Note also that, in conventional linear optics, the complex conjugation operation is missing.

We have completed the formalism necessary to evaluate the form of pulses emitted from a cw-pumped Kerr-like conjugator. Some examples will be presented in Section III.

III. Examples of Reshaping in cw-Pumped Conjugators

Now that we can evaluate the conjugate waveforms for arbitrary probe inputs, we shall examine a few interesting cases. For the sake of simplicity, we will restrict our attention to cases in which there is no background absorption ($\alpha = 0$). The stability criterion will then be cast in a simpler form, namely, that $\kappa L < \pi/2$.

A. The δ-Function Response

The response to the input of a temporal δ function can be obtained by evaluation of Eq. (8). The expression is given as Eq. (23) in Fisher *et al.* (1981). An alternative approach using the method of images was first developed by Bobroff and Haus (1967), and then specialized to this specific problem by Rigrod *et al.* (1980). Through either approach the expression for $H_0(t)$ is

$$\begin{aligned}\frac{2i}{\kappa v} H_0(t) &= [I_0(\kappa v t) - I_2(\kappa v t)]S(t)\\ &\quad - \sum_{n=1}^{\infty} [(r_n)^{n-1} I_{2n-2}(t_n)\\ &\quad -2(r_n)^n I_{2n}(t_n) + (r_n)^{n+1} I_{2n+2}(t_n)]S(t_n - n\tau), \qquad (13)\end{aligned}$$

where $r_n = (t - n\tau)/(t + n\tau)$, $t_n = \kappa v(t^2 - n^2\tau^2)^{1/2}$, the I_n are modified Bessel functions, $S(x)$ is the step function, and τ is given by the round-trip transit time $2L/v$. Figure 1 shows this expression evaluated for three cases: $\kappa L = 1$, $\pi/2$, and 2.

The δ-function responses exhibit many interesting features. Obviously, there is no response until the δ-function pulse strikes the entrance face of the conjugator. At that time the conjugate signal rises abruptly, and thereafter gradually rises until the round-trip time of the conjugator. At the time

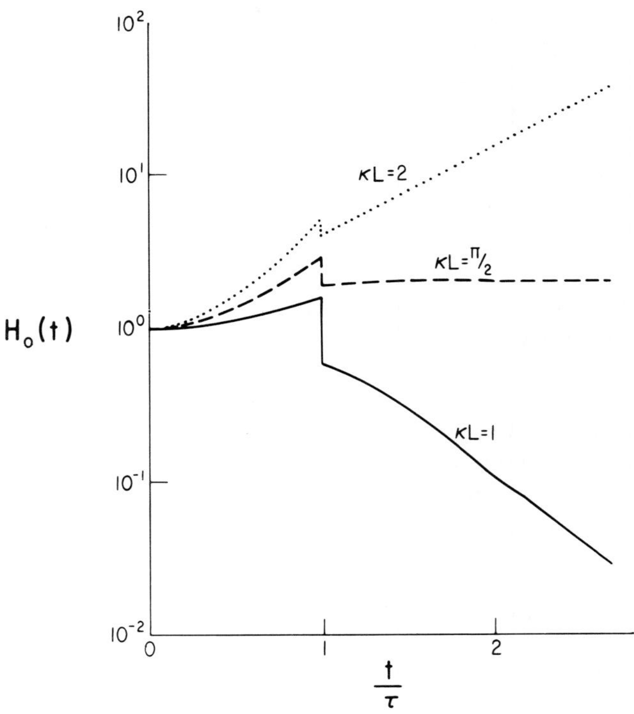

Fig. 1. Plot of backward-wave conjugate impulse response from eq. (13). The factor $-\frac{1}{2}i\kappa v$ has been normalized out so that all curves rise to unity at $t = 0$. The solid curve illustrates $H_0(t)$ for stable conjugation, with $\kappa L = 1$; the dashed curve, the critical case of oscillation threshold, with $\kappa L = \pi/2$; and the dotted curve, an unstable case, with $\kappa L = 2$. In all cases, the normalized curve exhibits a discontinuous drop of value 1 at the end of the first round-trip transit time. [Reproduced with permission from Rigrod *et al.* (1980).]

$2L/v$, the conjugate signal drops abruptly to a lower value, and thereafter rapidly approaches exponential decay for $\kappa L < \pi/2$, converges to a steady-state value at twice its initial value for $\kappa L = \pi/2$, or increases without limit for $\kappa L > \pi/2$. The explanation for the gentle rise during the first round-trip transit time is simple: as the δ-function pulse passes through the conjugator it generates a backward conjugate wave. Because of the narrow-band reflectivity of the conjugator, the backward-wave spectrum is peaked at the pump-wave frequency of the conjugator (in contrast to the flat spectrum associated with the forward-traveling δ function). As the backward wave travels through the conjugator on its way out, it, too, becomes conjugated into a forward-going wave. This conjugation and reconjugation continuously couples forward- and backward-going waves, resulting in a smooth tail on the forward-going δ function. It is the growth of this smooth tail that accounts for the increase in conjugate intensity as the δ-function pulse traverses the device.

As the δ-function pulse leaves the conjugator, only the temporally smooth backward- and forward-going waves remain inside; the δ function no longer generates a "new" backward wave. This accounts for the sudden discontinuous decrease in the conjugate signal at the time when an observer at the entrance face could first learn that the δ function has left the sample. For $\kappa L < \pi/2$, the coupling is inadequate for the remaining waves to subsequently generate stronger partial waves, and therefore the waves (both forward and backward) rapidly approach an exponential decay. For $\kappa L \geq \pi/2$, the regeneration continues even after the δ function leaves the conjugator, giving a steady-state ($\kappa L = \pi/2$) or exponentially growing ($\kappa L > \pi/2$) output.

For $\kappa L \ll 1$, the δ-function response is merely a flat-topped function that turns on at $t = 0$ and off at the round-trip transit time. This can also be easily understood in the partial-wave picture. The coupling is so weak that the primary backward-going wave is not reconjugated, and therefore no tail grows on the forward-going δ function. Thus the radiated intensity remains constant until the round-trip transit time, and thereafter the output is zero because the reconjugation of the partial waves can be neglected. This point was clearly appreciated by Marburger (1978).

B. Temporal Spreading in Conjugate Reflection

Because a temporal δ function is converted into a conjugate pulse with a duration longer than the round-trip transit time of the conjugator, it is interesting to examine the influence of this temporal spreading upon physical pulses. Figure 2 shows the evaluation of Eq. (12) when a 0.5-nsec (full width at $1/e$ of maximum) temporally Gaussian pulse is conjugated by a conjugator with a linear index of 4.0 (as would be the case for germanium). The expression κL has been set equal to $\pi/4$, but L (and, correspondingly, κ) is variable. Thus the conjugate reflectivity *on resonance* is unity, but the effective bandwidth of the conjugator is varied. In the figure the input pulse (with carrier frequency chosen equal to the pump-wave frequency) is shown for comparison. The peak intensity decreases and the temporal duration increases as the physical thickness of the conjugator is varied from 1 to 5 cm. These results can be explained by noting that the induced gratings consist of more and more lines as the conjugator becomes longer. Hence, the bandwidth of the conjugator is reduced to such an extent that it cannot efficiently conjugate all the frequency components

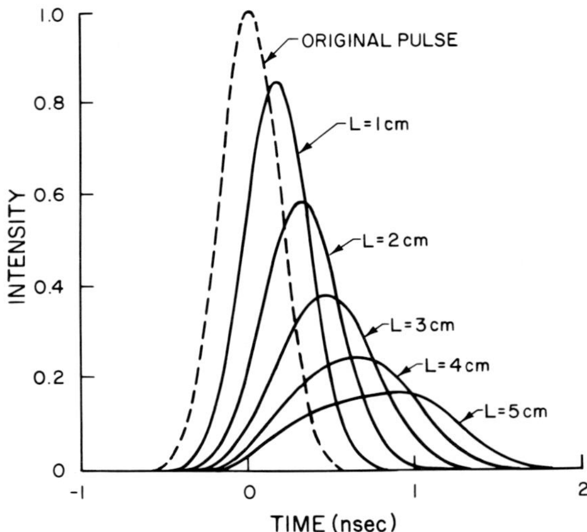

Fig. 2. Temporal spreading of the conjugate pulse. This calculation was performed by keeping κL equal to $\pi/4$ (so that the conjugate reflectivity on resonance is unity), but with L (and, correspondingly, κ) variable. This procedure varies the effective bandwidth of the conjugator. The original pulse (0.5 nsec FW$(1/e)$M is shown by the dashed line. The conjugator material has a refractive index of 4.0, as is the case for germanium. [Reproduced with permission from Fisher *et al.* (1981).]

in the input pulse. Note that the delay is approximately half the conjugator round-trip transit time.

The results of Fig. 2 are recast in Fig. 3. Peak intensity and total energy are evaluated for each conjugate waveform. Both quantities are plotted as functions of the ratio of the conjugator round-trip transit time to the duration of the pulse. The intensity curve falls off much faster than the energy curve; this is a manifestation of the temporal broadening process.

The results indicate that even if a conjugator has a cw on-resonance reflectivity of unity, the practical reflectivity will be greatly reduced unless the duration of the input pulse is far longer than the round-trip transit time of the conjugator.

C. Chirp Reversal

For a broadband conjugator, the implications for chirp reversal are clear: if $h(z, -i\Omega)$ is taken to be constant in Eq. (12), the chirp (or rate of change of instantaneous frequency with time) of the conjugate pulse is precisely opposite to that of the input pulse. This was first pointed out by Mar-

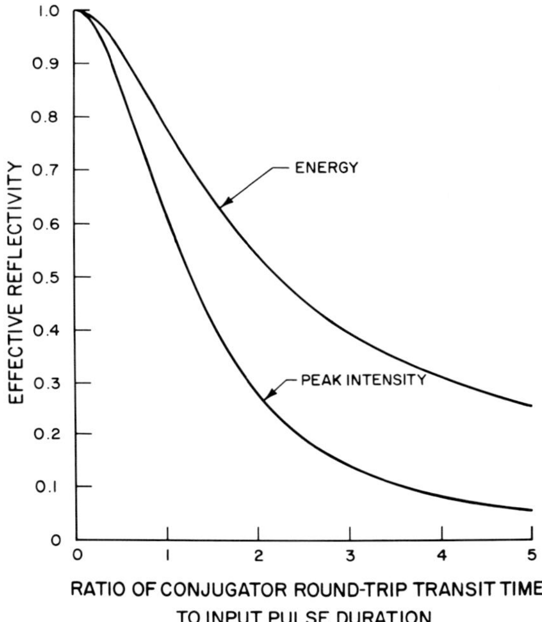

Fig. 3. The calculated degradation in effective reflectivity when the probe-pulse duration becomes shorter than the conjugator thickness. The integrated conjugate energy and intensity at the peak of the conjugate pulse are both plotted. All other conditions are the same as in Fig. 2. [Reproduced with permission from Fisher *et al.* (1981).]

burger (1978). Yariv *et al.* (1979) showed that a pulse that had undergone dispersive spreading could be conjugated in a suitably broadband conjugator so that the chirp reversal would cause subsequent dispersive narrowing upon retraversal of the dispersive element.

Figure 4 shows the results of a chirp reversal calculation for the conditions of Section II.B. The conjugate reflectivity on resonance is again set equal to unity, and the physical thickness of the germanium conjugator is varied (as an adjustment of its bandwidth). Just as in Section II.B, a Gaussian probe pulse is used, but this time a positive linear chirp is impressed upon it, so that the bandwidth is twice that of an unmodulated pulse. At the peak of the input pulse, the instantaneous frequency is that of the pump waves. In Fig. 4, the instantaneous frequency shift is plotted as a function of time for both the input pulse and the conjugate pulses. Three different cases were considered, with conjugator thicknesses of 0.1, 1.0, and 2.0 cm. As in the previous figures, $t = 0$ is the time at which the peak

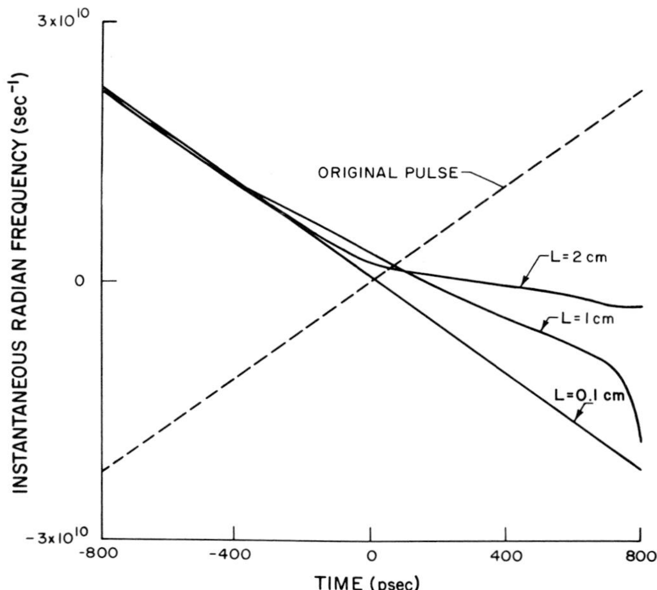

Fig. 4. Study of the chirp reversal process. The instantaneous frequency shift is plotted as a function of time for the positively chirped input pulse. As in Figs. 2 and 3, the cw on-resonance conjugate reflectivity is set equal to unity, but the physical thickness of the conjugator is varied. For a physically thin (i.e., sufficiently broadband) conjugator, chirp reversal is nearly perfect, but for thicker conjugators the chirp at the peak of the pulse is significantly reduced. [Reproduced with permission from Fisher *et al.* (1981).]

of the probe pulse strikes the input face of the conjugator. For the thinnest conjugator ($L = 0.1$ cm), the bandwidth of the device is adequate to produce nearly perfect chirp reversal, but for greater thicknesses the chirp reversal is clearly incomplete. The chirp shows a rather distinct disappearance at the temporal peak of the conjugate pulse (as evidenced by a flattening of the frequency-versus-time curves). Thus, a chirped pulse is conjugated as a relatively chirp-free pulse in this narrow-bandwidth limit.

In a related calculation, κL was set equal to $\pi/4$, and the conjugator thickness was set equal to 2 cm. Increasing the chirp on the input pulse resulted in decreasing the duration of the conjugate pulse. This can be easily understood by considering that as the chirp becomes more severe, the pulse sweeps more quickly through the high-reflectivity center frequency of the conjugator.

Decreasing κ at fixed large L does not produce perfect chirp reversal. This can be understood as follows. As one decreases κL, the δ-function response becomes a flat-topped function for the round-trip duration of the conjugator. The duration of the function depends only upon L, and the height of the function only upon κ. Thus the inverse Fourier transform of this limiting function has a spectral width that depends only upon L, as clearly shown in Eq. (8) of Pepper and Abrams (1978). This inability of a physically thick conjugator to reverse the input pulse chirp can be easily understood in terms of the impressed gratings; the resolution is determined merely by the number of lines in the grating. Thus weak coupling alone ($\kappa L \ll \pi/2$) is insufficient to guarantee faithful chirp reversal; the transfer function must be sufficiently broad.

Pepper and Yariv (1980) have shown that, in the steady state, a conjugator with a reflectivity of unity will compensate for the interposition of a weakly nonlinear aberrator as long as catastrophic self-focusing does not occur. Clearly, considerations in this section would restrict the transit time of the conjugator to far less than the duration of the input pulse; otherwise the time-varying divergence would not be faithfully reversed.

D. Concluding Comments

For the first time, a formalism is available to rapidly numerically evaluate a cw-pumped conjugator's response to an arbitrarily shaped input pulse. This predictive capability has implications for, e.g., signal processing, chirp reversal, and pulse spreading.

IV. The Inverse Problem for cw-Pumped Conjugators

In Sections II and III we discussed how, for cw-pumped Kerr-like conjugators, one evaluates the conjugate waveform when given the probe input waveform. In this section, we turn the problem around and ask how one predicts the input necessary to produce a specified conjugate waveform. We shall show that it is possible to generate a conjugate waveform shorter than the round-trip transit time of the conjugator by properly programing the probe. As we shall see, to make an arbitrarily short conjugate pulse the conjugator must have some residual nonsaturable absorption ($\alpha > 0$), and the probe programing is simplest if the conjugator is infinitely long.

Formally the problem of determining the probe envelope required to produce a specified conjugate pulse is trivially solved. Setting $z = 0$ into Eq. (6), solving for $f^*(s)$, and transforming gives

$$F^*(t) = -\frac{i}{\sqrt{2\pi}} \int_{\gamma-i\infty}^{\gamma+i\infty} [h(0, \sigma)]^{-1} e_c(0, s) e^{st}\, ds. \tag{14}$$

We are interested, of course, in the convergence of the integral. Note, first, that $[h(0, \sigma)]^{-1}$ has simple poles at the points

$$s = s_n = -\tfrac{1}{2}\alpha v \pm i(n\pi v/L)[1 - (\kappa L/n\pi)^2]^{1/2},$$
$$n = \pm 1, \pm 2, \ldots, \tag{15}$$

and no other singularities. If $\kappa L < \pi$, all these poles lie on the line $\mathrm{Re}(s) = -\tfrac{1}{2}\alpha v$. If $\kappa L > \pi$ but

$$\kappa L < \pi[1 + (\tfrac{1}{2}\alpha L)^2]^{1/2}, \tag{16}$$

there is on real pole, but it lies to the left of the origin. So if condition (16) is satisfied, $\mathrm{Re}(s) < 0$ at every singularity of $[h(0, \sigma)]^{-1}$. In this respect $[h(0, \sigma)]^{-1}$ is well behaved as a transfer function provided there is absorption ($\alpha > 0$). However, as $|s| \to \infty$, $[h(0, \sigma)]^{-1}$, is of the order of s, so that $[h(0, \sigma)]^{-1}$ does not belong to L_2 on any line $s = \gamma - i\Omega$, $\gamma \geq 0$.

Suydam (1981) solves the general problem, and shows that an arbitrarily short conjugate pulse can be formed if and only if that following conditions are satisfied.

1. $\alpha > 0$.
2. Condition (16) must be satisfied.
3. Both the desired envelope function $\mathscr{E}_c(0, t)$ and its time derivative must belong to L_2.

When these three conditions are satisfied, we can set $\gamma = 0$ in Eq. (14), which then expresses the required probe as an ordinary Fourier transform. We refer to Suydam (1981) for the full proof of these assertions, and shall give here a discussion of a simplified model for which a simple analytic solution can be obtained.

Let us imagine that κL is so small that we may neglect the reconjugation of the conjugate wave. Marburger (1978) was the first to use this model. We can obtain the transfer function and δ-function response by setting $z = 0$, allowing $\kappa \to 0$ in eq. (6), and renormalizing:

$$h^{(0)}(\sigma) = (1 - e^{-\sigma\tau})/(\sigma\tau), \tag{17}$$

where $\tau = 2L/v$, and

$$H_{\alpha}^{(0)}(t) = \begin{cases} 0 & \text{for} \quad t < 0 \\ (1/\tau)\exp(-\tfrac{1}{2}\alpha v t) & \text{for} \quad 0 < t < \tau \\ 0 & \text{for} \quad T > \tau. \end{cases} \tag{18}$$

Rewriting Eq. (7) as a convolution, we have

$$\mathscr{E}_c(0, t) = \frac{1}{\tau\sqrt{2\pi}} \int_{t-\tau}^{t} \exp\left[-\frac{1}{2}\alpha v(t - t')\right] F^*(t')\, dt'. \tag{19}$$

Differentiating Eq. (19) and solving for $F^*(t)$, we find

$$F^*(t) = \tau\sqrt{2\pi} \sum_{n=0}^{\infty} \exp(-\tfrac{1}{2}n\alpha v t)[\mathscr{E}_c'(0, t - n\tau) + \tfrac{1}{2}\alpha v \mathscr{E}_c(0, t - n\tau)]. \tag{20}$$

This is the exact expression for the desired input pulse. When the desired $\mathscr{E}_c(t)$ is precisely zero outside some interval of duration less than τ, and has a derivative within that same interval, the right-hand side of Eq. (20) consists of an infinite string of ever-decreasing nonoverlapping similar pulses, and we have

$$\int_{-\infty}^{-\infty} |F(t)|^2\, dt = 2\pi\tau^2[1 - e^{-\alpha v \tau}]^{-1} \int_{-\infty}^{-\infty} |\mathscr{E}_c'(0, t) + \alpha v \mathscr{E}_c(0, t)|^2\, dt. \tag{21}$$

Thus, $F(t)$ belongs to L_2 if and only if $\alpha > 0$ and $\mathscr{E}_c(0, t)$, $\mathscr{E}_c'(0, t)$ both belong to L_2. As $\alpha \to 0$, the required input energy unfortunately increases without limit.

Although the total probe energy is finite if $\alpha > 0$, Eq. (21) still requires an inconveniently long train for the input pulse. Let us therefore see what will happen if we truncate the series, Eq. (20), letting the sum run from only $n = 0$ to $n = N - 1$. We then obtain the approximate response $\hat{\mathscr{E}}_c$ given by

$$\hat{\mathscr{E}}_c(0, t) = \mathscr{E}_c(0, t) - [\exp(\tfrac{1}{2}N\alpha v\tau)]\mathscr{E}_c(0, t - N\tau). \tag{22}$$

This is the desired pulse followed at a time $N\tau$ later by a diminished and reversed replica of itself. If $\frac{1}{2}N\alpha v\tau$ and $N\tau$ are large enough, this second pulse is so small and arrives so late that it is of little consequence. Even for $N = 1$, provided L is sufficiently large, we have

$$F^*(t) \simeq \sqrt{2\pi}[\mathscr{E}_c'(0, t) + \tfrac{1}{2}\alpha v \mathscr{E}_c(0, t)]. \tag{23}$$

This becomes exact in the limit $L \to \infty$. Thus we find that, for an infinitely long conjugator with finite nonsaturable absorption, Eq. (23) shows the recipe for determining the input necessary to obtain the predetermined output pulse $\mathscr{E}_c(t)$. When the attenuation is modest, the first term on the right-hand side dominates, showing that one inputs the derivative of the desired output pulse. If a symmetric output pulse is desired, then the required input pulse will be an antisymmetric zero-pi pulse.

A simple physical picture explains why the antisymmetric zero-*pi* pulse generates an ultrashort conjugate pulse. Consider the input as a sequence of two closely spaced pulses of opposite phase. The first pulse strikes the sample, causing the emission of the δ-function response of the conjugator. The second pulse arrives immediately thereafter, causing the superposed emission of an oppositely phased δ-function response. These two responses very nearly cancel, except for the very short period prior to the arrival of the second pulse. Note that this is *not* a pulse compression scheme. For the generation of a symmetric conjugate pulse of short duration, we require (approximately) an antisymmetric zero-*pi* pulse of equally short duration. In fact, the input must have roughly twice the bandwidth of the desired output.

The inverse problem is more complicated in the absence of absorption. The interested reader is referred to Suydam (1981).

In this section we have shown how one deals with the inverse problem, and have obtained a prescription showing how one must program the input pulse to produce a desired output from a cw-pumped Kerr-like phase conjugator.

V. Transient Pumping

We now turn to the situation in which the pump waves may also vary in time. Clearly, this is an exceedingly complex situation, in which *either* temporal *or* spatial variations in the pump waves can lead to *both* temporal and spatial variations in the conjugate wave. Application of these effects will be discussed in Chapter 12.

For expedience, we limit ourselves to the simplest case. We assume that the probe and conjugate waves propagate in the z direction, that the conjugator is of length L in the z direction, that L/c is far greater than the duration of the input probe pulse, that $\kappa L \ll 1$, that the pumps propagate in the y direction (perpendicular to x), that the thickness of the conjugator in the y direction is far less than the speed of light times the duration of the input probe pulse, that all pumping occurs at a time when the probe pulse is entirely within the conjugator, and that there is no background absorption ($\alpha = 0$). This case was first considered by Miller (1980). Any relaxation of the above assumptions profoundly complicates the simple picture presented here.

For small κL, we retain from Eq. (2) only terms of order $|\kappa|$, to find

$$\mathscr{E}_{\mathrm{p}}^*(z, t) = \mathscr{E}_{\mathrm{p}}^*\left(t - \frac{z}{v}\right), \qquad \frac{1}{v}\frac{\partial \mathscr{E}_{\mathrm{c}}}{\partial t} - \frac{\partial \mathscr{E}_{\mathrm{c}}}{\partial z} = -i\kappa(t)\mathscr{E}_{\mathrm{p}}^*\left(t - \frac{z}{v}\right). \tag{24}$$

The coupling constant is given by

$$\kappa(t) = (\pi\omega_0 n_2/c)\mathscr{E}_1(t)\mathscr{E}_2(t). \tag{25}$$

Because the conjugator is very thin in the y direction, we neglect the dependence of κ upon y. Fourier transformed in time, Eq. (24) becomes

$$\frac{d\tilde{\mathscr{E}}_c(\Omega)}{dz} + \frac{i\Omega}{v}\tilde{\mathscr{E}}_c(\Omega) = \frac{i}{\sqrt{2\pi}}\int_{-\infty}^{\infty} e^{i\overline{\Omega}z/v}\tilde{\mathscr{E}}_p(\overline{\Omega})\tilde{\kappa}(\Omega - \overline{\Omega})\, d\overline{\Omega}. \tag{26}$$

To calculate the emerging conjugate wave at the conjugator entrance, $z = 0$, we must integrate Eq. (26) from $z = L$ to $z = 0$. Doing so and inverting the order of integration gives

$$\mathscr{E}_c(0, \Omega) = \frac{iv}{\sqrt{2\pi}}\int_{-\infty}^{\infty} \tilde{\mathscr{E}}_p^*\left(\frac{uv}{L} - \Omega\right)\tilde{\kappa}\left(2\Omega - \frac{uv}{L}\right)\frac{1 - e^{iu}}{iu}\, du, \tag{27}$$

where $u = (\Omega + \overline{\Omega})L/v$. We now take the limit as $L \to \infty$ and inverse Fourier transform, to get

$$\mathscr{E}_c(0, t) = \tfrac{1}{2}iv\int_{-\infty}^{\infty} \tilde{\mathscr{E}}_p^*(-\Omega)\tilde{\kappa}(2\Omega)e^{-i\Omega t}\, d\Omega. \tag{28}$$

Equation (28) is the response of a very long, weak conjugator to a short (but not necessarily δ-function) pump. In deriving it we assumed $L \to \infty$, but actually we need only require that the conjugator be so long that all of the probe pulse remains within the conjugator for the duration of the pump. Formally, Eq. (28) looks very much like our earlier result, Eq. (12), with $\tilde{\kappa}(2\Omega)$, the Fourier transform of $\kappa(t/2)$, serving as the transfer function. The one difference is that in Eq. (28) we have $\tilde{\mathscr{E}}_p^*(-\Omega)$, whereas in Eq. (12) we have $\tilde{\mathscr{E}}_p(-\Omega)$. This difference [the complex conjugate of $\tilde{\mathscr{E}}_p(-\Omega)$] leads to time reversal of the envelope when the pump durtion is very short. For short enough pumps, Eq. (28) reduces to

$$\mathscr{E}_c(0, t) = iv(\sqrt{\pi}/2)\tilde{\kappa}(2\Omega_0)\tilde{\mathscr{E}}_p^*(-t), \tag{29}$$

where Ω_0 is the frequency for which $|\tilde{\mathscr{E}}_p^*(\Omega)|$ is maximum.

Equation (29) gives us the basic time-reversal result. An arbitrary pulse is precisely time reversed if the pump is a true δ-function. Note that there are several differences between this behavior and more conventional phase conjugation from cw pumps. First, the input and output pulses in the case are now centered at the same frequency. The lack of shift can be explained by the fact that the pump-wave spectrum is flat, and therefore there is no pump-wave frequency about which the spectrum could invert. Second, if we write the probe pulse as

$$\mathscr{E}_p(t) = A(t)\exp[i\phi(t)], \tag{30}$$

where A is real, Eq. (29) shows that the conjugate pulse would be

$$\mathscr{E}_c(t) \propto A(-t) \exp[-i\phi(-t)]. \tag{31}$$

If $\phi(t)$ is written as a power series, the even and odd terms are treated oppositely. For example, a linear chirp is reversed in the sense that a positive linear chirp becomes a negative linear chirp, but a frequency offset is not reversed.

This ends our discussion of the simplest case in transient pumping, that of a short-pulse-pumped conjugator. The problem becomes much more complicated when, for example, we consider nonorthogonal propagation of pump and probe, spatially varying pumps, and chirped pumps. In general, information from the pumps can be transferred to the conjugate wave, but a complete description of the process has yet to be made.

VI. Concluding Remarks

In this chapter we have shown how, for a Kerr-like conjugator, one obtains expressions for the conjugate wave when either the probe or the pumps may vary in time. The complicated mathematical formalism is necessary because of the need to solve both forward- and backward-going wave equations consistently.

References

Abramowitz, M., and Stegun, I. A. (eds.) (1972). "Handbook of Mathematical Functions," p. 224. Dover, New York.

Bobroff, D. L. (1965). Coupled modes analysis of the phonon-photon parametric backward-wave oscillator, *J. Appl. Phys.* **36,** 1760.

Bobroff, D. L., and Haus, H. A. (1967). Impulse response of active coupled wave systems, *J. Appl. Phys.* **38,** 390.

Fisher, R. A., Suydam, B. R., and Feldman, B. J. (1981). Transient analysis of Kerr-like phase conjugators using frequency-domain techniques, *Phys. Rev. A* **23,** 3071.

Marburger, J. H. (1978). Optical-pulse integration and chirp reversal in degenerate four-wave mixing, *Appl. Phys. Lett.* **32,** 372.

Miller, D. A. B. (1980). Time reversal of optical pulses by four-wave mixing, *Opt. Lett.* **5,** 300.

Pepper, D. M., and Abrams, R. L. (1978). Narrow optical bandpass filter via nearly degenerate four-wave mixing, *Opt. Lett.* **3,** 212.

Pepper, D. M., and Yariv, A. (1980). Compensation for phase distortion in nonlinear media by phase conjugation, *Opt. Lett.* **5,** 59.

Rigrod, W. W., Fisher, R. A., and Feldman, B. J. (1980). Transient analysis of nearly degenerate four-wave mixing, *Opt. Lett.* **5,** 105.

Shakir, S. A. (1981). Zero area optical pulse processing by degenerate four-wave mixing, *Opt. Commun.* **40,** 151.
Shockley, R. C. (1981). Simplified theory of the impulse response of an optical degenerate four-wave mixing cell, *Opt. Commun.* **38,** 221.
Suydam, B. R. (1981). Can optical phase conjugation produce very short conjugate pulses? *Proc. Lasers '80 Meeting, New Orleans,* p. 319. STS Press, McLean, Virginia.
Yariv, A., Fekete, D., and Pepper, D. M. (1979). Compensation for channel dispersion by nonlinear optical phase conjugation, *Opt. Lett.* **4,** 52.
Zel'dovich, B. Ya., Orlova, M. A., and Shkunov, V. V. (1980). Nonstationary theory and calculation of the time of establishment of four-wave wavefront reversal, *Dokl. Akad. Nauk SSSR* **252,** 592 [*English transl.: Sov. Phys. Dokl.* **25,** 390 (1980)].

4 Improvements upon the Simple Theory of Degenerate Four-Wave Mixing

John H. Marburger

State University of New York at Stony Brook
Stony Brook, New York

I. Introduction

By the *simple* theory of four-wave mixing we mean the usual slowly varying envelope theory, linearized in the probe- and signal-beam strengths and ignoring pump depletion effects as in the preceding chapters. The need for improvements is suggested primarily by the development or discovery of materials with such large third-order susceptibilities that induced index changes approach unity, violating the slowly varying envelope approximation (SVEA) (see Smith and Miller, 1980). In practice, pump depletion may also influence the magnitude of the signal. The effects of pump depletion have been considered by a number of

Optical Phase Conjugation

ISBN 0-12-257740-X

authors in various degrees of approximation. We shall touch upon these effects later in this chapter, but devote most of the discussion to the linearized theory *without* making the slowly varying envelope approximation. Only the steady-state theory will be discussed.

Our aim is to construct a linearized theory that resembles the familiar optical theory of anisotropic media. The pumped nonlinear medium is regarded as a substance in which weak probe and signal beams propagate. It is inhomogeneous (because the refractive index is modulated by the pumps) and usually also anisotropic (if, for example, the pumps are both linearly polarized along the same direction, the only case we consider here). In anisotropic optics one proceeds by finding the normal (polarization) modes of the medium, and then constructing the desired solution by superposing them with amplitudes and phases chosen to satisfy specified boundary conditions (Born and Wolf, 1970). We set up the appropriate equations in the next section, and then find their normal modes and match boundary conditions in subsequent sections.

The analysis reveals an interesting complexity of behavior. The pumped medium (in the geometry considered) sustains 16 normal modes, 8 of which may be excited appreciably in practical cases by a probe beam entering from only one side of the medium. Except for special cases, the modes have different wave vectors, and therefore display "walkoff" phenomena familiar in anisotropic optics. The angles among modes depend upon angles of incidence and upon the parameter $(\delta n/n)^{1/2}$, where δn is the induced index change. When $\delta n/n$ is small, angles of incidence larger than this parameter cause decoupling of four of the relevant normal modes, and the remaining four suffice to describe the solutions accurately. Analytical expressions exist for the gain in each case, if nonsaturable linear loss may be ignored.

It is also of some interest to see how the absolute instability long recognized in DFWM appears in this more accurate theory: the instability still occurs, and we shall see that it occurs in an important geometry even in the presence of nonsaturable linear absorption; i.e. absorption does not prevent spontaneous oscillation. The thresholds of oscillation are of course accurately predicted by the linearized theory, although the magnitude of the oscillation is not.

Other features of the usual SVEA theory, including sensitivity of the frequency dependence of gain to forward-to-backward-pump asymmetry (Marburger and Lam, 1979b), are naturally embedded in the more general theory, but in a somewhat surprising way. Unless $\delta n/n$ is small, for example, it is not possible to "scale out" pump flux by introducing a critical length parameter. Increases in medium length L and pump flux do not have equivalent effects in the general case.

II. Linearized Equations for DFWM in Isotropic Media

A. Polarization Density

Hellwarth (1977) has discussed the third-order nonlinear polarization density $P(t)$ in the Born–Oppenheimer approximation. Our starting point is his Eq. (4.9), which may be written for isotropic media as

$$\mathbf{P}(t) = \tfrac{1}{2}\sigma(\mathbf{E} \cdot \mathbf{E})\mathbf{E} + \mathbf{E}A(id_t)(\mathbf{E} \cdot \mathbf{E}) + [\mathbf{E} \cdot B(id_t)\mathbf{E}]\mathbf{E}. \tag{1}$$

We are interested in the general case in which the total electric field $E(t)$ includes three monochromatic contributions: an intense pump field at ω and weaker fields at $\omega_{\pm} = \omega \pm \delta$. Thus, in place of $\mathbf{E}$ in Eq. (1) we insert

$$\mathbf{E} \to \mathrm{Re}[\mathbf{E}e^{-i\omega t} + \mathbf{F}e^{-i(\omega-\delta)t} + \mathbf{G}e^{-i(\omega+\delta)t}]. \tag{2}$$

Then P has a similar form, with the following coefficients replacing $\mathbf{E}$, $\mathbf{F}$, and $\mathbf{G}$, respectively:

$$P_i^0 = a_{ijk\ell}E_jE_k^*E_\ell, \tag{3}$$

$$P_i^- = 2a_{ijk\ell}(F_jE_k^*E_\ell + \tfrac{1}{2}G_k^*E_jE_\ell), \qquad P_i^+ = P_i^-(F \to G,\ G \to F). \tag{4}$$

Repeated indices indicate a sum, and only terms linear in the weak $\mathbf{F}$ and $\mathbf{G}$ are retained. P^+ is obtained from P^- by switching $\mathbf{F}$ and $\mathbf{G}$. The susceptibility tensor is

$$a_{ijk\ell} = \tfrac{1}{2}A(\delta_{ij}\delta_{k\ell} + \delta_{i\ell}\delta_{jk}) + \tfrac{1}{2}B\ \delta_{ik}\delta_{j\ell}, \tag{5}$$

$$A \equiv \tfrac{1}{4}[\sigma + 2A(0) + B(0)], \qquad B \equiv \tfrac{1}{4}[\sigma + 2\ B(0)]. \tag{6}$$

A and B are the nonlinear coefficients defined by Maker and Terhune (1965), and $A(0)$ and $B(0)$ are the zero-frequency values of $A(\omega)$ and $B(\omega)$ that appear in Eq. (1) with argument $id_t = id/dt$. We explicitly ignore dispersion in the nonlinear susceptibilities, so that $A(\delta)$ and $B(\delta)$ are nearly $A(0)$ and $B(0)$. [If dispersion were included, we would need $A(\omega)$, $B(\omega)$ at the frequencies $\omega = 0, \pm 2\delta$.] We also define

$$C = (2A - B)/(2A + B). \tag{7}$$

B. Pump Waves

We shall assume that the pump waves travel along the z axis with amplitude

$$E_j(z) = \mathscr{E}_j \exp(ik_f z) + \mathscr{E}_j' \exp(-ik_b z). \tag{8}$$

The wave equation for the positive-frequency component of the field at

ω is

$$(\nabla^2 + k_0^2 + ik_0\gamma)\mathbf{E} = -(4\pi k_0^2/n_0^2)\mathbf{P}^0, \tag{9}$$

where $k_0 = n_0\,\omega/c$, and $n_0 = n(\omega)$ is the linear refractive index at frequency ω; γ is the linear absorption coefficient for the pump waves. Assuming that the steady-state amplitudes are well behaved, we use Eq. (8) and obtain the Fourier transform of Eq. (9),

$$\begin{aligned}(-k^2 + k_0^2 + ik_0\gamma)\tilde{E}_i(\mathbf{k})\\ = -(4\pi k_0^2/n_0^2)(2\pi)^{-6}a_{ijk\ell}\int dq \int dq'\ \tilde{E}_j(\mathbf{k} - \mathbf{q} - \mathbf{q}')\tilde{E}_k^*(-\mathbf{q})\tilde{E}_\ell(\mathbf{q}').\end{aligned} \tag{10}$$

This may be evaluated explicitly by using Eq. (8), with the result

$$\begin{aligned}(-k_f^2 + k_0^2 + ik_0\gamma)\vec{\mathscr{E}} = -(4\pi k_0^2/n_0^2)\,\{A[|\mathscr{E}|^2\vec{\mathscr{E}} + |\mathscr{E}'|^2\vec{\mathscr{E}} + (\vec{\mathscr{E}}\cdot\vec{\mathscr{E}}'^*)\vec{\mathscr{E}}']\\ + \tfrac{1}{2}B[(\vec{\mathscr{E}}\cdot\vec{\mathscr{E}})\vec{\mathscr{E}}^* + 2(\vec{\mathscr{E}}\cdot\vec{\mathscr{E}}')\vec{\mathscr{E}}'^*]\},\end{aligned} \tag{11}$$

$$(-k_b^2 + k_0^2 + ik_0\gamma)\vec{\mathscr{E}}' = [\mathscr{E}' \leftrightarrow \mathscr{E}], \tag{12}$$

where the right-hand side of Eq. (12) is obtained by interchanging $\mathscr{E}$ and $\mathscr{E}'$ in Eq. (11). These equations admit any choice of pump polarization, but we shall continue with the case in which *both pumps are linearly polarized along the x axis*. Then,

$$(-k_f^2 + k_0^2 + ik_0\gamma)\mathscr{E} = -(4\pi k_0^2/n_0^2)(A + \tfrac{1}{2}B)(I + 2I')\mathscr{E},$$

where $I = |\mathscr{E}|^2$, $I' = |\mathscr{E}'|^2$. This implies

$$\begin{aligned}k_f &= k_0[1 + \alpha(I + 2I') + i\gamma k_0^{-1}]^{1/2}\\ &\approx k_0[1 + \tfrac{1}{2}\alpha(I + 2I')] + \tfrac{1}{2}i\gamma,\end{aligned} \tag{13}$$

where $\alpha = 2\pi(2A + B)/n_0^2$. Although it is not necessary, we have shown in Eq. (13) the approximate form with small loss and ignoring $(\alpha I)^2$ compared with unity. The expression $\alpha|\mathscr{E}|^2$ is twice the ratio of the nonlinear change in refractive index induced by a field $\mathscr{E}$ to the linear index, and a convenient dimensionless flux:

$$J \equiv 2\delta n/n_0 = \alpha|\mathscr{E}|^2 = 2\pi(2A + B)|\mathscr{E}|^2/n_0^2. \tag{14}$$

The backward-pump k vector is obtained from k_f by interchanging I and I'. It is convenient to define

$$\begin{aligned}g &\equiv \tfrac{1}{2}(k_f - k_b) \approx \tfrac{1}{4}k_0\alpha(I' - I),\\ h &\equiv \tfrac{1}{2}(k_f + k_b) \approx k_0 + \tfrac{3}{4}k_0\alpha(I + I') + \tfrac{1}{2}i\gamma,\end{aligned} \tag{15}$$

so that

$$E = \exp(igz)[\mathscr{E}\exp(ihz) + \mathscr{E}'\exp(-ihz - \tfrac{1}{2}\gamma L)]. \tag{16}$$

The second term is consistent with a backward amplitude of $\mathscr{E}'$ at $z = L$.

C. Signal Waves

We shall follow the procedure of the previous section to determine coupled-wave equations for the amplitudes F and G in Eq. (2). In order not to introduce too much complexity at once, we ignore the effect of non-saturable linear absorption, which will be examined in Section III E. The Fourier transforms of $P^{\pm}$ in Eq. (4) are easily evaluated once E is specified as in Eq. (16). One finds three kinds of terms.

1. Terms with **k**-vectors that are spatial harmonics of $k_{\pm} = n_{\pm}\omega_{\pm}/c$. These never phase-match with the probe or signal waves and will be ignored.
2. Terms with **k**-vectors close to $k_{\pm}$ for small angles between probe and pump beams, but increasingly mismatched at large angles. These will be called *cross-coupled waves* and must be retained at small angles.
3. Terms with **k**-vectors always near $k_{\pm}$. These give rise to the usual *direct-coupled waves:* the probe and phase-conjugate signal waves.

To be more precise, without giving all the algebra, each Fourier component $\tilde{F}_j(\mathbf{k})$ is coupled by its wave equation to itself and to three other Fourier amplitudes, one direct coupled and two cross coupled. The four modes

$$\tilde{F}_j(\mathbf{k}), \quad \tilde{G}_j^*(-\mathbf{p}), \quad \tilde{F}_j(-\mathbf{q}), \quad \tilde{G}_j^*(\mathbf{s}), \tag{17}$$

where

$$\mathbf{q} = 2h\hat{z} - \mathbf{k}, \qquad \mathbf{p} = -2g\hat{z} + \mathbf{k}, \qquad \mathbf{s} = 2(h + g)\hat{z} - \mathbf{k} \tag{18}$$

(see Fig. 1), form a closed set: the four equations they satisfy do not contain any new modes. Under the influence of the pump waves, they all couple together to form new normal modes of the radiation field. [The cross-coupled waves are $\tilde{F}_j(-\mathbf{q})$ and $\tilde{G}_j^*(\mathbf{s})$.]

To save space, we write the modes, in the order given in (17), as components of a four-vector Ψ. Then the Fourier-transformed wave equations for each component, analogous to Eqs. (11) and (12), may be written in the matrix form $M(\mathbf{k}) \cdot \Psi(\mathbf{k}) = 0$, where M is a 4×4 matrix

$$M(\mathbf{k}) = \begin{bmatrix} \mathcal{A}(\mathbf{k}) & \mathcal{B} \\ \mathcal{B}' & \mathcal{A}(-\mathbf{q}) \end{bmatrix}, \tag{19}$$

and $\mathcal{A}$, $\mathcal{B}$, $\mathcal{B}'$ are 2×2 matrices

$$\mathcal{A}(\mathbf{k}) = \begin{bmatrix} k_ik_j - (k^2 - k_-^2)\delta_{ij} + 2a_{ij}^-(I + I') & 2b_{ij}^-\mathcal{E}\mathcal{E}' \\ 2b_{ij}^+\mathcal{E}^*\mathcal{E}'^* & p_ip_j - (p^2 - k_+^2)\delta_{ij} + 2a_{ij}^+(I + I') \end{bmatrix}, \tag{20}$$

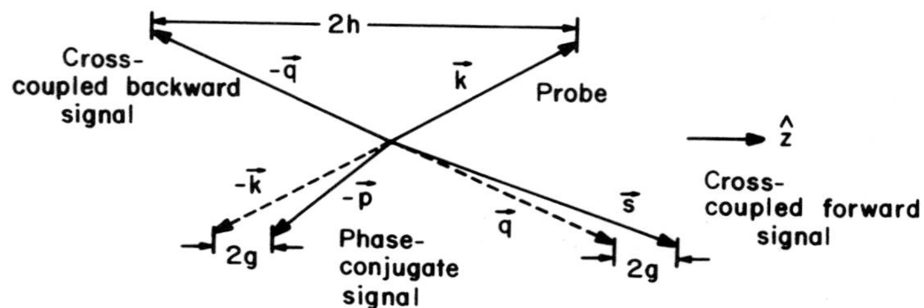

Fig. 1. Wave vectors for the four interacting modes in phase conjugation. g and h are related to the propagation vectors of the pump beams as modified by nonlinear effects [see Eq. (15)]. The small quantity g is proportional to the difference in intensity between the foward and backward beams; this difference establishes a preferred direction in the medium and thus breaks longitudinal wave-vector conservation.

$$\mathscr{B} = \begin{bmatrix} 2a_{ij}^{-}\mathscr{E}\mathscr{E}'^{*} & b_{ij}^{-}\mathscr{E}^{2} \\ b_{ij}^{+}\mathscr{E}'^{*2} & 2a_{ij}^{+}\mathscr{E}\mathscr{E}'^{*} \end{bmatrix}, \tag{21}$$

with $\mathscr{B}'$ obtained from $\mathscr{B}$ by interchanging $\mathscr{E}$ and $\mathscr{E}'$. The diagonal terms $k_i k_j$, etc., come from the $\nabla\nabla$ term in the vector wave equation, which ordinarily provides no contribution in an isotropic medium; but the pump beam induces an anisotropy that leads to polarizations of the fields that have components along the wave vectors. The constants in Eqs. (20) and (21) are

$$\begin{aligned} a_{ij}^{\pm} &= (2\pi k_{\pm}^{2}/n_{\pm}^{2})[A\delta_{ij} + (A + B)\delta_{ix}\delta_{jx}], \\ b_{ij}^{\pm} &= (2\pi k_{\pm}^{2})[2A\delta_{ix}\delta_{jx} + B\delta_{ij}]. \end{aligned} \tag{22}$$

To proceed, one fixes the component $\mathbf{k}_{\mathrm{T}}$ of $\mathbf{k}$ parallel to the entrance face by the boundary conditions on F. This is equivalent to choosing the angle of incidence of the probe wave. Then the condition $\det M(k_{\mathrm{L}}, \mathbf{k}_{\mathrm{T}}) = 0$ fixes the eigenvalues k_{L} of the component normal to the boundary for each of the normal modes $\Psi^{(n)}$. There are, in general, eight eigenvalues for each choice of polarization, but only four are consistent with an inward-traveling probe signal at the entrance face.

D. Geometry and Boundary Conditions

We are interested primarily in the signals generated by a probe beam at frequency ω_{-} traveling obliquely to the pump beams. When the angle between pump and probe is less than about 45°, boundary conditions are naturally specified on planes perpendicular to the pump-beam direction e.g., at $z = 0$ and $z = L$; however, when the incident angle is closer to 90°, it is natural to specify the signals on planes parallel to the pump beam,

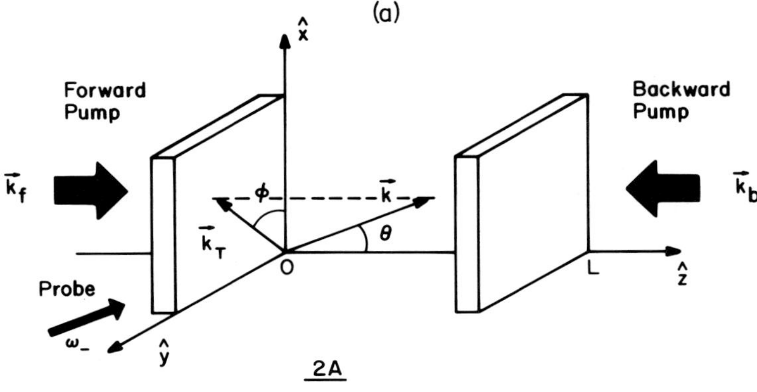

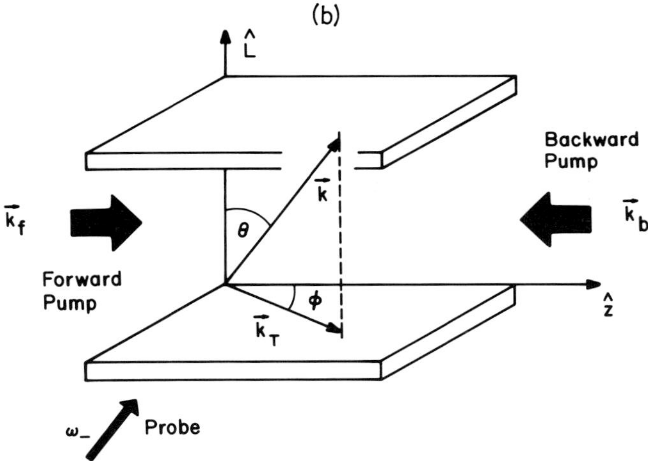

Fig. 2. Beam geometry for (a) collateral and (b) transverse boundary conditions. The entrance and exit planes are drawn with thickness to emphasize that the probe-wave vector **k** is measured inside the nonlinear medium.

e.g., at $y = 0$ and $y = L$. The two cases will be said to possess *collateral* and *transverse* boundary conditions, respectively. The geometry of the boundary conditions is shown in Fig. 2; the four waves coupled by the pumps are shown in Fig. 1. The wave vectors will always be specified by polar angles θ, ϕ. With collateral boundary conditions, θ and ϕ are defined with respect to the triplet (**x**, **y**, **z**) as usual. In the transverse case, the appropriate triplet is (**z**, $\hat{L} \times \hat{z}$, $\hat{L}$), where $\hat{L}$ is the unit vector normal

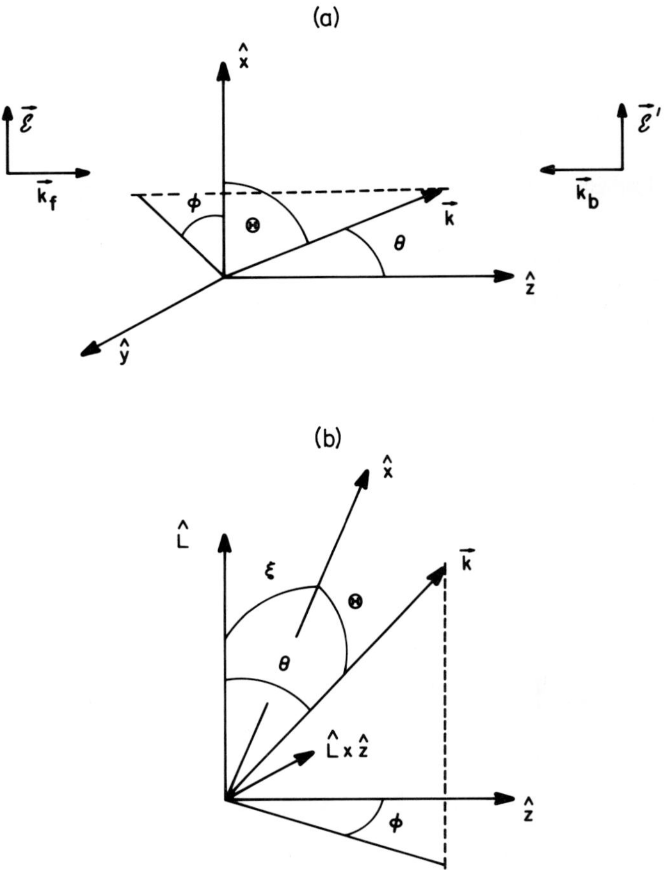

Fig. 3. Polarization geometry for (a) collateral and (b) transverse boundary conditions. Both pump beams are assumed to be linearly polarized along the x axis, which may be inclined, in the transverse case, to the normal vector L of the entrance face.

to the plane of incidence. Thus, in either case, θ is the angle of incidence of the probe-wave vector.

The pumped medium is both inhomogeneous (because the refractive index is modulated by the quasi-standing wave formed by the pumps) and anisotropic. For any probe direction $\mathbf{k}$, the probe polarization may be resolved into components perpendicular and parallel to $\hat{x}$. Contemplation of Eq. (22) convinces us that the former case corresponds to a polarization eigenstate of the system, and a mode with polarization orthogonal to $\hat{x}$ will be an *ordinary mode*, similar to the ordinary ray in anisotropic propagation. The other possibility gives *extraordinary modes*. The polarization geometry is shown in Fig. 3.

III. Approximate Theory for the Direct-Coupled Waves

A. General

Previous theories of degenerate four-wave mixing have included only two weak waves coupled by the pump wave. This is valid at large angles of incidence because the cross-coupled wave vectors **q** and **s** have z components that differ from the probe z component of roughly $k \cos \theta$ by about $k(1 - \cos \theta)$ [see Eq. (18)]. The difference can be compensated for by nonlinear index contributions to k, which are of order $k\, \delta n/n$. Thus the maximum angle at which cross-coupled waves can still be expected to be significant satisfies roughly $k(1 - \cos \theta_c) \approx k\, \delta n/n$, or $\theta_c^2 \approx \delta n/n$. For $\delta n/n \sim 10^{-3}$, which is larger than is achieved in most nonresonant experiments, the maximum angle is about 2°. An exact analysis is performed in Section IV.

Before developing the full theory in Section IV, we shall find it instructive to examine the approximate theory ignoring the cross-coupled waves. Our results will be accurate as long as $\theta > \theta_c$ and the amplitudes of the *direct-coupled* waves $\tilde{F}(\mathbf{k})$ and $\tilde{G}^*(-\mathbf{p})$ are not too large. When these amplitudes become very large, they can drive appreciable excitation of the cross-coupled modes despite the lack of phase matching.

B. Ordinary-Mode Solutions in the Fully Degenerate Case

The cross-coupled amplitudes form the lower two components of Ψ [see Eq. (17)], and neglecting them reduces our eigenvector equation to

$$\mathcal{A}_\nu(\mathbf{k}) \cdot U_\nu = 0, \qquad \nu = \mathrm{o, e},$$

where $U_\nu = [\tilde{F}_\nu(\mathbf{k}), \tilde{G}_\nu^*(-\mathbf{p})]$ is the upper two-vector in Ψ, and ν refers to one of the principal polarizations. In this section we take $\nu = \mathrm{o}$, for the ordinary modes: polarization perpendicular to the pump polarization **x**. The solutions $\mathbf{k}_n$ of

$$\det \mathcal{A}_\mathrm{o}(\mathbf{k}_n) = 0$$

are simply obtained when the frequency detuning vanishes ($\delta = 0$) and the forward- and backward-pump fluxes are equal ($|\mathscr{E}| = |\mathscr{E}'|$). These two conditions define the *fully degenerate* case. In this case, $\mathbf{p} = \mathbf{k}$, and $\mathscr{A}_o$ may be written

$$\mathscr{A}_o(\mathbf{k}) = \begin{bmatrix} \sigma & 2b \\ 2b^* & \sigma \end{bmatrix}. \tag{23}$$

where

$$\begin{gathered} \sigma = -k^2 + k_0^2 + 4a, \\ a = (2\pi k_0^2/n_0^2)A|\mathscr{E}|^2, \qquad b = (2\pi k_0^2/n_0^2)B\mathscr{E}^2. \end{gathered} \tag{24}$$

The four eigenvalues k_n are [see Eqs. (7) and (14) for C and J]

$$k_1 = -k_3 = k_0(1 + 2J)^{1/2}, \qquad k_2 = -k_4 = k_0(1 + 2CJ)^{1/2}, \tag{25}$$

and the corresponding eigenvectors are

$$U_o^{(1)} = U_o^{(3)} = (1, e^{-i\psi}), \qquad U_o^{(2)} = U_o^{(4)} = (1, -e^{-i\psi}), \tag{26}$$

where ψ is the phase of the product of the forward- and backward-pump amplitudes. Only $U_o^{(1)}$ and $U_o^{(2)}$ are required to match boundary conditions that specify an ingoing probe beam at $z = 0$, because k_3 and k_4 are oriented in the wrong direction.

Note that in each eigenmode the weights of the probe signal $F(\mathbf{k})$ and the backscattered signal $G(\mathbf{k})$ are equal. *An incoming probe thus excites equally two modes with different* **k** *vectors*. (This leads to double refraction, to be discussed in Section III C.) In the steady state, the amplitudes of these modes must be selected to obey the boundary conditions at the extremes of the nonlinear medium. If the forward probe is launched at $z = 0$ with amplitude F_0, and no other waves enter the medium, these conditions fix the normal coordinates A_1, A_2 in the general solution

$$U_o(\mathbf{r}) = A_1 U_o^{(1)} \exp i\mathbf{k}_1 \cdot \mathbf{r} + A_2 U_o^{(2)} \exp i\mathbf{k}_2 \cdot \mathbf{r}. \tag{27}$$

This is the case of collateral boundary conditions. The transverse part of both $\mathbf{k}_1$ and $\mathbf{k}_2$ is $\mathbf{k}_\mathrm{T}$, where $k_\mathrm{T} = k_0 \sin\theta$, with θ the angle of incidence of the probe-wave vector. Thus $k_{n\mathrm{L}}^2 = k_n^2 - k_\mathrm{T}^2$, and the angle of incidence of the nth-mode vector is θ_n, where $\sin\theta_n = k_\mathrm{T}/k_n$, $n = 1, 2$. The boundary conditions are explicitly

$$A_1 + A_2 = F_0, \qquad A_1 e^{i\phi_1} - A_2 e^{i\phi_2} = 0, \tag{28}$$

where $\phi_n = k_{n\mathrm{L}}L - \psi$. We define $\phi_- = \phi_1 - \phi_2$; then

$$A_1/F_0 = (A_2/F_0)^* = (1 + e^{i\phi_-})^{-1}. \tag{29}$$

The backscattered signal emergent at $z = 0$ is obtained from Eq. (27):

$$\overline{G}(0) = ie^{i\psi}F_0^* e^{-i\mathbf{k}_T\cdot\mathbf{r}} \tan \tfrac{1}{2}\phi_- . \tag{30}$$

The important feature of this result is that the signal is proportional to the conjugate of the incident probe F_0. The simple tangent gain formula is correct only at angles of incidence larger than about θ_c in the collateral case, but we shall find in Section IV that even at small angles there is a phase-conjugate wave, although with a more complicated gain formula owing to the cross-coupled waves.

At oblique angles, the double refraction implied by the difference between θ_1 and θ_2 causes a finite probe beam to split into two beams as it traverses the medium. Thus, one should observe the transmitted probe beam coming from two spots on the exit plane of the medium. Roughly speaking, the two modes correspond to in-phase and out-of-phase combinations of coupled probe and signal waves. The walkoff angle is greatest for large angles of incidence in collateral boundary conditions and approaches $(2J|1 - C|)^{1/2}$. This is about 5° for $J = 10^{-3}$ and C typical of CS_2.

Cross-coupled waves and double refraction can be reduced by operating with transverse boundary conditions as in Fig. 2. If L measures the distance between entrance and exit faces parallel to the z axis, Eqs. (29) and (30) are correct if θ is the angle of incidence for transverse boundary conditions. At normal incidence (e.g., along the y axis with polarization along $\hat{z}$), $k_T = 0$ and $\theta_n = 0$ for both modes, so there is no walkoff.

Equation (30) shows that the phase-conjugate signal will grow without bound compared with the probe amplitude when ϕ_- approaches an odd multiple of π. This condition is

$$(\cos^2\theta + 2J)^{1/2} - (\cos^2\theta + 2CJ)^{1/2} = \pi/k_0L. \tag{31}$$

For θ not too close to 90°,

$$\phi_- \approx (1 - C)Jk_0L/\cos\theta = \pi L/L_c \cos\theta, \tag{32}$$

where $L_c^{-1} = 4\omega B|\mathscr{E}|^2/n_0c$ is identical to the critical length in Marburger and Lam (1979a), and equal to $2|\kappa|/\pi$ in the notation of Bloom and Bjorklund (1977) and Yariv and Pepper (1977). (See also Chapters 2 and 3.)

C. Extraordinary-Mode Solutions in the Fully Degenerate Case

The modes $U_0^{(n)}$ with ordinary polarization (perpendicular to the polarization of the pump waves) have wave vectors k_n whose magnitudes do not depend on their directions. In the extraordinary case this is no longer possible, and we expect to find directionally dependent refractive indices, and polarizations not orthogonal to the wave vectors. Both conditions are of course features of extraordinary waves in linear uniaxial media.

We seek modes $U_e^{(n)} = U_o^{(n)}\hat{u}^{(n)}$, both of whose components have the same polarization direction, where $U_o^{(n)}$ is the same as in Eq. (26). Then, from Eq. (20)

$$\sum_j [k_i k_j - k^2\delta_{ij} + (k_0^2\delta_{ij} + 4a_{ij} + 2|b_{ij}|)]u_j^{(n)} = 0, \qquad n = 1,3, \tag{33}$$

$$\sum_j [k_i k_j - k^2\delta_{ij} + (k_0^2\delta_{ij} + 4a_{ij} - 2|b_{ij}|)]u_j^{(n)} = 0, \qquad n = 2,4, \tag{34}$$

by analogy to the notation of Eqs. (23) and (24). The quantities in parentheses play the role of dielectric tensors (times k_0^2) in the familiar theory of propagation in anisotropic media. Equations (20) and (22) give for the tensors of Eqs. (33) and (34), respectively,

$$\boldsymbol{\epsilon}^n \equiv \mathsf{I} + 2J\begin{bmatrix} 3 & 0 & 0 \\ 0 & 1 & 0 \\ 0 & 0 & 1 \end{bmatrix}, \qquad n = 1, 3,$$

$$\boldsymbol{\epsilon}^n \equiv \mathsf{I} + 2J\begin{bmatrix} 1 & 0 & 0 \\ 0 & C & 0 \\ 0 & 0 & C \end{bmatrix}, \qquad n = 2, 4.$$

These should be compared with Eqs. (25), which give the ordinary refractive indices as $|k_n|/k_0$. The usual formula for the refractive index of a wave whose **k** vector makes an angle Θ with the optical axis $\hat{x}$,

$$k_0^2/k_n^2 = (\epsilon_{11}^n)^{-1}\sin^2\Theta + (\epsilon_{22}^n)^{-1}\cos^2\Theta,$$

gives, to lowest order in J (the exact expression is cumbersome),

$$k_n^2 = k_0^2[1 + 2J(3 - 2\cos^2\Theta)], \qquad n = 1, 3, \tag{35}$$

$$k_n^2 = k_0^2[1 + 2J(\sin^2\Theta + C\cos^2\Theta)], \qquad n = 2, 4. \tag{36}$$

For collateral boundary conditions $\cos\Theta = \sin\theta\cos\phi$, whereas in the transverse case $\cos\Theta = \cos\theta\cos\xi + \sin\theta\sin\phi\sin\xi$, where ξ is the inclination of the x axis from the incident surface normal (see Fig. 3b). The polarizations corresponding to Eqs. (35) and (36) make an angle $\pi/2 - \Theta'$ with the optical axis, where $\Theta_n' = \Theta + \alpha_n$,

$$\tan\Theta_n' = (\epsilon_{22}^n/\epsilon_{11}^n)\tan\Theta,$$

or

$$\alpha_n \approx -2J\sin 2\Theta, \qquad n = 1, 3,$$

$$\alpha_n \approx -(1 - C)J\sin 2\Theta, \qquad n = 2, 4.$$

TABLE I

Refractive Index and Critical Lengths[a]

Mode	Refractive index[b]	Critical length[c]
Ordinary	$n_{o1}^2 = 1 + 2J$	$\pi/L_{oc} \simeq (1 - C)k_o J$
	$n_{o2}^2 = 1 + 2CJ$	
Extraordinary[d]	$n_{e1}^2 \simeq 1 + 2(3 + 2\cos^2\Theta)J$	$\pi L_{oc}/L_{ec} \simeq 1 + (2 + C)(1 - C)^{-1}\sin^2\Theta$
	$n_{e2}^2 \simeq 1 + 2(\sin^2\Theta + C\cos^2\Theta)J$	

[a] θ is the angle of incidence of the probe **k** vector, θ_n the angle of incidence of the nth normal-mode vector $\mathbf{k}_n$, and $L_{\nu c}$ the critical length corresponding to polarization type ν = o, e. The phase-conjugate small-signal gain for normal incidence ($\theta = 0$) is infinite when the medium length L equals the critical length $L_{\nu c}$.

[b] $\sin\theta_n = n_{\nu n}\sin\theta$.

[c] The small-signal gain formula for the completely degenerate case is

$$g = \tan^2(\pi L/2L_{\nu c}\cos\theta).$$

[d] $\cos\Theta = \sin\theta\sin\phi$ for collateral boundary conditions; $\cos\Theta = \cos\theta\cos\xi + \sin\theta\sin\phi\sin\xi$ for transverse conditions.

Equations (27)–(30) are valid for this case if the values of k_1 and k_2 given by Eqs. (35) and (36) are used. The maximum angle between the two **k** vectors is again of order $2J^{1/2}$.

The condition corresponding to Eq. (31) for infinite gain is

$$[\cos^2\theta + 2J(3 - 2\cos^2\Theta)]^{1/2} - \{\cos^2\theta + 2J[1 - (1 - C)\cos^2\Theta]\}^{1/2} \approx [2 - (1 + C)\cos^2\Theta]J/\cos\theta = \pi/k_0 L.$$

Thus,

$$\phi_- \approx (\pi L/L_c\cos\theta)[2 - (1 + C)\cos^2\Theta](1 - C)^{-1}. \tag{37}$$

Refractive indices and critical lengths for the various cases of polarization and boundary conditions are summarized in Table I. Note that the length L always appears multiplied by $J = 2\delta n/n$ in the lowest-order expression for the gain, Eq. (30), but this is not a feature of the exact expressions.

D. Effects of Detuning and Pump Asymmetry with Ordinary Polarization

The effect of detuning may be adequately understood in the context of ordinary polarization. For this and subsequent effects, motivated readers may analyze the extraordinary polarization cases themselves, using the method shown in the previous section.

Following the approach of Section III.B, we require the eigenvalues and eigenvectors of $\mathscr{A}(\mathbf{k})$ in Eq. (20). Dividing the top row by $(\omega_-/\omega)^2$ and the bottom row by $(\omega_+/\omega)^2$, we find the modified $\mathscr{A}(\mathbf{k})$ to be

$$\begin{bmatrix} -(\omega/\omega_-)^2k^2 + (n_-/n_0)^2k_0^2 + 4a & 2b \\ 2b^* & -(\omega/\omega_+)^2p^2 + (n_+/n_0)^2k_0^2 + 4a \end{bmatrix}, \tag{38}$$

where

$$\mathbf{p} = \mathbf{k} - 2g\mathbf{z} \qquad \text{[see Eqs. (15) and (18)]}$$

and

$$a = (2\pi k_0^2/n_0^2)A\tfrac{1}{2}(I + I'), \qquad b = (2\pi k_0^2/n_0^2)B\mathscr{E}\mathscr{E}'.$$

This will be similar to the matrix equation (23) if the frequency detuning $\delta = \omega_+ - \omega$ and pump reflection coefficient $R = I'/I$ are chosen to make the diagonal terms equal. That is, the effect of detuning can be compensated for by changing the forward-to-backward-pump ratio. This can be seen as follows.

The condition for equal diagonal terms may be written

$$(\omega_+^2 - \omega_-^2)(k^2/\omega_-^2) + (n_+^2 - n_-^2)(\omega_+^2/c^2) = -4gk_z + 4g^2. \tag{39}$$

The first term arises from detuning of the incident probe from the pump frequency, and the second term is the corresponding dispersion-induced phase mismatch. On the right-hand side are phase-mismatch terms arising from the inequality of the pump beams. Computation is easier when we keep only terms of lowest order in δ and J. Then Eq. (39) becomes

$$4(\delta/\omega)(1 + n'n_0^{-1}\omega) \approx J(1 - R)\cos\zeta, \tag{40}$$

where $n' = dn(\omega)/d\omega$, and ζ is the angle between the mode $\mathbf{k}$ vector and the z axis. If Eq. (39) is satisfied, Eq. (38) looks like Eq. (23), with

$$\sigma = -(\omega k/\omega_-)^2 + (n_-k_0/n_0)^2 + 4a.$$

The eigenvalues corresponding to Eq. (25) are

$$k_1^2/k_-^2 = 1 + 2J_-[\tfrac{1}{4}(1 + C)(1 - R^{1/2})^2 + R^{1/2}], \tag{41}$$

$$k_2^2/k_-^2 = 1 + 2J_-[\tfrac{1}{4}(1 + C)(1 - R^{1/2})^2 + CR^{1/2}], \tag{42}$$

where

$$J_- = (n_0/n_-)^2\,\mathrm{J}.$$

Equations (41) and (42) give refractive indices for the probe-wave vectors. The backscattered signals have different wave vectors:

$$\mathbf{p}_n = \mathbf{k}_n - 2g\hat{z} = \mathbf{k}_n + \tfrac{1}{2}k_0J(1 - R)\hat{z},$$

so the signals are not in the plane of incidence of the probe beam unless $\hat{z}$ is in that plane. (It need not be for transverse boundary conditions.)

Now that we have different wave vectors for probe and signal, we must write the general solution as

$$U(\mathbf{r}) = A_1 \begin{bmatrix} U_1^{(1)} e^{i\mathbf{k}_1\cdot\mathbf{r}} \\ U_2^{(1)} e^{i\mathbf{p}_1\cdot\mathbf{r}} \end{bmatrix} + A_2 \begin{bmatrix} U_1^{(2)} e^{i\mathbf{k}_2\cdot\mathbf{r}} \\ U_2^{(2)} e^{i\mathbf{p}_2\cdot\mathbf{r}} \end{bmatrix}. \tag{43}$$

For the special case we are considering, when detuning is just compensated for by pump asymmetry, the backward signal is given precisely by the tangent gain equation (30), with

$$\begin{aligned} \phi_- &= (p_{1L} - p_{2L})L = (k_{1L} - k_{2L})L \\ &\approx \pi R^{1/2} L/L_c \cos\theta \qquad \text{(compensated detuning).} \end{aligned} \tag{44}$$

That is, the pump flux in the gain expression for equal pumps is replaced with the geometric mean of the forward and backward pumps. Otherwise the formulas are the same in the small-J limit. Equation (40) gives the relation between detuning and pump reflection coefficient for this special case. Because of the factor $\cos\zeta = \hat{z}\cdot\mathbf{k}$ in Eq. (40), detuning is more difficult to compensate for as the modes point farther away from the pump directions; or, to put it more postively, the effect of pump asymmetry is weaker as ζ increases. If $\zeta = \frac{1}{2}\pi$ as with normal incidence in the transverse boundary condition, then the probe and signal cannot sample the relative pump phase at different z values, and the phase mismatch introduced by pump asymmetry can have no effect. Note that Eq. (40) is satisfied for $\delta = 0$, $\zeta = \frac{1}{2}\pi$, so that the gain equation for *normal incidence,*

$$|\overline{G}(0)/F_0| = \tan(\tfrac{1}{2}\pi R^{1/2} L/L_c) \qquad \text{(transverse boundary condition),} \tag{45}$$

is valid for arbitrary pump asymmetry. This is the gain expression first derived by Bloom and Bjorklund (1977) and Yariv and Pepper (1977). It is gratifying to see that it is valid for such an important special case.

At general angles, pump asymmetry uncompensated for by detuning prevents the gain from growing too large. This is most easily seen by ignoring terms of order J^2 in the matrix equation (38) for $\delta = 0$:

$$\mathscr{A}(\mathbf{k}) \approx \begin{bmatrix} \sigma & 2b \\ 2b^* & \sigma + k_0^2 J(1 - R)\cos\zeta \end{bmatrix}. \tag{46}$$

The extra contribution to the lower diagonal element spoils the symmetry that caused the two components of the mode eigenvectors to have equal weight. Not having equal weight, they can no longer interfere with each other to provide the vanishing denominators as in Eq. (29). Letting

$$\Delta = [k_0^2 J(R - 1)\cos\zeta](4|b|)^{-1} = \tfrac{1}{2}(R - 1)R^{-1/2}(1 - C)^{-1}\cos\zeta,$$

one finds eigenvectors

$$U^{(1)} = \{1, e^{-i\psi}[\Delta + (1 + \Delta^2)^{1/2}]\},$$
$$U^{(2)} = \{1, e^{-i\psi}[\Delta - (1 + \Delta^2)^{1/2}]\}.$$

Equation (43) still gives the general solution. The solution for the emergent signal at the entrance face yields the gain expression

$$|\overline{G}(0)/F_0|^2 = (\sin^2 \tfrac{1}{2}\phi_-)/(\Delta^2 + \cos^2 \tfrac{1}{2}\phi_-), \tag{47}$$

where Δ is the ratio of two small quantities, and may be large or small. For $C = -0.4$ (the value for CS_2) and $\zeta = 45°$, $R = \frac{1}{2}$, one finds $\Delta \approx -0.63$. The angle ϕ_- in Eq. (47) depends upon p_n as in Eq. (44). [See Eq. (49).]

Detuning can be treated just as was pump asymmetry. Setting the pumps equal, so $R = 1$, one finds that Eq. (38) may be written in the form (46), with a correction $4|b|\Delta'$ to the lower diagonal. In this case, however, σ and Δ' are given by

$$\sigma = -(\omega k/\omega_-)^2 + (n_- k_0/n_0)^2 + 4a,$$
$$(\omega_+/\omega)^2 4|b|\Delta' = (\omega_+^2 - \omega_-^2)(k/\omega_-)^2 + (n_+^2 - n_-^2)(\omega_+/c)^2.$$

Thus the gain function is the same as Eq. (47), with Δ' replacing Δ and an appropriate redefinition of ϕ_-.

When both detuning and pump asymmetry are present, the gain looks like

$$|\overline{G}(0)/F_0|^2 = (\sin^2 \tfrac{1}{2}\phi_-)/[(\Delta + \Delta')^2 + \cos^2 \tfrac{1}{2}\phi_-], \tag{48}$$

with

$$\phi_- \approx R^{1/2}\phi_{-0}[1 + (\Delta + \Delta')^2]^{1/2}, \tag{49}$$

where ϕ_{-0} is the degenerate phase difference given by Eq. (32).

The frequency sensitivity of degenerate four-wave mixing has been studied by Pepper and Abrams (1978), who proposed its use as a frequency filter; this frequency sensitivity is discussed in greater detail in Chapters 2 and 3. Equation (49) shows that the central passband of the filter may be tuned by changing the pump reflection coefficient. For comparison with the formulas of Marburger and Lam (1979b), Δ and Δ' here correspond to $-\Delta k_m/2\alpha_\perp$ and $\Delta k/2\alpha_\perp$ there.

E. Effect of Loss

Linear nonsaturable absorption affects the weak waves in DFWM in two ways. It attenuates them, and it modifies the pump intensities spatially so that the induced nonlinear index changes are inhomogeneous.

The latter effect is awkward to investigate analytically unless the pump absorption is very small. Unfortunately, the pump absorption is large in some of the most interesting nonlinear materials. There is one geometry, however, in which the inhomogeneous phase effect is absent: normal incidence with transverse boundary conditions. We have seen that the detuning effect of pump asymmetry is absent in the transverse normal geometry [Eq. (45)]. Thus the z-dependent pump asymmetry caused by pump loss should also be tractable in this geometry. This is the only lossy case we shall examine.

Let the probe and signal beams travel along the y axis and have polarizations along z. This is the ordinary polarization case. The weak waves F and G depend upon z only as a parameter, and we may Fourier-transform their wave equations on the y variable only, using Eq. (16) for $E(z)$. One finds that the Fourier amplitude $\tilde{F}(k, z)$ is coupled only to $\tilde{G}^*(-k, z)$, so there are no cross-coupled waves, as expected in this geometry. The remaining wave equations may be written as in Section II.C with

$$\Psi = [\tilde{F}(k, z), \tilde{G}^*(-k, z)],$$
$$M = \begin{bmatrix} -k^2 + k_-^2 + 2a^-|E(z)|^2 + ik_-\gamma_- & b^-E^2(z) \\ b^+E^{*2}(z) & -k^2 + k_+^2 + 2a^+|E(z)|^2 - ik_+\gamma_+ \end{bmatrix}, \tag{50}$$

where

$$a^\pm = 2\pi(\omega_\pm/c)^2 A, \qquad b^\pm = 2\pi(\omega_\pm/c)^2 B.$$

In the degenerate case $\omega_+ = \omega_- = \omega$, M may be written

$$M = \begin{bmatrix} \sigma + ik_0\gamma & bE^2 \\ bE^{*2} & \sigma - ik_0\gamma \end{bmatrix},$$

where $\sigma = -k^2 + k_0^2 + 2a|E|^2$. Defining

$$\sigma_0^2 = b^2|E|^4 - k_0^2\gamma^2, \tag{51}$$

one finds eigenvalues

$$k_1 = -k_3 = [k_0^2 + 2a|E(z)|^2 - \sigma_0]^{1/2},$$
$$k_2 = -k_4 = [k_0^2 + 2a|E(z)|^2 + \sigma_0]^{1/2}$$

and eigenvectors

$$U^{(1)} = U^{(3)} = \{1, -\exp[i(\Phi - \psi)]\},$$
$$U^{(2)} = U^{(4)} = \{1, \exp[-i(\Phi + \psi)]\}.$$

Here $\psi = \psi(z)$ is the phase of E^2 at z (assumed independent of y) and

$$\tan \Phi = k_0\gamma/\sigma_0, \qquad \sigma_0^2 > 0,$$
$$\Phi = \tfrac{1}{2}\pi - i \ln[(|\sigma_0| + k_0\gamma)/b|E|^2], \qquad \sigma_0^2 < 0.$$

The phase Φ is real or complex depending upon the sign of σ_0^2.

When σ_0^2 is positive, the weights of the probe and signal beams in the normal modes are equal, and infinite gains are possible, as in the previous cases. This somewhat surprising result is a consequence of the possibility that the nonlinear gain for F and G may be large enough to overcome the loss. The condition $\sigma_0^2 \geq 0$ may be written, by using Eqs. (51) and (32) with $|\mathscr{E}|^2 \rightarrow |E|^2$, as

$$(2/\pi)L_c \leq \gamma^{-1}, \tag{52}$$

which shows that the absorption length must exceed $2/\pi$ times the critical gain length before the loss can be compensated for. Note that $|E(z)|^2$ is the sum of the local forward- and backward-pump intensities plus an interference term [see Eq. (16)].

When Eq. (52) is satisfied and the loss is compensated for by the gain, the backward phase-conjugate signal at $y = 0$ is conveniently written

$$\overline{G}(0) = ie^{i\psi}F_0^*(\sin \tfrac{1}{2}\phi_-)/\cos(\tfrac{1}{2}\phi_- - \Phi) \qquad \text{(compensated loss)}, \tag{53}$$

where $\phi_- = (k_1 - k_2)L$. This reduces to Eq. (30) when $\gamma \rightarrow 0$, and shows that the gain reaches infinity at a critical transverse length increased by the presence of absorption. Both ϕ_- and Φ depend upon z through the magnitude of the pump field, and the gain reflects this dependence.

When loss dominates gain, Eq. (53) is still correct, but now ϕ_- and Φ are complex. The expression for the gain is not too complicated when the index change is small and the characteristic absorption and critical lengths are much longer than a wavelength (so far the analysis has been *exact* for the case we are considering). The result may be written

$$\overline{G}(0) = ie^{i\psi}F_0^*[\sinh(\tfrac{1}{2}|\phi_-|)/\sinh(\tfrac{1}{2}|\phi_-| + \ln x)], \tag{54}$$

where $x = (|\sigma_0| + k_0\gamma)/b|E|^2$. The expression in brackets is a monotonic function of the medium length L with maximum limiting magnitude $1/x$, which approaches $\pi/4\gamma L_c$ for large loss. Figure 4 shows how the gain changes with the length of the medium for various values of the loss. When the critical length L_c and the absorption length are both much longer than a wavelength of the radiation, ϕ_- may be written

$$\phi_- \approx \tfrac{1}{2}\pi(L/L_c)[1 - (2\gamma L_c/\pi)^2]^{1/2},$$

and L_c once again provides a scale of length for the process.

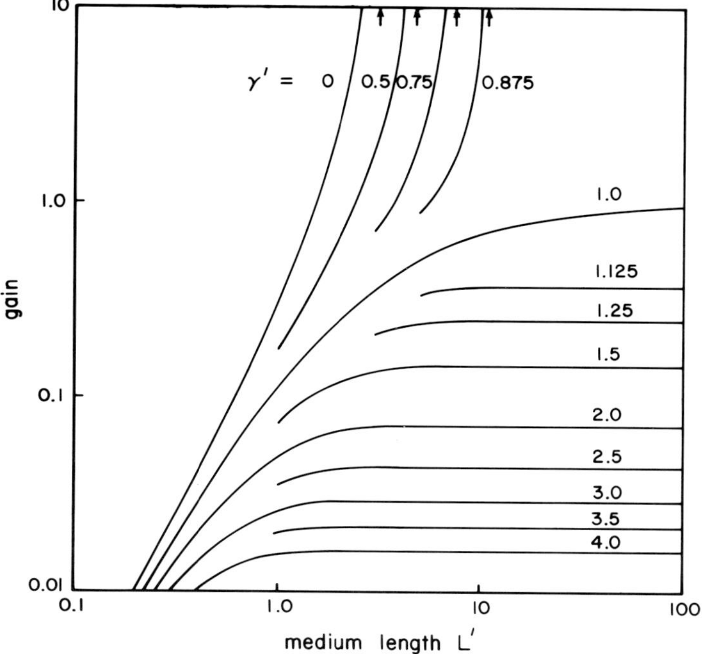

Fig. 4. Variation of small-signal phase-conjugate gain with dimensionless medium length $L' = \pi L/2L_c$ for different values of dimensionless loss $\gamma' = 2L_c\gamma/\pi$, according to Eqs. (53) and (54). Infinite-gain points are indicated at the top for $\gamma' < 1$. For $\gamma' \geq 1$, the gain saturates at finite values as shown. This situation applies for normal incidence and transverse boundary conditions.

IV. DFWM with Cross-Coupled Waves

We shall investigate the solutions of the full set of coupled-mode equations only in the fully degenerate case, with no detuning and equal forward- and backward-pump intensities. Our aim is to discover the influence of the cross-coupled waves on the gain of the phase-conjugate signal. As indicated at the beginning of Section III, we expect to find little influence except at small angles of incidence in the collateral geometry. We have already examined the transverse-geometry normal-incidence case exactly (in the linear approximation) in Section III.E, and found no cross-coupled waves.

Considering, for simplicity, only the case of ordinary polarization, we find that full degeneracy implies that $\mathcal{B} = \mathcal{B}'$ in the fundamental matrix M

of Eq. (19) whose eigenvectors are the normal modes. To summarize,

$$M = \begin{bmatrix} \mathscr{A}(\mathbf{k}) & \mathscr{B} \\ \mathscr{B} & \mathscr{A}(-\mathbf{q}) \end{bmatrix}, \tag{19$'$}$$

$$\mathscr{A}(\mathbf{k}) = \mathscr{A}_0\mathbf{k}) = \begin{bmatrix} \sigma(k) & 2b \\ 2b^* & \sigma(k) \end{bmatrix}, \tag{23$'$}$$

$$\mathscr{B} = \begin{bmatrix} 2a & b \\ b^* & 2a \end{bmatrix},$$

$$\mathbf{q} = 2h\hat{z} - \mathbf{k}, \qquad h = k_0(1 + \tfrac{3}{2}J), \tag{55}$$

and a, b, and $\sigma(k)$ are defined in Eqs. (24). The matrix M is block-diagonalized by the unitary transformation

$$U = 2^{-1/2}\begin{bmatrix} 1 & 0 & e^{i\psi} & 0 \\ e^{-i\psi} & 0 & -1 & 0 \\ 0 & 1 & 0 & e^{i\psi} \\ 0 & e^{-i\psi} & 0 & -1 \end{bmatrix}, \tag{56}$$

where

$$M' = U^+MU = \begin{bmatrix} N_+ & 0 \\ 0 & N_- \end{bmatrix},$$

$$N_\pm = \begin{bmatrix} \sigma(k) & 2a \\ 2a & \sigma(q) \end{bmatrix} \pm |b| \begin{bmatrix} 2 & 1 \\ 1 & 2 \end{bmatrix}$$

and ψ is the phase of the product of forward- and backward-pump amplitudes, as in Eq. (26). An eigenvector of M will be written as before:

$$\begin{aligned} \Psi(k) &= [\tilde{F}(\mathbf{k}), \tilde{G}^*(-\mathbf{k}), \tilde{F}(-\mathbf{q}), \tilde{G}^*(\mathbf{q})] \\ &\equiv [F, \overline{G}^*, \overline{F}, G^*]. \end{aligned} \tag{57}$$

Note that the columns of U in Eq. (56) are proportional to the eigenvectors (26) of the direct-coupled modes. There will be eight eigenvalues, four each of N_+ and N_-, and corresponding eigenvectors. Four of the eigenvectors have upper components, which are eigenvectors of N_+ with vanishing lower components, and similarly for N_-.

The characteristic equations

$$\det N_\pm(k) = 0 \tag{58}$$

are conveniently written in the notation $J_+ = J$, $J_- = CJ$, and

$$\mathbf{k} = (h + f)\hat{z} + \mathbf{k}_\mathrm{T}, \qquad H_\pm = k_0^2(\cos^2\theta + 2J_\pm),$$

where f is fixed by Eq. (58), and $k_\mathrm{T} = k_0 \sin\theta$, $\mathbf{k}_\mathrm{T} \cdot \hat{z} = 0$ as in Eq. (27) *et seq*. Then Eq. (58) may be written

$$(h^2 - f^2)^2 - 2H_\pm(h^2 + f^2) + H_\pm^2 - J_\pm^2 k_0^4 = 0,$$

which is a quadratic in f^2, with solutions

$$\begin{aligned} f_\pm^2 &= (h^2 + H_\pm) - (4h^2H_\pm + J_\pm^2 k_0^4)^{1/2}, \\ \hat{f}_\pm^2 &= (h^2 + H_\pm) + (4h^2H_\pm + J_\pm^2 k_0^4)^{1/2}. \end{aligned} \tag{59}$$

Examining the $J \to 0$ limit, we see that only the first four eigenvalues correspond to forward-traveling probe beams, so we may disregard the $\hat{f}$ solutions. We shall label the remaining eigenvalues f_n in the sequence

$$\{f_n\} = (|f_+|, -|f_+|, |f_-|, -|f_-|). \tag{60}$$

The corresponding eigenvectors are easily found to be

$$\begin{aligned} V'^{(1)} &= (1, e^{\Phi+}, 0, 0), \\ V'^{(2)} &= (e^{\Phi+}, 1, 0, 0), \\ V'^{(3)} &= (0, 0, 1, e^{\Phi-}), \\ V'^{(4)} &= (0, 0, e^{\Phi-}, 1), \end{aligned} \tag{61}$$

where

$$\tanh \Phi_\pm = 2h|f_\pm|/[2h^2 - (4h^2H_\pm + J_\pm^2 k_0^4)^{1/2}]. \tag{62}$$

Matching boundary conditions is now a straightforward computation, best performed in the unprimed basis: $V^{(n)} = UV'^{(n)}$. The general solution may be written

$$\Psi(\mathbf{r}) = \sum_{n=1}^{4} A_n \Psi^{(n)}(z) \exp[if_n z] \exp[i\mathbf{k}_T \cdot \mathbf{r}], \tag{63}$$

where $\Psi^{(n)}(z)$ is a four-vector whose components are related to those of $V^{(n)}$ through

$$\Psi^{(n)}(z) = [V_1^{(n)} e^{ihz}, V_2^{(n)} e^{ihz}, V_3^{(n)} e^{-ihz}, V_4^{(n)} e^{-ihz}]. \tag{64}$$

[All we are doing is multiplying each Fourier amplitude by the appropriate exponential, using Eqs. (55) and (57).] The dependence on transverse variables factors out, and we may ignore it in what follows. At the entrance plane $z = 0$ and exit plane $z = L$ we have

$$\begin{aligned} \Psi(0) &= [F_0, \overline{G}^*(0), \overline{F}(0), 0], \\ \Psi(L) &= [F(L), 0, 0, G^*(L)], \end{aligned} \tag{65}$$

where F_0 is regarded as the known probe amplitude, and all other quantities are to be determined. This can be done by eliminating A_n between Eqs. (63) and (65), by exact analogy to the method of Section III.B. As before, there are two characteristic path lengths like ϕ_1 and ϕ_2 in Eq. (28), but we shall call them ϕ_+ and ϕ_- here: $\phi_\pm = |f_\pm|L$. Then the coefficients A_n in Eq. (63) may be written

$$\begin{aligned} A_1(\Phi_\pm, \phi_\pm) &= SF_0/D, \qquad & A_2(\Phi_\pm, \phi_\pm) &= TF_0/D, \\ A_3(\Phi_\pm, \phi_\pm) &= A_1(\Phi_\mp, \phi_\mp), \qquad & A_4(\Phi_\pm, \phi_\pm) &= A_2(\Phi_\mp, \phi_\mp), \\ S(\Phi_\pm, \phi_\pm) &= e^{-\Phi_+}T(\mp\Phi_\pm, \mp\phi_\pm), \end{aligned} \tag{66}$$

$$e^{-\phi_+}T(\Phi_\pm, \phi_\pm) = e^{-i\phi_+}\sinh\Phi_- + \sinh(\Phi_- - i\phi_-) - ie^{-\Phi_+}\sin\phi_-, \tag{67}$$

$$\begin{aligned} \tfrac{1}{4}D = &-\sinh\Phi_+\sinh\Phi_-(1 + \cos\phi_+\cos\phi_-) \\ &-\sin\phi_+\sin\phi_-(1 + \cosh\Phi_+\cosh\Phi_-). \end{aligned} \tag{68}$$

With these coefficients, the solutions for the backward probe and signal waves emerging from the medium at $z = 0$, subject to boundary conditions (65), are

$$\overline{F}(0)/F_0 = 2(T - Se^{\Phi_+})/D, \tag{69}$$

$$e^{i\psi}\overline{G}^*(0)/F_0 = 2(S - Te^{\Phi_+} - \tfrac{1}{2}D)/D. \tag{70}$$

Similar formulas for $F(L)$ and $G^*(L)$ may be easily obtained from Eqs. (63), (64), and (66). These are complicated expressions, but we are chiefly interested in only two aspects of the solutions: (1) Can the denominator D vanish, and, if so, what is the ratio of the signal to the probe intensity? (2) What is the small-angle behavior of the intensities?

Inspection of Eq. (68) shows that D does vanish for certain values of the path length L and, in particular, it vanishes when

$$\begin{aligned} \tan\tfrac{1}{2}\phi_+\tan\tfrac{1}{2}\phi_- &= -\tanh\tfrac{1}{2}\Phi_+\tanh\tfrac{1}{2}\Phi_- \\ &= -\coth\tfrac{1}{2}\Phi_+\coth\tfrac{1}{2}\Phi_-. \end{aligned} \tag{71}$$

So *the presence of cross-coupled waves does not prevent infinite gains.* Further examination of Eqs. (67) and (68) shows that

$$|S|^2 - |T|^2 = e^{-\Phi_+}\sinh\Phi_- D, \tag{72}$$

which implies that S and T have the same magnitude for $D = 0$. With Eqs. (69) and (70) this implies that the ratio $|\overline{F}(0)/\overline{G}(0)|^2$ approaches unity when D vanishes, and therefore *the cross-coupled signal approaches the direct (phase-conjugate) signal in magnitude near a point of infinite gain.* A further important result may be inferred from the relation

$$\mathrm{Re}(S - Te^{\Phi_+}) = \tfrac{1}{2}D, \tag{73}$$

which, with Eq. (70), implies that the *signal $\overline{G}(0)$ always has its phase conjugate to that of the probe F_0, despite the presence of cross-coupled*

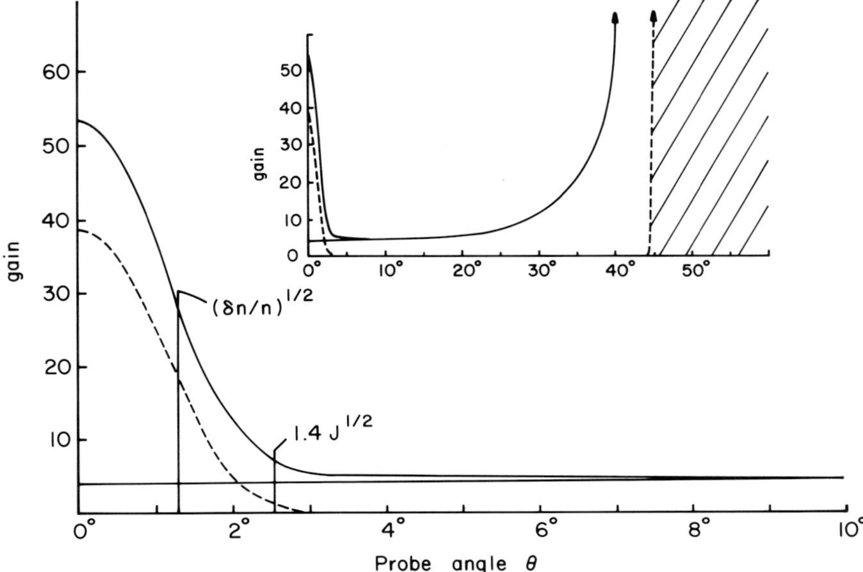

Fig. 5. Small-signal gain for phase-conjugate wave (—) and cross-conjugate reflected wave (---) versus probe angle of incidence for collateral boundary conditions in a slab of nonlinear medium. The length of the medium has been selected to give infinite gain at 45° owing to the effective lengthening of the path by tilting. The theory without cross-coupled waves gives the lower solid curve, which shows infinite gain at 45° (inset) but no enhancement at small angles. The crude estimate of the width of the enhanced region, $\theta_c \approx (\delta n/n)^{1/2}$, is seen to be useful. Inset: the non-phase-matched cross-coupled signal is large only in a region very near the infinite-gain angle. At greater angles, the linear steady-state theory is not valid. Other parameters for this case: $C = -0.425$, $J = 0.001$, $k_0L = 1556.7$.

waves. An exception should be made at normal incidence, where the cross-coupled mode becomes indistinguishable from the complex conjugate of the phase-conjugate signal. That is, *the cross-coupled mode at normal incidence becomes a nonconjugate contribution to the reflected signal.* The solution for this case was obtained by Marburger and Lam (1979b) in the slowly varying envelope approximation.

Turning now to the question of angular dependence, we see in Fig. 5 how the gains of direct and cross-coupled signals vary in a typical case. Note that the cross-coupled signal is small compared with the direct signal except at small angles and near the infinite-gain point, as anticipated. More general analytical progress is possible by expanding S, T, and D in small powers of $J = 2\delta n/n$. This is tedious, and must be done carefully because of many cancellations, but the work is straightforward and we reproduce here only the results of greatest utility.

The condition (71) for infinite gain becomes, at normal incidence ($\theta = 0$),

$$\cot[(9J/8)^{1/2}\tfrac{1}{2}JK_0L]\cot[3^{1/2}(1-C)^{1/2}(3-C)^{1/2}\tfrac{1}{2}JK_0L] \approx -[27J(1-C)/8(3-C)]^{1/2}$$

or

$$\tan[3^{1/2}(3-C)^{1/2}(1-C)^{-1/2}\pi L/2L_c] \to -\infty, \tag{74}$$

where the second expression is the limit of the first as $J \to 0$. This coincides with the SVEA result of Marburger and Lam (1979b). When the argument of Eq. (74) goes to $\pi/2$, L will by definition equal L_0, the critical length at $\theta = 0$. At angles that are not too close to $\theta = 0$, the critical length is $L_\theta \approx L_c \cos\theta$ as in Eq. (32). As θ approaches zero, the ratio

$$L_\theta/L_0 \cos\theta \equiv \tfrac{1}{2}3^{1/2}(3-C)^{1/2}(1-C)^{-1/2} \qquad \text{(ordinary polarization)} \tag{75}$$

approaches unity in a characteristic angular range

$$\Delta\theta \approx [3(1-C)(3-C)/2(3-2C)]^{1/2}J^{1/2} \qquad \text{(ordinary polarization)}, \tag{76}$$

which is about 1.4 $J^{1/2}$ rad for CS_2. This range is indicated in Fig. 5. Formulas for the angular dependence at small J within $\Delta\theta$ of normal incidence are so clumsy that calculations are as easily performed using the exact results (69) and (70).

As before, the two **k** vectors implied by Eq. (59) lead to double refraction even with ordinary polarization.

V. Difficulties with the Nonlinear Theory

Having gained some confidence with the exact linearized theory, we are now in a position to discuss approaches to the nonlinear theory. A moment's reflection convinces us that only in a collinear configuration is the problem of mutual four-wave interaction likely to be tractable. The spatial dependence of intensities in four strongly interacting beams of finite transverse extent traveling in four different directions has not to date been explored either numerically or experimentally. There may indeed be some highly symmetrical noncollinear geometries in which progress might be made, but nothing has as yet appeared in the literature. We shall therefore confine our attention to the collinear geometry, or what we have called *collateral boundary conditions with normal or nearly normal incidence.*

Unfortunately, this is precisely the case in which we have found that cross-coupled waves cannot be ignored in computing the strength of the phase conjugate signal. The stronger the nonlinear interaction, the larger the deviation from collinearity required to avoid the influence of cross-coupled modes. To date, no analysis of strongly interacting DFWM (i.e., including pump depletion) has been reported in which cross-coupled waves are included. The first question, then, is: What progress can be made with the full collinear problem? One would like analytical solutions, because numerical analysis of nonlinear boundary-value problems, even in one dimension, is extremely awkward and time consuming.

Analytical solutions for many lossless nonlinear optical processes in one dimension are indeed available, even when several interacting modes are present. The tractable situations are characterized by the existence of enough Manley–Rowe conservation equations to eliminate all but the intensity of a single mode from the problem (Armstrong *et al.*, 1962). In this chapter, and in other published works on the nonlinear theory, the pump and signal waves are assumed to be linearly polarized. For this case, there are not enough Manley–Rowe relations to reduce the problem to a single variable, and the prospect for analytical solutions is therefore not hopeful. The situation is different for circularly polarized waves, as subsequently described. The linearly polarized problem also becomes tractable if the cross-coupled waves are ignored, and several authors have reported solutions (Hsu, 1979; Marburger and Lam, 1979a; Soskin and Khizhnyak, 1980). The second question, then, is: What is the value of the solutions that ignore cross-coupled waves?

There are two important questions that can be answered only by the nonlinear theory: How does pump depletion affect the magnitude of the phase-conjugate signal (saturation behavior); and, How large is the spontaneous signal that appears when the process becomes unstable? The nonlinear theory without cross coupling answers both questions, but gives no clue as to how the answers should be modified to account for cross coupling. At best, the theory provides a model that can be manipulated to discover new qualitative features that might be present in nature. Winful and Marburger (1980) have used it, for example, to show how DFWM could exhibit bistable behavior, and Marburger and Lam (1979a) have shown that two types of qualitative behavior may be expected depending upon whether the material parameter $|4B(2A + B)^{-1}|$ is greater than or less than 2.

Although the full nonlinear problem is intractable for linearly polarized pump and signal waves, it becomes simplified once again if the interacting waves are circularly polarized. In this case, the collinear configuration yields enough Manley–Rowe relations to eliminate all but one intensity even when cross-coupled waves are included. The solutions made possible by this circumstance have not been reported elsewhere in the literature, but they resemble qualitatively the approximate solutions obtained for linear polarizations.

Much work remains to be done to complete our understanding of four-wave mixing in the strongly interacting regime. The problem is greatly exacerbated by its sensitivity to transverse-mode effects. The nonlinearity that causes DFWM also causes self-focusing. If the transverse scale a_T of intensity variation in any of the interacting beams is such that the self-focusing length [Marburger (1975)]

$$L_{sf} \approx J^{-1/2} a_T$$

is comparable with the shorter of L or L_c, then self-focusing definitely influences the process. The ratio

$$L_{sf}/L_c \approx (1 - C)J^{1/2} k_0 a$$

is the product of the generally small quantity $J^{1/2}$ and the generally large one $k_0 a$, so there are lengths and fluxes for which self-focusing must be considered.

VI. Concluding Remarks

The theory described in this chapter includes the effect of nonlinear index change, which leads to an asymmetry in the dependence upon forward- and backward-pump fluxes, and also takes into account the cross-coupled waves, which influence the gain at small angles of incidence. Our approach also emphasizes that the phase-conjugate signal is a superposition of two normal modes of the pumped medium, each of which propagates with a different **k** vector. This leads to double refraction in both the ordinary and the extraordinary polarization modes. As long as loss may be ignored, the theory remains analytically tractable when cross-coupled waves are included, and allows an expression for the phase-conjugate gain at small incidence angles. Loss may be analyzed easily only when the probe beam is at right angles to the pump beams, and in this case it is always possible to overcome the effect of loss by increasing the pump fluxes. The most important work still to be done in the theory is the general treatment of loss, and the inclusion of cross-coupled waves (for linear polarization) and transverse effects in the nonlinear theory.

References

Armstrong, J. A., Bloembergen, N., Ducuing, J., and Pershan, P. S. (1962). Interactions between light waves in a nonlinear dielectric, *Phys. Rev.* **127,** 1918.

Bloom, D. M., and Bjorklund, G. C. (1977). Conjugate wave front generation and image reconstruction by four wave mixing, *Appl. Phys. Lett.* **31,** 592.

Born, M., and Wolf, E. (1970). "Principles of Optics." Pergamon, Oxford.

Hellwarth, R. W. (1977). Theory of third order nonlinear susceptibilities, *Prog. Quantum Electron* **5.**

Hsu, H. (1979). Large signal theory of phase conjugate backscattering, *Appl. Phys. Lett,* **34,** 855.

Maker, P. D., and Terhune, R. W. (1965). Study of optical effects due to an induced polarization third order in electric field strength, *Phys. Rev.* **137A,** 801.

Marburger, J. H. (1975). Self focussing: Theory, *Prog. Quant. Electron.* **4.**

Marburger, J. H., and Lam, J. F. (1979a). Nonlinear theory of degenerate four wave mixing, *Appl. Phys. Lett.* **34,** 389.

Marburger, J. H., and Lam, J. F. (1979b). Effect of nonlinear index changes on degenerate four wave mixing, *Appl. Phys. Lett.* **35,** 249.

Pepper, D. M., and Abrams, R. L. (1978). Narrow optical bandpass filter via nearly degenerate four wave mixing, *Opt. Lett.* **3,** 212.

Smith, S. D., and Miller, D. A. B. (1980). Giant third order nonlinearities in semiconductors and application in bistability, transphasor action, and phase conjugation, *Int. Quantum Electron. Conf. Digest J. Opt. Soc. Am.* **70,** 653.

Soskin, M. S., and Khizhnyak, A. I. (1980). Interaction of four counter-propagating plane waves in a medium with an instantaneous cubic nonlinearity, *Kvantovaya Elektron,* **7,** 42 [*English transl.: Sov. J. Quantum Electron.* **10,** 21 (1980)].

Winful, H. G., and Marburger, J. H. (1980). Hysteresis and optical bistability in degenerate four wave mixing, *Appl. Phys. Lett.* **36,** 613.

Yariv, A., and Pepper, D. M. Amplified reflection, phase conjugation, and oscillation in degenerate four wave mixing, *Opt. Lett.* **1,** 16.

5 Phase Conjugation by Four-Wave Mixing in a Waveguide*

R. W. Hellwarth†
University of Southern California
Los Angeles, California

I. Introduction

In this chapter we review experiments and theory on phase conjugation by four-wave mixing in multimode optical waveguides. We assume that the reader is familiar with the concepts of phase conjugation and wave mixing. Yariv *et al.* (1978) pointed out that phase conjugation would occur by degenerate four-wave mixing in a multimode waveguide if the two counterpropagating pump waves were in the same single transverse mode. However, the first experiments of Jensen and Hellwarth (1978a,b) studied only the guided conjugation process with multimode pump waves, verifying the prediction of Hellwarth (1979) that high-fidelity phase conjugation will result even when the two counterpropagating pump waves are multimode and each has a different transverse-mode structure. These experiments also verified the prediction that the image wave being conjugated can serve simultaneously as its own pump beam. Subsequently, AuYeung *et al.* (1979) succeeded in launching a single-mode cw pump wave in a guide and generating a wave whose power dependence and polarization indicated that it was generated by four-wave mixing. However, the phase-conjugate nature of the signal was not ascertained. Before

* Some of the material in this chapter has been previously published with the permission of Academic Press.

† Work supported by the National Science Foundation under Grant No. ENG 78-04774 and the U.S. Air Force Office of Scientific Research under Grant No. 78-3478.

Optical Phase Conjugation

ISBN 0-12-257740-X

describing the experiments, we review the main results of the multimode pump theory. We concentrate on the degenerate case (i.e., where all beams are of the same frequency), and consider only briefly some multi-frequency experiments and theory.

II. Theory of Degenerate Mixing

The theory of the mixing in a waveguide of a multimode input image beam F with counterpropagating multimode pump beams G and H has been examined only in the small mixing limit, that is, where it may be assumed that the three input beams are negligibly altered by the mixing process (Hellwarth, 1979). In this case, a fourth beam E may be generated that is phase conjugate inside the guide (and also in interesting regions outside) to the image-bearing beam F.

The conditions for this to happen are as follows. For the degenerate case, phase conjugation results if there is negligible coupling among mode components of the four beams E, F, G, and H when at least three of the four components have different (unperturbed) transverse-mode patterns. Suppression of coupling will result from phase mismatch alone if the magnitudes of the propagation constants of all excitable guide modes are fairly uniformly distributed, provided that the guide is long enough so that the beams are well guided. However, in a multimode dielectric guide of isotropic material there are generally two sets of nearly degenerate modes (i.e., different modes having the same propagation constant). These two sets have orthogonal polarization. However, the nonlinear susceptibility tensor of isotropic media is such that, by launching only beams of the same linear polarization, the wave E is predicted to be generated also with the same polarization, thus effectively eliminating all coupling to the set of orthogonally polarized degenerate modes. This has been borne out by experiment. Intercoupling to degenerate modes can also be prevented by polarization selection and by eliminating degeneracies by using birefringent media. (The latter method holds promise for complete vector-wave conjugation.) In addition to phase mismatch, the four-mode overlap integral that occurs as a factor in the coupling coefficient is often zero, or at least small, when three or four different modes are involved. If many propagation constants are of closer than average separation, a longer guide will again produce phase mismatch. In any case, the elimination of three- and four-mode interaction terms would seem to present no essential difficulties in practice.

Even when coupling is limited to wave components consisting of two pairs of congruent transverse modes, the backward-generated wave E will still have a non-phase-conjugate component. However, theory predicts that the power fraction f in this unwanted component is small when the image-bearing beam excites a large number N of guide modes. It is estimated that f is of order N^{-1} times the fractional rms variation in the energies in the N modes excited. It is therefore expected, and found experimentally, that an input image of many elements will be phase conjugated with high fidelity by parallel polarized pump beams in a guide whose length is many diffraction lengths.

The fraction R of the power in the input image beam F that is backscattered as a (nearly) phase-conjugate wave E in the presence of counterpropagating pump waves G and H, under conditions described above in a waveguide, was found to be (Hellwarth, 1979)

$$R = \beta^2 P_G P_H \theta_{GH} S^{-2} e^{-\alpha L}(1 - e^{-\alpha L})^2 \alpha^{-2}, \qquad (1)$$

where P_G and P_H are the entering powers of the pump beams, S is the guide area, α is the guide (intensity) attenuation coefficient, L is the guide length, β is a coupling coefficient related to the nonlinear susceptibility coefficient $c_{1111}(-\omega, \omega, -\omega, \omega)$ defined by Maker and Terhune (1965) by

$$\beta = |96\pi^2 \omega c_{1111} \xi / n^2 c^2|, \qquad (2)$$

and θ_{GH} is the fraction of the power in one pump beam that is phase conjugate to the other. If g_b and h_b are the complex amplitudes of excitation of the unperturbed guide modes (labeled by subscripts a, b, c, etc.) defining the pump beams G and H, respectively,

$$\theta_{GH} = \left|\sum_b g_b h_b\right|^2 \bigg/ \left(\sum_c |g_c|^2 \sum_d |h_d|^2\right). \qquad (3)$$

In Eq. (2), ω is the angular frequency of all waves, c the velocity of light in a vacuum, n the refractive index at the guide axis, and ξ a dimensionless average of mode-overlap functions that is nearly unity.

The relation (1) also holds when the image to be conjugated is impressed on the pump beam G itself, in which case R of Eq. (1) represents the ratio of phase-conjugate power to G-beam power. In this case R becomes very dependent on the image itself, because θ_{GH} varies rapidly as image features are changed.

III. Degenerate Experiments

Experiments to date in which phase conjugation has been produced by four-wave mixing in a waveguide have taken the two forms described in Section II and illustrated schematically in Figs. 1a and 1b. The first exper-

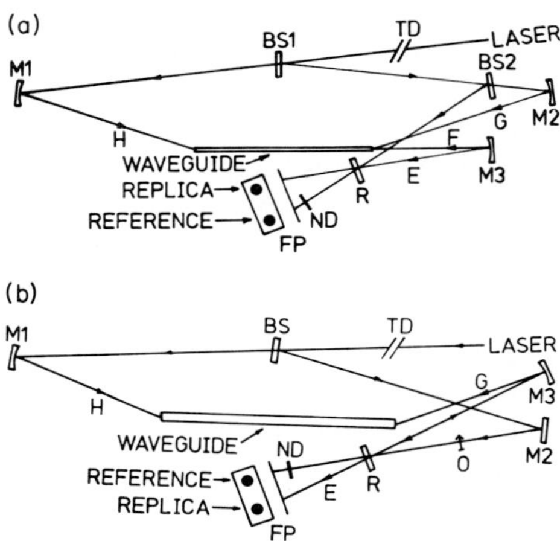

Fig. 1. Schematic of apparatus discussed in the text. (a) Three-beam arrangement whereby pump beams G and H interact in a CS_2-filled hollow glass waveguide with beam F to generate its time-reversed replica E. Isolation is supplied by a 30-nsec time delay TD. Beam splitter BS1 reflects 55%. Mirrors M1, M2, and M3 have radii of 1, 1, and 0.5 m, respectively, and focus the beams onto the ends of the 80-cm-long guide. About 8% of beam G is split by BS2 to form F. (b) Two-beam arrangement whereby image-bearing beam G interacts in the aforementioned guide with beam H to generate its time-reversed replica E. Mirrors M1 and M3 have radii of 36 cm and focus the beams onto the ends of the waveguide. Beam splitter BS reflects 55%, and R reflects 75% while transmitting 24%. (After Jensen and Hellwarth, 1978b.)

imental arrangement, shown in Fig. 1a, employed two pump beams G and H focused into opposite ends of a dielectric waveguide consisting of a 0.4-mm-i.d. × 80-cm-long glass tube filled with liquid CS_2. An input (image-bearing) beam F was focused into the same end of the guide as beam G. All beams originated from the same *Q*-switched ruby laser. The beam splitter R directed a beam E backscattered along F onto photographic plate FP, where it was compared with a reference fraction of beam F (as shown in the upper plate of Fig. 2a) to verify that it was a time-reversed replica of beam F. A critical test was the replication, recorded in Fig. 2a, of the fringes formed from the interference between reflections from parallel front and back surfaces of beam splitter BS2. The absence of polarization scrambling (checked also with a polarizer) and of phase distortions in the replicated beam were reflected in the fidelity and high contrast ratio of the fringes replicated. In Fig. 2a the reference-beam energy was 0.42 times that of the input beam F, showing that the energy in the replica E was about 0.4 that in F. Here a 20-nsec laser pulse supplied about 35 kW peak power to each pump wave. The replica beam E disappeared when either pump wave was absent, as shown in the lower plate of

Fig. 2a. Beams F, G, and H were focused to spots at the guide entrance that were much smaller than the inside diameter of the guide, so that nearly all of each beam (except for ~2% dielectric reflection) was coupled into the guide. Beams G and H entered at about 10° to the guide axis. They emerged from the guide in hollow conical beams whose dimensions showed that about 10^4 transverse modes had been excited by each beam. (The total number of bound transverse modes was ~10^6.) Efficiency of replication was greatest when the forward pump beam G was adjusted to enter the guide within the cone of the exiting beam H.

In the second experimental arrangement, shown in Fig. 1b, there was no third beam F; the pump beam G itself was observed to be replicated (phase conjugated) by its nonlinear interaction with a counterpropagating beam H when both were focused into the guide. A beam E backscattered along G was directed by partial reflector R onto film plate FP, where it was compared with a reference fraction of beam G, as in the upper plate

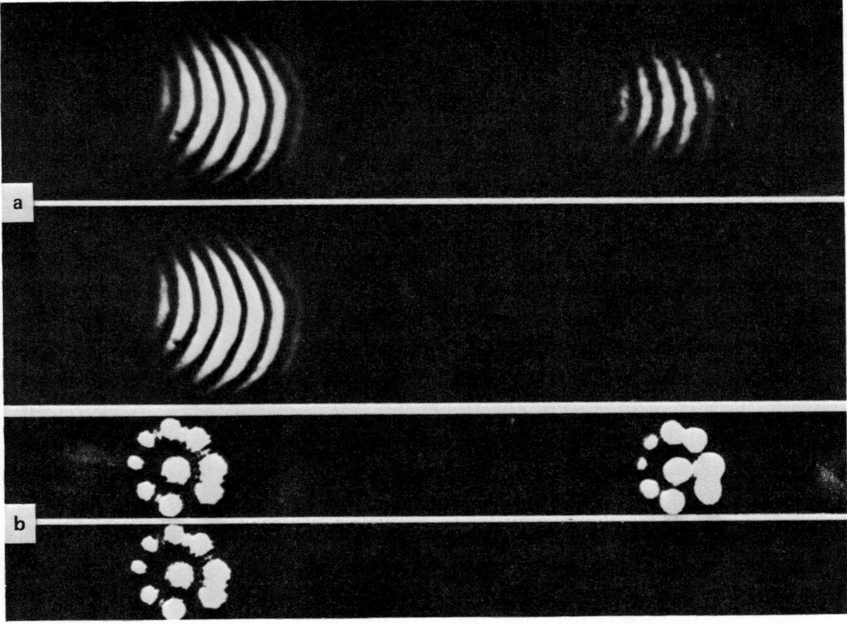

Fig. 2. Photographic images produced on Polaroid 410 film at film plate FP (see Fig. 1) by a single laser pulse. (a) The upper plate is from the arrangement of Fig. 1a, the lower plate from the same arrangement but with beam H blocked. (b) The upper plate is from the arrangement of Fig. 1b, the lower plate from the same arrangement but with beam H blocked. Reference images are shown on the left side, the time-reversed replicas on the right side. (After Jensen and Hellwarth, 1978b.)

of Fig. 2b. That the backscattered beam E was essentially a phase conjugate, or time-reversed replica, of beam G was verified by placing various objects in beam G (at position O) and observing their replicas in the E-beam patterns.

The reference-beam patterns differed somewhat from the replica patterns because the film plate FP was not the same effective distance from the object O for each beam. For example, one can see diffraction fringes in the reference-beam pattern in Fig. 2b where none exist in the replica (E-beam) pattern, which replicates beam G very near object O. The reference-beam fraction was 0.21 in this instance, showing that the replica energy was ~0.2 of that in beam G. Here beams G and H were 20-nsec pulses of about 25 kW peak power.

For the CS_2 medium at 6943 Å, $\beta \sim 6$ cm/GW, neglecting acoustic contributions (Hellwarth, 1977). It was therefore estimated that $\theta_{GH} \sim 10^{-3}$ in these experiments in order that Eq. (1) give the observed efficiency.

The pump powers required were nearly an order of magnitude less than were required for comparable replication efficiency R for free (unguided) waves in CS_2 (Bloom and Bjorklund, 1977; Jensen and Hellwarth, 1978). Furthermore, the number of resolution elements in a beam that could be replicated in the guide was almost the number of guide modes ($\sim 10^5$). This is orders of magnitude greater than the number of free-space modes replicated at similar power levels in CS_2.

IV. Nondegenerate Case

There are several applications of phase conjugation that can best be achieved by guided four-wave mixing of different frequencies. For example, if certain conditions on beam geometries are met, phase conjugation of multifrequency beams can be achieved in a guide without intermixing among the frequency components (Hellwarth, 1979). This is possible because of the special selectivity of phase matching in a guide. Another example is laser–Raman spectroscopy of small samples, which may be better achieved in a guide where fields are more concentrated. [Resonances in phase-conjugation efficiency will occur when appropriate beam frequencies differ by a Raman (Brillouin, etc.) frequency.] A third example is the realization of a wide-angle, narrow-band optical filter. This application is possible because the acceptance angle of a multimode guide can be large, while only probe-beam frequencies in a small range ($\sim c/L$) are efficiently conjugated. Experiments demonstrating such a filter have been reported by Lam and Hellwarth (1980). Many other applications of guided phase conjugation can be envisioned. Further experimentation is necessary to evaluate them.

V. Summary

Phase conjugation of monochromatic image-bearing optical beams by the process of nonlinear refractive mixing in a waveguide with multimode counterpropagating pump waves has been demonstrated. With this process, much less pump power is required, and the alignment and mode quality of the pump beams are much less critical, than in the corresponding process with unguided beams. Application of this process to narrow-band, wide-angle optical filtering has been demonstrated. Phase conjugation in a waveguide also promises to be of use in spectroscopy, multifrequency conjugation, and a variety of other applications.

Acknowledgment

The author would like to thank the Aspen Center for Physics for hospitality during the preparation of this chapter.

References

AuYeung, J., Fekete, D., Pepper, D. M., Yariv, A., and Jain, R. K. (1979). Continuous backward-wave generation by degenerate four-wave mixing in optical fibers, *Opt. Lett.* **4,** 42.

Bloom, D. M., and Bjorklund, G. C. (1977). Conjugate wave generation and image reconstruction by four-wave mixing, *Appl. Phys. Lett.* **31,** 592.

Hellwarth, R. W. (1977). Third-order optical susceptibilities of liquids and solids, *Prog. Quantum Electron.* **5,** Part 1.

Hellwarth, R. W. (1979). Theory of phase-conjugation by four-wave mixing in a waveguide, *IEEE J. Quantum Electron.* **QE-15,** 101.

Jensen, S. M., and Hellwarth, R. W. (1978a). Observation of the time-reversed replica of a monochromatic optical wave, *Appl. Phys. Lett.* **32,** 166.

Jensen, S. M., and Hellwarth, R. W. (1978b). Generation of time-reversed waves by nonlinear refraction in a waveguide, *Appl. Phys. Lett.* **35,** 404.

Lam, L. K., and Hellwarth, R. W. (1980). Wide-angle narrowband optical filter using phase conjugation by four-wave mixing in a waveguide, Abstract E.10, *Int. Quantum Electron. Conf. Digest Tech. Papers* Abstr. E.10, IEEE Catalog No. 80CH 1561-0.

Maker, P. D., and Terhune, R. W. (1965). Study of optical effects due to an induced polarization third order in electric field strength, *Phys. Rev.* **137A,** 801.

Yariv, A., AuYeung, J., Fekete, D., and Pepper, D. M. (1978). Image phase compensation and real-time holography by four-wave mixing in optical fibers, *Appl. Phys. Lett.* **32,** 635.

6 *Experimental Investigation of Wave-Front Reversal under Stimulated Scattering*

B. Ya. Zel'dovich
N. F. Pilipetskii
V. V. Shkunov

Institute for Problems in Mechanics
Academy of Sciences
Moscow, USSR

I. Introduction

Several contributors to this book have presented holographic-type techniques for wave-front reversal (WFR). These techniques employ ref-

Optical Phase Conjugation

ISBN 0-12-257740-X

erence waves of various kinds—of the same or of double frequency, either simultaneously present or separated in time. Stimulated scattering however, does not require reference waves. Experimental and theoretical investigation of self-reversal which takes place under stimulated scattering (SS) began in the USSR in 1971 (Zel'dovich *et al.*, 1972; Nosach, 1972). In this chapter some fundamental features of self-reversal are discussed, based primarily on our own experimental WFR results under-stimulated scattering (WFR–SS). We first present, in Section II, basic information on the physical mechanism of WFR–SS; this information is necessary to interpret the experimental results. Conventional methods of quantitative experiments of WFR are described in Section III, and major results are reported in Sections IV–VII.

For reading materials in this field important for the history of optical phase conjugation by the methods of static and dynamic holography, see the following:

W. L. Bragg, Nature (*London*) **166** (4218), 399 (1950).
D. Gabor, U.S. Patent 2,770,116 (July 6, 1951).
H. W. Kogelnik, U.S. Patent 3,449,577 (October 23, 1965).
G. T. Sincerbox, British Patent 1,218,331 (April 9, 1968).

II. Physical Mechanism of WFR–SS

We first review the basic concepts of SS [see, e.g., Bloembergen (1965), Hellwarth (1977), and Yariv (1975)]. Let two light waves—an exciting wave $\mathscr{E}_0(\mathbf{r})e^{-i\omega_0 t}$ and a signal wave $\mathscr{E}_s(\mathbf{r})e^{-i\omega_s t}$—propagate in a medium, the difference of their frequencies $\Omega = \omega_0 - \omega_s$ being close to the frequency Ω_0 of characteristic oscillations of the medium. The interference term $\mathscr{E}_0^*\mathscr{E}_s e^{i\Omega t}$ in the total light field intensity leads to these characteristic oscillations. As a result, a traveling spatial–temporal grating of the dielectric permittivity forms in the bulk of the medium; at acoustic resonance it has a phase shift of 90°: $\delta\epsilon(\mathbf{r}, t) = -iA\mathscr{E}_0^*\mathscr{E}_s e^{i\Omega t}$. The scattering of light by the grating leads to a term in the electric displacement $\delta D_s = -iA|\mathscr{E}_0(\mathbf{r})|^2\mathscr{E}_s(\mathbf{r})e^{-i\omega_s t}$, which produces a light wave whose frequency and propagation direction coincide with those of the signal wave. Thus, the signal wave is characterized by a negative imaginary part in the permittivity. The corresponding amplification $g(\mathbf{r})$ (in reciprocal centimeters) is proportional to the local wave intensity $g = G|\mathscr{E}_0(\mathbf{r})|^2$.

Stimulated scattering normally increases, starting from a very low level of spontaneous scattering. Under typical conditions, SS is observed in the saturation regime that occurs for the total amplification $\exp(G|\mathscr{E}_0|^2L) \sim$

$e^{30} \approx 10^{13}$. If the corresponding threshold intensity $|\mathscr{E}_0|^2$ is expressed in megawatts per square centimeter and the length L of the amplification region in centimeters, G is expressed in centimeters per megawatt. When the threshold condition is attained, the intensity of exponentially amplified spontaneous noise becomes comparable, by an order of magnitude, with the scattered wave intensity. Backward stimulated Brillouin scattering (SBS), which has fairly high constant $G \sim 10^{-1}$–10^{-2} cm/MW, small damping time $\tau_s \sim 10^{-8}$–10^{-9} sec, and small frequency shift $\Omega_0 \lesssim 1-10^{-2}$ cm^{-1}, is the most widely used stimulated process for wave-front reversal.

Two basic properties of SS—high total amplification and the relationship between the local amplification coefficient and the local intensity of the exciting field—underlie the physical mechanism of WFR–SS. Owing to exponential amplification, even a moderate increase in the effective increment $\Delta g/g \sim 1$ may lead to a radical change in the process. WFR–SS takes place when the spatial distribution of the intensity $|\mathscr{E}_0(\mathbf{r})|^2$ and of the local amplification coefficient of the Stokes wave, $g(\mathbf{r}) = G|\mathscr{E}_0(\mathbf{r})|^2$, both possess considerable spatial inhomogeneities (Zel'dovich *et al.*, 1972). Exciting radiation having an irregular divergence θ_0 has a speckle structure, with a characteristic transverse dimension of the inhomogeneities of the order of $\Delta r_\perp \sim (k\theta_0)^{-1}$ and a characteristic length $\Delta z \sim (k\theta_0^2)^{-1}$, where the radiation diffracts from one inhomogeneity to another. Under the conditions necessary for WFR, these dimensions must be much smaller than those of the interaction region. The effective amplification coefficient for a given configuration of the scattered field $\mathscr{E}_s(\mathbf{r}, z)$ is determined in every section $z =$ const by the intensity overlapping integral (Zel'dovich *et al.*, 1972):

$$g_{\text{eff}} = \frac{\int G|\mathscr{E}_0(\mathbf{r}, z)|^2|\mathscr{E}_s(\mathbf{r}, z)|^2\, d^2\mathbf{r}}{\int |\mathscr{E}_s(\mathbf{r}, z)|^2\, d^2\mathbf{r}}. \tag{1}$$

Therefore, maximum amplification is achieved for a wave $\mathscr{E}_s$ whose local maxima coincide everywhere in space with the maxima of $\mathscr{E}_0$. A scattered wave with a profile of the form $|\mathscr{E}_s(\mathbf{r}, z)| = \text{const}|\mathscr{E}_0(\mathbf{r}, z)|$, whose local maxima in every section $z =$ const coincide with the maxima of the amplification coefficient, has an amplification coefficient about twice that of any other configuration of $\mathscr{E}_s$ that does not correlate with the configuration of $\mathscr{E}_0$. Under diffraction, both fields $\mathscr{E}_0(\mathbf{r}, z)$ and $\mathscr{E}_s(\mathbf{r}, z)$ change their structure. To maintain the correlation of two inhomogeneous fields throughout the interaction volume, the scattered field $\mathscr{E}_s(\mathbf{r}, z)$ must be the conjugate of $\mathscr{E}_0(\mathbf{r}, z)$, that is, $\mathscr{E}_s(\mathbf{r}, z) \sim \mathscr{E}_0^*(\mathbf{r}, z)$.

The term *specklon* is used for a speckle-inhomogeneous field configuration correlated with the inhomogeneity of the medium. An essential feature of the specklon is that its fine inhomogeneities diffract according to the laws of free propagation in a homogeneous medium, and its evolution in space is reduced to a change in a smooth envelope. The term *mode*, introduced in the pioneering work of Sidorovich (1976, 1977), is used to denote a related concept of speckle-inhomogeneous solutions with an exponential change in the envelope along the longitudinal coordinate. There is an ongoing debate about the appropriateness of these terms.

Among the original spontaneous sources of the scattered field are both the conjugate (reversed) configuration and a large number of uncorrelated waves. When the WFR configuration $\mathscr{E}_s \sim \mathscr{E}_0^*$ is amplified by a factor of $\sim e^{30}$, the uncorrelated wave traveling the same distance is amplified only by a factor of $\sim e^{15}$. This discrimination of uncorrelated solutions accounts for the fact that the WFR component of the scattered field is sharply pronounced upon exiting the nonlinear medium.

Thus, the physical mechanism of WFR–SS is based on the predominant amplification of the reversed configuration when the scattered field local maxima coincide with those of the amplification coefficient throughout the interaction volume.

III. Quantitative Experimental Techniques

To describe WFR quantitatively, one must know the ratio of the energy of the scattered light to the incident energy (the reflectance) and the quality of reversal characterized by the fraction of energy of returned light corresponding to exactly reversed field. The techniques described in Sections III.A and III.B are used to measure these parameters (they are equally applicable to the study of all other types of WFR).

A. Phase Plate (Aberrator)

Etching a glass plate in hydrofluoric acid produces random variations in thickness. When the initial plane wave passes through such a plate, it acquires a phase increment $2\pi(n - 1)h(\mathbf{r})/\lambda$, where n is the refraction index of the glass, $h(\mathbf{r})$ the height of the inhomogeneities, and λ the wavelength of light, giving an irregular divergence $\delta\theta \sim \lambda/a$ for $h(n - 1) \lesssim \lambda$ or $\delta\theta \sim 2\pi(n - 1)h/a$ for $h(n - 1) \gtrsim \lambda$, where a is the transverse size of the inhomogeneities.[1] Plates with $\delta\theta \approx 10^{-2}$ rad are normally used. In all experiments on WFR, the reflected field contains the exactly reversed component and an uncorrelated component. To determine the fraction of

[1] If the height of the inhomogeneities is small, the wave passed through the plate also contains an intensity component with equivalent divergence.

WFR, these components must be spatially separated. When the reversed component passes back through the phase plate, it is transformed into a plane wave, yielding therefore, a bright diffraction-limited spot in the far field. The uncorrelated waves passed through the phase plate yield a beam of divergence $> \delta\theta$. Elements with phase inhomogeneities have been used in holographic studies of WFR for quite a while (see, for example, Cathey, 1968; Kogelnik, 1965), however, but only the technique of illuminating a phase plate by a diffraction-limited plane wave, used for the first time by Zel'dovich *et al.* (1972), permits spatial localization of the exactly reversed component and, therefore, quantitative measurement of its characteristics.

B. Measurement of the Reversed-Component Energy (Aperture, Mirror Wedge, WFR Interferometer)

The energy of the exactly reversed component can be measured in the far-field zone, behind the phase plate, by placing in the zone an aperture the size of the diffraction-limited divergence of the beam. This technique was used by Bespalov *et al.* (1977b), Mays and Lysiak (1979), and Vasilyev *et al.* (1981). The advantage of the technique is that it permits a simple interpretation of experimental results; its disadvantage is that it requires that an aperture of comparatively small size be placed exactly within the reversed laser beam. To reduce the sensitivity of the experimental setup to fluctuations of the beam propagation direction, one must increase the aperture diameter, which reduces the accuracy of measurement.

A mirror wedge technique, widely used by Zel'dovich *et al.*, 1972, compares the energies of the laser pumping and the reversed wave as well as the brightness of the pumping radiation and the scattered wave in the far-field zone. The energies are measured in the same units by two calorimeters, and it is essential that the reflected-wave calorimeter collect light propagating within a fairly wide solid angle. The brightness of both beams—the pumping with diffraction-limited divergence and the scattered wave transformed by the phase plate—is measured with the help of photomaterials. To overcome the limited dynamic range of these materials, either of the beams is split into a fan of spatially similar beams whose intensities are successively reduced by the same factor (twofold reduction is most convenient). This splitting is performed by a mirror wedge (Ragul'skii, 1976) formed by two semitransparent flat mirrors (Fig. 1a) or by one totally reflecting and one semitransparent mirror (Fig. 1b). As a result, among the various beams of the fan will be several beams whose en-

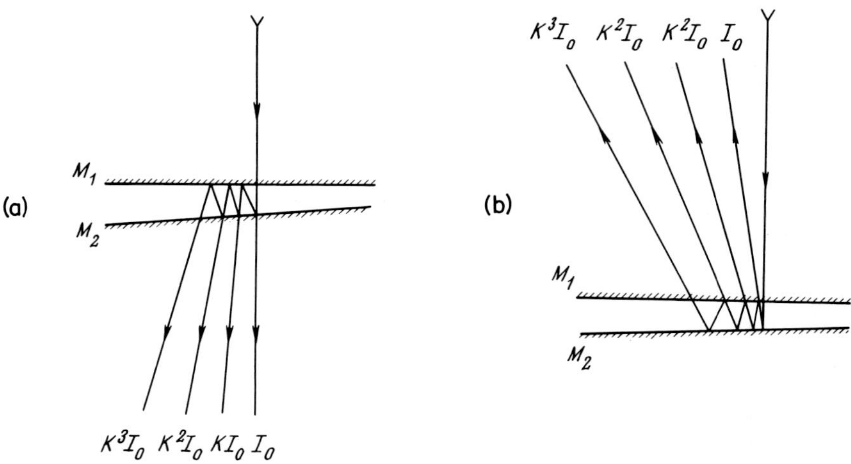

Fig. 1. (a) Transmission mirror wedge and (b) reflection mirror wedge, transforming a beam into a fan of spatially similar beams with a successive K-fold attenuation of their intensity. $\mathbf{M}_1$ and $\mathbf{M}_2$ are mirrors with reflection coefficients R_1 and R_2, respectively; $K = R_1R_2$.

ergies lie within the dynamic range of the film. Therefore, the brightness and angular distributions of any beam can be compared quantitatively during a single laser pulse. If the same photoplate is used in an experiment, variations in the properties and development conditions of photomaterials will not affect the accuracy of measurement. Moreover, this experimental scheme automatically gives the characteristic curve of the photomaterial.

The wave transformed by the phase plate contains a bright diffraction-limited central spot in the far-field. If the scattered wave consisted only of the reversed component (with respect to the exciting wave), the ratio of the brightness of the scattered and initial waves, B_s/B_0, at the diffraction-limited spot centers would coincide with the ratio of energies W_s/W_0 recorded by the calorimeters. Any distortion of the scattered field, as compared with the exactly reversed field, leads to a decrease in B_s/B_0 as compared with W_s/W_0. The fraction of reversal, H, is thus

$$H = (B_s/B_0)(W_0/W_s). \tag{2}$$

The expression for the fraction of WFR expressed in terms of the fields $\mathscr{E}_0(\mathbf{r})$ and $\mathscr{E}_s(\mathbf{r})$ being compared is Zel'dovich and Shkunov (1977)

$$H = \frac{|\int \mathscr{E}_0(\mathbf{r})\mathscr{E}_s(\mathbf{r})\, d^2\mathbf{r}|^2}{\int |\mathscr{E}_0|^2\, d^2\mathbf{r} \int |\mathscr{E}_s|^2\, d^2\mathbf{r}}. \tag{3}$$

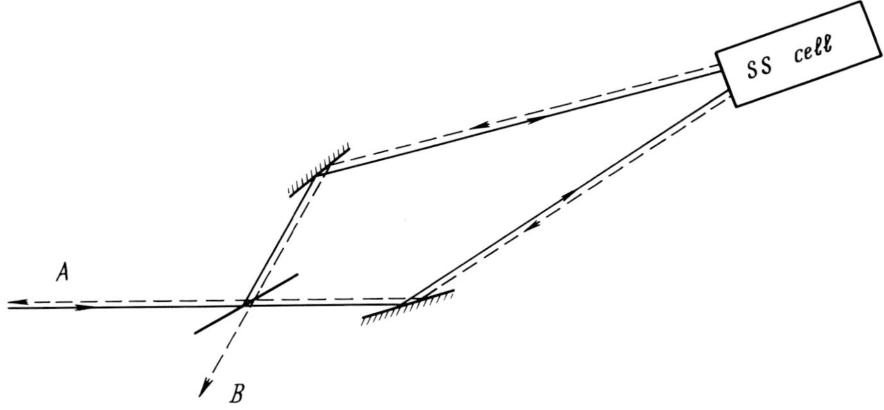

Fig. 2. Scheme for measuring the quality of WFR with the help of an interferometer. The exactly reversed portion of the total (for the two channels) field returns in the direction A of the input beam. The noise component is equally distributed between channels A and B.

Another fairly sensitive technique for recording the nonreversed part of the scattered field has been suggested (Basov *et al.*, 1980; Ragul'skii, 1980). It uses a scheme in which the initial beam is divided by a light-splitting surface into two beams of equal intensity before being directed into a nonlinear medium (see Fig. 2). If the reflected radiation is the reversed solution of the wave equation, the waves propagating from the light-splitting plate in direction B will completely cancel each other. The energy of the nonreversed part of the reflected wave will, on the average, be divided equally between directions A and B, owing to random phase differences. For a more detailed description of such an interferometer with a WFR mirror, see Ragul'skii (1980), Vasilyev *et al.* (1980), Bespalov *et al.* (1980), and Basov *et al.* (1980a,b).

C. Oscillographic Method for Recording Nonlinear Processes

In a number of WFR experiments, a simple and reliable method for recording nonlinear processes has been used (Popovichev *et al.*, 1974; Ragul'skii, 1976; Blashchuk *et al.*, 1978a,b). This method permits, in particular, the recording of the *steady-state* threshold value of the SS saturation and, owing to the threshold condition $gL = \text{const} \approx 30$, determination of the *steady-state* gain g. Oscillographic recording is necessary be-

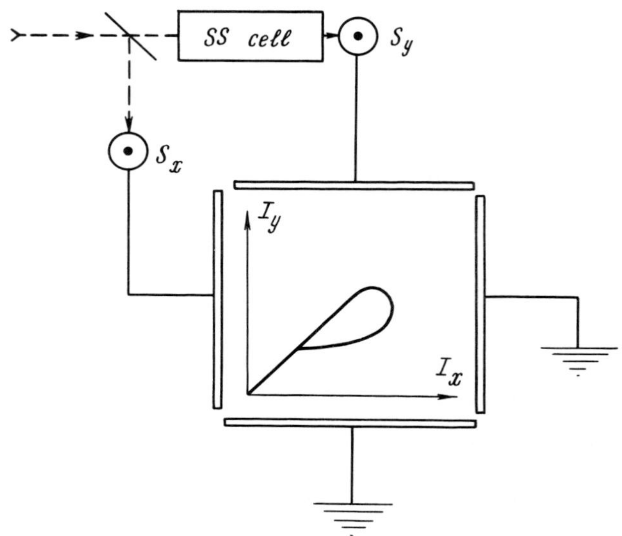

Fig. 3. Loop scheme for recording nonstationary processes. The power incident upon the cell is recorded by measuring device S_x, the transmitted power by S_y. For details see the text.

cause, in most experiments on WFR–SS, the length of the pumping pulses is of the same order of magnitude as the buildup time of SS.

The SS threshold is determined with the help of an oscillographic system as follows. The signal from a photodiode recording the exciting intensity I_x is applied to the horizontal deflection plates of the oscilloscope (see Fig. 3). The signal proportional to the intensity I_y of the radiation passed through the cell is simultaneously applied to the vertical deflection plates. In the absence of SS, $I_x(t) = I_y(t)$, so the electron beam displays a straight line on the screen, "drawing" it twice as the intensity $I_x(t)$ first increases and then decreases. In the presence of SS, the straight line on the screen is bent, owing to the nonlinear scattering of a certain fraction of the incident light, and in the quasi-stationary case each point of the curve is again displayed twice on the screen. In actual experiments, a loop, shown in Fig. 3, is displayed on the screen. Light with increasing intensity $I_x(t)$ does not excite SS immediately, because the process is nonstationary, so the oscilloscope beam displays the upper part of the loop $I_y(I_x)$. SS then attains a quasi-stationary level, and as the intensity $I_x(t)$ decreases, $I_y(t)$ drops gradually (the lower part of the loop). Finally, at the point where the loop is closed, the exciting intensity becomes such that only a small fraction of I_x is backscattered. The value of I_x at this point gives the saturation threshold for the steady-state regime of SS.

IV. Investigation of Reversed Waves

Because the presence of an exactly reversed component in the reflected field is of utmost importance for the practical application of WFR, in this section we discuss experimental results concerning just the WFR component. Amplification of waves uncorrelated with the pumping has been considered in detail in a number of papers, to be discussed in Section V.B.

A. Discovery of WFR–SS

Almost immediately after the discovery of stimulated scattering, it was found that the directivity of scattered radiation was rather high (see, e.g., Bespalov and Kubarev, 1966; Brewer, 1965; Dietz *et al.*, 1969; Kudriavtseva *et al.*, 1978; Rank *et al.*, 1967; Wiggins *et al.*, 1966; Zverev and Martynov, 1967). This was attributed only to geometrical factors, however, and the fine structure of the scattered wave front was not investigated at all for some time. This problem was first discussed by Zel'dovich *et al.* (1972).

Figure 4 is a diagram of the experimental setup used to discover self-reversal of the wave front. The radiation of the exciting ruby laser ($\tau = 110$ nsec, $W \approx 0.14$ J) had a vertical divergence of about 10^{-4} rad; its horizontal divergence was slightly greater. This radiation was recorded by the mirror wedge technique (see Section III.B), as shown in the photograph in Fig. 5a. Transmission of the radiation through the phase plate resulted in a large irregular divergence $\delta\theta \sim 3 \times 10^{-2}$ rad; see Fig. 5b, in which is shown the angular distribution of the pumping radiation passed through the phase plate a second time after reflection from an ordinary mirror. The phase plate Ph produced by a lens with $f = 1$ m was imaged onto the entrance of a hollow rectangular lightguide (internal dimensions 4×4 mm^2, length $L \approx 1$ m). As the radiation propagates along the lightguide, different angular components interfere with one another, producing a large number of maxima and minima. In this case, the energy practically does not leave the lightguide through the walls, owing to the high Fresnel reflection coefficient under glancing incidence.[2]

Compressed methane in the lightguide produces backward stimulated Brillouin scattering. In the experiment described, the energy reflection coefficient was $W_S/W_0 \approx 0.25$. The backscattered radiation passed through the elements of the optical system in the reverse direction and

[2] There are no strict requirements for the geometry and quality of the lightguide. Its main role is to provide for a great length of nonlinear interaction of light with the medium. We therefore prefer the term *lightguide* to *waveguide*.

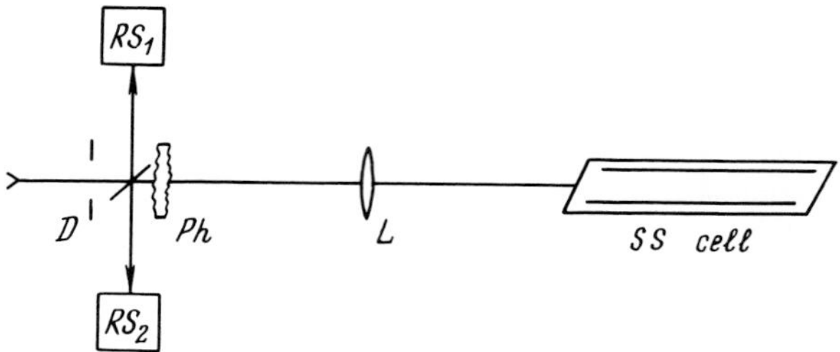

Fig. 4. Diagram of the experimental setup used to discover WFR–SS. The scattering is excited by a 1-MW ruby laser. D is a 6 × 6-mm aperture, Ph a phase plate. The lens L projected the image of the illuminated Ph onto the input face of a hollow lightguide placed inside a cell filled with gaseous methane under 125 atm. The cell length is 1 m, the lightguide cross section 4 × 4 mm. RS_1 and RS_2 are systems for recording the energies and angular distributions of the incident and reflected beams, respectively.

produced in the far-field zone (behind the phase plate) an angular distribution that reproduced the diffraction quality of the laser light along the vertical coordinate and all the details of the angular structure of this light along the horizontal coordinate (see Fig. 5c). Comparison of the brightness of the angular distributions of the laser light and the scattered wave passed through the phase plate yields the ratio $B_s/B_0 \approx 0.25$, which corresponds to the reversal fraction $H = 1$. Thus, within experimental error ($\sim$15%), *all* the energy of the reflected light was *exactly* reversed.

To elucidate the role of the pumping spatial inhomogeneity, the same experiment was performed without the phase plate. An almost ideally divergent spherical wave, just past the focus of a lens, was applied to the lightguide. The intensity inhomogeneities in this case are much larger than when a phase plate is used. The angular distribution of backward-reflected radiation is shown in Fig. 5d. Note that here, too, the directivity of the reflected radiation is sufficiently high, $\delta\theta \sim 5 \times 10^{-3}$ rad, yet WFR does not take place. Thus, if pumping inhomogeneities are not sufficiently fine, there is no correlated WFR solution (specklon) in the interaction volume. Theoretical considerations [see Sidorovich (1976) and, for quantitative theory Zel'dovich and Shkunov, (1978a)] show that the necessary condition for the existence of a specklon is a weak amplification over the longitudinal diffraction length of the inhomogeneities:

$$g\,\Delta z \equiv g/k\theta_0^2 \lesssim 1. \tag{4a}$$

Because in the regime of WFR recording $2gL \sim 30$, we obtain the condition

$$L/\Delta z \gtrsim 15, \tag{4b}$$

where L is the total interaction length.

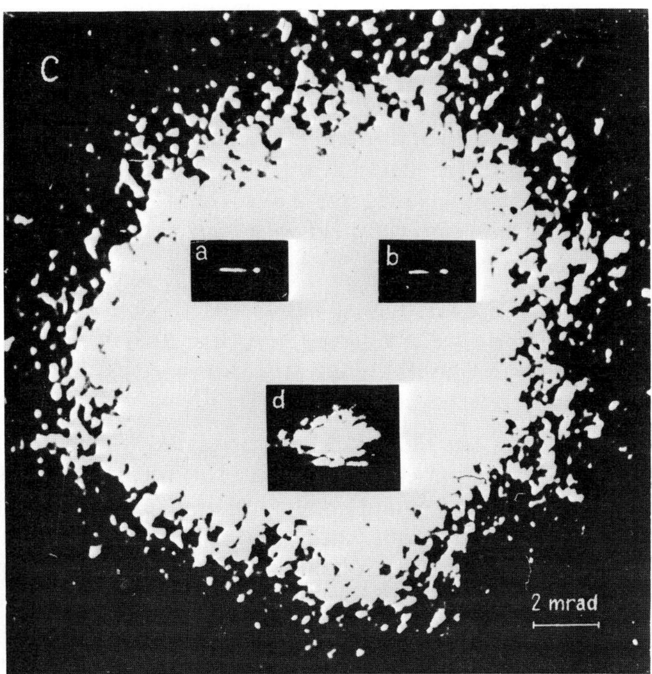

Fig. 5. Angular spectra: (a) incident radiation, (b) radiation in front of the phase plate and after double passage through it, (c) backscattered radiation after correction by the phase plate, (d) scattered radiation without a phase plate.

B. Effect of Nonuniformity of Polarization States on WFR-SS

The above situation with only the correlated solution—which reverses by a "specklon" (or, alternatively, "mode")—is characteristic only of the simplest case, where the monochromatic pumping radiation has a uniform polarization state throughout the interaction volume: $\vec{\mathscr{E}}_0(\mathbf{r}, z) = \mathbf{e}_0\mathscr{E}_0(\mathbf{r}, z)$, where the unit vector $\mathbf{e}_0$ does not depend on the coordinates. In this case the WFR specklon has the form $\vec{\mathscr{E}}_s(\mathbf{r}, z) = \mathbf{e}_s\mathscr{E}_0^*(\mathbf{r}, z)$, where the unit vector of the Stokes wave polarization reproduces the pumping unit vector *without conjugation*, $\mathbf{e}_s = \mathbf{e}_0$, owing to the scalar character of SBS. For pumping with incomplete spatial polarization,

$$\vec{\mathscr{E}}_0(\mathbf{r}, z) = \mathbf{e}_1\mathscr{E}_1(\mathbf{r}, z) + \mathbf{e}_2\mathscr{E}_2(\mathbf{r}, z), \tag{5}$$

there are four specklons of the scattered wave (Zel'dovich and Shkunov,

1978b). To characterize these specklons, it will be convenient if we choose in Eq. (5) the orthogonal vectors $\mathbf{e}_1$ and $\mathbf{e}_2$ so that the speckle structures $\mathscr{E}_1(\mathbf{r}, z)$ and $\mathscr{E}_2(\mathbf{r}, z)$ are uncorrelated over the space, i.e., $\int \mathscr{E}_1\mathscr{E}_2^* \, d^2\mathbf{r} = 0$. Let us also denote $\langle |\mathscr{E}_1|^2\rangle = I(1 + p)$ and $\langle |\mathscr{E}_2|^2\rangle = I(1 - p)$, where $|\mathscr{E}_1|^2 \geq |\mathscr{E}_2|^2$, and $p \geq 0$ is the degree of pumping polarization. Then, according to the detailed quantitative theory developed by Zel'dovich and Shkunov, 1978b, we have the following set of four specklons $\mathscr{S}_i(\mathbf{r}, z)$ of the scattered field with gain values g_i:

$$\begin{aligned}
\mathscr{S}_1(\mathbf{r}, z) &= \mathbf{e}_1\mathscr{E}_1^*(\mathbf{r}, z),\\
\mathscr{S}_2(\mathbf{r}, z) &= \mathbf{e}_2\mathscr{E}_2^*(\mathbf{r}, z),\\
\mathscr{S}_3(\mathbf{r}, z) &= \mathbf{e}_1(1 + p)\mathscr{E}_2^*(\mathbf{r}, z) + \mathbf{e}_2(1 - p)\mathscr{E}_1^*(\mathbf{r}, z),\\
\mathscr{S}_4(\mathbf{r}, z) &= \mathbf{e}_1\mathscr{E}_2^*(\mathbf{r}, z) - \mathbf{e}_2\mathscr{E}_1^*(\mathbf{r}, z),\\
g_{1,2} &= GI(1 \pm p), \qquad g_3 = GI, \qquad g_4 \equiv 0.
\end{aligned} \tag{6}$$

Obviously, there are, in addition, uncorrelated waves with gain $0.5GI(1 + p)$ for polarization $\mathbf{e}_1$ and $0.5GI(1 - p)$ for polarization $\mathbf{e}_2$.

In the experimental arrangement shown in Fig. 6, each specklon can be recorded separately and its energy measured (Blashchuk *et al.*, 1978a,b). The essential feature of this scheme is a special phase plate—a depolarizer—that makes nonuniform not only the phase of the wave but also its polarization state. The depolarizer is a calcite plate cut parallel to the optical axis and etched in nitric acid. The etching resulted in pits with an average depth of $\sim 8\ \mu$m and a transverse size of $\sim 250\ \mu$m. This produced a path difference $\gtrsim \lambda$ for the orthogonal polarizations ($n_0 - n_e = 0.16$). The plate was placed in a liquid with a refraction index $n \simeq \frac{1}{2}(n_0 + n_e)$. Therefore, the orthogonally polarized beams had the same divergence after leaving the depolarizer.

The optical axis was aligned vertically. To vary the degree of pumping polarization, the polarization of laser radiation was rotated by the Faraday cell F and then the beam was passed through the polarizer P_1 inclined at an angle α to the axis of the phase plate. The degree of polarization of the beam in the cell is $p = |\cos 2\alpha|$. The backward-scattered radiation passed through the depolarizer, and was then directed to a system for recording its energy and angular distribution. If the vertical polarization is denoted by $\mathbf{e}_1$ and the horizontal polarization by $\mathbf{e}_2$, then the specklon $\mathscr{S}_1(\mathbf{r}, t)$ passing through the depolarizer in the reverse direction is transformed into a vertically polarized plane wave and the specklon $\mathscr{S}_2(\mathbf{r}, t)$ into a horizontally polarized plane wave. Thereafter, these waves were spatially separated by the calcite double-refracting prism and directed to systems for recording the angular distribution and energy (see Section III) for each polarization. The specklons $\mathscr{S}_3$ and $\mathscr{S}_4$ are not transformed into plane waves in the arrangement shown in Fig. 6, and therefore only their

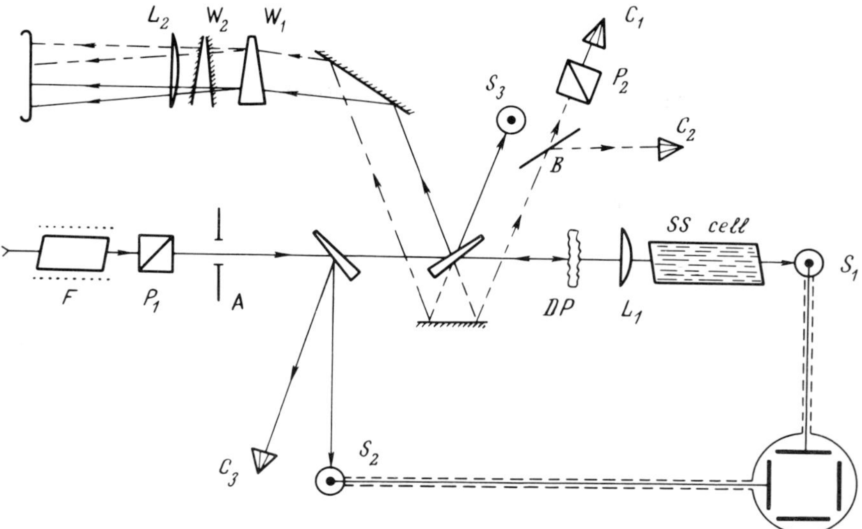

Fig. 6. Block diagram of an experimental setup used to study WFR of spatially depolarized radiation. Faraday cell F and polarizer P_1 govern the direction of the unit vector of linear polarization of Nd-laser radiation incident upon aperture A. C_3 measures the energy of laser radiation, S_1 the pulse shape. S_1 and S_2 are the measuring devices of the loop scheme for determining the steady-state threshold of SS. DP is the depolarizer. L_1 focuses the radiation into the cell containing a nonlinear substance. C_2 and C_1 measure the energy of the total reflected pulse and its vertical polarization component, respectively; double-refracting wedge W_1 split (in direction) the vertical and horizontal components of laser and scattered radiations. W_2 is a mirror wedge. Lens L_2 focuses four fans of beams onto the photoplate.

energy is recorded by the calorimeters [their brightness in the far-field zone is too small to be detected under these experimental conditions (Blashchuk *et al.*, 1978a,b)]. The arrangement can be slightly modified. A uniform polarization plate $\lambda/4$ with its axis at 45° to the vertical direction can be placed between the depolarizer and the scattering medium. Then the eigenvectors $\mathbf{e}_1$ and $\mathbf{e}_2$ in the medium will correspond to circular polarizations, and only the specklons $\mathcal{S}_3$ and $\mathcal{S}_4$ will be transformed into plane waves after backward propagation.

In this experiment, the brightness, energy, and reversal fraction were measured simultaneously in each of the polarization channels; moreover, a loop scheme like that in Fig. 3 was used to record the steady-state threshold of SBS. The results are presented in Figs. 7 and 8. The fraction of energy in the specklons $\mathcal{S}_1$ and $\mathcal{S}_2$ is denoted by η_1 and η_2, respectively. The reciprocal threshold pumping power, proportional to the effec-

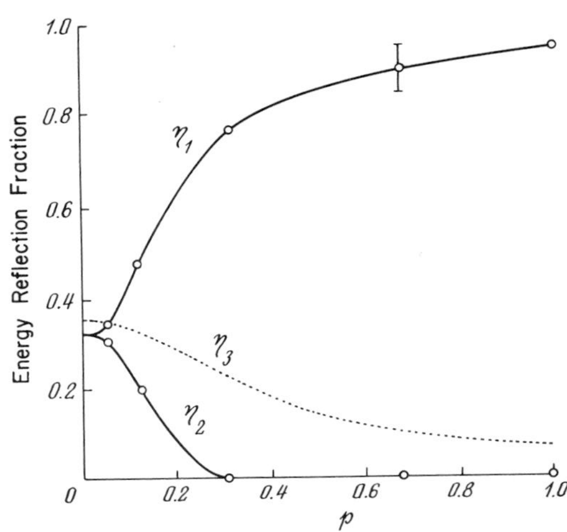

Fig. 7. The fraction of energy of reflected radiation corresponding to specklons $\mathscr{S}_1$, $\mathscr{S}_2$, and $\mathscr{S}_3$ as a function of the degree of polarization p of exciting radiation.

tive gain constant $g_1 \sim 1 + p$, is shown in Fig. 8 for the specklon $\mathscr{S}_1$ with maximal gain as a function of the degree of polarization p.

For completely polarized radiation ($p = 1$), only the WFR specklon $\mathscr{S}_1$ is excited, and it is to this specklon that almost all the backward-scattered energy [(95 ± 10)%] is transferred. Thus, uncorrelated waves appear to be almost completely suppressed owing to the discrimination mechanism. The same is true for the specklon $\mathscr{S}_4$, for which the gain is zero. For $p \lesssim 0.5$ the presence of the specklon $\mathscr{S}_3$ with $\eta_3 \equiv 1 - \eta_1 - \eta_2$ becomes noticeable, and for $p \lesssim 0.2$ also the presence of the specklon $\mathscr{S}_2$. If the pumping is completely depolarized in the bulk of the medium ($p = 45°$), the energy is shared equally, on the average, among all the specklons, i.e., $\eta_1 \approx \eta_2 \approx \eta_3 \approx \frac{1}{3}$. This result is in a good agreement with the theoretical prediction that the increments of all three specklons $\mathscr{S}_1$, $\mathscr{S}_2$, and $\mathscr{S}_3$ are equal for $p = 0$.

In experiments, the specklons $\mathscr{S}_2$ and $\mathscr{S}_3$ are present at a higher degree of polarization p than is predicted by theory according to the simple formula $\exp(g_i L)$, with g_i given by Eqs. (6). This descrepancy can probably be explained by the effects of SS saturation: a more intensive polarization component $\mathbf{e}_1\mathscr{E}_1$ is more attenuated owing to the transfer of energy to the specklon $\mathscr{S}_1$, so the effective degree of polarization in the interaction volume decreases.

An important advantage of the loop scheme is that it permits the re-

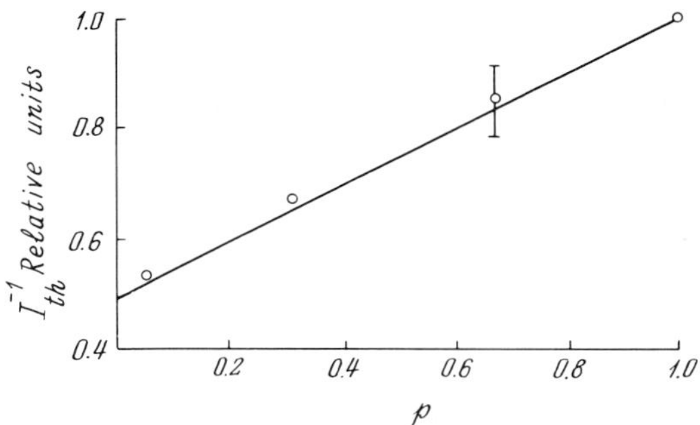

Fig. 8. Experimental values of the reciprocal SS threshold and theoretical increment of specklon $\mathcal{S}_1$ with a small gain as a function of degree of polarization p.

cording of the *threshold* of SS, where saturation effects were previously not revealed, and our theoretical calculations are completely applicable. The experimental data on the reciprocal threshold power fit well the theoretical dependence $g_1 \sim 1 + p$ for the specklon $\mathcal{S}_1$.

Thus, the experimental results (Blashchuk *et al.*, 1978a,b) confirm in detail the specklon concept of WFR–SS.

Two independent experiments (Blashchuk *et al.*, 1978a,b; Basov *et al.*, 1978, 1979b) followed the theoretical work on WFR–SS of depolarized beams (Sokolovskaya *et al.*, 1978b; Basov *et al.*, 1978, 1979b) deal with practical aspects of WFR–SS, in particular, with the depolarization of radiation as it passes through rods of neodymium amplifiers; they have also realized a scheme of spatial-polarization WFR based on the reversible transformation of a depolarized beam into a completely polarized beam of larger divergence.

Note that four-wave WFR permits complete spatial-polarization WFR without preliminary transformation of the depolarized beam into a polarized beam (Zel'dovich and Shkunov, 1979a,b). Various methods suggested by Zel'dovich and Shkunov (1979a,b) have been realized experimentally [see Blashchuk *et al.* (1980) and Martin *et al.* (1980)] in determining the fraction of WFR to be nearly 100%.

C. Fluctuations of the Specklon Total Phase

Both in laser and in SS-active media, stimulated radiation increases, starting from the level of spontaneous noise. In SBS, this spontaneous noise corresponds to the scattering of the exciting field from fluctuating hypersonic waves. Because the amplification is exponential, $\sim\exp(gz)$, a comparatively thin layer of the medium located at the far end of the cell ($\Delta L \sim g^{-1}$) contributes mainly to the amplified Stokes field. The characteristic correlation time of the initiating fluctuations is $\tau_s \sim \Gamma^{-1}$, where Γ is the scattering linewidth; τ_s coincides with the attenuation time of a hypersonic phonon. Under typical conditions, $\tau_s \sim 10^{-8}-10^{-9}$ sec.

Excitation of SS due to spontaneous noise leads to a number of important features of WFR–SS. For instance, the WFR of extremely small signals, to be considered in Section V, is limited by the level of amplified spontaneous uncorrelated noise. In this section we consider another consequence of this excitation—fluctuation of the complex amplitude of WFR specklons. The amplitude factor of the gain of a specklon is given in the linear approximation by $\exp[g(\omega)z]$, where $g(\omega)$ coincides, in general, with the spectral shape associated with the spontaneous scattering linewidth Γ. Simple arguments (Dyakov, 1969; Vasilyev *et al.*, 1981) show, therefore, that the correlation time of fluctuations of the amplified radiation increases, and appears to be of the order of $\tau_{cor} \sim \Gamma^{-1}(2gL)^{1/2}$, where $2gL \approx 30$. A fivefold reduction in the SS linewidth, as compared with the spontaneous linewidth, corresponds to this correlation time.

If the Stokes pulse length T_s exceeds $\tau_{cor} \sim 5/\Gamma$, the complex amplitude of the Stokes wave can experience $N \sim T_s/\tau_{cor}$ fluctuations. If, moreover, the effects of SS saturation are involved, the fluctuations of the scattered light intensity are almost completely suppressed (Tang, 1965; Dyakov, 1969), but the phase fluctuations remain unchanged.

Phase fluctuations of specklons under SS were first observed by Blashchuk *et al.* (1978a,b), using the scheme shown in Fig. 6; the optical axis of the depolarizer was rotated 45° relative to the vertical direction, and the light incident upon the depolarizer was vertically polarized. The pumping radiation in the measuring cell became completely depolarized ($p = 0$), and the unit vectors of polarization could be chosen arbitrarily. For the sake of further discussion, let us choose the unit vectors $\mathbf{e}_1$ and $\mathbf{e}_2$ coinciding with the new axes of the depolarizer. For $p = 0$ the scattered radiation contains the three specklons $\mathcal{S}_1$, $\mathcal{S}_2$, and $\mathcal{S}_3$, given by Eqs. (6), with approximately equal amplitudes. Owing to independent excitation and amplification of each of the specklons, their phases φ_i also fluctuate independently. Propagation of the specklons $\mathcal{S}_1$ and $\mathcal{S}_2$ through the depolarizer results in a plane wave with polarization $\mathbf{e}_1 e^{i\varphi_1} + \mathbf{e}_2 e^{i\varphi_2}$; the specklon $\mathcal{S}_3$ falls completely into the halo and is not recorded.

The intensities of diffraction-limited spots for the vertical and horizontal polarization components were measured. They are proportional to $\cos^2[\frac{1}{2}(\varphi_1 - \varphi_2)]$ and $\sin^2[\frac{1}{2}(\varphi_1 - \varphi_2)]$, respectively, and fluctuate because of fluctuations of the phases φ_1 and φ_2. Diffraction-limited spots were observed in the scattered wave for both horizontal and vertical polarizations.

From pulse to pulse the ratio of the energies of these spots fluctuated significantly (by up to a factor of 10), and so the independence of the excitation and amplification of specklons is confirmed experimentally.

Thus, the great energy fluctuations in both types of polarization can be attributed to the fact that phase correlation time for SBS in acetone (Blashchuk *et al.*, 1978a,b), $\tau_{cor} \sim 5/\Gamma \approx 20$ nsec, is of the order of the pumping pulse length $T_0 \approx 17$ nsec.

Phase fluctuations of two waves scattered with WFR from two different regions of the same cell were observed by Basov *et al.* (1979b). The fluctuations resulted in variations in the shape of the far-field spot.

Detailed quantitative investigations of temporal fluctuation of the complex amplitude of specklons were performed simultaneously by (Basov *et al.*, 1980a; Bespalov *et al.*, 1980b; and Vasilyev *et al.*, 1980). The results agree well, even though the recording techniques differed. We shall describe the results obtained by (Vasilyev *et al.*, 1980). The laser radiation with a flat wave front was split by a semitransparent mirror M into two beams of equal intensity, which passed through identical phase plates and were directed into measuring cells filled with the same substance (see Fig. 9). The radiation scattered with WFR was transformed by the plates into

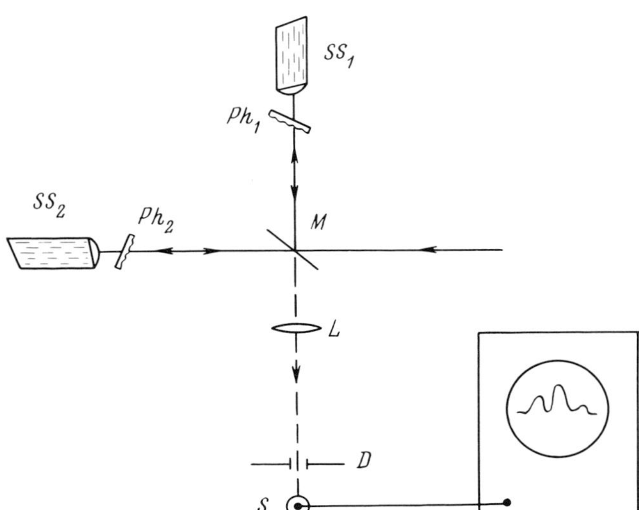

Fig. 9. Block diagram of an experimental setup used to phase fluctuations of the reversed wave. M is a flat, semitransparent mirror. Ph_1 and Ph_2 are identical phase plates, SS_1 and SS_2 identical cells filled with a liquid. L is a lens with a long focal length. D is an aperture for the separation of the reversed component of scattered light.

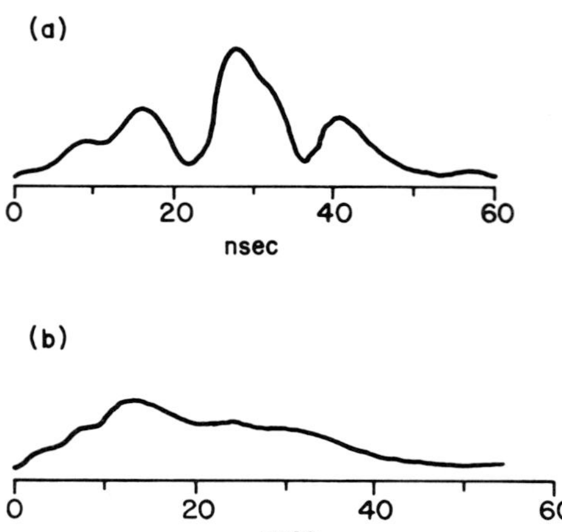

Fig. 10. Photocurrent oscillograms for scattering excited (a) in CCl_4, and (b) in acetone.

plane waves, which were then superimposed on mirror M to from a single plane wave with fluctuating amplitude. The temporal behavior of the intensity of this wave was recorded by a photodetector and displayed on the oscilloscope screen. A number of precautions were taken [for details see (Vasilyev *et al.*, 1980)] to avoid the possible coupling of these waves and nonequivalent conditions of their excitation.

The process of WFR–SS proceeded independently in each cell, and fluctuations of the phase difference of the reconstructed specklons resulted in the intensity fluctuations recorded by the photodetector. Typical intensity oscillograms for SS in CCl_4 and acetone are shown in Figs. 10a and 10b. The correlation time of the fluctuations agreed, within experimental accuracy, with the theoretical estimate $5/\Gamma$, where Γ is the linewidth of spontaneous Brillouin scattering in the substance.

V. WFR of Weak Signals against the Noise Background and Amplification of Uncorrelated Waves

It would seem to be impossible to perform WFR–SS of signals A whose power is lower than the threshold of stimulated scattering. To overcome this difficulty, one can use the following procedure. Let another beam B, with a power exceeding the threshold, excite backward SS with WFR in the medium. If the weak beam A to be reversed is mixed at some angle with the strong beam B (within the interaction volume), WFR of the resultant field $B + A$ will occur because its power exceeds the threshold. The reflected field will be of the form $\mathscr{E}_s = r(A^* + B^*)$, the reversed component of the weak signal being separated (with respect to direction) from the reversed reference wave B^*. Such a scheme has been studied in detail by Basov *et al.* (1979).

In this section, we deal with studies of stimulated scattering of waves uncorrelated with the exciting wave. The presence of such waves is an essential property of WFR–SS; the experimental study of these waves is therefore essential to an understanding of the physics of the process. The presence of uncorrelated waves decreases the fraction (quality) of reversal.

This is especially important for WFR of weak signals (Ragul'skii, 1979). Experimental data show that the brightness of uncorrelated waves is almost constant within a wide solid angle, and hence is the same in the directions of both the powerful reference wave and the signal. The relative fraction of noise for a weak signal is much greater than for a strong signal (for the same angle of detection).

A. Differential Efficiency of WFR of Weak Signals

To investigate WFR–SS of extremely weak signals and determine the brightness of the uncorrelated component of the scattered radiation, a special experiment was devised (Pilipetskii *et al.*, 1979b,c). A schematic diagram of the experimental setup is shown in Fig. 11. The radiation of a ruby laser was split by a mirror M_1 into a powerful reference beam, and about ten weak beams. The reference beam was fed through a phase plate into a lightguide containing liquid carbon disulfide. The weak beams were fed directly into the lightguide without first passing through the phase plate. The mirror wedge technique (see 2.2) with the successive attenuation factor $K = 9.7$ was used to produce the weak beams. The angle between the adjacent beams was 2.4 mrad and their divergence 0.3 mrad. The reference beam was confined at the lightguide input, within a cone having an apex angle of 24 mrad. The angle between the first weak beam and the axis of the reference beam was 26 mrad.

Both the reference beam and the fan of weak beams were backward scattered with WFR. Let us call the differential scattering efficiency η_i, the ratio of the energy of the ith component of the scattered field to the energy of the corresponding component of the exciting field. To compare the differential efficiencies of the scattered waves of sharply differing intensities, the weak reflected beams were transmitted through the same

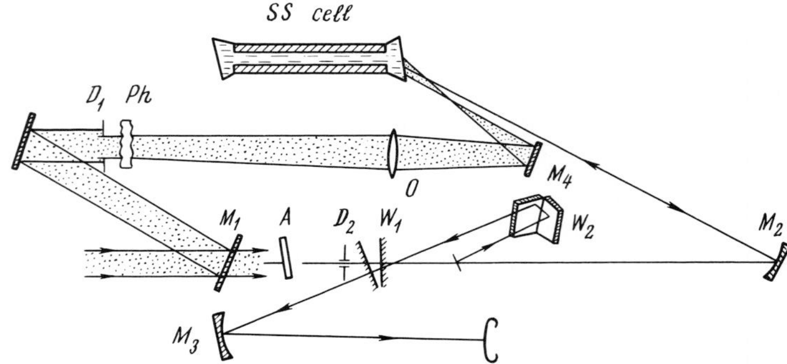

Fig. 11. Block diagram of an experimental setup used to study the quality of reversal of a light field. M_1 is a semitransparent dielectric mirror. D_1 and D_2 are apertures. Ph is a phase plate. O, M_2, and M_3 are objective lenses. *A* is an attenuator. W_1 is a mirror wedge with $K \approx 10$, W_2 a corner-type reflector with the edge rotated 25° relative to W_1. The cell containing a hollow lightguide is filled with carbon dioxide.

mirror wedge, but with a preliminary rotation of the fan direction of 50°. As a result, *each* of the scattered beams produced its own series of successively attenuated spots, so that the total pattern on the film was that shown in Fig. 12.

Suppose that a given spot is produced by *m*fold attenuation by the wedge in the wave incident upon the measuring cell and by *n*fold attenuation by the wedge in the recording process. The energy of the spot is const $\eta_m K^{-(m+n)}$, where the constant value is the same for all beams. Thus, in comparing spots with a fixed value of $m + n$, one can compare the values of η_m for various m. In Fig. 12 the spots with a fixed $m + n$ form columns. Photometric analysis shows that the spots of a column are identical in both angular dimension and brightness. Thus, the differential efficiency of the WFR reflection, η_m, does not depend on the number m if the attenuation factor k^m varies from 1 to 5×10^{-6}. In the experiment, η_m was $\approx 50\%$ and coincided with the reflectance of the powerful reference wave. The peak power of the weakest exciting beam whose scattering was recorded was ≈ 3 W and its energy ≈ 1 erg.

As the power of a weak beam is further reduced, it is lost in the noise field. A photograph of this is shown at the bottom of Fig. 12; greater exposure was needed for this photograph. The speckle structure of the noise is rather inhomogeneous. The photograph is positioned so that its top edge approximately corresponds (along the vertical axis) to the power of the weak scattered beams whose brightness coincides with that of the most intensive spots of the noise field.

Thus, under stimulated scattering of a spatially inhomogeneous field,

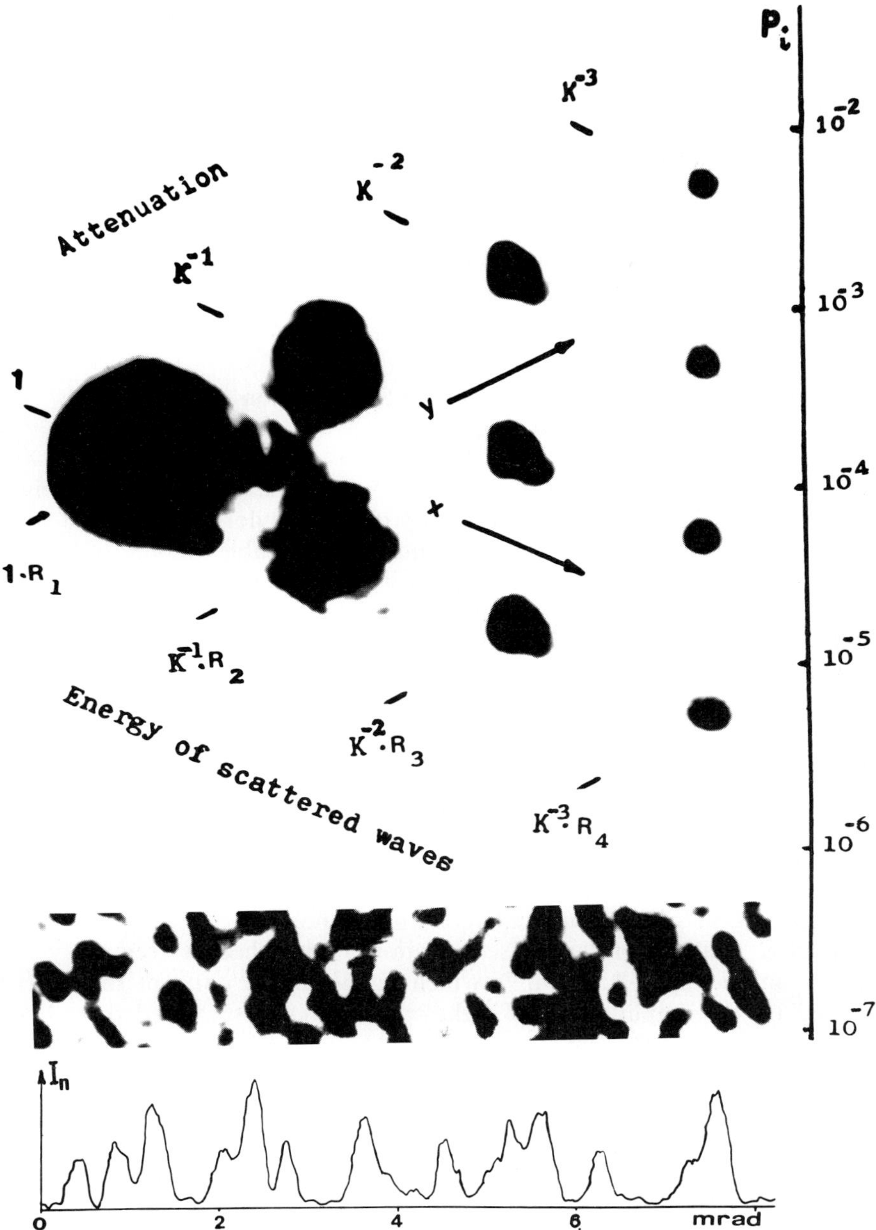

Fig. 12. Image of the far-field zone of the scattered beams and noise. The brightness of the scattered beams decreases in direction x, that of the secondary beams in direction y. The energy of the first scattered beam is assumed to be unity. R_{1-4} is the scattering efficiency for different components of the exciting field, P_i the relative power of the ith component. The vertical scale refers to the right series of beams. The photograph at the bottom is the noise field densitogram.

the information on its very weak details is not distorted if the brightness exceeds the level of noise-scattered radiation.

These results permit us to delineate the limits of possible wave-front reversal of a weak light field mixed with radiation of high power (exceeding the threshold of stimulated scattering). The relative power of the weak field, corresponding to the solid angle of diffraction, must be greater than $\sim 10^{-6}$, in which case the brightness of the weak field after scattering will exceed that of the noise field, and the former can be separated from the noise by means of angular selection.

B. Investigation of the Component of SS Radiation Uncorrelated with Pumping[3]

As has been shown, the noise component determines the minimal brightness of the weak field details reproduced under reversal. On the other hand, this component may contain a considerable amount of backscattered light, because the solid angle within which the component propagates can be fairly large. Experiments on the direct determination of the brightness and angular distribution of the noise have been performed (Pilipetskii *et al.*, 1979a), using an experimental setup similar to one used previously (Pilipetskii *et al.*, 1979b,c). A photograph of the noise field is shown at the bottom of Fig. 12. The minimal angular dimension of the recorded spots of the microstructure is ~ 0.2 mrad, which corresponds to diffraction at the output of the lightguide containing the scattering medium. Thus, the noise waves, in general, uniformly fill the output face of the lightguide. The high contrast of the speckle structure observed is due to the fact that the length of the scattered pulse is of the order of the correlation time of the amplified spontaneous noise.

Spectral analysis of the noise radiation shows that its frequency is Stokes shifted by $\sim 0.19\ \text{cm}^{-1}$, which supports the concept that it is formed by the amplification of light spontaneously scattered from hypersonic waves in the carbon disulfide. The noise had the same linear polarization as the pumping.

Photographic measurements of the mean noise intensity show that under the experimental conditions of (Pilipetskii *et al.*, 1979c) the fraction of the energy of the exciting radiation that is transformed into the diffraction solid angle of noise, in the vicinity of the central direction of the pumping beam, is, on the average, $\sim 5 \times 10^{-8}$. This result was verified (Pilipetskii *et al.*, 1979a) by direct measurement of the intensity of light

[3] The results of this section were obtained by Dr. V. V. Ragulskiy.

collected by a condenser within a large solid angle and recorded by a calorimeter (see the scheme in Fig. 13). It is convenient to interpret this result in terms of the reflectance per unit solid angle of the scattered noise component: $R_{\text{noise}} = 1.55 \pm 0.23\ \text{sr}^{-1}$.

These measurements are for a direction near the central direction and for a solid angle measured within the medium. Special measurements of the mean brightness as a function of the solid angle have been performed with the same experimental arrangement (see Fig. 13).

In Fig. 14 the dependence of R_{noise} (sr^{-1}) on the angle β measured from the central direction of the pumping in the scattering medium is shown. The distribution $R_{\text{noise}}(\beta)$ is almost independent of the pumping divergence, which was varied from 10 to 40 mrad (in air), and of the pumping power, when operating under saturation conditions. Thus, the noise measured does not correlate with the pumping. The results obtained by using lightguides of various lengths L (0.33–1 m) and diameters d (0.2–0.6 cm) fit well the empirical relation $R_{\text{noise}}(\beta) = R_0 \exp[-(n - 1)L\beta^2/d]$. Here β is the angle in the medium measured in radians and n the ratio of the refractive indices of carbon disulfide and quartz, the materials of the core and walls of the lightguide.

The fit is achieved for the following reason. The lightguide used has a circular cross section. Because the angle of total internal reflection at the carbon disulfide–quartz interface, $\beta_c \approx 0.5$ rad, is considerably greater than the angles of interest, the reflection of uncorrelated waves from the walls of the lightguide does not produce energy losses; it depolarizes the scattered beam. This leads, owing to the scalar character of SBS, to a reduction in the effective amplification. The number of reflections of the beam with a given β, governing the rate of depolarization, is proportional to $L\beta/d$, and this is revealed in the angular dependence observed.

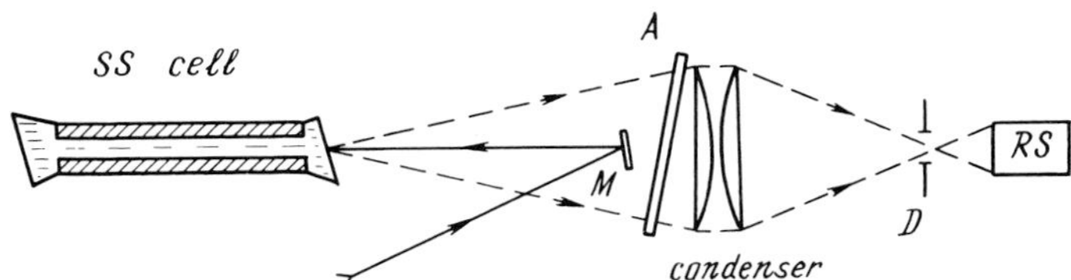

Fig. 13. Diagram of an experimental setup for recording noise radiation. M is a totally reflecting mirror which directs the exciting radiation into the lightguide containing a scattering liquid. The condenser collects the noise component of backscattered light and concentrates it at the plane of the aperture. RS is a system for recording the energy and angular distribution of scattered light. *A* is a glass attenuator, which eliminates the effects of the recording instruments of the propagation of radiation in the SS cell.

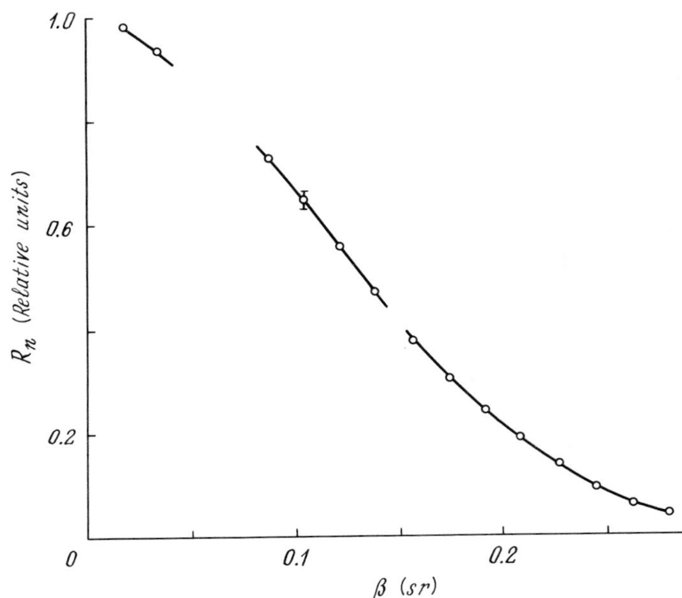

Fig. 14. Experimental angular dependence of the reversal efficiency R_{noise} into the noise component. β is the angle between the directions of propagation of reversed and nonreversed waves measured in the scattering medium.

Integration of $R_{\text{noise}}(\beta)$ over the solid angle permits estimation of the total energy of noise radiation, which, under typical experimental conditions (Pilipetskii *et al.*, 1979a,b,c), was ~13% of the incident energy. 50% of the pumping energy was reflected into the WFR component. The fraction of WFR determined by a formal approach was $H \approx 0.79$. Actually, however, the operating solid angle of most of the elements of laser devices is not large, $\lesssim (10^{-1}\ \text{rad})^2$. About $\frac{1}{10}$ of the total noise strikes the aperture, so that the effective fraction of WFR is almost unity.

Thus, under WFR–SS in a lightguide, uncorrelated noise has little effect on the fraction of WFR important in practice, but limits the energy level of the weak signals to be reversed. Although under spontaneous initiation of SS this noise cannot be eliminated, it can be slightly reduced by choosing an appropriate scattering medium and varying the temperature and the geometry.

VI. WFR in Focused Beams

A lightguide affords high power density over large interaction length L. Therefore, the threshold of the pumping power density $|\mathscr{E}_0|^2$, as determined from the relation $2G|\mathscr{E}_0|^2L \approx 30$, is relatively low. Another important advantage of the scheme with a lightguide is that the different angular components of pumping are well mixed, so the mean intensity is nearly constant throughout the interaction volume.

A pumping beam focused into the medium without a lightguide is frequently used for WFR–SBS (see, for instance, Basov *et al.*, 1978, 1979a,b, 1980a,b; Bespalov *et al.*, 1977a,b, 1980a; Blashchuk *et al.*, 1977, 1978a,b; Koptev *et al.*, 1978; Krivoshchekov *et al.*, 1980; Kudriavtseva *et al.*, 1970, 1978; Mays and Lysiak, 1979; Sokolovskaya *et al.*, 1977 and 1978; Sokolovskaya and Brekhovskikh, 1978; Vasilyev *et al.*, 1980). A large number of speckle inhomogeneities must be present in the interaction volume (see Section II) for the discrimination mechanism to separate the reversed configuration.

Let the divergence $\Delta\theta_0$ of a speckle-inhomogeneous pump beam with diameter d be ξ times greater than the diffraction limit: $k\,\Delta\theta_0 d \gg 1$. Then, behind the lens of focal length f, the beam is localized in a focal region with diameter $d_f \sim 2f\,\Delta\theta_0$ and length $L_f \sim f^2\,\Delta\theta_0/d$. The corresponding dimensions of the speckle pattern in the region are much smaller: $\Delta r_\perp \sim d_f/\xi$ and $\Delta z \sim L_f/\xi$. Because the total amplification near the SS threshold in the interaction length $\sim L_f$ is about e^{30}, the condition (4b) gives an estimate of the lowest divergence of the beam that could be reversed with good quality: $\xi \gtrsim 15$.

Amplification of uncorrelated noise sets an upper limit to the possible values of ξ for which a noticeable fraction of WFR occurs. Under WFR of a focused beam, the number of independent noise components with significant gain grows as ξ^2 with increasing irregular pumping divergence.

Uncorrelated waves are amplified primarily within an area $S = (0.2d)^2$ and solid angle $(\Delta\theta)^2 = (0.2d/L_f)^2$. The ratio of these dimensions and angles to the dimensions and angles for pumping, d and d/L_f, is $1/\sqrt{gL_f} \approx \frac{1}{30}$. Such a ratio occurs owing to a large total gain in the tube of an effective lightguide formed by the pumping envelope; it is typical of such an amplification geometry and quite similar to the narrowing of the SS spectral line considered in Section IV. The number of modes over the area S and within the solid angle $(\Delta\theta)^2$ is $N = S(\Delta\theta)^2/\lambda^2$, from which one obtains (Bespalov *et al.*, 1977b) the estimate $N_{\text{noise}} \approx \xi^2/30^2$ for the total number of excited uncorrelated waves.

In the absence of saturation in the field of focused pumping, WFR occurs in an amplification profile that depends upon the radius from the axis as $\sim f(\mathbf{r})\mathscr{E}_0^*(\mathbf{r})$, and the envelope $f(\mathbf{r})$ has its maximum on the axis $|\mathbf{r}| = 0$ (Baranova *et al.*, 1978, 1980; Bespalov *et al.*, 1978). This leads to some mismatching of the specklon $\mathscr{S}_s$ with the amplification profile, resulting not only in a deterioration of the quality of small- and large-scale WFR in the specklon but also in a reduction of its amplification rate. A detailed calculation (Bespalov *et al.*, 1978), shows that the mean increment in the gain of a specklon is about 1.3–1.4 times the mean increment for uncorrelated waves. Therefore, the discrimination of noise components appears to be less efficient without a lightguide. Thus, for an individual noise component the gain (in the unsaturated regime) is $1/Q \approx 10^{-3}$, the value under saturation conditions. As the irregular divergence grows, the process of WFR becomes ineffective when the number of noise components N_{noise} exceeds the discrimination factor Q corresponding to $\xi \approx 900$.

WFR–SBS in a focused pumping beam (i.e., without a lightguide) was first detected by Blashchuk *et al.* (1977) and Bespalov *et al.* (1977a). A detailed experimental study of WFR–SBS in a focused beam was performed by Bespalov *et al.* (1977b). The threshold of SS and the quality of WFR were investigated as a function of ξ. In the experiment, the threshold of SS, as determined from the integral pumping power, grew linearly with ξ, because the effective length of the interaction region, (the length of the focal region L_f) was proportional to ξ, whereas the power density fell as $a_\perp^{-2} \sim \xi^{-2}$ for a given integral pumping power. The quality of WFR was determined by the aperture technique with the use of phase plates, the latter permitting variation of the irregular pumping divergence $\Delta\theta_0 \approx \xi\lambda/d$.

WFR was observed in the range $100 \lesssim \xi \lesssim 1000$, in good agreement with theory (Bespalov *et al.*, 1978). The fraction of WFR reached the maximal value 0.8 for $100 \lesssim \xi \lesssim 500$.

The theory of WFR–SS in a focused beam presently available (Bespalov *et al.*, 1978; Baronova *et al.*, 1978) does not take into account saturation effects almost always encountered. It can be argued that saturation levels the pumping intensity envelope, increasing both the quality WFR and the degree of discrimination. Moreover, the effects on WFR of such factors as the focal length of the lens and the position of focus in the medium, as well as nonstationary effects, have not yet been studied in detail.

The presence of a large number of parameters makes it possible to optimize the regime of WFR of focused beams. This is confirmed by experimental results. For instance, Blashchuk *et al.*, 1977, 1978a, have achieved unity WFR (within an experimental accuracy of 10–20%).

VII. WFR for Other Mechanisms of Amplification in the Field of Nonuniform Pumping

The discrimination mechanism of WFR, first discovered for SBS, leads to a predominance of the gain of a WFR specklon, correlated with pumping, under a physical process that provides for a local increase in the gain in places with maximal pumping intensity. So far, WFR has been studied, to various extents, under the following similar physical processes: stimulated Raman scattering (SRS) (Kudriavtseva *et al.*, 1978; Sololovskaya *et al.*, 1977, 1978; Sololovskaya and Brekhovskikh, 1978; Zel'dovich *et al.*, 1977), stimulated Rayleigh wing scattering (SRWS) (Kudriavtseva *et al.*, 1978), stimulated temperature scattering (STS–II) (Krivoshchekov *et al.*, 1980), and superluminescence (SL) (Koptev *et al.*, 1978).

SRS and SL are characterized by a relatively large frequency shift of the amplified radiation (up to tens of percent) (Yariv, 1965). This hinders their practical use in laser systems, because reversed radiation with a large frequency shift does not remain in the laser gain profile. Nevertheless, for an understanding of the physics of WFR as a whole, the study of WFR–SRS and WFR–SL is of considerable interest. The specific features of WFR with a large frequency shift arise from the differences in the laws of diffraction and propagation of fields of various wavelengths.

Suppose that two fields of different frequencies, $\mathscr{E}_0$ and $\mathscr{E}_s$, are exactly reversed in some section z = const. Then, in the paraxial case, under free propagation the speckle inhomogeneities of these fields are similar in structure but differ in longitudinal scales, $\Delta z_0/\Delta z_s = k_0/k_s$, where $k_{0,s}$ is the wave number. Mismatching of the speckle patterns in free space takes place at a length $L_{mis} \sim \Delta z_0 k_0/|k_0 - k_s|$.

Nevertheless, the wave $\mathscr{E}_s$ can acquire the WFR structure and longitudinal scale Δz_0 throughout the amplification volume owing to spatially inhomogeneous amplification. The condition (Zel'dovich and Shkunov, 1977) $(g_{cor} - g_{uncor})L_{mis} \gtrsim 1$ must be fulfilled. Because, as a rule, $g_{cor} - g_{uncor} \sim g$, the condition for the existence of a WFR specklon in the interaction volume is

$$|k_0 - k_s|\theta_0^2 \lesssim g \approx 15/L.$$

Once the wave $\mathscr{E}_s$ leaves the nonlinear medium, it propagates according to the laws of diffraction in free space.

The first investigations of WFR–SRS were performed by Zel'dovich *et al.* (1977) and Sokolovskaya *et al.* (1977). A great deal of experimental work followed [see, for instance, Kudriavtseva *et al.* (1978), Sokolovskaya and Brekhovskikh (1978), and Sokolovskaya *et al.* (1978)]. The phase plate technique described above was used by Zel'dovich *et al.*, 1977 for unambiguous detection of WFR. Among the features of the experiments of Kudriavtseva *et al.* (1978), Sokolovskaya *et al.* (1977, 1978), and Sokolovskaya and Brekhovskikh (1978) are the use of amplitude (rather than phase) aberrators and the focusing of radiation into the medium without the use of a lightguide. The pumping passed through the amplitude test pattern contains, in addition to a nonuniform portion, a considerable portion of the initial plane wave. Therefore, the subsequent focusing of radiation generates a diffraction-limited bright spot in the focal region. As one moves away from the focal point, components distorted by the test pattern are mixed with the regular component. As a result, SRS is first excited in the central spot (Sokolovskaya *et al.*, 1978; Sokolovskaya and Brekhovskikh, 1978), and the discrimination mechanism starts to work only in the vicinity of the input window of the sample where all the pumping components are mixed. In these experiments, WFR was detected by the appearance of the test image formed by waves of Stokes frequency. A novel result is the detection and quantitative study of the shift in position of the reconstructed image and of the change in its transverse scales (Sokolovskaya and Brekhovskikh, 1978; Sokolovskaya *et al.*, 1978). The values of the shift and scale changes correspond to the departure of the wave front of the Stokes wave from the pumping just at the input window of the sample.

WFR–SRS was also studied by Mays and Lysiak (1979); it was recorded with the help of a phase plate and an aperture in the far-field zone.

The experiment of Kudriavtseva *et al.* (1978) is the only attempt so far to excite WFR–SS of ultrashort pulses ($T_0 \sim 25$ psec) by using stimulated Rayleigh wing scattering. SRWS has a relatively low frequency shift and a short buildup time, so it is of considerable interest for practical applications, including the possibility of achieving complete spatial-polarization WFR predicted by theory (Zel'dovich and Yakovleva, 1980).

Stimulated temperature scattering, caused by absorption, has a low frequency shift but, unfortunately, a relatively long buildup time. Experimental observation of WFR–STS–II was reported by Krivoshchekov *et al.* (1980).

In the experiment of Koptev *et al.* (1978), pulsed radiation ($\lambda = 532$ nm) was passed through a phase plate and then used as spatially inhomogeneous pumping for a dye. The backward superluminescence of the dye was observed in a wide spectral range (545–565 nm). The backward radiation of the dye occupied approximately the same solid angle as the pumping. The radiation contained the reversed component, though the fraction of reversal was not large (10–20%). The authors attribute the poor quality of WFR to a small degree of discrimination of a WFR specklon owing to effects of saturation of the local gain by the pumping (Lazaruk, 1979).

VIII. Conclusion

Space limitations have not allowed us to discuss all the work on wave-front reversal under stimulated scattering. Thus, we have discussed only work concerning the physical mechanism of WFR–SS. Investigations relating to practical applications have been excluded, in particular, work reporting the possibility (Wang and Guiliano, 1978; Zakharov, 1977a,b) and realization (Peregudov *et al.*, 1979; Pilipetskii *et al.*, 1978; Dolgopolov *et al.*, 1979) of autofocusing of radiation onto objects, and experiments on the compensation for distortions introduced by laser amplifiers (Dolgopolov *et al.*, 1979; Efimkov *et al.*, 1980; Nosach *et al.*, 1972; Zubarev *et al.*, 1980). Self-reversal under stimulated scattering holds great promise for these applications, because it permits operation with high-power radiation and does not require appropriately oriented additional waves or mirror elements.

Acknowledgments

Our research on WFR and the preparation of this chapter was carried out in close collaboration with Dr. V. V. Ragulskiy. We are deeply grateful to him. We are also greatly indebted to A. Talashev for help in translating the manuscript.

References

Baranova, N. B., and Zel'dovich, B. Ya. (1980). Wavefront reversal of focused beams (Theory of Brillouin backscattering), *Kvantovaya Elektron.* **7,** 973 [*English transl.: Sov. J. Quantum Electron.* **10,** 555 (1980)].

Baranova, N. B., Zel'dovich, B. Ya., and Shkunov, V. V. (1978). Wavefront reversal in stimulated light scattering in a focused spatially inhomogeneous pump beam, *Kvantovaya Elektron.* **5,** 973 [*English transl.: Sov. J. Quantum Electron.* **8,** 559 (1978)].

Basov, N. G., Efimkov, V. F., Zubarev, I. G., Kotov, A. V., Mikhailov, S. I., and Smirnov, M. G. (1978). Inversion of wavefront of SMBS of a depoparized pump, *Pis'ma Zh. Eksp. Teor. Fiz.* **28,** 215 [*English transl.: JETP Lett.* **28,** 197 (1978)].

Basov, N. G., Zubarev, I. G., Kotov, A. V., Mikhailov, S. I., and Smirnov, M. G. (1979a). Small-signal wavefront reversal in nonthreshold reflection from a Brillouin mirror, *Kvantovaya Elektron.* **6,** 394 [*English transl.: Sov. J. Quantum Electron.* **9,** 237 (1979)].

Basov, N. G., Efimkov, V. F., Zubarev, I. G., Kotov, A. V., Mikhailov, S. I., and Smirnov, M. G. (1979b). Influence of certain radiation parameters on wavefront reversal in a Brillouin mirror, *Kvantovaya Elektron.* **6,** 765 [*English transl.: Sov. J. Quantum Electron.* **9,** 455 (1979)].

Basov, N. G., Zubarev, I. G., Mironov, A. B., Mikhailov, S. I., and Okulov, A. Yu. (1980a). Phase fluctuations of the Stokes wave produced as a result of stimulated scattering of light, *Pis'ma Zh. Eksp. Teor. Fiz.* **31,** 685 [*English transl.: JETP Lett.* **31,** 645 (1980)].

Basov, N. G., Zubarev, I. G., Mironov, A. B., Mikhailov, S. I., and Okulov, A. Yu. (1980b). Laser interferometer with wavefront-reversing mirrors, *Zh. Eksp. Teor. Fiz.* **79,** 1678 [*English transl.: Sov. Phys. JETP* **5,** 847 (1980)].

Bespalov, V. I., and Kubarev, A. N. (1966). *Proc. All-Un. Symp., Nonlinear Opt., 2nd,* p. 247. Novosibirsk (in Russian).

Bespalov, V. I., Betin, A. A., and Pasmanik, G. A. (1977a). Reconstruction effects in stimulated scattering, *Pis'ma Zh. Tekh. Fiz.* **3,** 215 [*English transl.: Sov. Tech. Phys. Lett.* **3,** 85 (1977)].

Bespalov, V. I., Betin, A. A., and Pasmanik, G. A. (1977b). Experimental investigation of the threshold of the stimulated scattering (SS) of multimode light beams and the degree of regeneration of the pumping in scattered radiation, *Izv. Vuz. Radiofiz.* **20,** 791 [*English transl.: Radiophys. Quantum Electron.* **20,** 544 (1977)].

Bespalov, V. I., Betin, A. A., and Pasmanik, G. A. (1978). Reproduction of the pumping wave in stimulated-scattering radiation, *Izv. Vuz. Radiofiz.* **21,** 961 [*English transl.: Radiophys. Quantum Electron.* **21,** 675 (1979)].

Bespalov, V. I., Betin, A. A., Dyatlov, A. I., Kulagina, S. N., Manishin, V. G., Pasmanik, G. A., and Shilov, A. A. (1980a). Reversal of wave front in four-photon processes under conditions of two-quantum resonance, *Zh. Eksp. Teor. Fiz.* **79,** 378 [*English transl.: Sov. Phys. JETP* **52,** 190 (1980)].

Bespalov, V. I., Betin, A. A., Pasmanik, G. A., and Shilov, A. A. (1980b). Observation of transient field oscillations in the radiation of stimulated Mandel'shtam–Brillouin scattering, *Pis'ma Zh. Eksp. Teor. Fiz. Zhetf* **31,** 668 [*English transl.: JETP Lett.* **31,** 630 (1980)].

Blashchuk, V. N., Zel'dovich, B. Ya., Mel'nikov, N. A., Pilipetskii, N. F., Popovichev, V. I., and Ragul'skii, V. V. (1977). Wavefront inversion in stimulated scattering of focused light beams, *Pis'ma Zh. Tekh. Fiz.* **3,** 211 [*English transl.: Sov. Tech. Phys. Lett.* **3,** 1977)].

Blashchuk, V. N., Zel'dovich, B. Ya., Krasheninnikov, V. N., Mel'nikov, N. A., Pilipetskii, N. F., Ragul'skii, V. V., and Shkunov, V. V. (1978a). Stimulated scattering of depolarized radiation, *Dokl. Akad. Nauk SSSR* **241,** 1322 [*English transl.: Sov. Phys. Dokl.* **23,** 588 (1978)].

Blashchuk, V. N., Krasheninnikov, V. N., Melnikov, N. A., Pilipetskii, N. F., Ragulskii, V. V., Shkunov, V. V., and Zel'dovich, B. Ya. (1978b). SBS wavefront reversal for the depolarized light-theory and experiment, *Opt. Commun.* **27,** 137.

Blashchuk, V. N., Zel'dovich, B. Ya., Mamaev, A. V., Pilipetskii, N. F., and Shkunov, V. V. (1980). Complete wavefront reversal of depolarized radiation under degenerate four-photon interaction conditions (theory and experiment), *Kvantovaya Elektron.* **7,** 627 [*English transl.: Sov. J. Quantum Electron.* **10,** 356 (1980)].

Bloembergen, N. (1965). "Nonlinear Optics." Benjamin, New York.

Brewer, R. G. (1965). Growth of optical plane waves in stimulated Brillouin scattering, *Phys. Rev.* **140A,** 800.

Cathey, W. T. (1968). Holographic simulation of compensation for atmospheric wavefront distortion, *Proc. IEEE* **56,** 340.

Dietz, D. R., Cho, C. W., Rank, D. H., and Wiggins, T. A. (1969). Stimulated scattering in argon, *Appl. Opt.* **8,** 1248.

Dolgopolov, Yu. V., Jomarevskii, V. A., Kormer, S. B., Kochemasov, G. G., Kulikov, S. M., Murugov, V. M., Nikolaev, V. D., and Sukharev, S. A. (1979). Experimental investigation of the feasibility of application of the wavefront reversal phenomenon in stimulated Mandel'shtam–Brillouin scattering, *Zh. Eksp. Teor. Fiz.* **76,** 908 [*English transl.: Sov. Phys. JETP* **49,** 458 (1979)].

Dyakov, Yu. E. (1969). Estimate of line width of stimulated Mandel'stam–Brillouin and Raman scattering of light in saturation, *Pis'ma Zh. Eksp. Teor. Fiz.* **10,** 545 (1969).

Efimkov, V. F., Zubarev, I. G., Kotov, A. V., Mironov, A. B., Mikhailov, S. I., and Smirnov, M. G. (1980). Investigation of systems for obtaining short high-power pulses by wavefront reversal of the radiation in stimulated Brillouin scattering mirror, *Kvantovaya Elektron.* **7,** 372 [*English transl.: Sov. J. Quantum Electron.* **10,** 211 (1980)].

Hellwarth, R. W. (1977). Third-order optical susceptibilities of liquids and solids, *Prog. Quantum Electron.* **5,** No. 1 (1977).

Kogelnik, H. (1965). Holographic image projection through inhomogeneous media, *Bell Syst. Tech. J.* **44,** 2451.

Koptev, V. G., Lazaruk, A. M., Petrovich, I. P., and Rubanov, A. S. (1978). Wavefront inversion in superradiance, *Pis'ma Zh. Eksp. Teor. Fiz.* **28,** 468 [*English transl.: JETP Lett.* **28,** 434 (1979)].

Krivoshchekov, G. V., Struts, S. G., and Stupak, M. F. (1980). Stimulated thermal-scattering spectrum with wavefront inversion, *Pis'ma Zh. Tekh. Fiz.* **6,** 428 [*English transl.: Sov. Tech. Phys. Lett.* **6,** 184 (1980)].

Kudriavtseva, A. D., Sokolovskaya, A. I., and Sushchinskii, M. M. (1970). Stimulated Raman scattering and self-focusing of light in liquid nitrogen, *Zh. Eksp. Teor. Fiz.* **59,** 1556.

Kudriavtseva, A. D., Sokolovskaya, A. I., Gazengel, J., Xuan, N. P., and Rivoire, G. (1978). Reconstruction of the laser wave-front by stimulated scatterings in the picosecond range, *Opt. Commun.* **26,** 446.

Lazaruk, A. M. (1979). Wavefront reversal in amplifying dynamic dye-solution holograms, *Kvant. Elektron.* **6,** 1770. [*English transl.: Sov. J. Quantum Electron.* **9,** 1041 (1979)].

Martin, G., Lam, L. K., and Hellwarth, R. W. (1980). Generation of a time-reversed replica of a nonuniformly polarized image-bearing optical beam, *Opt. Lett.* **5,** 185.

Mays, R., and Lysiak, R. J. (1979). Phase conjugated wavefronts by stimulated Brillouin and Raman scattering, *Opt. Commun.* **31,** 89.

Nosach, O. Yu., Popovichev, V. I., Ragul'skii, V. V., and Faizullov, F. S. (1972). Cancellation of phase distortions in an amplifying medium with a "Brillouin mirror," *Zh. Eksp. Teor. Fiz. Pisma Red.* **16,** 617 [*English transl.: Sov. Phys. JETP* **16,** 435 (1972)].

Peregudov, G. V., Plotkin, M. E., and Ragozin, E. N. (1979). Use of wave-front reversal effect to investigate a jet formed by focusing laser radiation on a plane target, *Kvantovaya Elektron.* **6,** 2401 [*English transl.: Sov. J. Quantum Electron.* **9,** 1413 (1979)].

Pilipetskii, N. F., Popovichev, V. I., and Ragul'skii, V. V. (1978). Concentration of light by inverting Its wavefront, *Pis'ma Zh. Eksp. Teor. Fiz.* **27,** 619 [*English transl.: JETP Lett.* **27,** 595 (1978)].

Pilipetskii, N. F., Popovichev, V. I., and Ragul'skii, V. V. (1979a). The precision of the reproduction of a light field at stimulated scattering, *Proc. Int. Conf. Lasers '79, December 17–21* p. 673.

Pilipetskii, N. F., Popovichev, V. I., and Ragul'skii, V. V. (1979b). Accuracy of reconstruction of a light field after stimulated scattering, *Dokl. Akad. Nauk SSR* **248,** 1097 [*English transl.: Sov. Phys. Dokl.* **24,** 845 (1979)].

Pilipetskii, N. F., Popovichev, V. I., and Ragul'skii, V. V. (1979c). The reproduction of weak components of a light field at stimulated scattering, *Opt. Commun.* **31,** 97.

Popovichev, V. I., Ragul'skii, V. V., and Faizullov, F. S. (1974). Stimulated Mandel'shtam–Brillouin scattering excited by radiation with a broad spectrum, *Zh. Eksp. Teor. Fiz. Pis'ma Red.* **19,** 350 [*English transl.: JETP Lett.* **19,** 196. (1974)].

Ragul'skii, V. V. (1976). Stimulated Mandel'shtam–Brillouin scattering, *Lebedev Tr.* **85,** p. 1.

Ragul'skii, V. V. (1979). Wavefront inversion of weak beams in stimulated scattering, *Pis'ma Zh. Tekh. Fiz.* **5,** 251 [*English transl.: Sov. Tech. Phys. Lett.* **5,** 100 (1979)].

Ragul'skii, V. V. (1980). Detection of slight absorption of light through wavefront inversion, *Pis-ma Zh. Tekh. Phys.* **6,** 687 [*English transl.: Sov. Tech. Phys. Lett.* **6,** 297 (1980)].

Ragul'skii, V. V., and Faizullov, F. S. (1969). A simple method of measuring laser Radiation divergence, *Opt. Spektrosk.* **27,** 707.

Rank, D. H., Cho, C. W., Foltz, N. D., and Wiggins, T. A. (1967). Stimulated thermal Rayleigh scattering, *Phys. Rev. Lett.* **19,** 828.

Sidorovich, V. G. (1976). Theory of the "Brillouin Mirror," *Zh. Tekh. Fiz.* **46,** 2168 [*English transl.: Sov. Phys. Tech. Phys.* **21,** 1270 (1976)].

Sidorovich, V. G. (1977). Theory of the transformation of light fields by amplitude three-dimensional holograms recorded in amplifying media, *Opt. Spektrosk.* **42,** 693 [*English transl.: Opt. Spectrosc.* **42,** 395 (1977)].

Sokolovskaya, A. I., and Brekhovskikh, G. L. (1978). Dynamic holograms under stimulated light scattering, *Dokl. Akad. Nauk SSSR* **243,** 630 (in Russian).

Sokolovskaya, A. I., Brekhovskikh, G. L., and Kudryavtseva, A. D. (1977). Restoration of wave front of light beams during stimulated Raman scattering, *Dokl. Akad. Nauk SSSR* **233,** 356 [*English transl.: Sov. Phys. Dokl.* **22,** 156 (1977)].

Sokolovskaya, A. I., Brekhovskikh, G. L., and Kudryavtseva, A. D. (1978). Light beam wavefront reconstruction and real volume image reconstruction of the object at the stimulated Raman scattering, *Opt. Commun.* **24,** 74.

Tang, C. L. (1965). Saturation and spectral characteristics of the Stokes emission in the stimulated Brillouin process, *J. Appl. Phys.* **37,** 2945.

Vasilyev, M. V., Gyulameryan, A. L., Mamaev, A. V., Ragul'skii, V. V., Semenov, M., and Sidorovich, V. G. (1980). Recording of phase fluctuations of stimulated scattered light, *Pis'ma Zh. Eksp. Teor. Fiz.* **31,** 673 [*English transl.: JETP Lett.* **31,** 634 (1980)].

Vasilyev, A. A. *et al.* (1981). Wave front reversal by an optically controlled liquid-crystal transparency at microwatt power of reference waves (accepted for publication).

Wang, V., and Guiliano, C. R. (1978). Correction of phase aberrations via stimulated Brillouin scattering, *Opt. Lett.* **2,** 4.

Wiggins, T. A., Wick, R. V., and Rank, D. H. (1966). Stimulated effects in N_2 and CH_4 gases, *Appl. Opt.* **5,** 1069.

Yariv, A. (1975). "Quantum Electronics." Wiley, New York.

Zakharov, S. D. (1977a). "New Potentialities of Laser Optics," No. 12, p. 116. Priroda, "Nauka," Moscow (in Russian).

Zakharov, S. D. (1977b). A Method of Light Autofocusing in the Study of Interaction between Laser Radiation and Matter. Preprint FIAN, No. 210, Moscow (in Russian).

Zel'dovich, B. Ya., and Shkunov, V. V. (1977). Wavefront reproduction in stimulated Raman scattering, *Kvantovaya Elektron.* **4,** 1090 [*English transl.: Sov. J. Quantum Electron.* **7,** 610 (1977)].

Zel'dovich, B. Ya., and Shkunov, V. V. (1978a). Limits of existence of wavefront reversal in stimulated light scattering, *Kvantovaya Elektron.* **5,** 36 [*English transl.: Sov. J. Quantum Electron.* **8,** 15 (1978)].

Zel'dovich, B. Ya., and Shkunov, V. V. (1978b). Reversal of the wave Front of light in the case of depolarized pumping, *Zk. Eksp. Teor. Fiz.* **75,** 428 [*English transl.: Sov. Phys. JETP* **48,** 214 (1978)].

Zel'dovich, B. Ya., and Shkunov, V. V. (1979a). Spatial-polarization wavefront reversal in four-photon interaction, *Kvantovaya Elektron.* **6,** 629 [*English transl.: Sov. J. Quantum Electron.* **9,** 379 (1979)].

Zel'dovich, B. Ya., and Shkunov, V. V. (1979b). Spatial-polarization wave front reversal under four-photon interaction, In "Reversal of Wave Front of Optical Radiation in Nonlinear Media," pp. 23–43. Gorky (in Russian).

Zel'dovich, B. Ya., and Yakovleva, T. V. (1980). Spatial-Polarization wavefront reversal in stimulated scattering of the Rayleigh wing, *Kvantovaya Elektron.* **7,** 880 [*English transl.: Sov. J. Quantum Electron.* **10,** 501 (1980)].

Zel'dovich, B. Ya., Popovichev, V. I., Ragul'skii, V. V., and Faizullov, F. S. (1972). Connection between the wave fronts of the reflected and exciting light in stimulated

Mandel'shtam–Brillouin scattering, *Zh. Eksp. Teor. Fiz. Pis'ma Red.* **15,** 160 [*English transl.: Sov. Phys. JETP* **15,** 109 (1972)].

Zel'dovich, B. Ya., Melnikov, N. A., Pilipetskii, N. F., and Ragul'skii, V. V. (1977). Observation of wave-front inversion in stimulated Raman scattering of light, *Pis'ma Zh. Eksp. Teor. Fiz.* **25,** 41 [*English transl.: JETP Lett.* **25,** 36 (1977)].

Zubarev, I. G., Mironov, A. B., and Mikhailov, S. I. (1980). Single-mode pulse-periodic oscillator-amplifer system with wave front reversal, *Kvantovaya Elektron.* **7,** 2035 [*English transl.: Sov. J. Quantum Electron.* **10,** 1179 (1980)].

Zverev, G. M., and Martynov, A. D. (1967). Investigation of stimulated Mandel'shtam–Brillouin scattering thresholds for different media at wavelengths 0.35, 0.69, and 1.06 μ," *Pis'ma Zh. Eksp. Teor. Fiz.* [*English transl.: JETP Lett.* **6,** 351 (1967)].

7 Phase Conjugation by Stimulated Backscattering

R. W. Hellwarth†
University of Southern California
Los Angeles, California

I. Introduction

When a strong monochromatic wave is incident on a transparent medium, it causes waves at lower frequencies to experience exponential gain if their frequency offset corresponds to the frequency of some excitation in the medium. This effect was discovered by Eckhardt *et al.* (1962) using Raman vibrations in liquids. They called it "stimulated (Raman) scattering," by analogy to Einstein's "stimulated emission." Regardless of the nature of the mediating excitation, the stimulated gain has a relation to the ordinary spontaneous inelastic scattering from the excitation, and this relation has a close analogy to the Einstein relation between stimulated emission and spontaneous emission of radiation (Hellwarth, 1963). The same physical process, but with nonpropagating radio-frequency fields interacting via spin magnetism, was envisaged earlier by Javan (1958a,b). Chaio *et al.* (1964) soon demonstrated stimulated Brillouin scattering (stimulated "Mandelstam–Brillouin" scattering in Soviet liter-

† Work supported by the U. S. Air Force Office of Scientific Research under Grant No. 78-3479 and the National Science Foundation under Grant No. ENG78-04774.

Optical Phase Conjugation

ISBN 0-12-257740-X

ature), in which the mediating excitations are acoustic waves. Mash *et al.* (1966) identified stimulated Rayleigh wing scattering, in which the mediating excitations involve molecular reorientation or rotation. Zaitsev *et al.* (1967), Rank *et al.* (1967), and Fabelinskii *et al.* (1968) discovered stimulated Rayleigh scattering, in which the mediating excitations are entropy fluctuations. This effect is most easily observed when the light couples by absorption; hence the effect is called stimulated thermal Rayleigh scattering. Stimulated polariton scattering was observed by Kurtz and Giordmaine (1969). These stimulated effects, and many more have been investigated in hundreds of articles, to which this brief introduction cannot do justice.

The important new aspect of stimulated scattering of prime interest here is the discovery by Zel'dovich *et al.* (1972) that in some circumstances the stimulated wave will be nearly phase conjugate to the incident wave. Phase conjugation has so far been verified for stimulated Brillouin scattering (acoustic-wave scattering) in gases, liquids, and plasma, for stimulated Raman scattering in liquids, and for stimulated Rayleigh wing scattering in liquids, and it is likely that phase conjugation will be observed in all the aforementioned stimulated processes, at least under certain experimental conditions. A review of published experimental results is given in Section II. The conditions that permit phase conjugation by stimulated scattering are a major concern of the theory of Section III. A discussion of the main conclusions of Sections II and III is given in Section IV. Needless to say, the wedding of stimulated scattering to phase conjugation of optical (or other) waves is a recent union. A review at this point is incomplete at best; the full range of practical applications that will be realized can hardly be envisioned at this time. It is hoped, however, that this review will help to speed new developments.

II. Phase Conjugation by Stimulated Scattering: Experiment[1]

The discovery fact that stimulated Brillouin scattering (SBS) in the backward direction could be phase conjugate to the incident beam was first reported by Zel'dovich *et al.* (1972), who scattered a ruby laser from methane gas in an optical waveguide. They also sketched the theoretical ideas behind this effect, the elaboration of which is the main intent of this chapter. In essence, the nonlinear polarization density, mediated by some

[1] This section was written when the completion of Chapter 6 was in doubt, to fill partially that void. Although Chapter 6 did indeed arrive, the present section is left unaltered because it gives a complementary view.

driven Raman- (Brillouin-, etc.) active excitation of the optical medium, couples normal-mode solutions of the linear Maxwell equations so as to create a new set of normal modes in backscattering with, in effect, complex propagation constants representing index change and gain. Under conditions to be discussed in later sections, the gain of one of these new modes is nearly twice that of the nearest competing mode, and this predominant mode has an electric field amplitude at the guide entrance that is nearly the complex conjugate of the incident wave amplitude in the same plane. That is, the predominant mode is "phase conjugate" to the incident wave. To confirm this ocurrence, Zel'dovich *et al.* (1972) demonstrated, as did Kogelnik (1965) in static holographic phase conjugation, that the backscattered wave would restore itself to a coherent wave front in passing back through an aberrating plate placed in the path of the incident beam. The experimental setup used by Zel'dovich *et al.* is shown in Fig. 1. They verified that the backscattered radiation was frequency-shifted by the appropriate Brillouin shift by the Fabry–Perot interferograms of Fig. 2. Their published beam restoration results were not clearly reproduced, but appeared to be essentially the same as the results of Nosach *et al.* (1972) shown in Fig. 3.

Nosach *et al.* (1972) demonstrated an important practical application of phase conjugation by backward SBS. They used a "Brillouin mirror" to cause wave distortions introduced by a ruby laser amplifier to be canceled upon reverse passage through the amplifier. The amplifier, in effect, replaced the aberrator in Fig. 1, its gain property having negligible effect on propagation of the phase front. (See Fig. 3.)

Curiously, it was five years before the next experimental articles on the subject appeared. Following the suggestion of Zel'dovich and Shkunov (1977), Zel'dovich *et al.* (1977) and Sokolovskaya *et al.* (1977) independ-

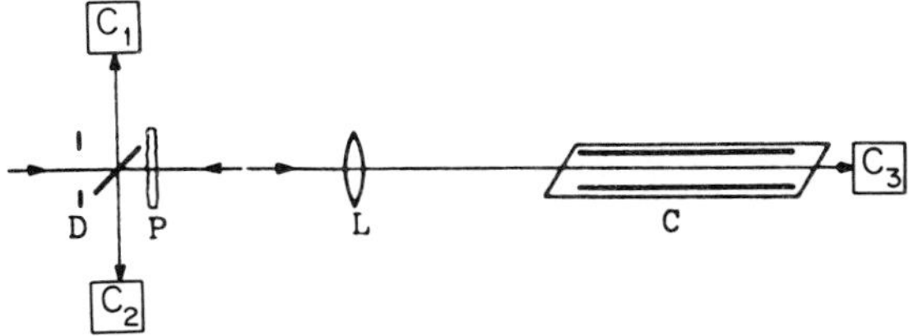

Fig. 1. Schematic diagram of apparatus for veryifying the phase-conjugate character of stimulated backscattering from a medium (in a waveguide) in cell C. Camera C_1 records the profile of the incident beam. Camera C_2 records the profile of the backscattered beam after retracing phase-distorting plate P. Lens L is adjusted for optimum conjugation. Camera C_3 records the distorted-wave profile when lens L and cell C are removed. [After Zel'dovich *et al.* (1972).]

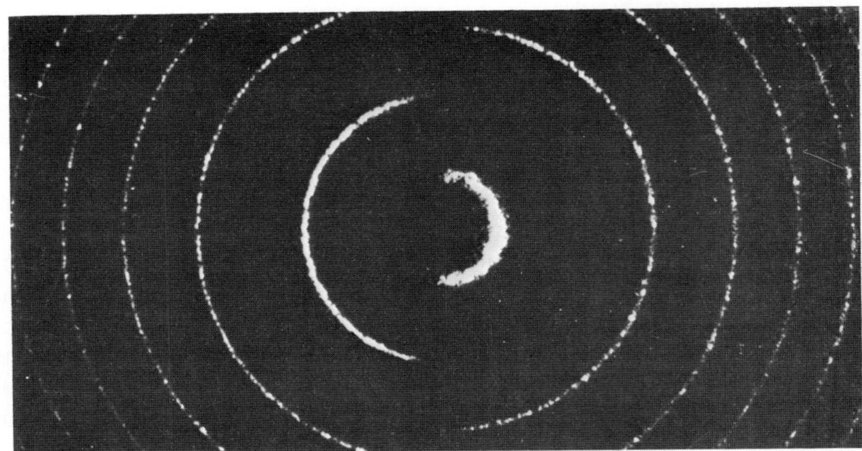

Fig. 2. Fabry–Perot interferograms of incident and backscattered phase-conjugate (Brillouin-shifted) beams. [After Zel'dovich *et al.* (1972).]

ently demonstrated phase conjugation by stimulated backward Raman scattering. In this case, although the scattered field amplitude was conjugate to the incident amplitude at the entrance to the interaction region, the large frequency difference caused the backward wavefront to be a laterally magnified version of the incident wave front (as in two-color holography) rather than to be a time-reversed replica of the incident wave front. Nevertheless, excellent restoration of a beam retracing an aberrator is still possible if the aberrator is close enough to the conjugator, as is seen in the results of Zel'dovich *et al.* (1977) reproduced in Fig. 4.

Sokolvskaya *et al.* (1977, 1978) observed that forward stimulated Raman scattering (SRS) by a multimode beam (not far from steady state) was weaker, had higher threshold, and contained very much less image-preserving component than backward SRS.

Also in 1977, Blaschuk *et al.* (1977) and Bespalov *et al.* (1977a,b) showed that phase conjugation could be produced in SBS from unguided beams in certain media (ether, acetone, and CCl_4) if the power was held near threshold. This and many subsequent experiments were probably beset by self-focusing, simultaneous SRS and SBS, and transient effects that confused attempts at quantitative interpretation.

High image fidelity of phase conjugation in stimulated backscattering was confirmed in an important way by Basov *et al.* (1979a), Ragulskiy (1979), and Pilipetskii *et al.* (1979a,b), all of whom studied the reproduction of weak details in the incident image. Pilipetskii *et al.* (1979a,b) used

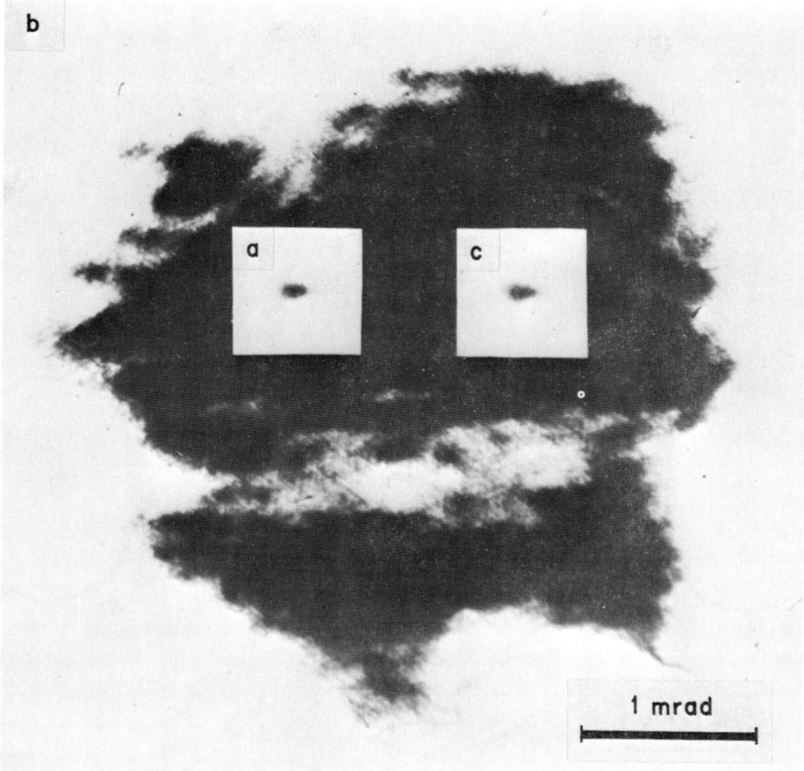

Fig. 3. Beam restoration results obtained by using the apparatus of Fig. 1 at 6943 Å. Beam profile (a) was recorded in the far field at C_1 in Fig. 1. Large distorted profile (b) was recroded equidistant at C_3 with L and C removed. Reconstructed profile (c) was recorded at C_2. Phase-distorting plate P was a ruby laser amplifier. The Brillouin-active medium in C was liquid CS_2. [After Nosach *et al.* (1972).]

a guided interaction geometry that clearly reproduced weak side-lobe patterns in a ruby laser input beam that were attenuated by as much as 10^6 below the intensity of the brightest beam spots.

Image-bearing stimulated Brillouin backscatter from laser-produced plasma was recognized to have a "ray-retracing" component by Eidmann and Sigel (1974). This phase-conjugation behavior has also been studied in laser-produced plasma by Basov *et al.* (1977) and Ripin *et al.* (1977), and identified in extended plasma by Herbst *et al.* (1979).

Studies by Basov *et al.* (1978) and Blaschuk *et al.* (1978a,b) make it clear that high-fidelity phase conjugation occurs in stimulated backscat-

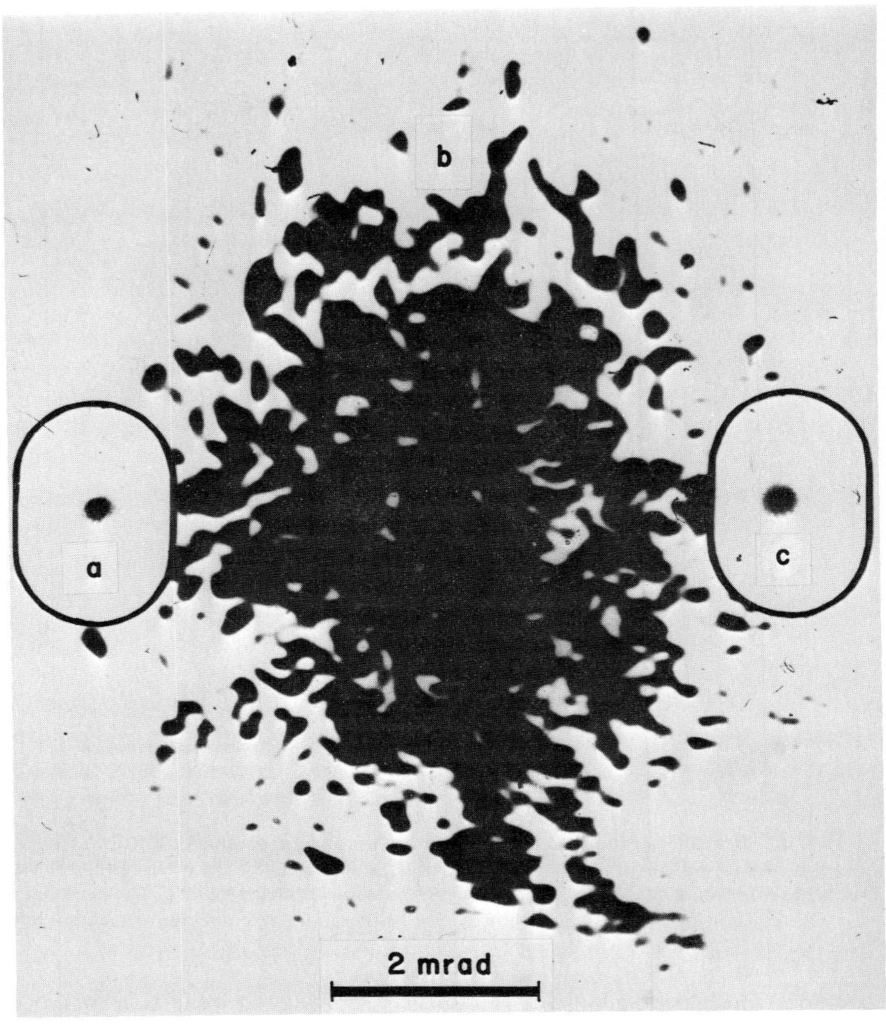

Fig. 4. Demonstration of the restoration of an aberrated beam by stimulated Raman scattering from the 656-cm^{-1} excitation of liquid CS_2. The results were obtained by using the apparatus of Fig. 1. The input beam as recorded by camera C_1 is shown in inset (a). Photograph (b) is of the aberrated beam as recorded by camera C_3. The restored beam as recorded by camera C_2 is shown in inset (c). [After Zel'dovich *et al.* (1977).]

tering only when the incident image-bearing beam is uniformly polarized in a sense that preserves polarization while traversing the interaction region (guided or unguided) at low power.

Kurdriavtseva *et al.* (1978) found that phase conjugation occurred in stimulated Raman backscattering (from benzene and cyclohexane) by using pulses having 25 psec duration. With such short pulses they were also able to observe image reconstruction at a Stokes frequency shift of less than 5 cm^{-1}. Because the pulses were presumed to be too short to excite SBS, this was reasonably assumed to be the first observation of phase conjugation by stimulated Rayleigh-wing scattering. In the nanosecond-pulse regime, Basov *et al.* (1979b) found that phase conjugation by SBS was not degraded even when the optical pulse length was less than the phonon lifetime. More remarkable for applications was the discovery by Hon (1980b) that phase conjugation can occur in transient stimulated Brillouin backscattering so strong that the reflected pulse is compressed from 20 to 2 nsec in length.

Quantitative measurements of phase-conjugation fidelity and efficiency as functions of the geometry and medium of the stimulated backscattering have been reported by Wang and Giuliano (1978), Basov *et al.* (1979b), Blekhovskikh *et al.* (1978), Borisov *et al.* (1977, 1979); Fan *et al.* (1980), Gerasimov *et al.* (1977a), and Mays and Lysiak (1979, 1980).

Applications of stimulated phase conjugation by backscattering to realizing high-brightness laser sources, performing diagnostics, etc., have been explored in experiments reported by Lesnik *et al.* (1979), Dolgopolov *et al.* (1979), Peregudov *et al.* (1979), Efimkov *et al.* (1979a,b, 1980), and Erosenko *et al.* (1980).

Perhaps the most striking feature of all the experiments on phase conjugation by stimulated backscattering is the clear persistence of the phase-conjugate nature of a significant portion of the backscattered beam in the face of simultaneous extraneous nonlinear effects, especially self-focusing, and in the presence of inhomogeneities, strong transients, absorption, and other complicating features. Nature surely loves the phase-conjugate beam.

III. Theory

The theoretical basis for the existence of a significant component in stimulated backscattering that is phase conjugate to the incident stimulating beam was outlined in the article of Zel'dovich *et al.* (1972) in which the effect was first reported. The idea is essentially the same whether the complex incident optical field $\mathrm{Re}\mathbf{E}_\nu(\mathbf{r})\exp(-i\nu t)$ is launched into a waveguide containing the nonlinear medium or into a tranversely infinite medium. In the limit of low scattering, the nonlinear Maxwell equations become linear in the backscattered optical field $\mathrm{Re}\mathbf{E}_\omega(\mathbf{r})\exp(-i\omega t)$ (the nonlinear polarization density at ω being proportional to $\mathbf{E}_\nu\mathbf{E}_\nu^*\mathbf{E}_\omega$). These linear equations for $\mathbf{E}_\omega(\mathbf{r})$ have many solutions that are "modes" in the same sense as are the usual solutions to the linear Maxwell equations (i.e., Gaussian–Hermite modes in free space or waveguide modes in a guide). However, the new modes have complex propagation vectors such that the solutions will have gain, provided that $\nu > \omega$ (i.e., for Stokes scattering). Zel'dovich *et al.* (1972) suggested that, under certain conditions, (a) one mode must have a significantly (two to three times) higher gain coefficient than any of the other solutions, (b) this mode would have nearly

$$\mathbf{E}_{\omega}(\mathbf{r}) \propto \mathbf{E}_{\nu}^{*}(\mathbf{r}) \tag{1}$$

at the entrance plane to the interaction region, i.e., be nearly phase conjugate to the incident beam, and (c) this mode predominates when the backscatter grows in one pass from noise (whence the total energy gain must be high, $\sim 10^{9}-10^{11}$).

In the first article to pursue this idea, Sidorovich (1976) assumed that the incident wave satisfied periodic boundary conditions in the transverse beam directions and found that, after certain approximations, there was indeed a high-gain, predominant solution that was phase conjugate to the incident beam. Since then, dozens of theoretical articles have expanded on this idea, including those of Baranova *et al.* (1978), Baranova and Zel'dovich (1980a,b), Beldyugin *et al.* (1976, 1979a), Beldyugin and Zemskov (1978, 1980), Bespalov *et al.* (1977a, 1979a,b), Fan *et al.* (1980), Gerasimov *et al.* (1977a), Gerasimov and Orlov (1977), Shkunov (1977, 1978a,b), Hellwarth (1978), Kochemasov and Nikolaev (1977, 1978, 1979), Ostrovskii (1979), Pasmanik (1979), Sidorovich and Shkunov (1979), and Zel'dovich and Yakovleva (1980). A different approach to a related problem has been tried by Lehmberg (1978) and Lehmberg and Holder (1980), who performed numerical simulations of ray retracing, incorporating self-focusing due to pondermotive forces, to treat stimulated backscatter from laser-produced plasma.

A critical comparison of the foregoing works has not yet been made and is beyond the scope of this review. These works are not always mutually consistent, and the reader is urged to examine the particular articles of interest.

The theory of steady-state scattering presented in Section III.A is based on the author's work (Hellwarth, 1978), the only work to treat scattering from a beam that has been launched into a real waveguide in which the nonlinear interaction takes place. Section III.B is devoted to transient phase conjugation. Although the results presented here display all the major features of experiments where stimulated backscattering is the predominant effect, there is still much room for quantitative verification. We treat unguided-wave interactions approximately as the limit of a short guide. Complications from competing nonlinear effects, such as self-focusing, are often in evidence in experiments but are considered here only qualitatively. Although the framework in which to treat broadband pump radiation and pump depletion is available, much remains to be explored in these areas.

A. Steady-State Scattering from Nondepleted, Monochromatic Incident Waves

In this section, we consider the stimulated backscattering of monochromatic, nondepleted incident waves in a waveguide. Our major conclusions for multimode incident beams are as follows:

(a) A backscattered solution that is nearly phase conjugate predomites (because it has gain nearly twice that of any other mode) provided that the interaction length is not too long; the Stokes shift $\nu - \omega$ is not too large; the number N of guide modes excited falls within a certain large range; and the total number of guided modes is not too large.

(b) The phase-conjugate wave has an exponential gain factor that is nearly twice normal, i.e., twice what would be experienced by a backward plane wave growing in the presence of an incident plane wave whose intensity equals the guided incident power divided by the guide area.

(c) Multimode forward-scattered solutions exist which have gains proportional to the gains of the corresponding backward wave solutions but scaled by the peak spontaneous differential scattering cross section for forward scattering from the same excitation. There is no forward SBS, but anomalous forward Raman gain is possible.

(d) The nonconjugate backscattered background can have an intensity per unit solid angle orders of magnitude lower than the phase-conjugate wave at high scattering efficiency.

(e) Large differences in the amplitudes of modes excited do not degrade phase conjugation significantly if a large number ($>10^2$) of modes are excited.

(f) The gain length need not exceed the diffraction length of the entering beam.

(g) The solid angle of the entering beam has no restriction other than that it allow the beam to be uniformly guided (or entirely unguided).

This theory of stimulated backscattering in a waveguide follows.

In the following sections, we first discuss the form of the nonlinear polarization density (Section III.A.1), and then use it in Maxwell's equations to obtain the couplings among unperturbed guide modes (Section III. A.2). A perturbation method of solution is developed (Section III.A.3) that divides the problem into an exactly soluble part (Section III.A.4) and a part that gives a small correction (Section III.A.5) if certain conditions are met (Sections III.A–III.A.8). The part of the backscattering that is not phase conjugate sets limits on resolution of detail (Section III.A.9). Forward stimulated scattering gains are discussed (Section III.A.10). Extensions to high conversion and time-dependent pump will be outlined in Section III.B.

1. Form of the Nonlinear Polarization

It is well known that, for $\nu - \omega$ near a Raman-active resonance frequency, the nonlinear polarization density

$$\text{Re } \mathbf{P}^{\text{NL}}(\mathbf{r}) \exp(-i\omega t)$$

at the scattered wave frequency ω has an amplitude whose leading term may be expressed by

$$\mathbf{P}^{\text{NL}}(\mathbf{r}) = -i\mathbf{G}(\nu - \omega, \nu) \vdots \mathbf{E}_\nu(\mathbf{r})\mathbf{E}_\nu^*(\mathbf{r})\mathbf{E}_\omega(\mathbf{r}). \tag{2}$$

Here $\mathbf{G}$ is an appropriate fourth-rank susceptibility tensor.

One can show that the spatially local relation (2) holds also for stimulated Brillouin backscattering (SBS) despite sound dispersion, provided that the entrance beam divergence (convergence) angle θ_i obeys

$$\theta_\text{i} < (\omega_\text{B}/\Gamma_\text{B})^{1/2}, \tag{3}$$

where ω_B is the Brillouin shift and Γ_B the damping rate of the mediating acoustic phonon of wave vector $\hat{z}(\nu + \omega)n/c$. Here n is the linear refractive index of the nonlinear medium, c the velocity of light in vacuum, and $\hat{z}$ a unit vector along the guide axis (Hellwarth, 1978). The condition (3) is not restrictive in practice, and it is likely that even were it violated, phase conjugation would occur by SBS anyway, mediated more or less independently by different sets of phonons.

When the scattering excitation is of the pure scalar or symmetric type (as for longitudinal acoustic waves, pure symmetric Raman vibrations, or plasmons), the nonlinear susceptibility tensor takes the simple form $G\mathbf{II}$, so that Eq. (3) becomes

$$\mathbf{P}^{\text{NL}} \to -iG\mathbf{E}_\nu\mathbf{E}_\nu^* \cdot \mathbf{E}_\omega. \tag{4}$$

We will use only the form (4) in our calculations; the full generalization to Eq. (2) is straightforward but cumbersome, and changes the results very little.

2. *The Coupled-Mode Equations*

In this section we derive from Maxwell's equations the coupled-mode equations with which we will analyze the stimulated backscattering from an incident optical beam at frequency ν in a waveguide. In Sections III.A.3–III.A.10 we will discuss the weak backscattering regime, in which the electric field amplitude $\mathbf{E}_\nu(\mathbf{r})$ of the incident beam is negligibly altered by the backscattering or by other nonlinear effects. In this regime we will find that the mode amplitudes of the stimulated backscattered waves obey a large number N of coupled linear equations (one equation for each mode).

The number N of guide modes involved is large because we are interested in generating the phase conjugate of an image-bearing beam containing a large number N_p ($\sim 10^6$) of resolution elements (i.e., pixels). Clearly N must be at least as large as N_p. It is also in the limit of large N that we will find that the emerging stimulated backscattered wave approaches the phase conjugate of the incident beam.

To proceed, we first express the amplitude $\mathbf{E}_\nu(\mathbf{r})$ of the given electric field of the incident wave in terms of the (complex) amplitudes A_m for this wave to be in the N excitable normal modes of a cylindrical guide:

$$\mathbf{E}_\nu = \sum_{m=1}^{N} A_m \hat{e}_m(x, y) e^{ik_{m\nu} z - \mu z/2}. \tag{5}$$

Here μ is the linear attenuation coefficient for waves in the guide. The transverse-mode patterns $\hat{e}_m(x, y)$ are normalized so that

$$\int dx\, dy\, \hat{e}_m^* \cdot \hat{e}_n = \delta_{mn}. \tag{6}$$

We will assume that the transverse refractive index variations, or the metallic walls that form the guide, are such that the transverse functions $\hat{e}_m$ do not differ significantly for the incident and backscattered waves of index m. However, the propagation constants $k_{m\nu}$ do vary with wave angular frequency ν according to

$$k_{m\nu}^2 + u_m^2 = n_\nu^2 \nu^2 / c^2, \tag{7}$$

where u_m^2 is the (frequency-independent) eigenvalue associated with the function $\hat{e}_m(x, y)$, and n_ν^2 the real part of the optical dielectric constant at the guide axis. We need not digress here to discuss the characteristics of waveguide modes. Our results will be largely independent of the detailed nature of the waveguide, provided that it propagates forward and backward modes at ν and ω whose transverse-mode patterns $\hat{e}_m$ are congruent.

For the amplitude $\mathbf{E}_\omega(\mathbf{r})$ of the electric field of the backscattered wave, we will try a solution of the form

$$\mathbf{E}_\omega = \sum_{n=1}^{N} B_n \hat{e}_n(x, y) \exp(-ik_{n\omega}z + \mu z/2 - \gamma z/2). \tag{8}$$

If $\gamma = 0$, Eq. (8) gives the solutions of the usual linear Maxwell equations for the guide. However, if the nonlinear polarization density of Eq. (4) is added to Maxwell's equations, we will find that there are N solutions or "modes" of the form (8) that replace the N old normal-mode solutions for the unperturbed waveguide. Each new mode solution has a characteristic eigenvalue γ_a ($a = 1, 2, \ldots, N$) and a particular set of coefficients $B_n^{(a)}$. [We will omit the solution label (a) if no ambiguity results.] More important, we will find certain conditions under which the γ of one of these new modes has (1) a positive real part significantly greater than that of any other solution and (2) a set of B_n nearly proportional to A_n^*. The first property enables this mode to predominate over all others after growing from noise over a sufficiently long path L. The second property implies that the phase-conjugate property (1) obtains for this predominant backscattered wave. To establish these properties, we first substitute Eqs. (5) and (8) in Eq. (1) to obtain a polarization density, linear in the backscattered mode amplitudes B_n, that must be balanced by the extra terms in $(\nabla\times\nabla\times - n^2\omega^2/c^2)\mathbf{E}_\omega$ arising from $\exp(-\frac{1}{2}\gamma z)$. That is, Maxwell's equations will still be satisfied if

$$\gamma k_{m\omega} B_m = 4\pi\omega^2 c^{-2} G \sum_{ijn} \int dx \int dy\, \hat{e}_m^* \cdot \hat{e}_i \hat{e}_j^* \cdot \hat{e}_n K_{mijn} A_i A_j^* B_n, \tag{9}$$

where

$$K_{mijn} \equiv \frac{1}{L} \int_0^L e^{i\Delta kz - \mu z}\, dz \tag{10a}$$

and

$$\Delta k \equiv k_{m\omega} + k_{i\nu} - k_{j\nu} - k_{n\omega}. \tag{10b}$$

Note that $|K_{mijn}|$ is less than or equal to unity, being largest (unity) when the wave-vector mismatch Δk is zero and the loss μ is much less than L^{-1}.

It is not practical or useful to seek an exact closed-form solution of (9) for a specific waveguide type (rectangular, circular, etc.) and N arbitrary amplitudes A_n of a large number N of modes. However, much can be learned from the wide variety of special cases that can be solved either exactly or quite accurately by the perturbation theory that we now develop.

3. *Perturbation Method for Solution*

Inspection of the N coupled equations (9) shows that their solution is equivalent to finding the eigenvectors $B_n^{(a)}$ and eigenvalues γ_a of an $N \times N$ square "Hamiltonian" matrix operator. We now rewrite Eq. (9) as an eigenvalue equation, separating the equivalent Hamiltonian into three terms: an unperturbed Hamiltonian $\mathcal{H}_{mn}$ that is exactly soluble, a term V_{mn} that can be treated by second-order perturbation theory when $N \gg 1$ (at least for nonpathological input beams), and a term U_{mn} that can be neglected because the average of K_{mijn} is small. That is, Eq. (9) is equivalent to

$$\sum_n (\mathcal{H}_{mn} + V_{mn} + U_{mn})B_n = \Upsilon B_m, \tag{11}$$

where

$$\mathcal{H}_{mn} \equiv (\overline{\alpha} - \overline{w})\,\delta_{mn} + \overline{\beta} a_m^* a_n, \tag{12}$$

$$V_{mn} \equiv \sum_\ell (\alpha_{n\ell} - \overline{\alpha} - w_{n\ell})|a_\ell|^2\,\delta_{nm} + (\beta_{mn} - \overline{\beta})a_m^* a_n, \tag{13}$$

and

$$U_{mn} \equiv {\sum_{ij}}' a_i a_j^* \int dx \int dy\; \hat{e}_m^* \cdot \hat{e}_i \hat{e}_j^* \cdot \hat{e}_n K_{mijn}. \tag{14}$$

In Eq. (14) the prime on the summation means to omit terms where $i = j$ and $m = n$, and also terms where $i = n$ and $j = m$. Normalized incident beam amplitudes a_m are defined as

$$a_m \equiv A_m \left(\sum_n A_n^* A_n\right)^{-1/2}, \tag{15a}$$

so that

$$\sum_{m=1}^{N} |a_m|^2 = 1 \tag{15b}$$

and

$$|a_m| \sim N^{-1/2} \tag{15c}$$

The matrices in Eqs. (12)–(14) are defined by

$$\alpha_{mn} \equiv \int dx \int dy\; |e_m^* \cdot e_n|^2 K_{mnnm}, \tag{16}$$

$$\beta_{mn} \equiv \int dx \int dy\; (e_m^* \cdot e_n)^2 K_{mnmn}, \tag{17}$$

and

$$w_{n\ell} \equiv \alpha_{nn}|a_n|^2 - \alpha_{\ell\ell}|a_\ell|^2. \tag{18}$$

The averaged matrices are defined as

$$\overline{\alpha} \equiv \sum_{mn} \alpha_{mn}|a_m|^2|a_n|^2, \tag{19}$$

$$\overline{\beta} \equiv \sum_{mn} \beta_{mn}|a_m|^2|a_n|^2, \tag{20}$$

and

$$\overline{w} \equiv \sum_{m} \alpha_{mm}|a_m|^4. \tag{21}$$

The renormalized eigenvalue Υ in Eq. (11) is proportional to γ in an obvious way. More useful is its relationship to the usually defined stimulated gain coefficient g_s (cm/MW) for plane pump and stimulated waves. This free-space coefficient g_s is the same as that $(\gamma S/P)$ in a guide excited in a single mode $\hat{e}_n$ with a uniform intensity profile ($\hat{e}_n$ = const). (Here S is the guide area and P the incident wave power.) For the single-mode case with no attenuation ($\mu = 0$), Eqs. (19)–(21) give $\overline{\alpha} = \overline{\beta} = \overline{w} = 1/S$ and $V_{nm} = U_{nm} = 0$. Then $B_m = a_m^*$ is the solution of Eq. (11) with eigenvalue

$$\Upsilon_{\mathrm{PW}} = 1/S. \tag{22}$$

For this case, and therefore for any case, we have

$$\gamma\ (\mathrm{cm}^{-1}) = \Upsilon\ (\mathrm{cm}^{-2}) \cdot g_s(\mathrm{cm/MW}) \cdot P\ (\mathrm{MW}), \tag{23}$$

which allows the immediate conversion of the eigenvalues Υ that we will derive to the spatial gain constant γ of any solution (9) in terms of the commonly tabulated (or calculable) parameter g_s.

In the analysis of the following sections, we will find that $\mathcal{H}_{mn}$ acting alone produces a dominant backscattered mode that is perfectly phase conjugate to the incident beam, and that (for $\omega - \nu$ smaller than a critical value) the V_{nm} produce a nonconjugate fraction of backscattered power that is of order $N^{-3/2}$, and is hence unimportant for large N. The U_{mn} are assumed to be zero in this analysis, an assumption that we now show can be justified only for input beam polarizations that make $\mathbf{P}$ parallel to $\mathbf{E}_\nu$ in Eq. (2). We cannot prove that it is always necessary that $U_{mn} \ll V_{mn}$ for phase conjugation to occur. However, experiments suggest that phase conjugation is spoiled, even for large N and $\omega \cong \nu$, when the input beam excites modes of mixed polarization. In this case many U_{mn} are comparable with V_{mn}. To see this, note that, even though $i \neq n$ and $j \neq m$ in Eq. (18), degeneracies make it possible that $k_{i\nu} = k_{n\omega}$ and $k_{j\nu} = k_{m\omega}$ (so that

$\Delta k = 0$) while at the same time $\int dx \int dy\, e_m^* \cdot e_i e_j^* \cdot e_n$ is of the usual magnitude ($\sim S^{-1}$). For example, this happens when m and i refer to x-polarized modes and j and n refer to y-polarized modes, mode m having the same spatial dependence as mode j, and mode i the same spatial dependence as mode n. The sum over i and j in Eq. (18) compounds such terms, so U_{mn} may even be much larger than V_{mn} in some cases. However, if only one linear (or circular) polarization is excited in a guide filled with isotropic material, such terms do not occur. Neither do terms with $k_{i\nu} = k_{j\nu}$ and $k_{m\omega} = k_{n\omega}$ (while $i \neq j$ and $m \neq n$), because this sort of degeneracy can be eliminated by proper choice of basis functions for the modes. "Accidental" degeneracies can occur where $\Delta k = 0$ without any pair of wave vectors canceling in Eq. (10b), but model calculations show that in this case the x–y overlap integral either vanishes (as for a rectangular waveguide) or is very small.

It remains for us to show that phase mismatch ($\Delta k \neq 0$) makes any typical U_{mn} negligibly small. It does so because the K_{mijn} are negligible when

$$\Delta k \gg L_{\text{eff}}^{-1}, \tag{24a}$$

where the effective guide length is defined as

$$L_{\text{eff}} \equiv (1 - e^{-\mu L})/\mu. \tag{24b}$$

To estimate Δk, we note that the transverse eigenvalues of Eq. (7) are of magnitude

$$u_n^2 \sim N/S, \tag{25}$$

so that

$$\Delta k \sim N\lambda/S, \tag{26}$$

where λ is a typical longitudinal wavelength in the guide. Now L_{eff} is at least as long as the diffraction length L_{d} for an N-mode beam whose beam waist area is of the order of the guide area S (assuming $\mu L_{\text{d}} \ll 1$). Even unguided incident beams (focused into an infinite medium) will have a stimulated backscatter interaction path at least as long as L_{d}. Because

$$L_{\text{d}} \sim S/N^{1/2}\lambda, \tag{27}$$

we conclude that for guided, or for focused unguided, stimulated backscattering

$$\Delta k\, L_{\text{eff}} \gtrsim (N\lambda/S)(S/N^{1/2}\lambda) \sim N^{1/2} \gg 1, \tag{28}$$

and the criterion (24a) is satisfied. Therefore, we are justified in proceeding with the calculation of the new mode functions and their growth rate (eigenvalues Y) by assuming $U_{mn} = 0$ in Eq. (11). Having noted that even for unguided (free) waves, the troublesome U_{mn} can be neglected when the input is uniformly polarized, let us consider how the analysis which follows of the interactions in a guide can be used to treat stimulated backscattering of free waves, at least qualitatively. A waveguide that imitates well the interaction of focused unguided waves may be imagined as follows. Construct the complete orthonormal set of free-space Gaussian-beam modes whose parameters are such that the smallest number N of modes need be superposed to give a good representation of the (focused) incident beam. Then let the waveguide axis coincide with the z axis of these modes, and let it barely encompass the beam waist over the length where the waist size does not change appreciably. This length is generally of order $S\lambda^{-1}N^{-1/2}$. Calculate the backscattered wave by using the mode decompositions in Eqs. (5) and (8) as we have prescribed. The field patterns calculated at the entrance to the waveguide should approximate those in the same plane in free space, because guiding is minimal.

4. *Character of Unperturbed Solutions*

The unperturbed matrix $\mathcal{H}_{mn}$ of Eq. (11) is a weighted superposition of the unit matrix and a projection operator $P_{\text{inc}} = a_m^* a_n$ onto the N-dimensional vector whose components are the mode amplitudes a_n of the incident beam.

It follows immediately that one eigenvector $b_n^{(0)}$ of $\mathcal{H}_{mn}$ is the vector a_n^* itself with eigenvalue

$$h_0 = \overline{\alpha} - \overline{w} + \overline{\beta}, \tag{29}$$

and that all $N - 1$ of the other mutually orthogonal vectors $b_n^{(a)}$ orthogonal to a_n^* ($\Sigma_n\, b_n^* a_n^* = 0$) are also eigenstates of $\mathcal{H}_{mn}$ with eigenvalue

$$h_a = \overline{\alpha} - \overline{w}. \tag{30}$$

It is common in experiments that ω is close enough to ν that

$$K_{mnnm} \to K_{mnmn} \to L_{\text{eff}}/L. \tag{31}$$

If incident modes are also all of the same linear polarization,

$$\overline{\alpha} \to \overline{\beta} \sim L_{\text{eff}}/LS. \tag{32}$$

Further, because the average $|a_n|^2$ is of order N^{-1}, Eq. (21) shows that $\overline{w} \sim \overline{\alpha}/N$, so that for $N \gg 1$

$$\overline{w} \ll \overline{\alpha}. \tag{33}$$

Therefore, with the foregoing assumptions, $\mathcal{H}_{mn}$ acting alone results in one backscattered mode that is a perfect phase conjugate of the incident wave (i.e., has $b_n^{(0)} \propto a_n^*$) and has a gain (i.e., eigenvalue) twice that of any other solution:

$$h_0 \rightarrow 2h_a, \qquad a = 1, 2, \ldots, N - 1. \tag{34}$$

The experimental conditions under which this phase-conjugate character persists along with the gain "gap" (34) when V_{mn} is also considered in Eq. (11) will be the main concern of the following four sections.

5. *A Class of Exactly Soluble Incident Images*

When ω differs from ν by less than a critical value to be derived in Section III A 8, Eq. 31 obtains, and there is an important class of incident mode patterns in a rectangular waveguide for which the eigenvectors of $\mathcal{H}_{mn}$ and V_{nm} are the same set, and exactly soluble along with their eigenvalues Υ. This is the class of incident waves that excite N of the total N_g of guide modes with equal energy but arbitrary phase ϕ_n, so that

$$a_n = N^{-1/2}e^{-i\phi_n}, \tag{35}$$

and that leave the other $N_g - N$ modes totally unexcited. The mode index n in a rectangular guide can be expressed by a pair of integers (n_x, n_y) labeling the number of nodes in the x and y directions, respectively. For simplicity, we study the case where the first $N = M^2$ modes of the guide are excited, i.e., where

$$n_x, n_y = 0, 1, 2, \ldots, M - 1. \tag{36}$$

The overlap integrals in Eqs. (16) and (17) are easily evaluated, by assuming that the mode functions are sine waves vanishing at the boundaries, to obtain [for use in Eqs. (11)–(13)]

$$\alpha_{n\ell} = \beta_{n\ell} = [2 + \delta(n_x, \ell_x)][2 + \delta(n_y, \ell_y)]/4\bar{S}, \tag{37}$$

where the δ function is unity for equal arguments and zero otherwise; $\bar{S} \equiv SL/L_{\text{eff}}$.

Let us relabel the N sets of solutions (i.e., eigenvectors) $B_n^{(a)}$ of Eq. (11) by representing the label (a) $(a = 0, 1, \ldots, N)$ by the pair of integers (a_x, a_y). Each integer can be any integer from 0 to $M - 1$:

$$a_x, a_y = 0, 1, 2, \ldots, M - 1. \tag{38}$$

One can easily verify that the following N normalized functions (i.e., N-dimensional vectors) are solutions of Eq. (11) (with $U_{mn} = 0$):

$$B_n(a_x, a_y) = M^{-1} \exp[i\phi_n + 2\pi i(n_x a_x + n_y a_y)/M], \tag{39}$$

for the mode integers of Eq. (36).

Clearly the eigenvector with $a_x = a_y = 0$ is the desired phase-conjugate solution $[B_n(0, 0) \propto a_n^*]$, and has the largest eigenvalue:

$$\Upsilon_{00} = \bar{\alpha} + \bar{\beta} - 9/4\bar{S}N. \tag{40}$$

Then there are $2M - 2$ eigenvectors with either $a_x = 0$ or $a_y = 0$ (but not both) that all have the eigenvalue

$$\Upsilon_{0\nu} = \bar{\alpha} + (M - 4)/2\bar{S}N. \tag{41}$$

The $(M - 1)^2$ states for which $a_x \neq 0$ and $a_y \neq 0$ have the eigenvalue

$$\Upsilon_{\mu\nu} = \bar{\alpha} - 2/\bar{S}N. \tag{42}$$

Finally, every unexcited, unperturbed guide mode, whose mode integer n' is not included in Eq. (36), is a solution of Eq. (11) with eigenvalue $\Upsilon_{n'} = \bar{\alpha}(1 + O[M^{-1}])$.

Inserting Eq. (37) in Eqs. (19) and (20), we see that

$$\bar{\alpha} = \bar{\beta} = (1 + M^{-1} + M^{-2}/4)/\bar{S}. \tag{43}$$

The gap in gain between Υ_{00} of the conjugate-wave solutions and the next nearest gain $\Upsilon_{0\nu}$ is

$$\Upsilon_{00} - \Upsilon_{0\nu} = (1 - 1/2M)/\bar{S}. \tag{44}$$

We see that, for a large number N of modes excited, each eigenvalue approaches its unperturbed value (i.e., that for $V_{mn} = 0$). We shall argue that this is generally the case, by solving Eq. (11) approximately for a larger class of incident waves.

6. *Perturbation Calculation of the Non-Phase-Conjugate Fraction of the Dominant Backscattered Wave*

Having seen from the exact solutions for B_n in the special case of the previous section that the effect of V_{mn} in Eq. (11) is small when the number N of pump-wave modes excited is large, we are encouraged to treat V_{mn} as a small perturbation on $\mathcal{H}_{mn}$ in solving the eigenvalue equation (11) (in which we continue to assume $U_{mn} = 0$ for reasons stated). Let use denote the exact eigenvector of $\mathcal{H}_{mn} + V_{mn}$ with largest eigenvalue Υ_0 by B_{on}. Because we have shown in Eq. (29) that the eigenvector $b_n^{(0)}$ of $\mathcal{H}_{mn}$ with highest eigenvalue $h_0 = \bar{\alpha} - \bar{w} + \bar{\beta}$ is a_n^*, let us write the exact solution as

$$B_{0n} = a_n^* + c_n. \tag{45}$$

Then standard perturbation theory gives for the correction c_n (repeated mode indices are to be summed henceforth)

$$c_n = \sum_{a \neq 0} \frac{b_n^{(a)} b_\ell^{(a)*} V_{\ell m} a_m^*}{h_0 - h_a} \tag{46}$$

to lowest order in the perturbation $V_{\ell m}$. Recall that each $b_n^{(a)}$ is an eigenvector of $\mathcal{H}_{mn}$ with eigenvalue h_a. From Eqs. (29) and (30) we recall that the energy denominator in Eq. (46) is a constant $\bar{\beta}$. Because, from its definition in Eqs. (13)–(17), V_{mn} has the property

$$a_m V_{mn} a_n^* = 0, \tag{47}$$

the sum in Eq. (46) may be extended over all a (with constant denominator), and closure invoked, to obtain

$$c_n = V_{nm} a_m^* / \bar{\beta}, \tag{48}$$

which is convenient for calculation. The most important quantity to calculate for our purposes is the fraction f_1 of power in the highest-gain solution B_{0n} that is not in the phase-conjugate wave a_n^*. Because $a_n^* a_n = 1$ by Eq. (15),

$$f_1 = r/(1 - r), \tag{49}$$

where $r \equiv c_n^* c_n$. In Section III A 9, we will estimate the fraction f_2 of backscattered power in the lower-gain solutions. Both contributions make a similar background "noise" fraction that must be kept smaller than unity for useful phase conjugation.

From Eqs. (48) and (13) we have, to lowest order in $V_{\ell m}$,

$$r = |a_\ell|^2 \theta_{\ell m} |a_m|^2 \theta_{mn} |a_n|^2 / \bar{\beta}^2, \tag{50}$$

where $\theta_{\ell m} \equiv \alpha_{\ell m} - \alpha_{\ell\ell}|a_\ell|^2 + a_{mm}|a_m|^2 + \beta_{\ell m} - \bar{\alpha} - \bar{\beta}$. This is the most important result of perturbation theory for assessing the fraction f of stimulated backward scattering that is not phase conjugate to the pump wave. If r is not much less than unity, the process is useless, and it is pointless to calculate further corrections to Eq. (46). If $r \ll 1$, Eqs. (46) and (50) are accurate enough. Note that the gain Υ_0 of the important wave is somewhat larger than h_0. From nondegenerate perturbation theory for eigenvalues,

$$\Upsilon_0 = h_0 + r\bar{\beta} \tag{51}$$

to second order in V_{mm}. [The correction linear in V_{mn} vanishes because of Eq. (47).] That is, the gain is increased by the fraction r of the gain gap $\bar{\beta}$.

We next use Eq. (50) to calculate r for a fairly general class of pump beams. If this incident beam excites only two or three modes with arbitrary amplitudes a_n, it is easy to see from Eq. (50) that the backscattered wave with largest gain is not nearly phase conjugate to the input. However, as the number N of guide modes excited becomes larger, r becomes smaller. To see why this should be, we rewrite Eq. (50) as the mean-square deviation among modes of a dimensionless quantity K_ℓ, defined for guide mode ℓ as

$$K_\ell \equiv (\alpha_{\ell n} + \beta_{\ell n} - \alpha_{\ell\ell}|a_\ell|^2 + \alpha_{nn}|a_n|^2)|a_n|^2/\bar{\beta}. \tag{52}$$

Then Eq. (50) becomes

$$r = \langle |K_\ell|^2 \rangle - |\langle K_\ell \rangle|^2, \tag{53}$$

where the average $\langle f_\ell \rangle$ of any quantity depending on the mode label ℓ is defined as

$$\langle f_\ell \rangle \equiv \sum_\ell f_\ell |a_\ell|^2. \tag{54}$$

From Eq. (52) it is evident that K_ℓ is itself an average of another quantity over all modes, and hence its rms fluctuation decreases as the number of modes increases. We will study how r depends on the number N of modes excited in greater detail in the next section.

7. *Nonconjugate Fraction of Stimulated Backscattering in a Rectangular Guide with Small Stokes Shift*

The effects of the perturbation V_{mn} in Eq. (11) on the degree of phase conjugation [as expressed in Eq. (53)] for multimode excitation can be seen by studying the case where the Stokes shift $\nu - \omega$ is not so large that Eq. (31) is violated. (The exact limit will be determined in the next section.) We again study interactions in the rectangular waveguide described in Section III.A.5. Using Eqs. (37) and (52), we see that we must calculate the mean-square fluctuation from mode to mode of

$$K_{\ell_0} = \sum_{m_x, m_y} \frac{[4 + 2\delta(\ell_x, m_x) + 2\delta(\ell_y, m_y) + \delta_{\ell m}]|a_m|^2}{2\bar{S}\bar{\beta}}, \tag{55}$$

using the average defined in Eq. (54). Clearly the term 4 within the brackets contributes nothing to the fluctuation of Eq. (53). The last term ($\delta_{\ell m}$) fluctuates as $|\alpha_\ell|^2$ itself. If the $|a_\ell|^2$ were of purely random magnitude over N modes, their mean would be of order N^{-1} and mean-square deviation of order N^{-2}. So this last term alone would contribute a term to r of order N^{-2} for large N.

The second term within the brackets in Eq. (55) requires a sum over energies $|a_m|^2$ of modes with a fixed value m_x of the x index. If the $|a_m|^2$ are randomly distributed (and uncorrelated), this sum has an average value $\sim M_y/N$ and a variance $\sim M_y^{1/2}/N$, where M_y is the number of y indices excited. This term and the third term therefore contribute terms in Eq. (53) of order $(M_y + M_x)/N^2$, which is larger than the contribution of the other terms. For nearly equal numbers of modes excited in the x and y directions, $M_x \sim M_y \sim N^{1/2}$. We conclude, therefore, that if N guide modes are excited by the incident wave with randomly distributed mode energies, the nonconjugate fraction of the stimulated backscatter (with largest gain) may be estimated by

$$r \sim N^{-3/2}. \tag{56}$$

Hellwarth (1978) considered a model distribution for $|a_m|^2$ in which the various $|a_m|^2$ at fixed n_x were highly correlated. There the sum over n_y had a variance of order M_y/N (rather than of order $M_y^{1/2}/N$, as for the case of random amplitudes discussed above). Therefore, a more pessimistic estimate ($r \sim M_y/N^2 \sim N^{-1}$) was found. However, intuitively it would seem that a random distribution of mode energies would better approximate the typical situation where the image itself is of random composition, and the coupling of unaligned input beams to waveguide modes is also random. We believe, therefore, that Eq. (56) is the better estimate for most experiments.

Alternative ways to understand why an image with large amplitude variations can be well conjugated by stimulated Brillouin backscattering have been given by Bespalov, *et al.* (1978) and by Abrams, *et al.* (1981).

8. *Effects of Finite Frequency Shift on the Nonconjugate Fraction of the Predominant Backscattered Mode*

So far, we have ignored the contribution of finite Stokes shift $\nu - \omega$ to the phase mismatch Δk of Eqs. (10). The effect of this shift is to alter $\beta_{\ell n}$ everywhere. We proceed to estimate its effect on the nonconjugate fraction r by using the altered $\beta_{\ell n}$ in Eqs. (52) and (53). The effect will place limits on the interaction length for given shift, guide area, number of modes excited, and desired r.

To calculate Δk in Eqs. (10) and (17) for $\beta_{\ell n}$ we use for the transverse eigenvalue in Eq. (7) the value u_n for a square waveguide; the results obtained will apply qualitatively to a guide of any cross section:

$$u_n^2 = \pi^2(n_x^2 + n_y^2)/S \tag{57}$$

for a mode whose indices n_x and n_y are integers as in Eq. (36). Inserting Eqs. (7) and (57) into Eq. (10b), we have

$$\Delta k \cong q(m_x^2 - n_x^2 + m_y^2 - n_y^2), \tag{58}$$

where

$$q \equiv \pi\,\Delta\lambda/4S, \tag{59}$$

in which

$$\Delta\lambda \equiv 2\pi c(\omega^{-1} - \nu^{-1})/n \tag{60}$$

is the difference between scattered and input wavelengths in the medium.

Using Eqs. (58) and (10) for K_{mnmn} and substituting the result in Eq. (17) for β_{mn} gives

$$4SL\beta_{mn} = \int_0^L dz[2\exp(iqz(m_x^2 - n_x^2)) + \delta(m_x, n_x)] \times [2\exp(iqz(m_y^2 - n_y^2)) + \delta(m_y, n_y)]\exp(-\mu z). \tag{61}$$

This creates an obvious change from the $q = 0$ case to K_ℓ in Eq. (52).

We consider first the use of an axial input beam for which the significant mode integers m_x and m_y run from 0 to $N^{1/2}$. To see the effect of the change in K_ℓ when $q \neq 0$, we need only use the average value N^{-1} for $|a_n|^2$ in Eq. (52). We try to express the change in Eq. (61) only to lowest (first order in q, and find below that this approximation is good if r is to be less than unity. These assumptions give a K_ℓ in Eq. (53) that differs from our previous ($q = 0$) value $K_{\ell 0}$ in Eq. (55) by

$$\Delta K_{\mathrm{m}} = (iq\tilde{L}/2S\bar{\beta})(m_x^2 + m_y^2), \tag{62}$$

where $\tilde{L} \equiv 2L(1 - e^{-v} - ve^{-v})/v^2$, $v \equiv \mu L$, is the result of the z integral in Eq. (61). The fact that $\bar{\beta}$ is also a function of q creates higher-order corrections that are negligible when $r \ll 1$.

The contribution of Stokes shift q to the nonconjugate fraction r that we seek is now obtained by using ΔK_ℓ of Eq. (62) for K_ℓ in Eq. (53) and performing the indicated sums with $|a_\ell|^2 = N^{-1}$. The result, for $N \gg 1$, is [with $\bar{\beta}$ from Eq. (32)]

$$\Delta r = \tfrac{2}{45}(q\eta LN)^2. \tag{63}$$

The dimensionless loss function

$$\eta \equiv 2(1 - e^{-v} - ve^{-v})/v(1 - e^{-v}), \qquad v \equiv \mu L, \tag{64}$$

is unity for lossless media ($\mu = 0$) and less otherwise.

In Eq. (63) we see that, for $\Delta r \ll 1$, $qL\eta N$ is small compared with unity, and because m_x^2 is never larger than N, the exponents within the parentheses of Eq. (61) are also small compared with unity. Therefore, our expansion of the right-hand side of Eq. (61) to first order in q is accurate in the regime of interest, and Eq. (62) is appropriate.

From Eq. (63) we have estimated the nonconjugate fraction for published experiments in which the input beam is axial, and found consistency. For example, Wang and Giuliano (1978) obtained 70% conversion to high-fidelity phase conjugation of a ruby laser pulse (694 nm) launched into a cell of CS_2 ($n = 1.62$) with $L = 1$ m and $S = 5 \times 10^{-2}$ cm². The aberrated beam entered the guide with a divergence angle $\theta \sim 25$ mrad. Because $N \sim \theta^2 S/\lambda^2$, we estimate $N \sim 10^4$. The Brillouin shift was $\Delta\lambda \sim 3 \times 10^{-9}$ cm. From Eq. (63) we expect, therefore, $\Delta r \sim 10^{-4}$, which is larger than the $r \sim 10^{-5}$–10^{-6} expected from Eq. (56) to arise from the mode amplitude fluctuations. Wang and Giuliano observed about 1% of the backscatter to appear in some nonconjugate side lobes after passage back through an aberrator. This probably occurs because not all of the aberrated beam is launched into the waveguide, owing to side lobes of diffraction by the aberrator.

Another interesting example is the unguided conjugate stimulated Raman backscatter from a ruby laser observed by Zel'dovich *et al.* (1977) from the 656-cm^{-1} vibration of CS_2 (simultaneously with SBS). The divergence angle θ_i of the axial input beam was of order $\lambda p^{1/2}/\theta_{d0} f$, where θ_{d0} is the unaberrated laser divergence (0.14 mrad) and f the focal length (7 cm) of the input focusing lens. The effective guide area S was of order the diffraction spot area λ^2/θ_i^2 times the number p ($\sim$300) by which the aberrator increased the beam solid angle (p the number of image elements). The effective guide length L was of order $S^{1/2}/\theta_i$. The number N of modes excited in this equivalent guide is p. Combining these approximations gives $NL/S \sim p^{1/2}/\lambda \sim 2.5 \times 10^5 \text{ cm}^{-1}$. Because $\Delta\lambda \sim 2 \times 10^{-6}$ cm and $\eta \sim 1$, we expect from Eq. (63) a nonconjugate fraction $\Delta r \sim 0.7\%$. This is consistent with the unquantified but "excellent" degree of phase conjugation reported. We have found no inconsistencies with these results or with other experiments, although many experiments are complicated by self-focusing.

Unfortunately, our calculations for Δr as a function of $\Delta\lambda$ have not yet been extended to off-axis beams, which have been employed in some experiments. This can be done by using the full expression (61) for β_{mn} (with the small-q approximation) in Eqs. (52) and (53).

9. *Contribution of Low-Gain Noise to Nonconjugate Backscatter*

We have noted that even when all of an incident beam is launched into a guide or interaction region, there is an irreducible background to the stimulated phase-conjugate wave, and this background backscattering arises from two sources: (1) the nonconjugate fraction f_1 [Eq. (49)] of the predominant (high-gain) backscattered mode and (2) the backscattered energy in all of the other (low-gain) modes, which we define to be the fraction f_2 of the total backscattering. Both backgrounds are expected to be fairly uniformly (if not randomly) distributed among all unperturbed guide modes. (In a dielectric guide there must also be some scattering out of the sides, which, however, experiences so little gain that we assume it to be negligible.) Clearly, when the total nonconjugate fraction

$$f = f_1 + f_2$$

of backscattering approaches unity, phase conjugation becomes inefficient, if not spoiled, for practical purposes.

To estimate the fraction f_2 we assume that the initiating noise power in any mode is proportional to the gain coefficient γ for that mode. [This proportionality is analogous to the proportionality between the Einstein A and B coefficients describing spontaneous and stimulated emission (Hellwarth, 1963).] We will write the gain factor by which the noise power in the predominant mode becomes amplified as e^{2G}. In terms of the stimulated gain coefficient g_s (cm/W) for the medium, incident power P (W), guide length L (cm), guide area S, and linear absorption coefficient μ, we have from Eqs. (23), (29), and (32)

$$G \sim (1 - e^{-\mu L}) P g_s / \mu S \tag{65}$$

in steady state.

In the analysis of Section III.A.4–III.A.8 we found that when the waveguide is much longer than the diffraction length of the incident beam, essentially all the $N_g - 1$ low-gain modes of the guide experience a gain factor of order e^G. For backscatter from an unguided focal region, the effective number N_g of guided modes is close to the number p of image elements. In either case we estimate that the ratio of powers in the phase conjugate and non-phase conjugate modes is approximately

$$\bar{f}_2 \sim (N_g - 1)e^G/2e^{2G} \sim \tfrac{1}{2}N_g e^{-G}. \tag{66}$$

The factor $\frac{1}{2}$ arises because the noise power in each low-gain mode is about half that in the predominant mode. $f_2 = \bar{f}_2/(1 + \bar{f}_2)$.

It is useful to infer the gain G from the observed backscattered power P_{pc} in the predominant (nearly phase-conjugate) mode as follows. Assume that the linear loss μ of the medium is zero and that the incident wave power P is essentially uniform along the length L of the guide, creating a power gain γ (cm^{-1}) for the predominant mode. A small fraction of the incident wave is backscattered into this predominant mode, creating in it a constant noise power p_N per unit length along the guide. Therefore

$$P_{pc} = \int_0^L dz\, p_N e^{\gamma z} \cong \frac{L p_N e^{2G}}{2G}. \tag{67}$$

Using $\gamma \sim 2Pg_s/S$ from Eq. (65) and relation (2) of Hellwarth (1963) between γ (multiplied by c/n) and p_N, we obtain from Eq. (67)

$$G^{-1}e^{2G} = \frac{(1 - e^{-\hbar(\nu-\omega)/kT})Sn}{\hbar\omega g_s c}\frac{P_{pc}}{P}. \tag{68}$$

Let us consider the example of Brillouin scattering of a ruby laser (at 6943 Å) by liquid CCl_4 in a guide of area 10^{-4} cm^{-2} at $T = 300$ K, for which

$n = 1.454$, $\nu - \omega = 2\pi$ (4.39 GHz), and $g_s \sim 8$ cm/GW (Kaiser and Maier, 1972), so that $P_{pc} \sim P$. Then Eq. (68) gives $e^{2G} \sim 4 \times 10^{10}$.

We may estimate the number N_g of guided propagating modes by using the approximation for a dielectric waveguide whose cladding material has refractive index n_c:

$$N_g \sim S(n^2 - n_c^2)\omega^2/\pi^2 c^2. \tag{69}$$

In the above example, with CCl_4 in a glass tube having $n_c \sim n - 0.05$, the number N_g of modes would be $\sim 10^4$. Therefore, from Eq. (66) we expect the low-gain modes would contribute $f_2 \sim 2 \times 10^{-2}$ in this case. This is a satisfactory situation for most purposes.

Note that, in a transient situation, e^G can be much larger, momentarily, than is given by the steady-state formula (68), because it takes time for the backscattering to develop from a situation of sudden extreme gain G. In this transient regime Pilipetskii *et al.* (1979a) and Ragulskiy (1979) produced a phase-conjugate image with a small noise power per unit solid angle: of order 10^{-7} of the average image power per unit solid angle. They used a large-diameter (3 mm) waveguide, 100 cm long, filled with CS_2, in which there were about 10^7 guided modes. They used a ruby laser pulse of 30 nsec duration and 0.6 MW peak power. At this power level the steady-state phase-conjugate-mode coefficient $2G$ is of order $2(0.6\text{ MW}) \times (0.2\text{ cm/MW}) \times (100\text{ cm}) \div (0.07\text{ cm}^2) \sim 340$. This gives $e^{2G} \sim 10^{145}$. A calculation of f_1 in the transient regime has been given by Zel'dovich and Yakovleva (1979) who found it to be reduced by $\sim\frac{1}{2}$. A complete calculation of f_1 and f_2 in a highly transient situation would certainly have to take into account the finite bandwidth (probably not transform limited) of the laser pulse, self-phase modulation etc. A step in the transient analysis will be described in Section III.B.

In summary, the noise backscattered into low-gain, non-phase-conjugate modes radiates diffusely from the interaction region and should pose no problem for phase-conjugating images of up to 10^6 pixels in steady-state SBS and even larger images in SRS and transient regimes.

10. Forward Stimulated Scattering

The coupled-mode analysis above may also be applied to stimulated forward scattering in the guide. The analysis and results are formally almost identical, predicting one solution with anomalously high gain (but with $B_n \propto A_n$ rather than A_n^*) under essentially the same circumstances as the phase-conjugate solution. The gains γ of the forward-scattered solutions in steady state are proportional to the coupling coefficient G

in Eq. (4) by the same formulas that applied to backward-scattering. Of course there is no stimulated forward Brillouin scattering for which the phonon frequency vanishes. Whether forward or backward stimulated Raman scattering predominates in steady state depends the relative values of G, small differences in G making a large difference in scattered powers. In practice transient effects, self-focusing, and other nonlinear effects affect the ratio of forward-to-backward scattering. A large predominance of backward SRS over forward SRS has been reported, without explanation, by Maier *et al.* (1969) for 30-nsec pulses focused into cells of CS_2. Although SBS, self-focusing, and pump depletion were present, clearly phase conjugation may also have played a role. The predominance over forward SRS of a backscattered phase-conjugate wave has been reported by Sokolovskaya *et al.* (1977, 1978) and Blekhovskikh *et al.* (1979). Here transient and pump-depletion effects may also have played a role.

Our backward scattering analysis may be adapted to forward scattering by simply replacing z by $-z$ in Eq. (8) and $k_{m\omega}$ by $-k_{m\omega}$ in Eqs. (9) and (10). This requires the effective Hermitian Hamiltonian in (11) to be replaced by its transpose. Therefore, the new eigenvalues remain the same as for those for the backward-scattered solutions, and the new eigenvectors are the complex conjugates of the corresponding eigenvectors which described the backward-wave solutions. The actual wave gains γ per cm are easily seen to be equal to these eigenvalues times a constant times the coefficient G of Eq. (4). For resonant scattering, G is itself proportional to the total spontaneous (Raman, Brillouin, etc.) scattering intensity divided by the linewidth.

B. Pulse Compression with Stimulated Phase Conjugation

Pulse compression by efficient backward SRS has been applied to pulse compression for laser fusion by Jacobs *et al.* (1980). Background references for pulse compression by stimulated backscattering are given by Murray *et al.* (1979) and Kaiser and Maier (1972). Pulse compression by backward SBS was observed by Maier and Renner (1971a,b). Transient backward SBS in fibers is discussed by Ippen and Stolen (1972) and Johnson and Marburger (1971). However, only recently (Hon, 1980b) was it demonstrated that an image-bearing laser pulse can be compressed (from 20 to 2 nsec) while being simultaneously phase conjugated by stimulated backscattering (SBS). See Fig. 5.

Although exact conditions have not been derived for the simultaneous occurrence of pulse compression and phase conjugation by stimulated

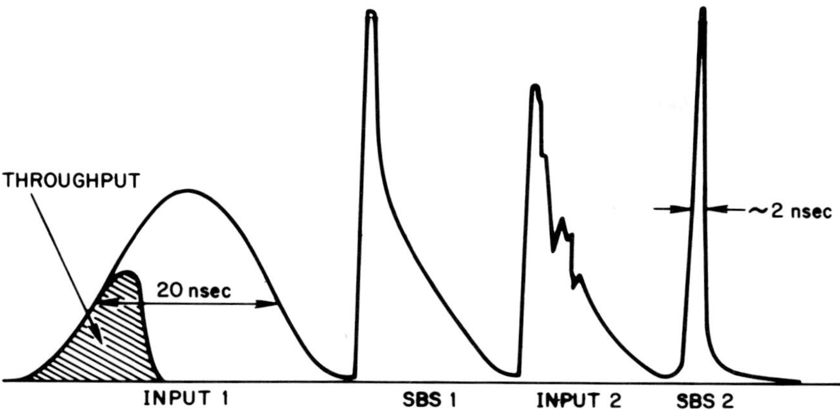

Fig. 5. Results on pulse compression by backward SBS. A 200-mJ pulse from a Nd:YAG laser, recorded as input 1, was launched into a 1.3-m optical guide with inner diameter tapered from 4 mm i.d. at the entrance face to 2 mm at the end. The Brillouin scattering medium in the guide was methane gas at 130 atm. The compressed, phase-conjugate pulse labeled SBS 1 contained about 70% of the incident energy while the "throughput" was reduced to 30% as shown. This backscattered pulse retraversed the Nd:YAG laser, where it was slightly amplified and distorted, to reemerge as the pulse labeled input 2, which was also backscattered to produce the additionally compressed pulse labeled SBS 2. The vertical scale is the same for each pulse. [After Hon (1980).]

backscattering, it is likely that widely useful predictions can be obtained by calculating pulse-shortening and phase-conjugation characteristics separately.

Pulse compression can be treated by considering forward- and backward-propagating plane waves for which the incident field may be

$$\mathbf{E}_{\mathrm{i}}(\mathbf{x}, t) = \mathrm{Re}\ \hat{e}_{\mathrm{i}} A(z, t) e^{i\nu(nz/c - t)} \tag{70}$$

and the backscattered field

$$\mathbf{E}_{\mathrm{b}}(\mathbf{x}, t) = \mathrm{Re}\ \hat{e}_{\mathrm{b}} B(z, t) e^{-i\omega(nz/c + t)}. \tag{71}$$

The matter excitation that mediates the scattering may be described by

$$U(\mathbf{x}, t) = \mathrm{Re}\ \mathbf{t} Q(z, t) e^{ikz - i\Delta t}, \tag{72}$$

where $k \equiv (\nu + \omega)n/c$ and $\Delta \equiv \nu - \omega$ are the wave vector and frequency of the excitation. The symbol $\mathbf{t}$ indicates that the excitation may be scalar, vector, or tensor. We will assume that, whatever the nature of $\mathbf{t}$, the

Raman-active excitation does not couple other polarizations to e_i and e_b, but couples these to each other. Then the excitation creates a change in the local optical susceptibility for each wave proportional to U. Using this modulated susceptibility in Maxwell's equations, and assuming that fields do not change much during an optical period (the "slowly varying envelope approximation"), one arrives at the well-known set of coupled equations for A, B, and Q (Kaiser and Maier, 1972),

$$\frac{\partial A}{\partial z} + \frac{n}{c}\frac{\partial A}{\partial t} = -\frac{1}{2} g_A^* B Q - \frac{1}{2}\mu A, \tag{73}$$

$$\frac{\partial B}{\partial z} - \frac{n}{c}\frac{\partial B}{\partial t} = -\frac{1}{2} g_B^* A Q^* + \frac{1}{2}\mu B, \tag{74}$$

$$\left(\frac{i}{2\omega_B}\frac{\partial^2}{\partial t^2} + \frac{\partial}{\partial t} + \frac{1}{2}\Gamma\right) Q = \frac{1}{2}\Gamma A B^* + S(z, t). \tag{75}$$

Here g_A and g_B are coupling coefficients, which for nonthermally coupled stimulated scattering are real and related by $g_A = (\nu/\omega) g_B$; Γ is the relaxation rate of the excitation, $S(z, t)$ the noise source responsible for initiating undriven pulse compression, and μ the linear attenuation coefficient for the guided optical waves.

It has recently been shown (Hon and Hellwarth, 1981) that Eqs. (73)–(75) are valid even for stimulated Brillouin scattering provided that Eq. (3) is satisfied (with $\omega_B = \Delta$ and $\Gamma_B = \Gamma$) and

$$A^{-1}|\partial A/\partial t|,\ B^{-1}|\partial B/\partial t| \ll \nu(F/\Delta). \tag{76}$$

Under these conditions the optical momenta do not vary enough to excite more than one phonon of a given wave vector, to within the natural uncertainty of the phonon wave vector caused by its damping rate.

To date, analyses of Eqs. (73)–(75) have been carried out by computing the evolution of an initial pulse existing at $t = 0$ without noise $S(z, t)$.

Two distinct regimes have been probed experimentally. The first, and most studied, is where the medium excitation responds essentially instantaneously to changes in the optical fields; see, for example, Maier *et al.* (1968) and Jacobs *et al.* (1980). In the second regime the optical field amplitudes change rapidly compared with medium response times. See, for example, Efimkov *et al.* (1979c), Korner *et al.* (1979), and Hon (1980a,b). The crossover between these regimes has not yet been studied experimentally. Because both regimes have been observed to produce phase-conjugate backscatter, we consider them separately in the following sections.

1. *Regime of Low-Gain and Fast-Responding Excitations*

When the medium excitation that mediates the stimulated backscattering responds nearly instantaneously to driving optical fields, the time derivatives in Eq. (75) can be neglected (being small compared with the excitation damping rate Γ). In this case, the coupled equations (73)–(75) may be rewritten in a dimensionless form in which the only adjustable parameter represents linear loss, which has been neglectible in experiments. Specific solutions take on a physical universality. The important scaling parameter is the characteristic steady-state gain length ℓ_0 for a particular unperturbed real pump field amplitude A_0, which we take to be the maximum input pulse amplitude at the initial time. Specifically, we define

$$\ell_0 = (A_0^2 \operatorname{Re} g_B)^{-1}. \tag{77a}$$

In this section we measure lengths in terms of ℓ_0, so that $z/\ell_0 \to z'$ in Eqs. (73)–(75). We measure time in units of

$$t_0 = c\ell_0/n, \tag{77b}$$

the time for light to travel a distance ℓ_0 in the medium. Therefore, $t/t_0 \to t'$ in Eqs. (73)–(75). In addition, we measure all optical field amplitudes in terms of the maximum magnitude A_0 of the forward wave at the initial time according to

$$a(z, t) \equiv A(z, t)/A_0 \tag{78a}$$

and

$$b(z, t) \equiv [B(z, t)/A_0](\operatorname{Re} g_A/\operatorname{Re} g_B)^{1/2}. \tag{78b}$$

The possible advantages of using an absorbing medium for phase conjugation have not yet been explored, so we consider only the limit $\mu = 0$. We also consider the common case where the coupling coefficients g_A and g_B are real. Because linear gain creates a lower bound on the rate of change of weak signals, a necessary condition for the neglect of $\partial/\partial t$ in Eq. (73) is clearly that

$$\Gamma t_0 \gg 1. \tag{79}$$

This condition will in fact define the low-gain regime of this section. With our new variables and the neglect of $\partial/\partial t$ in Eq. (75), Eqs. (73)–(75) immediately become

$$\frac{\partial a}{\partial z'} + \frac{\partial a}{\partial t'} = -\frac{1}{2}|b|^2 a - bs \tag{80a}$$

and

$$\frac{\partial b}{\partial z'} - \frac{\partial b}{\partial t'} = -\frac{1}{2}\,|a|^2 b - as^*, \tag{80b}$$

where s is a renormalized noise source.

These two coupled equations have been studied extensively by Maier *et al.* (1969) for $s = 0$. They found that a weak initial backscattered pulse, of length τ_0 much longer than $c/\Gamma n$, will be linearly amplified until its intensity is comparable with the incident intensity ($b \sim a$). Then pump depletion becomes strong, and the pulse length τ begins to shorten exponentially with further travel of distance x (in a region of constant incident amplitude):

$$\tau(x) \sim \tau_0 e^{-x/\ell_0}, \tag{81}$$

until the condition

$$\tau(x) > c/\Gamma n \tag{82}$$

is violated and the medium can no longer respond instantaneously. After this point, τ approaches a value near $c/\Gamma n$.

The noise sources proportional to s and s^* in Eqs. (80) generate spontaneous Raman (Brillouin, etc.) scattering, which serves to initiate strong stimulated backscattering even when there is no injected backscattered pulse. Johnson and Marburger (1971) have used an approximate form of s to calculate the evolution of the expected values of $|b|^2$ and $|a|^2$ when the backscattering is initiated by this spontaneous scattering term for a step-function pump incident at the front face ($x = 0$) of the interaction region ($0 \leq x \leq L_{\text{eff}}$). They found that a nearly square backscattered pulse emerged from the front face after a time ηt_0. Smith (1972) estimates η to be 20–24, in agreement with experiments with Raman and Brillouin oscillators. The length of the pulse was predicted by Johnson and Marburger to be $2L_{\text{eff}} - \eta\ell_0$. (Subsequent pulses were also predicted to occur at intervals of $\eta L_{\text{eff}}/c$; we do not consider these here.) Most important, the peak pulse amplitude b_m was essentially unity ($b_m \cong a \cong 1$). Therefore, amplified spontaneous backscattering of a square pulse was predicted never to produce efficient pulse compression, but only pulse clipping (or chopping) at best. Even allowing for the inexact treatment of the initiating noise source, the author feels that this conclusion is widely useful as a rule of thumb.

To obtain a backscattered phase-conjugate pulse that is shorter than the input pulse and that contains a significant fraction of its energy, one must *inject* a small backward-traveling pulse of length much less than L_{eff} that contains a phase-conjugate component. To see how this works in the regime of Eqns. (80), assume that the injected pulse length τ_0 is longer than $c/\Gamma n$, and assume an efficient interaction geometry such that the leading edge of the input pulse reaches the end of the interaction region ($z = L$) at the moment the backscattered pulse is injected (see Fig. 6, trace (a)). The backscattered pulse is linearly amplified over a distance z_1 before reaching saturation ($|b_m|^2 \sim \ell_0/\tau_0$) as depicted in Fig. 6, trace (b). The backscattered pulse then depletes the forward wave almost entirely while traveling the remainder x of the distance to the entrance face, and while experiencing shortening according to Eq. (81), as depicted in Fig. 6, trace (c). If the incident pulse ends as the backscattered pulse reaches the entrance face, then the fraction ϵ of the incident pulse that is backscattered is of order

$$\epsilon \sim x/L. \tag{83}$$

This can be seen from Fig. 6, trace (d). The span $z_1 \sim L - x$ over which the injected pulse is linearly amplified depends on its initial amplitude b_0 according to

$$|b_0|^2 e^{z_1/\ell_0} \sim \ell_0/\tau_0, \tag{84}$$

b_0 and the gain length ℓ_0 being adjusted to achieve the desired conversion efficiency ϵ. The length τ_f of the backscattered pulse as it emerges from the interaction region is given approximately by Eq. (83) with Eq. (81):

$$\tau_f \sim \tau_0 e^{-\epsilon L/\ell_0}. \tag{85}$$

Because amplified spontaneous backscattering will produce a spurious unshortened pulse if $L \gtrsim 22\ell_0$, the pulse compression of Eq. (85) is limited to be of order 10^{-4} of the original τ_0. Furthermore, when $\eta\tau_0 c^{-1} \exp(-\epsilon L/\ell_0)$ becomes as short as the medium response time Γ^{-1}, the pulse stops compressing and

$$\tau_f \to c/n\Gamma. \tag{86}$$

One can achieve the conditions of Fig. 6 by the following scheme. After reaching the end of the main section of the waveguide ($0 < z < L$), the incident pulse is conducted through a short taper to a smaller-diameter section of guide spanning $L < z < L + L_i$ and containing the same scattering medium. The diameter of this added section is so small that the leading-edge intensity of the incident pulse causes L_i to be greater than $\sim$22 gain lengths, and a strong ($b_{max} \sim a$) pulse is generated spontaneously and injected back into the main section, just after the incident pulse has filled it. The injected pulse is nearly phase conjugate at L for the same reasons as in cw stimulated backscattering. The length of the injected pulse τ_0 is close to the Johnson–Marburger estimate of $L_i - \eta\ell_{0i}$, where

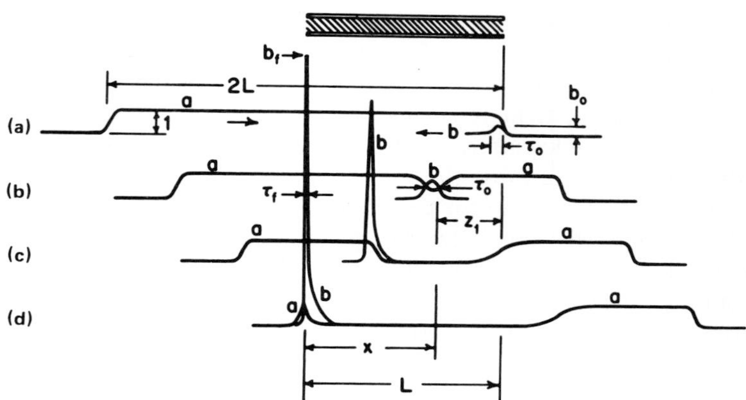

Fig. 6. Schematic diagram of salient temporal phases in the compression of an injected backward-traveling pulse b by stimulated scattering from a forward-traveling pulse a. The spatial distribution of pulse energies is represented at four succeeding characteristic times labeled (a), (b), (c), and (d), as described in the text.

$\overline{\eta \ell_{0i}}$ is the appropriate average over the time-varying leading edge of the critical gain factor $\eta \sim 22$ and the gain length ℓ_{0i} appropriate to the instantaneous pump intensity.

The pulse compression of Eq. (85) depends sensitively on the injected power levels. For example, at $L \sim 10\ell_0$ and $\epsilon \sim 1$, a 10% reduction in incident pulse power causes an e-fold increase in the backscattered pulse length. If this sensitivity is unsatisfactory, one may wish to work to the limit placed by the medium response time, as expressed in Eq. (86). However, if the response time of useful media does not correspond to the desired backscattered pulse length, one may wish to use the high-gain, transient-medium-response regime, discussed in the next section, to obtain stable pulse compression.

2. *Regime of High-Gain and Transient Scattering Excitations*

It is often of interest to allow the incident pulse to be so strong that the gain length ℓ_0 is much shorter than the length $c/\Gamma n$ that light travels in one medium response time:

$$\ell_0 \ll c/\Gamma n. \tag{87}$$

In this case, the evolution of the backscattered pulse will usually be found to drive the scattering excitations in such a manner that

$$\partial Q/\partial t \gg \tfrac{1}{2}\Gamma Q \tag{88}$$

in Eq. (72). An exception occurs when both $|a(t)|^2$ and $|b(t)|^2$ are less than the small dimensionless parameter γ defined by

$$\gamma \equiv (\ell_0 \Gamma n/c)^{1/2}. \tag{89}$$

This exception applies to points where neither wave sees large stimulated gain or loss because of the other, that is, in overlapping pulse tails. Computer solutions show that only minor alterations are produced by approximating the left-hand side of Eq. (75) by $\partial Q/\partial t$ when $\gamma \ll 1$ in all cases examined (Hon and Hellwarth, unpublished). This fact motivates us to rewrite Eqs. (72)–(74) in a dimensionless form so that, when Eq. (88) holds, there are no adjustable parameters in the coupled equations (at least when $L\mu \ll 1$).

This form is obtained by the following transformations.

(1) Reexpress the optical field amplitudes as in Eq. (78).
(2) Measure lengths in units of a characteristic length

$$\ell_1 \equiv (\ell_0 c/\Gamma n)^{1/2} = \ell_0/\gamma, \tag{90}$$

the geometric mean of the steady-state gain length ℓ_0 and the length ℓ_{p} that light travels in a phonon decay time Γ^{-1}. That is, use the dimensionless length variable u defined by

$$u \equiv z/\ell_1. \tag{91}$$

(3) Measure time in units of

$$t_1 \equiv \ell_1 n/c, \tag{92}$$

the time for light to travel one characteristic length. For this, use the dimensionless time variable s defined by

$$s \equiv t/t_1. \tag{93}$$

(4) Reexpress the excitation coordinate amplitude Q in terms of the variable

$$q \equiv Q\Gamma t_1 A_0^2(\mathrm{Re}\ g_B/\mathrm{Re}\ g_A)^{1/2}. \tag{94}$$

With these substitutions the coupled matter–Maxwell equations (71)–(74) become

$$\left(\frac{i}{2\Omega}\frac{\partial^2}{\partial s^2} + \frac{1}{2}\gamma + \frac{\partial}{\partial s}\right) q = \frac{1}{2}ab^* + \sigma, \tag{95}$$

$$\frac{\partial a}{\partial u} + \frac{\partial a}{\partial s} = -\frac{1}{2}bq(1 - i\rho_A) - \frac{1}{2}\mu\ell_1 a, \tag{96}$$

$$\frac{\partial b}{\partial u} - \frac{\partial b}{\partial s} = -\frac{1}{2} a q^*(1 + i\rho_B) + \frac{1}{2}\mu \ell_1 b, \tag{97}$$

where σ is a renormalized noise source, $\Omega \equiv \omega_B t_1$, and

$$\rho_A \equiv \mathrm{Im}\, g_A/\mathrm{Re}\, g_A, \tag{98a}$$

$$\rho_B \equiv \mathrm{Im}\, g_B/\mathrm{Re}\, g_B. \tag{98b}$$

In the discussion that follows we will assume, as is commonly the case, that $\rho_A = \rho_B = 0$ and $\mu\ell_1$ can be neglected. We will also assume that $\gamma \ll 1$ because of Eq. (87) and that pulse times are longer than an acoustic cycle, so that the second derivative in Eq. (95) can be neglected.

An important case to consider is that of a backscattered pulse that (1) has a smooth symmetric initial shape $b(x, s = 0) = \beta(x)$ whose length τ_0 is much greater than ℓ_1, and (2) grows under the influence of a constant undepleted pump ($a = 1$). Equations (95) and (97) are satisfied for times less than $\tau_0^2 n/c\ell_1$ and $\gamma \ll 1$ by

$$b(x, s) \sim \beta(x + \tfrac{1}{2}s) \cosh \tfrac{1}{2}s, \qquad s < \tau_0^2/\ell_1^2. \tag{99}$$

This solution is essentially the same as for low gain, except that the pulse velocity is $\frac{1}{2}c/n$ and the gain length ℓ_1 rather than ℓ_0. When s exceeds $2\tau_0^2/\ell_1^2$ the pulse broadens, developes structure and gains energy less rapidly. However, an initial pulse several ℓ_1 long, and of typical noise amplitude, would grow to saturation approximately as described by Eq. (99).

The second important case to consider is like the first, except that the peak amplitude b_0 of the injected pulse initially is comparable to, or larger than the initial fixed pump amplitude, i.e., $|b_0| \gtrsim 1$. The computer solutions of Hon and Hellwarth (unpublished) show that complete pump depletion sets in immediately (nearly all subsequent pump photons are converted to backscattered photons) and the backscattered pulse length τ shortens approximately according to

$$\tau^{-1} \sim 10^{-2} s \ell_1^{-1} + \tau_0^{-1} \tag{100}$$

until $\tau \sim \ell_1$ after which no more shortening occurs.

Assuming that the driving spontaneous noise power is unchanged from the low-gain value, one sees from Eq. (99) that a backscattered signal will grow from noise in the presence of a uniform pump amplitude in a number η of characteristic times t_1, where η is again of order 20–24. However, because of the slowed propagation velocity in Eq. (99), one expects that spontaneous backscattering in steady state grows to depleting levels in only 10–12 characteristic gain lengths ℓ_1. An analysis of the time dependence of backscattering growing from noise in the presence of a step-function input pulse has not been made. One might suppose that it could produce results similar to those obtained by Johnson and Marburger (1971), however, with ℓ_0 replaced by ℓ_1 and η reduced from 22 to 11. Pulse compression when $\gamma \ll 1$ might also then be visualized as in Fig. 6, except that the greatly reduced narrowing rate of Eq. (100) occurs during the sweep of the injected pulse through the main interaction region.

Some specific solutions of Eqs. (95)–(97), with $\rho_A = \rho_B = 0$ and $\mu = 0$, have been given by Kormer *et al.* (1979) and Eroshenko *et al.* (1980) in conjunction with experiments in which an input iodine laser pulse was backscattered by a medium (SF_6 at 18 atm) whose acoustic decay time was about ten times the pulse length.

In the pulse-compression experiments reported by Hon (1980), stimulated Brillouin backscattering from CH_4 (130 atm) was generated from noise in a 1.3-m-long glass tube tapered from 4 to 2 mm i.d. This tapering acted somewhat like adding the small-diameter section of optical guide onto the main section (to supply an injected pulse near $z = L$) as described in the previous section. The value of γ at the front of the tapered guide was $\sim 10^{-4}$. (Here, $\Gamma/2\pi \sim 20$ MHz, and $\ell_0^{-1} \sim 0.1$ times the pump intensity in MW/cm^2.) Clearly much remains to be analyzed in this high-gain regime ($\gamma \ll 1$) before such experiments are well understood, and before optimum geometries for efficient pulse compression to desired pulse lengths can be reliably predicted.

IV. Summary

Stimulated backward scattering of a multimode optical beam at frequency ω in a waveguide, or in a focal region, can efficiently generate a beam that is phase conjugate to an incident beam, at a frequency ν differing from ω by the frequency of a Raman (Brillouin, etc.) excitation of the medium of propagation. This backscattering process can result from noise amplified by the stimulated Raman (Brillouin, etc.) scattering, and therefore requires no auxiliary beams or pump sources to achieve the phase-conjugate signal. In addition, if the incident beam is an optical pulse not longer than twice the length of the guide or interaction region, the backscattered pulse can be much shorter in length and still preserve its phase-conjugate nature; that is, efficient pulse compression is possible with phase conjugation by stimulated backscattering. The subject is not closed, and all the conditions governing successful phase conjugation by stimulated backscattering are not yet established. However, the theory presented here leads one to the following principal conclusions.

The main considerations for ensuring that the stimulated backscattering from an arbitrary multimode beam contains mainly a phase-conjugate component, in a quasi-steady-state regime, are as follows. First, there is a nonconjugate backscattered power fraction of the order of $N^{-3/2}$ if N is the number of modes excited by the incident beam and the amplitudes of these excited modes vary more or less randomly from zero to some maximum value. Therefore, the number of guide modes excited (or number of focused image elements) must be larger than ~ 10 before the phase-conjugate component predominates. On the other hand, the total number of guide modes available cannot be so large that competing (low-gain) backward and forward background scattering predominates over the (high-gain) phase-conjugate component. Depending on the stimulated cross section and frequency shift, this limits the maximum number of pixels that can be usefully conjugated to typically $10^4 - 10^7$ in steady state, and higher numbers in transient scattering. Also, the two independent states of polarization must not be mixed; only uniformly polarized beams, and often only linearly or circularly polarized beams, will be successfully conjugated. There are, however, no stringent conditions on guide uniformity; small imperfections can actually aid phase conjugation by removing unwanted mode degeneracies. An effective interaction length is set by the guide (or focal region) length or by the absorption length, whichever is shorter. The product of this effective length times the wavelength shift times the number of modes excited divided by the guide area cannot exceed a certain number without unwanted noise exceeding a certain level. This restriction is not important in practice for stimulated Brillouin processes (the wavelength shift is so small), but can further limit stimulated processes that produce a large Stokes wavelength shift. Although the conditions under which transient effects spoil phase conjugation are not known, experiments have shown that useful and efficient pulse compression can accompany wave-front conjugation in stimulated backscattering. Experiments have also shown excellent phase conjugation to be produced by stimulated backscattering when self-focusing and other nonlinear effects were certainly also present.

We conclude that phase conjugation by stimulated backscattering is an important and widely applicable effect, favored even by circumstances beyond our present understanding.

Acknowledgment

The author would like to thank the Aspen Center for Physics for hospitality during the preparation of this chapter.

References

Abrams, R. L., Giuliano, C. R., and Lam, J. F. (1981). On the equality of stimulated Brillouin scattering reflectivity to conjugate reflectivity of a weak probe beam, *Opt. Lett.* **9,** 131.

Baranova, N. B., and Zel'dovich, B. Ya. (1980a). Transverse enhancement of coherence of the scattered field in wavefront reversal, *Kvantovaya Elektron.* **7,** 299 [*English transl.: Sov. J. Quantum Electron.* **10,** 172 (1980)].

Baranova, N. B., and Zel'dovich, B. Ya. (1980b). Wavefront reversal of focused beams (theory of stimulated Brillouin backscattering), *Kvantovaya Elektron.* **7,** 973 [*English transl.: Sov. J. Quantum Electron.* **10,** 555 (1980)].

Baranova, N. B., Zel'dovich, B. Ya., and Shkunov, V. V. (1978). Wave front reversal in stimulated light scattering in a focused spatially inhomogeneous pump beam, *Kvantovaya Elektron.* **5,** 973 [*English transl.: Sov. J. Quantum Electron.* **8,** 559 (1978)].

Basov, N. G., Boiko, V. A., Danilychev, V. A., Zvorykin, V. D., Kholin, I. U., and Chugunev, A. Yu. (1977). Reflection of radiation from a plasma mirror of an electron-beam-controlled CO_2 laser, *Kvantovaya Elecktron.* **4,** 2268 [*English transl.: Sov. J. Quantum Elecktron.* **7,** 1300 (1977)].

Basov, N. G., Efimkov, V. F., Zubarev, I. G., Kotov, A. V., Mikhailov, S. I., and Smirnov, M. G. (1978). Inversion of wavefront of SMBS of a depolarized pump, *Pis'ma Zh. Eksp. Teor. Fiz.* **28,** 215 [*English transl.: JETP Lett.* **28,** 197 (1978)].

Basov, N. G., Zubarev, I. G., Kotov, A. V., Mikhailov, S. I., and Smirnov, M. G. (1979a). Small-signal wave front reversal in nonthreshold reflection from a Brillouin mirror, *Kvantovaya Elektron.* **6,** 394 [*English transl.: Sov. J. Quantum Electron.* **9,** 237 (1979)].

Basov, N. G. *et al.* (1979b). Influence of certain radiation parameters on wavefront reversal in a Brillouin mirror, *Kvantovaya Elektron.* **6,** 765 [*English transl.: Sov. J. Quantum Electron.* **9,** 455 (1979)].

Bel'dyugin, I. M., and Zemskov, E. M. (1978). Influence of changes in the pump field on the form of the field of a signal amplified under stimulated scattering conditions, *Kvantovaya Elektron.* **5,** 2055 [*English transl.: Sov. J. Quantum Electron.* **8,** 1163 (1978)].

Bel'dyugin, I. M., and Zemskov, E. M. (1980). Stimulated Raman scattering in a dispersion medium with nonmonochromatic broad-spectral-line pump, *Kvantovaya Elektron.* **7,** 2233.

Bel'dyugin, I. M., Galushkin, M. G., Zemskov, E. M., and Mandrosov, V. I. (1976). Complex conjugation of fields in stimulated Brillouin scattering, *Kvantovaya Elektron.* **3,** 2467 [*English transl.: Sov. J. Quantum Electron.* **6,** 1349 (1976)].

Bel'dyugin, I. M., Zemskov, E. N., and Klushin, V. M. (1979a). On a problem of wavefront reversal by means of stimulated Brillouin scattering, *Sov. J. of Quantum Electron.* **9,** 1200.

Bel'dyugin, I. M., Galushkin, M. G., and Zemskov, E. M. (1979b). Properties of resonators with wavefront-reversing mirrors, *Kvantovaya Elektron.* **6,** 38 [*English transl.: Sov. J. Quantum Electron.* **9,** 20 (1979)].

Bel'dyugin, I. M., Zubarev, I. G., and Mikhailov, S. I. (1979c). Analysis of conditions for stimulated Raman scattering of multimode pump radiation in dispersive media, *Sov. J. Quant. Electron.* **10,** 847.

Bespalov, V. I., Betin, A. A., and Pasmanik, G. A. (1977a). Reconstruction effects in stimulated scattering, *Pis'ma Zh. Tekh. Fiz.* **3,** 215 [*English transl.: Sov. Tech. Phys. Lett.* **8,** 85 (1977)].

Bespalov, V. I., Betin, A. A., and Pasmanik, G. A. (1977b). Experimental investigation of the threshold of the stimulated scattering (SS) of multimode light beams and the degree of regeneration of the pumping in scattered radiation, *Izv. Vuz. Radiofiz.* **20,** 791 (1977b) [*English transl.: Radiophys. Quantum Electron.* **20,** 544 (1977)].

Bespalov, V. I., Betin, A. A., and Pasmanik, G. A. (1978). Reproduction of the pumping wave in stimulated-scattering radiation, *Izv. Vuz. Radiofiz.* **21,** 961 [*English transl.: Radiophys. Quantum Electron.* **21,** 675 (1979)].

Bespalov, V. I., Manishin, V. G., and Pasmanik, G. A. (1979a). Nonlinear selection of optical radiation on reflection from a stimulated Mandel'shtam–Brillouin scattering mirror, *Zh. Eksp. Teor. Fiz.* **77,** 1756 [*English transl.: Sov. Phys. JETP* **50,** 879 (1979)].

Bespalov, V. I., Betin, A. A., Pasmanik, G. A., and Shilov, A. A. (1979b). Wavefront inversion in the Raman conversion of the Stokes wave in oppositely directed pump beams, *Pis'ma Zh. Tekh. Fiz.* **5,** 242 [*English transl.: Sov. Tech. Phys. Lett.* **5,** 97 (1979)].

Blashchuk, V. N., Zel'dovich, B. Ya., Mel'nikov, N. A., Pilipetskii, N. F., Popovichev, V. I., and Ragul'skii, V. V. (1977). Wave front inversion in stimulated scattering of focused light beams, *Pis'ma Zh. Tekh. Fiz.* **3,** 211 [*English transl.: Sov. Tech. Phys. Lett.* **3,** 83 (1977)].

Blashchuk, V. N. *et al.* (1978a). Stimulated scattering of depolarized radiation, *Dokl. Akad. Nauk SSSR* **241,** 1322 [*English transl.: Sov. Phys. Dokl.* **23,** 589 (1978)].

Blashchuk, V. N. *et al.* (1978b). SBS wave front reversal for the depolarized light—theory and experiment, *Opt. Commun.* **27,** 137.

Blekhovskikh, G. L., Kudryavsteva, A. D., and Sokolovskaya, A. I. (1978). Reconstruction of the wave front of light beams in stimulated Raman scattering, *Kvantovaya Elektron.* **5,** 1812 [*English transl.: Sov. J. Quantum Electron.* **8,** 1028 (1978)].

Blekhovskikh, G. L., Okladnikov, N. V., and Sokolovskaya, A. I. (1979). Experimental study of the effect of amplification saturation on wavefront establishment in SRS, *J. Appl. Spectr.* **32,** 15.

Borisov, B. N., Kruzhilin, Yu. J., and Shklyarik, S. V. (1977). Wavefront inversion of the output from a neodymium laser caused by a stimulated Brillouin scattering mirror, *Sov. Tech. Phys. Lett.* **4,** (2), 66.

Borisov, B. N., Kruzhilin, Yu. I., Nashchekin, S. A., Orlov, V. K., and Shklarik, S. V. (1979). Wavefront reversal of radiation during SBS in glass without distortion, *Zh. Tekh. Fiz.* **50** (5) 1073.

Chiao, R. Y., Townes, C. H., and Stoicheff, B. P. (1964). Stimulated Brillouin scattering and coherent generation of intense hypersonic waves, *Phys. Rev. Lett.* **12,** 592.

Dolgopolov, Yu. V., *et al.* (1979). Experimental investigation of the feasibility of application of the wavefront reversal phenomenon in stimulated Mandel'shtam–Brillouin scattering, *Zh. Eksp. Teor. Fiz.* **76,** 908 [*English transl.: Sov. Phys. JETP* **49,** 458 (1979)].

Eckhardt, G., Hellwarth, R. W., McClung, F. J., Schwarz, S. E., Weiner, D., and Woodbury, E. J. (1962). Stimulated Raman scattering from organic liquids, *Phys. Rev. Lett.* **9,** 455.

Efimkov, V. F., Zubarev, I. G., Kotov, A. V., Mironov, A. B., Mikhailov, S. I., and Smirnov, M. G. (1979a). Generation of high-power short pulses with wavefront reversal under steady stimulated Brillouin scattering, *Sov. J. Quantum Electron.* **9** (9), 1193.

Efimkov, V. F., Zubarev, I. G., Kotov, A. V., Mironov, A. B., Mikhailov, S. I., and Smirnov, M. G. (1979b). An investigation of schemes for emission of high-power short pulses with radiation wavefront reversal in a stimulated Brillouin scattering (SBS) mirror, *Sov. J. Quantum. Electron.* **10** (2), 211.

Efimkov, V. F., *et al.* (1979c). Inertial effects of Mandel'shtam–Brillouin stimulated scattering and nonthreshold reflection of short pulses with reversal of the wavefront, *Zh. Eksp. Teor. Fiz.* **77,** 2(8) 526 [*English transl.: Sov. Phys. JETP* **50,** 267 (1979)].

Eidman, K., and Sigel, R. (1974). Backscatter experiments, *in* "Laser Interaction and Related Plasma Phenomena'' (H. J. Schwarz and H. Mora, eds.), Vol. 38, p. 667. Plenum, New York.

Eroshenko, V. A., Kir'yanov, Yu. F., Kormer, S. B., Kochemasov, G. G., and Nikolaev, V. D. (1980). A numerical investigation into possible use of the stimulated Brillouin scattering in laser fusion facilities, *Kvantovaya Elektron.* **7** (12), 2536. [*English trans.: Sov. J. Quantum Electron.* **10,** 1481 (1980)].

Fabelinskii, I. L., Mash, D. I., Morozov, V. V., and Starunov, V. S. (1968). Stimulated scattering of light in hydrogen-gas at low pressures. *Phys. Lett.* **27A,** 253.

Fan, J. Y., Wu, C. K., and Wang, Z. Y. (1980). Nature of the phase conjugation in stimulated Raman scattering backward wave, *Laser. J.* **7** (3), 14.

Gerasimov, V. B., and Orlov, V. K. (1977). Influence of wavefront reversal on the operation of Brillouin lasers, *Kvantovaya Elektron.* **5,** 906 [*English transl.: Sov. J. Quantum Electron.* **8,** 517 (1978)].

Gerasimov, V. B., Gerasimova, S. A., and Orlov, V. K. (1977a). Wavefront of the Stokes component in stimulated Brillouin backscattering, *Kvantovaya Elektron.* **4,** 930 [*English transl.: Sov. J. Quantum Electron.* **9,** 352 (1979)].

Gerasimov, V. B., Gerasimova, S. A., and Orlov, V. K. (1977b). Substantial improvement in selective properties of a stimulated Brillouin scattering mirror resulting from use of wide-band pumping, *Kvantovaya Elektron.* **4,** 932 [*English transl.: Sov. J. Quantum Electron.* **7,** 528 (1977)].

Hellwarth, R. W. (1963). Theory of stimulated Raman scattering, *Phys. Rev.* **130,** 1850–1852.

Hellwarth, R. W. (1978). Theory of phase conjugation by stimulated scattering in a waveguide, *J. Opt. Soc. Am.* **68,** 1050.

Herbst, M. J., Clayton, C. E., and Chen, F. F. (1979). Saturation of Brillouin backscatter. *Phys. Rev. Lett.* **43,** 1591.

Hon, D. T. (1980a). High-brightness ND:YAG laser using SBS phase conjugation, *11th Int. Quantum Electron. Conf. Digest Tech. Papers. J. Opt. Soc. Am.* **70,** 635.

Hon, D. T. (1980b). Pulse compression by stimulated Brillouin scattering, *Opt. Lett.* **5,** 516.

Ippen, E. P., and Stolen, R. H. (1972). Stimulated Brillouin scattering in optical fibers, *Appl. Phys. Lett.* **21,** 539.

Jacobs, R. R., Goldhar, J., Eimerl, D., Brown, S. B., and Murray, J. R. (1980). High efficiency energy extraction in backward wave Raman scattering, *Appl. Phys. Lett.* **37,** 264.

Javan, A. (1958a). *Bull. Am. Phys. Soc.* **3,** 213.

Javan, A. (1958b). *J. Phys. Radium* **19,** 806.

Johnson, R. V., and Marburger, J. H. (1971). Relaxation oscillations in stimulated Raman and Brillouin scattering, *Phys. Rev. A* **4,** 1175.

Kaiser, W., and Maier, M. (1972). Stimulated Rayleigh, Brillouin, and Raman spectroscopy, *in* ''Laser Handbook'' (F. T. Arecchi and E. O. Schulz-Dubois, eds.), Vol. 2, pp. 1077–1150. North-Holland Publ., Amsterdam.

Kochemasov, G. G., and Nikolaev, V. D. (1977). Reproduction of the spatial amplitude and phase distributions of a pump beam in stimulated Brillouin scattering, *Kvantovaya Elektron.* **4,** 115 [*English transl.: Sov. J. Quantum Electron.* **7,** 60 (1977)].

Kochemasov, G. G., and Nikolaev, V. D. (1978). Wavefront reversal in stimulated scattering of two-frequency pump radiation, *Kvantovaya Elektron.* **5,** 1837 [*English transl.: Sov. J. Quantum Electron.* **8,** 1043 (1978)].

Kochemasov, G. G., and Nikolaev, V. D. (1979). Investigation of the spatial characteristics of Stokes radiation in stimulated scattering under saturation conditions, *Kyantovaya Elektron.* **6,** 1960 [*English transl.: Sov. J. Quantum Electron.* **9,** 1155 (1979)].

Kogelnik, H. (1965). Holographic image projection through inhomogeneous media. *Bell System Tech. J.* **44,** 2451.

Kormer, S. B., Kochemasov, G. G., Kulikov, S. M., Nikolaev, V. D., Nikolaev, V., and Sukharev, S. A. (1979). Use of SBS for sharpening pulses and intercascade insolation experiments in LTS, *Zh. Tekh. Fiz.* **50**(6), 1319. [*English trans.: Sov. Phys. Tech. Phys.* **25,** 757 (1980)].

Kudryavtseva, A. D., Sokolovskaia, A. I., Gazengel, J., Xuan, N. P., and Rivoire, G. (1978). Reconstruction of the laser wave-front by stimulated scatterings in the picosecond range, *Opt. Commun.* **26,** 446.

Kurtz, S. K., and Giordmaine, J. A. (1969). Stimulated scattering by polaritons. *Phys. Rev. Lett.* **22,** 192.

Lehmberg, R. H. (1978). Theory of optical ray retracing in laser-plasma backscatter, *Phys. Rev. Lett.* **41,** 863.

Lehmberg, R. H., and Holder, K. A. (1980). Numerical study of optical ray retracing in laser-plasma backscatter, *Phys. Rev. A* **22,** 2156.

Lesnik, S. A., Soskin, N. S., and Khizhnyak, A. I. (1978). Laser with a stimulated-Brillouin-scattering complex-conjugate mirror, *Zh. Tekh. Fiz.* **49,** 2257 [*English transl.: Sov. Phys. Tech. Phys.* **24,** 1249 (1978)].

Maier, M., and Renner, G. (1971a). Transient threshold power of stimulated Brillouin and Raman scattering, *Phys. Lett.* **34A,** 299.

Maier, M., and Renner, G. (1971b). Transient and quasistationary stimulated scattering of light, *Opt. Commun.* **3,** 301.

Maier, M., Kaiser, W., and Giordmaine, J. A. (1969). Backward stimulated Raman scattering. *Phys. Rev.* **177,** 580.

Mash, D., Morozov, V. V., Starunov, V. S., and Fabelinski, I. L. (1965). Stimulated scattering of light of the Rayleigh wing, *JETP Lett.* **2,** 25.

Mays, R., and Lysiak, R. J. (1979). Phase conjugated wavefronts by stimulated Brillouin and Raman scattering, *Opt. Commun.* **31,** 89.

Mays, R., and Lysiak, R. J. (1980). Observations of wavefront reproduction by stimulated Brillouin and Raman scattering as a function of pump power and waveguide dimensions, *Opt. Commun.* **32,** 334.

Murray, J. R., Goldhar, J., Eimerl, D., and Szöke, A. (1979). Raman pulse compression of excimer lasers for application to laser fusion, *IEEE J. Quantum Electron.* **QE-15,** 342.

Nosach, D. Yu., Popovichev, V. I., Ragul'skii, V. V., and Faizullov, F. S. (1972). Cancellation of phase distortions in an amplifying medium with a Brillouin mirror, *Zh. Eksp. Teor. Fiz. Pis'ma Red.* **16,** 617 [*English transl.: Sov. Phys. JETP* **16,** 435 (1972)].

Ostrovskii, Yu. L. (1979). Wavefront reconstruction in stimulated scattering of light, *Sov. Tech. Phys. Lett.* **5,** 315.

Pasmanik, G. A. (1979). Reconstruction of the wavefront of a complex signal in stimulated back scattering, *Sov. Tech. Phys. Lett.* **4** (5), 201.

Peregudov, G. V., Plotkin, M. E., and Ragozin, E. N. (1979). Use of wave-front reversal effect to investigate a jet formed by focusing laser radiation on a plane target, *Sov. J. Quantum Electron.* **9,** 1413.

Pilipetskii, N. F., Popovichev, V. I., and Ragul'skii, V. V. (1979a). Accuracy of reconstruction of a light field after stimulated scattering. *Dokl. Akad. Nauk SSSR* **248,** 1097 [*English transl.: Sov. Phys. Dokl.* **25,** 845 (1979)].

Pilipetskii, N. F., Popovichev, V. I., and Ragul'skii, V. V. (1979b). The reproduction of weak components of a light field at stimulated scattering, *Opt. Commun.* **31,** 97.

Ragul'skii, V. V. (1979). Wavefront inversion of weak beams in stimulated scattering, *Pis'ma Zh. Tekh. Fiz.* **5,** 251 [*English transl.: Sov. Tech. Phys. Lett.* **5,** 100 (1979)].

Rank, D. H., Cho, C. W., Foltz, N. D., and Wiggins, T. A. (1967). Stimulated thermal Rayleigh scattering. *Phys. Rev. Lett.* **19,** 828.

Ripin, B. H., *et al.* (1977). Enhanced backscatter with a structured laser pulse. *Phys. Rev. Lett.* **39,** 611.

Sidorovich, V. G. (1976). Theory of the Brillouin Mirror, *Zh. Tekh. Fiz.* **46,** 2158 [*English transl.: Sov. Phys. Tech. Phys.* **21,** 1270 (1976)].

Sidorovich, V. G., and Shkunov, V. V. (1979). Capture of a Stokes pump wave in a stimulated wave Raman scattering amplifier, *Zh. Tekh. Fiz.* **49,** 816 [*English transl.: Sov. Phys. Tech. Phys.* **24,** 472 (1979)].

Smith, R. G. (1972). Optical power handling capacity of low loss optical fibers as determined by stimulated Raman and Brillouin scattering, *Appl. Opt.* **11,** 2489.

Sokolovskaya, A. I., Brekhovskikh, G. L., and Kudryavtseva, A. D. (1977). Restoration of wave front of light beams during stimulated Raman scattering, *Dokl. Akad. Nauk SSSR* **233,** 356 [*English transl.: Sov. Phys. Dokl.* **22,** 156 (1977)].

Sokolovskaya, A. I., Brekhovskikh, G. L., and Kudryavtseva, A. D. (1978). Light beam wavefront reconstruction and real volume image reconstruction of the object at the stimulated Raman scattering, *Opt. Commun.* **24,** 74.

Wang, V., and Giuliano, C. R. (1978). Correction of phase aberrations via stimulated Brillouin scattering, *Opt. Lett.* **2,** 4.

Zaitzev, G. I., Kyzlasov, Yu. I., Starunov, V. S., and Fabelinskii, I. L. (1967). Stimulated temperature scattering of light in liquids. *Zh. Eksp. Teor. Fiz. Pis'ma Red.* **6,** 802. [*English trans.: JETP Lett.* **6,** 255 (1967)].

Zel'dovich, B. Ya., and Shkunov, V. V. (1977). Wavefront reproduction in stimulated Raman scattering, *Kvantovaya Elektron.* **4,** 1090 [*English transl.: Sov. J. Quantum Electron.* **7,** 610 (1977)].

Zel'dovich, B. Ya., and Shkunov, V. V. (1978a). Reversal of the wave front of light in the case of depolarized pumping, *Zh. Eksp. Teor. Fiz.* **75,** 428 [*English transl.: Sov. Phys. JETP* **48,** 214 (1978)].

Zel'dovich, B. Ya., and Shkunov, V. V. (1978b). Limits of existence of wavefront reversal in stimulated light scattering, *Kvantovaya Elektron.* **5,** 36 [*English transl.: Sov. J. Quantum Electron.* **8,** 15 (1978)].

Zel'dovich, B. Ya., and Yakovleva, T. V. (1979). Fine structure distortions in the problem of wavefront reversal-stimulated Brillouin scattering under nonsteady-state conditions, *Kvantovaya Elektron.* **7** (10), 2293. [*English trans.: Sov. J. Quantum Electron.* **10,** 1306 (1980)].

Zel'dovich, B. Ya., and Yakovleva, T. V. (1980). Small-scale distortions in wavefront reversal of a beam with incomplete spatial modulation (stimulated Brillouin backscattering, theory), *Kvantovaya Elektron.* **7,** 316 [*English transl.: Sov. J. Quantum Electron.* **9,** 351 (1979)].

Zel'dovich, B. Ya., Popovichev, V. I., Ragul'skii, V. V., and Faizullov, F. S. (1972). Connection between the wave fronts of the reflected and exciting light in stimulated Mandel'shtam–Brillouin scattering, *Zh. Eksp. Teor. Fiz. Pis'ma Red.* **15,** 160 [*English transl.: Sov. Phys. JETP* **15,** 109 (1972)].

Zel'dovich, B. Ya., Mel'nikov, N. A., Pilipetskii, N. F., and Ragul'skii, V. V. (1977). Observation of wave-front inversion in stimulated Raman scattering of light, *Pis'ma Zh. Eksp. Teor. Fiz.* **25,** 41 [*English transl.: JETP Lett.* **25,** 35 (1977)].

Zel'dovich, B. Ya., Pilipetskii, N. F., Ragul'skii, V. V., and Shkunov, V. V. (1978). Wavefront reversal of nonlinear optics methods, *Kvantovaya Elektron.* **5,** 1800 [*English transl.: Sov. J. Quantum Electron.* **8,** 1021 (1978)].

8 Phase Conjugation and High-Resolution Spectroscopy by Resonant Degenerate Four-Wave Mixing

R. L. Abrams
J. F. Lam
R. C. Lind
D. G. Steel
Hughes Research Laboratories
Malibu, California

P. F. Liao
Bell Telephone Labs
Holmdel, New Jersey

I. Introduction

A. Mechanism of Resonant Degenerate Four-Wave Mixing

Thus far, degenerate four-wave mixing in Kerr-like media has been discussed. In such media, the only nonlinear behavior is the linear dependence of the index of refraction on intensity. We now turn our attention to the use of the same four-wave mixing geometry in resonant or near-

Optical Phase Conjugation

ISBN 0-12-257740-X

resonant materials. Here much more complicated nonlinearities are involved.

In the most general sense, phase-conjugate optical waves are generated through the nonlinear response of a material to an optical stimulus. The most common examples of nonlinear processes are stimulated Brillouin scattering (Zel'dovich *et al.*, 1972), three-wave mixing (Yariv, 1976), and four-wave mixing (Hellwarth, 1977), but in fact even the nonlinearity exhibited by volume holograms can be used. In this case, the hologram (after development) is dependent on the integrated intensity of the light to which it was previously exposed. Owing to its slow response, traditional holography is impractical for real-time phase conjugation, and we are motivated to investigate materials whose properties are capable of responding rapidly to light with high sensitivity, i.e., highly nonlinear optical materials.

Most students of nonlinear optics are introduced to nonlinear polarization as a source term in Maxwell's equations, and it often remains a mysterious material parameter. In fact, these mechanisms are not so difficult to comprehend once one looks carefully at the nonlinear polarization generated by an atomic system (Bloembergen, 1965). Degenerate four-wave mixing (DFWM) is one of the most useful techniques for the generation of phase-conjugate waves. DFWM relies on the nonlinear polarization of the medium P, where all of the incident and generated waves are at the same frequency. In this chapter, we will concentrate on the nonlinear portions of the polarization P arising from resonances, i.e., where the optical signals are resonant or nearly resonant with atomic or molecular transitions in the nonlinear medium.

The physical significance of P in DFWM is that the refractive index and absorption coefficient of a material can change with light intensity. Saturated absorption and dispersion are obvious examples, and are well known to laser scientists as fundamental to the operating characteristics of lasers. One does not usually think of simple saturation in considering nonlinear optical phenomena, probably because such processes are not generally useful for the generation of new frequencies—which is the most common objective in nonlinear optics. DFWM, however, does not require the medium to respond to or generate new frequencies as does harmonic generation, so saturation of atomic and molecular resonances can be an effective mechanism.

The geometry and nomenclature to be used throughout this chapter are illustrated in Fig. 1. In an arrangement similar to that discussed in previous chapters, E_f and E_b are the forward and backward pumps, both plane waves at frequency ω and counterpropagating along the x axis. E_p is the probe wave, also at frequency ω, and is the wave front of which we

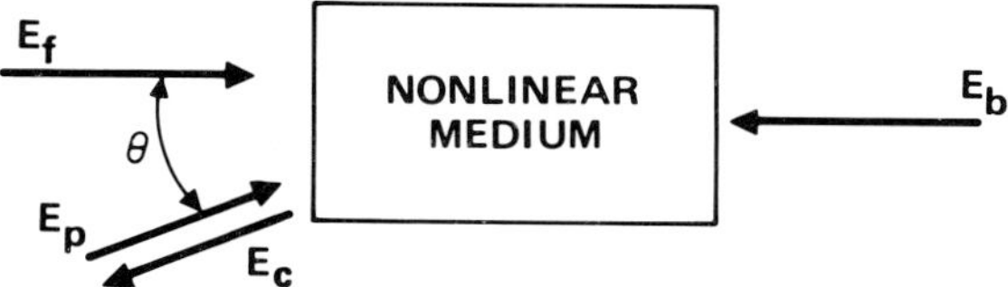

Fig. 1. Geometry for DFWM.

wish to generate the conjugate; it is incident at an angle θ to the x axis. E_c is the conjugate wave; it is the conjugate of E_p, also at frequency ω.

Consider the effects of saturated absorption and dispersion in the given geometry. Because E_p, E_f, and E_b all oscillate at the same frequency ω, they interfere to form a spatial intensity modulation pattern. Saturation of an atomic resonance then leads to spatial modulation of the complex re-

FORMATION

FORWARD PUMP, PROBE, θ; BACKWARD PUMP, PROBE

READOUT

BACKWARD PUMP, CONJUGATE; FORWARD PUMP, CONJUGATE

$D = \lambda/2 \sin(\theta/2)$ $D = \lambda/2 \cos(\theta/2)$

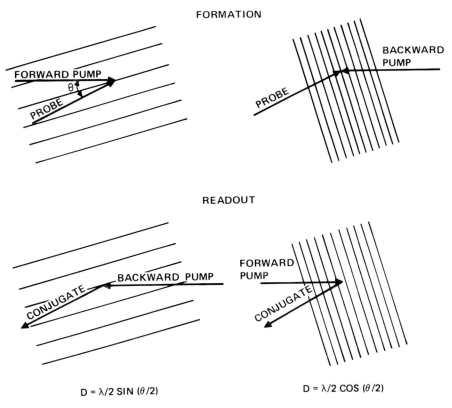

Fig. 2. Dual grating picture. [After Lind *et al.* (1982).]

fractive index of the material—either dispersive or absorptive, or both. Two important modulation patterns in the case where E_p is also a plane wave are illustrated in Fig. 2. E_p and E_f form a grating of relatively large period in which the important Fourier component is exactly phase matched to Bragg-scatter E_b into E_c, the phase conjugate of E_p. E_p and E_b, on the other hand, form a finer grating with spacing and orientation to similarly scatter E_f into E_c. A third grating, due to the interference of E_f and E_b, does not contribute to this process, because it is not phase matched for scattering of E_c or E_p.

In addition to the well-known phenomena of saturated absorption and dispersion, a number of more subtle effects in resonant systems can produce or modify the observed nonlinear polarization. These include quantum coherence (two-photon, Zeeman coherence, crossovers, etc.) and nonlinearities arising from optically induced level shifts (adiabatic following).

B. General Approach

To understand and distinguish these effects, we begin with the simplest possible system—steady-state solutions of a nondegenerate two-level atomic system in which all the atoms are at rest. Using a density matrix formulation, we calculate P for this simple system for pump, probe, and conjugate waves all at the same frequency, which can be varied through the atomic resonance. The nonlinear polarization term in this case will be seen to be due entirely to spatial grating formations, where the gratings are due to spatial modulation of the populations of levels 1 and 2. This model and experiments in which this approximation is valid are discussed in Section II.A.1.

The first complication we introduce is allowing the atoms random thermal motion, as in an atomic or molecular vapor (Section II.A.2). This generally results in the reduction of the effective nonlinearity, because the spatial population gratings become smeared out as the atoms drift through the grating maxima and minima. Atoms with zero velocities contribute most strongly to P, making this a very effective tool for eliminating Doppler broadening in high-resolution spectroscopic measurements.

In Section II.A.3 we evaluate the effects of frequency differences in the probe and pumps. In this case the standing wave that produces the grating becomes a traveling wave. These effects are very important for practical applications and spectroscopy.

Although the two-level model provides a very convenient and simple way of looking at the importance of resonance behavior in DFWM, nature is generally more complicated. In particular, atoms have angular momentum associated with their internal degrees of freedom, and their energy levels have associated rotational degeneracy. This allows additional mechanisms to contribute to DFWM—namely, spatially modulated coherent excitation of degenerate sublevels and optical pumping—resulting in new polarization effects unrealizable in a simple two-level system. We will treat this subject theoretically and describe some experimental confirmations in Section II.A.4.

Pulsed excitation, a major subject unto itself, is discussed only briefly in this chapter in Section II.B. Theory and experiments for two-level systems are described. Coherent transient effects include phase conjugation by photon echoes, treated in the following chapter.

Thus far, we have considered only single-photon resonances, where the incident radiation is nearly resonant with the atomic energy-level splitting. The case of two-photon resonances is also important. Strong nonlinearities can occur when twice the incident photon energy is resonant with the atomic system; this is treated in Section III with a three-level model. Closely related to this problem are nonlinearities arising from crossover resonances, which occur in three-level systems with two transitions very close to each other in wavelength and sharing a common level.

In Section IV we discuss potential applications of resonantly enhanced DFWM, including its importance for spectroscopic measurements. We also attempt to predict fruitful areas of future work.

To acquaint the reader with the analytical approach of this chapter, in Section II.A.1 we provide a detailed derivation of the basic DFWM nonlinear interaction (including saturation) leading to a phase-conjugate signal. In the remaining sections we discuss other nonlinearities that give rise to DFWM, but emphasize the physics of the interaction rather than the analysis. Hence we provide the reader with the calculational starting point and the basic analytical result. Discussions of experiments further illustrate the physical behavior.

II. Single-Photon Transitions

A. Near Resonance

The response of an isotropic medium to applied and stimulated fields is measured by the ensemble average of the expectation value of the induced dipole moment operator $\boldsymbol{\mu}$. The ensemble average takes into account the physical uncertainties associated with the random initial states of the macroscopic quantum-mechanical systems. An operator exists in quantum mechanics that allows a systematic treatment of the response of the macroscopic medium and takes into account the statistical nature of the quantum states of the system in question. This quantity is the density matrix $\rho(\mathbf{r}, \mathbf{v}, t)$, where $\mathbf{r}$ and $\mathbf{v}$ are macroscopic displacement and velocity of an atomic or molecular species. A knowledge of $\rho(\mathbf{r}, \mathbf{v}, t)$ provides a means of evaluating the expectation value of any Hermitian operator.

The ensemble average of the induced dipole moment—in short, the polarization—is given by

$$\mathbf{P}(\mathbf{r}, t) = \text{trace} \int_{-\infty}^{\infty} d\mathbf{v}\, \rho(\mathbf{r}, \mathbf{v}, t)\boldsymbol{\mu}, \tag{1}$$

where the integration over velocity takes into account the possibility of the random macroscopic motion of the atomic species. The density matrix $\rho(\mathbf{r}, \mathbf{v}, t)$ is defined by $\Sigma_n C_n |\Psi_n\rangle \langle\Psi_n|$, where C_n is the statistical weight of the state function $|\Psi_n\rangle$. Each $|\Psi_n\rangle$ describes a set of initial conditions of the medium.

The Schrödinger equation describes the evolution of the state function $|\Psi_n\rangle$ and yields the following space–time evolution for the density matrix $\rho(\mathbf{r}, \mathbf{v}, t)$:

$$i\hbar\left(\frac{\partial}{\partial t} + \mathbf{v}\cdot\boldsymbol{\nabla}\right)\rho = [H_0, \rho] + [V, \rho] - \frac{i\hbar}{2}\{\Gamma, \rho\} + i\hbar\Lambda, \tag{2a}$$

where $\mathbf{v}\cdot\boldsymbol{\nabla}\rho$ describes the macroscopic translational motion of the atomic species and accounts for such effects as Doppler shifts and population pulsations,

$$H_0|n\rangle = \hbar\omega_n|n\rangle \tag{2b}$$

is the unperturbed electronic Hamiltonian that describes the eigenstates and energy eigenvalues of the quantum system, and

$$V = -\mu\cdot\mathbf{E}(\mathbf{r}, t) \tag{2c}$$

is the electric-dipole-allowed coupling to the external and generated classical radiation fields $\mathbf{E}(\mathbf{r}, t)$. In Eq. (2a), brackets denote commutation and braces anticommutation. Relaxation processes such as natural or spontaneous decay to the reservoir (any other states of the system) are described by the operator Γ such that

$$\Gamma|n\rangle = \gamma_n|n\rangle. \tag{2d}$$

The excitation (or pumping) of the various energy levels from the reservoir is described by the operator Λ such that

$$\Lambda|n\rangle = \lambda_n|n\rangle. \tag{2e}$$

The solution of the density matrix equation (2a) assumes a knowledge of the initial condition for ρ. In the steady-state regime such conditions are taken to be at time $t \to -\infty$. Equation (2a) can be decomposed into two distinct contributions, the first given by the diagonal components of ρ, which describe the population of each eigenstate of the system, the second by the off-diagonal components of ρ, which describe the atomic coherence between two eigenstates of the system. In the cases we will consider, the atomic coherences are induced only by the radiation fields, and the initial populations are given in terms of the density of the atomic species. Hence, in the absence of radiation fields, diagonal density matrix elements ρ_{mm} are given quantities, and $\rho_{mn} = 0$, $m \neq n$. The density matrix equation (2a) and the initial conditions constitute the starting point in the analysis of resonantly enhanced four-wave mixing.

For two-level systems with ground state $|1\rangle$ and excited state $|2\rangle$, the polarization is reduced to

$$\mathbf{P}(\mathbf{r}, t) = \int_{-\infty}^{\infty} d\mathbf{v}\; \rho_{12}(\mathbf{r}, \mathbf{v}, t)\boldsymbol{\mu}_{21} + \text{c.c.} \tag{3}$$

Hence the dynamic response of the medium to the external and generated radiation fields involves the calculation of $\rho_{12}(\mathbf{r}, \mathbf{v}, t)$, one of the main subjects of this chapter.

1. Homogeneously Broadened: Without Motion

In this section we will discuss the solution of the simplest problem of a resonant system, that of a two-level saturable absorber (Abrams and Lind, 1978). We will assume that we are dealing with a homogeneously broadened medium and that atomic and molecular motion is neglected. The results for inhomogeneously broadened media and the effects of motion in homogeneously broadened media will be considered later. We will discuss the relevant equations, give a detailed discussion of their solutions, and finally describe selected experiments and compare the experimental results with the model.

A complete analytical description of DFWM in resonant materials begins with the solution of the quantum-mechanical transport equation (QMTE) given by Eq. (2). For this derivation we assume the simple two-level model shown in Fig. 3. The energy levels are nondegenerate and separated by energy $\hbar\omega_{21}$. If we eliminate the motion operation $\mathbf{v} \cdot \boldsymbol{\nabla}$, the simple QMTE becomes

$$i\hbar \frac{\partial \rho}{\partial t} = [H_0, \rho] + [V, \rho] - \frac{i\hbar}{2}\{\Gamma, \rho\} + ih\Lambda. \tag{4}$$

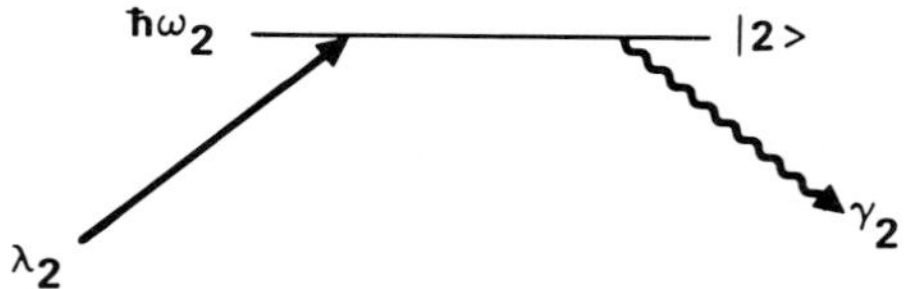

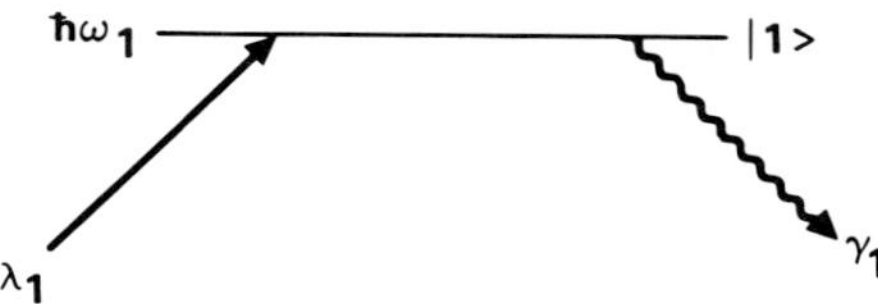

Fig. 3. Two-level model for DFWM. λ_1 and λ_2 are the incoherent pumping rates $|1\rangle$ and $|2\rangle$; γ_1 and γ_2 are the decay rates.

H_0 is the unperturbed Hamiltonian for the atom,

$$H_0 = \begin{bmatrix} \hbar\omega_1 & 0 \\ 0 & \hbar\omega_2 \end{bmatrix};$$

V is the interaction energy resulting from dipole coupling between the atom and the radiation field,

$$V = \begin{bmatrix} 0 & V_{12} \\ V_{21} & 0 \end{bmatrix},$$

where

$$V_{ij} = -\tfrac{1}{2}\,\boldsymbol{\mu}_{ij} \cdot (\mathbf{E}e^{i\omega t} + \text{c.c.});$$

Γ is the relaxation rate,

$$\Gamma = \begin{bmatrix} \gamma_1 & 0 \\ 0 & \gamma_2 \end{bmatrix};$$

and Λ is the incoherent pumping rate,

$$\Lambda = \begin{bmatrix} \lambda_1 & 0 \\ 0 & \lambda_2 \end{bmatrix}.$$

The geometry of interest is shown in Fig. 1. The pumps are assumed to be counterpropagating. Equation (4) is usually approached analytically via perturbation theory. However, the situation of interest involves the presence of arbitrary pump amplitudes E_f and E_b, and it is generally in the limit of strong pumps that the most interesting and useful behavior is observed. In contrast, the probe and conjugate amplitudes often remain small compared with the saturation intensity. That is, the probe and conjugate waves need not significantly affect the level populations. In this limit a full solution of the density matrix equation including arbitrary pump amplitudes is obtained, and perturbation theory is used to include the probe and conjugate waves E_p and E_c. Hence we can expand V and ρ as

$$V = V^{(0)} + q\,V^{(1)}, \qquad \rho = \rho^{(0)} + q\,\rho^{(1)},$$

where

$$\begin{aligned} V^{(0)} &= -\tfrac{1}{2}\boldsymbol{\mu}\cdot(\mathbf{E}_f + \mathbf{E}_b)e^{i\omega t} + \text{c.c.}, \\ V^{(1)} &= -\tfrac{1}{2}\boldsymbol{\mu}\cdot(\mathbf{E}_p + \mathbf{E}_c)e^{i\omega t} + \text{c.c.}, \end{aligned}$$

and q is a bookkeeping parameter to be set equal to unity. The QMTE may now be written by equating terms of equal powers of q. The zeroth- and first-order terms are

$$i\hbar\frac{\partial\rho^{(0)}}{\partial t} = [H_0, \rho^{(0)}] + [V^{(0)}, \rho^{(0)}] - \frac{i\hbar}{2}\{\Gamma, \rho^{(0)}\} + i\hbar\Lambda, \tag{5a}$$

$$i\hbar\frac{\partial\rho^{(1)}}{\partial t} = [H_0, \rho^{(1)}] + [V^{(0)}, \rho^{(1)}] + [V^{(1)}, \rho^{(0)}] - \frac{i\hbar}{2}\{\Gamma, \rho^{(1)}\}, \tag{5b}$$

respectively.

The zeroth-order equations are solved exactly in the rotating wave approximation (RWA). Assuming steady state and noting that the off-diagonal elements have the form

$$\rho_{ij} = \tilde{\rho}_{ij}e^{i\omega t}, \tag{6}$$

we obtain for the population difference and atomic coherence the expressions

$$\rho_{22}^{(0)} - \rho_{11}^{(0)} = \frac{-\Delta N_0}{1 + (\gamma_{12}/\Gamma_0)(\boldsymbol{\mu}_{12}\cdot\mathbf{E}_0/\hbar)^2/(\gamma_{12}^2 + \Delta^2)} \tag{7a}$$

and

$$\tilde{\rho}_{12}^{(0)} = -\frac{1}{2i}\frac{\boldsymbol{\mu}_{12}\cdot\mathbf{E}_0}{\hbar}\frac{\rho_{22}^{(0)} - \rho_{11}^{(0)}}{\gamma_{12} + i\Delta}, \tag{7b}$$

where $\Delta N_0 = \lambda_1/\gamma_1 - \lambda_2/\gamma_2$ is the population difference in the absence of applied fields, $\gamma_{12} = \frac{1}{2}(\gamma_1 + \gamma_2)$ is the atomic coherence decay rate, and $\Gamma_0^{-1} = \frac{1}{2}(\gamma_1^{-1} + \gamma_2^{-1})$ is the population decay rate. $\mathbf{E}_0 = \mathbf{E}_f + \mathbf{E}_b$ is the sum of both pump fields including the spatial dependence, and $\Delta = \omega - \omega_{21}$.

Note the expected field saturation of the population difference in Eq. (7a). In the limit of strong fields the population difference is driven to zero. The case $\Delta N_0 > 0$ corresponds to absorption, whereas $\Delta N_0 < 0$ corresponds to an inverted system exhibiting gain.

Having obtained the system response in the presence of the strong field $\mathbf{E}_0$, we can now proceed to obtain the first-order variation of $\rho^{(1)}$ described in Eq. (5b). It is this perturbation in ρ that gives rise to the polarization that generates phase-conjugate waves by DFWM. If we again assume the RWA and seek a steady-state solution, the first-order equations may be written

$$\begin{aligned} \rho_{22}^{(1)} - \rho_{11}^{(1)} &= -\frac{1}{i\Gamma_0}\left(\frac{\boldsymbol{\mu}_{12}\cdot\mathbf{E}_0^*}{\hbar}\,\tilde{\rho}_{12}^{(1)} - \frac{\boldsymbol{\mu}_{12}\cdot\mathbf{E}_0}{\hbar}\,\tilde{\rho}_{21}^{(1)}\right) \\ &\quad -\frac{1}{i\Gamma_0}\left(\frac{\boldsymbol{\mu}_{12}\cdot\mathbf{E}_1^*}{\hbar}\,\tilde{\rho}_{12}^{(0)} - \frac{\boldsymbol{\mu}_{12}\cdot\mathbf{E}_1}{\hbar}\,\tilde{\rho}_{21}^{(0)}\right), \end{aligned} \tag{8a}$$

$$\begin{aligned} \tilde{\rho}_{12}^{(1)} &= -\frac{1}{2i}\,\frac{\boldsymbol{\mu}_{12}\cdot\mathbf{E}_0}{\hbar}\,\frac{\rho_{22}^{(1)} - \rho_{11}^{(1)}}{\gamma_{12} + i\Delta} \\ &\quad -\frac{1}{2i}\,\frac{\boldsymbol{\mu}_{12}\cdot\mathbf{E}_1}{\hbar}\,\frac{\rho_{22}^{(0)} - \rho_{11}^{(0)}}{\gamma_{12} + i\Delta}, \end{aligned} \tag{8b}$$

where $\mathbf{E}_1 = \mathbf{E}_c + \mathbf{E}_p$ is the sum of the conjugate and probe waves, and $\tilde{\rho}_{12} = \tilde{\rho}_{21}^*$. Using Eqs. (7) and (8b), we may rewrite the first-order population difference (8a) as

$$\begin{aligned} \rho_{22}^{(1)} - \rho_{11}^{(1)} &= \Delta N_0\,\frac{\gamma_{12}/\Gamma_0}{\gamma_{12}^2 + \Delta^2} \\ &\quad \times \frac{(\boldsymbol{\mu}_{12}\cdot\mathbf{E}_0^*/\hbar)\boldsymbol{\mu}_{12}\cdot\mathbf{E}_1/\hbar + (\boldsymbol{\mu}_{12}\cdot\mathbf{E}_0/\hbar)\boldsymbol{\mu}_{12}\cdot\mathbf{E}_1^*/\hbar}{[1 + (\gamma_{12}/\Gamma_0)(\boldsymbol{\mu}_{12}\cdot\mathbf{E}_0/\hbar)^2/(\gamma_{12}^2 + \Delta^2)]^2}. \end{aligned} \tag{9}$$

Equation (9) represents the first-order population difference generated by *simultaneous* action of the strong and weak fields. Note that this population difference is time independent (as expected), but has a spatial dependence of $\exp[i(\mathbf{k}_1 - \mathbf{k}_0)\cdot\mathbf{r}]$. This property is commonly referred to as spatial hole burning.

Using Equation (9), we can now, with the aid of Eq. (8b), evaluate the off-diagonal matrix element (atomic coherence):

$$\begin{aligned} \tilde{\rho}_{12}^{(1)} &= \frac{\Delta N_0}{2i}\,\frac{\boldsymbol{\mu}_{12}\cdot\mathbf{E}_1/\hbar}{\gamma_{12} + i\Delta}\,\frac{1}{[1 + (\gamma_{12}/\Gamma_0)(\boldsymbol{\mu}_{12}\cdot\mathbf{E}_0/\hbar)^2/(\gamma_{12}^2 + \Delta^2)]^2} \\ &\quad - \frac{\Delta N_0}{2i}\,\frac{\gamma_{12}}{\Gamma_0}\,\frac{(\boldsymbol{\mu}_{12}\cdot\mathbf{E}_0/\hbar)^2\boldsymbol{\mu}_{12}\cdot\mathbf{E}_1^*/\hbar}{(\gamma_{12} + i\Delta)(\gamma_{12}^2 + \Delta^2)} \\ &\quad \times \frac{1}{[1 + (\gamma_{12}/\Gamma_0)(\boldsymbol{\mu}_{12}\cdot\mathbf{E}_0/\hbar)^2/(\gamma_{12}^2 + \Delta^2)]^2}. \end{aligned} \tag{10}$$

Using the expression $\mathbf{P} = \mathrm{Tr}\, \boldsymbol{\mu}\rho$ for the polarization, we find that the total polarization for use in Maxwell's equations can now be written

$$P = -\frac{\Delta N_0}{2\hbar} \frac{i\gamma_{12} + \Delta}{\gamma_{12}^2 + \Delta^2} |\mu_{12}|^2 \frac{E_1 e^{i\omega t}}{[1 + (\gamma_{12}/\Gamma_0)(\mu_{12}E_0/\hbar)^2(\gamma_{12}^2 + \Delta^2)^{-1}]^2}$$

$$+ \frac{\Delta N_0}{2\hbar^3} \frac{\gamma_{12}}{\Gamma_0} \frac{i\gamma_{12} + \Delta}{(\gamma_{12}^2 + \Delta^2)^2} |\mu_{12}|^4$$

$$\times \frac{E_0^2 E_1^* e^{i\omega t}}{[1 + (\gamma_{12}/\Gamma_0)(\mu_{12}E_0/\hbar)^2(\gamma_{12}^2 + \Delta^2)^{-1}]^2} + \text{c.c.} \tag{11}$$

This polarization constitutes the complete response of the medium for DFWM in the presence of arbitrarily strong pumps but weak signal and probe. The first term corresponds to the saturating absorption and nonlinear dispersion, and the second term is responsible for the phase-conjugate signal. Note, however, that the phase-conjugate term has both a real and an imaginary part, corresponding to dispersion and absorption, respectively.

We can now insert Eq. (11) into the complete set of Maxwell's equations to evaluate the response. The geometry for this nonlinear interaction was shown in Fig. 1. Recall that the two strong pumps E_f and E_b are counterpropagating and the probe E_p is incident at $x = 0$. The medium is of length L. The $\mathbf{k}$-vector phase-matching conditions for DFWM are once again

$$\mathbf{k}_f + \mathbf{k}_b - \mathbf{k}_p - \mathbf{k}_c = 0.$$

Because $\mathbf{k}_f + \mathbf{k}_b = 0$ and the frequencies are degenerate, the conjugate wave is counterpropagating to the probe wave as shown in Fig. 1. In general, the coupled-wave equations for all four waves must be solved. However, if we assume that there is no pump depletion (numerical solutions that relax this assumption will be described later), the coupled-wave equations for the probe E_p and signal E_c must be solved subject to the boundary conditions $E_p(x = 0) = E_p(0)$ and $E_c(x = L) = 0$.

The radiation field can be written

$$E(\mathbf{r}, t) = \tfrac{1}{2} \sum_n \mathscr{E}_n(\mathbf{r}) e^{i(\omega t - \mathbf{k}\cdot\mathbf{r})} + \text{c.c.}, \tag{12}$$

where the field envelope $\mathscr{E}_n(\mathbf{r})$ satisfies the slowly varying envelope approximation (SVEA)

$$|\nabla \mathscr{E}_n(\mathbf{r})| \ll k|\mathscr{E}_n(\mathbf{r})|. \tag{13}$$

Before we can proceed, the SVEA equations of evolution for the probe $\mathscr{E}_p$ and conjugate signal $\mathscr{E}_c$ must be developed. This entails choosing a volume of the nonlinear medium large enough that the radiation fields are distinguishable from each other but small enough that the SVEA is preserved. An appropriate choice of the dimension is the wavelength of light. The equations for $\mathscr{E}_p$ and $\mathscr{E}_c$ are generated by inserting Eqs. (11) and (12) into the wave equation, multiplying the wave equation by the complex conjugate of the corresponding phase modulation factor, and performing spatial integration over the wavelength. The SVEA equations may then be written

$$\frac{d\mathscr{E}_p}{dz} = -\alpha\mathscr{E}_p - i\beta\mathscr{E}_c^*, \qquad \frac{d\mathscr{E}_c}{dz} = \alpha\mathscr{E}_c + i\beta\mathscr{E}_p^*, \tag{14}$$

where α is the attenuation (or gain, if $\Delta N_0 < 0$) coefficient

$$\alpha = \alpha_0 \frac{1}{1 + (\Delta/\gamma_{12})^2} \frac{1 + (I_f + I_b)/I_{sat}}{\{[1 + (I_f + I_b)/I_{sat}]^2 - 4I_fI_b/I_{sat}^2\}^{3/2}} \tag{15}$$

and β the nonlinear coupling coefficient

$$\beta = \alpha_0 \frac{i + \Delta/\gamma_{12}}{1 + (\Delta/\gamma_{12})^2} \frac{2(I_fI_b/I_{sat})^{1/2}}{\{[1 + (I_f + I_b)/I_{sat}]^2 - 4I_fI_b/I_{sat}^2\}^{3/2}}. \tag{16}$$

α_0 is the familiar *field* absorption (or gain, if $\Delta N_0 < 0$) coefficient

$$\alpha_0 = (\omega/2nc)\, \Delta N_0|\mu_{12}|^2/\hbar E_0\gamma_{12} \tag{17}$$

and I_{sat} the frequency-dependent saturation intensity

$$I_{sat} = \tfrac{1}{2}\epsilon_0 c(\hbar^2\gamma_{12}\Gamma_0/|\mu_{12}|^2)[1 + (\Delta/\gamma_{12})^2] = I_{sat}^0(1 + \delta^2), \tag{18}$$

where $\delta = \Delta/\gamma_{12}$, and I_{sat}^0 is the line-center saturation intensity. The intensity-dependent index of refraction in Eq. (17) is given by

$$n^2 = 1 + \frac{\Delta N_0}{\hbar\gamma_{12}E_0}|\mu_{12}|^2 \frac{\delta}{1 + \delta^2} \times \frac{1 + (I_f + I_b)/I_{sat}}{\{[1 + (I_f + I_b)/I_{sat}]^2 - 4I_fI_b/I_{sat}^2\}^{3/2}}. \tag{19}$$

Note that in Eqs. (14) the conjugate and probe waves are related by their phase conjugates, clearly showing that this nonlinear interaction gives rise to phase conjugation of the incident wave. Equations (14) may now be solved for the reflectivity subject to the boundary conditions given above:

$$R = \left|\frac{\mathscr{E}_c(0)}{\mathscr{E}_p(0)}\right|^2 = \left(\frac{\beta \sin \gamma L}{\gamma \cos \gamma L + \alpha \sin \gamma L}\right)^2, \tag{20}$$

where $\gamma^2 = |\beta|^2 - \alpha^2$. If $n \cong 1$ (valid for most gaseous systems), the solutions given above are the same as those obtained by Abrams and Lind (1978).

In discussing these solutions we first note that if $|\beta|^2 > |\alpha|^2$, γ is a positive real number and the functions in Eq. (20) are sinusoidal. If $|\beta|^2 < |\alpha|^2$, the functions become hyperbolic. For $|\beta|^2 > \alpha^2$ the oscillation condition, $R \to \infty$, becomes

$$\tan \gamma L = -\gamma/\alpha, \tag{21}$$

where (a) if $\alpha_0 > 0$ (i.e., absorption), the oscillation condition requires $\gamma L > \pi/2$; (b) if $\alpha_0 < 0$ (i.e., gain), the oscillation condition requires $\gamma L < \pi/2$; and (c) if $\alpha_0 = 0$, the oscillation condition requires $\gamma L = \pi/2$. Condition (c) (no absorption) corresponds to the solution discussed in Chapter 2. For $|\beta|^2 < \alpha^2$ the oscillation condition is given by

$$\tanh \gamma L = -\gamma/\alpha. \tag{22}$$

In this case oscillation occurs only if $\alpha_0 < 0$.

Some examples of reflectivity R, determined from Eq. (20), are given in Figs. 4–12, plotted (except where noted) as a function of I/I_{sat}.

In Fig. 4 we show R for line-center operation, $\delta = 0$, for various values of $\alpha_0 L$. In this case the reflectivity is due to a purely absorptive grating contribution. In general, R peaks near $I = I_{sat}$. For small values of $\alpha_0 L$ the reflectivity increases linearly with $\alpha_0 L$. As $\alpha_0 L$ becomes large, the reflectivity saturates but never becomes greater than unity. Taking the limit of Eq. (20) for large $\alpha_0 L$ and large I/I_{sat}, we find that $R \to 1$. Physically, this saturation occurs because the increase in reflectivity with $\alpha_0 L$ due to the participation of more absorbers (i.e., $\alpha_0 \sim$ density) is offset by the increase in absorption of the probe and conjugate signals as they propagate through the medium. To achieve reflectivities larger than unity—to obtain large R in general—requires either operation off line center, as subsequently discussed, or the use of an inverted medium. The equations are equally valid for inverted media (gain) ($\alpha_0 > 0$), and the results for various values of small signal gain are given in Fig. 5. For comparable values of $|\alpha_0 L|$, much larger reflectivities ($R > 1$) are obtained in the gain case because of additional amplification of the signal. Also note that the peak returns occur for $I \ll I_{sat}$, in contrast to the absorption case. This is because one wants to maintain a large saturated gain coefficient, which would be reduced for $I/I_{sat} \geq 1$. We see that for a gain of -4 the solution approaches the oscillation condition.

In Fig. 6 we show the effect of detuning the laser off line center while maintaining the number density or $\alpha_0 L$ at a constant value. In general, the reflectivity decreases with detuning, owing to an effectively reduced absorption coefficient. However, the contribution of a dispersive grating (and reduced absorption) begins to show its importance if we detune the laser, by increasing $\alpha_0 L$ by an amount equal to $1 + \delta^2$. This keeps the

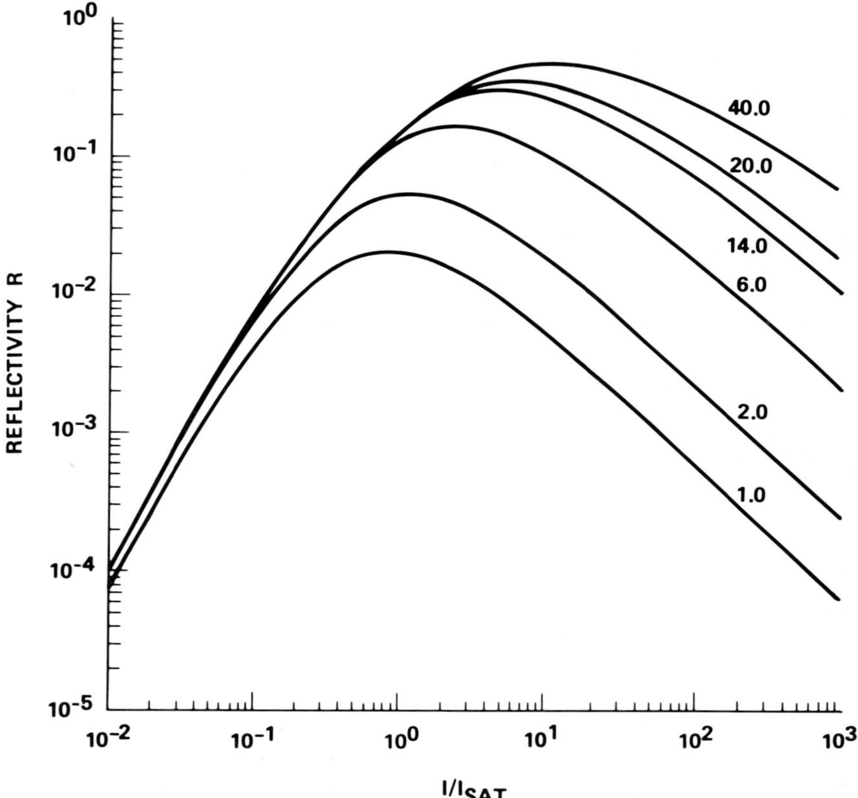

Fig. 4. Reflectivity R versus I/I_{sat} for various values of $\alpha_0 L$ at line center (α_0 is the field absorption coefficient). [After Lind *et al.* (1982).]

absorption coefficient constant at our operating frequency. In this case the reflectivity increases dramatically and values of $R > 1$ are easily achieved. This is shown in Fig. 7, where the quantity $\beta L \equiv \alpha_0 L/(1 + \delta^2)$ is maintained at a value of unity while $\alpha_0 L$ and δ are both increased. The price of this increased reflectivity is the requirement that the pump intensity be continually increased at the same rate, because I_{sat} increases as $1 + \delta^2$. This can be seen from Fig. 7, where, for $\delta = 8$, $R > 1$ requires $I_{pump} = 60 I_{sat}$. The above examples indicate what parameters must be varied in an experiment to achieve large reflectivities.

Modifications of the simple theory have been made to include the effects of pump absorption and depletion (Brown, 1981) and unequal pump fields (Dunning and Lam, 1981; Dunning and Steel, 1982). The numerical

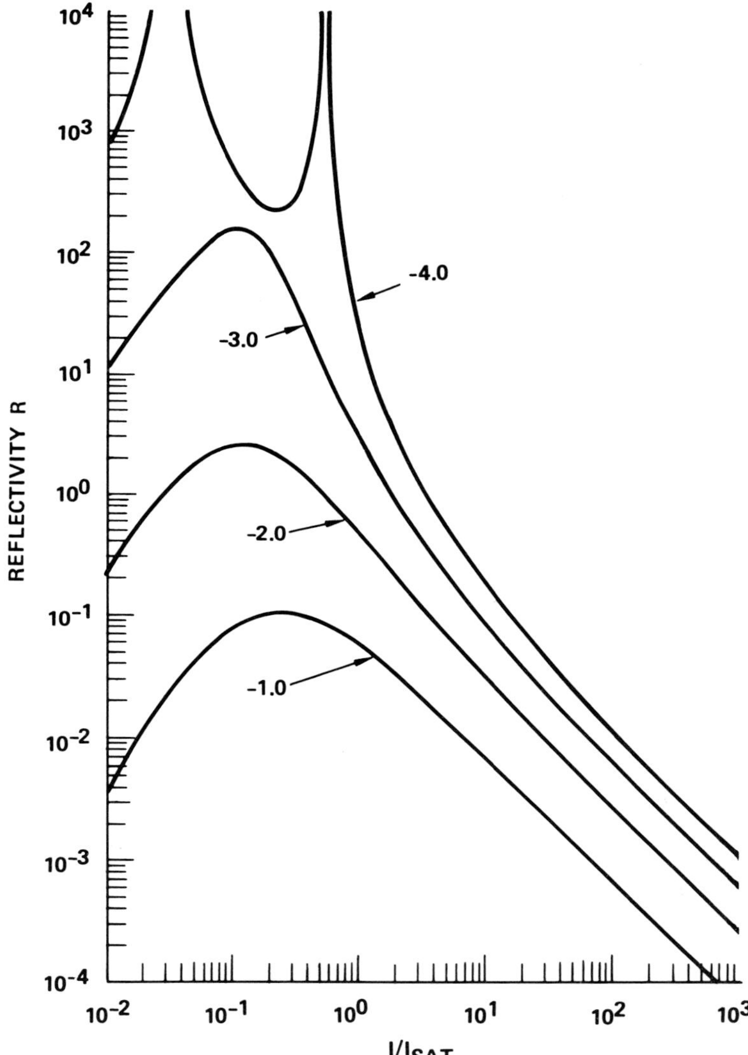

Fig. 5. Reflectivity R versus I/I_{sat} for various values of gain at line center. [After Lind *et al.* (1982).]

solution to the pump absorption problem obtained by Brown (1981) is compared with the prediction of Eq. (20) in Fig. 8 for $\alpha_0 L = 1.5$. For large I/I_{sat} the effects of pump absorption are reduced by bleaching, and both solutions give the same results. However, for small I/I_{sat} the effect of pump absorption is significant as expected.

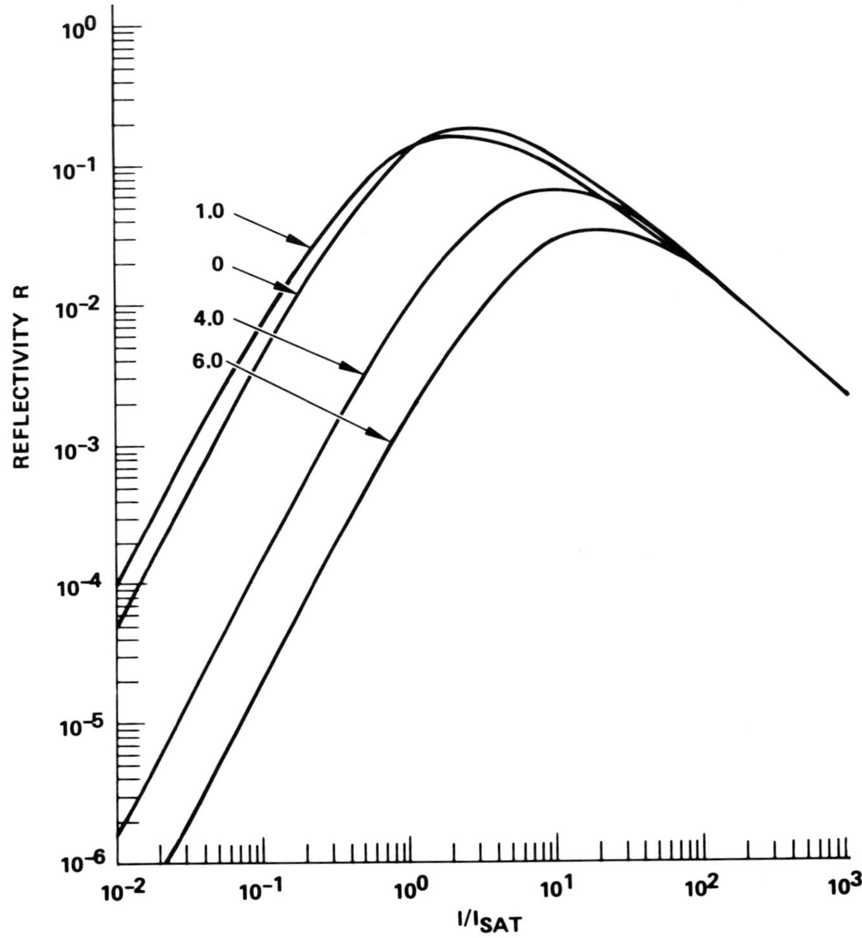

Fig. 6. Reflectivity R versus I/I_{sat} for $\alpha_0 L = 6$ and various detunings; δ values shown.

A quantitative comparison of the analytical expression (20) with experiment requires that the medium be a true two-level system. One such system is ReO_4-doped KCl, investigated by Watkins *et al.* (1980a,b). This system has been characterized as homogeneously broadened, and satisfies the assumptions of the theory presented in this section. Their experiments were performed for values of $2\alpha_0 L$ of 0.47 and 3.0. In Fig. 9a the experimental results are given for $2\alpha_0 L = 0.47$ together with a calculation of the reflectivity with and without pump absorption. The agreement

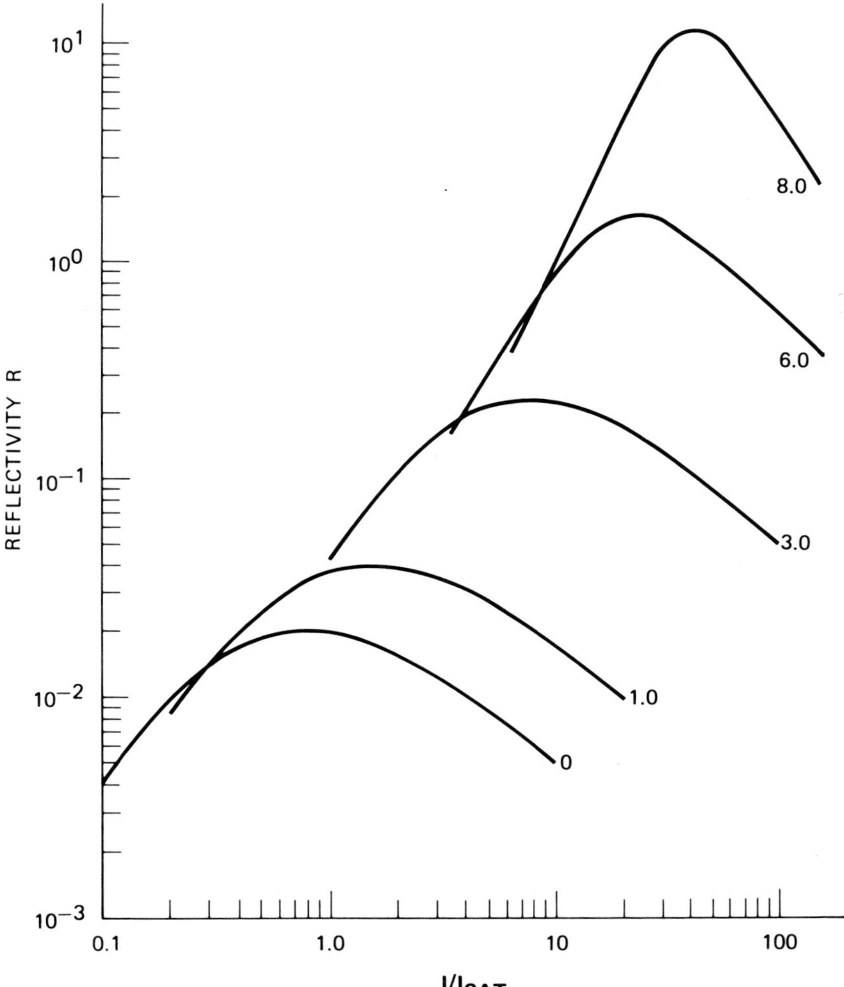

Fig. 7. Reflectivity R versus I/I_{sat} for $\beta L = \alpha_0 L/(1 + \delta^2) = 1.0$, with δ values shown. [After Lind *et al.* (1982).]

is excellent. In Fig. 9b the same results are given for $2\alpha_0 L = 3.0$. In this case, there is agreement only when pump absorption is included, as would be expected from our calculations. The increase in reflectivity at high pump powers is anticipated to be due to a contribution from excited states. We will discuss this later in connection with similar results seen in SF_6.

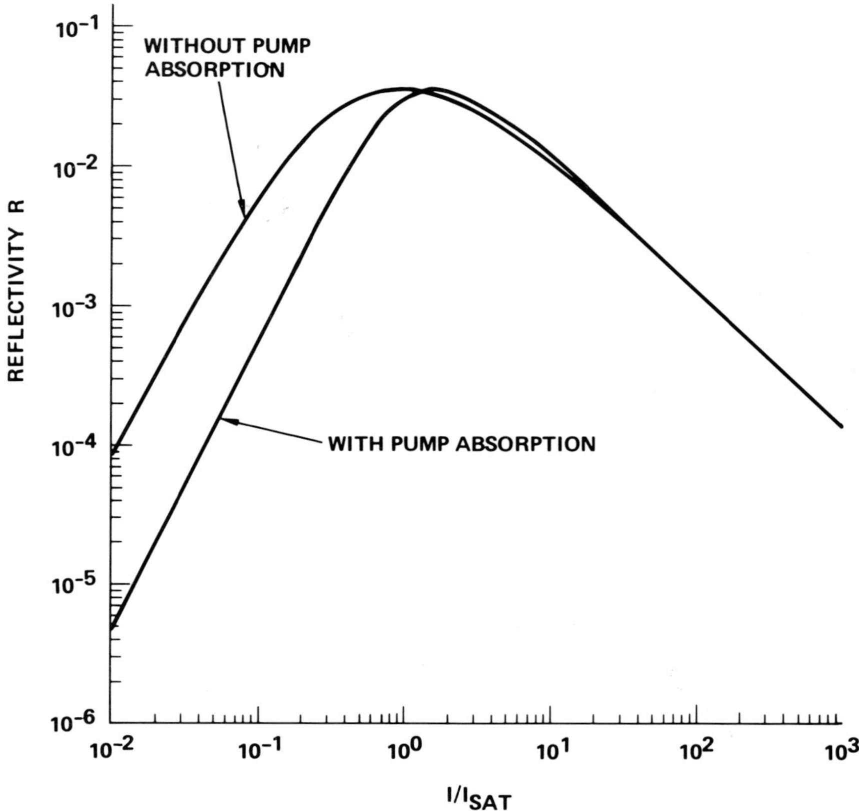

Fig. 8. Reflectivity R versus I/I_{sat} for $\alpha_0 L = 1.5$ and $\delta = 0$ at line center with and without pump absorption. [After Brown (1981); Lind *et al.* (1982).]

We now examine the situation where all four waves have equal magnitude; that is, $\mathscr{E}_p$ and $\mathscr{E}_c$ are not small compared with $\mathscr{E}_f$ and $\mathscr{E}_b$. In this case four coupled differential equations must be solved. Numerical techniques must be used, as has been done by Brown (1981). An example of this fully nonlinear solution is shown in Fig. 10, a calculation of the reflectivity near an oscillation point given by Eq. (21). The particular case shown is for an absorption of $\beta L = \alpha_0 L/(1 + \delta^2) = 1$ and $\delta = 15$. Curve A is the result obtained from Eq. (20) and predicts that the reflectivity diverges at two values of pump intensity. Curve B is the result of the numerical calculation including pump absorption but not depletion. Note that the numerical calculation predicts that the reflectivity diverges at only one value of pump intensity. The dashed curve C gives the results obtained when the

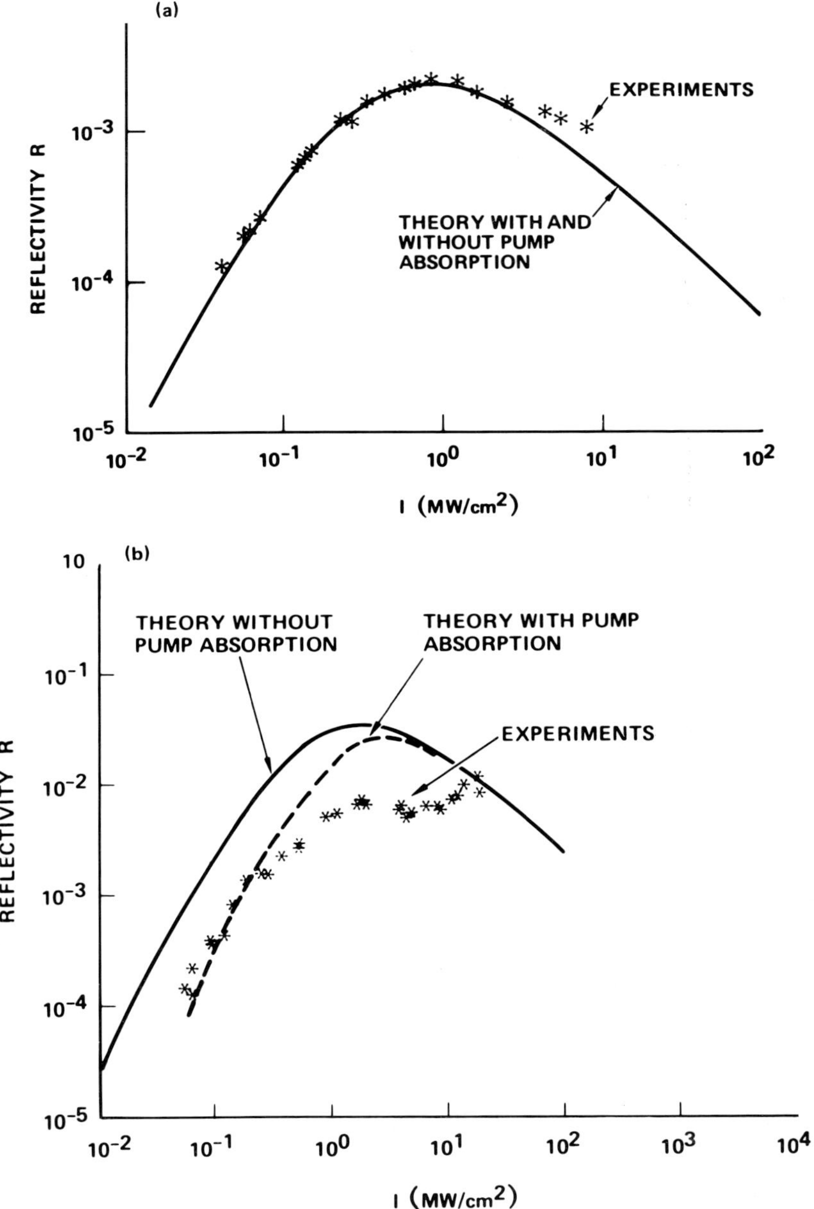

Fig. 9. Phase-conjugate reflectivity R versus average incident intensity in KCl:ReO_4 for (a) $2\alpha_0 L = 0.47$ and (b) $2\alpha_0 L = 3.0$ [theory with pump absorption from Brown (1981)]. [After Watkins *et al.* (1980a, 1980b).]

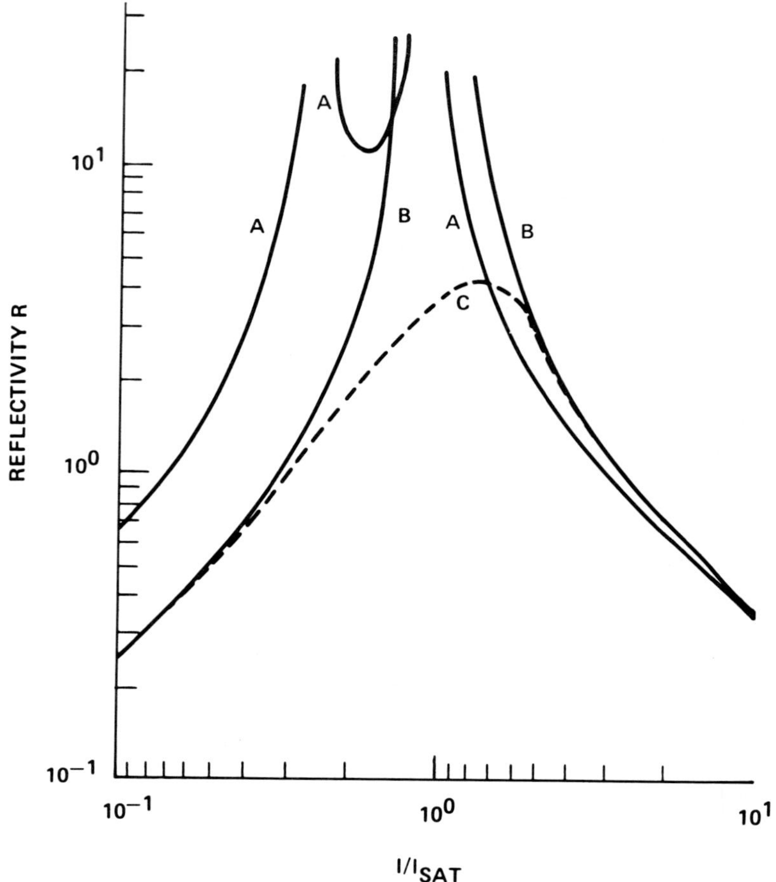

Fig. 10. Probe amplitude reflectivity R versus I/I_{sat} in the vicinity of a backward-wave resonance where $\beta L = 1$ and $\delta = 15$. Curve A is from Eq. (20); curve B, with pump absorption; curve C, with pump absorption and depletion ($I_p/I_{sat} = 0.01$). [After Brown (1981).]

effects of pump depletion are taken into account. In this case we must specify the magnitude of the probe intensity relative to that of the pump. As noted in the figure, the results shown apply to a case in which the probe irradiance is 1% that of the pump. It is obvious from these results that pump depletion effects are important in the vicinity of such resonances, because Eq. (20) predicts essentially unlimited power in the reflected field. In reality, the reflected power is finite, because the pump fields are strongly depleted under these conditions. Using these numerical calculations, we can examine the question of the overall efficiency of phase conjugation (including pump energy) when the reflectivity is large. The variation of the reflected (conjugate) wave as a function of the magnitude of the pump and probe fields is given in Fig. 11, again for $\beta L = 1$, $\delta = 15$, and $I_f = I_b = I_p$. The results indicate that roughly 50% of the probe energy is reflected under optimum conditions. This implies that the maximum overall efficiency is approximately 17%. No attempt was made to investigate the effect of distributing the input energy unequally among the pump and probe waves, and thus this result must be considered as indicative only of the attainable efficiency.

The analytical model has been compared with other experiments, both cw and pulsed, on materials that are not strictly two-level saturable absorbers. Modifications of the two-level theory to include a four-level saturable absorber have been made. With this modified form of the model, excellent quantitative agreement has been achieved with 1-nsec experiments performed in SF_6 at 10.6 μm (Dunning and Lam, 1981; Steel *et al.* 1981; Lind *et al.*, 1979). These results are shown in Fig. 12. The solid curve is a four-level model and the squares are the experimental results. A

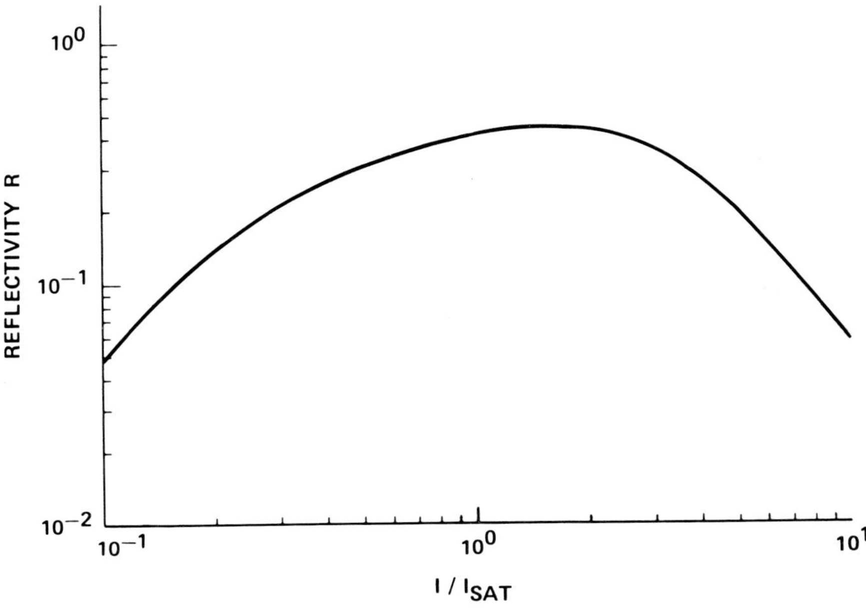

Fig. 11. DFWM reflectivity R versus I/I_{sat} for equal pump and probe amplitudes: $I = I_f = I_b = I_p$, where $\delta = 15$ and $\beta L = 1$. [After Brown (1981).]

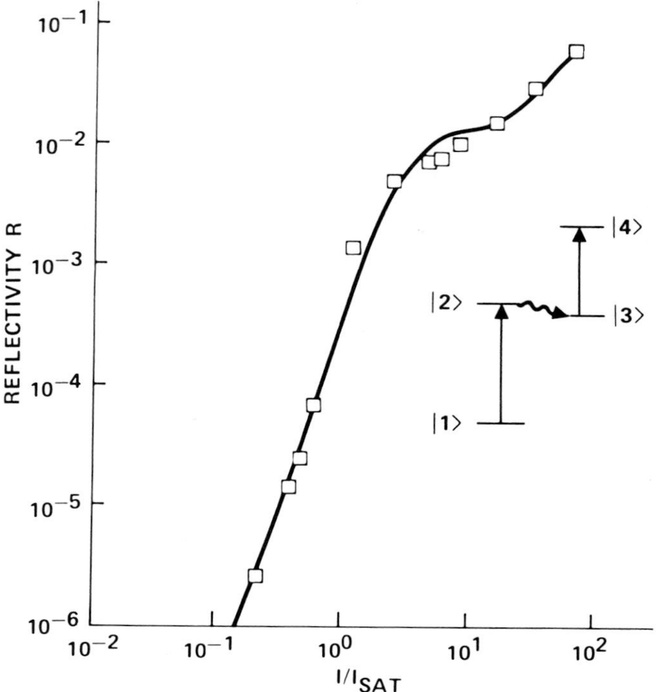

Fig. 12. Reflectivity R versus I/I_{sat} for SF_6. Measurements were made with $\tau_p \sim 1$ nsec and $P = 10$ Torr in a 2-cm cell. The solid line is a four-level model. Conditions: $\delta_{12} = \delta_{34} \sim 0$; $I_{34}/I_{12} = 55$; $\alpha_{12}L = 2.9$; and $\alpha_{34}L = 2.1$. [After Steel *et al.* (1981).]

best fit gave the values indicated for the relevant absorption coefficients and saturation intensities, which are in agreement with known parameters in SF_6.

Experiments performed in inverted media such as CO_2 (Fisher and Feldman, 1979) at 10.6 μm and Nd:YAG (Tomita, 1979) at 1.06 μm are consistent with the theory for appropriate values of the parameters $\alpha_0 L$ and I_{sat} ($\delta = 0$ for these cases). Continuous-wave experiments performed at 5145 Å with ruby as the nonlinear medium have been compared with a modified form of the solution (Liao and Bloom, 1978). In this case a three-level extension of the model was made. The comparison of experiment and theory is shown in Fig. 13. The deviation at high pump intensity is attributed to self-focusing. Experiments performed at 1.06 μm with the saturable absorber BDN have been performed by Moses and Wu (1980). Phase-conjugate reflectivities as high as 600% were achieved. The authors

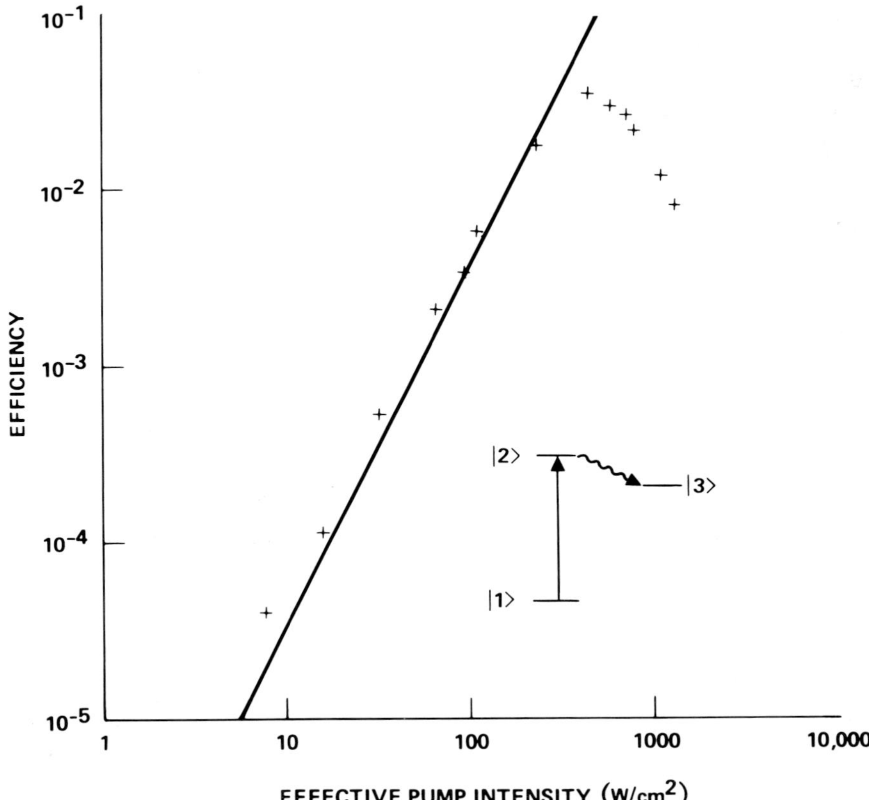

Fig. 13. Conversion efficiency versus effective power density for the 514.5-nm line. The solid line is a theoretical calculation from Eq. (2) of Liao and Bloom (1978).

attributed the high returns to excited-state contributions, and they modeled the system by a three-level formalism. Organic dye saturable absorbers have been used by Tocho *et al.* (1980) to generate phase-conjugate returns. They modeled the dye as a two-level saturable absorber, confirming the expected dependence of reflectivity on pump power. They were also the first to compare the measured reflectivity for various concentrations of dye with that predicted by the analytical model (modified to include pump absorption), as shown in Fig. 14. The experiments were performed at 605 nm with 5-psec mode-locked pulses. However, their comparison with the saturable absorber model was apparently fortuitous, because in a subsequent experiment (Tocho *et al.*, 1981) they

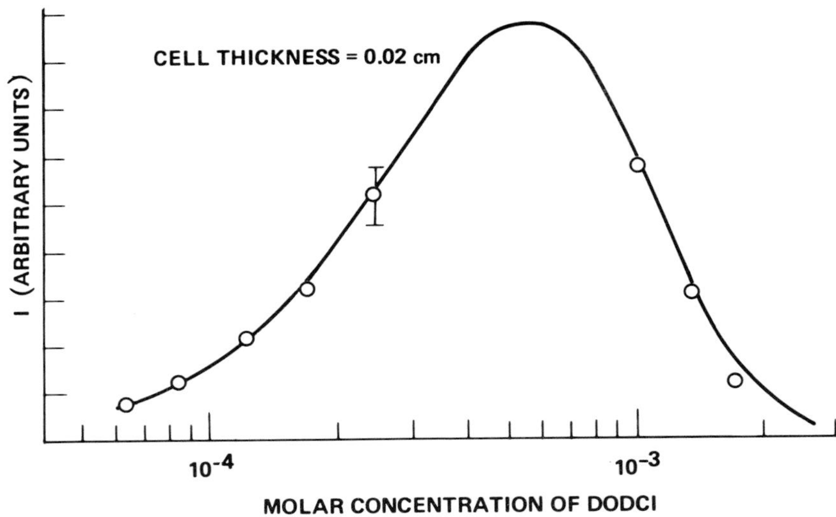

Fig. 14. Reflected intensity I versus DODCI molar concentration; cell thickness = 0.02 cm. [After Tocho *et al.* (1980).]

showed that phase conjugation in organic dyes arises from thermal gratings.

It is interesting that the model compares well with experiments with excitation times ranging from cw to short pulses over a wide wavelength range. The important common feature of all the experiments is that the medium is effectively homogeneously broadened and atomic motion is not important. For inhomogeneously broadened systems such as Na vapor, many new features appear that require the additional analysis given below. Numerous other experiments have been performed, and more complicated multiresonant theories developed; discussions of these can be found in Anan'ev *et al.* (1979), Apanasevich (1976), Carquille and Froehly (1980), Depatie *et al.* (1979), Elci and Rogovin (1979, 1980), Fan *et al.* (1980), Ivakin *et al.* (1975, 1979), Lazaruk (1979), Oraevskii (1979), Ostrovskii *et al.* (1975), Shtyrkov (1970), Steel and Lam (1980), Stepanov *et al.* (1971), Tan-No *et al.* (1980), Woerdman and Schurmans (1980), and Wu *et al.* (1980a,b).

2. *The Effect of Atomic Motion*

In the preceding section, we considered the regime in which the atoms are essentially stationary; i.e., the center-of-mass velocity $\mathbf{v} = 0$. However, degenerate four-wave mixing in gaseous systems has additional complications associated with random translational motion of the atomic or molecular species (Wandzura, 1979; and Bloch, 1980). In the inhomogeneously broadened regime, the natural linewidth γ_{12} is much smaller than the Doppler width ku_0, where u_0 is the thermal velocity. Hence only a certain number of atoms whose velocities are in the vicinity of a specific velocity group can interact simultaneously with the two counterpropagating pump waves, the input probe wave, and the generated conjugate wave. For collinear geometry (all wave vectors lie along a line), only those atoms whose velocities are centered on $\mathbf{v} = 0$ are effective in generating a conjugate signal. Furthermore, the effect of motion leads to a further reduction of the generated signal owing to the washout of the spatial grating. This effect arises from the competition between the mean lifetime $\Gamma_0^{-1} = T_1$ of the grating and the time τ it takes an atom to move from maximum to maximum in the periodic structure (see Fig. 2). The time τ is determined by the average speed u_0 of the atom and the grating period d: $\tau = d/u_0$, where $u_0 = (2k_BT/m)^{1/2}$, with k_B the Boltzmann constant, T the equilibrium temperature, and m the mass of the atom. For collinear geometry, the grating formed by the forward pump and the input probe waves has infinite periodicity. Hence, during the lifetime T_1 of the grating, the atoms do not have sufficient time to randomize the periodic structure. However, the spatial grating produced by the backward pump and input probe waves has a periodicity of $\lambda/2$ (λ is the optical wavelength). Here, under certain circumstances (low gas pressure) the atomic motion will lead to a complete washout within the lifetime T_1 of this grating, and one expects that the conjugate signal generated by the backward pump-probe contribution will be negligible compared with that generated by forward pump-probe contribution.

In the homogeneously broadened regime, the natural linewidth γ_{12} is much greater than the Doppler width ku_0, and all atoms participate effectively in generating the phase-conjugate signal. However, the problem of grating washout due to atomic motion still remains, and leads to the signal reductions just discussed for the case of inhomogeneously broadened media.

The starting point for a quantitative analysis is the full density matrix equation (2a) with initial conditions. The distribution of atomic velocities is described by the Maxwellian distribution function

$$W(\mathbf{v}) = (1/(\pi u_0^2)^{3/2} e^{-(v/u_0)^2}, \tag{23}$$

which represents thermal equilibrium. The response of the medium is obtained via a perturbation scheme of the form

$$\begin{Bmatrix} \rho_{11}^{(0)} \\ \rho_{22}^{(0)} \end{Bmatrix} \rightarrow \begin{Bmatrix} \rho_{12}^{(1)} \\ \rho_{21}^{(1)} \end{Bmatrix} \rightarrow \begin{Bmatrix} \rho_{22}^{(2)} \\ \rho_{11}^{(2)} \end{Bmatrix} \rightarrow \begin{Bmatrix} \rho_{12}^{(3)} \\ \rho_{21}^{(3)} \end{Bmatrix}, \tag{24}$$

where the superscript describes the order of the perturbation theory. The perturbation parameter is given by the Rabi flopping frequency $\boldsymbol{\mu}_{12} \cdot \vec{\mathscr{E}}/\hbar$, where $\boldsymbol{\mu}_{12}$ is the matrix element of the electric dipole moment operator, and $\vec{\mathscr{E}}$ the slowly varying envelope of the radiation field. Recall that in this section the superscript is such that each field envelope (regardless of whether it is the pump or probe) contributes to an order of the perturbation. This notation differs that of Section II A 1, where density matrix $\rho_{ij}^{(0)}$ referred to strong pump effects and $\rho_{ij}^{(1)}$ to the weak probe effects. The perturbation sequence (24) has a simple physical interpretation when read from left to right. Assume that initial populations $\rho_{22}^{(0)}$ and $\rho_{11}^{(0)}$ exist in the upper and lower states, respectively. The action of the radiation field, to first order in the applied field, generates a coherent superposition of upper and lower states ($\rho_{12}^{(1)}$ and $\rho_{21}^{(1)}$). To second order in the field, changes in the populations are produced ($\rho_{11}^{(2)}$ and $\rho_{22}^{(2)}$) that have a spatial dependence. The spatial modulation in the population leads to both absorbing and dispersive gratings. The uniform changes in the population due to the counterpropagating pump waves lead to a modified absorption and dispersion coefficient. To third order in the fields, an atomic coherence is generated between the upper and lower states, resulting in a coherent scattering of either of the counterpropagating pump waves. This scattering generates a phase-conjugate signal. A simple calculation yields the following expression for the polarization density giving rise to DFWM:

$$\begin{aligned}\mathbf{P}(\mathbf{r}, t) = \frac{\Delta N_0 \boldsymbol{\mu}_{21}}{(2i\hbar)^3} &(r_{\mathrm{fp}} \boldsymbol{\mu}_{12} \cdot \vec{\mathscr{E}}_{\mathrm{b}} \boldsymbol{\mu}_{12} \cdot \vec{\mathscr{E}}_{\mathrm{p}}^{*} \boldsymbol{\mu}_{21} \cdot \vec{\mathscr{E}}_{\mathrm{f}} \\ &+ r_{\mathrm{bp}} \boldsymbol{\mu}_{12} \cdot \vec{\mathscr{E}}_{\mathrm{f}} \boldsymbol{\mu}_{12} \cdot \vec{\mathscr{E}}_{\mathrm{p}}^{*} \boldsymbol{\mu}_{21} \cdot \vec{\mathscr{E}}_{\mathrm{b}}) \\ &\times e^{i\omega t - i(\mathbf{k}_{\mathrm{f}} + \mathbf{k}_{\mathrm{b}} - \mathbf{k}_{\mathrm{p}})\cdot\mathbf{r}},\end{aligned} \tag{25}$$

where

$$\begin{aligned} r_{\mathrm{fp}} = \int_{-\infty}^{\infty} d\mathbf{v}\, W(\mathbf{v}) &\frac{1}{\gamma_{12} + i(\Delta + \mathbf{k}_{\mathrm{p}} \cdot \mathbf{v})} \\ &\times \left(\frac{1}{\gamma_{12} + i(\Delta - \mathbf{k}_{\mathrm{f}} \cdot \mathbf{v})} + \frac{1}{\gamma_{12} - i(\Delta - \mathbf{k}_{\mathrm{p}} \cdot \mathbf{v})} \right) \\ &\times \left(\frac{1}{\gamma_1 + i(\mathbf{k}_{\mathrm{p}} - \mathbf{k}_{\mathrm{f}}) \cdot \mathbf{v}} + \frac{1}{\gamma_2 + i(\mathbf{k}_{\mathrm{p}} - \mathbf{k}_{\mathrm{f}}) \cdot \mathbf{v}} \right), \end{aligned} \tag{26}$$

$$\begin{aligned} r_{\mathrm{bp}} = \int_{-\infty}^{\infty} d\mathbf{v}\, W(\mathbf{v}) &\frac{1}{\gamma_{12} + i(\Delta + \mathbf{k}_{\mathrm{p}} \cdot \mathbf{v})} \\ &\times \left(\frac{1}{\gamma_{12} + i(\Delta - \mathbf{k}_{\mathrm{b}} \cdot \mathbf{v})} + \frac{1}{\gamma_{12} - i(\Delta - \mathbf{k}_{\mathrm{p}} \cdot \mathbf{v})} \right) \\ &\times \left(\frac{1}{\gamma_1 + i(\mathbf{k}_{\mathrm{p}} - \mathbf{k}_{\mathrm{b}}) \cdot \mathbf{v}} + \frac{1}{\gamma_2 + i(\mathbf{k}_{\mathrm{p}} - \mathbf{k}_{\mathrm{b}}) \cdot \mathbf{v}} \right). \end{aligned} \tag{27}$$

The dynamics of the formation of the spatial grating by means of the interference between the forward pump field $\mathscr{E}_f$ and input probe field $\mathscr{E}_p$, and the subsequent coherent scattering of the backward pump field $\mathscr{E}_b$ to generate a phase-conjugate signal $\mathscr{E}_c$, is described by the quantity r_{fp}. In the same manner, the dynamics of the formation of the spatial grating by means of the interference between the backward pump field $\mathscr{E}_b$ and the input probe field $\mathscr{E}_p$, and the subsequent coherent scattering of the forward pump field $\mathscr{E}_f$ to generate a phase-conjugate signal $\mathscr{E}_c$, is described by r_{bp}. The quantities r_{fp} and r_{bp} describe the behavior of the reflectivity for the various regimes of interest. For simplicity, we shall assume that the energy decay rates γ_1 and γ_2 are equal: $\gamma_1 = \gamma_2$.

a. The Homogeneously Broadened Regime. This regime is the simplest to treat, and is defined by the condition $\gamma_{12} \gg ku_0$, where γ_{12} is the natural linewidth and ku_0 the Doppler width. In this regime all atoms interact with the incoming and generated radiation fields, producing a resonant response identical to that of the atomic system. The Doppler shift terms in the Lorentzian denominators can therefore be neglected, and the velocity integration is carried out only for the term containing the energy relaxation rate γ_1:

$$r_{fp} = \frac{2}{\gamma_{12} + i\Delta} \frac{2\gamma_{12}}{\gamma_{12}^2 + \Delta^2} \int_{-\infty}^{\infty} d\mathbf{v}\, W(\mathbf{v}) \frac{1}{\gamma_1 + i(\mathbf{k}_p - \mathbf{k}_f) \cdot \mathbf{v}}, \tag{28a}$$

$$r_{bp} = \frac{2}{\gamma_{12} + i\Delta} \frac{2\gamma_{12}}{\gamma_{12}^2 + \Delta^2} \int_{-\infty}^{\infty} d\mathbf{v}\, W(\mathbf{v}) \frac{1}{\gamma_1 + i(\mathbf{k}_p - \mathbf{k}_b) \cdot \mathbf{v}}. \tag{28b}$$

Let θ be the angle between the forward pump and input probe propagation directions. The integration over velocities can be easily performed, to yield

$$r_{fp} = \frac{1}{\gamma_{12} + i\Delta} \frac{4\gamma_{12}}{\gamma_{12}^2 + \Delta^2} \frac{1}{i\gamma_1} S_+ Z(iS_+), \tag{29a}$$

$$r_{bp} = \frac{1}{\gamma_{12} + i\Delta} \frac{4\gamma_{12}}{\gamma_{12}^2 + \Delta^2} \frac{1}{i\gamma_1} S_- Z(iS_-), \tag{29b}$$

where $S_+ = \gamma_1/2ku_0 \sin(\theta/2)$ and $S_- = \gamma_2/2ku_0 \cos(\theta/2)$; $Z(iS_\pm)$ is the plasma dispersion function of argument $iS_\pm$ (Fried and Conte, 1961). These results are identical to those of Wandzura (1979), who described the behavior of DFWM processes in the presence of atomic motion. To study the degree of thermal washout, Wandzura plotted the parameter

$$m(\Gamma_1, \theta) = (1/2i)[S_+ Z(iS_+) + S_- Z(iS_-)] \tag{30}$$

as a function of the angle θ as shown in Fig. 15, where $\Gamma_1 = \gamma_1/ku_0$. For large values of Γ_1, the effect of atomic motion plays essentially no role in

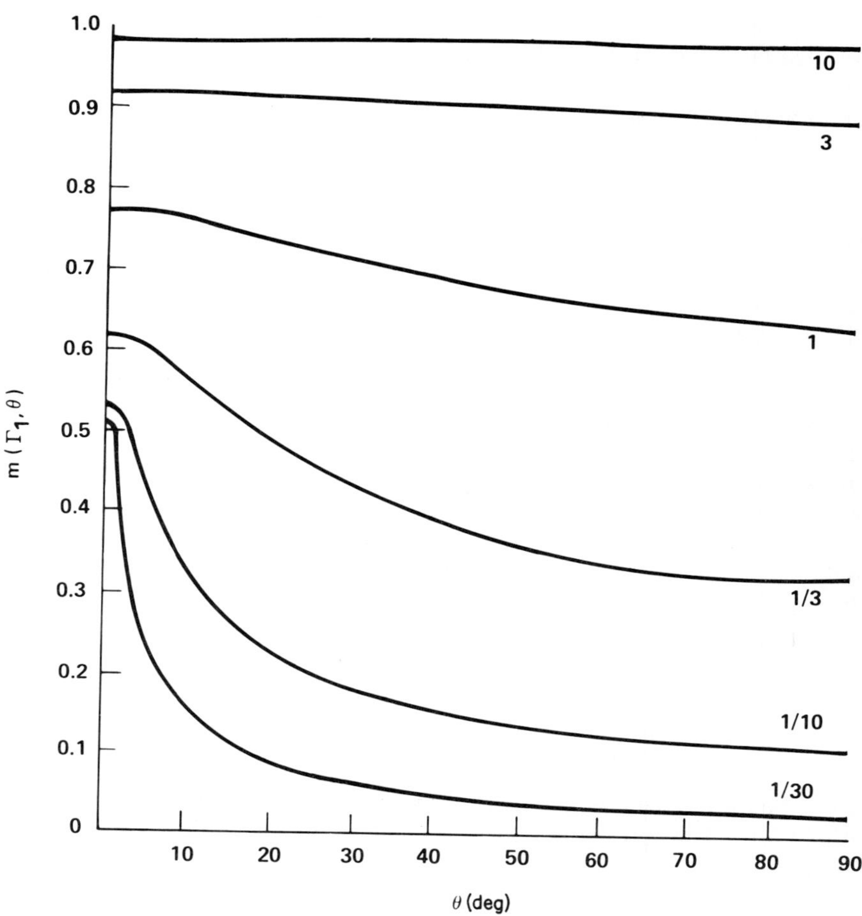

Fig. 15. Amplitude reduction factor $m(\Gamma_1, \theta)$ versus angle θ for a homogeneously broadened system. Values for Γ_1 given for each curve. [After Wandzura (1979).]

the determination of the response of the medium; i.e., the grating lifetime T_1 is made shorter than the time it takes the atomic motion to cause any washout. For very small angles θ and $\Gamma_1 \ll 1$, only the grating formed by the interference of the forward pump and input probe contributes significantly to the signal. This grating has a long wavelength, and atomic motion does not have time to lead to a complete washout. The calculation of the reflectivity under the conditions for Eq. (30) would proceed by evaluating Eq. (20) and multiplying the result by $m(\Gamma_1, \theta)$.

To examine the effects of motion in a homogeneously broadened system, experiments were performed at 10.6 μm in 10 Torr of SF_6 (Steel *et al.*, 1979). To interpret such measurements it is necessary to isolate the signal produced by the forward pump and probe from the signal produced by the backward pump and probe. This is accomplished by using polarization rotation. For an understanding of this concept, see Eq. (42b), to be discussed in detail in Section II A 4. In particular, using the cross-population coupling, we observe that if $\vec{\mathscr{E}}_f$ and $\vec{\mathscr{E}}_p$ are copolarized and $\vec{\mathscr{E}}_b$ is 90° cross polarized, the conjugate signal $\vec{\mathscr{E}}_c$ is polarized parallel to $\vec{\mathscr{E}}_b$ rather than $\vec{\mathscr{E}}_p$ and arises purely from the grating formed by $\vec{\mathscr{E}}_f$ and $\vec{\mathscr{E}}_p$. In this case the signal is proportional to $|r_{fp}|^2$. Likewise if $\vec{\mathscr{E}}_b$ and $\vec{\mathscr{E}}_p$ are copolarized and $\vec{\mathscr{E}}_f$ is 90° cross polarized, the signal is now polarized parallel to $\vec{\mathscr{E}}_f$ and arises purely from the grating formed by $\vec{\mathscr{E}}_b$ and $\vec{\mathscr{E}}_f$, and is proportional to $|r_{bp}|^2$. There are no contributions from Eqs. (42a) (normal population) and (42c) (Zeeman coherence) because of dipole selection rules. This provides a means to measure the ratio $|r_{bp}/r_{fp}|^2$. By cross-polarizing $\vec{\mathscr{E}}_f$ and $\vec{\mathscr{E}}_b$ we can study either the r_{fp} or the r_{bp} term by choosing $\vec{\mathscr{E}}_p$ parallel to either $\vec{\mathscr{E}}_f$ or $\vec{\mathscr{E}}_b$. Using this technique, we measured this ratio for various pulse widths. The important parameter is Γ_1, the ratio of the characteristic motion time for the system, T_0, to a characteristic longitudinal relaxation time T_1 or pulse length T_p, whichever is shorter. For the experiments in SF_6, where $T_0 = \lambda/u_0$, for pulse lengths of 180, 140, and 1 nsec, Γ_1 is 0.92, 1.17, and 140, respectively. The experimental results for the ratio $(r_{bp}/r_{fp})^2$ obtained for these values of Γ_1 are given in Fig. 16. As expected, for $\Gamma_1 < 1$ the signal from the large-period grating ($\vec{\mathscr{E}}_f \| \vec{\mathscr{E}}_p \perp \vec{\mathscr{E}}_b$) was significantly stronger than that from the small-period grating ($\vec{\mathscr{E}}_b \| \vec{\mathscr{E}}_p \perp \vec{\mathscr{E}}_f$). For $\Gamma_1 > 1$ the contributions from each grating were found to be equal as expected, indicating no motion effects. The solid curve gives the theory of Wandzura (1979), Eq. (30), with $\theta \sim 2°$ and $u_0 \sim 1.4 \times 10^4$ cm/sec.

b. Inhomogeneously Broadened Regime. This regime is defined by the condition $\gamma_{12} \ll ku_0$. In this case one must keep all Doppler-shifted terms. In the homogeneously broadened limit discussed above, all atoms interacted with the radiation field, producing a frequency response given by Eq. (11). In contrast, owing to the velocity-dependent resonant condition $\omega - \omega_0 + \mathbf{k}_n \cdot \mathbf{v} = 0$, the Doppler regime is characterized by individual velocity groups interacting with the incoming and generated radiation fields. The $\mathbf{k}_n \cdot \mathbf{v}$ term in the resonance condition results in four separate velocity groups corresponding to each of the four waves in Fig. 1; these must all be considered in the general solution to the noncollinear problem. Hence, because of the four resonance conditions, the spectral and angular response is not so simply described as in the homogeneously

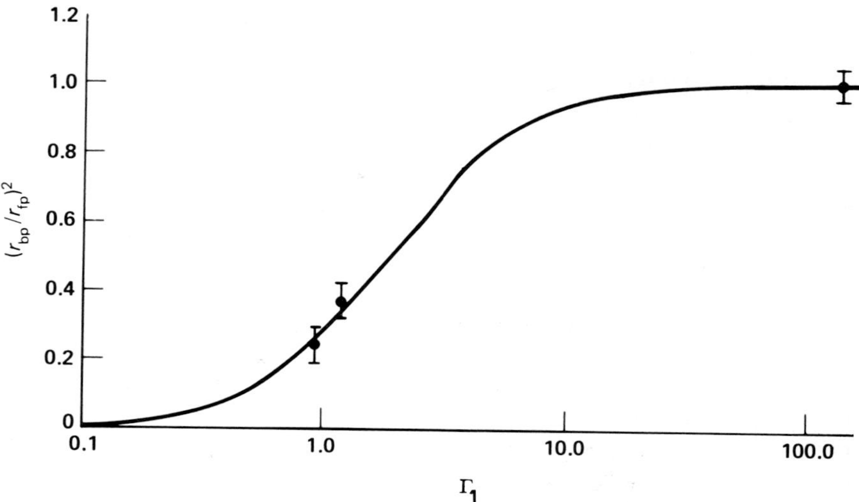

Fig. 16. Comparison of the density matrix calculation with experiment, where θ is 2°. For $\Gamma_1 > 1$, motion does not affect the polarization term. For $\Gamma_1 < 1$, the signal from the small-period grating is weaker than that from the large-period grating, as expected. [After Steel *et al.* (1979).]

broadened limit. This general problem has been examined in the limit of perturbation theory by Bloch (1980) and Nilsen and Yariv (1981). For collinear geometry, exact analytical solutions can be found for r_{fp} and r_{bp} of Eqs. (26) and (27). In this geometry, only the $\mathbf{v} = 0$ velocity group leads to an effective interaction process. This is similar to the case of saturated absorption spectroscopy. The expressions for r_{fp} and r_{bp} as given by Eqs. (26) and (27) are evaluated to yield

$$r_{fp} = \frac{2}{i\gamma_1 ku_0}\left\{\frac{1}{\gamma_{12}+i\Delta} Z\left(\frac{i\gamma_{12}-\Delta}{ku_0}\right) - \frac{1}{2i\Delta}\left[Z\left(\frac{i\gamma_{12}-\Delta}{ku_0}\right) - Z\left(\frac{i\gamma_{12}+\Delta}{ku_0}\right)\right]\right\}, \tag{31a}$$

$$\begin{aligned} r_{bp} = \frac{1}{iku_0}\Bigg\{&\frac{1}{\frac{1}{2}\gamma_1-\gamma_{12}-i\Delta}\,\frac{Z'((i\gamma_{12}-\Delta)/ku_0)}{iku_0} \\ &-\left(\frac{1}{(\frac{1}{2}\gamma_1-\gamma_{12}-i\Delta)^2} + \frac{1}{2i\Delta(\frac{1}{2}\gamma_1-\gamma_{12}-i\Delta)}\right) \\ &\times\left[Z\left(\frac{i\gamma_{12}-\Delta}{ku_0}\right) - Z\left(\frac{i\gamma_1}{2ku_0}\right)\right] \\ &+\frac{1}{2i\Delta(\frac{1}{2}\gamma_1-\gamma_{12}+i\Delta)}\left[Z\left(\frac{i\gamma_{12}+\Delta}{ku_0}\right) - Z\left(\frac{i\gamma_1}{2ku_0}\right)\right]\Bigg\}, \end{aligned} \tag{31b}$$

where $Z'(x + iy)$ is the derivative with respect to the complex argument $x + iy$ of the plasma dispersion function. Equations (31a) and (31b) hold for any value of γ_{12}/ku_0, γ_1/ku_0, and Δ/ku_0. Apparently complicated, these equations can be reduced to a simpler form in the case of the extreme Doppler limit, i.e., $\gamma_{12} \ll ku_0$, $\gamma_1 \ll ku_0$, and $\Delta \ll ku_0$. In this regime one obtains, by suitable expansion of the plasma dispersion function,

$$r_{\mathrm{fp}} \cong \frac{2\sqrt{\pi}}{ku_0}\frac{1}{\gamma_1}\frac{\gamma_{12} - i\Delta}{\gamma_{12}^2 + \Delta^2}, \tag{32a}$$

$$r_{\mathrm{bp}} \cong 2\sqrt{\pi}/(ku_0)^3. \tag{32b}$$

Note that to lowest order r_{bp} is a frequency-independent term. The frequency spectrum of r_{fp} resulting from interaction with only the $\mathbf{v} = 0$ velocity group has a Lorentzian profile centered at the resonance frequency ω_0 and with a linewidth γ_{12}. The ratio of the r_{bp} and r_{fp} contributions at line center is given by $\gamma_1\gamma_{12}/(ku_0)^2$, which is typically much less than unity, thus indicating that the grating described by r_{bp} is washed out. The frequency dependence is therefore given by r_{fp} [Eq. (32a)] and is Doppler-free. An immediate application of such behavior is in high-resolution spectroscopy[1] as indicated throughout the text. The technique is similar to saturated absorption spectroscopy, but with a distinct improvement in signal-to-noise ratio owing to the angular separation between the weak signal and strong pumps.

For arbitrarily increasing pump intensities, the DFWM reflectivity is expected to decrease (for constant $\alpha_0 L$ and δ) owing to saturation of the inhomogeneously Doppler-broadened system. Indeed, the analytical description for a homogeneously broadened system given in Section II.A.1 showed that the response did saturate as expected for large pumps. In a homogeneously broadened system, saturation of the traveling-wave response varies as $(1 + I/I_{\mathrm{sat}})^{-1}$, whereas in an inhomogeneously broadened system the response varies as $(1 + I/I_{\mathrm{sat}})^{-1/2}$. Hence, one would intuitively expect the DFWM response to saturate more slowly in an inhomogeneously broadened medium than in a homogeneously broadened medium. However, experiments in inhomogeneous gases show a saturation of the reflectivity consistent with the simple homogeneous theory, although the measured reflectivity is reduced to a value several orders of magnitude below the calculated value. This behavior can be intuitively understood by considering the line-center solution to the homogeneous problem in the two limits $I/I_{\mathrm{sat}} \ll 1$ and $I/I_{\mathrm{sat}} \gg 1$. In this case the reflectivity given by Eq. (20) (for homogeneous broadening without motion)

[1] When the probe is slightly angularly displaced from the pump.

reduces to $R \propto \alpha_0^2$, where α_0 is the small signal field absorption coefficient. If the system were inhomogeneously broadened, only that fraction of atoms whose Doppler shift was less than the natural width would contribute to the interaction. That is, if $\gamma_1 = \gamma_2$, the reflectivity would be reduced by $(\gamma_1/ku_0)^2$. Therefore R_{inh} would be expected to have the form $R_{\text{inh}} \simeq (\gamma_1/ku_0)^2 R$, where $R \propto \alpha_0^2$, and α_0 is determined by the individual atomic absorption cross section. In most gas phase experiments, the large-angle grating is completely washed out as previously discussed. However, Eq. (20) implicitly assumes the coherent addition of two gratings. Hence, to correct for the missing term we divide r by 4. The final expression for R_{inh} is

$$R_{\text{inh}} \simeq [\gamma_1^2/4(ku_0)^2]R. \tag{33}$$

An experimental study of this behavior was made by using 220 mTorr of SF_6 in a 10-cm cell (Steel and Lam, 1981). The laser was a tunable cw CO_2 laser operated on the P16 line of the 10.4-μm branch. The laser was tuned to the strongest transition of SF_6 by using a piezoelectric translator on the output coupler. The line center was determined for $I/I_{\text{sat}} \ll 1$ to avoid confusion due to power broadening. The measured intensity dependence of the reflectivity is shown in Fig. 17. The results of Eq. (33) are also

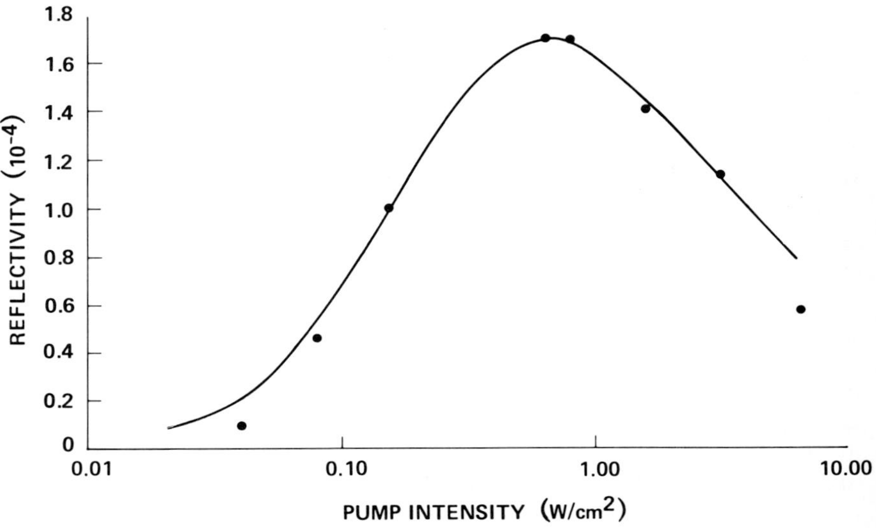

Fig. 17. Continuous-wave DFWM reflectivity versus pump intensity, measured in low-pressure inhomogeneously broadened SF_6. [After Steel and Lam (1981).]

shown, for a Doppler width of 30 MHz and a natural linewidth of 1 MHz. A correction factor of 0.85 was used to obtain agreement between theory and experiment at the peak reflectivity. Considering the simplistic nature of the model, the agreement is quite good.

c. Transient Grating Experiments: Diffusion Regime The possible washout of the gratings by atomic motion suggests a way of measuring rates of energy diffusion by using degenerate four-wave mixing processes. Pulsed laser excitation might be used to time-resolve the diffusion process. One of the earliest uses of the degenerate four-wave mixing configuration to measure diffusion processes was reported by Woerdman (1970), who measured the diffusion of free carriers excited in silicon by a *Q*-switched Nd:YAG laser. The experimental arrangement for these diffusion measurements was identical to that for a usual degenerate four-wave mixing experiment, with one exception. Only one pulsed pump beam is allowed to temporally overlap the pulsed object wave. Hence, only one interference grating is formed in the material. The second pump wave is brought into the sample at a later time, and produces the conjugate wave as it diffracts off the decaying grating of free carriers. If one assumes that the grating is washed out by a diffusion process that can be represented by a single diffusion constant, the intensity of the generated backward wave exhibits a time dependence given by

$$I(t) = I_0 \exp(-Kt), \tag{34a}$$

where

$$K = 2/\tau_0 + 2D[(4\pi/\lambda)\sin(\theta/2)]^2. \tag{34b}$$

The value τ_0 is the natural lifetime of the excitation of the carriers, λ the excitation laser wavelength in the nonlinear medium, θ the angle between the image wave and the first pump wave, which creates the grating, D the diffusion constant, and t the delay time between the two pump-wave pulses. By measuring the phase-conjugate wave intensity as a function of θ and t, it is possible to determine both τ_0 and D. For a nonresonant nonlinearity, the "readout" of the transient grating can be made with a laser beam of another wavelength (Eichler, 1978) providing one satisfies the Bragg condition, although the degenerate four-wave configuration has the advantage of simplicity and assures that all phase-matching conditions are satisfied. Diffusion measurement by transient gratings was recently performed by Salcedo *et al.* (1979), who studied the dynamics of energy transfer in molecular crystals. In Fig. 18 are shown the results of a series of measurements of the transient grating decay constant K versus angle θ for measurement of energy diffusion in single crystals of pentacene-doped

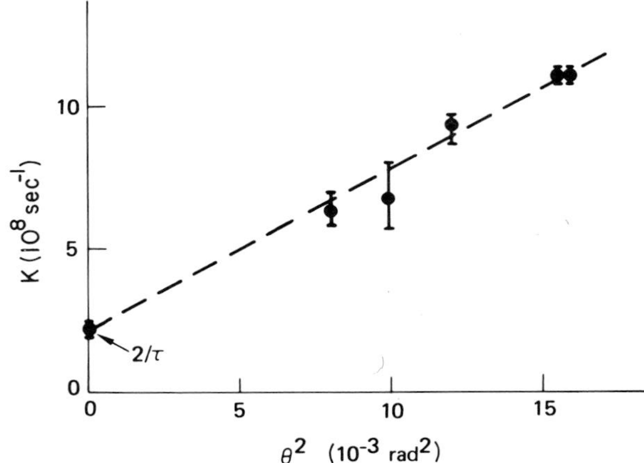

Fig. 18. Transient grating decay constant K versus θ^2. From the slope, the diffusion coefficient is $D = 2$ cm^2/sec. The predicted intercept $2/\tau$ is observed (τ is the independently measured lifetime). [After Salcedo *et al.* (1979).]

p-terphenyl (10^{-3} molar). Both τ_0 and the diffusion constant can be easily extracted from the data.

Note that this four-wave technique is a very useful method for measuring fluorescence lifetimes of materials exhibiting strong reabsorption, as long as the resonance reabsorption length is large compared with the grating period. Ordinary methods for measuring fluorescence lifetimes are subject to substantial errors owing to entrapment of the light within the sample.

3. *Nearly Degenerate Four-Wave Mixing*

The previous section was devoted to cases of resonant DFWM where all frequencies were equal. There are practical situations, however, where the input probe may be a wave oscillating at a slightly different frequency from the pump waves (or from a different laser). This occurs when, for example, Doppler shifts arise from the translational motion of a receiver or transmitter system, or the gain bandwidth of an external resonator differs from that of the phase-conjugate mirror, or the laser system drifts in frequency during operation. In addition, such a process has been proposed for use as a narrow-band tunable filter (Nilsen *et al.*, 1981). These practical considerations compel us to discuss the frequency response of a phase conjugator for the case where the frequency of the input probe differs from that of the counterpropagating pump waves.

The interaction process for nearly degenerate operation is shown in

Fig. 19. $\omega = \omega_0 + \Delta$, $\Omega = \omega + \delta$, and $2\omega - \Omega = \omega - \delta$ are the frequencies of the counterpropagating pump waves, input probe, and phase-conjugate signal, respectively where ω_0 is the atomic resonance frequency. Let us consider the interaction of the pump and probe waves with the two-level system. The interference of the forward pump wave $\mathscr{E}_f$ and input probe wave $\mathscr{E}_p$ no longer generates a spatial modulation in the population difference; rather, it produces a traveling-wave excitation in the population with frequency $\omega - \Omega$ and phase velocity $(\omega - \Omega)/|\mathbf{k}_p - \mathbf{k}_f|$. The direction of propagation of the traveling wave is given by $\mathbf{k}_p - \mathbf{k}_f$. The scattering of the backward pump wave $\mathscr{E}_b$ from this traveling-wave excitation yields a phase-conjugate signal. In a similar manner, the backward pump wave $\mathscr{E}_b$ and input probe wave $\mathscr{E}_p$ generate a traveling-wave excitation with frequency $\omega - \Omega$ and phase velocity $(\omega - \Omega)/|\mathbf{k}_p - \mathbf{k}_b|$. The direction of the traveling-wave excitation is given by $\mathbf{k}_p - \mathbf{k}_b$. The scattering of the forward pump wave $\mathscr{E}_f$ generates a phase-conjugate signal. The efficiency of the signal generation is determined in part by the amount of phase mismatch that arises in the four-wave mixing process; i.e., if a counterpropagating geometry is maintained, there is a **k**-vector mismatch

$$\Delta\mathbf{k} = \mathbf{k}_f + \mathbf{k}_b - \mathbf{k}_p - \mathbf{k}_c \neq 0, \tag{35}$$

which occurs because the frequency of the probe wave (p) differs from that of the generated conjugate signal (c). One defines a coherence length L_c by

$$L_c = \pi/|\Delta\mathbf{k}|, \tag{36}$$

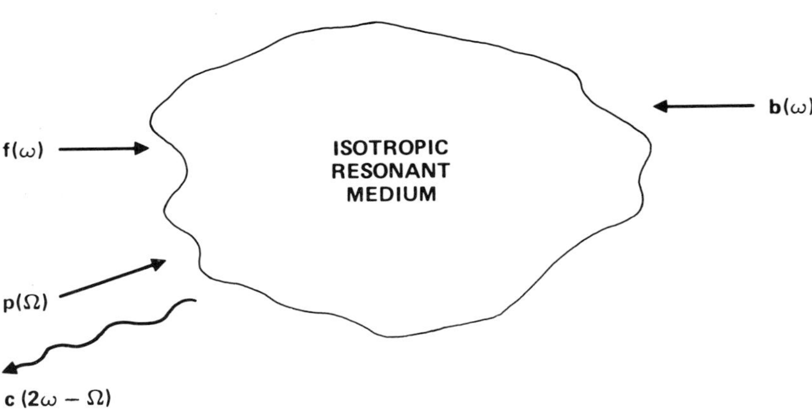

Fig. 19. Interaction geometry for nearly degenerate four-wave mixing.

the critical length of the interaction region L such that if L exceeds L_c, generation of a phase-conjugate signal becomes rather inefficient.

The frequency dependence of the phase-conjugate reflectivity (as a function of the difference between pump and probe frequencies) for the two-level homogeneously broadened system can be analyzed as follows. The phase-conjugate signal response in a stationary two-level system as a function of $\omega - \Omega$ is determined by the energy relaxation rate $1/T_1$. If $\omega = \Omega$, a standing-wave excitation is generated in the excited state of an atomic species. As Ω is tuned away from ω, the standing wave becomes a traveling wave. The time for the traveling wave to move one period is given by $(\omega - \Omega)^{-1}$, whereas the excited state has a finite lifetime given precisely by T_1. Hence, for $(\omega - \Omega)^{-1}$ much greater than T_1, the atom effectively returns to the ground state via spontaneous decay before the traveling wave moves one period, and the strength of the population modulation is maintained. However, for $(\omega - \Omega)^{-1}$ much less than T_1, the atom cannot relax fast enough. Hence the spatial modulation is washed out and drives the reflectivity to zero. Thus only for a pump–probe detuning of less than T_1^{-1} will the reflectivity be large.

The technique of nearly degenerate four-wave mixing has been used by Raj *et al.* (1980) to study the line shapes of atomic and molecular systems. They took advantage of the fact that in the collinear geometry two distinct signals, at frequencies $\omega + \Omega$ and $\omega - \Omega$, are generated inside the nonlinear medium if $E_f \sim E_p \sim E_b$. If the counterpropagating pump waves oscillate at frequency ω and the probe at frequency $\omega + \Omega$, then the interference of the forward pump and input probe yields two traveling-wave excitations oscillating at frequencies $\pm\Omega$. The scattering of the backward pump wave from the two excitations gives rise to two backward generated waves oscillating at frequencies $\omega \pm \Omega$. In the detector this results in a beat signal at 2Ω. Using phase-sensitive detection, Raj *et al.* were able to achieve shot-noise-limited detection, thus surpassing all other nonlinear spectroscopy techniques for sensitivity (Bloch *et al.*, 1981).

The Doppler-broadened problem has been solved only in the limit of low-intensity pumps by using third-order perturbation theory (Nilson and Yariv, 1979, 1981), and the frequency spectrum of the conjugate wave gives rise to two resonances. For a given pump detuning from resonance, Δ, it is possible to determine the resonance condition for the pump–probe detuning δ. The resonances are given by $\delta = 0$ ($\Omega = \omega_0 + \Delta$) and $\delta = 2\Delta$ ($\Omega = \omega_0 + 3\Delta$). Physically, the first resonance arises when the detuning between the two beams is less than the linewidth, which enables both the forward pump and the probe to interact simultaneously with the same group of atoms, producing a spatial modulation of the population from which the backward pump can scatter. The physical origin of the second resonance is somewhat more subtle; it arises from a combination of Doppler effects. The forward pump, at ω detuned from resonance by Δ, is Doppler shifted into resonance for a particular velocity group v. The resonance observed in tuning the probe to $\delta = 2\Delta$ occurs at a frequency when, in the rest frame of the atom, the atom sees the Doppler-shifted probe and backward pump at the same frequency.

The pump–probe detuning problem has also been examined in the presence of arbitrarily strong pump fields in homogeneously broadened Doppler-free media, by Fu and Sargent (1979) and Harter and Boyd 1980). A steady-state solution to the quantum-mechanical transport equation was found that is valid to all orders in the pump field and to first order in the probe and signal, and shows the multiresonant behavior illustrated for various pump intensities in Fig. 20 (Harter and Boyd, 1980). The qualitative solution to this problem may be anticipated by recalling the behavior of a two-level atom subject to a strong field at frequency ω tuned near resonance at ω_0. The population will undergo strong oscillation between the two levels at an angular frequency given by the generalized Rabi frequency $\omega_R' = [(\omega - \omega_0)^2 + \omega_R^2]^{1/2}$, where $\omega_R \equiv \mu\mathscr{E}/\hbar$, with μ the transition dipole moment and $\mathscr{E}$ the electric field envelope. This effect can be viewed as a splitting of each level into two levels separated by an energy

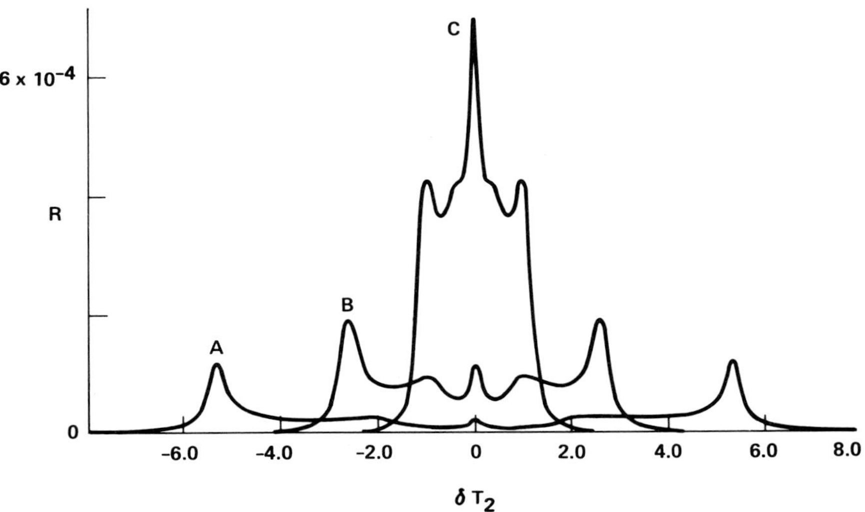

Fig. 20. Reflectivity versus pump–probe tuning for various pump intensities. Curve A: $(\omega - \omega_0)T_2 = 19$, $\omega_R'T_2 = 25$. Curve B: $(\omega - \omega_0)T_2 = 8$, $\omega_R'T_2 = 12.5$. Curve C: $(\omega - \omega_0)T_2 = 3$, $\omega_R'T_2 = 5$. [After Harter and Boyd (1980), © 1980 IEEE.]

$\hbar\omega'_R$, and is known as the ac Stark effect. Hence, one would expect such a system to be characterized by three resonances at frequencies $\delta = 0$ and $\delta = \pm\omega'_R$. However, in nearly degenerate FWM the presence of strong counterpropagating pumps slightly complicates this expectation. The simultaneous action of both pumps produces a standing-wave modulation of the net electric field. Hence, the Stark splitting of the levels is also spatially modulated, producing a macroscopic polarization with resonances anticipated by the presence of high- and low-intensity fields. Using perturbation theory in stationary media, one can show that, in the presence of low-intensity pumps, pump–probe detuning resonances occur at $\delta = 0$ and $\delta = \pm\Delta$. Hence a total of *five* resonances are expected. The observation of the low-intensity resonances in Doppler-broadened media might be expected only if the power broadening exceeds the pump detuning Δ, and because of velocity effects, quantitative comparison of existing theory and experiments in atomic sodium is not possible.

The experimentally measured reflectivity as a function of pump–probe

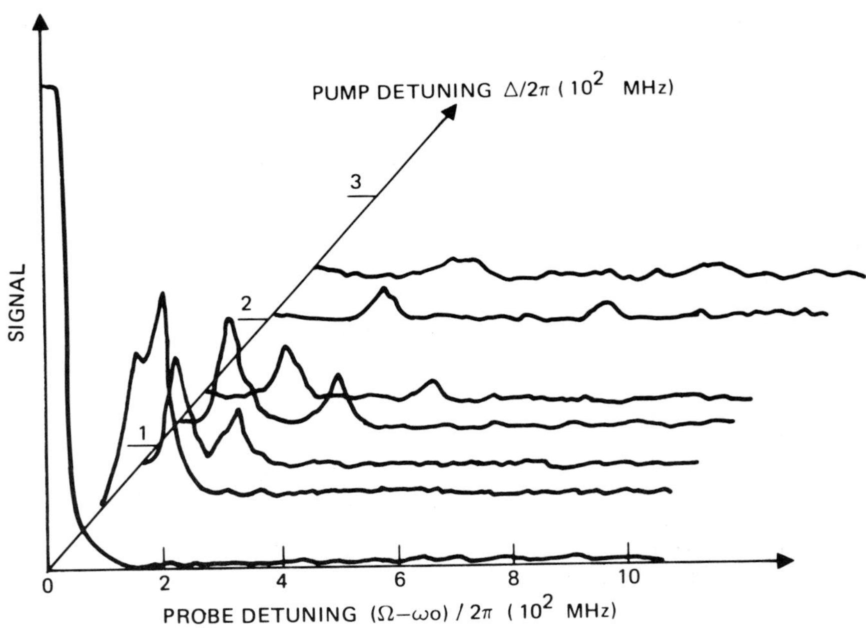

Fig. 21. Experimental measurement of reflectivity in nearly degenerate FWM for $I < I_{sat}$ in Doppler-broadened material. The reflectivity is plotted as a function of probe frequency for various pump frequencies. The atomic resonance is at ω_0. [After Steel and Lind (1981).]

detuning δ and pump detuning Δ is shown in Fig. 21 for $I \leq I_{\mathrm{sat}}$ (Steel and Lind, 1981) in inhomogeneously broadened atomic sodium. As anticipated, when $\Delta = 0$, there is a single strong resonance. However, as Δ is increased, two resonances form at $\delta = 0$ ($\Omega - \omega_0 = \Delta$) and $\delta = 2\Delta$ ($\Omega - \omega_0 = 3\Delta$). As indicated above, the bandwidth of the first resonance is determined by the longitudinal relaxation rate while the bandwidth of the second resonance is determined by the transverse relaxation rate (Lam, *et al.*, 1982). For a two-level problem with the lower state the ground state, that rate is given by the *A* coefficient. Hence, the bandwidth is twice the atomic linewidth. In Fig. 22 the measured bandwidth when $\Delta = 0$ is shown. The measured bandwidth is 25 MHz, in good agreement with the expected value of 20 MHz.

As expected at high intensities, the pump–probe detuning dependence is considerably more complicated. The interaction is further complicated in sodium by atomic motion and hyperfine structure. Nevertheless, by appropriate choices for the pump detuning Δ and pump intensity I, it appears possible to observe the multiresonant behavior qualitatively.

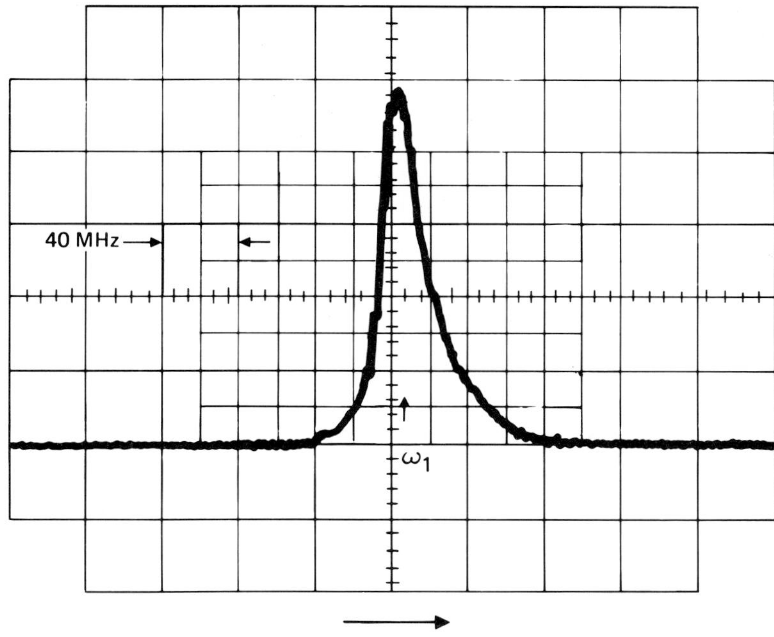

Fig. 22. Measurement of the nearly degenerate FWM bandwidth when $\Delta = 0$. The bandwidth is given (FWHM in M2) by $I/\pi T_1$. [After Steel and Lind (1981).]

Shown in Fig. 22 is the observed five-peak behavior. The pump intensity was 23 W/cm^2, corresponding to $\omega_R = 365$ MHz. The pump detuning was $\Delta = 210$ MHz, giving a generalized Rabi frequency of $\omega_R' = 421$ MHz. As shown in the figure, the location of the sidebands (structure A) agrees excellently with ω_R'. Structure B is the resonance at $\delta = \pm\Delta$ expected from the perturbation theory. The dependence of the measured ω_R' on the pump electric field is expected to be complicated by velocity effects in the presence of the standing-wave electric field. However, our measurements show that over a range of $0.1 < E < 0.45$ eV the measured ω_R' corresponds to the calculated value.

Not easily predicted or understood is the dip at $\delta = 0$ ($\Omega = \omega$). This structure has been observed to be very narrow (~30 MHz). Such an effect has been predicted to result from a constructive interference between the probe and signal at $\delta = 0$. This explanation is not easily applied to our data, because the microstructure is a dip for $\Delta < 0$ and a spike for $\Delta > 0$ for the 3s $^2S_{1/2}(F = 2) \rightarrow$ 3p $^2P_{3/2}(F = 3)$ transition. Further, for transitions out of the $F = 1$ ground state it is a dip for $\Delta > 0$ and a spike for $\Delta < 0$. Additional work is needed here.

The net bandwidth in Fig. 23 is given by $2\omega_R'$. However, as the absorption length product is increased, the bandwidth narrows significantly and

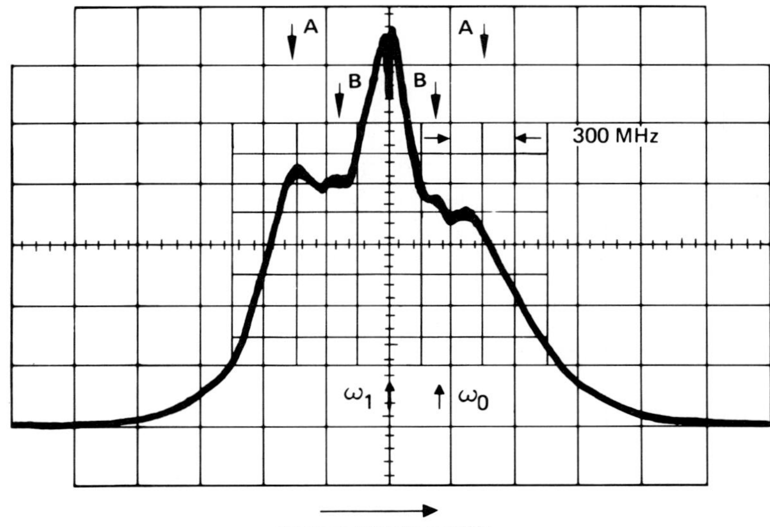

Fig. 23. Experimental measurement of multiresonant reflectivity in nearly degenerate FWM for $I \gg I_{sat}$. The central resonance occurs at $\delta = 0$. The two structures designated B occur at $\delta = \pm\Delta$. The two structures designated A are due to the ac Stark splitting of the atomic levels and occur at $\delta = \pm\omega_R'$. [After Steel and Lind (1981).]

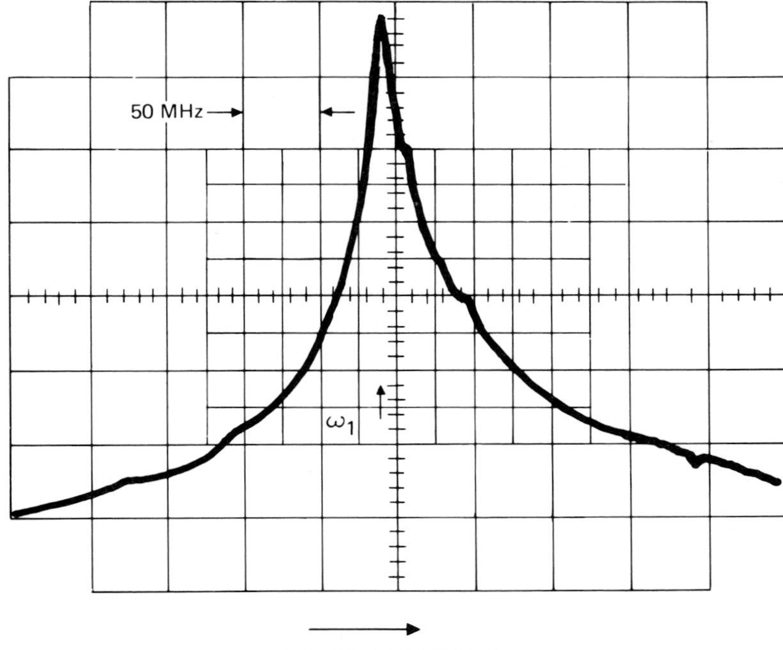

Fig. 24. Bandwidth of the pump–probe detuning signal under high-reflectivity conditions ($R \simeq 150\%$, $\alpha_0 L \simeq 30$). [After Steel and Lind (1981).]

secondary resonances are not observed. This is because the probe and signal are strongly absorbed as the detuning δ exceeds the frequency width of the hole burned by the strong saturating pumps. Furthermore, at high $\alpha_0 L$, the phase matching is no longer perfect, owing to a finite dispersion. Additional narrowing also results due to distributed feedback effects (Pepper and Abrams, 1978). However, under these conditions the reflectivity is over 150% for the 3s $^2S_{1/2}(F = 1) \rightarrow$ 3p $^2P_{3/2}(F = 0)$ transition. An example of such a bandwidth is shown in Fig. 24.

4. *Effects of Degeneracies*

So far, we have considered only a purely two-level model. However, in the absence of external forces an atom with a central potential is spherically symmetric. In such a system it can be shown that the minimum problem to be solved really has four levels owing to degenerate angular momentum states, with quantum members J and degeneracy $2J + 1$. The simplest case is that of an atom undergoing a transition between $J = 0$ and

$J = 1$ states, and involves solving a four-level problem or, equivalently, a two-level problem with degenerate states.

The purpose of this section is to elaborate on studies of DFWM associated with degenerate levels. In the typical laboratory situation, degenerate states are usually considered only by their contribution to some multiplicity. However, experiments and calculations have shown that many resonance properties described earlier are completely masked or complicated by the presence of degenerate states. The effects can be divided into two distinct types. The first is related to optical pumping: the redistribution of the ground-state population among the degenerate sublevels by the radiation field. These effects (including alignment and orientation) are polarization sensitive and affect the apparent saturation intensity and velocity distributions. The second involves the dependence of the phase-conjugate signals on the polarization states of the applied fields. In this case nonlinearities not expected in simple two level-theory are observed, and lead to unique spectroscopic information with potential practical applications.

We consider optical pumping first, and refer to extensive work in analyzing Lamb dip spectroscopy (Feld *et al.*, 1980; Pappas *et al.*, 1980). This is closely related to DFWM, because the basic nonlinearity in both interactions arises from population saturation, and both are characterized by velocity hole burning by the counterpropagating fields. The effect on saturation is easily seen by considering the level diagram shown in Fig. 25, in which levels 1 and 3 may be degenerate. We consider an electromagnetic field that can couple levels 1 and 2 but cannot interact with levels 2 and 3, either because it is nonresonant or because the polarization is not consistent with the change in the angular momentum of the transition. Note, however, that an atom excited from state $|1\rangle$ to $|2\rangle$ can decay to either $|1\rangle$ or $|3\rangle$. If the atom decays to level $|3\rangle$, it cannot interact with the field again until it relaxes to level $|1\rangle$ via some collision. If no collisions are present, the population of level $|1\rangle$ will be transferred to level $|3\rangle$, effectively reducing the saturation intensity necessary to drive the population difference between $|1\rangle$ and $|2\rangle$ to zero. We typically take I_{sat} to be $\hbar\omega/\sigma T_1$, where σ is the absorption cross section and T_1 the population decay time of the transition from $|1\rangle$ to $|2\rangle$. A heuristic derivation of the new I_{sat} in the presence of optical pumping can easily be made if we assume a transit time τ_{m} for motion across the field ($\tau_{\text{m}} \sim 2a/u_0$, where a is the beam radius and u_0 the thermal velocity). We consider the simple rate equation between levels $|1\rangle$ and $|3\rangle$ given by $dN_3/dt = \sigma_{12}IN_1/h\omega$, assuming $\gamma_3 \gg \gamma_1$, where γ_3 is the decay rate from level $|2\rangle$ to level $|3\rangle$. Recognizing that the atom can interact for a time τ_{m}, we find that significant population has been transferred to level $|3\rangle$ when $I = I'_{\text{sat}} =$

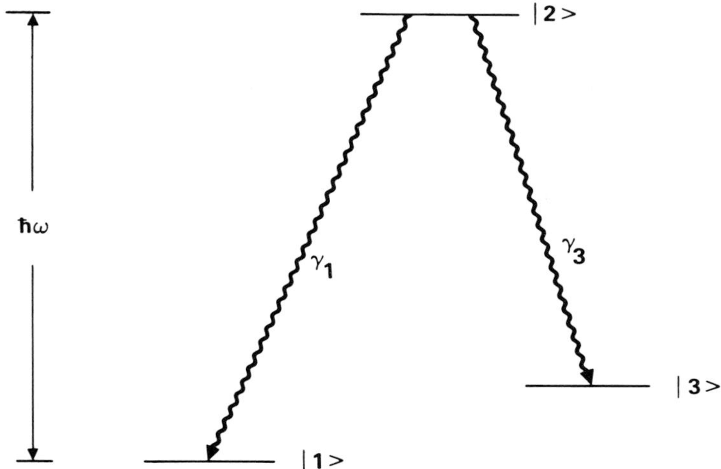

Fig. 25. Three-level system depicting the optical pumping scheme. A photon at $\hbar\omega$ excites state $|2\rangle$, which can decay to $|1\rangle$ or $|3\rangle$. Eventually all of level $|1\rangle$ is transferred to level $|3\rangle$. Levels $|1\rangle$ and $|3\rangle$ need not be degenerate to be affected by this kind of spatial pumping.

$\hbar\omega/\sigma\tau_{\mathrm{m}} = I_{\mathrm{sat}}T_1/\tau_{\mathrm{m}}$. A more precise derivation of the saturation parameter and its effect on line shape is given by Pappas *et al.* (1980). They show that

$$I'_{\mathrm{sat}} = \frac{\hbar\omega}{\sigma\tau}\frac{1 + \tau/\tau_{\mathrm{m}}}{2 + \gamma_3\tau_{\mathrm{m}}}, \tag{37}$$

where $\tau = (\gamma_1 + \gamma_3)^{-1}$ is the radiative lifetime of state $|2\rangle$. In the limits $\tau \ll \tau_{\mathrm{m}}$ and $\gamma_3\tau_{\mathrm{m}} \gg 1$, this expression reduces to the simple expression given above. Recalling now the power-broadened linewidth $\gamma'_{12} = \gamma_{12}(1 + I/I'_{\mathrm{sat}})^{1/2}$, we see that even for experiments conducted with $I \lesssim I_{\mathrm{sat}}$ significant power broadening may be observed in the line shape, because, typically, $I'_{\mathrm{sat}}/I_{\mathrm{sat}} \sim 10^{-2}$.

Optical pumping can also affect the velocity distribution, a result with impact on the spectral response of the reflectivity. In this case we consider a two-level system $J \rightarrow J + 1$, where the ground-state level J has at least three $2J + 1$ degenerate states and the upper level has degeneracy $2J + 1$. In the presence of right circularly polarized light, the entire population of the ground state, which initially was uniformly distributed over the $2J + 1$ levels, is now pumped into the magnetic substate $m = J$. Whereas, initially, the velocity distribution of the $m = J$ level was Maxwellian [see Eq. (23)], it now has an additional spike whose profile is

Lorentzian, centered at a velocity such that $\mathbf{k} \cdot \mathbf{v} = \omega - \omega_0$ and with a velocity width γ'_{12}/k, where γ'_{12} is the so-called optical pumping linewidth discussed above. The wave vector $\mathbf{k}$ is associated with the wave doing the pumping. In the presence of two counterpropagating pumps (at frequency ω), two spikes in the velocity distribution are formed symmetrically about $\mathbf{v} = 0$ and are associated with atoms with oppositely directed but equal-magnitude velocities. The effect is a large sudden increase in the number of atoms participating in the nonlinear interaction when $\omega = \omega_0$ and the spikes coincide at $\mathbf{v} = 0$.

When considering the effects of the velocity modification to DFWM, we recall that the two pumps are always collinear and strong, but the probe is relatively weak and at some finite angle with respect to the pumps. Hence, it is the dynamics induced by the pumps that determine the velocity modification with which the probe interacts. In this case only atoms with velocities about $\mathbf{v} = 0$ within a width γ'_{12}/k can participate in the DFWM interaction, which results in the now intuitively obvious Doppler-free behavior, which is independent of the angle between the pump and probe. In the limit of third-order perturbation theory with the usual approximations (RWA and SVEA) the reflectivity will be given by (Humphrey *et al,* 1980)

$$R = \frac{|\mathscr{E}_c|^2}{|\mathscr{E}_p|^2} = \left(\eta \frac{4\mu_{12}^4 \pi^4 N |\mathscr{E}_f|\,|\mathscr{E}_b| L}{\hbar^3 \lambda (ku)^2 \sin\theta}\right)^2 \left(\frac{\gamma'_{12}}{\gamma_{12} + \gamma'_{12}}\right)^2 \frac{1}{(\gamma_{12} + \gamma'_{12})^2 + 4\Delta^2}, \tag{38}$$

where $\Delta = \omega - \omega_0$, θ is the angle between $\mathbf{k}_f$ and $\mathbf{k}_p$, and η is a constant.

This behavior was demonstrated experimentally by using a cw dye laser tuned to the D_2 line in sodium vapor at 5890 Å. Both the pumps and the probe were linearly polarized. In Fig. 26 (from Lam *et al.*, 1981) the entire spectrum in the vicinity of the D_2 line is shown. For a better understanding of the data, we show in Fig. 27 the level structure of the D_2 line. The ground state is 3s $^2S_{1/2}$ and the upper level 3p $^2P_{3/2}$. The nuclear spin–orbit coupling ($I \cdot J$) splits these levels into the hyperfine (hf) components shown in the diagram. Because the laser bandwidth is small compared with the hf splitting, F is the quantum number of interest, with dipole selection rules $\Delta F = 0, \pm 1$. Hence, there are a total of six distinct transitions associated with the D_2 line shown in Fig. 27. However, four lines are clearly missing. This is easily understood as follows. When the laser is tuned to one of these missing transitions—say, $^2S_{1/2}(F = 2) \rightarrow {}^2P_{3/2}(F = 2)$—the population in the ground state ($F = 2$) is moved to the other level ($F = 1$). That is, any atoms excited from $F = 2$ to $F = 2$ end up in $F = 1$ after a while, and thus are no longer eligible for any subsequent interaction, because they are detuned from the laser frequency.

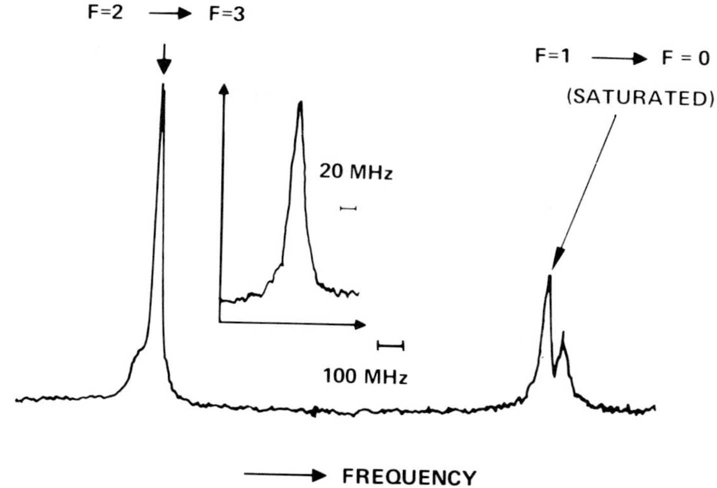

Fig. 26. DFWM reflectivity versus laser frequency on the D_2 line in sodium at λ = 589.0 nm. [After Lam *et al.* (1981).]

Hence only $F = 2 \rightarrow F = 3$ and $F = 1 \rightarrow F = 0$ transitions are observed. Note that this kind of optical pumping has just the opposite effect in saturation spectroscopy. Optical pumping here produces an increase in the signal for the optically pumped transitions. The geometry used to obtain these data (Fig. 26) is nearly collinear, and the data clearly show sub-Doppler profiles. While the laser intensity was below I_{sat}, the $F = 1 \rightarrow F = 0$ transition still exhibited a saturation dip, owing to the strong effect of optical pumping of the degenerate magnetic substates (the ground state $F = 1$ gives $2F + 1 = 3$ substates and the upper level $F = 0$ only one substrate). A closer examination of the line profile for the $F = 2 \rightarrow F = 3$ transition is shown in Fig. 28 (Humphrey *et al.*, 1980). Here the observed linewidth is of order 20 MHz. Recall that the laser intensity is less than I_{sat} and that the natural linewidth of sodium is 10 MHz. Using Eq. (38), Humphrey *et al.* (1980) found good agreement with their data, assuming $\gamma'_{12} = 3\gamma_{12} = 9.4 \times 10^7 \text{ sec}^{-1}$. They also varied the angle between the pump and probe, and showed that the linewidth is indeed dominated by optical pumping effects, that the linewidth is angle independent, and that R decreases as $(\sin \theta)^{-2}$ as shown in Fig. 29. The apparent singularity at $\theta = 0$ in Eq. (38) is not real, but the result of an approximation involved in the derivation.

The generalization of the two-level model to include degenerate states also leads to the vectorial description of the physics of resonant degen-

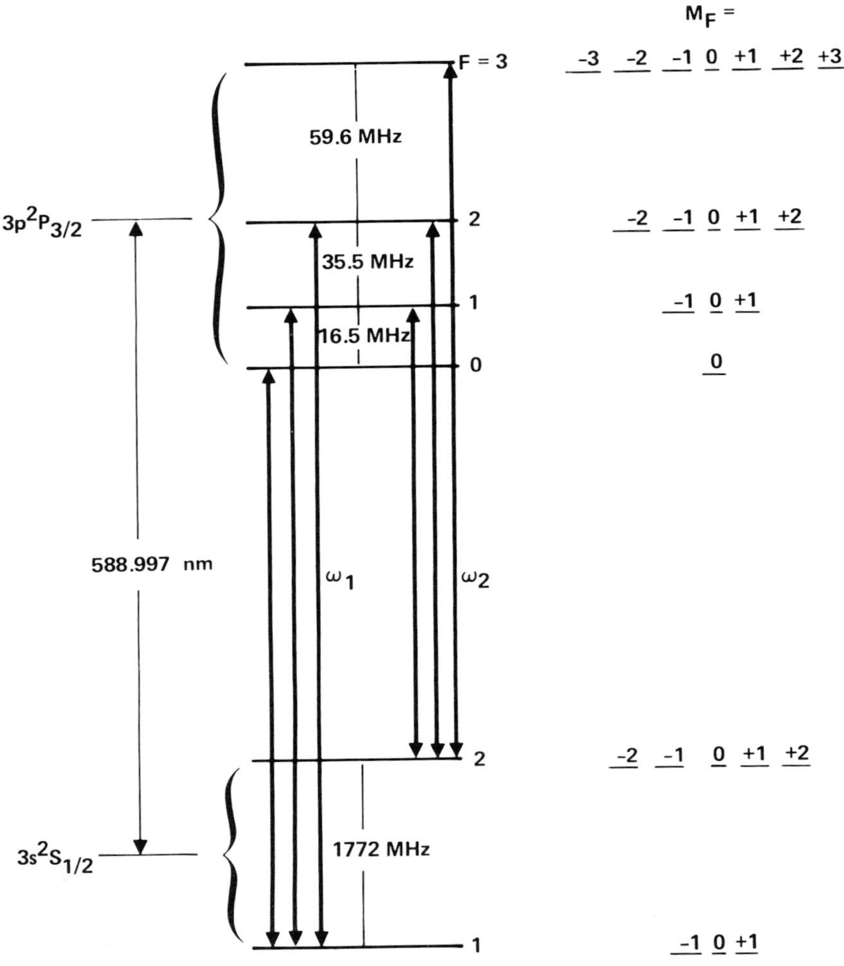

Fig. 27. Energy-level structure for the D_2 line in sodium.

erate four-wave mixing (Lam and Abrams, 1982). The relative orientation of the input electric field polarization plays an important role in determining the physical interactions.

To better understand the vectorial properties of DFWM, let us define a quantum state of the system given by the ket vector $|\gamma J_\gamma M_\gamma\rangle$, where J_γ and M_γ are the total and z component of the angular momentum, and γ denotes all other good quantum numbers (α and β below). Consider the

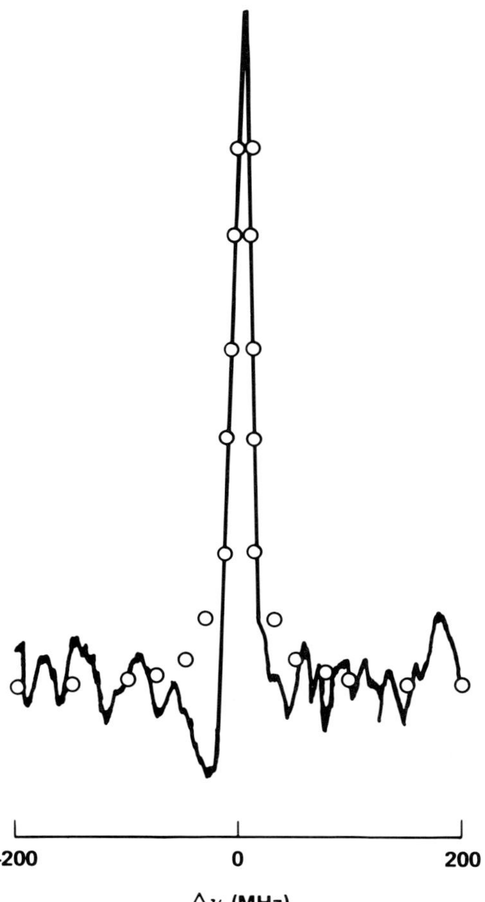

Fig. 28. DFWM intensity versus laser frequency for the 3s $^2S_{1/2}(F = 2) \rightarrow$ 3p $^2P_{3/2}(F = 3)$ transition in sodium. The pump intensity is 6 mW/cm². [After Humphrey *et al.* (1980).]

one-photon transition shown in Fig. 30. The frequency separation between the upper level $|\beta J_\beta\rangle$ and lower level $|\alpha J_\alpha\rangle$ is given by the transition frequency ω_0. The electric fields in a degenerate four-wave mixing scheme can have any electric field polarization. The polarization states of the radiation fields are defined in the laboratory frame by $\sigma_\pm = (\hat{x} \pm i\hat{y})/\sqrt{2}$, such that σ_+ fields induce $\Delta m = \pm 1$ transitions and σ_- fields $\Delta m = -1$ transitions.

Consider three examples. In the first example, the forward and back-

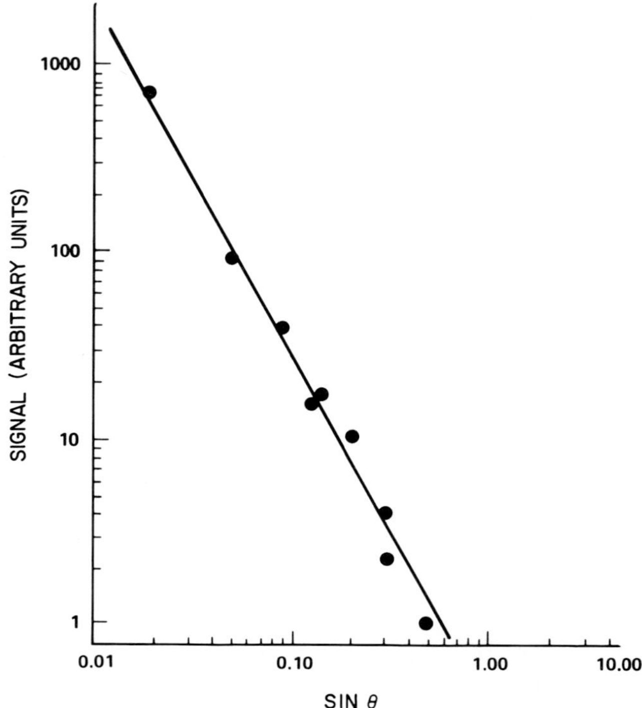

Fig. 29. Experimental measurement of DFWM reflectivity versus angle. The solid line is the theoretical prediction including the effects of optical pumping. [After Humphrey *et al.* (1980).]

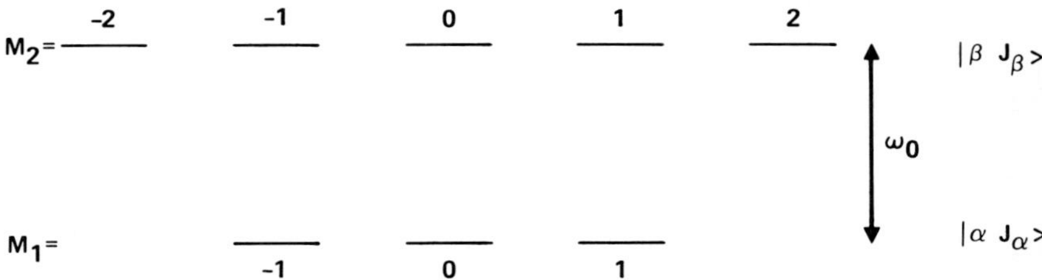

Fig. 30. One-photon transition showing degenerate states in both lower and upper states.

ward pumps and input probe are σ_+-polarized fields (Fig. 31a) in a collinear or nearly collinear geometry. A spatial modulation in the population of the upper state $|J_\beta\rangle$ is generated by the interference of the forward pump and input probe (a possible transition is between $M_1 = -1$ and $M_2 = 0$, shown in Fig. 30). The periodicity of the spatial modulation is given by $2\pi/|\mathbf{k}_f - \mathbf{k}_p|$, where $\mathbf{k}_f$ and $\mathbf{k}_p$ are the wave vectors of the forward pump and input probe, respectively. The coherent scattering of the backward pump wave, which couples the same transition (in this case $M_1 = -1 \to M_2 = 0$), yields a phase-conjugate wave with the same polarization as the backward pump wave. In the same manner, the interference of the backward pump and input probe waves generates a spatial modulation in the upper state of periodicity $2\pi/|\mathbf{k}_b - \mathbf{k}_p|$, where $\mathbf{k}_b$ is the wave vector of the backward pump. The scattering of the forward pump from the spatial grating yields a phase-conjugate wave with the same electric field polarization as the forward pump wave. We shall designate this contribution as due to *normal population* effects.

In the second example (Fig. 31b), the polarization of the backward pump is rotated to be cross polarized to the electric field polarization of the forward pump and input probe. The formation of the spatial grating by the interference of the forward pump and input probe is the same as before (say, $M_1 = -1$ with $M_2 = 0$). The coherent scattering proceeds in a different channel (in this case it couples $M_2 = 0$ to $M_1 = +1$). The scattering yields a phase-conjugate wave with the same polarization as the backward pump wave. We shall designate this contribution as due to *cross-population* effects.

In the third example, we consider the interaction of the forward pump and input probe with the electric field polarizations shown in Fig. 31c. The interference of the pump and input probe generates an electric quadrupole coupling between upper states $|M_2 = -1\rangle$ and $|M_2 = 1\rangle$. This is not a population effect, but rather a coherent coupling between states having the same energy. The scattering yields a phase-conjugate wave with an electric field polarization that is the conjugate of the input probe wave. We shall designate this contribution as due to *Zeeman coherence*.

The physical description of the effects of the relative orientation of the electric field polarization vectors on the resonant interaction can be quantified (Lam *et al.*, 1981; Lam and Abrams, 1982; Bloch, 1980; Bloch *et al.*, 1980; and Raj *et al.*, 1980) by using the quantum-mechanical transport equation (2). We shall denote the matrix element of an operator Θ between states $|\alpha J_\alpha M_\alpha\rangle$ and $|\beta J_\beta M_\beta\rangle$, in standard representation, as

$$\Theta_{J_\alpha M_\alpha\ J_\beta M_\beta} = \langle \alpha\ J_\alpha M_\alpha | \Theta | \beta\ J_\beta M_\beta \rangle. \tag{39}$$

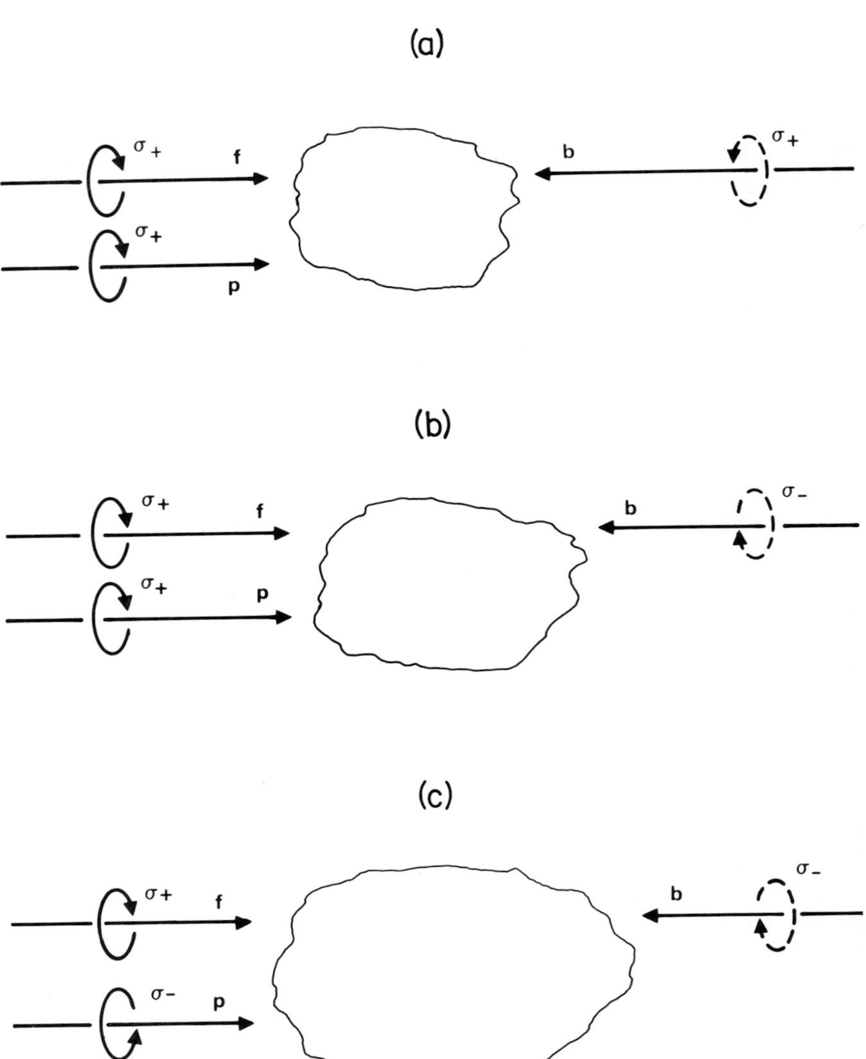

Fig. 31. Interaction configuration (a) giving rise to normal population processes, (b) for cross-population processes, and (c) for Zeeman coherence generated by the forward pump and input probe waves.

The perturbation schemes leading to the distinct contributions to the generation of the phase-conjugate signal are as follows.

Normal population:

$$\begin{pmatrix}\rho_{J_1M_1:J_1M_1}\\ \rho_{J_2M_1:J_2M_2}\end{pmatrix} \to \begin{pmatrix}\rho_{J_1M_1:J_2M_2}\\ \rho_{J_2M_2:J_1M_1}\end{pmatrix} \to \begin{pmatrix}\rho_{J_2M_2:J_2M_2}\\ \rho_{J_1M_1:J_1M_1}\end{pmatrix} \to \begin{pmatrix}\rho_{J_2M_2:J_1M_1}\\ \rho_{J_1M_1:J_2M_2}\end{pmatrix}. \tag{40a}$$

Cross population:

$$\begin{pmatrix}\rho_{J_1M_1:J_1M_1}\\ \rho_{J_2M_2:J_2M_2}\end{pmatrix} \to \begin{pmatrix}\rho_{J_1M_1:J_2M_2}\\ \rho_{J_2M_2:J_1M_1}\end{pmatrix} \to \begin{pmatrix}\rho_{J_2M_2:J_2M_2}\\ \rho_{J_1M_1:J_1M_1}\end{pmatrix} \to \begin{pmatrix}\rho_{J_2M_2:J_1M_1'}\\ \rho_{J_1M_1':J_2M_2}\end{pmatrix}. \tag{40b}$$

Zeeman coherence:

$$\begin{pmatrix}\rho_{J_1M_1:J_1M_1}\\ \rho_{J_2M_2:J_2M_2}\end{pmatrix} \to \begin{pmatrix}\rho_{J_1M_1:J_2M_2}\\ \rho_{J_1M_2:J_1M_1}\end{pmatrix} \to \begin{pmatrix}\rho_{J_2M_2':J_2M_2}\\ \rho_{J_1M_1':J_1M_1}\end{pmatrix} \to \begin{pmatrix}\rho_{J_2M_2':J_1M_1}\\ \rho_{J_1M_1':J_2M_2}\end{pmatrix}. \tag{40c}$$

The polarization density, to third order in the field strength, is given by

$$\mathbf{P}(\mathbf{r}, t) = \int_{-\infty}^{\infty} d\mathbf{v} \sum_{M_1}\sum_{M_2} \rho_{J_1M_1:J_2M_2}\boldsymbol{\mu}_{J_2M_2:J_1M_1}, \tag{41}$$

where the summation is over all possible values of the subscripts M_1 and M_2. The results can be summarized as follows.

Normal population:

$$\begin{aligned}\mathbf{P} = {} & R_{\mathrm{fp}}^{(2)} \sum_{M_1}\sum_{M_2} \boldsymbol{\mu}_{J_2M_2:J_1M_1}\boldsymbol{\mu}_{J_1M_1:J_2M_2}\cdot\mathbf{E}_{\mathrm{b}}\boldsymbol{\mu}_{J_2M_2:J_1M_1}\cdot\mathbf{E}_{\mathrm{p}}^{*}\boldsymbol{\mu}_{J_1M_1:J_2M_2}\cdot\mathbf{E}_{\mathrm{f}}\\ & + R_{\mathrm{fp}}^{(1)} \sum_{M_1}\sum_{M_2} \boldsymbol{\mu}_{J_1M_1:J_2M_2}\cdot\mathbf{E}_{\mathrm{f}}\boldsymbol{\mu}_{J_2M_2:J_1M_1}\cdot\mathbf{E}_{\mathrm{p}}^{*}\boldsymbol{\mu}_{J_1M_1:J_2M_2}\cdot\mathbf{E}_{\mathrm{b}}\boldsymbol{\mu}_{J_2M_2:J_1M_1}\\ & + (\mathrm{b} \rightleftarrows \mathrm{f}).\end{aligned} \tag{42a}$$

Cross population:

$$\begin{aligned}\mathbf{P} = {} & R_{\mathrm{fp}}^{(2)} \sum_{M_1\neq M_1'}\sum_{M_2} \boldsymbol{\mu}_{J_2M_2:J_1M_1'}\boldsymbol{\mu}_{j_1M_1':J_2M_2}\cdot\mathbf{E}_{\mathrm{b}}\boldsymbol{\mu}_{J_2M_2:J_1M_1}\\ & \cdot\mathbf{E}_{\mathrm{p}}^{*}\boldsymbol{\mu}_{J_1M_1:J_2M_2}\cdot\mathbf{E}_{\mathrm{f}}\\ & + R_{\mathrm{fp}}^{(1)} \sum_{M_2\neq M_2}\sum_{M_1} \boldsymbol{\mu}_{J_1M_1:J_2M_2}\cdot\mathbf{E}_{\mathrm{f}}\boldsymbol{\mu}_{J_2M_2:J_1M_1}\cdot\mathbf{E}_{\mathrm{p}}^{*}\boldsymbol{\mu}_{J_1M_1:J_2M_2}\\ & \cdot\mathbf{E}_{\mathrm{b}}\boldsymbol{\mu}_{J_2M_2:J_1M_1}\\ & + (\mathrm{b} \rightleftarrows \mathrm{f}).\end{aligned} \tag{42b}$$

Zeeman coherence:

$$\mathbf{P} = R_{\text{fp}}^{(2)} \sum_{M_1} \sum_{M_2' \neq M_2} \boldsymbol{\mu}_{J_2M_2':J_1M_1} \boldsymbol{\mu}_{J_1M_1:J_2M_2} \cdot \mathbf{E}_{\text{b}} \boldsymbol{\mu}_{J_2M_2:J_1M_1} \cdot \mathbf{E}_{\text{p}}^{*} \boldsymbol{\mu}_{J_1M_1:J_2M_2'} \cdot \mathbf{E}_{\text{f}} + R_{\text{fp}}^{(1)} \sum_{M_2} \sum_{M_1 \neq M_1'} \boldsymbol{\mu}_{J_1M_1:J_2M_2} \cdot \mathbf{E}_{\text{f}} \boldsymbol{\mu}_{J_2M_2:J_1M_1'} \cdot \mathbf{E}_{\text{p}}^{*} \boldsymbol{\mu}_{J_1M_1':J_2M_2} \cdot \mathbf{E}_{\text{b}} \boldsymbol{\mu}_{J_2M_2:J_1M_1} + (\text{b} \rightleftarrows \text{f}). \tag{42c}$$

where

$$R_{n\text{p}}^{(m)} = \frac{\Delta N_0}{(2i\hbar)^3} \int_{-\infty}^{\infty} d\mathbf{v}\, W(\mathbf{v}) \frac{1}{\gamma_{12} + i(\Delta + \mathbf{k}_{\text{p}} \cdot \mathbf{v})} \times \left(\frac{1}{\gamma_{12} + i(\Delta - \mathbf{k}_n \cdot \mathbf{v})} + \frac{1}{\gamma_{12} - i(\Delta - \mathbf{k}_{\text{p}} \cdot \mathbf{v})} \right) \times \frac{1}{\gamma_m + i(\mathbf{k}_{\text{p}} - \mathbf{k}_n) \cdot \mathbf{v}}, \qquad n = \text{f, b}, \quad m = 1, 2. \tag{43}$$

Note that the $R_{n\text{p}}^{(m)}$'s are identical to Eqs. (26) and (27). However, one must expect that, in the presence of collisions, the relaxation time for the populations is different from the coherence time for the magnetic state mixing.

Experimental studies of the polarization properties described above have been performed in molecular SF_6 (Steel *et al.*, 1979) and atomic sodium (Lam *et al.*, 1981). The effects are most dramatic in sodium. For these experiments a cw dye laser tuned to the D_2 line (5890 Å) was used. The pumps and probe were linearly polarized. The pumps were vertically polarized, and the probe polarization could be adjusted either parallel or perpendicular to the pumps. In Fig. 32a the spectrum recorded when the probe is parallel to the pumps is shown. The two lines correspond to the 3s $^2S_{1/2}(F = 2) \rightarrow$ 3p $^2P_{3/2}(F = 3)$ and 3s $^2S_{1/2}(F = 1) \rightarrow$ 3p $^2P_{3/2}(F = 0)$ transitions as shown in Fig. 26. The spectrum of the induced Zeeman coherence obtained by rotating the probe polarization 90° is shown in Fig. 32b. Note the clear absence of any signal for the high-frequency transition 3s $^2S_{1/2}(F = 1) \rightarrow$ 3p $^2P_{3/2}(F = 0)$. There is no induced coherence in the upper state of this transition, because the degeneracy $2F + 1$ is one. Thus the only signal observed is for the 3s $^2S_{1/2}(F = 2) \rightarrow$ 3p $^2P_{3/2}(F = 3)$ transition. The signal halfway between the two components is due to an inverted V-type crossover resonance, to be discussed in Section III.D. The polarization of the signal was identical to that of the probe as expected. Polarization rotation of the probe due to cross-population effects was also

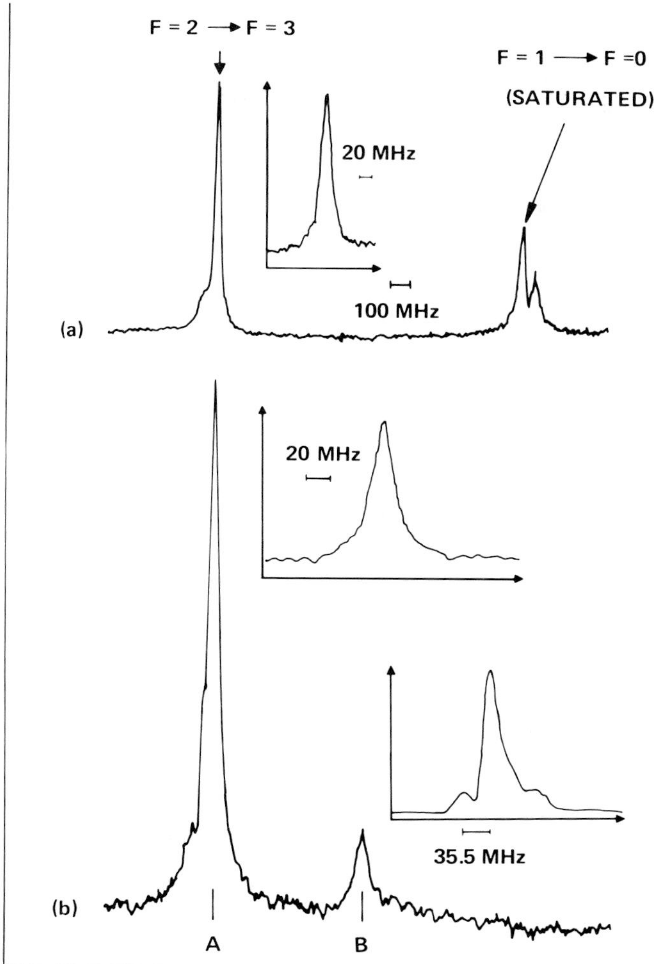

Fig. 32. DFWM frequency spectra of the sodium D_2 line ($\lambda = 5890$ Å) obtained by scanning the dye laser (the vertical scale is arbitrary). (a) Typical response when all beams are copolarized. Only the two hyperfine components are observed, owing to hf optical pumping. (b) Spectral response when the pump polarization is crossed with respect to the probe polarization. The detector is copolarized with the probe. The structure demonstrates the existence of magnetic coherence between the degenerate substates of the $F = 2 \rightarrow F = 3$ hf component. Note the absence of any signal due to magnetic coherence for the $F = 1 \rightarrow F = 0$ component, as expected. Structure B is the crossover signal mentioned in the text. The insets show the line shapes at reduced intensity and on an expanded scale. [After Lam *et al.* (1981).]

demonstrated by arranging the probe and one pump ($\mathbf{E}_f$) to be copolarized and rotating the linear polarization of the remaining pump ($\mathbf{E}_b$) by 90°. The signal had a polarization identical to that of E_b, and a spectral dependence similar to that shown in Fig. 32b.

B. Adiabatic-Following Limit

A useful technique for evaluating the nonlinear polarization of a two-level system is the adiabatic-following approximation. Developed by Grischkowsky (1970), this technique can readily give the saturation behavior of the nonlinearity evidenced in Eq. (11). More interesting, it provides a simple way to prove that atomic motion does *not* degrade DFWM efficiency if Δ, the detuning from resonance, exceeds the Doppler linewidth ku_0. This condition is fulfilled in many practical cases in which one attempts to maximize the DFWM efficiency by increasing the atomic density. One then operates slightly away from resonance to avoid problems due to absorption. Indeed, the first demonstration (Bloom *et al.*, 1978) of phase-conjugate gain was accomplished with fairly dense sodium vapor as the nonlinear medium and a detuning of 1.25 cm^{-1} from the D_1 resonance line. This detuning was well outside the Doppler width of $\sim$0.05 cm^{-1}. The data are illustrated in Fig. 33. A sodium density of $\sim 3 \times 10^{14}$ cm^{-3} was used. At this density the 5-cm-long path length of the vapor was completely opaque at line center, so that no generation was observed at resonance. However, at the detuning of 1.25 cm^{-1} the vapor was essentially transparent and had a linear absorption of less than 5%. Gains for both the generated conjugate wave and the transmitted probe wave approaching 100 were seen. Gain was limited by saturation at high intensities owing to pump depletion and saturation of the atomic nonlinearity.

A study (Grischkowsky *et al.*, 1978) of the saturation of the atomic nonlinearity appeared soon after the report of gain in sodium vapor. In Fig. 34 we show the comparison of theory with experimentally measured points as a function of $(\tan \Phi)^6$, $= |\mu_{12}\mathscr{E}_0/\hbar\Delta|.^6$ In this experiment the total laser intensity was varied so that $\mathscr{E}_p$, $\mathscr{E}_f$, and $\mathscr{E}_b$ varied proportionally. Hence the low-intensity theory would show that the signal should vary approximately as $(\tan \Phi).^6$ The deviation from this simple behavior and the good agreement with the full calculation [as obtained from Eq. (11) or the adiabatic-following calculation given by Grischkowsky *et al.* (1978)] is evident. The parameter Φ is a natural parameter of the adiabatic-following approximation.

The adiabatic-following approximation makes use of concepts first developed in nuclear magnetic resonance studies. It provides a relatively simple and useful way of obtaining the solution of the nonlinear polariza-

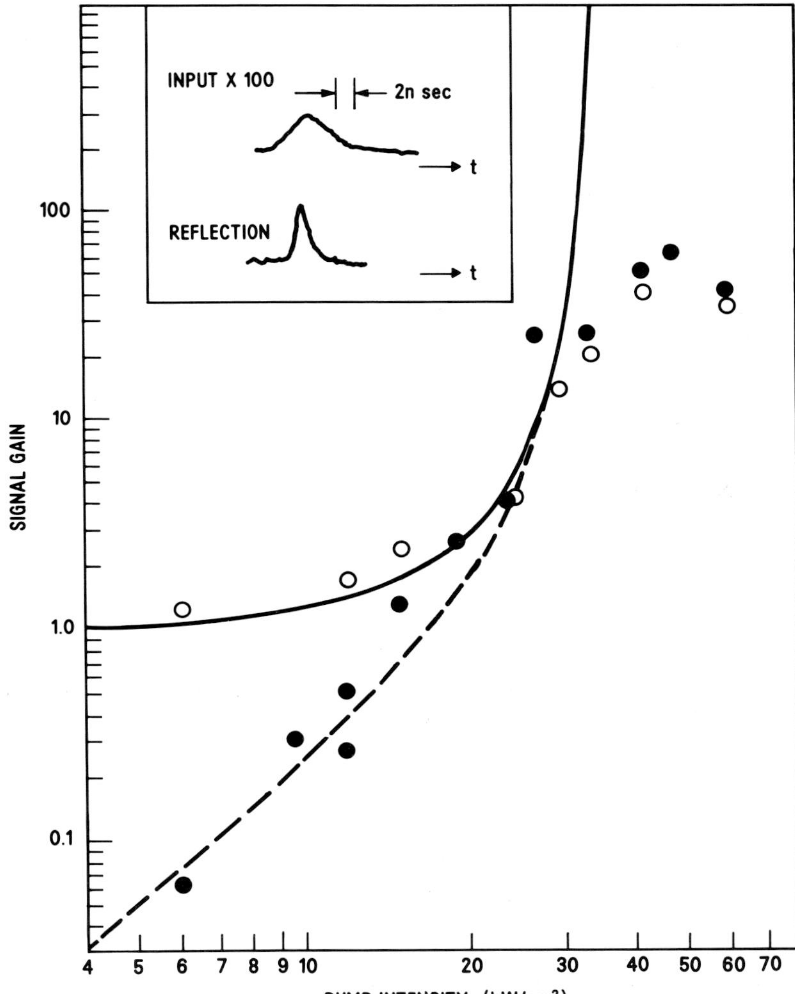

Fig. 33. Signal gain versus pump intensity. The measured gain for the backward wave (●), and the measured gain for the transmitted probe wave (○) are shown. The curves are theoretical. Inset: Oscilloscope traces of the input probe pulse and the backward-generated wave. [After Bloom *et al.* (1978).]

tion when the atom is subjected to time-varying optical fields. Because, as we shall see, the effects of atomic motion can be considered as producing time-varying fields, the adiabatic-following model will be useful in understanding atomic motion.

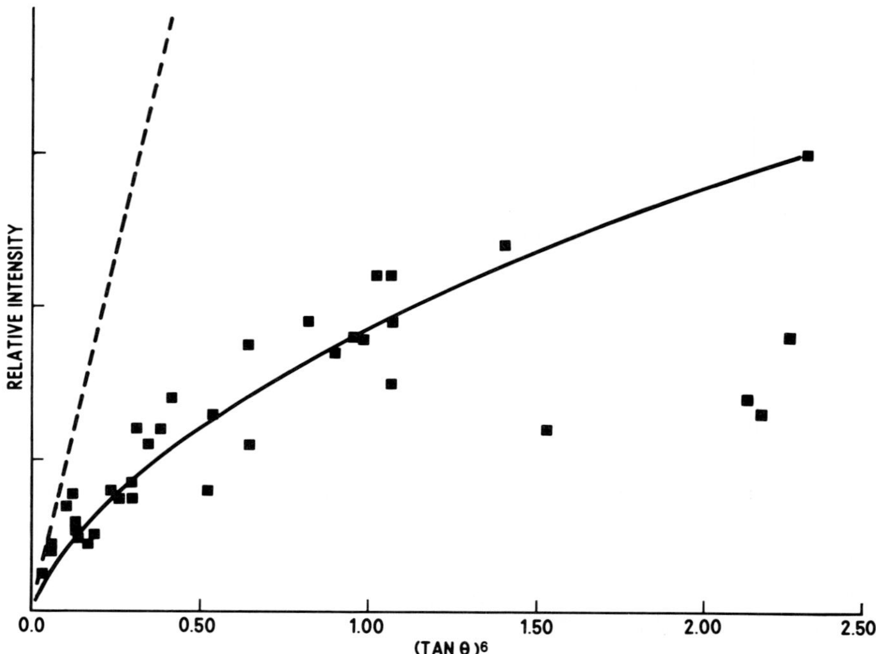

Fig. 34. Comparison of the low-intensity theory (---) and the complete theory (—) with experimental data (■). [After Grischkowsky *et al.* (1978).]

By defining

$$u = \tilde{\rho}_{12} + \tilde{\rho}_{21}, \tag{44a}$$

$$v = -i(\tilde{\rho}_{12} - \tilde{\rho}_{21}), \tag{44b}$$

$$w = \rho_{22} - \rho_{11}, \tag{44c}$$

we can rewrite the density matrix equation of motion

$$i\hbar \frac{d\rho}{dt} = [H, \rho], \tag{45}$$

where

$$H = \begin{pmatrix} \hbar\omega_1 & V_{12} \\ V_{21} & \hbar\omega_2 \end{pmatrix}, \tag{46}$$

with

$$V = -\boldsymbol{\mu} \cdot \vec{\mathscr{E}} \cos(\omega t + \phi). \tag{47}$$

For simplicity we shall assume that the states 1 and 2 are connected by a $\Delta M = 0$ transition, and that $\vec{\mathscr{E}}$ is linearly polarized, so that $\boldsymbol{\mu} \cdot \vec{\mathscr{E}} = \mu\mathscr{E}$. We shall also adjust arbitrary phases associated with the states so that $\phi = 0$.

Using the rotating wave approximation, we find that Eq. (45) can be written (Feynman *et al.*, 1957)

$$\hbar \frac{d\mathbf{S}}{dt} = \vec{\mathscr{E}}_{\text{eff}} \times \mathbf{S}, \tag{48}$$

where the pseudomoment $\mathbf{S}$ is given by the vector

$$\mathbf{S} = \mu_{12} \begin{pmatrix} u \\ v \\ w \end{pmatrix}, \tag{49}$$

and the effective field $\vec{\mathscr{E}}_{\text{eff}}$ is given by

$$\vec{\mathscr{E}}_{\text{eff}} = \frac{1}{\mu_{12}} \begin{pmatrix} -\mu_{12}\mathscr{E} \\ 0 \\ \hbar\Delta \end{pmatrix}. \tag{50}$$

Equation (48) has exactly the same form as the equation of motion of a pseudomoment of magnitude μ precessing about an effective field $\vec{\mathscr{E}}_{\text{eff}}$. This concept of a pseudomoment and effective field is extremely useful, because techniques such as adiabatic following, developed for magnetic spin resonance problems, can be used to evaluate the time dependence of the nonlinear susceptibility. In the absence of an applied optical field, the effective field $\vec{\mathscr{E}}_{\text{eff}}$ is given by

$$\vec{\mathscr{E}}_{\text{eff}} = \mu_{12} \begin{pmatrix} 0 \\ 0 \\ \hbar\Delta \end{pmatrix}, \tag{51}$$

and the pseudomoment is aligned exactly along $\vec{\mathscr{E}}_{\text{eff}}$. Applying an optical field of frequency ω causes $\vec{\mathscr{E}}_{\text{eff}}$ to change its direction by the angle $\Phi = \tan^{-1}(\mu_{12}\mathscr{E}/\hbar\Delta)$ and to change its magnitude to $(1/\mu_{12})[(\mu_{12}\mathscr{E})^2 + (\hbar\Delta)^2]^{1/2}$.

In the adiabatic-following approximation, the solution to Eq. (48) is for the pseudomoment $\mathbf{S}$ to stay closely aligned to the effective field. The pseudomoment is said to adiabatically follow the field as it changes direction. This approximation is valid if optical field amplitude $|\vec{\mathscr{E}}|$ varies slowly enough that the effective field $\vec{\mathscr{E}}_{\text{eff}}$ changes its direction slowly compared with the rate $\mu_{12}\mathscr{E}_{\text{eff}}$ at which $\mathbf{S}$ precesses about it. One can easily show that the maximum angle 2ψ at which the pseudomoment will deviate from the direction of the effective field is given by $\tan\psi = \hbar\Omega/\mu_{12}\mathscr{E}_{\text{eff}}$, where

Ω is the angular frequency associated with the rate at which $\vec{\mathscr{E}}_{\text{eff}}$ changes direction.

The expression for the accuracy of the adiabatic-following approximation is obtained as follows. One transforms to a new reference frame, which rotates with angular frequency Ω so that $\vec{\mathscr{E}}_{\text{eff}}$ is stationary. In this frame one obtains an equation of motion identical to Eq. (48), except that **S** precesses about the vector sum of $\vec{\mathscr{E}}_{\text{eff}}$ and $\vec{\mathscr{E}}_{\text{R}} = n\hbar\Omega/\mu_{12}$, where $\hat{n}$ is the unit direction axis vector around which $\vec{\mathscr{E}}_{\text{eff}}$ rotates. Because **S** is initially aligned along $\vec{\mathscr{E}}_{\text{eff}}$, the maximum angle that **S** will achieve from $\vec{\mathscr{E}}_{\text{eff}}$ as **S** precesses about $\vec{\mathscr{E}}_{\text{eff}} + \vec{\mathscr{E}}_{\text{R}}$ is $2\psi = \tan^{-1}(\hbar\Omega/\mu_{12}\mathscr{E}_{\text{eff}})$. Therefore, if

$$\Omega \leq \left|\frac{1}{\mathscr{E}}\frac{d\mathscr{E}}{dt}\right| \ll |\Delta| \leq \mu_{12}\mathscr{E}_{\text{eff}}/\hbar, \tag{52}$$

the adiabatic-following conditions are satisfied and the pseudomoment is quite accurately aligned along the effective field. The laser pulse width should also be short compared with the homogeneous relaxation times, although this restriction can be relaxed in the limit of low intensities if Δ is much larger than the homogeneous widths.

The atomic polarization is determined by $\tilde{\rho}_{12}$, and therefore by the transverse components of the pseudomoment. Using the adiabatic-following approximation, we obtain

$$|P| = \mu_{12} \sin \Phi. \tag{53}$$

Although this expression can be used to obtain the same results as Eq. (11), the algebra is somewhat tedious. We shall not derive the result here, but refer the reader to the work of Grischkowsky *et al.* (1978). Instead, we shall address the problem of atomic motion, because the adiabatic-following theory answers one question often posed about atomic motion.

As discussed in Section I.A, the four-wave mixing process is most easily visualized as a formation of thick holographic phase gratings (see Fig. 2). The incident probe wave interferes with either pump waves to form a spatial modulation of intensity, which through the nonlinear phase shift leads to a modulation of refractive index. This phase grating diffracts the counterpropagating pump wave to produce the phase-conjugate wave. The question immediately arises as to the effect of atomic motion when one uses an atomic or molecular vapor as the nonlinear medium. Motion would cause atoms to traverse the grating fringes and hence see a time-varying optical field. If the time to traverse the fringe is less than an atomic lifetime, perhaps the resulting phase holograms would wash out, and effectively reduce the fringe contrast ratio and hence diffraction efficiency as shown in Section II.A.2. The fringe spacing is given by $\Lambda = \lambda/\sin(\theta/2)$, where θ is the angle between the pump beam and probe wave. The typical inverse transit time between fringes is therefore of the order of the Doppler width, $\Delta\nu_{\text{D}} = U_0/\lambda$, where U_0 is essentially the average atomic thermal velocity. Hence, whenever the Doppler width exceeds the natural width of a transition, one might expect substantial washout effects. However, if the adiabatic-following conditions are satisfied, the pseudomoment of each atom maintains its alignment with the effective field and instantaneously responds to the changing field amplitudes, and in particular to the change in field amplitude that at atom experiences as it moves across the intensity fringes. Hence, the refractive index changes will exactly mirror the intensity fringes regardless of motion; i.e., the nonlinear polarization is unaffected by atomic motion. In terms of the Doppler width, the adiabatic-following condition (52) is given by

$$\left|\frac{1}{\mathscr{E}}\frac{d\mathscr{E}}{dt}\right| \simeq \Delta\nu_{\mathrm{D}} \ll |\Delta|. \tag{54}$$

Hence, as long as the detuning exceeds the Doppler width, the nonlinear polarization will respond to the motion-produced time-varying fields and atomic motion can be ignored. Only if the detuning is very small will the adiabatic-following condition be violated and the holographic phase grating wash out. For such small detunings, absorption also becomes an important factor. The effects of absorption are particularly complicated, because the probe wave, the generated conjugate wave, and both pump waves will be absorbed.

This rather surprising result explains why the experimental results shown in Fig. 33 were *not* degraded by atomic motion, and why the low signal theory without motion represented by the solid and dashed curves gave such a good fit.

III. Two-Photon Resonances

A. Weak-Pump Theory

Thus far we have confined our attention to atoms described in terms of two-level systems with generalizations to include the effects of angular momentum. The process of degenerate four-wave mixing can also occur in media modeled by three-level systems. In Fig. 35 we show a cascade-up three-level system that serves to describe two-photon-type processes. The physics of four-wave mixing for such a system can be described as follows. The action of the counterpropagating pump waves ($\mathbf{k}_{\mathrm{f}} + \mathbf{k}_{\mathrm{b}} = 0$) generates a coherence between states $|1\rangle$ and $|3\rangle$. This coherent excitation oscillates at frequency 2ω and is spatially uniform. Furthermore, all the atoms participate in the interaction process, because the two-photon excitation is Doppler-free:

$$2\omega - \omega_{31} - (\mathbf{k}_{\mathrm{f}} + \mathbf{k}_{\mathrm{b}}) \cdot \mathbf{v} = 0, \tag{55}$$

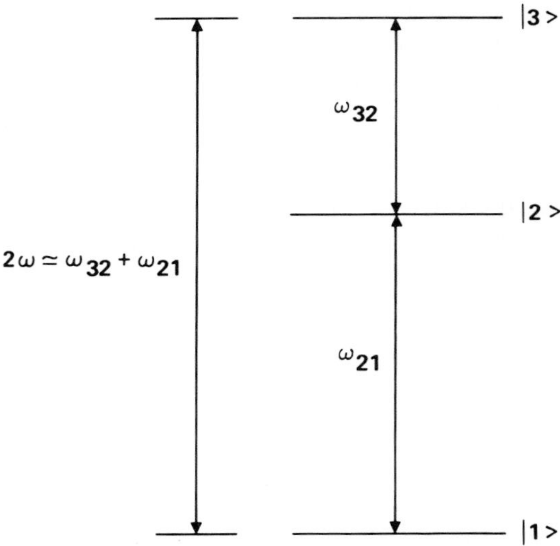

Fig. 35. Cascade-up three-level system.

because $\mathbf{k}_f + \mathbf{k}_b = 0$. The action of the probe wave induces a stimulated two-photon process that generates a phase-conjugate signal.

The starting point in the formulation of the process of degenerate four-wave mixing in three-level systems is the density matrix equations. We shall consider the perturbation regime where $I \ll I_{sat}$ (in Section III.B we consider the saturated solution). As before, there are two distinct perturbation chains for a cascade-up three-level system.

Two-quantum:

$$\rho_{11}^{(0)} \to \begin{pmatrix} \rho_{12}^{(1)} \\ \rho_{21}^{(1)} \end{pmatrix} \to \begin{pmatrix} \rho_{13}^{(2)} \\ \rho_{31}^{(2)} \end{pmatrix} \begin{matrix} \nearrow \begin{pmatrix} \rho_{12}^{(3)} \\ \rho_{21}^{(3)} \end{pmatrix} \\ \searrow \begin{pmatrix} \rho_{23}^{(3)} \\ \rho_{32}^{(3)} \end{pmatrix} \end{matrix} \tag{56a}$$

Stepwise:

$$\rho_{11}^{(0)} \to \begin{pmatrix} \rho_{12}^{(1)} \\ \rho_{21}^{(1)} \end{pmatrix} \to \begin{pmatrix} \rho_{11}^{(2)} \\ \rho_{22}^{(2)} \end{pmatrix} \begin{matrix} \nearrow \begin{pmatrix} \rho_{12}^{(3)} \\ \rho_{21}^{(3)} \end{pmatrix} \\ \searrow \begin{pmatrix} \rho_{23}^{(3)} \\ \rho_{32}^{(3)} \end{pmatrix} \end{matrix} \tag{56b}$$

The first (two-quantum) chain does not include the excitation of population in the intermediate state $|2\rangle$; it is a pure coherent two-photon process. The second (stepwise) chain involves the generation of a spatial modulation in the intermediate state $|2\rangle$, and is analogous to degenerate four-wave mixing in two two-level systems.

The polarization density for degenerate four-wave mixing via two-photon transitions is given by

$$\mathbf{P}(\mathbf{r}, t) = \int_{-\infty}^{\infty} d\mathbf{v}\, [\rho_{12}(\mathbf{r}, \mathbf{v}, t)\boldsymbol{\mu}_{21} + \rho_{23}(\mathbf{r}, \mathbf{v}, t)\boldsymbol{\mu}_{32} + \text{c.c.}], \tag{57}$$

where the integration over velocities takes into account the random motion of the atoms. If all fields are copolarized, the perturbation calculation yields the following expressions.

Two-quantum:

$$\begin{aligned} P(\mathbf{r}, t) &= \frac{N_0}{(2i\hbar)^3} \frac{|\mu_{12}|^2|\mu_{23}|^2 \mathscr{E}_\mathrm{f}\mathscr{E}_\mathrm{b}\mathscr{E}_\mathrm{p}^*}{\gamma_{13} + i(2\omega - \omega_{31})} \int_{-\infty}^{\infty} d\mathbf{v}\, W(\mathbf{v}) \\ &\times \left(\frac{1}{\gamma_{12} + i(\Delta_{21} - \mathbf{k}_\mathrm{f} \cdot \mathbf{v})} + \frac{1}{\gamma_{12} + i(\Delta_{21} - \mathbf{k}_\mathrm{b} \cdot \mathbf{v})} \right) \\ &\times \left(\frac{1}{\gamma_{12} + i(\Delta_{21} + \mathbf{k}_\mathrm{p} \cdot \mathbf{v})} - \frac{1}{\gamma_{23} + i(\Delta_{32} + \mathbf{k}_\mathrm{p} \cdot \mathbf{v})} \right) \\ &\times e^{i(\omega t + \mathbf{k}_\mathrm{p} \cdot \mathbf{r})}. \end{aligned} \tag{58a}$$

Stepwise:

$$\begin{aligned} P(\mathbf{r}, t) &= \frac{N_0}{(2i\hbar)^3} |\mu_{12}|^2\, \mathscr{E}_\mathrm{f}\mathscr{E}_\mathrm{b}\mathscr{E}_\mathrm{p}^* \int_{-\infty}^{\infty} d\mathbf{v}\, W(\mathbf{v}) \\ &\times \left(\frac{1}{\gamma_{12} + i(\Delta_{21} - \mathbf{k}_\mathrm{f} \cdot \mathbf{v})} + \frac{1}{\gamma_{12} - i(\Delta_{21} - \mathbf{k}_\mathrm{p} \cdot \mathbf{v})} \right) \\ &\times \left[\left(\frac{1}{\gamma_1 + i(\mathbf{k}_\mathrm{p} - \mathbf{k}_\mathrm{f}) \cdot \mathbf{v}} + \frac{1}{\gamma_2 + i(\mathbf{k}_\mathrm{p} - \mathbf{k}_\mathrm{f}) \cdot \mathbf{v}} \right) \right. \\ &\times \frac{|\mu_{12}|^2}{\gamma_{12} + i(\Delta_{21} + \mathbf{k}_\mathrm{p} \cdot \mathbf{v})} - \frac{1}{\gamma_2 + i(\mathbf{k}_\mathrm{p} - \mathbf{k}_\mathrm{f}) \cdot \mathbf{v}} \\ &\times \left. \frac{|\mu_{23}|^2}{\gamma_{23} + i(\Delta_{32} + \mathbf{k}_\mathrm{p} \cdot \mathbf{v})} \right] e^{i(\omega t + \mathbf{k}_\mathrm{p} \cdot \mathbf{r})} \\ &+ (\mathrm{f} \rightarrow \mathrm{b}). \end{aligned} \tag{58b}$$

The notation $(\mathrm{f} \rightarrow \mathrm{b})$ means that all subscripts established by f should be changed to subscripts established by b.

In the collinear geometry, the integration over velocities leads to expressions given in terms of the plasma dispersion function. To deduce the spectrum near the two-photon resonance defined by $2\omega - \omega_{31} < \gamma_{13}$, observe that in any real atomic or molecular system, the detuning from the intermediate states Δ_{21} and Δ_{32} must necessarily be large compared with both the natural linewidth and the Doppler width. In this limit we neglect all velocity terms $\mathbf{k}_n \cdot \mathbf{v}$ and apply the normalization condition $\int_{\infty}^{\infty} d\mathbf{v}\, W(\mathbf{v}) = 1$, to obtain the following expressions.

Two-quantum:

$$P(\mathbf{r}, t) = -\frac{N_0}{(2i\hbar)^3} \frac{|\mu_{12}|^2 |\mu_{23}|^2 \mathscr{E}_{\mathrm{f}} \mathscr{E}_{\mathrm{b}} \mathscr{E}_{\mathrm{p}}^*}{\gamma_{13} + i(2\omega - \omega_{31})} \times \frac{2}{\Delta_{21}} \left(\frac{1}{\Delta_{21}} - \frac{1}{\Delta_{32}} \right) e^{i(\omega t + \mathbf{k}_{\mathrm{p}} \cdot \mathbf{r})}. \tag{59a}$$

Stepwise:

$$P(\mathbf{r}, t) = -\frac{2N_0}{(2i\hbar)^3} |\mu_{12}|^2 \mathscr{E}_{\mathrm{f}} \mathscr{E}_{\mathrm{b}} \mathscr{E}_{\mathrm{p}}^* \frac{2\gamma_{12}}{\Delta_{21}^2} \times \left(\frac{2}{\gamma_0} \frac{|\mu_{12}|^2}{\Delta_{21}} - \frac{1}{\gamma_2} \frac{|\mu_{23}|^2}{\Delta_{32}} \right) e^{i(\omega t + \mathbf{k}_{\mathrm{p}} \cdot \mathbf{r})}. \tag{59b}$$

At the two-photon resonance $\Delta_{32} = -\Delta_{21} = -\Delta$, the ratio of the stepwise to the two-quantum contributions is given approximately by

$$\frac{\gamma_{12}}{\Delta} \left(\frac{2\gamma_{13}}{\gamma_0} \frac{|\mu_{12}|^2}{|\mu_{23}|^2} + \frac{\gamma_{13}}{\gamma_2} \right). \tag{60}$$

Typically, $\gamma_{12} \ll \Delta$, and the quantity in parentheses is of order unity. Hence the ratio is much less than unity, and the stepwise contribution is negligible compared with the two-quantum contribution. In our discussion of the stepwise process we have neglected the spatial grating formed by the backward pump and probe waves, because in the collinear geometry the effect of atomic motion leads to a washout of the grating.

Finally, the frequency spectrum of the phase-conjugate signal generated via DFWM in three-level systems shows a Lorentzian profile of linewidth γ_{13} centered at $\omega = \omega_{31}/2$. It is the Doppler-free nature of the two-photon coherence generated by the counterpropagating pump waves that provides a sensitive technique of nonlinear spectroscopy via DFWM. The importance of these results was pointed out by Matsuoka (1975), Haueisen (1979), and Popov and Shalaev (1980).

B. Saturation Effects

The fundamental physics of DFWM via two-photon processes was discussed in the last section in the limit of weak counterpropagating pump waves. However, in most cases of interest, the pump intensity exceeds the two-photon saturation intensity (Larsen and Bloembergen, 1976). We have solved this problem in a self-consistent manner by using the density matrix approach of the previous sections. The presentation of this solution is beyond the scope of this chapter, owing to the complexity of the problem. However, using the phenomenological approach of Narducci *et al.* (1977), Fu and Sargent (1980) have derived an expression for the two-photon resonant DFWM polarization including the effects of saturation.

The polarization giving rise to the signal is expressed as (after expansion about strong pumps)

$$
\begin{aligned}
P(\mathbf{r}) = \frac{N_0}{2}\Bigg[K_{11} + K_{33} &+ \frac{K_{11} - K_{33}}{1 + I_0^2 L}\left(1 - \frac{2LI_0^2}{1 + I_0^2 L}\right) \\
&+ \frac{2K_{13}}{\sqrt{T_1 T_2}}\, D\, \frac{I_0}{1 + I_0^2 L}\Bigg] E_c \\
&- N_0(K_{11} - K_{33})L\, \frac{I_0 E_0^2}{(1 + I_0^2 L)^2}\, E_p^* \\
&+ N_0 \frac{K_{13}}{\sqrt{T_1 T_2}}\, D\, \frac{E_0^2}{1 + I_0^2 L}\, E_p^*,
\end{aligned}
\tag{61}
$$

where N_0 is the population difference between states $|1\rangle$ and $|3\rangle$. The constants K_{11}, K_{33}, and K_{13} are defined as

$$
\begin{aligned}
K_{\alpha\alpha} &= \frac{2}{\hbar}\, |\mu_{2\alpha}|^2 \frac{\omega_{2\alpha}}{\omega_{2\alpha}^2 - \omega_0^2}, \qquad \alpha = 1 \text{ or } 3, \\
K_{13} &= \frac{1}{\hbar}\, \frac{\mu_{21}\mu_{23}}{\omega_{21} - \omega},
\end{aligned}
\tag{62}
$$

where $\mu_{2\alpha}$ is the electric dipole moment between the intermediate state $|2\rangle$ and state $|\alpha\rangle$. $I_0 = K_{13}(T_1 T_2)^{1/2}|E_0|^2/2\hbar$ is the dimensionless pump intensity. $E_0 = \mathscr{E}_f e^{-ikx} + \mathscr{E}_b e^{ikx}$, where $\mathscr{E}_f$ and $\mathscr{E}_b$ are the slowly varying envelopes of the forward and backward pump waves, respectively. $D = T_2/(1 + i\delta)$ is the complex Lorentzian and $L = 1/(1 + \delta^2)$ the real Lorentzian profile. The effective two-photon detuning is given by $\delta = (\omega_{\text{eff}} - 2\omega)/T_2$. The effective line center $\omega_{\text{eff}} = \omega_0 + \omega_s I_0$, where $\omega_s = (K_{11} - K_{33})/2K_{13}(T_1 T_2)^{1/2}$, with T_1 and T_2 the two-photon coherence and dipole decay times, respectively.

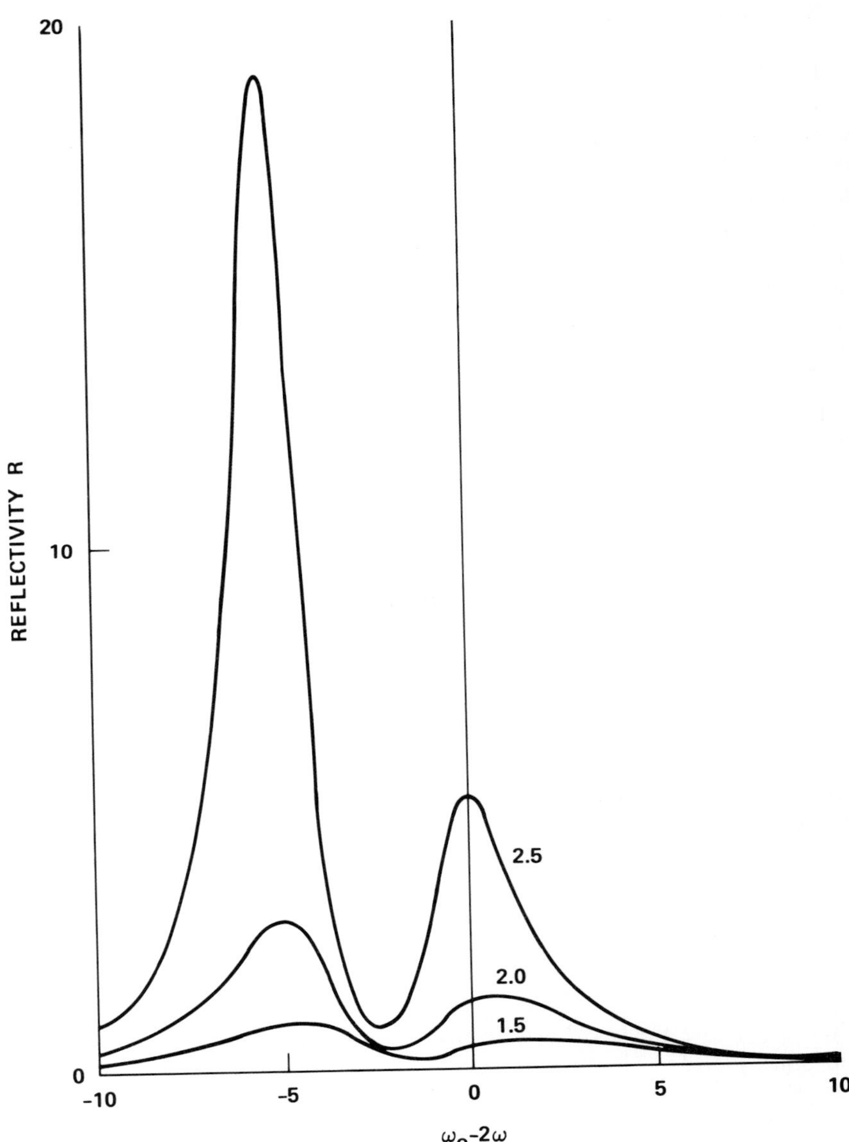

Fig. 36. Reflectivity R versus frequency detuning, normalized to T_2. T_2 and T_1 are chosen to be equal, and pump intensity is set equal to unity. Curves are shown for 3 choices of the product $\omega_s T_1$. [After Fu and Sargent (1980).]

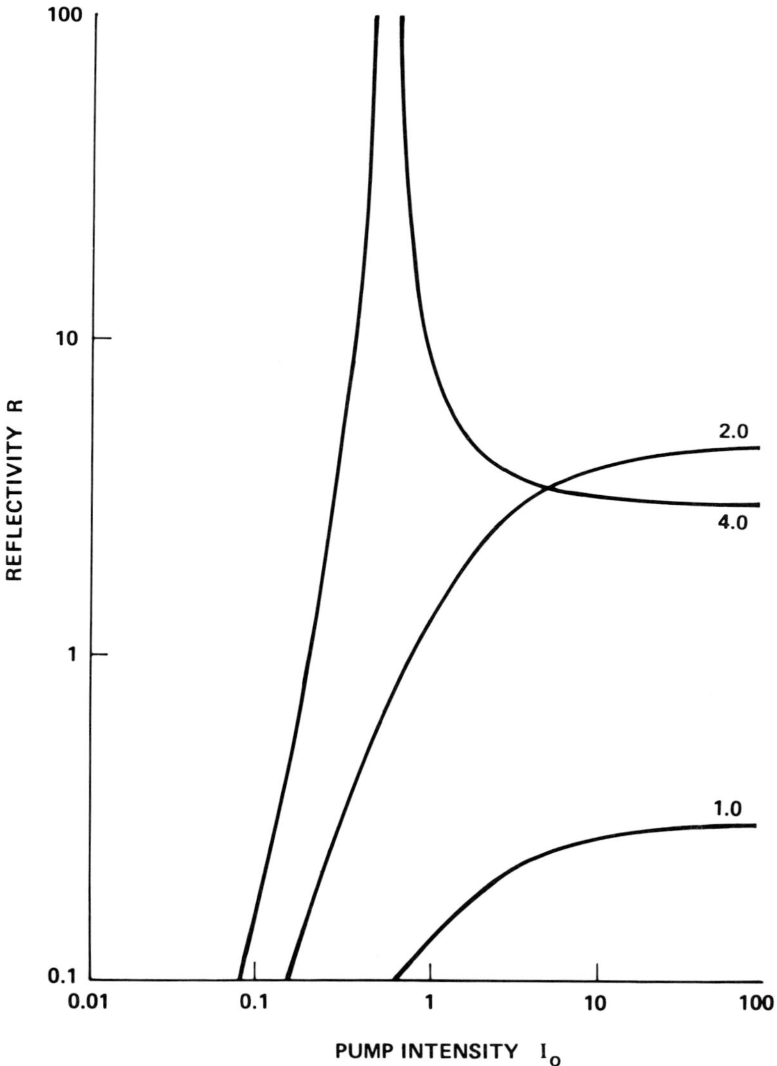

Fig. 37. Reflectivity R versus pump intensity I_0 in an absorbing medium. $T_1 = T_2$, $KN_0K_{13}L/4\mathscr{E}_0 = 1$. Curves shown for $\omega_s T_1$; coupled mode oscillation occurs at $\omega_s T_1 = 4$. [After Fu and Sargent (1980).]

A physical interpretation of the three terms in the polarization expression (61) is in order. The first term leads to linear dispersion phenomena modified by pump-induced saturation effects. The second term arises from the formation of a spatial grating and yields phase conjugation and is the strong-signal analog of the stepwise path. The third term is due to the excitation of a two-photon coherence by the pump waves, and is the strong-signal analog of the two-quantum path.

The appropriate components of the polarization are inserted into Maxwell's equations, and the reflectivity is obtained by solving the boundary-value problem of DFWM. There are three distinctive features of the reflectivity not found in two-level theory. First, the contribution to the phase-conjugate signal due to both the spatial grating (second term) and the two-photon coherence (third term) gives rise to an interference term that yields a double-peak feature in the reflectivity as a function of two-photon detuning, as shown in Fig. 36. Second, as shown in Fig. 37, for the case where $\omega_s T_1 = 4$ a singularity exists in the reflectivity in the same manner as in the two-level case in Eq. (20). Third, the reflectivity does not saturate as the pump intensity increases, because the ac Stark shift causes the two-photon resonance to shift out of resonance with applied field.

C. Transient Effects

Owing to the Doppler-free properties of two-photon resonantly enhanced DFWM, experimental work has emphasized spectroscopic applications. These properties were used to measure the collisional dephasing rate of the 4d state in sodium (Liao *et al.*, 1977). A similar experiment was performed by Steel and Lam (1979) to measure the collisionless dephasing rate in SF_6. In principle, these measurements could have been made by observing the line shape obtained by scanning a narrow-band laser through the resonance. However, detuning of the intermediate state resulted in a signal too small to be observed with available cw sources. Hence, with the aid of a powerful pulsed laser, a coherent transient technique was used to make the measurement. As discussed above, the two counterpropagating pump beams E_f and E_b generate the two-photon coherence. When these beams are turned off, the coherence decays at a rate γ_{13}, the two-photon dephasing rate. The decay rate (in the time domain) is the same as the rate that gives rise to the finite linewidth (in the frequency domain) obtained if a cw scanning technique is used. In the time domain, this decay of the coherence may be directly observed by delaying the probe beam by an amount τ and observing the signal level. The decay rate is measured by recording the signal as a function of τ. The polarization induced at a time τ is given by (Steel and Lam, 1979)

$$P(r, t, \tau) = \frac{-N_0}{4i\hbar^3} \frac{|\mu_{12}|^2|\mu_{23}|^2\mathscr{E}_f\mathscr{E}_b\mathscr{E}_p^* \exp(i[\mathbf{k}_p \cdot \mathbf{r} + (\omega_{31} - \omega)t] - \gamma_{13}t)}{\gamma_{13} + i(\Delta_{32} + \Delta_{21})} \times \left(\frac{1}{\gamma_{12} + i\Delta_{21}} - \frac{1}{\gamma_{32} + \Delta_{32}}\right) F(t), \tag{63}$$

where $F(t)$ arises from the initial conditions and describes the pulse shape of the signal, and $\Delta_{ij} = \omega - \omega_{ij}$. The signal is observed to decay at a rate $2\gamma_{13}$ as a function of τ (assuming that the pulse width is less than τ). In Fig. 38 we show the signal intensity as a function of τ in Na using a pulsed dye laser tuned to the 3s–4d two-photon resonance ($\lambda = 5787$ Å) in the pres-

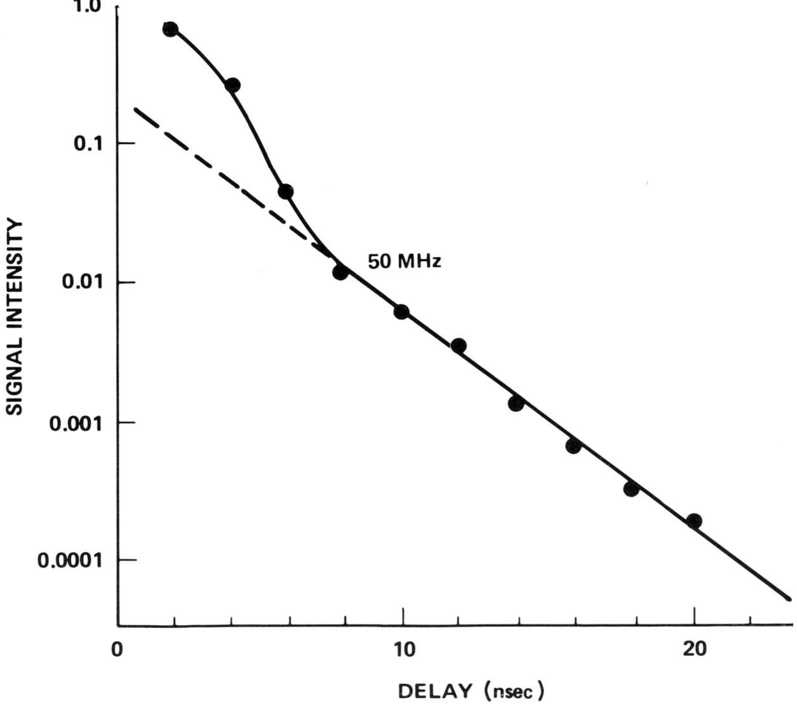

Fig. 38. Two-photon enhanced DFWM signal intensity versus time delay between pumps and probe. The slope yields the decay rate. The experiments were performed in sodium with 0.8 Torr neon. [After Liao *et al.* (1977).]

ence of 0.8 Torr Ne buffer gas. Similar results were obtained in SF_6 (Steel and Lam, 1979) for the collisionless dephasing rate. However, owing to the dense number of resonances, the SF_6 data must be interpreted with care. In particular, in the SF_6 experiments the decay rate was observed to be nearly equal to the pulse width. Thus the dephasing is probably due to an ensemble of oscillators being excited by the finite bandwidth of the pulse, giving the dephasing rate equal to the reciprocal of the pulse width (in the field). Hence the two-photon coherent transient technique provides a powerful spectroscopic tool, but the interpretation depends quite sensitively on the system studied.

D. Crossover Resonances

Closely related to three-level behavior characterized by a two-photon resonance described above is three-level behavior characterized by a crossover resonance. These V-type transitions (both normal and inverted), shown schematically in Fig. 39, are characterized by two dipole-allowed transitions sharing a common level, with the remaining two levels having only a slight energy separation. The dynamics of the interaction process can be understood as follows. An atom can interact resonantly with counterpropagating waves provided the left-going wave is Doppler shifted up into resonance with the transition $|3\rangle \leftrightarrow |2\rangle$ while the right-going wave is Doppler shifted down into resonance with the transition $|1\rangle \leftrightarrow |2\rangle$. That is, the laser frequency must satisfy two velocity-dependent resonance conditions simultaneously:

$$\omega + kv = \omega_{32}, \qquad \omega - kv = \omega_{12}. \tag{64}$$

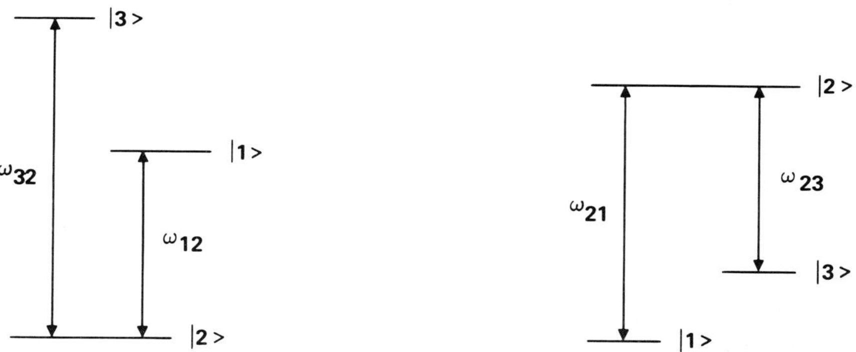

Fig. 39. Three-level model for crossover resonances: normal (left diagram) and inverted (right diagram) V-type three-level systems.

Hence ω must satisfy the condition

$$\omega = (\omega_{32} + \omega_{12})/2. \tag{65}$$

Such crossover resonances will only occur in inhomogeneously broadened gases. The perturbation schemes used to calculate the polarization are similar to those discussed in Section III.A. The polarization is again composed of both a two-quantum coherent contribution and a stepwise population-dependent contribution. Consider the case of inverted V-type three-level systems. The polarization for these two contributions can be expressed as follows.

Two-quantum:

$$\begin{aligned} P(\mathbf{r}, t) = &-\frac{|\mu_{12}|^2|\mu_{23}|^2}{(2i\hbar)^3} \mathscr{E}_f \mathscr{E}_b \mathscr{E}_p^* e^{i(\omega t + \mathbf{k}_p \cdot \mathbf{r})} \\ &\times \int_{-\infty}^{\infty} d\mathbf{v}\, W(\mathbf{v}) \left[\frac{1}{\gamma_{12} + i\Delta_{21} + i\mathbf{k}_p \cdot \mathbf{v}} \frac{1}{\gamma_{13} - i\omega_{31} + i(\mathbf{k}_p - \mathbf{k}_f) \cdot \mathbf{v}} \right. \\ &\times \left(\frac{N_3}{\gamma_{23} - i\Delta_{32} + i\mathbf{k}_p \cdot \mathbf{v}} + \frac{N_1}{\gamma_{12} + i\Delta_{21} - i\mathbf{k}_f \cdot \mathbf{v}} \right) \\ &+ \frac{1}{\gamma_{23} + i\Delta_{32} + i\mathbf{k}_p \cdot \mathbf{v}} \frac{1}{\gamma_{13} + i\omega_{31} + i(\mathbf{k}_p - \mathbf{k}_f) \cdot \mathbf{v}} \\ &\left. \times \left(\frac{N_3}{\gamma_{23} + i\Delta_{32} - i\mathbf{k}_f \cdot \mathbf{v}} + \frac{N_1}{\gamma_{12} - i\Delta_{21} + i\mathbf{k}_p \cdot \mathbf{v}} \right) \right]. \end{aligned} \tag{66a}$$

Stepwise:

$$\begin{aligned} P(\mathbf{r}, t) = &\frac{1}{(2i\hbar)^3} \frac{|\mu_{12}|^2|\mu_{23}|^2}{\gamma_2 + i(\mathbf{k}_p - \mathbf{k}_f) \cdot \mathbf{v}} \mathscr{E}_f \mathscr{E}_b \mathscr{E}_p^* e^{i(\omega t + \mathbf{k}_p \cdot \mathbf{r})} \\ &\times \int_{-\infty}^{\infty} d\mathbf{v}\, W(\mathbf{v}) \left[\frac{N_3}{\gamma_{12} + i\Delta_{21} + i\mathbf{k}_p \cdot \mathbf{v}} \right. \\ &\times \left(\frac{1}{\gamma_{23} + i\Delta_{32} - i\mathbf{k}_f \cdot \mathbf{v}} + \frac{1}{\gamma_{23} - i\Delta_{32} + i\mathbf{k}_p \cdot \mathbf{v}} \right) \\ &+ \frac{N_1}{\gamma_{23} + i\Delta_{32} + i\mathbf{k}_p \cdot \mathbf{v}} \\ &\left. \times \left(\frac{1}{\gamma_{12} + i\Delta_{21} - i\mathbf{k}_f \cdot \mathbf{v}} + \frac{1}{\gamma_{12} - i\Delta_{21} + i\mathbf{k}_p \cdot \mathbf{v}} \right) \right]. \end{aligned} \tag{66b}$$

N_1 and N_3 are the initial populations of states $|1\rangle$ and $|3\rangle$, respectively.

Detailed experimental studies of this effect and the important lineshape properties of the resonances are currently under way. However, clear evidence of both crossovers is shown in Fig. 32. The shoulder on the low-frequency side of the 3s $^2S_{1/2}(F = 2) \rightarrow$ 3p $^2P_{3/2}(F = 3)$ transition shown in Fig. 32a is due to two *V*-type crossovers, and the signal shown in Fig. 32b located centrally between the two principal transitions (labeled B) is due to an inverted *V*-type crossover (Schlossberg and Javan, 1966).

IV. Conclusions

We have discussed the physics of DFWM in media where the nonlinear polarization arises from saturation of resonant transitions. Starting with the density matrix formulation, we derived the nonlinear polarization for a simple two-level system, which provided a physical understanding of the mechanisms underlying the nonlinear response of resonant systems. Successive complications such as atomic motion, saturation, and multi-level effects were then added, along with examples of experimental observations.

In Chapters 13 and 14, applications of phase conjugation will be extensively discussed. Here, we wish to point out where *resonant* effects in DFWM can be used to advantage. DFWM in Kerr-like media (see Chapters 2–5) generally allows for very fast (subnanosecond) time response of the nonlinear medium but relatively modest reflectivities. For this reason, high-power pulsed lasers or very long interaction lengths (in fibers) are required to achieve reasonable phase-conjugate returns. Resonant DFWM, however, allows one to trade off response time (~nsec) for large nonlinear response. In effect, one can "engineer" a nonlinear material by choosing a resonance interaction for the largest conjugate reflectivity *R* giving the required response time. One example is the recent demonstration by Lind and Steel, (1981) of a cw laser where one of the laser "mirrors" is formed by DFWM in atomic sodium vapor. Continuous-wave phase-conjugate reflectivity greater than 100% was achieved with fast response time (<20 nsec), providing the first demonstration of a real-time, cw phase-conjugate laser resonator.

To date, spectroscopic applications of resonant DFWM have received the most attention. The Doppler-free nature of the interaction and the large signal-to-noise ratio (achieved by separating the probe and return waves and using heterodyne detection) have resulted in numerous sub-Doppler resolution experiments, discussed in this chapter. Multilevel effects and degeneracies have allowed additional polarization-dependent interactions, leading to a new understanding of phenomena in resonantly excited systems. The dependence of the DFWM signal strength and bandwidth on atomic motion has also provided a new tool for investigation of collision phenomena in vapors and of diffusion of resonance excitation in solids.

Resonant DFWM promises to provide efficient phase conjugation for such diverse applications as laser fusion, photolithography, real-time holography, optical data processing, high-energy laser aberration correction, and narrow-band optical filtering.

Undoubtedly, many additional experiments will evolve from DFWM research. For example, DFWM is currently being used at Hughes Research Laboratories to study atomic Rydberg levels near the ionization limit in atomic species. The applications and scientific benefits arising from the study of resonant DFWM have already been numerous, and will undoubtedly continue to grow in the future.

Acknowledgments

We would like to thank Dr. R. A. McFarlane for a critical reading of this chapter, and G. J. Dunning for computational assistance. We would also like to express our appreciation to Mrs. Cheryl Krieger for her excellent typing of the manuscript.

References

Abrams, R. L., and Lind, R. C. (1978). Degenerate four-wave mixing in absorbing media, *Opt. Lett.* **2,** 94; **3,** 203 (1978).

Anan'ev, Yu. A., Gorlanov, A. V., Grishmanova, N. I., Sventsiskaya, N. A., and Solov'ev, V. D. (1979). Transient self-diffraction of coherent light beams in an absorbing liquid, *Kvantovaya Elektron.* **6,** 1813 [*English transl.: Sov. J. Quantum Electron.* **9,** 1072 (1979)].

Apanasevich, P. A., Afanas'ev, A. A., and Urbanovich, A. I. (1975). Mechanism of diffraction of light by optically induced gratings in absorbing media. *Kvantovaya Elektron.* **2,** 2423 [*English transl.: Sov. J. Quantum Electron.* **5,** 1320 (1976)].

Bloch, D. (1980). Conjugaison de phase dans les milieux gazeux. Spectroscopic de saturation heterodyne. These de Troisiene Cycle, Univ. de Paris-Nord.

Bloch, D., Raj, R. K., Snyder, J. J., and Ducloy, M. (1980). High-frequency optically heterodyned saturation spectroscopy via resonant degenerate four-wave mixing, *Int. Quantum Electron. Conf. Digest Tech. Papers J. Opt. Soc. Am., 11th* **70,** 624.

Bloch, D., Raj, R. K., and Ducloy, M. (1981). Doppler-free heterodyne spectroscopy of H_a. Measurement of the $2S_{1/2}$ collisional quenching in gas cell, *Opt. Commun.* **37,** 183.

Bloembergen, N., (1965). "Nonlinear Optics," Benjamin, New York.

Bloom, D. M., Liao, P. T., and Economou, N. P. (1978). Observation of amplified reflection by degenerate four-wave mixing in atomic sodium vapor, *Opt. Lett.* **2,** 58.

Brown, W. P. (1981). Pump Attenuation and Depletion Effects in Degenerate Four-Wave Mixing, CLEO '81, Washington D.C., June 10.

Carquille, B., and Froehly, C. (1980). Real-time high-resolvance image correlation by Bragg diffraction in saturable absorbers, *Appl. Opt.* **19,** 207.

Depatie, D., Haueisen, D., Elci, A., and Rogovin, D. (1979). Observation of infrared phase conjugation in molecular gases, *Proc. Int. Conf. Lasers '78* (V. J. Corcoran, ed.), p. 525. STS Press, McLean, Virginia.

Dunning, G. J., and Lam, J. F. (1981). Optical Phase Conjugation in Saturable Absorbers. CLEO, Washington D.C. 1981.

Dunning, G. J., and Steel, D. G. (1982). Effects of unequal pump intensity in resonantly enhanced degenerate four-wave mixing, *IEEE J. Quant. Elec.*, **QE-18,** 3.

Eichler, H. J. (1978). Festkörperproblems, *Adv. Solid State Phys.* **18,** 241.

Elci, A., and Rovovin, D. (1979). Phase conjugation in nonlinear molecular gases, *Chem. Phys. Lett.* **61,** 407.

Elci, A., and Rogovin, D. (1980). Phase conjugation in an inhomogeneously broadened medium, *Opt. Lett.* **5,** 255.

Fan, J. Y., Wu, C. K., Wang, Z. Y. (1980). The theory of resonantly enhanced four-wave mixing in absorbing media, *Acta Phys. Sinica* **29,** 879 [*English trans.: Chinese Phys.* **1,** 570 (1981)].

Feld, M. S., Burns, M. M., Kühl, T. U., and Pappas, P. G. (1980). Laser-saturation spectroscopy with optical pumping, *Opt. Lett.* **5,** 79.

Feynman, R. P., Vernon, F. L., and Hellwarth, R. W. (1957). Geometrical representation of the Schrodinger equation for solving maser problems, *J. Appl. Phys.* **28,** 49.

Fisher, R. A., and Feldman, B. J. (1979). On-resonant phase-conjugate reflection and amplification at 10.6 μm in inverted CO_2, *Opt. Lett.* **4,** 140.

Fried, B. D., and Conte, S. D. (1961). "The Plasma Dispersion Function." Academic Press, New York.

Fu, Y. Y., and Sargent, M. III (1979). Effects of signal detuning on phase conjugation, *Opt. Lett.* **4,** 366.

Fu, T. Y., and Sargent, M. III (1980). Theory of two-photon phase conjugation, *Opt. Lett.* **5,** 433.

Grischkowsky, D. (1970). Self-focusing of light by potasium vapor, *Phys. Rev. Lett.* **24,** 866.

Grischkowsky, D., Shiren, N. S., and Bennett, R. J. (1978). Generation of time-reversed wavefronts using a resonantly enhanced electronic nonlinearity, *Appl. Phys. Lett.* **33,** 805.

Harter, D. J., and Boyd, R. W. (1980). Nearly degenerate four-wave mixing enhanced by the A. C. Stark effect, *IEEE J. Quantum Electron.* **QE-16** (10), 1126.

Haueisen, D. C. (1979). Doppler-free two-photon spectroscopy using degenerate four-wave mixing, *Opt. Commun.* **28,** 183.

Hellwarth, R. W. (1977). Generation of time-reversed wave fronts by nonlinear refraction, *J. Opt. Soc. Am.* **67,** 1.

Humphrey, L. M., Gordon, J. P., and Liao, P. F. (1980). Angular dependence of line shape and strength of degenerate four-wave mixing in a Doppler-broadened system with optical pumping, *Opt. Lett.* **5,** 56.

Ivakin, E. V., Petrovich, I. P., Rubinov, A. S., and Stepanov, B. I. (1975). Dynamic holograms in an amplifying medium, *Kvantovaya Elektron.* **2,** 1556 [*English transl.: Sov. J. Quantum Electron.* **5,** 840 (1975)].

Ivakin, E. V., Koptev, V. G., Lazaruk, A. M., Petrovich, I. P., and Rubanov, A. S. (1979). Phase conjugation of light fields as a result of nonlinear interaction in saturable media, *Zh. Eksp. Teor. Fiz. Pis'ma Red.* **30,** 648 [*English transl.: JETP Lett.* **30,** 613 (1979)].

Lam, J. F., and Abrams, R. L. (1982). Theory of nonlinear optical coherences in resonant degenerate four-wave mixing, *Phys. Rev. A.* **26,** 1539.

Lam, J. F., Steel, D. G., McFarlane, R. A., and Lind, R. C. (1981). Atomic coherence effects in resonant degenerate four-wave mixing, *Appl. Phys. Lett.* **38,** 977.

Lam, J. F., Steel, D. G., and McFarlane, R. A. (1982). Collisionally induced narrowing of the longitudinal relaxation linewidth in nearly degenerate four-wave mixing. *Phys. Rev. Lett.* **49,** 1628.

Larsen, D. M., and Bloembergen, N. (1976). Excitation of polyatomic molecules by radiation, *Opt. Commun.* **17,** 254.

Lazaruk, A. M. (1979). Wavefront reversal in amplifying dynamic dye-solution holograms, *Kvantovaya Elektron.* **6,** 770 [*English transl.: Sov. J. Quantum Elecron.* **9,** 1041 (1979)].

Liao, P. F., and Bloom, D. M. (1978). Continuous-wave backward-wave generation by degenerate four-wave mixing in ruby, *Opt. Lett.* **3,** 4.

Liao, P. F., Economou, N. P., and Freeman, R. R. (1977). Two-photon coherent transient measurements of Doppler-free linewidths with broadband excitation, *Phys. Rev. Lett.* **39,** 1473.

Lind, R. C., and Steel, D. G. (1981). A CW Dye Laser with a Phase Conjugate Mirror. CLEO, Washington D.C.; Demonstration of the longitudinal modes and aberration correction properties of a cw dye laser with a phase conjugate mirror, *Opt. Lett.* **6,** 554 (1981).

Lind, R. C., Steel, D. G., Klein, M. B., Abrams, R. L., Giuliano, C. R., and Jain, R. K. (1979). Phase conjugation at 10.6 μm by resonantly enhanced degenerate four-wave mixing, *Appl. Phys. Lett.* **34,** 457.

Lind, R. C., Steel, D. G., and Dunning, G. J. (1982). Phase conjugation by resonantly enhanced degenerate four-wave mixing, *Opt. Eng.* **21,** 190.

Matsuoka, M., (1975). Doppler-free two-photon induced coherence and emission, *Opt. Commun.* **15,** 84.

Moses, E. I., and Wu, F. Y. (1980). Amplification and phase conjugation by degenerate four-wave mixing in a saturable absorber, *Opt. Lett.* **5,** 64.

Narducci, L. M., Eidson, W. W., Furcinitti, P., and Eteson, D. C. (1977). Theory of a 2-photon laser amplifier, *Phys. Rev. A* **16,** 1665.

Nilsen, J., and Yariv, A. (1979). Nearly degenerate four-wave mixing applied to optical filters, *Appl. Opt.* **18,** 143.

Nilsen, J., and Yariv, A. (1981). Nondegenerate four-wave mixing in a doppler broadened resonant medium, *J. Opt. Soc. Am.* **71,** 180.

Nilsen, J., Gluck, N. S., and Yariv, A. (1981). Narrow-band optical filter through phase conjugation by nondegenerate four-wave mixing in sodium vapor, *Opt. Lett.* **6,** 380.

Oraevskii, A. N. (1979). Possible use of resonantly excited media in phase conjugation, *Kvantovaya Elektron.* **6,** 218 [*English transl.: Sov. J. Quantum Electron.* **9,** 119 (1979)].

Ostrovskii, Yu. I., Sidorovich, V. G., Stastl'ko, D. I., and Tanin, L. V. (1975). Dynamic holograms in sodium vapor, *Zh. Eksp. Teor. Fiz. Pis'ma Red.* **1,** 1030 [*English transl.: Sov. Tech. Phys. Lett.* **1,** 442 (1975)].

Pappas, P. G., Burns, M. M., Hinshelwood, D. D., Feld, M. S., and Murnick, D. E. (1982). Saturation spectroscopy with laser optical pumping in atomic barium, *Phys. Rev. A* **21,** 1955.

Pepper, D. M., and Abrams, R. L. (1978). Narrow optical bandpass filter via nearly degenerate four-wave mixing, *Opt. Lett.* **3,** 212.

Popov, A. K., and Shalev, V. M. (1980). Doppler-free spectroscopy and wavefront conjugation by four-wave mixing of non-monochromatic waves, *Appl. Phys.* **21,** 93.

Raj, R. K., Bloch, D., Snyder, J. J., Camy, G., and Ducloy, M. (1980). High frequency optically heterodyned saturation spectroscopy via resonant degenerate four-wave mixing, *Phys. Rev. Lett.* **44,** 1251.

Salcedo, J. R., Siegman, A. E., Dlott, D. D., and Fayer, M. D. (1978). Dynamics of energy transport in molecular crystals: The picosecond transient grating method, *Phys. Rev. Lett.* **41,** 131.

Schlossberg, H., and Javan, A. (1966). Saturation behavior of a Doppler-broadened transition involving levels with closed spaced structure, *Phys. Rev.* **150,** 267.

Shtyrkov, E. I. (1970). Scattering of light by a periodic structure of excited and unexcited atoms, *Zh. Eksp. Teor. Fiz. Pis'ma Red.* **12,** 134 [*English transl.: JETP Lett.* **12,** 92 (1970)].

Steel, D. G., and Lam, J. F. (1979). Two-photon coherent transient measurements of the nonradiative collisionless dephasing rate in SF_6 via Doppler-free degenerate four-wave mixing, *Phys. Rev. Lett.* **43,** 1588.

Steel, D. G., and Lam, J. F. (1980). Multiline phase conjugation in resonant materials, *Opt. Lett.* **5,** 297.

Steel, D. G., and Lam, J. F. (1981). Saturation effects in inhomogeneous broadening in Doppler-free degenerate four-wave mixing, *Opt. Commun.* **6,** 587.

Steel, D. G., and Lind, R. C. (1981). Multiresonant behavior in nearly degenerate four-wave mixing, *Opt. Lett.* **6,** 587.

Steel, D. G., Lind, R. C., Lam, J. F., and Giuliano, C. R. (1979). Polarization-rotation and thermal-motion studies via resonant degenerate four-wave mixing, *Appl. Phys. Lett.* **35,** 376.

Steel, D. G., Lind, R. C., and Lam, J. F. (1981). Degenerate four-wave mixing in a resonant homogeneously broadened system, *Phys. Rev. A* **23,** 2513.

Stepanov, B. I., Ivakin, E. V., Rubanov, A. S. (1971). Recording two-dimensional and three-dimensional dynamic holograms in bleachable substances, *Dokl. Akad. Nauk SSSR* **196,** 567 [*English transl.: Sov. Phys. Dokl.-Tech. Phys.* **16,** 46 (1971)].

Tan-No, N., Hishimiya, T., and Inaba, H. (1980). Dispersion-free amplification and oscillation in phase-conjugate four-wave mixing in an atomic vapor doublet, *IEEE J. Quantum Electron.* **QE-16,** 147.

Tocho, J. D., Sibbett, W., and Bradley, D. J. (1980). Picosecond phase-conjugate reflection from organic dye saturable absorbers, *Opt. Commun.* **34,** 122.

Tocho, J. D., Sibbett, W., and Bradley, D. J. (1981). Thermal effects in phase conjugation in saturable absorbers with picosecond pulses, *Opt. Commun.* **37,** 67.

Tomita, A., (1979). Phase conjugation using gain saturation of a ND:YAG laser, *Appl. Phys. Lett.* **34,** 463.

Wandzura, S. M., (1979). Effects of atomic motion on wavefront conjugation by resonantly enhanced degenerate four-wave mixing, *Opt. Lett.* **4,** 208.

Watkins, D. E., Thomas, S. J., and Figueira, J. F. (1980a). Phase conjugation via resonantly enhanced degenerate four-wave mixing in a doped alkali halide, *Int. Quantum Electron. Conf. Digest Tech. Papers J. Opt. Soc. Am., 11th* **70,** 600.

Watkins, D. E., Figueira, J. F., and Thomas, S. J. (1980b). Observation of resonantly enhanced degenerate four-wave mixing in doped alkali-halides, *Opt. Lett.* **5,** 169.

Woerdman, J. P. (1970). Formation of a transient free carrier hologram in Si, *Opt. Commun.* **2,** 212.

Woerdman, J. P., and Schuurmans, M. F. H. (1980). Wave-front conjugation in atomic sodium vapor: Rabi regime, *Int. Quantum Electron. Conf. Digest Tech. Papers J. Opt. Soc. Am.* **70,** 598.

Wu, C. K., Fan, J. Y., and Wang, Z. Y. (1980a). Generation of phase conjugated backward wave by degenerate four-wave mixing in chlorophyll solution, *Acta Phys. Sinica* **29,** 305.

Wu, C. K., Cui, Y. Z., and Wang, Z. Y. (1980b). Degenerate four-wave mixing and phase conjugation in organic dye solutions, *Acta Phys. Sinica* **29,** 937.

Yariv, A. (1976). Three-dimensional pictorial transmission in optical fibers, *Appl. Phys. Lett.* **28,** 88.

Zel'dovich, B. Ya, Popovichev, V. I., Ragul'skii, V. V., and Faizullov, F. S. (1972). Connection between the wave fronts of the reflected and exciting light in stimulated Mandel'Shtam-Brillouin scattering, *Zh. Eksp. Teor. Fiz Pis'ma Red.* **15,** 160 [*English transl.: Sov. Phys. JETP* **15,** 109 (1972)].

9 Phase Conjugation from Nonlinear Photon Echoes*

John C. AuYeung†
California Institute of Technology
Pasadena, California

I. Introduction

As discussed in Chapter 8, a two-level medium is a simple system in which the mechanism of conjugate wave-front generation can be easily understood. A theoretical treatment of optical phase conjugation in a two-level system under a steady-state condition was made by Abrams and Lind (1978). Experiments on time-reversed wave-front generation in such media have been reported by Bloom *et al.* (1978), Liao and Bloom (1978), Liao *et al.* (1978), Fisher and Feldman (1979), Lind *et al.* (1979), Tomita (1979), Steel and Lam (1980) and Watkins *et al.* (1980). In addition to the model described by Abrams and Lind, Grischkowsky *et al.* (1978) have observed in rubidium vapor phase conjugation due to nonlinear mixing described by the adiabatic-following model of a two-level medium. Typically, a medium interacts with cw optical fields or optical pulses such that continuous or quasi-continuous absorption and phase gratings are formed in it. With a counterpropagating pump beam geometry, the phase-matching condition dictates that a wave diffracted by these gratings be directed in a direction opposite to that of a probe wave. The duration of

* Work supported by the National Aeronautics and Space Administration under Contract No. NAS7-100.

† Present address: Newport Corporation, Fountain Valley, California 92708.

Optical Phase Conjugation

ISBN 0-12-257740-X

the field–matter interaction is much longer than the dephasing time of the two-level system. The population grating induced by the optical fields is maintained in space for a "wash-out" period determined by the relaxation time T_1, by diffusion, by charge migration, or by other mechanisms depending on the specific properties of the medium. It is therefore possible in an experiment to allow a readout pump pulse (e.g., E_2) to be separated from a reference pump pulse (e.g., E_1) and a probe pulse E_p by a time delay less than this period. However, the grating must be formed by the interference of two simultaneously incident pulses (e.g., E_1 and E_p).

In this chapter we shall discuss optical phase conjugation in the opposite regime. The behavior of a two-level system exposed to pulses on an ultrashort time scale is far different from that observed in the cw or quasi-steady-state situation. When the optical fields are pulses with pulse durations shorter than the washout period of the phase memory of the individual atomic quantum states, a transient population grating can be induced in the medium on this time scale by nonsimultaneously incident optical pulses. A conjugate optical pulse is spontaneously radiated at a later time by the medium in the form of a photon echo. The echo wave front propagates in either the forward or the backward direction, depending on the directions of the wave vectors of the exciting pulses. Experimental observations of both types of echoes with phase-reversal properties have been reported by Griffen and Heer, (1978), Shtyrkov *et al.* (1978), and Fujita *et al.* (1979). Despite some unique features of this special phase-conjugation technique, a close analogy can still be drawn between this transient nonlinear process and conventional degenerate four-wave mixing (Shiren, 1978; and Yariv and AuYeung, 1979). In the next section, the properties of a two-level medium and its interaction with ultrashort optical pulses will be described. The vector model of the Schrödinger equation will be used to describe qualitatively the coherent–cooperative phenomenon of the photon echo. A mathematical derivation of the photon echo with its phase-conjugate properties will be presented in Section III, followed by a comparison with degenerate four-wave mixing. Finally, in Section IV the influence of atomic motion on the echo-conjugation effect will be discussed.

II. Photon Echo in a Two-Level System

A photon echo is an optical version of the nuclear-spin echo discovered by Hahn (1950). When illuminated by a sequence of two pulses separated in time by τ, a medium reradiates a third pulse—the photon echo—at a time τ after the arrival of the second pulse. It was first observed in ruby by

Kurnit *et al.* (1964). The echo phenomenon can be produced in inhomogeneously broadened systems with a finite number of quantum states coupled by resonant interactions. A nuclear-spin system is a two-level system with a spin-up ($\uparrow$) state and a spin-down ($\downarrow$) state, which can be coupled by an external transverse rf magnetic field. Protons and electrons in magnetic fields also have two spin states and can produce spin echoes. An isolated atom in a gas or an impurity ion in a solid such as chromium in ruby crystal can also be considered as two-level systems capable of being excited through an interaction of the electric dipole moment with an external optical field. A medium is inhomogeneously broadened when the individual atoms in it can be distinguished from one another because each has a slightly different resonant frequency. The inhomogneous distribution in the resonant frequencies is brought about by a Doppler shift due to the thermal motion in an atomic gas, or by a Stark shift due to the local crystalline field in the case of impurity ions in a solid. In a small volume $\sim\lambda^3$ at a point $\mathbf{r}$ inside the medium, the number of atoms is sufficiently large that a continuous distribution function $g(\Delta\omega)$ can be used to characterize the frequency variation among the atoms. $\Delta\omega$ is the deviation in resonant frequency from a central frequency value ω_0. The value of a macroscopic quantity at the point $\mathbf{r}$ inside the medium is the result of an average of the corresponding microscopic quantities over the distribution $g(\Delta\omega)$. For example, the optical radiation that originates at $\mathbf{r}$ is due to the macroscopic polarization contributed by all the atoms in the volume λ^3 at $\mathbf{r}$. It is precisely the summation of the dipole moments weighted by the distribution of the resonant frequencies among the atoms within the small volume that gives rise to the coherent–cooperative effect of the photon echo.

There are numerous theoretical discussions of the photon echo in the literature (Abella *et al.*, 1966; Scully *et al.*, 1968; Alekseev and Evseev, 1969; Gordon *et al.*, 1969; Alekseev and Evseev, 1975; and Mossberg *et al.*, 1979). The discussion here will be confined to the photon echoes from two-level media. The properties of a two-level system are completely specified by its wave function $\psi(t)$, which consists of three variables: the probability amplitudes of the states and the relative phase between them. When a single two-level atom (as shown in Fig. 1), initially in a pure quantum state, is irradiated by a resonant or near-resonant electromagnetic pulse, its atomic wave function evolves after the interaction as a linear superposition of the two atomic eigenstates, with a definite phase relationship maintained between the probability amplitudes $a(t)$ and $b(t)$ of the two states:

$$\psi(t) = a(t)|a\rangle + b(t)|b\rangle. \tag{1}$$

Evolution of the wave function with time both with and without external

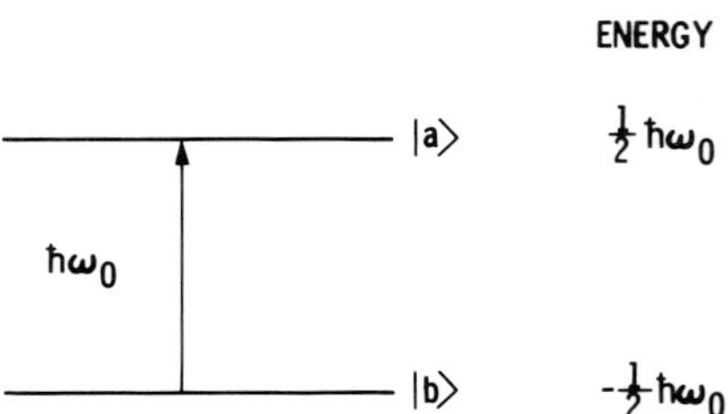

Fig. 1. Two-level system with energy separation $\hbar\omega_0$.

excitation is governed by the Schrödinger equation

$$i\hbar \frac{d}{dt} \psi(t) = (\mathcal{H}_0 + V)\psi(t), \tag{2}$$

where $\mathcal{H}_0$ is the field-free Hamiltonian and V the interaction due to the external field. The phase of the incident optical field is retained in the wave function until it is erased by some irreversible-dephasing mechanism such as atomic collisions. Therefore, during the transient period of coherent superposition of its two basis states, an atom acts as a storage medium capable of memorizing the phase distribution of an external field. As a result, a grating can be established in the medium by the interference of two separate optical pulses, provided that the interpulse time interval is shorter than the irreversible-dephasing time T_2. The diffraction of a third optical field by a grating formed under such a condition was observed experimentally by Shtyrkov and Samartsev (1978), and Shtyrkov *et al.* (1978, 1980).

The spin or photon echo results when an inhomogeneously broadened two-level system interacts with a sequence of exciting pulses. Its generation mechanism is most easily visualized by using the vector model of the Schrödinger equation proposed by Feynman *et al.* (1957) to describe geometrically the behavior of a two-level system with a three-vector **P**. The coordinate axes in the three-dimensional space are labeled 1, 2, and 3. The components of **P** are real functions of the probability amplitudes, with

$$P_1 = ab^* + ba^*, \tag{3}$$

$$P_2 = i(ab^* - ba^*), \tag{4}$$

$$P_3 = aa^* - bb^*, \tag{5}$$

where $|\mathbf{P}|^2 = |a|^2 + |b|^2 = 1$. An asterisk denotes complex conjugation. The equation of motion of **P** can be obtained from the Schrödinger equation as a vector cross product:

$$d\mathbf{P}/dt = \mathbf{\Omega} \times \mathbf{P}. \tag{6}$$

The motion of **P** is that of a precession about a vector **Ω** in the same space. **Ω** is also a three-vector, with

$$\Omega_1 = (V_{ab} + V_{ba})/\hbar, \tag{7}$$

$$\Omega_2 = i(V_{ab} - V_{ba})/\hbar, \tag{8}$$

$$\Omega_3 = \omega_0. \tag{9}$$

V_{ij} is the transition energy $\langle i|V|j\rangle$, and ω_0 the resonant frequency.

The formalism applies to any two-level system. We shall use the example of a two-level atom subject to a resonant external circularly polarized pulse

$$\mathbf{E}(t) = \mathscr{E}_0(t)(\mathbf{x} \cos \omega_0 t + \mathbf{y} \sin \omega_0 t). \tag{10}$$

The applied field stimulates an electric dipole transition through the interaction Hamiltonian $V = -\mu_x E_x - \mu_y E_y$. In a coordinate system rotating about the **3** axis at an angular frequency ω_0, the equation of motion of **P** (**P**′ in the new frame) given by Eq. (6) is transformed to

$$d\mathbf{P}'/dt = (-2\mu\mathscr{E}_0/\hbar)(\mathbf{1}' \times \mathbf{P}'). \tag{11}$$

The axes in the rotating frame are labeled 1′, 2′, and 3′.

P′ is constant in the absence of an applied field. When the circularly polarized field is applied, **P**′ precesses around a stationary vector of magnitude $-2\mu\mathscr{E}_0/\hbar$ and parallel to the **1**′ axis. An atom in its lower energy state at time $t = 0$ has $\mathbf{P}' = (0, 0, -1)$ as shown in Fig. 2a. Upon the incidence of the pulse given by Eq. (10), **P**′ rotates about the **1**′ axis starting from its initial position. In Fig. 2b is shown the situation where **P**′ has rotated to a final position along the negative **2**′ axis when the field is turned off. The applied pulse in this case is referred to as a $\pi/2$ pulse, because it causes **P**′ to rotate 90°. All the atoms in the medium with the same initial states have their **P**′ vectors aligned along the same direction, and a large macroscopic polarization is created in the sample. If all atoms had identical resonant frequencies, their **P**′ vectors would remain parallel to one another, because with no external field present, **P** precesses about the **3** axis at the resonant frequency of the atom. Because our sample is inhomogeneously broadened, the variation in resonant frequency from atom to atom causes the precession rate of **P** for each atom to be sightly different. In the primed coordinate system rotating at fixed angular frequency ω_0, the **P**′ of atoms with resonant frequencies larger than ω_0 will move ahead and those of atoms with smaller resonant frequencies will lag behind. This situation is shown in Fig. 2c, where $\omega_{\text{i}} > \omega_{\text{ii}} = \omega_0 > \omega_{\text{iii}} > \omega_{\text{iv}}$. The fanning out of

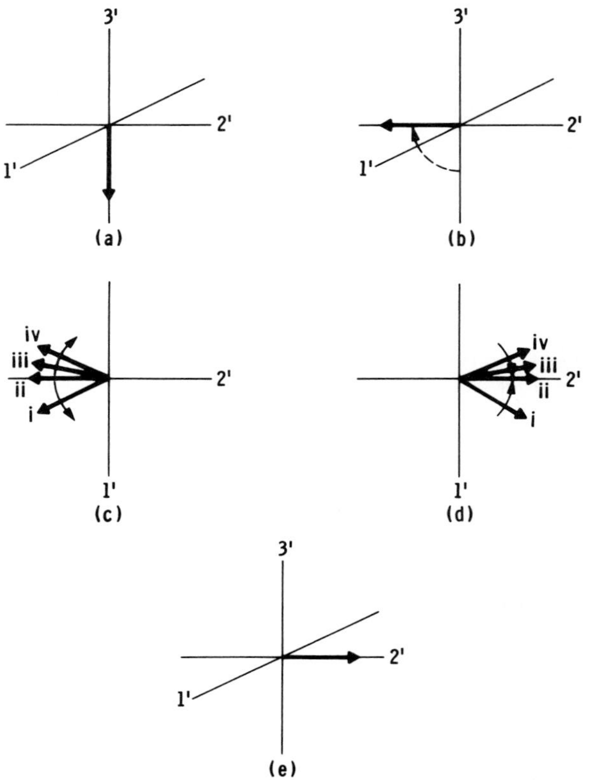

Fig. 2. $\mathbf{P}'$ in the rotating coordinate system (a) at $t = 0$ before the incidence of a $\pi/2$ pulse and (b) immediately after the $\pi/2$ pulse, (c) at $t = \tau$ before the incidence of a π pulse, (d) immediately after the π pulse, and (e) at $t = 2\tau$.

the $\mathbf{P}'$ vectors in the transverse plane implies that the induced microscopic dipole moments of the individual atoms begin to fall out of step with one another, and consequently cannot contribute constructively to maintain the macroscopic polarization. The optical radiation emitted by this polarization source decays at a rate $1/T_2^*$, the inhomogeneous linewidth. This phenomenon is referred to as free-induction decay. The dephasing of the microscopic dipole moments among the atoms happens even though each atom is still in a coherent superposition of its basis states, the inverse inhomogeneous linewidth T_2^* being shorter than T_2. Hence, the phase information of the external field is not lost, because it is still retained by each atom.

The dephasing of the dipole moments via T_2^* is reversible. When a second optical pulse is applied at a time τ after the first pulse, the $\mathbf{P}'$ vector of each atom rotates again about the $\mathbf{1}'$ axis. Specifically, the $\mathbf{P}'$ vectors can be made to rotate exactly 180° about the $\mathbf{1}'$ axis, such that they all end up back in the transverse plane and with the relative phase angle between the vectors of any two atoms in the medium exactly the negative of what it was before the incidence of the second pulse. This second pulse is called a π pulse because it rotates the vectors π radians about a particular axis. In Fig. 2d the final positions of the vectors after the π pulse are shown. Vectors which were ahead before the arrival of the π pulse are now behind and those behind are now ahead, because of the 180° rotation. After the π pulse is over, the $\mathbf{P}$ vectors freely precess again in the transverse plane, each at its own resonant frequency. It takes exactly the same time interval τ in the rotating frame for the $\mathbf{P}'$ vectors that were lagging behind to catch up because of their larger resonant frequencies and for those that were ahead to fall behind because of their smaller precession rates. Thus, at time $t = 2\tau$, a large macroscopic polarization is once again formed in the medium. The individual dipole moments are all in step and again add coherently as shown in Fig. 2e. This delayed macroscopic polarization radiates the photon echo. When the dipole moments are randomly phased, they still radiate at the normal spontaneous emission rate, with the radiated energy proportional to N, the number density of the atoms. When their phases become equal, the collective mode of radiation has a N^2 dependence, which indicates a superradiance phenomenon. As the precession in the transverse plane continues, the vectors again dephase and the echo disappears. The echo pulse duration is therefore approximately equal to T_2^*. The entire process can take place within a time shorter than or comparable with the irreversible-dephasing time (i.e., $T_2^* < 2\tau < T_2$). Within this interval, the phase information of the external fields is retained in the atomic wave functions. The spatial variation of any externally applied optical field of interest is slight within one wavelength λ. From a small volume $\sim\lambda^3$ centered at a point $\mathbf{r}$ around which the electromagnetic field remains fairly constant, the radiated echo field bears amplitude and phase information related to that of the incident fields at this spatial point. As the incident fields vary over the medium, their entire spatial variations are stored and subsequently reconstructed by the echo emitted from the medium.

III. Phase Conjugation by Photon Echoes

Since its first observation, the photon echo has been extensively used as a tool to provide information on relaxation processes and to determine atomic lifetimes and scattering cross sections. Shtyrkov and Samartsev

(1976) pointed out that a system of atoms excited into a superposition state can retain the spatial phase distribution of the excited field. Such a medium can therefore serve as the recording material of a dynamic hologram. They also pointed out that a conventional two-pulse sequence can be used to produce a forward-propagating echo that is a phase-conjugate version of the first pulse in the sequence. Heer and McManamon (1977) later suggested the use of a two-pulse photon echo to correct wave-front distortion. Shiren (1978) and Fujita *et al.* (1979) described the generation of a backward echo by using a standing-wave second pulse or a three-pulse sequence. An analogous situation had actually been observed earlier in phonon echoes. Backward ultrasonic spin echoes were first reported by Shiren and Kazyaka (1972), and the time-reversal property of the backward phonon echoes was recognized by Shiren and Melcher (1976).

In the previous section, a qualitative description of the photon-echo phenomenon in an atomic medium was presented. The special example of a $\pi/2$ and π pulse sequence was discussed. In this section, phase-conjugate echoes will be shown to be generated by using several different pulse sequences and geometries. A quantitative expression for the radiated echo wave front for arbitrary pulse areas will be derived. To simplify the derivation, we shall assume an optically thin sample. We shall first derive the macroscopic polarization of the medium subject to a sequence of exciting electromagnetic pulses, and then treat the polarization as a source of the echo radiation.

Consider a sample consisting of stationary atoms with finite energy levels, two of which can be coupled by an applied monochromatic field through a dipole transition. When an external optical pulse

$$\mathbf{E}_\alpha(\mathbf{r}, t) = \mathbf{x}\tfrac{1}{2}[\mathscr{E}_\alpha(\mathbf{r}, t)\exp\{i[\phi_\alpha(\mathbf{r}) + \mathbf{k}_\alpha \cdot \mathbf{r} - \omega t]\} + \text{c.c.}] \tag{12}$$

is incident onto the sample at the time t_α, the interaction Hamiltonian of Eq. (2) is given by $V_{ij} = -(\boldsymbol{\mu} \cdot \mathbf{E}_\alpha)_{ij}$. We have assumed a linearly polarized field with spatial amplitude and phase variations. $\mathscr{E}_\alpha(\mathbf{r}, t)$ is a real function, varying slowly in space and time, and the phase information is carried in $\phi_\alpha(\mathbf{r})$. The applied optical pulse is in resonance with the atomic medium, with $\omega \sim \omega_0$. The pulse duration is typically so short that we can neglect T_2^* dephasing during the interaction. That is, $|\omega - \omega_0|\,\Delta t \ll 1$, where Δt is the pulse width. This condition is equivalent to saying that the Rabi flopping frequency $\mu\mathscr{E}_\alpha/\hbar$ is much greater than $|\omega - \omega_0|$. In a gaseous medium with moving atoms, this condition implies that the pulse duration is sufficiently short that the atoms can be considered stationary during the pulse. As a result of this simplification, a small delay in the precise time of the echo radiation will be dropped. The wave function is solved from Eq.

(2), to give

$$\psi(\mathbf{r}, t) = \exp[-i\mathcal{H}_0(t - t_\alpha)]U_\alpha(\mathbf{r})\psi(\mathbf{r}, t_\alpha), \tag{13}$$

where $U_\alpha(\mathbf{r})$ can be represented as

$$U_\alpha(\mathbf{r}) = \begin{bmatrix} \cos[\theta_\alpha(\mathbf{r})/2] & i\sin[\theta_\alpha(\mathbf{r})/2] \\ & \times \exp\{i[\phi_\alpha(\mathbf{r}) + \mathbf{k}_\alpha \cdot \mathbf{r} - \omega t_\alpha]\} \\ i\sin[\theta_\alpha(\mathbf{r})/2] & \cos[\theta_\alpha(\mathbf{r})/2] \\ \times \exp\{-i[\phi_\alpha(\mathbf{r}) + \mathbf{k}_\alpha \cdot \mathbf{r} - \omega t_\alpha]\} & \end{bmatrix}. \tag{14}$$

$\theta_\alpha(\mathbf{r})$ is the pulse area defined as $\int (\mu/\hbar)\mathscr{E}_\alpha(\mathbf{r}, t')\, dt'$. For example, a $\pi/2$ pulse has a $\mathscr{E}_\alpha(t)$ such that $\theta_\alpha(\mathbf{r}) = \pi/2$. The phase information of the field is contained in the off-diagonal elements of $U_\alpha(\mathbf{r})$, and hence in the probability amplitude of the final state to which the atom makes a transition when stimulated by the incident field.

A. Two-Pulse Echo

We begin by discussing the conventional two-pulse echo in a geometry first used by Kurnit *et al.* (1964) and shown in Fig. 3. Consider an atom located at $\mathbf{r}$ in its lower energy level at time $t = t_p$, that is, $a(\mathbf{r}, t = t_p) = 0$ and $b(\mathbf{r}, t = t_p) = 1$. It is irradiated by two short and intense optical pulses $\mathbf{E}_p(\mathbf{r}, t)$ and $\mathbf{E}_1(\mathbf{r}, t)$ given by Eq. (12) at t_p and t_1, respectively. The wave vectors $\mathbf{k}_p$ and $\mathbf{k}_1$ need not be parallel in general. In fact, in an experiment it is desirable for the case of echo observation that the two wave vectors not be parallel. The wave function of the atom at $t > t_1$ is given by

$$\psi(\mathbf{r}, t) = \exp[-i\mathcal{H}_0(t - t_1)U_1(\mathbf{r}) \exp[-i\mathcal{H}_0(t_1 - t_p)]U_p(\mathbf{r})\psi(\mathbf{r}, t_p). \tag{15}$$

Because the pulse widths and interpulse time intervals are much shorter than T_1 and T_2, the dependence of the echo intensity on these two lifetimes is small. The incorporation of the lifetimes is quite straightforward and will be neglected here. The expectation value $\langle\psi^*(\mathbf{r}, t)|\boldsymbol{\mu}|\psi(\mathbf{r}, t)\rangle$ gives the induced dipole moment at time t. Of the four terms contributing to the atomic dipole moment, the term of interest is

$$\begin{aligned}\langle\boldsymbol{\mu}\rangle =& (-i\boldsymbol{\mu}_{ba}/2) \sin\theta_p(\mathbf{r}) \sin^2[\theta_1(\mathbf{r})/2] \\ &\times \exp\{i[2\phi_1(\mathbf{r}) - \phi_p(\mathbf{r}) + (2\mathbf{k}_1 - \mathbf{k}_p)\cdot\mathbf{r} - \omega t - \Delta\omega(t - 2t_1 + t_p)]\} \\ &+ \text{c.c.,}\end{aligned} \tag{16}$$

where $\Delta\omega = \omega_0 - \omega$. Equation (16) is a general expression for arbitrary pulse areas $\theta_p(\mathbf{r})$ and $\theta_1(\mathbf{r})$. It can be seen that the maximum dipole moment is induced by a sequence of $\pi/2$ and π pulses. To obtain the polarization density at $\mathbf{r}$, the dipole moment given by Eq. (16) is summed over all the

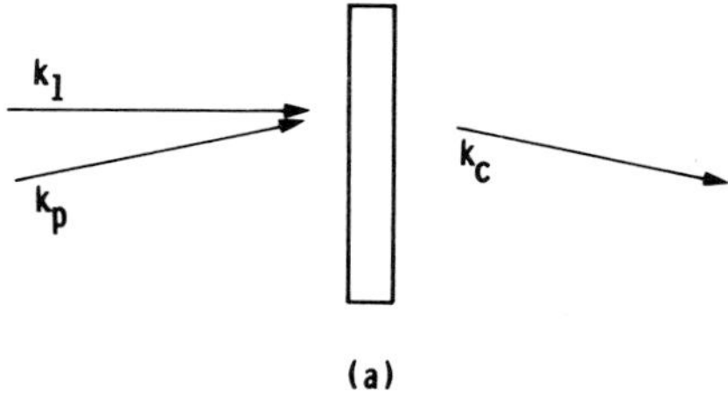

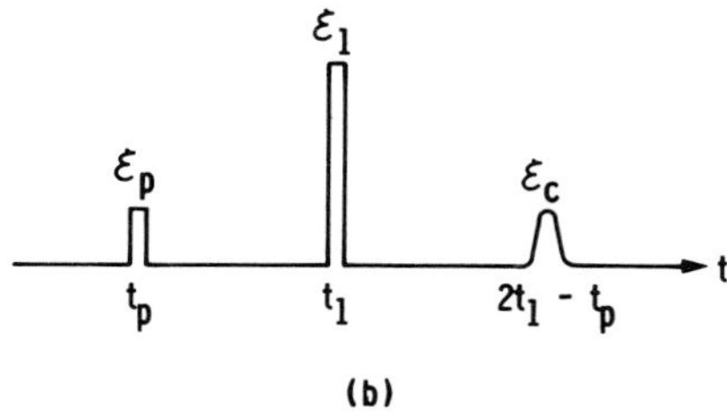

Fig. 3. (a) Propagation directions of the fields in the conventional two-pulse echo. (b) Temporal sequence of the two-pulse echo.

atoms at $\mathbf{r}$ by averaging over $g(\Delta\omega)$. Statistically, there are initially a finite number of atoms in the upper and lower energy states. If we assume that the total number density of atoms in the two states is N, with N_a in state $|a\rangle$ and N_b in state $|b\rangle$ at $t = t_p$, then $N_a - N_b = -N \tanh(\hbar\omega_0/2k_B T_{th})$, where k_B is the Boltzmann constant and T_{th} the temperature at thermal equilibrium. the polarization density giving rise to the echo is therefore

$$\begin{aligned}\mathbf{P}_{echo} = {} & [-iN \tanh(\hbar\omega_0/2k_B T_{th})\boldsymbol{\mu}_{ba}/2] \sin\theta_p(\mathbf{r}) \sin^2[\theta_1(\mathbf{r})/2] \\ & \times \exp\{i[2\phi_1(\mathbf{r}) - \phi_p(\mathbf{r}) + 2(\mathbf{k}_1 - \mathbf{k}_p)\cdot\mathbf{r} - \omega t]\} \\ & \times G(t - 2t_1 + t_p) + \text{c.c.},\end{aligned} \tag{17}$$

where

$$G(t) = \int_{-\infty}^{\infty} d(\Delta\omega)\, g(\Delta\omega) \exp(-\Delta\omega t). \tag{18}$$

The averaging over the resonant frequencies results in a temporal behavior given by the Fourier transform of the inhomogenous distribution function. It is negligible for all $t > t_1$ except at $t = 2t_1 - t_p$. Note that

$$t_p = \mathbf{n}_p \cdot \mathbf{r}/c, \tag{19}$$

$$t_1 = t_{1p} + \mathbf{n}_1 \cdot \mathbf{r}/c, \tag{20}$$

where $\mathbf{n}_p$ and $\mathbf{n}_1$ are the unit vectors normal to the phase fronts of the first and second pulses, respectively. t_{1p} is the time interval between the instants when the two pulses pass through the origin of $\mathbf{r}$. The first pulse is assumed to arrive at the origin at $t = 0$. After the incidence of the second pulse, it takes exactly the interpulse separation time for the microscopic dipole moments to rephase again. The time when the macroscopic polarization density is at a maximum and an intense burst of photons is emitted by the volume of atoms located at $\mathbf{r}$ is therefore given by

$$t = 2t_{1p} + (2\mathbf{n}_1 - \mathbf{n}_p) \cdot \mathbf{r}/c. \tag{21}$$

The total electric field $\mathbf{E}_c$ of the radiated echo with a wave vector $\mathbf{k}_c$ at a point $\mathbf{R}$ outside the medium is obtained from the polarization density (17), integrated over the entire medium and evaluated at the retarded time $t = t' + |\mathbf{R} - \mathbf{r}|/c$:

$$\mathbf{E}_c(\mathbf{R}, t) = \frac{1}{4\pi\epsilon_0} \nabla \times \nabla \times \int \frac{[\mathbf{P}_{\text{echo}}(\mathbf{r}, t')]_{\text{ret}}}{|\mathbf{R} - \mathbf{r}|} d^3r, \tag{22}$$

where ϵ_0 is the free-space permittivity. Substituting Eq. (17) into Eq. (22), we obtain

$$\begin{aligned}\mathbf{E}_c(\mathbf{R}, t) \propto \int d^3r\, G\left(t - \frac{R}{c} - 2t_{1p} + \frac{\mathbf{r}}{c} \cdot (\mathbf{n}_c - 2\mathbf{n}_1 + \mathbf{n}_p)\right) \\ \times \sin\theta_p(\mathbf{r}) \sin^2[\theta_1(\mathbf{r})/2] \\ \times \exp\{i[2\phi_1(\mathbf{r}) - \phi_p(\mathbf{r}) + (2\mathbf{k}_1 - \mathbf{k}_p - \mathbf{k}_c) \cdot \mathbf{r} - \omega(t - R/c)]\} \\ + \text{c.c.},\end{aligned} \tag{23}$$

where $\mathbf{n}_c = \mathbf{R}/R = \mathbf{k}_c/k_c$. We have assumed that R is much larger than r. Equation (23) gives a peak when all the atoms in the sample radiate in phase, namely, when

$$t = 2t_{1p} + R/c, \tag{24}$$

$$\mathbf{n}_c = 2\mathbf{n}_1 - \mathbf{n}_p. \tag{25}$$

Equation (24) gives the time when the echo pulse reaches $\mathbf{R}$. Equation (25) is the phase-matching condition, similar to what is commonly seen in nonlinear mixing. It requires the radiated field to have the same phase as the atomic dipole radiators along its propagation path. Transfer of power from the induced polarization to the echo field can be achieved only when they have the same wave vectors. The echo pulse propagates in the forward direction. Of particular interest is the case when the second pulse is a plane wave with uniform $\mathscr{E}_1$ and ϕ_1. $\mathbf{E}_c(\mathbf{R}, t)$ is recognized as the phase-conjugate version of the first pulse $\mathbf{E}_p$ if $\mathbf{E}_p$ has only spatial phase variation with uniform amplitude. A plane wave passing through a phase-distorting medium will satisfy this condition. The restriction arises because of the sin $\theta_p(\mathbf{r})$ dependence of $\mathbf{E}_c$. Note that, for a small pulse area, sin $\theta_p(\mathbf{r})$ can be approximated by $\theta_p(\mathbf{r})$, and $\mathbf{E}_c$ is a phase-conjugate replica of $\mathbf{E}_p$ even when the amplitude of $\mathbf{E}_p$ varies spatially.

The phase-matching condition (25) requires $\mathbf{k}_c = \mathbf{k}_1 = \mathbf{k}_p$. However, the configuration shown in Fig. 3 is usually preferred for easier detection of the echo. When a small angle β exists between $\mathbf{k}_1$ and $\mathbf{k}_p$, the echo is emitted at an angle 2β with respect to $\mathbf{k}_p$, and can thus be spatially separated from the preceding laser pulses. Owing to the violation of the phase-matching requirement, the echo amplitude decreases with increasing sample thickness l or larger β. The limitations on both parameters can be expressed as $\beta^2 l/\lambda \ll 1$. Another problem with the forward echo arises from the finite sample thickness, which can cause a wave-front distortion. For example, Heer and McManamon (1977) pointed out that if $\mathbf{E}_p$ is a divergent spherical wave front, the conjugate wave front generated by the front and rear parts of the sample will focus at different positions in space. These complications associated with the forward echo do not arise for a backward echo.

Shiren (1978) proposed the generation of a backward optical echo by using a standing wave as the second pulse in a two-pulse sequence. Such a configuration is shown in Fig. 4. Shiren's treatment will be followed here. In our notation, the standing wave is

$$\mathbf{E}_1(\mathbf{r}, t) = \mathbf{x}\ \tfrac{1}{2}\{\mathscr{E}_1(t) \cos \mathbf{k}_1 \cdot \mathbf{r} \exp[i(\phi_1 - \omega t)] + \text{c.c.}\} \tag{26}$$

$\mathscr{E}_1(t)$ and ϕ_1 are uniform in space. The standing wave is applied to the entire sample at the same time. $U_1(r)$ has the same form as Eq. (14), except that $\theta_1(\mathbf{r})$ in this case is $\theta_{10} \cos \mathbf{k}_1 \cdot \mathbf{r}$, where θ_{10} is $\int (\mu/\hbar)\mathscr{E}_1(t')\, dt'$. The ensuing derivation of the echo wave front is identical to that of the for-

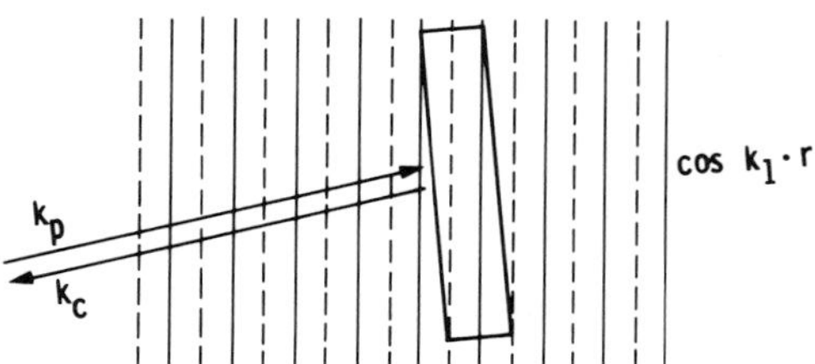

Fig. 4. Configuration of backward-echo generation by using a standing wave as the second pulse in the two-pulse sequence.

ward echo. The echo pulse is

$$E_c(\mathbf{R}, t) \propto \int d^3r\, G\left(t - \frac{R}{c} - 2t_{1p} + \frac{\mathbf{r}}{c} \cdot (\mathbf{n}_c + \mathbf{n}_p)\right)$$
$$\times \sin \theta_p(\mathbf{r}) \sin^2[\theta_1(\mathbf{r})/2]$$
$$\times \exp\{i[2\phi_1 - \phi_p(\mathbf{r}) - (\mathbf{k}_p + \mathbf{k}_c) \cdot \mathbf{r} - \omega(t - R/c)]\}$$
$$+ \text{c.c.} \tag{27}$$

To investigate the dependence of the radiation field on $\mathbf{k}_1$, we expand $\sin^2[\theta_1(\mathbf{r})/2]$ as

$$\sin^2[(\theta_{10} \cos \mathbf{k}_1 \cdot \mathbf{r})/2] = \tfrac{1}{2}\left(1 - \sum_{m=-\infty}^{\infty} (-1)^m J_{2m}(\theta_{10}) \exp(i2m\mathbf{k}_1 \cdot \mathbf{r})\right). \tag{28}$$

The term J_{2m} is the Bessel function. Note that for the $m = 0$ term all the atoms radiate in phase, with

$$t = 2t_{1p} + R/c, \tag{29}$$

$$\mathbf{n}_c = -\mathbf{n}_p. \tag{30}$$

This is a backward-propagating echo with an amplitude proportional to $1 - J_0(\theta_{10})$ and $\mathbf{k}_c = -\mathbf{k}_p$ for any $\mathbf{k}_1$. As we have shown for a forward echo, a backward echo will be a true conjugate replica of $\mathbf{E}_p(\mathbf{r}, t)$ only for small $\theta_p(\mathbf{r})$. When $\mathbf{k}_1$ is parallel to $\mathbf{k}_p$, the $m = +1$ term also gives rise to a forward echo. However, the echo amplitude is small, owing to the imperfect phase matching for G in Eq. (27).

B. Three-Pulse Echo

A photon echo induced by a sequence of three exciting pulses is sometimes referred to as a stimulated echo. The stimulated echo is a more interesting case because of an additional wave vector of the third pulse that can be varied to create new situations. Consider three pulses $\mathbf{E}_p(\mathbf{r}, t)$, $\mathbf{E}_1(\mathbf{r}, t)$, and $\mathbf{E}_2(\mathbf{r}, t)$ incident onto the medium at time $t = t_p$, t_1, and t_2, respectively. The optical fields have the same form as Eq. (12), and the evolution operators acting on the atomic wave function the same form as Eq. (14). It will be shown that, depending on the directions of the wave vectors $\mathbf{k}_p$, $\mathbf{k}_1$, and $\mathbf{k}_2$ of the three pulses, either a forward or a backward echo can be produced. The wave function at $t > t_2$ is

$$\psi(\mathbf{r}, t) = \exp[-i\mathcal{H}_0(t - t_2)]U_2(\mathbf{r}) \exp[-i\mathcal{H}_0(t_2 - t_1)]U_1(\mathbf{r})$$
$$\times \exp[-i\mathcal{H}_0(t_1 - t_p)]U_p(\mathbf{r})\psi(\mathbf{r}, t_p). \tag{31}$$

Again assuming that the atom is at the lower state $|b\rangle$ at $t = t_p$, we calculate the induced dipole moment $\langle \psi^*(\mathbf{r}, t)|\boldsymbol{\mu}|\psi(\mathbf{r}, t)\rangle$ for $t > t_2$. The terms that will give rise to echoes are identified to be

$$\begin{aligned}\langle \boldsymbol{\mu} \rangle = (-i\boldsymbol{\mu}_{ba}/8) \sin \theta_p(\mathbf{r}) \sin \theta_1(\mathbf{r}) \sin \theta_2(\mathbf{r}) \\ \times (\exp\{i[\phi_2(\mathbf{r}) + \phi_1(\mathbf{r}) - \phi_p(\mathbf{r}) + (\mathbf{k}_2 + \mathbf{k}_1 - \mathbf{k}_p) \cdot \mathbf{r} \\ - \omega t - \Delta\omega(t - t_2 - t_1 + t_p)]\} \\ + \exp\{i[\phi_2(\mathbf{r}) + \phi_p(\mathbf{r}) - \phi_1(\mathbf{r}) + (\mathbf{k}_2 + \mathbf{k}_p - \mathbf{k}_1) \cdot \mathbf{r} \\ - \omega t - \Delta\omega(t - t_2 + t_1 - t_p)]\}) + \text{c.c.}\end{aligned} \tag{32}$$

Only the first term will rephase to cause the emission of a photon echo for a sample of stationary atoms. The second term is included because we will show in the next section that in a medium of moving atoms a backward echo can be emitted from the sample only because of this term. By averaging over the resonant frequencies and adding the contributions from the atoms that can be in the initial state $|a\rangle$ at t_p, the polarization density is given by

$$\begin{aligned}\mathbf{P}_{\text{echo}} = [-iN \tanh(\hbar\omega_0/2k_B T_{\text{th}})\boldsymbol{\mu}_{ba}/8] \sin \theta_p(\mathbf{r}) \sin \theta_1(\mathbf{r}) \sin \theta_2(\mathbf{r}) \\ \times (\exp\{i[\phi_2(\mathbf{r}) + \phi_1(\mathbf{r}) - \phi_p(\mathbf{r}) + (\mathbf{k}_2 + \mathbf{k}_1 - \mathbf{k}_p) \cdot \mathbf{r} - \omega t]\} \\ \times G(t - t_2 - t_1 + t_p) \\ + \exp\{i[\phi_2(\mathbf{r}) + \phi_p(\mathbf{r}) - \phi_1(\mathbf{r}) + (\mathbf{k}_2 + \mathbf{k}_p - \mathbf{k}_1) \cdot \mathbf{r} - \omega t]\} \\ \times G(t - t_2 + t_1 - t_p)]) + \text{c.c.}\end{aligned} \tag{33}$$

where t_p, t_1, and t_2 are the time instants when the atoms located at $\mathbf{r}$ see the three pulses. The first term contains $G(t - t_2 - t_1 + t_p)$, which has a maximum at $t = t_2 + (t_1 - t_p)$. That is, the rephasing after the third pulse is completed when the elapsed time equals the interval between the first and second pulses. The second term contains $G(t - t_2 + t_1 - t_p)$, which peaks at $t = t_2 - (t_1 - t_p)$. This time occurs before the incidence of the third pulse, and therefore the second term cannot emit an echo. Note that the ordinary two-pulse echo can be suppressed by a proper choice of $\mathbf{k}_p$ and $\mathbf{k}_1$ such that the phase-matching condition for echo radiation is not satisfied. The terms t_p and t_1 are given by Eqs. (19) and (20), and

$$t_2 = t_{2p} + \mathbf{n}_2 \cdot \mathbf{r}/c, \tag{34}$$

where t_{2p} is the interval separating the arrival times of the first and last pulses at the origin. When the second and third pulses are plane waves analogous to the pump waves in degenerate four-wave mixing, the echo radiation field is

$$\mathbf{E}_c(\mathbf{R}, t) \propto \int d^3r \sin \theta_p(\mathbf{r}) \times \exp\{i[-\phi_p(\mathbf{r}) + (\mathbf{k}_2 + \mathbf{k}_1 - \mathbf{k}_p - \mathbf{k}_c) \cdot \mathbf{r} - \omega(t - R/c)]\} \times G\left(t - t_{2p} - t_{1p} - \frac{R}{c} + \frac{\mathbf{r}}{c} \cdot (\mathbf{n}_c - \mathbf{n}_2 - \mathbf{n}_1 + \mathbf{n}_p)\right) + \text{c.c.} \tag{35}$$

The conditions for echo observation at $\mathbf{R}$ are

$$t = t_{2p} + t_{1p} + R/c, \tag{36}$$

$$\mathbf{n}_c = \mathbf{n}_2 + \mathbf{n}_1 - \mathbf{n}_p. \tag{37}$$

A forward-propagating conjugate echo is observed when $\mathbf{k}_p = \mathbf{k}_1 = \mathbf{k}_2$. The interesting case is when $\mathbf{k}_1 = -\mathbf{k}_2$; then $\mathbf{k}_c = -\mathbf{k}_p$. The echo that propagates in the opposite direction of the first pulse is a conjugate replica of $\mathbf{E}_p(\mathbf{r}, t)$ for small or spatially uniform $\theta_p(\mathbf{r})$. In Fig. 5 we show schematically the directions of the wave vectors for producing forward and backward stimulated echoes.

We have shown that both forward and backward echoes can be produced by a two- or three-pulse sequence. The echo is always the con-

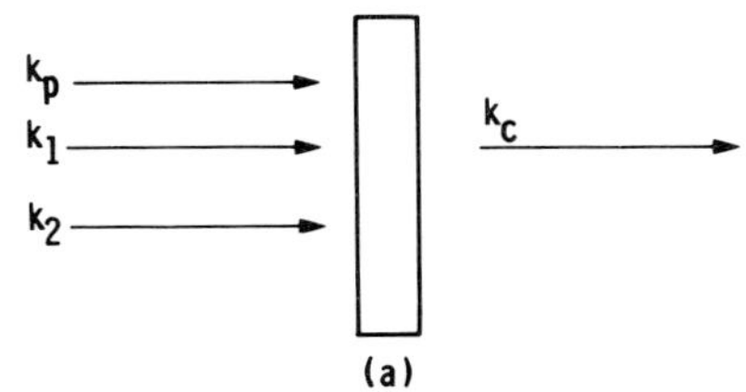

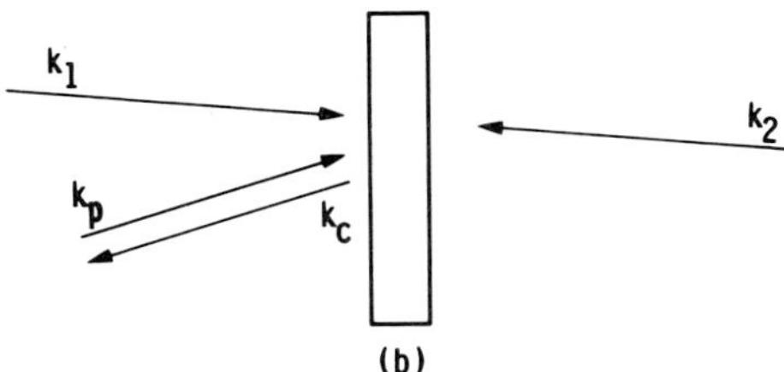

Fig. 5. Configurations of (a) forward and (b) backward stimulated echoes.

jugate replica of the first pulse in the sequence for weak excitation ($\sin \theta_p \sim \theta_p$). Note that the backward echo is superior to the forward echo in producing a conjugate wave front. The difference lies in the phase-matching conditions that the echo wave vector $\mathbf{k}_c$ must satisfy for the two cases. A forward echo is radiated when $\mathbf{k}_c = 2\mathbf{k}_1 - \mathbf{k}_p$ for the two-pulse echo or $\mathbf{k}_1 + \mathbf{k}_2 - \mathbf{k}_p$ for the three-pulse echo. The phase-matching condition is strictly satisfied only when all the wave vectors are equal and point in the same direction. Hence the first pulse is restricted to a plane wave. If the first pulse has spatially varying amplitude and phase, it can be viewed as a superposition of plane waves each with a different wave vector. The phase-matching condition is met for only one of these plane waves, with substantially lower echo amplitude for the non-phase-matched components, thus degrading the fidelity of the conjugation. In the case of a backward echo, $\mathbf{k}_c = -\mathbf{k}_p$ independent of the wave vectors of the other pulses. The phase-matching condition is satisfied for all the Fourier components of $\mathbf{E}_p$, giving rise to a true conjugate echo.

Both the photon echo and degenerate four-wave mixing can be described as real-time holographic processes. For example, the three-pulse echo can be viewed as a wave diffracted by a grating formed by the first two pulses and subsequently read out by the third. The main difference is that simultaneity of the pulses is not required. The phase-matching condition resembles that of four-wave mixing, with $\mathbf{E}_1$ and $\mathbf{E}_2$ playing the role of the counterpropagating pump waves and $\mathbf{E}_p$ being the object wave. When the pump pulses coincide in time, the two-pulse echo with a standing-wave second pulse results. In conventional four-wave mixing with optical pulses, the conjugate pulse is typically shorter than the pump and probe pulses because it results from a multiplication of the latter pulses. In the photon echo process, the conjugate pulse width is independent of the widths of the exciting pulses; it is determined by the Fourier transform of the inhomogeneous line shape. Conjugation via echoes also has the advantage that the echo pulse duration is independent of the transit time of the pulses through the sample. A drawback of the photon echo process is that the phase-conjugation property is adversely affected by atomic motion, as will be shown in the next section. Finally, we have shown that true phase conjugation from a photon echo occurs only for weak excitation ($\sin \theta_p \sim \theta_p$). The same restriction applies to degenerate four-wave mixing that arises from the third term in a perturbation series. The limit on the perturbation strength is of course different in each case. In a two-level resonant medium, the steady-state interaction among the pump waves, object wave, and conjugate wave originates from the saturation effect. The coupling constant responsible for backward-wave generation derived by Abrams and Lind (1978) is a first-order term when the nonlinear effect is expressed as an infinite series. Only when the effects of the remaining terms in the series can be neglected is a true conjugate wave front possible. There is therefore a close analogy between the photon echo process and the steady-state nonlinear effect in a two-level medium in producing phase conjugation (Shiren, 1978; and Yariv and AuYeung, 1979).

IV. Conjugate Echoes in Gaseous Media

In a gaseous medium, all the atoms have the same resonant frequency ω_0. The dephasing among the microscopic dipole moments after the first pulse is brought about by the rapid motion of the atoms. Even though the atomic velocities are not reversed by the subsequent pulses, it can be shown (Scully *et al.*, 1968) that a photon echo is still radiated from the medium. In this section, the influence of atomic motions on the phase-conjugation property of the photon echoes is examined. The atoms are assumed to be moving with uniform velocities, and the effect of velocity-changing collisions is not considered. Only the interesting case of the stimulated backward echo will be discussed in detail. Conditions for echo generation in the simple case when the three exciting pulses are plane waves will be derived. Extension to the case of a probe pulse with many Fourier components can then be easily applied to examine the phase-conjugation capability of the process.

Consider a gaseous medium that interacts with three plane waves $\mathbf{E}_p$, $\mathbf{E}_1$, and $\mathbf{E}_2$. A moving atom interacts with each optical pulse in the series at a different position in space $\mathbf{r}_p$, $\mathbf{r}_1$, and $\mathbf{r}_2$ at t_p, t_1, and t_2 related by

$$\mathbf{r}_i = \mathbf{r}_j + \mathbf{v}(t_i - t_j), \tag{38}$$

where $\mathbf{v}$ is the atomic velocity. The atom that subsequently moves to the point $\mathbf{r}$ at $t > t_2$ has a wave function

$$\begin{aligned}\psi(\mathbf{r}, t) = {} & \exp[-i\mathcal{H}_0(t - t_2)]U_2(\mathbf{r}_2) \exp[-i\mathcal{H}_0(t_2 - t_1)]U_1(\mathbf{r}_1) \\ & \times \exp[-i\mathcal{H}_0(t_1 - t_p)]U_p(\mathbf{r}_p)\psi(\mathbf{r}_p, t_p).\end{aligned} \tag{39}$$

The matrix elements of $U_p(\mathbf{r}_p)$, $U_1(\mathbf{r}_1)$, and $U_2(\mathbf{r}_2)$ contain phase information of the electric fields at the spatial points where the atom happens to be when it encounters the optical pulses. The precise instants of the field–atom interactions are

$$t_p = \mathbf{n}_p \cdot \mathbf{r}_p/c, \tag{40}$$

$$t_1 = t_{1p} + \mathbf{n}_1 \cdot \mathbf{r}_1/c, \tag{41}$$

$$t_2 = t_{2p} + \mathbf{n}_2 \cdot \mathbf{r}_2/c. \tag{42}$$

The induced dipole moment (32) is modified to become

$$\begin{aligned}\langle \boldsymbol{\mu} \rangle = (-i\boldsymbol{\mu}_{ba}/8) \sin\theta_p \sin\theta_1 \sin\theta_2 \\ \times \{\exp[i(\phi_2 + \phi_1 - \phi_p + \mathbf{k}_2 \cdot \mathbf{r}_2 + \mathbf{k}_1 \cdot \mathbf{r}_1 - \mathbf{k}_p \cdot \mathbf{r}_p - \omega t)] \\ + \exp[i(\phi_2 + \phi_p - \phi_1 + \mathbf{k}_2 \cdot \mathbf{r}_2 + \mathbf{k}_p \cdot \mathbf{r}_p - \mathbf{k}_1 \cdot \mathbf{r}_1 - \omega t)]\} + \text{c.c.}\end{aligned} \tag{43}$$

The polarization density at $\mathbf{r}$ is obtained by averaging over the velocity distribution $g(\mathbf{v})$ of all the atoms at $\mathbf{r}$, yielding a total echo field

$$\begin{aligned}\mathbf{E}_c(\mathbf{R}, t) \propto \int d^3r\, d^3v\, g(\mathbf{v}) \sin\theta_p \sin\theta_1 \sin\theta_2 \\ \times [\exp(i\{\phi_2 + \phi_1 - \phi_p + (\mathbf{k}_2 + \mathbf{k}_1 - \mathbf{k}_p - \mathbf{k}_c) \\ \cdot [\mathbf{r} - \mathbf{v}(t - R/c)] - \mathbf{v} \cdot [(t - R/c)\mathbf{k}_c - t_{2p}\mathbf{k}_2 - t_{1p}\mathbf{k}_1] \\ - \omega(t - R/c)\}) + \exp(i\{\phi_2 + \phi_p - \phi_1 + (\mathbf{k}_2 + \mathbf{k}_p - \mathbf{k}_1 - \mathbf{k}_c) \\ \cdot [\mathbf{r} - \mathbf{v}(t - R/c)] - \mathbf{v} \cdot [(t - R/c)\mathbf{k}_c - t_{2p}\mathbf{k}_2 + t_{1p}\mathbf{k}_1] \\ - \omega(t - R/c)\})] + \text{c.c.}\end{aligned} \tag{44}$$

Terms of second and higher order in $1/c^2$ in the exponentials are neglected. The first term peaks when

$$\mathbf{k}_c = \mathbf{k}_2 + \mathbf{k}_1 - \mathbf{k}_p, \tag{45}$$

$$(t - R/c)\mathbf{k}_c = t_{2p}\mathbf{k}_2 + t_{1p}\mathbf{k}_1. \tag{46}$$

Only a forward echo is possible. A backward echo with $\mathbf{k}_2 = -\mathbf{k}_1$ and $\mathbf{k}_c = -\mathbf{k}_p$ cannot satisfy condition (46). The second term of $\mathbf{E}_c(\mathbf{R}, t)$ peaks when

$$\mathbf{k}_c = \mathbf{k}_2 + \mathbf{k}_p - \mathbf{k}_1, \tag{47}$$

$$(t - R/c)\mathbf{k}_c = t_{2p}\mathbf{k}_2 - t_{1p}\mathbf{k}_1. \tag{48}$$

A backward echo is generated for counterpropagating $\mathbf{E}_p$ and $\mathbf{E}_2$ ($\mathbf{k}_p = -\mathbf{k}_2$) at

$$t = R/c + t_{2p}(\mathbf{n}_c \cdot \mathbf{n}_2) + t_{1p}. \tag{49}$$

It propagates in the opposite direction of the second pulse $\mathbf{E}_1$. In Fig. 6 a configuration that satisfies Eqs. (47) and (48) is shown. Observation of a backward echo under such a condition was reported by Fujita *et al.* (1979).

Unfortunately, although $\mathbf{k}_c = -\mathbf{k}_1$ independently of $\mathbf{k}_p$ and $\mathbf{k}_2$ as long as $\mathbf{k}_p + \mathbf{k}_2 = 0$, an echo that is a conjugate replica of an $\mathbf{E}_1$ with many Fourier components has a very small amplitude, because the time of the echo radiation given by Eq. (49) depends on the direction of $\mathbf{k}_c$ and is different for each component. The average over the atomic velocities when Eq. (48) is not exactly satisfied results in a weak signal with a dependence

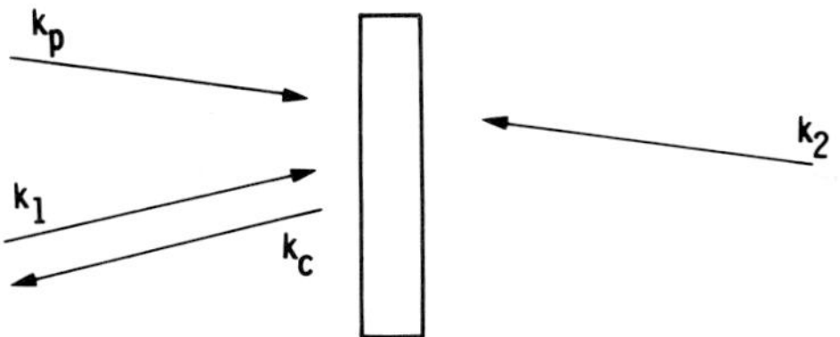

Fig. 6. Configurations of a backward echo generated in a gaseous medium.

of the amplitude on the angle between the pump pulses and the conjugate pulse.

A qualitatively similar behavior is found (Wandzura, 1979) in degenerate four-wave mixing. Conjugate-signal reflectivity is reduced when atomic motion washes out the gratings formed in a gaseous medium. An angular dependence of the reflection coefficient limits the spatial bandwidth of the object wave, in a manner similar to that demonstrated in the photon echo process.

V. Conclusion

We have discussed optical phase conjugation by photon echoes. In a medium of stationary atoms, both the forward- and the backward-propagating echoes are the phase conjugate of the first pulse in the exciting sequence, with the backward echo being the better technique in producing a large-spatial-bandwidth conjugate replica. In a gaseous medium, the echo propagates in a direction opposite to that of the second pulse in a three-pulse sequence when the first and third pulses are counterpropagating. However, atomic motion has a serious adverse effect on phase conjugation. Finally, we have compared four-wave mixing and photon echoes in generating time-reversed wave fronts, and found a close analogy between them.

References

Abella, I. D., Kurnit, N. A., and Hartmann, S. R. (1966). Photon echoes, *Phys. Rev.* **141,** 391.

Abrams, R. L., and Lind, R. C. (1978). Degenerate four-wave mixing in absorbing media, *Opt. Lett.* **2,** 94; **3,** 205.

Alekseev, A. I., and Eveseev, I. V. (1969). Photon echo polarization in a gas medium, *Zh. Eksp. Teor. Fiz.* **56,** 2118. [*English transl.: Sov. Phys. JETP* **29,** 1139 (1969)].

Alekseev, A. I., and Eveseev, I. V. (1975). Properties of photon echo on broad spectral lines, *Zh. Eksp. Teor. Fiz.* **68,** 456 [*English transl.: Sov. Phys. JETP* **41,** 222 (1975)].

Bloom, D. M., Liao, P. F., and Economou, N. P. (1978). Observation of amplified reflection by degenerate four-wave mixing in atomic sodium vapor, *Opt. Lett.* **2,** 58.

Feynman, R. P., Vernon, F. L., and Hellwarth, R. W. (1957). Geometrical representation of the Schrödinger equation for solving maser problems, *J. Appl. Phys.* **28,** 49.

Fisher, R. A., and Feldman, B. J. (1979). On-resonant phase-conjugate reflection and amplification at 10.6 μm in inverted CO_2, *Opt. Lett.* **4,** 140.

Fujita, M., Nakatsuka, H., Nakanishi, H., and Matsuoka, M. (1979). Backward echo in two-level systems, *Phys. Rev. Lett.* **42,** 974.

Gordon, J. P., Wang, C. H., Patel, C. K. N., Slusher, R. E., and Tomlinson, W. J. Photon echoes in gases, *Phys. Rev.* **179,** 294.

Griffen, N. C., and Heer, C. V. (1978). Focusing and phase conjugation of photon echoes in Na vapor, *Appl. Phys. Lett.* **33,** 865.

Grischkowsky, D., Shiren, N. S., and Bennett, R. J. (1978). Generation of time-reversed wave fronts using a resonantly enhanced electronic nonlinearity, *Appl. Phys. Lett.* **33,** 805.

Hahn, E. L. (1950). Spin Echoes, *Phys. Rev.* **80,** 580.

Heer, C. V., and McManamon, P. F. (1977). Wavefront correction with photon echoes, *Opt. Communi.* **23,** 49.

Kurnit, N. A., Abella, I. D., and Hartmann, S. R., (1964). Observation of a photon echo, *Phys. Rev. Lett.* **13,** 567.

Liao, P. F., and Bloom, D. M. (1978). Continuous-wave backward wave generation by degenerate four-wave mixing in ruby, *Opt. Lett.* **3,** 4.

Liao, P. F., Bloom, D. M., and Economou, N. P. (1978). CW optical wavefront conjugation by saturated absorption in atomic sodium vapor. *Appl. Phys. Lett.* **32,** 813.

Lind, R. C., Steel, D. G., Klein, M. B., Abrams, R. L., Giuliano, C. R., and Jain, R. K. (1979). Phase conjugation at 10.6 μm by resonantly enhanced degenerate four-wave mixing, *Appl. Phys. Lett.* **34,** 457.

Mossberg, T. W., Kachru, R., Hartmann, S. R., and Flusberg, A. M. (1979). Echoes in gaseous media: A generalised theory of rephasing phenomena, *Phys. Rev. A* **20,** 1976.

Scully, M., Stephen, M. J., and Burnham, D. C. (1968). Photo echo in gaseous media, *Phys. Rev.* **171,** 213.

Shiren, N. S. (1978). Generation of time-reversed optical wave fronts by backward-wave photon echoes, *Appl. Phys. Lett.* **33,** 299.

Shiren, N. S., and Kazyaka, T. G. (1972). Ultrasonic spin echoes, *Phys. Rev. Lett.* **28,** 1304.

Shiren, N. S., and Melcher, R. L. (1976). Time reversed ultrasonic echoes as an experimental tool, *in* "Phonon Scattering In Solids" (L. J. Challis, V. W. Rampton, and A. F. G. Wyatt, eds., p. 405. Plenum Press, New York.

Shtyrkov, E. I., and Samartsev, V. V. (1976). Imaging properties of dynamic echo holograms in resonant media, *Opt. I. Spektrosk.* **40,** 392 [*English transl: Opt. Spectrosc.* **40,** 224 (1976)].

Shtyrkov, E. I., and Samartsev, V. V. (1978). Dynamic holograms on the superposition states of atoms, *Phys. Status Solidi* (*A*) **45,** 647.

Shtyrkov, E. I., Lobkov, V. S., and Yarmukhametov, N. G. (1978). Grating induced in ruby by interference of atomic states, *Pis'ma Zh. Eksp. Teor. Fiz.* **27,** 685 [*English transl.: JETP Lett.* **27,** 648 (1978)].

Shtyrkov, E. I., Nevelskaya, N. L., Lobkov, V. S., and Yarmukhametov, N. G. (1980). Transient light-induced spatial gratings by successive optical coherent pulses, *Phys. Status Solidi* (*B*) **98,** 472.

Steel, D. G., and Lam, J. F. (1980). Multiline phase conjugation in resonant materials, *Opt. Lett.* **5,** 297.

Tomita, A. (1979). Phase conjugation using gain saturation of a Nd:YAG laser, *Appl. Phys. Lett.* **34,** 463.

Wandzura, S. M. (1979). Effects of atomic motion on wavefront conjugation by resonantly enhanced degenerate four-wave mixing, *Opt. Lett.* **4,** 208.

Watkins, D. E., Figueira, J. F., and Thomas, S. J. (1980). Observation of resonantly enhanced degenerate four-wave mixing in doped alkali-halides, *Opt. Lett.* **5,** 169.

Yariv, A., and AuYeung, J. (1979). Transient four-wave mixing and real time holography in atomic systems, *IEEE J. Quantum Electron.* **QE-15,** 224.

10 Degenerate Four-Wave Mixing in Semiconductors*

R. K. Jain
M. B. Klein

Hughes Research Laboratories
Malibu, California

I. Introduction

The relatively recent discovery of various applications of degenerate four-wave mixing (DFWM), most notably the generation of a wave whose phase over the entire wave front is exactly reversed with respect to that of a wave incident on the medium, has stimulated interest in the search for

* Some of the material in this chapter has been previously published with permission of Academic Press.

Optical Phase Conjugation

ISBN 0-12-257740-X

suitable nonlinear materials for DFWM. Semiconductors are unique as a class of materials for DFWM and other self-interaction types of experiments, because they possess an unusually large variety of nonlinear optical mechanisms that may be used for this purpose. The abundance of such mechanisms in semiconductors is largely due to the presence of free-carrier states (in addition to the bound-electron states present in all optical materials) and to the ease with which free carriers may be photogenerated in them (even with optical wavelengths as long as those in the middle infrared).

From a historical perspective, it is interesting to note that one of the earliest nonlinear optics experiments in semiconductors (Patel *et al.*, 1966) involved the use of free carriers in a study of four-wave mixing with frequencies that are nearly degenerate. The third-order nonlinear susceptibilities of $\sim 10^{-10}$ esu reported for these experiments in highly doped n-type semiconductors (InSb, InAs, and GaAs) were higher than those reported for any material at that time. Nonlinearities due to the anharmonic motion of bound electrons far from their resonant frequencies were subsequently studied in Si and Ge by Wynne and Boyd (1968). This nonresonant nonlinearity was the first semiconductor nonlinearity with which phase-conjugation experiments were performed (in Ge at 10.6 μm) under the nomenclature of DFWM (Bergmann *et al.*, 1978). Note, however, that earlier transient holography experiments performed by Woerdman (1971) and related studies of carrier dynamics via transient gratings (e.g., Jarasiunas and Vaitkus, 1977) were essentially the first DFWM experiments in semiconductors. These early transient grating experiments invoked relatively large nonlinearities ($\chi^{(3)} \sim 10^{-8}$ esu) due to the generation of a free-carrier plasma.

In recent years, numerous other experiments have been performed on DFWM and closely related nonlinear optical phenomena (such as intrinsic optical bistability), using a large variety of semiconductors, wavelengths, and nonlinear optical mechanisms. For instance, nonlinearities as high as 10^{-1} esu have been demonstrated (Miller *et al.*, 1980b; Jain and Steel, 1981) in InSb and HgCdTe via slow ($\sim 10^{-7}$ sec) accumulation of the optically induced free-carrier plasma, and DFWM reflectivities as high as 800% have been measured at high intensities ($\sim$100 MW/cm^2) via nonlinear generation of a free-carrier plasma in Ge (Watkins *et al.*, 1981).

In this chapter our primary emphasis will be on the review and discussion of the principal nonlinear optical mechanisms that may be used for DFWM in semiconductors, and a secondary emphasis will be on the review of recent experiments on DFWM and phase conjugation in which these nonlinear optical mechanisms have been used. The chapter is organized as follows. In Section II, we discuss a few broad issues relating to DFWM in semiconductors. Our discussion includes a description of the relevant terms in the nonlinear polarization density for backward DFWM in isotropic and anisotropic media, and for single-photon and multiphoton transitions between the valence and conduction bands. In Section III, we discuss the nonlinear optical mechanisms in some detail, and present derivations of expressions of third-order susceptibilities for DFWM based on the simplest models. The concepts of "effective" and "steady-state" susceptibilities are used to describe the nonlinear polarization density for slowly responding media. Magnitudes and speeds of the third-order susceptibilities are estimated for a representative choice of semiconductors and wavelengths throughout Section III. In Section IV, we describe specific details of recent DFWM experiments with various semiconductors, with an emphasis on novel features in the physics and in the experimental techniques used.

We will assume that the reader has some familiarity with the band structure of solids and with elementary semiconductor concepts, but will briefly review some relatively simple properties and definitions (such as density-of-states effective masses and exciton binding energies) whenever they can be easily incorporated into the discussion. For more detailed information on semiconductors and solid-state physics, we refer the reader to standard textbooks, such as those by Ziman (1964), McKelvey (1966), Smith (1968), Harrison (1970), Kittel (1971), and Ashcroft and Mermin (1976). For additional background material on nonlinear optics in solids, we recommend the treatises by Bloembergen (1965), Butcher (1965), and Flytzanis (1970), and recent comprehensive reviews by Chemla and Jerphagnon (1980) and Chemla (1980).

II. DFWM in Semiconductors: General Considerations

A. Description of the Nonlinear Polarization

The general formulation of the problem of the coupling of the various waves in the degenerate four-wave mixing interaction has been discussed amply in earlier chapters. In essence, the solution of the DFWM problem involves estimation of the appropriate polarization density $\mathbf{P}^{NL}(\mathbf{r}, t)$ and its incorporation into a complete set of Maxwell's equations, resulting in coupled differential equations between the various waves involved, which are usually solved numerically or with the use of simplifying approximations.

The configuration of most interest to us is the backward DFWM configuration, depicted schematically in Fig. 1a for the purpose of defining the nomenclature used in this chapter. All four waves are at the same optical

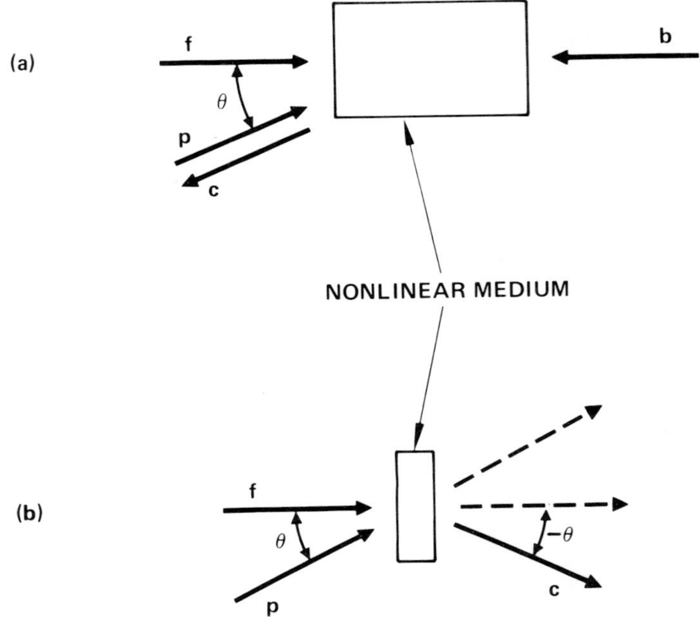

Fig. 1. Schematic illustration of the two common configurations for the DFWM interaction, involving two combinations of four waves f, p, b, c, whose frequencies are equal: $\omega_f = \omega_p = \omega_b = \omega$ and $\omega_c = \omega_f + \omega_b - \omega_p = \omega$ where f is the forward pump wave, b the backward pump wave, p the probe wave, and c the conjugate signal wave. (a) The backward configuration, with $\theta \ll 1$, is the principal DFWM interaction considered in this chapter. (b) The forward configuration is used principally for highly absorbing thin samples. In the nomenclature of four-wave mixing, the forward pump beam constitutes *two* input waves and the probe the third input wave.

frequency ω (wavelength λ), and we use $\mathbf{k}_i$ and $\vec{\mathscr{E}}_i$ to denote their wave vectors and electric field amplitudes,[1] respectively, with i = f, p, b, or c. The letters f and b designate the exactly counterpropagating ($\mathbf{k}_f + \mathbf{k}_b = 0$) forward and backward pump waves, respectively, and p designates the probe wave, which makes a small nonzero angle θ ($0 < \theta \ll 1$) with respect to the forward pump wave. Phase matching ($\mathbf{k}_c = \mathbf{k}_f + \mathbf{k}_b - \mathbf{k}_p$) then requires that the phase-conjugate signal wave (designated by c) be radiated in a direction backward to the probe wave. Note that our four-wave definition of the DFWM interaction specifies only the number of

[1] The complex electric field amplitude $\vec{\mathscr{E}}(\mathbf{r}, \omega)$ at frequency ω is related to the real electric field $\mathbf{E}(\mathbf{r}, t)$ by $\mathbf{E}(\mathbf{r}, t) = \frac{1}{2}\vec{\mathscr{E}}(\mathbf{r}, \omega)\exp(i\omega t)$ + c.c. For fields that are not monochromatic, the real electric field is simply $\mathbf{E}(\mathbf{r}, t) = \int_{-\infty}^{\infty} \vec{\mathscr{E}}'(\mathbf{r}, \omega)\exp(i\omega t)\,d\omega$.

waves involved, and poses no restriction on the order of the nonlinearity. Thus, even higher-order interactions involving four waves (three input, one output) with the geometrical constraints of Fig. 1a are covered by our definition as a DFWM interaction.

For the backward DFWM interaction, the coupled-wave equations for the four waves are usually solved approximately by perturbative expansions (with $|\vec{\mathscr{E}}_p|, |\vec{\mathscr{E}}_c| \ll |\vec{\mathscr{E}}_f|, |\vec{\mathscr{E}}_b|$). In the simplest cases (third-order interaction only, low linear attenuation, equal pump intensities, negligible pump depletion, etc.; see Chapter 2), simple expressions can be obtained (Yariv and Pepper, 1977) for the DFWM reflectivity $R = |\vec{\mathscr{E}}_c|^2/|\vec{\mathscr{E}}_p|^2$ in terms of any Kerr-like third-order susceptibility $\chi^{(3)}$. However, if higher-order susceptibilities are invoked, or $\chi^{(3)}$ is dependent on other parameters (such as the absorption coefficient α), the above approach has only a limited validity. More general solutions are then obtained by inserting the functional form of the nonlinear polarization density $\mathbf{P}^{NL}(\mathbf{r}, t)$ directly into the propagation equations, as has been done for a simple two-level saturable absorber by Abrams and Lind (1978) and for plasma generation via single-photon valence-to-conduction-band transitions by Kukhtarev and Kovalenko (1980).

In Fig. 1b is depicted a second DFWM configuration, often called forward degenerate four-wave mixing or forward phase conjugation, which is used principally for thin, highly absorbing samples. For this configuration, the four-wave nomenclature is explained by the consideration that for the chosen interaction the strong pump beam may be assumed to constitute two forward waves. The weak probe beam then provides a third input wave at a small input angle θ, so that for thin samples the four-wave interaction is nearly phase matched in the $-\theta$ direction (i.e., $\mathbf{k}_c = 2\mathbf{k}_f - \mathbf{k}_p$). This configuration has also been called three-wave mixing or a two-beam interaction. An interesting aspect of this geometry is that the coupled propagation equations can be solved analytically (Maruani, 1980).

We will confine our discussion in this section to the backward DFWM configuration, giving particular attention to the general form and magnitude of the nonlinear polarization density for this specific interaction. Most of the general conclusions and specific comments about the backward DFWM geometry may then be easily extended to forward DFWM by analogy.

A rigorous derivation of the nonlinear polarization density leading to the DFWM signal would involve the use of a perturbative approach in a density matrix formulation similar to that outlined in Chapter 8. However, in a semiconductor, the presence of a quasi-continuum of states and the lack of detailed information on the matrix elements and dephasing rates makes such an approach very cumbersome. When the radiation-induced changes (if any) in the populations of the energy levels are small, a much simpler and more effective approach involves estimating the magnitudes and signs of the various orders of the nonlinear susceptibility invoked. The nonlinear polarization density leading to the DFWM signal can then be written

$$\mathcal{P}^{\mathrm{NL}}(\mathbf{r}, \omega) = \overleftrightarrow{\chi}^{(3)} \vdots \vec{\mathscr{E}}\vec{\mathscr{E}}\vec{\mathscr{E}} + \overleftrightarrow{\chi}^{(5)} \vdots \vec{\mathscr{E}}\vec{\mathscr{E}}\vec{\mathscr{E}}\vec{\mathscr{E}}\vec{\mathscr{E}} + (\text{terms of higher odd orders in } \vec{\mathscr{E}}). \quad (1)$$

We remind the reader that the *four*-wave description of the DFWM interaction specifies only the *number* of beams involved, and does not constrain the interaction to a third-order nonlinearity. Odd-order terms are chosen in Eq. (1) because only such terms may be phase matched to radiate in the backward direction. The order of the susceptibilities and the specific terms that contribute significantly to the DFWM signal depend on the specifics of the nonlinear interaction. For instance, when the nonlinear mechanism is due to two-photon transitions from the valence to the conduction band, terms involving both the third- and the fifth-order nonlinear susceptibilities will generally be significant, and the dominance of one over the other will depend on such details as the two-photon transition probability, the dephasing rate of the two-photon coherence, and the lifetime of the plasma. By the same token, *all* higher-order terms become important when significant population changes are caused by the radiation, such as near the saturation intensity of a resonant transition; in this case, a more general form for the nonlinear susceptibility or polarization density, such as that used by Abrams and Lind (1978), becomes *necessary*. Note, however, that for many of the nonlinearities of interest, or if only low light intensities are involved, the $\chi^{(3)}$ terms are dominant. Consequently, we will concentrate principally on third-order terms in the polarization density in our discussion (except in Section II.C).

In isotropic media, the most general form of the third-order polarization density for the backward DFWM geometry of Fig. 1a can be expressed as

$$\vec{\mathcal{P}}^{(3)} = \chi_{\mathrm{fp}}^{(3)}(\vec{\mathscr{E}}_{\mathrm{f}} \cdot \vec{\mathscr{E}}_{\mathrm{p}}^{*})\vec{\mathscr{E}}_{\mathrm{b}} + \chi_{\mathrm{bp}}^{(3)}(\vec{\mathscr{E}}_{\mathrm{b}} \cdot \vec{\mathscr{E}}_{\mathrm{p}}^{*})\vec{\mathscr{E}}_{\mathrm{f}} + 2C(\vec{\mathscr{E}}_{\mathrm{f}} \cdot \vec{\mathscr{E}}_{\mathrm{b}})\vec{\mathscr{E}}_{\mathrm{p}}^{*}, \quad (2)$$

where $\vec{\mathcal{P}}^{(3)}$ is the complex polarization density amplitude and is related to the real polarization density $\mathbf{P}^{(3)}$ by $\mathbf{P}^{(3)}(r, t) = \frac{1}{2}\vec{\mathcal{P}}^{(3)}(r, \omega)\exp(i\omega t)$ + c.c. In Eq. (2), the first two terms are identified as spatial grating terms (described in earlier chapters), and the third term corresponds to the scattering of the probe wave from a spatially uniform two-photon coherence. If the nonlinearity is strongly affected by motion, such as that due to the diffusion of carriers (see e.g., Jain and Klein, 1979), the susceptibilities $\chi_{\mathrm{fp}}^{(3)}$

and $\chi_{\mathrm{bp}}^{(3)}$ become functions of the grating spacing, and thus strong functions of the angle θ between the forward and probe waves (especially near $\theta = 0$). The susceptibility C is unaffected by motion, and is usually negligible unless there is a strong resonant enhancement from a two-photon transition.

B. Crystal Symmetry and Anisotropy Considerations

The general form of the third-order polarization density for DFWM in anisotropic media can be expressed as (Maker and Terhune, 1965)

$$\mathcal{P}_i^{(3)}(\omega) = 3\chi_{ijk\ell}^{(3)}(-\omega;\ \omega,\ \omega,\ -\omega)\mathscr{E}_j\mathscr{E}_k\mathscr{E}_\ell^*, \tag{3}$$

where a sum is taken over all repeated indices, and $\mathscr{E}_i$ are the complex amplitudes of the electric field. Note that, in contrast to the subscripts in Eq. (2), which designate the different waves involved in the nonlinear interaction, the subscripts in Eq. (3) designate the Cartesian components of the various electric fields and polarization densities. The number of independent components of $\chi_{ijk\ell}^{(3)}$ is determined by the symmetry properties of the nonlinear medium. The lattice structure for most of the semiconductors of interest here is cubic, of point group m3m or $\bar{4}$3m (Long, 1968; Kittel, 1971). In these point groups, the tensor elements of $\chi_{ijk\ell}^{(3)}$ reduce to four independent components $\chi_{xxxx}^{(3)}$, $\chi_{xyxy}^{(3)}$, $\chi_{xyyx}^{(3)}$, and $\chi_{xxyy}^{(3)}$ (Butcher, 1965), where the subscripts x and y designate principal axes of the crystal. When all frequencies are far from any material resonances, the susceptibility is a real quantity and the number of independent components reduces to two: $\chi_{xxxx}^{(3)}$ and $\chi_{xyxy}^{(3)} = \chi_{xyyx}^{(3)} = \chi_{xxyy}^{(3)}$.

For such cubic crystals, it is conventional to express the anisotropy in $\chi^{(3)}$ in terms of the factor σ (Flytzanis, 1975)

$$\sigma = (3\chi_{xxyy}^{(3)} - \chi_{xxxx}^{(3)})/\chi_{xxxx}^{(3)}. \tag{4}$$

In semiconductors with cubic crystal symmetry, nonlinearities involving free carriers are isotropic, owing to the isotropic behavior of the net carrier transport properties (McKelvey, 1966). Although there are generally three independent components of $\chi^{(3)}$ even for isotropic media, in the absence of dispersion the number of independent components reduces to one, with $\chi_{xyxy}^{(3)} = \chi_{xyyx}^{(3)} = \chi_{xxyy}^{(3)} = \chi_{xxxx}^{(3)}/3$, resulting in nonlinear polarization terms similar to those in Eq. (2). Note that, for isotropic media, σ is equal to zero.

For anisotropic nonlinearities (such as the bound-electron nonlinearity) in cubic media (crystal symmetry classes 432, m3m, and $\bar{4}$3m), the third-order polarization density can be shown to be of the form

$$\vec{\mathscr{P}}^{(3)} = 3(\chi_{xxyy} + \chi_{xyxy})(\vec{\mathscr{E}} \cdot \vec{\mathscr{E}}^*)\vec{\mathscr{E}} + 3\chi_{xyyx}(\vec{\mathscr{E}} \cdot \vec{\mathscr{E}})\vec{\mathscr{E}}^* + 3(\chi_{xxxx} - \chi_{xxyy} - \chi_{xyxy} - \chi_{xyyx})\,\mathrm{tr}(\vec{\mathscr{E}}\vec{\mathscr{E}}) \cdot \vec{\mathscr{E}}^*, \tag{5}$$

where all susceptibilities are defined with relation to the principal crystal directions, and diffusion is neglected. For the geometry of Fig. 1a, the polarization density leading to the conjugate wave can be written

$$\vec{\mathscr{P}}_{\mathrm{c}}^{(3)} = A[\vec{\mathscr{E}}_{\mathrm{f}} \cdot \vec{\mathscr{E}}_{\mathrm{p}}^*)\vec{\mathscr{E}}_{\mathrm{b}} + (\vec{\mathscr{E}}_{\mathrm{b}} \cdot \vec{\mathscr{E}}_{\mathrm{p}}^*)\vec{\mathscr{E}}_{\mathrm{f}}] + 2C(\vec{\mathscr{E}}_{\mathrm{f}} \cdot \vec{\mathscr{E}}_{\mathrm{b}})\vec{\mathscr{E}}_{\mathrm{p}}^* + (B - C)[\mathrm{tr}(\vec{\mathscr{E}}_{\mathrm{f}}\vec{\mathscr{E}}_{\mathrm{b}}) \cdot \vec{\mathscr{E}}_{\mathrm{p}}^* + \mathrm{tr}(\vec{\mathscr{E}}_{\mathrm{b}}\vec{\mathscr{E}}_{\mathrm{f}}) \cdot \vec{\mathscr{E}}_{\mathrm{p}}^*], \tag{6}$$

where $A = 3(\chi_{xxyy} + \chi_{xyxy})$, $B = 3(\chi_{xxxx} - \chi_{xxyy} - \chi_{xyxy})$, and $C = 3\chi_{xyyx}$. The contribution of each term in Eq. (6) to the polarization density can be determined by proper selection of the polarization vectors of the input waves. In Table I we give the polarization of the conjugate wave and the coefficient(s) of the terms in Eq. (6) that contribute to the polarization density, for all independent combinations of polarization vectors of the input waves. Note that certain cases (e.g., combinations 1–3) allow independent measurement of C and A. The coefficient B can then be obtained through measurement of $B - C$ (case 5). Such experiments might provide a convenient and independent technique for measuring the Born–Oppenheimer coefficients (Hellwarth, 1977) in cubic and isotropic media. Note that, in the absence of dispersion, only two of the three coefficients

TABLE I

Terms Contributing to $\mathscr{P}^{(3)}$ in Cubic Crystals and Polarization Vector Dependence of $\mathscr{P}^{(3)}$ on Combinations of Polarization Vectors Associated with the Three Input Waves

Combination	Polarization state	Wave				Terms contributing to $\mathscr{P}^{(3)}$
		f	b	p	c	
1	Linear[a]	↑	↑	→	→	$2C$
2	Linear[a]	↑	→	↑	→	A
3	Linear[a]	↑	→	→	↑	A
4	Linear[a]	↑	↑	↑	↑	A, $2C$
5	Circular[b]	R	R	L	L	$B - C$
6	Circular[b]	R	L	R	L	A, $2C$, $B - C$
7	Circular[b]	R	L	L	R	A, $2C$, $B - C$
8	Circular[b]	R	R	R	R	A, $B - C$

[a] These polarization vectors are parallel to the principal axes of the crystal.
[b] Propagation is along a principal axis of the crystal.

are independent, and that the anisotropy factor σ reduces to

$$\sigma = [(C - B)/(2C + B)]. \tag{7}$$

Thus, polarization studies in DFWM experiments provide a relatively simple technique for the determination of σ in such media. Similar techniques should be applicable to the study of symmetry properties in other anisotropic crystal structures.

C. Multiphoton Transitions and Higher-Order Susceptibilities

We have seen that the most general forms [Eqs. (2) and (6)] for the third-order nonlinear polarization density at frequency ω contain terms in which a 2ω oscillation occurs as an intermediate condition. Instead of the spatially modulated variations in population (or in the electron–hole plasma density), the physical process involves the creation of a spatially uniform coherence at 2ω by the counterpropagating pumps, which then scatters the probe wave in the backward direction. As mentioned in Section II.A, the contribution from this term is usually negligible unless strongly enhanced by a two-photon transition (and, ideally, a near-resonant intermediate single-photon resonance). Also, as mentioned earlier, even for the case of a single-photon transition, terms of higher than third order may contribute, but such contributions are usually negligible unless significant population changes are involved.

Several interesting questions arise when the single-photon absorption is negligible and overshadowed (at appropriate light intensities) by multiphoton transitions. For example, what is the nature of the phase-matched terms in the polarization density, and of their spatial and temporal dependence? What is the dependence of the spatial terms, if any, on the intensities of the three waves? For simplicity, let us consider the case where a two-photon transition is dominant, as occurs when the band-gap energy $E_g \gg \hbar\omega > E_g/2$ or when $2\hbar\omega$ closely matches the energy of the biexciton transition (Hanamura and Haug, 1977). As in the general case of Section II A, the polarization density terms leading to DFWM can still be written as:

$$\mathcal{P}^{\mathrm{NL}}(\omega) = \mathcal{P}^{(3)} + \mathcal{P}^{(5)} + (\text{higher-order terms}).$$

In contrast to the single-photon resonant terms in $\mathcal{P}^{\mathrm{NL}}$, in two-photon resonant media a much more significant contribution to the net polarization density $\mathcal{P}^{\mathrm{NL}}$ may occur from $\mathcal{P}^{(5)}$ and higher-order terms, even when the population changes are negligible. For the four-wave interaction of Fig.

1a, some of the fifth-order terms that are phase matched to radiate in the conjugate-wave direction are shown in Table II, along with the corresponding phase-matching diagrams. Numerous other similar terms are possible, including those obtained simply by interchanging the subscripts f and b. Although not normal convention, parentheses are inserted in these terms to facilitate visualization of one perturbational sequence that results in such a term; a description of the corresponding physical sequence in the nonlinear polarization density is given in the last column. Note that in the presence of an intermediate state one can expect term c to dominate, owing to the formation of a relatively long-lived intermediate-state population. If this intermediate state is already well saturated at the light intensities of interest, this term reduces essentially

TABLE II

Nature of Significant $\mathscr{P}^{(5)}$ Terms in Two-Photon Resonantly Enhanced Materials

Term	Typical term in $\mathscr{P}^{(5)}(\omega)$	Phase-matching diagram	Comments
a	$\chi_a^{(5)}[(\bar{\mathscr{E}}_f \cdot \bar{\mathscr{E}}_f)\bar{\mathscr{E}}_f^* \cdot \bar{\mathscr{E}}_p^*]\bar{\mathscr{E}}_b$	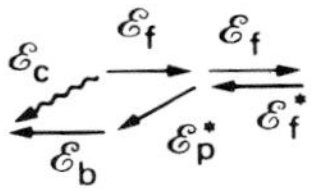 $\mathscr{E}_f$, $\mathscr{E}_f$, $\mathscr{E}_c$, $\mathscr{E}_f^*$, $\mathscr{E}_p^*$, $\mathscr{E}_b$	2ω Coherence → spatial grating
b	$\chi_b^{(5)}[(\bar{\mathscr{E}}_f \cdot \bar{\mathscr{E}}_b)\bar{\mathscr{E}}_b^* \cdot \bar{\mathscr{E}}_p^*]\bar{\mathscr{E}}_b$	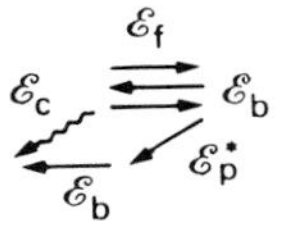 $\mathscr{E}_f$, $\mathscr{E}_c$, $\mathscr{E}_b$, $\mathscr{E}_p^*$, $\mathscr{E}_b$	2ω Coherence → spatial grating
c	$\chi_c^{(5)}[(\bar{\mathscr{E}}_f \cdot \bar{\mathscr{E}}_f^*)\bar{\mathscr{E}}_f \cdot \bar{\mathscr{E}}_p^*]\bar{\mathscr{E}}_b$	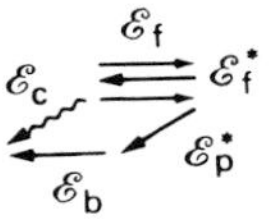 $\mathscr{E}_f$, $\mathscr{E}_c$, $\mathscr{E}_f^*$, $\mathscr{E}_p^*$, $\mathscr{E}_b$	Population of intermediate state → spatial grating
d	$\chi_d^{(5)}[(\bar{\mathscr{E}}_f \cdot \bar{\mathscr{E}}_b)\bar{\mathscr{E}}_f^* \cdot \bar{\mathscr{E}}_f]\bar{\mathscr{E}}_p^*$	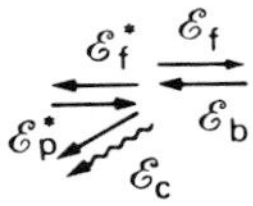 $\mathscr{E}_f^*$, $\mathscr{E}_f$, $\mathscr{E}_b$, $\mathscr{E}_p^*$, $\mathscr{E}_c$	2ω Coherence only, no spatial grating
e	$\chi_e^{(5)}[(\bar{\mathscr{E}}_f \cdot \bar{\mathscr{E}}_f^*)\bar{\mathscr{E}}_f \cdot \bar{\mathscr{E}}_b]\bar{\mathscr{E}}_p^*$	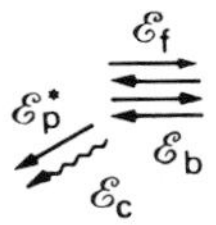 $\mathscr{E}_f$, $\mathscr{E}_p^*$, $\mathscr{E}_b$, $\mathscr{E}_c$	2ω Coherence only, no spatial grating

to a third-order term. Also, terms a and b contain population redistributions (manifested by the spatial modulation), and because population relaxation times for states in semiconductors are generally much longer than the dephasing times of the coherences, one can expect $\chi_{\mathrm{a}}^{(5)}$ and $\chi_{\mathrm{b}}^{(5)}$ to be significantly larger than $\chi_{\mathrm{d}}^{(5)}$ and $\chi_{\mathrm{e}}^{(5)}$, and the corresponding polarization terms may be expected to result in DFWM signals comparable to the third-order [see Eq. (2)] 2ω coherence term in such a semiconductor. Note also the different functional dependence of each term in $\mathcal{P}^{(5)}$ on the amplitudes of the three input waves. (For instance, term b has a cubic dependence on the intensity of the backward wave.) With appropriate polarization selection, and by intensity and angular dependence experiments, it may be possible to study the individual contributions of these terms to the DFWM signal. Note, finally, that in the present discussion we have assumed negligible contributions of the 3ω terms to $\mathcal{P}^{(5)}$.

The above discussion can easily be extended to higher-order transitions, but the bookkeeping is fairly tedious. Note, however, that for a medium that is m-photon resonant ($m > 1$) the lowest-order term that may contribute significantly to a DFWM signal is of order $2m - 1$, whereas the lowest-order term with a spatial modulation is of order $2m + 1$.

III. Nonlinear Mechanisms for DFWM in Semiconductors

As is perhaps obvious from the description of the nonlinear polarization density terms in Section II, the DFWM signal is generated either via coherences at harmonics of the input wave or via a simpler nonlinearity, viz., intensity-dependent changes in the absorption coefficient or in the refractive index of the medium. The latter type of nonlinearity has been variously called a self-interaction effect, a Kerr-like effect, or a self-induced change in the optical properties of the medium; it is the type of nonlinearity responsible for self-focusing and intrinsic optical bistability, and also participates in self-Q-switching, self-mode-locking, and regenerative pulsations in lasers. Apart from the two-photon coherences that may be created in biexciton states, all of the third-order nonlinearities that have been used for DFWM in semiconductors fall into this category. Note also that self-induced changes in the *refractive index* are generally much more important for DFWM than self-induced changes in the absorption coefficient.

The semiconductor nonlinearities can be further divided into two principal categories: resonant and nonresonant, viz., those involving real or virtual transitions. The first two nonlinearities described below—

anharmonic motion of bound electrons and nonlinear motion of free carriers—apparently have physical origins in valence-to-conduction-band transitions; nevertheless, these nonlinearities are by and large nonresonant, and are characterized by very fast speeds. Most of the remaining nonlinearities described are resonant, i.e., involve redistribution of populations, and are consequently limited by speeds corresponding to energy buildup and relaxation times, as will be discussed subsequently. Nevertheless, a major advantage of the resonant nonlinearities is the large enhancements in the values of $\chi^{(3)}$ that may be possible, e.g., from $\sim 10^{11}$ esu for the bound-electron nonlinearity in Ge to $\sim 10^{-1}$ esu for interband transitions in InSb and HgCdTe. We will elaborate on the general tradeoff between the speed τ and the magnitude of the nonlinearity at the end of this section.

An important general point about slow nonlinearities, however, is that the optically induced polarization densities are not dependent upon the value of the instantaneous fields, but are transient functions. For times $t \ll \tau$, the self-induced nonlinearities are essentially proportional to the integrated *energy* from the incident waves. Thus, for short laser pulses (pulse duration $\tau_L \ll \tau$), if one were to estimate the magnitude of the nonlinear effect in terms of nonlinear susceptibilities, one could define the concept of *effective* third-order susceptibilities that are essentially proportional to the pulse energy, and thus to the laser pulse duration τ_L. Likewise, when steady-state values of the induced nonlinearity are reached, as for cw laser radiation or extremely long laser pulses, at $t \gg \tau$, one could talk of *steady-state* third-order susceptibilities. In this regime, the nonlinear polarization is again proportional to the instantaneous fields, and as such we may deal with the steady-state third-order susceptibility as a conventional nonlinear susceptibility. The concepts of effective and steady-state susceptibilities are discussed in more detail for the case of the nonlinearity induced by the plasma generated by valence-to-conduction-band transitions, which serves as an illustrative example in the discussion of all slow nonlinearities.

A. Anharmonic Response of Bound Electrons

The third-order nonlinearity due to bound electrons results from anharmonic terms in the potential-energy well of each electron. This nonlinearity is present in all crystalline solids, and is the dominant nonlinearity in intrinsic semiconductors at frequencies well below the band gap. A useful model for this nonlinearity can be developed from the equation of motion of a one-dimensional anharmonic oscillator. The driving fields are all assumed to be at frequency ω, as is appropriate for DFWM. The equation of motion is given by

$$\ddot{x} + \gamma\dot{x} + \omega_0^2 x + vx^2 + \mu x^3 = (e/m)\mathscr{E}e^{i\omega t} + \text{c.c.}, \tag{8}$$

where x is the displacement, γ is the damping constant, ω_0 is the resonant frequency, and v and μ are anharmonic force constants. The nonlinear restoring force includes a term that is second order in displacement, as appropriate for noncentrosymmetric materials (Bloembergen, 1965), and a term that is third order in displacement (Wynne and Boyd, 1968). The solution to Eq. (8) for the linear susceptibility is

$$\chi^{(1)}(\omega) = Ne^2/m(\omega_0^2 - \omega^2 - i\gamma\omega), \tag{9}$$

where N is the number of oscillators per unit volume. If we assume $\omega \ll \omega_0$, Eq. (8) may be solved for a third-order susceptibility when one nonlinear term dominates (Jamroz, 1980):

$$\chi^{(3)}(\omega;\, \omega, \omega, -\omega) = (m\mu/N^3e^4)[\chi^{(1)}(\omega)]^4 = \delta_\mu[\chi^{(1)}(\omega)]^4 \tag{10a}$$

for centrosymmetric materials, and

$$\chi^{(3)}(\omega;\, \omega, \omega, -\omega) = (m^2v^2/N^4e^6)[\chi^{(1)}(\omega)]^5 = \delta_v[\chi^{(1)}(\omega)]^5 \tag{10b}$$

for noncentrosymmetric materials. Accurate calculation of the anharmonic force constants v and μ requires detailed knowledge of electronic wave functions. However, the constants can be estimated (Bloembergen, 1965) by assuming that the linear and nonlinear restoring forces are equal when x equals one lattice spacing. The resulting values of δ_μ and δ_v are nearly constant for a wide variety of crystals, and the values of nonlinear susceptibility estimated from these are within an order of magnitude of the corresponding experimentally measured values. The relationships for $\chi^{(3)}$ given above are directly analogous to Miller's rule (Miller, 1964) for second-order nonlinear susceptibilities, and Eqs. (10) are thus a modified form of Miller's rule. Using known values of the linear susceptibilities, Wynne and Boyd (1968) and Wang (1970) have used the modified Miller's rule to estimate third-order nonlinear susceptibilities due to bound electrons in several semiconductors (at frequencies in the region of transparency). The estimated values of $\chi^{(3)}$ for several semiconductors, scaled from experimental values of $\chi^{(3)}$ in Ge or GaAs (the choice being determined by the inversion symmetry of the material), are listed in column 5 of Table III.

Wang (1970) has developed an alternative relationship to Eqs. (10), obtained from a perturbative expression for $\chi^{(3)}$ in hydrogen, with modification for local-field effects. He showed that

$$\chi^{(3)} \approx 1.2(\chi^{(1)})^2 f^3/N_{\text{eff}}\hbar\omega_0, \tag{11}$$

where f is a local-field correction factor, N_{eff} the number of effective

TABLE III

Third-Order Susceptibility Due to Anharmonic Motion of Bound Electrons

	Band Gap		Susceptibility $\chi^{(3)}_{xxxx}$ (10^{-12} esu)			Anisotropy σ	
Material	E_g (eV)	λ_g (μm)	Calculated (low-frequency limit)	Estimated (low-frequency limit)	Measured at 10.6 μm	Calculated (low-frequency limit)	Measured at 10.6 μm
Ge	0.67	1.85	−88,[a] 30,[b] 51[c]	25,[d] 41[e]	100,[f] 25[g]	1.4,[b] 1.0[c]	0.83,[b] 0.56[g]
Si	1.11	1.12	−63,[a] 4.4,[b] 5.1[c]	7.8,[d] 9.6[e]	6[f]	1.1,[b] 0.98[c]	0.44[f]
GaAs	1.43	0.87	−12.5,[a] 6,[b] 7.8[c]	12,[h] 11[e]	12[i]	0.50,[b] 1.0[c]	−0.25[i]
InSb	0.17	7.3	−1250,[a] 37.8,[b] 4300[c]	98.7[h]	—	0.56[b]	—
InAs	0.36	3.4	−150,[a] 15.5,[b] 370[c]	19.4[h]	—	0.44[b]	—
CdS	2.42	0.51	—	0.18,[h] 3.0[e]	1.6[j] (0.694 μm)	—	—

[a] Jha and Bloembergen (1968a,b).
[b] Flytzanis (1970).
[c] Van Vechten *et al.* (1970).
[d] Anharmonic oscillator model, $\chi^{(3)} = 4.9 \times 10^{-16}(\chi^{(1)})^4$.
[e] Wang (1970).
[f] Wynne and Boyd (1968).
[g] Watkins *et al.* (1980b).
[h] Anharmonic oscillator model, $\chi^{(3)} = 1.3 \times 10^{-16}(\chi^{(1)})^5$.
[i] Wynne (1969).
[j] Maker and Terhune (1965).

charges, and ω_0 a dispersion frequency. Estimated values of $\chi^{(3)}$ using Eq. (11) (Wang, 1970) are also presented in column 5 of Table III.

More precise calculations of $\chi^{(3)}$ can be performed by using time-independent perturbation theory, but these are difficult because they depend on accurate knowledge of the band structure throughout the Brillouin zone. A good approximation for the contribution of all bound electrons may be obtained by considering tetrahedral molecular bonding orbitals for the ground-state wave functions. Jha and Bloembergen (1968a,b) used the molecular-orbital technique to calculate susceptibilities for a number of semiconductors (see Table III, column 4). Their model assumed isotropic bands, so that $\sigma = 0$. The magnitudes of their estimated bound-electron susceptibilities are in good agreement with experiment, but the negative signs calculated are in disagreement with those measured experimentally, perhaps owing to the simplified molecular-orbital wave functions used. Flytzanis (1970) used improved wave functions and added local-field correction factors in the molecular bonding model to obtain improved values of susceptibility and values of σ in good agreement with experiment. Using a different approach, Van Vechten *et al.* (1970) considered the contribution to the bound-electron $\chi^{(3)}$ due to both interband and intraband transitions, and calculated the latter from the Franz–Keldysh effect. In contrast to the negative sign of the interband contribution, the intraband contribution is of positive sign and is always larger by a factor of $\frac{5}{3}$. The resulting sign and magnitudes of the net bound-electron susceptibilities are in good agreement with the values measured in Ge, Si, and GaAs at 10.6 μm (see Table III).

We caution the reader that experimental values for $\chi^{(3)}$ due to bound electrons in Table III were obtained principally via nondegenerate four-wave mixing, or difference frequency mixing (Maker and Terhune, 1965; Wynne and Boyd, 1968; Wynne, 1970). As such, the measured values are only approximately valid for DFWM, assuming that no resonances or anomalous dispersion terms exist at any of the frequencies of interest. In most reported experiments the susceptibility was obtained from measurements of power conversion in samples shorter than one coherence length. Proper account must also be taken of bulk absorption, reflection from the sample surfaces, and the focusing conditions. Studies have concentrated primarily on measurements in the 10-μm spectral region, because of the wider range of semiconductors that are transparent at this wavelength and the ease in generating separate CO_2 laser wavelengths for difference frequency mixing. Using difference frequency mixing near 10 μm, Wynne and Boyd (1968) measured $\chi^{(3)}_{xxxx}$ and $\chi^{(3)}_{xxyy}$ for silicon and germanium as a function of doping and temperature. The measured susceptibilities were found to be independent of doping and temperature (in the range 80–300 K), as expected for the bound-electron nonlinearity. In a similar experiment, Wynne (1969) measured $\chi^{(3)}_{xxxx}$ and $\chi^{(3)}_{xxyy}$ in high-resistivity GaAs, and confirmed that the signs for $\chi^{(3)}$ are positive in GaAs, Ge, and Si. Watkins *et al*. (1980b) measured self-interaction values $\chi^{(3)}_{xxxx}$ and $\chi^{(3)}_{xxyy}$ in germanium using an ellipse rotation technique at 10.6 μm. The values obtained by this technique are expected to be more accurate than those measured earlier, because this technique circumvents the need for absolute power measurements. The values of susceptibility measured by Watkins *et al*. are approximately four times smaller than those of Wynne and Boyd, and are in better agreement with the values calculated by Flytzanis (1970) and Wang (1970) (see Table III).

B. Nonlinear Motion of Carriers

1. Expressions for $\chi^{(3)}$ Due to Nonlinear Carrier Motion

In doped semiconductors with moderately large free carrier concentrations ($\gtrsim 10^{15}/\text{cm}^3$), significant third-order (and higher) susceptibilities may occur via the nonlinear motion and energy relaxation of the free carriers in response to the driving optical fields (Wolff and Pearson, 1966; Kaw, 1968; Krishnamurthy and Paranjape, 1969). Such nonlinearities can be particularly large ($\gtrsim 10^{-9}$ esu) in narrow gap semiconductors (such as InSb, InAs, and HgCdTe), because of the large nonparabolicity of the conduction bands, typified by an energy-or momentum-dependent electron effective mass, m_e^*, and a momentum-dependent electron velocity $\mathbf{v}(\mathbf{p})$.

Using Kane's band structure calculations for InSb (Kane, 1957), the conduction band energy, effective mass, and momentum-dependent electron velocity can be approximated by

$$E_c = \frac{p^2}{2m} + \frac{E_g}{2} + \left[\frac{E_g^2}{4} + E_g\,\frac{p^2}{2m_e^*}\right]^{1/2}, \tag{12a}$$

$$m^*(p) \approx m_e^*\left(1 + \frac{2p^2}{m_e^* E_g}\right), \tag{12b}$$

and

$$\mathbf{v}(\mathbf{p}) = \frac{\partial E_c}{\partial \mathbf{p}} \approx \frac{\mathbf{p}/m_e^*}{(1 + 2p^2/m_e^* E_g)^{1/2}}, \tag{12c}$$

where m is the mass of a free electron, m_e^* is the effective mass of the electron at the bottom of the conduction band, E_g is the band-gap energy, and we have assumed $p^2/2m_e^* \ll E_g$, $m_e^* \ll m$.

a. Nondegenerate Frequencies ($\omega \approx \omega_1 \approx \omega_2 \ll \omega_g$; $(\omega_1 - \omega_2) \gg 1/T_{2m}$) The first derivation of third-order susceptibilities due to nonlinear motion of conduction band electrons was performed by Wolff and Pearson (1966), with an emphasis on explaining the experimental observations (Patel, *et al.* 1966) of relatively large nonlinearities for nearly degenerate difference frequency mixing ($\omega_4 = 2\omega_1 - \omega_2$). The derivation of Wolff and Pearson is strictly valid only for nondegenerate four-wave mixing, for the case where $\omega \approx \omega_1 \approx \omega_2 \ll E_g/\hbar$, and $(\omega_1 - \omega_2) \gg 1/T_{2m}$, where T_{2m} is the momentum relaxation time. Nevertheless, we will review this derivation as the starting point of our discussion, and then discuss the enhancement factors that must be considered for degenerate four-wave mixing ($\omega_1 \to \omega_2$) and for $\omega_1 \approx \omega_2 \gtrsim E_g/2\hbar$.

In the presence of two optical fields, the classical equation of motion for the conduction band electrons is simply:

$$\frac{d\mathbf{p}}{dt} + \frac{\mathbf{p}}{T_{2m}} = e\mathbf{E} = \frac{1}{2} e(\vec{\mathscr{E}}_1 e^{i\omega_1 t} + \vec{\mathscr{E}}_2 e^{i\omega_2 t} + \text{c.c.}), \tag{13a}$$

with the solution

$$\mathbf{p} = \frac{e\vec{\mathscr{E}}_1 e^{i\omega_1 t}}{2(i\omega_1 + 1/T_{2m})} + \frac{e\vec{\mathscr{E}}_2 e^{i\omega_2 t}}{2(i\omega_2 + 1/T_{2m})} + \text{c.c.} \tag{13b}$$

For most cases of interest $T_{2m} \ll \omega_1^{-1}, \omega_2^{-1}$, so that Eq. (13b) reduces to

$$\mathbf{p} = \frac{e\vec{\mathscr{E}}_1 e^{i\omega_1 t}}{2i\omega_1} + \frac{e\vec{\mathscr{E}}_2 e^{i\omega_2 t}}{2i\omega_2} + \text{c.c.} \tag{13c}$$

If we neglect the energy distribution of the conduction band electrons, then the induced current density is

$$\mathbf{J} = Ne\mathbf{v} = \frac{Ne}{m_e^*} \frac{\mathbf{p}}{(1 + 2p^2/m_e^* E_g)^{1/2}}, \tag{14a}$$

where N is the carrier density. Expanding Eq. (14a) in a Taylor series in p^2 and noting the linear dependence of $\mathbf{p}$ on $\vec{\mathscr{E}}$, we have

$$\mathbf{J}^{(3)} = -\frac{Nep^2}{(m_e^*)^2 E_g} \mathbf{p} \tag{14b}$$

for the third-order current density. By combining with Eq. (13c), and retaining only terms which oscillate at $\omega_4 = 2\omega_1 - \omega_2$, we obtain

$$\mathbf{J}^{(3)}(\omega_4) = \frac{iNe^4|\vec{\mathscr{E}}_1|^2}{8(m_e^*)^2 E_g \omega_1^2 \omega_2} \vec{\mathscr{E}}_2 e^{i\omega_4 t} + \text{c.c.} \tag{14c}$$

Finally, by using

$$\mathbf{P}^{(3)}(\omega_4) = \int \mathbf{J}^{(3)}(\omega_4)\,dt = \tfrac{1}{2}\chi_{\rm NP}^{(3)}|\vec{\mathscr{E}}_1|^2\vec{\mathscr{E}}_2 e^{i\omega_4 t} + \text{c.c.} \tag{15}$$

we obtain from Eqs. (14c) and (15):

$$\chi_{\rm NP}^{(3)}(\omega_4;\,\omega_1,\,\omega_1,\,-\omega_2) = \frac{Ne^4}{4(m_{\rm e}^*)^2 E_{\rm g}\omega_1^2\omega_2\omega_4}. \tag{16}$$

An alternative derivation for Eq. (16) has been obtained by Yuen and Wolff (1982), who consider the kinetic energy $\Delta W = p^2/2m_{\rm e}^*$ of the carriers and the net carrier energy fluctuations at the beat frequency $\delta W_{\rm e}(\omega_1 - \omega_2)$ from the cross-term of $|\mathbf{p}|^2$ [see Eq. (13b)]:

$$\delta W_{\rm e}(\omega_1 - \omega_2) = \frac{e^2}{4m^*}\left(\frac{\vec{\mathscr{E}}_1 \cdot \vec{\mathscr{E}}_2^{\,*} e^{i(\omega_1-\omega_2)t}}{(\omega_1 + i/T_{2m})(\omega_2 - i/T_{2m})} + \text{c.c.}\right) \tag{17}$$

and deduce therefrom the modulation of the effective mass due to the nonparabolicity [see Eq. (12b)]:

$$\delta(1/m_{\rm e}^*) \simeq -\frac{4\delta W}{m_{\rm e}^* E_{\rm g}}. \tag{18}$$

The nonlinear index change can then be easily estimated from the change in the index contribution of the Drude plasma:

$$\Delta n_{\rm NP} = -\frac{2\pi e^2}{n\delta(1/m_{\rm e}^*)\omega^2}. \tag{19}$$

By combining Eqs. (17), (18), and (19) and by considering the polarization induced by a third light field (see Section III.C), we get essentially the same expression for $\chi^{(3)}$ as in Eq. (16).

In the above derivations for $\chi_{\rm NP}^{(3)}$, the electron energy distribution has been ignored. Inclusion of the energy distribution function in the calculation of the third order current [Eq. (14c)] yields an expression for $\chi_{\rm NP}^{(3)}$ given by (Wolff and Pearson, 1966):

$$\chi_{\rm NP}^{(3)} = \frac{Ne^4}{4(m_{\rm e}^*)^2 E_{\rm g}\omega_1\omega_2\omega_4}\left[\frac{1 + 8E_{\rm F}/5E_{\rm g}}{(1 + 4E_{\rm F}/E_{\rm g})^{5/2}}\right], \tag{20}$$

where $E_{\rm F}$ is the Fermi energy measured from the bottom of the conduction band, and is given by the classical expression

$$E_{\rm F} = \frac{\hbar^2}{2m_{\rm e}^*}(3\pi^2 N)^{2/3}. \tag{21}$$

Eq. (21) is clearly valid only for semiconductors that are doped heavily enough so that the Fermi level lies within the conduction band. For more lightly doped samples a more exact expression for E_F is required, and the averaging over the energy distribution (Wang and Ressler, 1969) gives a modified version of Eq. (20):

$$\chi_{NP}^{(3)} = \frac{Ne^4}{4(m_e^*)^2 E_g \omega_1 \omega_2 \omega_4} \left[1 + \frac{6kT}{E_g}\right]^{-7/2}. \tag{22}$$

b. Degenerate Frequencies ($\omega = \omega_1 = \omega_2 \ll \omega_g$) In the case considered above, we have ignored the dissipation of energy via thermal relaxation of the carriers via inelastic collisions. However, as first elucidated by Kaw (1968) and by Krishnamurthy and Paranjape (1969), who concentrated on the effect of the energy dependence of the momentum relaxation time, and clarified very recently by Yuen and Wolff (1982) for the case of nonparabolic bands, the energy dissipation can provide a dominant contribution to the free electron susceptibility, $\chi_F^{(3)}$. Using a basic formulation similar to that of Kaw (1968) and Krishnamurthy and Paranjape (1969), which considers the kinetic equation for the average energy of the carriers:

$$\frac{d(\delta W_i)}{dt} + \frac{\delta W_i}{\tau_{th}} = \frac{e(\mathbf{E} \cdot \mathbf{p})}{m^*}. \tag{23}$$

Yuen and Wolff (1982) have derived an expression for the energy fluctuations for the case of inelastic electron scattering:

$$\delta W_i(\omega_1 - \omega_2) = \delta W_e \left(1 + \frac{2/T_{2m} - 1/\tau_{th}}{i(\omega_2 - \omega_1) + 1/\tau_{th}}\right) + \text{c.c.} \tag{24}$$

Once again using the effective mass modulation due to nonparabolicity [Eq. (18)], we have a more generalized expression for the third-order susceptibility due to inelastic free carrier motion (Yuen and Wolff, 1982):

$$\chi_F^{(3)} = \chi_{NP}^{(3)} \left(1 + \frac{2\tau_{th} - T_{2m}}{T_{2m}[1 - i(\omega_2 - \omega_1)\tau_{th}]}\right). \tag{25}$$

Equation (25) shows that inelastic collisions result in a significant enhancement of the third-order susceptibility $\chi_F^{(3)}$, especially when $(\omega_1 - \omega_2) \to 0$. Note that the energy loss to the lattice by the electrons results in an energy loss from the fields themselves; thus, this enhancement in the nonlinear susceptibility can also be understood as the nonlinear dispersion associated with such free carrier absorption.

Note that susceptibility enhancements for the case when $\omega_1 \to \omega_2$ are also predicted by the calculations of Kaw (1968) and Krishnamurthy and Paranjape (1969). These earlier calculations also include the effect of the energy dependence of the momentum relaxation time. For most practical cases, the latter effect results in only small fractional changes in $\chi_F^{(3)}$, and we will ignore this contribution in the present discussion.

c. High-Frequency Case ($\omega \lesssim \omega_g$) The expressions for $\chi_{NP}^{(3)}$ given earlier [Eqs. (20) and (22)] are valid only for photon energies much less than the bandgap, where the dominant contributions come from the conduction band.[2] For $\hbar\omega \gtrsim E_g/2$ the effect of all bands must be considered. Jha and Bloembergen (1968a, 1968b) derive an expression for $\chi_{NP}^{(3)}$ in this limit using time-dependent perturbation theory:

$$\chi_{NP}^{(3)} = -\frac{e^4}{4m^4\omega_1^2\omega_2\omega_4} \sum_{\mathbf{k}} \sum_{s,t,u} f_0(E_{ck}) Q(c, s, t, u, \omega, \mathbf{k}). \tag{26}$$

In this relation, m is the free electron mass, V is the volume of the solid, $f_0(E_{ck})$ is the Fermi distribution function, E_{ck} is the energy of state $\mathbf{k}$ in the conduction band, and s, t, u are band indices. The factor Q contains products of momentum matrix elements and resonant denominators of the type $(E_g - \hbar\omega)$. Damping is not included in this model, so the range of validity extends only to within $\sim\hbar/T_2$ of the bandgap energy, where T_2 is the momentum relaxation time of states near $\mathbf{k} = 0$. In the limit of long wavelengths ($\hbar\omega \ll E_g$), Eq. (26) reduces simply to

$$\chi_{NP}^{(3)} = -\frac{e^4}{24\hbar^4\omega_1^2\omega_2\omega_4 V} \sum_{\mathbf{k}} f_0(E_{ck}) \frac{\partial^4 E_{ck}}{\partial k^4}. \tag{27}$$

We note that the magnitude of $\chi^{(3)}$ provides information about the nonparabolicity of the conduction band. In particular, if we expand E_{ck} in a power series in a we find (using Eq. (12a))

$$E_{ck} \approx E_g + \frac{\hbar^2 k^2}{2m_e^*} - \frac{\hbar^4 k^4}{4(m_e^*)^2 E_g} + \cdots. \tag{28}$$

Thus only terms in k^4 (and higher) contribute to $\chi^{(3)}$, independent of the parabolic k^2 term. If we evaluate $\partial^4 E_{ck}/\partial k^4$ using Eq. (28) and neglect details of the distribution of conduction band electrons, then Eq. (27) reduces to Eq. (16) derived earlier.

2. *Calculated Values and Measurements*

Several calculations and measurements of $\chi^{(3)}$ due to conduction band electrons have been performed; the results (scaled to a carrier concentra-

[2] Note that the effect of the valence bands is also implied via the effective mass m_e^*, which itself is strongly related to valence-to-conduction band transitions via the Kane momentum matrix element (Kane, 1957).

tion of $2 \times 10^{16}/cm^3$) are collected in Table IV. The first measurements of $\chi^{(3)}$ were performed by Patel, *et al.* (1966). The semiconductors InSb, InAs, and GaAs were studied using difference frequency mixing at 10 μm. The largest susceptibilities were observed in InSb, because of its small bandgap and small effective mass. For each material the measured values of susceptibility were compared with values calculated on the basis of conduction band nonparabolity [Eq. (20)] and on the basis of anharmonicity of the bound carriers [Eq. (10)]. The measured values of $\chi^{(3)}$ for InSb and InAs were in order-of-magnitude agreement with values determined from Eq. (20), whereas the calculated contribution from bound electrons was ~100 times smaller. Thus, for InSb and InAs, nonparabolicity is the dominant contributing mechanism. For GaAs the calculated contributions from free and bound electrons were of comparable magnitude, so that both mechanisms contribute at 10 μm for the chosen doping levels.

The model of Jha and Bloembergen (1968a,b) provides improved accuracy in the calculation of the separate contribution of bound and free electrons to $\chi^{(3)}$ in InSb and InAs, especially when $\hbar\omega \gtrsim E_g/2$. Using the model band structures of Kane (1957), Jha and Bloembergen calculated $\chi^{(3)}$ at 10 μm due to conduction band nonparabolicity using the exact expression [Eq. (26)] and the long wavelength approximation of Wolff and Pearson [Eq. (20)]. In each case, values were obtained for carrier concentrations in the range 10^{15}–$10^{18}/cm^3$. For InAs the two models yield similar values, since $\hbar\omega \approx \frac{1}{3}E_g$. For InSb (with $\hbar\omega \sim \frac{1}{2}E_g$) the exact values are ~50% larger than the values from Eq. (20). The bound electron contribution to $\chi^{(3)}$ was determined using the tetrahedral bonding orbital approximation for the ground-state wave function. The bound and free electron contributions are comparable in InSb and InAs for carrier concentrations in the range 2×10^{15} to 4×10^{15} cm^{-3}.

The third-order nonlinear properties of GaAs at 10 μm were studied by Wynne (1969). Difference frequency mixing was used to measure $\chi^{(3)}$ in a high resistivity sample ($N < 10^{10}/cm^3$) and in samples with carrier densities in the range $10^{15}/cm^3$ to $5 \times 10^{16}/cm^3$. The contributions to $\chi^{(3)}$ from bound and free electrons were found to be equal at $N \approx 3 \times 10^{15}/cm^3$, and the sign of each contribution was observed to be positive. At high carrier densities no anisotropy in $\chi^{(3)}$ was observed, indicating that the conduction band is spherical near $\mathbf{k} = 0$. The measured susceptibility was observed to vary linearly with the carrier density; the value at a density of $2 \times 10^{16}/cm^3$ is given in Table IV. Susceptibilities were also calculated from the theory of Jha and Bloembergen [Eq. (26)] using Kane's model band structure [Eq. (12a)]. Wynne concluded that the Kane model is inaccurate for GaAs, because of the subsidiary conduction band minima which lie energetically close to the minimum at $\mathbf{k} = 0$. A modified expression for E_{ck} was given, resulting in much better agreement between the measured and calculated values of the third order susceptibilities in GaAs.

The nonlinearity due to conduction band electron in n-type germanium at 10 μm was studied by Wang and Ressler (1970) for samples with carrier densities in the range $10^{14}/cm^3$ to $2 \times 10^{17}/cm^3$. Germanium is an indirect bandgap material with a relatively large direct bandgap, so that the contribution of conduction band electrons to $\chi^{(3)}$ is smaller than in the earlier materials studied. The contributions from bound and free electrons were found to be equal only at the highest donor concentration ($2 \times 10^{17}/cm^3$); no absolute values of the measured susceptibilities were given. In order to estimate the nonlinearity due to free electrons, the $\mathbf{k} \cdot \mathbf{p}$ approach was used to derive $E_c(\mathbf{k})$ at the L_1 direct band edge in germanium. The susceptibility $\chi^{(3)}$ was then obtained by averaging the induced polarization over the four tetrahedrally coordinated conduction band ellipsoids. The calculated result for a carrier density of $2 \times 10^{16}/cm^3$ is given in Table IV.

The large increase in the third order susceptibility with reduction in the frequency detuning ($\omega_1 - \omega_2 \rightarrow 0$), as predicted by Eq. (25) for the case that includes inelastic collisions, has also been observed recently by

TABLE IV

Third-Order Susceptibility Due to Conduction-Band Electrons Scaled to a Carrier Concentration $N = 2 \times 10^{16}/cm^3$ [a]

	Band gap		$\chi^{(3)}$ (10^{-10} esu)	
Semiconductor	E_g (eV)	λ_g (μm)	Calculated	Measured
InAs	0.36	3.4	9.4,[b] 11[c]	1.8[b]
InSb	0.17	7.3	50,[b] 40[c]	8,[b] 500 ($\Delta\omega$ = 1.4 cm^{-1})[d]
GaAs	1.43	0.87	0.38,[b] 0.58[e]	0.07,[b] 0.88[e]
Ge	0.67	1.85	0.065[f]	—
$Hg_{0.784}Cd_{0.216}Te$	0.19	6.8	420[g]	3800[g] ($\Delta\omega$ = 0)

[a] For all cases, λ = 10.6 μm and T = 300 K. Unless noted, the measurements are based on difference frequency measurements, with $\Delta\omega$ between 10 and 100 cm^{-1}.

[b] Patel *et al.* (1966).

[c] Jha and Bloembergen (1968a,b).

[d] Yuen and Wolff (1982). The magnitude of $\chi^{(3)}$ was obtained by comparison with the four-wave mixing signal from a Ge reference sample. For the value listed here, we chose to use the Ge measurements of Watkins *et al.* (1980b) as the reference value.

[e] Wynne (1969).

[f] Wang and Ressler (1970).

[g] Khan *et al.* (1980).

Yuen and Wolff (1982). In these measurements, which are particularly relevant to estimation of values of $\chi^{(3)}$ for DFWM, Yuen and Wolff measured the dependence of $\chi^{(3)}$ in InSb near 10 μm as a function of $(\omega_1 - \omega_2)$ over the range 1.4–100 cm^{-1}; they found that $\chi^{(3)}$ for $\omega_1 - \omega_2 = 1.4$ cm^{-1} exceeded that for $\omega_1 - \omega_2 = 100$ cm^{-1} by a factor of 10^2–10^3, depending on the doping density. The absolute magnitude of $\chi^{(3)}$ was determined by comparing the difference frequency signal with that from a reference sample of Ge. The assumed reference value of $\chi^{(3)}$ for Ge under these conditions ($\chi^{(3)} = 1.0 \times 10^{-10}$ esu; Wynne and Boyd, 1969) is perhaps too large by a factor of 4, in light of the recent measurements of Watkins, *et al.* (1980); in our listing of values of $\chi^{(3)}$ in Table IV, we have modified the value reported by Yuen and Wolff accordingly.

For comparison of the exact dispersion of the $\chi^{(3)}$ data with the functional form predicted by Eq. (25), the values τ_{th} used by Yuen and Wolff were chosen on the basis of curve fitting and not on the basis of previously published values. Thus, accuracy of the function form of Eq. (25) will be clear only when additional independent measurements are performed in order to determine values of T_{2m} and τ_{th} for samples and excitation conditions similar to those used by Yuen and Wolff.

The data of HgCdTe in Table IV are due to Khan, *et al.* (1980), who have attempted to both calculate and measure $\chi^{(3)}$ due to conduction band nonparabolicity under conditions of degenerate four-wave mixing. The calculations are based on the theory of Wolff and Pearson [Eq. (20)], and thus do not account for the case of degenerate frequencies [Eq. (25)]. Two other factors influence the authors' calculation of $\chi^{(3)}$ using the Wolff and Pearson model. First, the experimental samples are not degenerately doped for all samples and temperatures used; in such cases, the model of Wang and Ressler [Eq. (22)] is more appropriate. Second, the photon energy is close to the bandgap energy for all the compositions and temperatures studied. To account for this, Khan *et al.* include a multiplicative "resonant" factor $E_g^2/[E_g^2 - (\hbar\omega)^2]$ in Eq. (20), by analogy with expressions for $\chi^{(3)}$ due to spin-flip Raman scattering (Wolff, 1966; Kruse *et al.* 1979). The applicability of such a factor is generally inconsistent with other related derivations for $\chi^{(3)}$. Furthermore, the theory of Jha and Bloembergen (Eq. 26) provides a more reliable basis for calculating $\chi^{(3)}$, since it incorporates both bandgap proximity and nondegenerate doping. On the other hand, the high values of $\chi^{(3)}$ measured at Khan *et al.* in several of their HgCdTe samples are very possibly due to the inelastic collision factor for degenerate frequencies [Eq. (25)] as well as the electron-hole plasma (see Section III.C) that is generated in these absorbing samples. Thus, the specific contribution of non-parabolicity in the measurements of Khan *et al.* (1980) is difficult to assess, and the interpretation of nonparabolicity in these measurements must be thus treated with caution.

C. Nonlinearities Associated with Real Transitions

A large variety of energy states may occur in semiconductors, both in the form of bands and in the form of discrete (but generally broadened) energy levels. Pure semiconductors of perfect crystallinity are characterized by a set of broad valence and conduction electron bands and an abun-

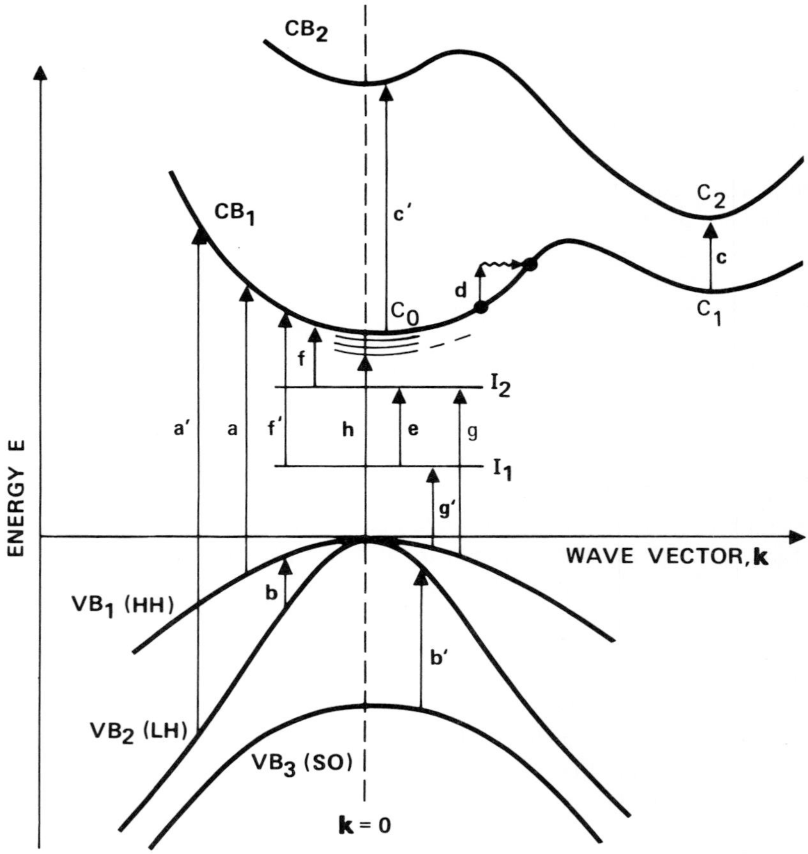

Fig. 2. Schematic representation of several of the most prominent types of transitions that may be present in a semiconductor. The energy-band structure and impurity levels (I_1, I_2) are hypothetical; all the features shown are typically not present simultaneously in a given semiconductor. CB_i are conduction bands and VB_i valence bands, with HH, LH, and SO the heavy-hole, light-hole, and splitoff bands, respectively. The vertical arrows indicate a variety of transitions that may be induced by optical radiation. Among the transitions shown, only d is phonon assisted or *indirect:* the remaining are all *direct* transitions.

dance of phonon modes (lattice vibrations) at modest temperatures; discrete states occur in the presence of doped impurities and lattice defects, and at low temperatures, owing to the possible binding of electrons and holes into excitons and excitonic complexes. Several of these states are depicted schematically in the energy-level diagram of Fig. 2, in which energy is plotted as a function of wave vector for a hypothetical direct-band-gap semiconductor. Various types of transitions are thus possible, as depicted by the vertical arrows. These include valence-to-conduction-band transitions (a) (which also lead to free-carrier generation and photoconductivity), free-carrier transitions [inter-valence-band (b), inter-conduction-band (c), and intraband (d)], and transitions between the electronic levels of impurities (e), between impurity levels and conduction bands (f), and between valence bands and impurity (g) or excitonic levels

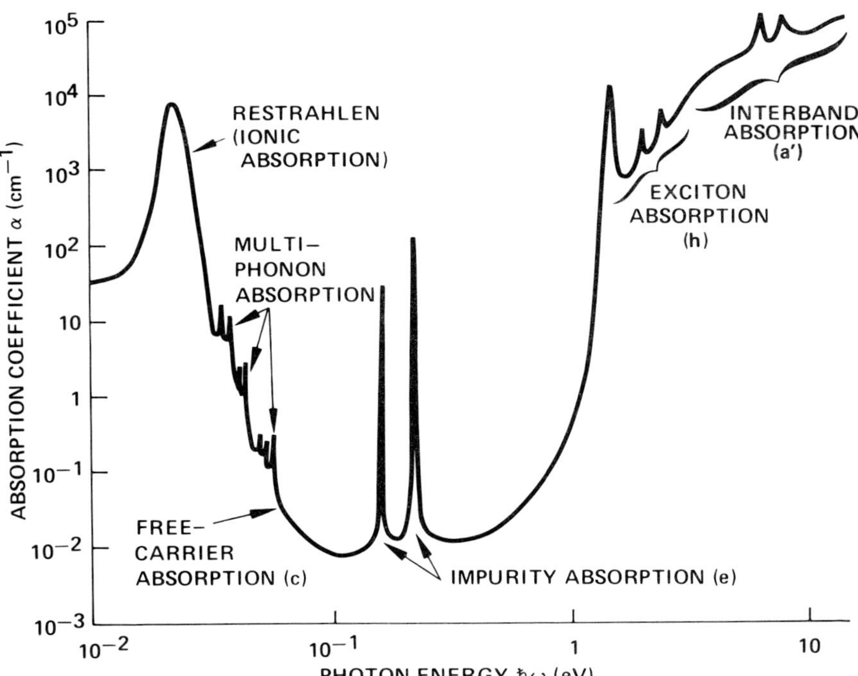

Fig. 3. Typical low-intensity absorption spectrum for the hypothetical direct-band-gap compound semiconductor ($E_g \sim 1$ eV) whose band structure is represented schematically in Fig. 2. The semiconductor is assumed to be at low temperatures and doped with deep-level impurities. The letters in parentheses designate transitions shown in Fig. 2.

(h). Note that near **k** = 0 the band diagram of Fig. 2 is quantitatively similar to that of most III–V semiconductors. The absorption spectrum of a hypothetical direct-gap semiconductor with an energy-level diagram similar to that of Fig. 2 is illustrated in Fig. 3 so as to depict *qualitatively* the spectral location and relative magnitude of the low-intensity absorption coefficients associated with some of these transitions. The value of these absorption coefficients is often a strong function of the light intensity. In particular, some of these transitions may show saturation behavior similar to that of a two-level system, as discussed in Chapter 8; the associated dispersive nonlinearities may be similar in character, as will be discussed in greater detail below. Nevertheless, the band character of semiconductors, and the possibility of interacting with free carriers, result in several unique nonlinear mechanisms, such as the large nonlinear index change due to the generation of a free-carrier plasma and the nonlinearity due to resonant interaction with the cyclotron or Landau levels that may be created in a free-carrier plasma in the presence of an external (large) magnetic field (Lax, 1963; Dresselhaus and Dresselhaus, 1966; Scott, 1975; MacKenzie *et al.*, 1981).

As opposed to the "fast" ($\tau_{NL} \lesssim 10^{-12}$ sec) nonresonant nonlinearities (discussed in Sections III. A and III. B) that are observed in the transparency regions of a semiconductor and are associated only with virtual transitions, optical nonlinearities associated with real transitions are generally much slower, because they depend upon population changes incurred during the transitions, and thus on the buildup and decay time τ of these populations. This decay time is typically of the order of nanoseconds or longer. Thus, with the use of long optical pulses or cw radiation, large steady-state susceptibilities proportional to the medium response time τ may be observed with several of these nonlinear mechanisms. In this section, we will consider self-induced nonlinearities associated with the following transitions.

1. valence-to-conduction-band transitions (free-carrier generation)
2. free-carrier transitions ($VB_i \rightarrow VB_j$, $CB_i \rightarrow CB_j$, intraband)
3. impurity transitions ($I_i \rightarrow I_j$, $VB \rightarrow I$, $I \rightarrow CB$)
4. transitions in excitons and exitonic complexes

1. Valence-Band-to-Conduction-Band Transitions (Free-Carrier Generation)

a. Modification of the Dielectric Function via Plasma Generation It is well known that the generation of an optical plasma can result in a strong modification of the dielectric function $\epsilon(\omega)$ in a semiconductor, either by single-electron effects or by correlated many-body effects (Haug *et al.*, 1981). For our discussion here, we will assume that the plasma densities are low enough that many-body effects are negligible, and we will estimate the change Δn in the refractive index $n(\omega)$ due to such an optically generated plasma in a direct-gap semiconductor with cubic symmetry. In addition, we will present the functional form of the band-edge absorption coefficients $\alpha(\omega)$, because the density of the generated plasma, and consequently the nonlinear index change $\Delta n(\omega)$ itself, is directly related to $\alpha(\omega)$.

In a cubic direct gap semiconductor without any impurity or exciton levels, the real part of the dielectric function (in the single electron approximation) is given by (Ehrenreich, 1966; Ashcroft and Mermin, 1976; Auston *et al.*, 1978)

$$\epsilon_1(\omega) - 1 = -\frac{4\pi e^2}{\omega^2}\int \frac{d^3k}{4\pi^3}\sum_n f(E_n)\frac{\nabla_k^2 E_n(\mathbf{k})}{3\hbar^2} + \frac{4\pi\hbar^2 e^2}{m^2}\sum_{n,n'}\int \frac{d^3k}{4\pi^3}\frac{[f(E_{n'}) - f(E_n)]|p_{nn'}|^2}{(E_n - E_{n'})[(\hbar\omega)^2 - (E_n - E_{n'})^2]} \tag{29}$$

where n and n' are the electron states in the conduction and valence bands, $p_{nn'}$ is the momentum matrix element for a direct transition between these states, $E_n(\mathbf{k})$, $E_{n'}(\mathbf{k})$ are the energies of the electron states in these bands, $f(E_n)$, $f(E_{n'})$ are the Fermi carrier distribution functions, and m is the mass of a *free* electron. In Eq. (29), the first term corresponds to intraband contributions, whereas the second term is associated with interband (valence-to-conduction band) transitions.

The imaginary part of the dielectric function for transitions between the valence and conduction bands, and the corresponding absorption coefficient $\alpha(\omega)$, are described by (Ehrenreich, 1966; Harbeke, 1972; Bassani, 1969; Stern, 1963; Casey and Stern, 1976):

$$\epsilon_2(\omega) = \frac{2\pi e^2\hbar^2}{m^2\omega^2}\sum_{n,n'}\int \frac{d^3\mathbf{k}}{4\pi^3}[f(E_{n'}) - f(E_n)]|p_{nn'}|^2\,\delta(\hbar\omega - E_n + E_{n'}), \tag{30a}$$

and

$$\alpha(\omega) = \omega\epsilon_2(\omega)/n(\omega)c. \tag{30b}$$

For allowed direct-gap transitions in the vicinity of the band gap E_g, it can be shown (Harbeke, 1972) that Eq. (30) results in an absorption coefficient given by

$$\alpha(\omega) = \frac{1}{nc}\frac{8\hbar^2 e^2}{m^2\omega}\left(\frac{2m_{\mathrm{eh}}^*}{\hbar^2}\right)^{3/2}|p_{\mathrm{vc}}|^2(\hbar\omega - E_g)^{1/2}, \tag{31}$$

where $m_{\mathrm{eh}}^* = m_e m_h/(m_e + m_h)$ is the reduced effective mass, p_{vc} is the momentum matrix element, assumed to be independent of $\mathbf{k}$ and ω, and $n(\omega)$ is the refractive index, determined from Eq. (29). It is seen from Eq.

(31) that in a direct-gap semiconductor, for $\hbar\omega > E_g$, the primary dependence of the absorption coefficient on the photon energy is of the form $(\hbar\omega - E_g)^{1/2}$, a result that can also be deduced from simpler arguments [see Smith (1968) or Pankove (1971)] on the joint density of states in a semiconductor with isotropic parabolic conduction- and valence-band edges.

Similarly, if the direct transition at the band edge is forbidden in the electric dipole approximation, one can make use of the $\mathbf{k}$ dependence of the matrix element $p_{vc}(\mathbf{k})$, and expand around the critical point at $\mathbf{k}_0$:

$$p_{vc}(\mathbf{k}) = p_{vc}(\mathbf{k}_0) + \sum_{i=1}^{3} \frac{\partial p_{vc}(\mathbf{k})}{\partial \mathbf{k}_i}(k_i - k_{0i}) + \cdots. \tag{32}$$

Keeping only the linear term in $\mathbf{k}$, and assuming an isotropic dependence of $p_{vc}(\mathbf{k})$, we obtain the absorption coefficient near the band gap for forbidden direct transitions as

$$\alpha(\omega) = \frac{1}{nc}\frac{8\hbar^2 e^2}{m^2\omega}\left(\frac{2m_{eh}^*}{\hbar^2}\right)^{5/2}\left|\frac{\partial p_{vc}}{\partial|\mathbf{k}|}\right|^2 (\hbar\omega - E_g)^{3/2}. \tag{33}$$

Note that the absolute values for $\alpha(\omega)$ are much lower for the "forbidden" transition case of Eq. (33) compared with the "allowed" case of Eq. (31), because the magnitude of the first derivative of $\mathbf{k}$ in the matrix element p_{vc} is usually small near the critical point ($\mathbf{k} = \mathbf{k}_0$).

Returning to Eq. (29), let us estimate the change in the real part of the dielectric function, and thereby estimate the nonlinear refractive index, due to the generation of carriers near the band edge via absorption processes described by Eq. (31). For simplicity, we will assume that only two bands (i.e., one conduction and one valence band) exist, that the temperature is 0° K, (so that in the absence of light, $f(E_n) = 1$ and $f(E_n') = -1$), and that the conduction and valence bands are isotropic and described by density-of-states effective masses

$$1/m_e^* = (1/3\hbar^2)\nabla_k^2 E_n(\mathbf{k})|_c \quad \text{and} \quad 1/m_h^* = -(1/3\hbar^2)\nabla_k^2 E_n(\mathbf{k})|_v.$$

Then, the change in $\epsilon_1(\omega)$ is estimated to be

$$\Delta\epsilon_1(\omega) = -\frac{4\pi e^2}{\omega^2}\left(\frac{N_e}{m_e^*} + \frac{N_h}{m_h^*}\right) + \frac{4\pi\hbar^2 e^2}{m^2}\frac{2(N_e + N_h)}{E_g}\frac{|p_{vc}|^2}{(\hbar\omega)^2 - E_g^2}, \tag{34}$$

where it is assumed that the relevant energy states E_n and $E_{n'}$ are described by $E_n - E_{n'} = E_g$, and that

$$N_e = \int \frac{d^3\mathbf{k}}{4\pi} f(E_n) \quad \text{and} \quad N_h = -\int \frac{d^3\mathbf{k}}{4\pi} f(E_{n'})$$

are the conduction- and the valence-electron densities. Assuming that the momentum matrix element can be estimated by Kane's **k** · **p** perturbation model (Kane, 1957), i.e., $|p_{vc}|^2 = \frac{1}{4}E_g(m^2/m^*_{eh})$, and defining $N \equiv N_e = N_h$ as the carrier number density, we have (Auston *et al.*, 1978)

$$\Delta\epsilon_1(\omega) = -\frac{4\pi Ne^2}{m^*_{eh}}\left(\frac{1}{\omega^2} - \frac{1}{\omega^2 - \omega_g^2}\right) \tag{35a}$$

or

$$\Delta\epsilon_1(\omega) = -\frac{4\pi Ne^2}{m^*_{eh}\omega^2}\frac{\omega_g^2}{\omega_g^2 - \omega^2}. \tag{35b}$$

In Eq. (35a), the first and second terms in large parentheses can be identified as intraband and interband contributions, respectively. If $\epsilon_2(\omega)$ is small, $\epsilon_1 + \Delta\epsilon_1 \approx (n + \Delta n)^2$, where Δn is the optically induced change in the refractive index, deduced to be

$$\Delta n = -\frac{2\pi Ne^2}{nm^*_{eh}\omega^2}\frac{\omega_g^2}{\omega_g^2 - \omega^2}. \tag{36}$$

Note that if $\omega \ll \omega_g$, as may be true when carriers are at frequencies far from the direct gap (e.g., near the indirect gap of Si), Eqs. (35) and (36) reduce to the familiar Drude result. Here $\Delta n = -(2\pi e^2/nm^*_{eh}\omega^2)N$. Note also that, in the light of the classical Drude and Lorentz models, Eq. (35) may be interpreted as the change in the $\epsilon_1(\omega)$ due to the generation of a density of N free carriers with the simultaneous elimination of a density of N bound oscillators resonant at ω_g.

The optically generated carrier density N in Eqs. (35) and (36) is in the general case a function of position **r** and time t, determined by the generation of carriers by the local flux density $\Phi(\mathbf{r}, t)$ and by their decay via diffusion and recombination. For optically thin samples, under the assumption of intensity variations in only one transverse direction x, the spatial and temporal dependence of N is given by the one-dimensional diffusion equation

$$\frac{\partial N(x, t)}{\partial t} = \eta\alpha\Phi(x, t) - \frac{N(x, t)}{\tau_R} + D_a\frac{\partial^2 N(x, t)}{\partial x^2}. \tag{37a}$$

Here, $\Phi(x, t)$ is the optical flux density, η is the quantum efficiency of carrier generation, and is nearly unity in intrinsic semiconductors at low light intensities, τ_R is the electron–hole recombination time, and D_a is the ambipolar diffusion coefficient. For the case of spatially interfering light fields, $\Phi(x, t)$ varies sinusoidally with x. Assuming that the spatial variation of N is also a sinusoid, of period Λ, the time dependence $N(t)$ of the carriers near the locations of the peak carrier density is given by

$$\frac{dN}{dt} = \eta\alpha\Phi(x, t) - \frac{N}{\tau}, \tag{37b}$$

where

$$\tau = \tau_D\tau_R/(\tau_D + \tau_R) \tag{38a}$$

is the total lifetime of the carriers, with τ_D the ambipolar diffusional decay time given by

$$\tau_D = \Lambda^2/4\pi^2 D_a. \tag{38b}$$

For simplicity, we ignore the dependence of τ_R and τ_D on the carrier density in this discussion. The solution to Eq. (37b) is simply

$$N(t) = \eta\alpha e^{-t/\tau} \int_0^t e^{t'/\tau}\Phi(t')\, dt', \tag{39}$$

and is of particular interest in two time regimes, $t \gg \tau$ and $t \ll \tau$. The relevant time regimes are determined by the duration and shape of the laser pulses.

For long pulses (pulse duration $\tau_L \gg \tau$) of slowly varying intensity ($dI/dt < I/\tau$), the steady-state carrier density (i.e., $dN/dt \approx 0$) is given by

$$N(x, t) = \eta\alpha\tau\Phi(x, t) = \frac{\eta\alpha\tau}{\hbar\omega} I(x, t).$$

For the sinusoidal variation in the carrier density generated by the interference of the probe wave with either of the pump waves (see Fig. 1),

$$N_{ip}(x, t) = \frac{\eta\alpha n c}{8\pi\hbar\omega} \tau_{ip} \vec{\mathscr{E}}_i(t) \cdot \vec{\mathscr{E}}_p^*(t), \tag{40a}$$

where $i = \text{f}$ or b, corresponding to the forward and backward pump waves, respectively. To calculate the appropriate third-order susceptibilities for this process, we associate each grating term in Eq. (2) with its corresponding index modulation from Eq. (36):

$$\frac{2\pi}{n} \chi_{ip}^{(3)} \vec{\mathscr{E}}_i \cdot \vec{\mathscr{E}}_p^* = \Delta n_{ip} = -\frac{2\pi e^2 N_{ip}(x, t)}{n m_{eh}^* \omega^2} \left(\frac{\omega_g^2}{\omega_g^2 - \omega^2}\right). \tag{40b}$$

Then, from Eqs. (40a) and (40b), we obtain

$$\chi_{ip}^{(3)} = -\frac{\eta\alpha n c e^2}{8\pi m_{eh}^* \hbar\omega^3} \tau_{ip} \left(\frac{\omega_g^2}{\omega_g^2 - \omega^2}\right) \tag{41}$$

as the steady-state third-order nonlinear susceptibilities invoked for

DFWM via free-carrier generation by valence-to-conduction-band transitions near the direct band gap of a semiconductor. Note that the term in parentheses may be dropped when the transitions occur near an indirect band gap for which $\omega^2 \ll \omega_g^2$, as in the case of Si at 1.06 μm (Jain and Klein, 1979). The subscripts correspond to the two spatial interference terms [see Eq. (2)] characteristic of DFWM. The occurrence of τ_{ip} in Eq. (41) is a direct consequence of the slow integrating character [Eq. (39)] of such a nonlinearity, dependent on both the recombination and the diffusion of the carriers. These may differ significantly for the two spatial gratings, especially if the grating periods are vastly different and one (or both) of these decay times is diffusion dominated (see, e.g., Jain and Klein, 1979).

In the opposite limit of short laser pulses ($\tau_L \ll \tau$), one has a *time dependent transient susceptibility,* determined by the carrier buildup described by Eq. (39), which in turn is related simply to the integrated photon flux. Thus

$$N(t) = \eta\alpha \int_0^t \Phi(t')\, dt' = \frac{\eta\alpha nc}{8\pi\hbar\omega} \int_0^t \vec{\mathscr{E}}_i(t') \cdot \vec{\mathscr{E}}_p^*(t')\, dt'. \tag{42}$$

For cases of practical interest one can compute the carrier density and effective third-order susceptibility $\chi_{\text{eff}}^{(3)}$ at the temporal peak of the short laser pulse. For symmetric pulses of peak electric field $\mathscr{E}_0$, the carrier density at the pulse peak is approximately $N_0 = (\eta\alpha nc\tau_L/16\pi\hbar\omega)\mathscr{E}_0^2$, resulting in an effective susceptibility (at the pulse peak) described by

$$\chi_{\text{eff}}^{(3)} = -\frac{\eta\alpha nce^3\tau_L}{16\pi m_{\text{eh}}^{*3}\hbar\omega^3} \frac{\omega_g^2}{\omega_g^2 - \omega^2}. \tag{43}$$

It has been assumed that the semiconductor is isotropic, which is essentially true for both the absorption coefficient and the transport properties of cubic semiconductors. Although both of the energy versus wave vector surfaces near the conduction-band valleys are highly nonisotropic for indirect-band-gap semiconductors (such as Si and Ge), the net contribution of all the different valleys to overall transport properties of the electrons results in an isotropic effective mass (McKelvey, 1966)

$$1/m_e^* = \tfrac{1}{3}(1/m_L + 2/m_T), \tag{44}$$

where m_L and m_T are the longitudinal and transverse effective masses, respectively. Note that in a semiconductor with an indirect band gap it is the conductivity effective mass of Eq. (44) that enters into this computation of the nonlinear index change and susceptibility expressions of Eqs. (35), (41), and (43).

For semiconductors in which the hole density is partitioned between two isotropic hole bands (heavy- and light-hole bands) such that the hole populations in the heavy- and light-hole bands are given by N_H and N_L respectively, we have

$$N_H/N_L = (m_{HH}/m_{LH})^{3/2} \tag{45}$$

and

$$N_H + N_L = p \equiv N. \tag{46}$$

Equation (45) is a direct consequence of the fact that the densities of states near the band edge of semiconductors are proportional to $m_{LH}^{3/2}$ (see, e.g., Moss *et al.,* 1973). From Eqs. (45) and (46) it follows that the effective hole mass that contributes to m_{eh}^* in, e.g., Eqs. (35), (41), and (43) is given by

$$1/m_h^* = (m_{HH}^{1/2} + m_{LH}^{1/2})/(m_{HH}^{3/2} + m_{LH}^{3/2}). \tag{47}$$

Estimates of the third-order susceptibility based on Eqs. (41) and (43) have been made for several semiconductors and found to be in good agreement with those measured experimentally (Jain and Klein, 1979; Jain and Steel, 1980, 1981; Miller *et al.,* 1980a; Lind and Jain, 1981). These will be discussed later. Note that with the use of extremely long carrier lifetimes ($\sim 10^{-7}$–10^{-6} sec) and correspondingly long laser pulses (or cw radiation) extremely high third-order susceptibilities have been predicted and measured on the basis of this model. Third-order susceptibilities as high as 10^{-1} esu (Miller *et al.,* 1980a; Jain and Steel, 1981) have been estimated, and, with an appropriate choice of semiconductor and experimental parameters, values of $\chi^{(3)}$ of over 10^3 esu should be obtainable. (See the discussion on HgCdTe in Section IV.)

Although simplifying assumptions were made [prior to Eq. (34)] to obtain our approximate expressions for Δn [Eq. (36)] and $\chi^{(3)}$ [Eqs. (41) and (43)], the model is based on a quantum-mechanically rigorous expression for the dielectric function. Thus, if one were to use better values for the electron energy distributions, band structures, and momentum matrix elements in Eq. (29)—say, from specific band-structure calculations—one could obtain more precise values for the third-order optical susceptibility (and nonlinear index change) by numerical integration. Note, finally, that the formalism described above, relating the nonlinear index change to the third-order susceptibility and defining the concepts of transient, effective, and steady-state susceptibilities, is valid for all nonlinearities that involve relatively slow nonlinear index changes. In the following sections our discussion will be confined to either the nonlinear refractive index changes or the third-order susceptibility, depending on the nature of the derivation; also, we will ignore the temporal behavior of the third-order susceptibility and simply use the term "third-order susceptibility" to designate the steady-state or effective values as appropriate.

b. Alternative Models for Nonlinear Refraction via Valence-to-Conduction-Band Transitions Several alternative models for calculating the nonlinear index change (and thus the third-order susceptibility) have recently been proposed (Miller *et al.*, 1980a, 1981; Moss, 1980; Wherrett and Higgins, 1981; Elci and Rogovin, 1981). The important physics invoked by these models is well contained by the "modification of the dielectric function" approach of Eqs. (29) and (34). Nevertheless, these alternative models represent different approaches that occasionally provide additional insight into the nature of this nonlinearity; as elaborated below, calculations based on such simplified models often lend themselves to a more natural inclusion of different parameters and approximations, such as the momentum relaxation times and carrier distribution functions, and thus to alternative simplified expressions for the third-order susceptibility.

Two basic approaches have been used in these alternative models for calculating the nonlinear index change near the band edge of a direct-gap semiconductor: a phenomenological "blocking" approach (the dynamic Burstein–Moss model) and an "atomic" or "two-level" approach (also called a direct saturation model).

Let us first consider the blocking approach, discussed by Miller *et al.* (1980a, 1981), Moss (1980), and Wherrett and Higgins (1981). In this approach, the intensity-dependent (i.e., nonlinear) refractive index is explained in terms of a three-stage process: (1) generation of a moderate-density (10^{14}–10^{16} cm^{-3}) electron–hole plasma by the incident radiation; (2) filling of states in the conduction band (due to low conduction-band effective masses, i.e., low conduction-band densities of states, as occur in InSb and HgCdTe), resulting in the "blocking" or unavailability of these states for further transitions, manifested by a saturation of absorption; and (3) a change in the refractive index due to the loss of the filled states. The "loss" in the refractive index due to the blocking of states in the conduction band (the dynamic Burstein–Moss effect) can be estimated simply by integrating the refractive index contribution from each of these states over the number of lost states. Such calculations have been performed by Miller *et al.* (1980a) and Wherrett and Higgins (1981) and is outlined below.

In the rotating wave approximation, the refractive index contribution from an isolated pair of homogeneously broadened states separated by $\hbar\omega_{12}$ and with a dephasing time T_2 may be approximated by

$$\delta n = 2\pi|er_{12}|^2/\hbar(\omega_{12} - \omega - iT_2^{-1}). \qquad (48a)$$

In this spirit of approximation, the refractive index change due to removal of a set of valence-to-conduction-band level pairs of the same **k** value is

$$\delta n \approx -\frac{2\pi}{3n}\left|\frac{ep_{\text{vc}}}{m\omega}\right|^2 \frac{1}{\hbar}\sum_{\text{c,v}} [\omega_{\text{cv}}(\mathbf{k}) - \omega]^{-1}, \tag{48b}$$

where we have neglected T_2 broadening, and replaced the dipole matrix element $|er_{12}|^2$ with the momentum matrix element $\frac{1}{3}|ep_{\text{vc}}/m\omega|^2$ as appropriate for transitions between the valence and conduction bands. Then, for conduction-band states occupied up to $E_{\text{F}} = h^2k_{\text{F}}^2/2m_{\text{eh}}^*$,

$$\Delta n = -\frac{2\pi}{3n}\left|\frac{ep_{\text{vc}}}{m\omega}\right|^2 \frac{1}{h\pi^2}\int_0^{k_{\text{F}}} k^2\, dk \left(\omega_{\text{g}} + \frac{\hbar k^2}{2m_{\text{eh}}^*} - \omega\right)^{-1}. \tag{49}$$

For $\omega < \omega_{\text{g}}$,

$$\begin{aligned}\int_0^{k_{\text{F}}} k^2\, dk \left((\omega_{\text{g}} - \omega) + \frac{\hbar k^2}{2m_{\text{eh}}^*}\right)^{-1} &= \left(\frac{2m_{\text{eh}}^*}{\hbar}\right)^{3/2} (\omega_{\text{g}} - \omega)^{1/2}(X_{\text{F}} - \tan^{-1} X_{\text{F}}) \\ &\approx \left(\frac{2m_{\text{eh}}^*}{\hbar}\right)^{3/2} (\omega_{\text{g}} - \omega)^{1/2}\, X_{\text{F}}^3,\end{aligned}$$

where

$$X_{\text{F}} = k_{\text{F}}\left(\frac{\hbar/2m_{\text{eh}}^*}{\omega_{\text{g}} - \omega}\right)^{1/2} \ll 1.$$

Equation (49) leads to a saturation-induced nonlinear index change (Miller *et al.*, 1980a)

$$[n_{\text{I}}]_{\text{sat}} = \frac{\Delta n_{\text{sat}}}{I} = -\frac{2\pi}{3n}\left(\frac{eP}{\hbar\omega}\right)^2 \frac{\alpha\tau}{\hbar(\omega_{\text{g}} - \omega)\hbar\omega}, \tag{50}$$

where I is the intensity, and n_{I} is defined by $\Delta n = n_{\text{I}}I$. This is to be compared with the plasma-induced index change Δn_{p} computed directly [see Eqs. (36) and (38)] from the estimated change in the dielectric function (assuming $\eta = 1$):

$$[n_{\text{I}}]_{\text{p}} = \frac{\Delta n_{\text{p}}}{I} = -\frac{2\pi e^2\alpha\tau}{nm_{\text{eh}}^*\hbar\omega^3}\frac{\omega_{\text{g}}^2}{\omega_{\text{g}}^2 - \omega^2}. \tag{51}$$

By using $1/m_{\text{eh}}^* \approx 4P^2/3\hbar^3\omega_{\text{g}}$, as expected from Kane's model [$P^2 = (3\hbar^2/2m^2)|p_{\text{vc}}|^2$], and assuming that $\omega \approx \omega_{\text{g}}$, so that $\omega_{\text{g}}^2 - \omega^2 \approx 2\omega_{\text{g}}(\omega_{\text{g}} - \omega)$, the two different models result in expressions for n with an identical functional form. The apparent difference of a factor of 4 is due to factor-of-2 simplifications associated with the use of the two-band model of Eq. (50). The conclusions of the two models thus agree remarkably well, especially considering the different simplifying assumptions and approximations used to obtain the final expressions [Eqs. (50) and (51)] for the two calculations.

The essential difference between the two approaches is that in the plasma-modified dielectric function model [Eq. (51)] one first computes the dielectric function integrated over the entire set of states in the valence and conduction bands, and then estimates the modification of this overall dielectric function by the transitions, whereas in the saturation model one starts with an estimate for the index change due to transitions in each of these discrete states, and then integrates over all the states affected by these transitions. Note that, in the simplifying approximations for both of these calculations, thermal equilibration of the carriers is implicit, but the temperature T is assumed to be 0 K for Eq. (50), and no energy distributions are accounted for in Eq. (51).

In a more recent calculation, Miller *et al.* (1981) have included the effect of a Boltzmann distribution (valid only for low carrier densities) of carriers in the conduction and valence bands. The change in the absorption coefficient [Eqs. (30) and (31)] due to this light-induced Boltzmann distribution is then used to estimate the change in the refractive index from the Kramers–Kronig relations, from which the nonlinear index is estimated to be

$$[n_{\rm I}]_{\rm B} = -\frac{8\sqrt{\pi}}{3}\frac{e^2h^2}{m_{\rm e}}\frac{m_{\rm eh}^*}{m_{\rm e}}\frac{mP^2}{\hbar^2}\frac{1}{n}\frac{1}{kT}\frac{\alpha(\omega)\tau}{(\hbar\omega)^3}J(a), \tag{52}$$

where

$$a = m_{\rm eh}^*(\hbar\omega - E_{\rm g})/m_{\rm e}kT$$

and

$$J(a) = \int_0^\infty \frac{x^{-1/2}e^{-x}}{x-a}\,dx.$$

Because of the Boltzmann approximation, Eq. (52) is valid only at low carrier densities (and thus at low light intensities). The values of n_I calculated on the basis of the above theory were compared with experimental data (Miller *et al.*, 1980a) on the frequency dependence of n_I, and a reasonable fit was obtained when the experimentally measured frequency dependence of the absorption coefficient was included in the model, with significant errors occurring mainly at frequencies close to the band gap. Unfortunately, however, the limited spectral range of the data does not allow discrimination of the accuracy of the Boltzmann model with respect to the two models discussed earlier [Eqs. (36) and (50)]; in particular, the low-temperature plasma model of Eq. (51) agrees at least as well with the data when the measured frequency dependence of the absorption coefficient is included in the model. Also, surprisingly, considering the attempted accuracy of this semiempirical Boltzmann model [Eq. (52)], an independent measurement of the recombination time τ_R at the intensities of interest was *not* performed in these experiments, and τ_R effectively represents a free parameter chosen within the range of typically reported (10^{-7}–10^{-6} sec) values.

Further experimental information about the near-band-edge frequency dependence of the free-carrier nonlinear index n_I has recently been obtained by Nurmikko and co-workers (1982), who measured the free-electron–hole contribution to the dielectric constant by picosecond pump-and-probe methods. In their measurements of optically-induced changes in the index of refraction as a function of photon energy, these workers have observed a distinct sign change at photon energies near the band gap, similar to that described earlier by Auston *et al.* (1978). The experimentally observed sign reversal is found to be consistent with numerical calculations of Nurmikko (1982), which are based on the change in the absorption coefficient due to a Maxwellian distribution of the excess electron–hole plasma, and the Kramers–Kronig relations.

We note here for completeness that Moss (1980) has also estimated an expression for the nonlinear index change, based more explicitly on the dynamic Burstein–Moss effect. This calculation represents a simple alternative approach to that of Miller *et al.* (1980a, 1981), but the final expression for Δn (and n_I) on the basis of this model is expected to be less accurate than those in Eqs. (50)–(52). In this calculation the dynamic Burstein–Moss shift in the absorption edge, $\Delta E = (\hbar^2/2m_e^*)(3N/8\pi)^{2/3}$, is first estimated from the corresponding change in the absorption spectrum; the refractive index change is then calculated (by using the Kramers–Kronig relationships) to be $\Delta n = \Delta\lambda/4\pi^2(\lambda - \lambda_g)$, where $\Delta\lambda$ represents the wavelength shift in the absorption edge due to the dynamic Burstein–Moss shift.

Let us now briefly consider the other alternative approach to the estimation of $\chi^{(3)}$: the "atomic" or "two-level" approach, introduced by Miller *et al.* (1980a), and discussed in detail by Wherrett and Higgins (1981) and very recently by Elci and Rogovin (1981). Simply stated, these models start with atomic or two-level expressions for the nonlinear susceptibility, and integrate these microscopic contributions over all possible sets of two-level transitions (at $\hbar\omega$) between states in the valence and conduction bands. The atomic or two-level expressions for the nonlinear susceptibility are based on a full density matrix approach (for instance, see Chapter 8) or on a perturbative expansion for the third-order susceptibility (Miller *et al.*, 1980a; Elci and Rogovin, 1981) valid only for low intensities.

Using an atomic expression for a saturation susceptibility and integrating over states in the valence and conduction bands, Wherrett and Higgins (1981) obtain

$$\chi(\omega, \vec{\mathscr{E}}) = \left|\frac{ep_{\mathrm{vc}}}{m\omega}\right|^2 \frac{1}{\hbar\pi^2} \int_0^\infty k^2\, dk[\omega_{\mathrm{cv}}(\mathbf{k}) - \omega - iT_2^{-1}]$$
$$\times \left(1 + \frac{4\tau T_2^{-1}|e\vec{\mathscr{E}} \cdot \mathbf{p}_{\mathrm{vc}}/m\hbar^2|}{[\omega_{\mathrm{cv}}(\mathbf{k}) - \omega]^2 + T_2^{-2}}\right)^{-1}, \tag{53}$$

where $T_2(\mathbf{k}, \vec{\mathscr{E}})$ corresponds to the momentum relaxation times of the free carriers. Then, using a perturbative expansion for $\chi(\omega, \vec{\mathscr{E}})$ (valid only at low intensities) and

$$n_{\mathrm{I}} = (4\pi^2/n^2c)\,\mathrm{Re}[\chi^{(3)}], \tag{54}$$

Wherrett and Higgins obtain the following expression for the saturation-associated nonlinear index:

$$(n_{\mathrm{I}})_{\mathrm{sat}} = -\frac{1}{\hbar^3}\left(\frac{ep_{\mathrm{vc}}}{\hbar\omega}\right)^4 \frac{\tau}{T_2}\frac{2\pi}{15n^2c}\left(\frac{2m_{\mathrm{eh}}^*}{\hbar}\right)^{3/2}(\omega_{\mathrm{g}} - \omega)^{-3/2}. \tag{55}$$

By deriving an expression for absorption in the band tail (i.e., for $\omega < \omega_{\mathrm{g}}$) via the same atomic saturation model,

$$\alpha(\omega) = \frac{1}{3nc}\left(\frac{ep_{\mathrm{vc}}}{\hbar\omega}\right)^2 \left(\frac{2m_{\mathrm{eh}}^*}{\hbar}\right)^{3/2}(\omega_{\mathrm{g}} - \omega)^{-1/2}, \tag{56}$$

they obtain excellent agreement in functional form and order of magnitude between the expressions for the saturation-induced refractive index changes given by Eqs. (50) and (55). Note, however, that the simplified expression (56) for band tails does not agree well with experimental data on band-tail absorption, which show a near-exponential behavior; this is in contrast with the impurity band-tail models of Kane (1966) and Halperin and Lax (1966) which are in much better agreement with the near-exponential behavior of typical data (Halperin and Lax, 1966; Casey and Panish, 1978). The inclusion of the wave-vector and energy dependence of T_2 in the theoretical model, and the use of ultrapure samples in the experiments, will perhaps yield better agreement.

The expressions for the nonlinear index changes due to interband transitions are applicable to a large variety of semiconductors, particularly to those in the III–V family. Note also that an assumption of low intensities ($I < I_{\mathrm{sat}}$) is implicit in all of the models used. The expressions are also applicable to most indirect-band-gap semiconductors if one associates $\hbar\omega_{\mathrm{g}}$ with the energy of the direct gap closest to the photon energy $\hbar\omega$. Besides the values shown in Table V, order-of-magnitude estimates for $\chi^{(3)}$ or n_2 can easily be made for a large variety of semiconductors on the basis of

TABLE V

Third-Order Nonlinear Susceptibilities Obtained by Single-Photon Valence-to-Conduction-Band Transitions[a]

Semiconductor	Band gap E_g (eV)	Band gap λ_g (μm)	Wavelength λ (μm)	Absorption coefficient α (cm^{-1})	Laser pulse width τ_L (nsec)	Estimated[b] $\chi^{(3)}_{eff}$ (esu)	Measured[b] $\chi^{(3)}_{m}$ (esu)	Recombination time τ_R (nsec)	Estimated[b] steady-state $\chi^{(3)}$ (esu)
Si[c]	1.11	1.12	1.06	10	15	$\approx 8 \times 10^{-8}$	1.1×10^{-7}	$\sim 10^3$	$\approx 1.7 \times 10^{-5}$
InSb[d] (5 K)	0.235	5.28	5.3	1	cw	$\approx 10^{-1}$	$\approx 10^{-1}$	~ 300	$\approx 10^{-1}$
CdS[e]	2.42	0.51	0.53	10	15	$\approx 4 \times 10^{-9}$	3.4×10^{-9}	$\sim 10^3$	$\approx 5.3 \times 10^{-7}$
$Hg_{0.8}Cd_{0.2}Te$[f]	0.17	7.5	10.6	26	180	$\approx 8 \times 10^{-5}$	5.4×10^{-6}	~ 0.5	$\approx 8 \times 10^{-5}$
$Hg_{0.78}Cd_{0.22}Te$[g] (180 K)	0.138	9.0	10.6	35	cw	$\approx 9 \times 10^{-2}$	$\gtrsim 5 \times 10^{-2}$	$\sim 10^3$	$\approx 9 \times 10^{-2}$

[a] Unless otherwise noted, $T = 300$ K.
[b] All values quoted for $\chi^{(3)}$ are of *negative* sign.
[c] Jain and Klein (1979).
[d] Miller *et al.* (1980a,b).
[e] Lind and Jain (1981).
[f] Jain and Steel (1980).
[g] Jain and Steel (1981).

any of the above models. The relationship between n_I and $\chi^{(3)}$ (both in esu) is given in Eq. (54). To convert n_I to units of cm^2/W, one simply uses n_I (cm^2/W) $= 10^7 n_I$ (esu).

From Eqs. (50) and (51) we see that the magnitude of $\chi^{(3)}$ increases with both α and τ, and decreases inversely as ω^3. However, in practice an increase in α also has detrimental effects on the magnitude of the DFWM signal (AuYeung *et al.*, 1979; Jain *et al.*, 1979b). For a sample length L, a good compromise is a value of $\alpha = 1/L$. For purposes of comparison we will attempt to choose an α of 10 cm^{-1} (L = 1 mm), and select the wavelength λ accordingly. Note that an alternative "figure of merit" for comparison is $\chi^{(3)}/\alpha$ or even $\chi^{(3)}/\alpha\tau$, when the tradeoff between bandwidth and speed is significant (see Section III.E). Because of the difficulty in obtaining precise values of α and τ, the calculated values of $\chi^{(3)}$ in Table V are typically accurate only to within an order of magnitude.

For values of $\chi^{(3)}_{\text{eff}}$ [see Eq. (43)], τ_L corresponds to the experimental laser pulse width; note that $\chi^{(3)}_{\text{eff}} = \chi^{(3)}$ in Table V if $\tau_L \gg \tau_R$. The values of third-order susceptibility $\chi^{(3)}_m$ measured in the experiments should thus be compared with $\chi^{(3)}_{\text{eff}}$. In the last column is given the magnitude of the steady-state third-order susceptibility ($\chi^{(3)}$) that may be obtainable in this material if laser pulse widths much greater than the recombination time are used, and if the angle θ is chosen to be small enough that the diffusional decay time τ_D of the large-period grating is much greater than the plasma recombination time τ_R.

2. *Saturation of Free-Carrier Transitions*

a. Valence-Band-to-Valence-Band Transitions As illustrated schematically in Fig. 2 and repeated for clarity in Fig. 4, the valence bands of several semiconductors consist of three subbands, one of which—the split-off (SO) band—is separated by the spin–orbit interaction energy. In heavily doped p-type semiconductors, the availability of holes may extend to energies considerably below the top of the valence band, allowing the possibility of various valence-to-valence-band direct transitions, particularly in the long-wavelength spectral region. Three possible transitions exist: from the LH to the HH band, from the SO to the HH band, and from the SO to the LH band. Transitions have been observed in a large number of semiconductors, including Ge at 10.6 μm, and are characterized by an absorption edge that shifts towards higher energies with increasing hole density; likewise, for a specific wavelength, intervalence-band absorption increases with hole density and is negligible if the material is near intrinsic or n type. In Fig. 5 is shown the absorption spectrum of p-type Ge, which is very well understood in light of the

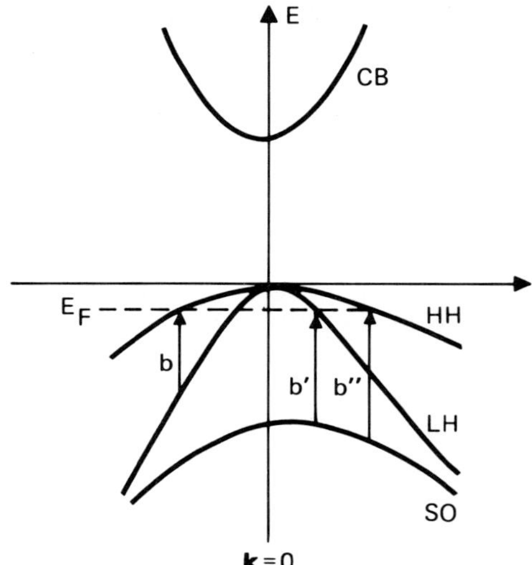

Fig. 4. Schematic depiction of the typical valence-band structure near $\mathbf{k} = 0$ and of the various inter-valence-band transitions, designated by b (LH → HH), b′ (SO → LH), and b″ (SO → HH), where LH, HH, and SO designate the light-hole, heavy-hole, and split-off bands, respectively.

theory of inter-valence-band absorption (Kane, 1957). In this figure, the 3300-cm^{-1} ($\sim$0.4 eV) peak is attributed to the SO → HH transition (labeled b″ in Fig. 4) and the 2400-cm^{-1} ($\sim$0.3 eV) peak to the SO → LH absorption (b′), whereas the broad peak at lower energies is attributed to the LH → HH absorption (b). The peaks b′ and b″ coincide at low temperatures because the Fermi level moves close to the top of the band, where the LH and HH bands coincide.

Saturation behavior of the inter-valence-band absorption in various semiconductors has been easily observed experimentally, and shown (James and Smith, 1980a) to possess the functional form of an inhomogeneously broadened line:

$$\alpha(I, \omega) = \frac{\alpha_0(\omega)}{[1 + I/I_{\text{sat}}(\omega, T)]^{1/2}}, \tag{57}$$

where $\alpha_0(\omega)$ is the low-intensity absorption coefficient and $I_{\text{sat}}(\omega, T)$ the frequency- and temperature-dependent saturation intensity.

Keilmann (1976) has successfully used a phenomenological saturation model, based on homogeneous broadening, to explain this saturation of inter-valence-band absorption. A more rigorous expression, based on the

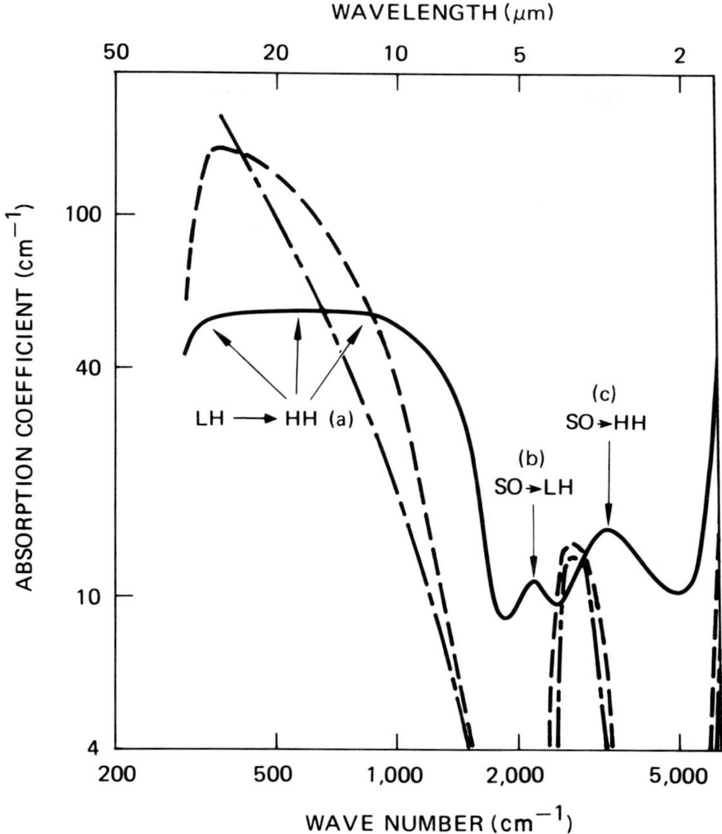

Fig. 5. Absorption due to inter-valence-band transitions in p-type germanium. The transitions are identified for the 300 K data (—); peaks (b) and (c) coincide at low temperatures 15 K(– —), and 77 K(– –) because the Fermi level moves closer to the top of the valence band, reducing the energy spread between the holes in the light- and heavy-hole bands. (After Kaiser, *et al.*, 1953, used with permission.)

intensity dependence of the imaginary part of the dielectric function in a microscopic form similar to Eq. (30), has been derived by James and Smith (1980a,b). They show that the absorption coefficient as a function of light intensity I can be written

$$\alpha(I, \omega) = \frac{4\pi^2}{n_\infty m^2 \omega c} \frac{N_{\mathrm{H}} e^2}{3\hbar} \left(\frac{1}{2\pi}\right)^3 \int_{\Omega(\mathbf{k})=\omega} \frac{dS}{|\nabla_{\mathrm{c}} \Omega(\mathbf{k})|} \frac{|p_{\mathrm{HL}}(\mathbf{k})|^2 [f_{\mathrm{H}}(\mathbf{k}) - f_{\mathrm{L}}(\mathbf{k})]}{[1 + I/I(\mathbf{k})]^{1/2}}, \tag{58}$$

where the subscripts L and H refer to states in the LH and HH bands, n_∞ is the high-frequency refractive index, N_H is the density of the holes (i.e., the doping density), $p_{HL}(\mathbf{k})$ is the momentum matrix element, f_H and f_L are equilibrium distribution functions in the HH and LH bands, and the integral is calculated over a surface of constant $\Omega(\mathbf{k})$, where $\Omega(\mathbf{k}) = E_H(\mathbf{k}) - E_L(\mathbf{k})$ is the energy difference between states in the HH and LH bands.

The essential difference between the *functional form* of Eq. (58) and that due to $\alpha(\omega)$ for interband transitions based on Eq. (30) is the inclusion of the saturation-type intensity-dependent denominator. By comparison of Eq. (58) with the macroscopic expression (57) for saturation in the presence of inhomogeneous broadening with a phenomenological saturation intensity I_{sat}, the quantity $l(\mathbf{k})$ can be loosely described as a "microscopic" saturation parameter, which is directly related to basic parameters in the semiconductor. This relationship has been derived by James and Smith (1980b):

$$l(\mathbf{k}) = \frac{3h^2(n_\infty m^2\omega^2)}{[T_H(\mathbf{k}) + T_L(\mathbf{k})]T_2(\mathbf{k})2\pi e^2|p_{HL}(\mathbf{k})|^2}, \tag{59}$$

where $T_H(\mathbf{k})$ and $T_L(\mathbf{k})$ are the overall dephasing times for the scattering of holes (originally in the HH and LH bands, respectively) between these two hole bands, and $T_2^{-1} = (T_H^{-1} + T_L^{-1})/2$.

By using the momentum matrix elements from Kane's $\mathbf{k} \cdot \mathbf{p}$ perturbation theory, relating $T_H(\mathbf{k})$ and $T_L(\mathbf{k})$ to hole–phonon scattering matrix elements, and then integrating Eq. (58) numerically, James and Smith (1980b) were able to calculate both the absolute values of the saturation intensity and the dependence of the saturation intensity on such parameters as photon energy and temperature. For p-type Ge at 10.6 μm, in which numerous experiments have been performed, the calculated values (James and Smith, 1979) are found to be in good agreement with the experimental data. These authors also studied the dependence of the saturation intensity on impurity concentration and residual absorption of p-Ge at 10.6 μm, and the laser-induced changes in the dispersive properties and photoconductivity (via hole redistribution) in p-Ge (James and Smith, 1980c, 1981a,b).

In Table VI we show the transition wavelengths and saturation intensities for a large number of p-type semiconductors at a fixed temperature (295 K) and specified photon energies. The calculated values are either based on James and Smith's rigorous calculations or, in the absence of such calculations, estimated from experimental absorption cross sections σ and relaxation lifetimes τ using $I_{sat} \propto 1/\lambda\sigma\tau$. Measured values of I_{sat} are also given when available.

Once the saturation intensity and the inhomogeneous nature of the broadening mechanism are established, the nonlinear index changes may easily be estimated for $I \ll I_{sat}$ by means of a power-series expansion of the overall susceptibility:

$$\chi(I, \omega) = \frac{\chi_0(\omega)}{(1 + I/I_{sat})^{1/2}}. \tag{60}$$

Thus, using

$$\chi(I, \omega)\Big|_{I<<I_{sat}} = \chi_0(\omega)(1 - I/2I_{sat} + I^2/4I_{sat}^2 + \cdots) \tag{61a}$$

and noting that

$$\mathcal{P} = \chi\mathcal{E} = \chi_0\mathcal{E} - \chi_0(I/2I_{sat})\mathcal{E} + \chi_0(I^2/4I_{sat}^2)\mathcal{E} + \cdots, \tag{61b}$$

we have, using I (W/cm^2) $= (nc/8\pi) \times 10^{-7}|\mathcal{E}|^2$ (esu),

$$\chi^{(3)} = -\frac{\chi_0}{2}\left(\frac{nc}{8\pi} \times 10^{-7}\right)\frac{1}{I_{sat}} \tag{62a}$$

and

$$\chi^{(5)} = \frac{\chi_0}{4}\left(\frac{nc}{8\pi} \times 10^{-7}\right)^2\frac{1}{I_{sat}^2}, \tag{62b}$$

and similar expressions for all higher-order odd terms. Values of $\chi^{(3)}$ estimated from Eq. (62a) are also tabulated in Table VI for several of the semiconductors. Note that p-GaSb at 10.6 μm with its relatively low saturation intensity appears to be a particularly promising candidate for the study of this nonlinearity. Note also that at intensities near the saturation intensity the power-series expansion of Eq. (61) is not vaid, and one must use the functional form of the overall susceptibility [Eq. (60)] in the coupled propagation equations for DFWM. Using an approach similar to that of Abrams and Lind (1978), one may then solve directly for DFWM signals and the DFWM reflectivity, without separately invoking any particular order of the nonlinearity. These considerations apply for all nonlinearities derived from saturation behavior, as may be the case for various other transitions observed in semiconductors, such as the conduction band-to-conduction-band, impurity-level, and excitonic transitions.

b. Conduction-Band-to-Conduction-Band Transitions The valence-to-valence-band (V-V) transitions discussed above (for heavily doped p-type semiconductors) occur as direct transitions at $\mathbf{k} = 0$ at energies well below the band gap. Thus there is no competition between these transitions and the fundamental (V–C) absorption, so that at sufficiently high intensities the filling of states causes the inter-valence-band (V-V) absorption to saturate. The conduction-to-conduction-band transitions at $\mathbf{k} = 0$ (from the lowest filled conduction band at the edge of the band gap) gener-

TABLE VI

Transition Wavelengths, Saturation Intensities, and $\chi^{(3)}$ for Inter-Valence-Band Transitions in p-Type Semiconductors at 295 K

Semiconductor	λ_Δ[a] (μm)	λ (μm)	Transition[b]	Saturation intensity I_{sat} (MW/cm^2) Calculated	Measured	$\chi^{(3)}$ (10^{-9} esu) Calculated	Measured	Ref.
Ge	4.20	10.6	LH→HH	3.5[c]	3.2,[e] 3.5,[f] 4.3[g]	~1.2	~0.3[i]	[j]
Ge	4.20	9.4	LH→HH	6.2[c]	6.5[e]	—	—	—
Ge	4.20	3.4	SO→HH	~100	—	—	—	[k,l,m]
GaAs	3.65	10.6	LH→HH	18[c]	20[h]	~0.24	—	[n]
GaAs	3.65	8.9	LH→HH	42[c]	—	—	—	—
GaAs	3.65	3.0	SO→HH	~200	—	—	—	[m,o,p]
GaSb	1.61	10.6	LH→HH	1.8[d]	—		—	[n]
AlAs	4.43	10.6	LH→HH	400[d]	—	~0.012	—	[q]
AlSb	1.65	10.6	LH→HH	24[d]	—	~0.18	—	[o]
InAs	3.26	10.6	LH→HH	5.1[d]	—	~1.78	—	[r]
ZnTe	1.17	1.17	SO→HH	≳500	—	—	—	[s,t]

[a] Wavelength corresponding to spin–orbit splitting.
[b] LH = light-hole band, HH = heavy-hole band, SO = split-off band.
[c] James and Smith (1980a).
[d] James and Smith (1980b).
[e] Phipps and Thomas (1977).
[f] Keilmann (1976).
[g] Carlson *et al.* (1977).
[h] Gibson *et al.* (1972).
[i] Watkins *et al.* (1981).
[j] Arthur *et al.* (1967).
[k] Briggs and Fletcher (1953).
[l] Kaiser *et al.* (1953).
[m] Chang (1981).
[n] Cardona *et al.* (1967).
[o] Braunstein and Kane (1962).
[p] Picus *et al.* (1959).
[q] Onton (1970).
[r] Pidgeon *et al.* (1967).
[s] Watanabe and Usui (1966).
[t] Watanabe (1966).

ally occur at energies higher than the band gap (see transition c′ in Fig. 2), so that the above condition is not easily satisfied. However, the band structure of several semiconductors allows the possibility of a low-energy direct transition from a neighboring (**k** $\neq$ 0) conduction valley, also depicted schematically (as transition c) in Fig. 2. Thus, at sufficiently high donor doping densities, state C_1 may have a high enough population for a distinct absorption peak corresponding to the $C_1 \rightarrow C_2$ transition to be observed. This is particularly true (and much smaller doping densities are required) if state C_1 occurs at a lower energy than C_0, i.e., if the lower state of this transition occurs as the *conduction-band-minimum,* as occurs for the X_1 valley in several III–V compounds. Notable among these is GaP, the band structure of which is shown schematically in Fig. 6. The absorption spectrum for n-type GaP with a donor concentration of 10^{18} cm^{-3} is shown in Fig. 7. The distinct peak near 3 μm corresponds to the $X_1 \rightarrow X_3$ transition. Although no saturation measurements have apparently been performed yet for this transition, the large absorption cross section ($\sim 2 \times 10^{-16}$ cm^2) implies that this transition should saturate at relatively low intensities, making GaP an excellent candidate for DFWM at 3 μm. Note also that the location and intensity of this absorption peak can be tuned to some extent by using $GaAs_{1-x}P_x$ alloys with $x > 0.8$ (Allen and Hodby, 1962).

3. *Transitions Involving Impurities*

From the standpoint of the absorption spectrum, transitions involving impurities in semiconductors may result either in mere modification of the

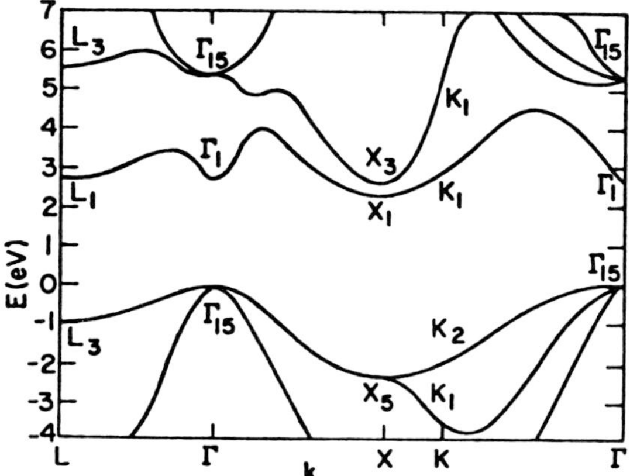

Fig. 6. Band structure of GaP. Of particular relevance are the conduction valleys at the symmetry point X. (After Cohen and Bergstrasser, 1966, used with permission.)

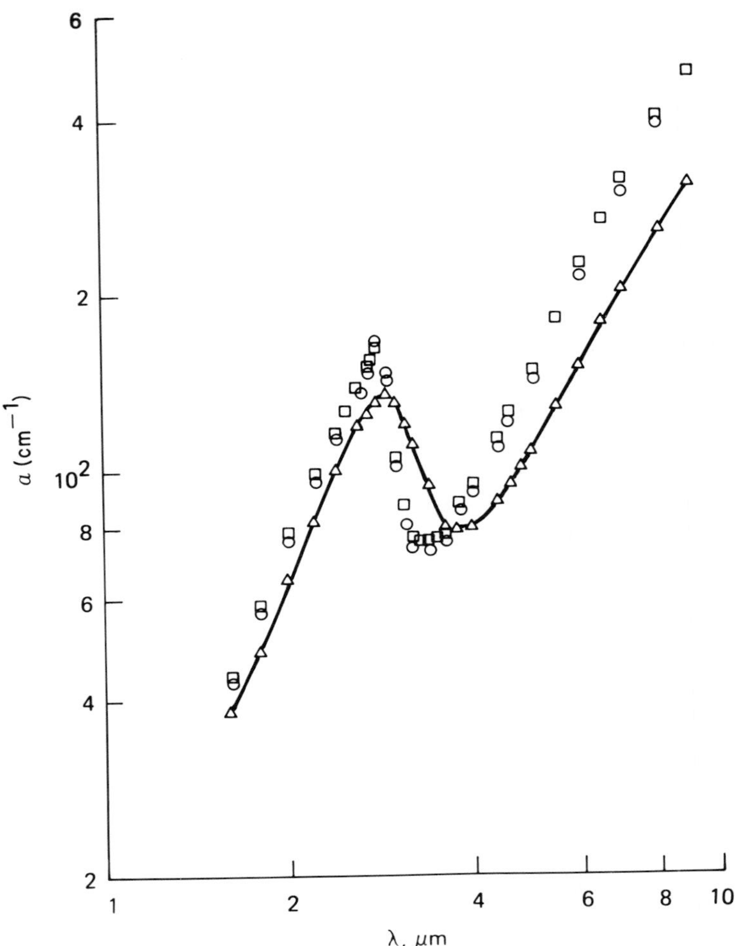

Fig. 7. Absorption spectrum of n-type GaP ($N_e = 1 \times 10^{18}$ cm^{-3}), at 295 K(△), 80 K (□), and ~5 K (○). The peak near 3 μm is due to the conduction-to-conduction-band transition indicated in Fig. 6. (After Spitzer *et al.*, 1959, used with permission.)

edges of the fundamental absorption or in the introduction of discrete spectral lines, at energies well below the band gap (see Fig. 3). The impurities may be present either as replacements of constituent atoms in the semiconductor crystal or as interstitials. When the impurities replacing the constituent atoms have the same valence as (i.e., are isoelectronic with) the replaced atom, they have only a second-order effect on the energy levels and the general properties of the semiconductor (the dominant effect is that of lattice distortion). However, when the impurities have a different valence, they provide an excess or deficit of nearly free electrons, thereby altering the free-carrier density (in a donor- or acceptor-type behavior), to result in n- or p-type semiconductors. We note that impurities at interstitial locations usually act as donors, and that "vacancies" (i.e., absence of atoms from their normal lattice locations) usually result in acceptorlike behavior.

Apart from altering the free-carrier density, impurities provide a new set of energy levels within the band gap of the semiconductors; these levels are usually sharp and distinct at low impurity concentrations. As an example, consider the case of a single-electron donor, which provides one extra unbonded electron besides its "bonding" electrons. This extra electron will be attracted most strongly to the positive charge of the impurity nucleus, which is effectively immersed in a crystalline "sea," described by the dielectric constant ϵ. Thus, the electron-ion system is roughly equivalent to a one-electron hydrogenic atom, whose energy levels are thus "modified" Rydberg levels:

$$E_n = \frac{m_e^* e^4}{2\hbar^2\epsilon^2 n^2} = \frac{m_e^*}{m\epsilon^2 n^2}\,13.6\text{ eV}, \tag{63}$$

where m_e^* is the effective mass of the electron and n is an integer. Because the energy at the bottom of the conduction band corresponds to the ionization of the electron, and because ϵ is typically of the order of 10 and m_e^* typically of the order of m, the "ground state" of the Rydberg-series energy levels of a single-electron donor is typically approximately 0.1 eV below the bottom of the conduction band.

If the impurity atom is of still higher valence, it can contribute more than one electron (multiple donor) or be multiply ionized. When it is multiply ionized, the higher charge on the impurity nucleus results in binding energies successively greater than that for the singly ionized impurity. Thus, as the degree of ionization increases, the various donor levels go deeper below the conduction-band edge, to form "deep" levels that often result in serious modification of the properties of the semiconductor. Note that deep levels may also be caused by other sources (such as transition element impurities), especially in semiconductors with wide band gaps. (The energy-level scheme resulting from the transition element impurities is not hydrogenic and is not completely understood.) Note also that whereas the discussion preceding Eq. (63) relates to donor-type impurities, acceptor-type impurities also result in similar energy states just above the valence band.

The energy levels due to various impurities have been measured and

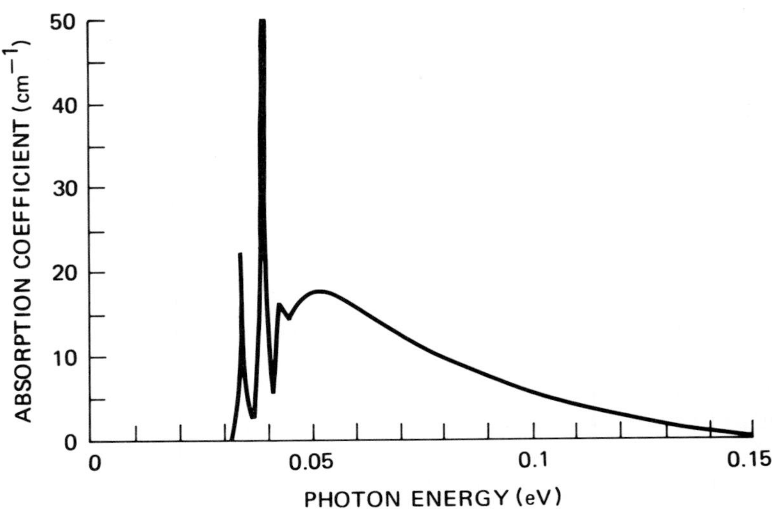

Fig. 8. Absorption spectrum of a boron-doped sample of silicon. (After Burstein, *et al.*, 1956, used with permission.)

tabulated for several semiconductors, notably Ge, Si, and GaAs, and data on these are given in several books and review articles (Conwell, 1967; Sze and Irwin, 1968; Sze, 1969). If transitions occur between the various energy levels of the impurity, sharp absorption lines are usually observed, as seen in Fig. 8. For the case of boron-doped silicon, the sharp absorption lines correspond to Rydberg-like series between the hydrogenic energy levels of the boron impurity, whereas the broad peak corresponds to absorption into the band. The dropoff in the absorption coefficient at higher photon energies, in spite of the increased density of participating states, is due to the rapid decrease in this transition probability away from the band edge (Kohn, 1957). We note that at high impurity concentrations the increased concentration of carriers results in stronger screening of the nucleus [i.e., a smaller value of ϵ in Eq. (63)] and a reduction of the ionization energies E_n (Pearson and Bardeen, 1949; Ermanis and Wolfsturn, 1966). Thus, at very large impurity concentrations a smearing of the impurity energy levels will usually occur, resulting in band tails near the fundamental absorption edge, and the usual localized level concepts implied by Eq. (63) are no longer valid.

Unlike the sharp absorption lines of transitions between impurity levels, transitions between impurity levels and the bands result in broad asymmetric bands in the absorption spectrum or in shoulders at the edge of the fundamental absorption, due to the possibility of transitions into band states at all photon energies above the energy required for a transition to the band edge. The spectrum for the absorption between the impurity levels and the band edge of Ge (Johnson and Levinstein, 1960) is shown in Fig. 9. More specifically, this spectrum corresponds to vertical transitions between the valence band and acceptor levels 0.15 eV above the valence-band edge.

In each of the above-mentioned examples of transitions involving impurities, saturation of absorption may easily occur, resulting in still another set of candidates for DFWM. On either side of the spectral lines of Fig. 8, and near the low-energy edge of the impurity-level-to-band-edge absorption (Fig. 9), these impurity transitions represent new examples of resonantly enhanced nonlinearities in saturable absorbers, and the DFWM analysis of Abrams and Lind (1978) is roughly applicable to both of these cases. Saturation intensities can be estimated from a knowledge of the absorption cross sections σ and relaxation times τ. Pertinent data on transition wavelengths, saturation intensities, and speeds of several of these

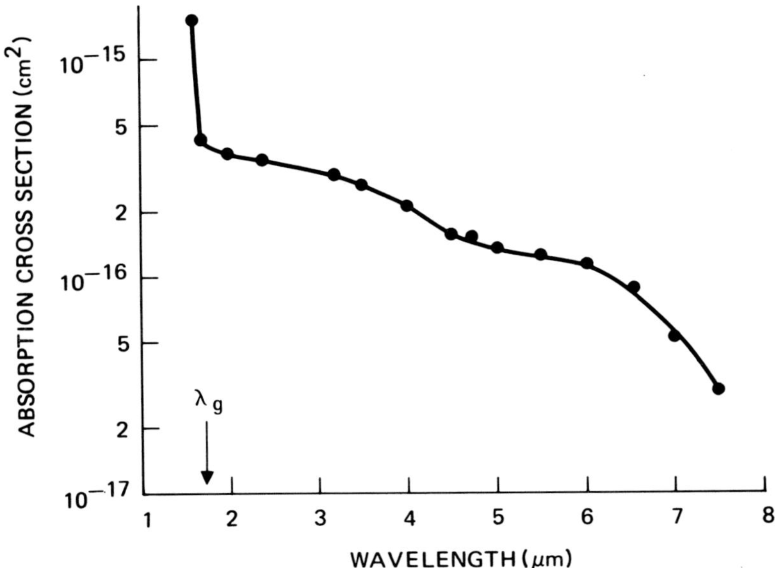

Fig. 9. Absorption spectrum for a Au-doped sample of germanium ($T = 5$ K). The absorption corresponds to transitions from the valence-band edge of Ge to the Au acceptor levels. (After Johnson and Levinstein, 1960, used with permission.)

transitions are listed in Table VII, which is essentially identical to a table compiled by Chang (1981). We remind the reader that for laser pulse widths smaller than the relaxation time of the states participating in the nonlinearity, the saturation *energy* density $I_{sat}\tau$ is a more relevant parameter than I_{sat} for the description of saturation behavior. The saturation energy density is proportional to $1/\lambda\sigma$, and may be estimated easily from experimental data on the absorption cross section σ. Values of $I_{sat}\tau$ are plotted in units of Joules per square centimeter in column 6 of Table VII, and the saturation intensities can be estimated therefrom. However, the lack of information on the relaxation time τ precludes estimation of the saturation intensities for several of these semiconductor impurity transitions.

To our knowledge, DFWM via impurity transitions has not yet been reported in a bona fide semiconductor. Nevertheless, assuming that an insulator is essentially a semiconductor with a very wide band gap, we note that Watkins *et al.* (1980a) have demonstrated DFWM via transitions in ReO_4 doped in KCl ($E_g = 6$ eV). Using a transition resonant at 10.6 μm, they observed saturation intensities of $\sim$1 MW/cm^2, and obtained DFWM results in good agreement with the Abrams–Lind model for DFWM in homogeneously broadband systems. The latter provides indirect evidence that homogeneous broadening may be the dominant line-broadening mechanism in the ReO_4–KCl system.

4. *Transitions in Excitons and Excitonic Complexes*

If an electron and hole are present in reasonable proximity and relatively immobile, or traveling at similar velocities with respect to the crystal lattice, their Coulomb attraction may result in their being bound to each other. Such a bound state is called an exciton. Similar to the ionized impurity discussed in the previous section, the energy levels of the exciton represent a set of hydrogenlike states at energies E_{xn} below the conduction band, given by

$$E_{xn} = -m_{eh}^* e^4/2\hbar^2\epsilon^2 n^2, \tag{64}$$

where m_{eh}^* is the reduced effective mass of the electron–hole pair. These bound electron–hole pair states are the energetically lowest electronic excitations in weakly excited pure semiconductors. The binding energy $|E_{x1}| \simeq m_{eh}^* e^4 2\hbar^2\epsilon^2$ for an exciton is lower than typical donor and acceptor energies, because $m_{eh}^* < m_e^*$, m_h^*, and is typically of the order of 20 meV in a bulk semiconductor. Because $kT \simeq 26$ meV at room temperature, low temperatures are usually necessary to prevent thermal dissociation of excitons.

Low temperature absorption spectra of the direct-gap semiconductor

TABLE VII

Transition Wavelengths, Absorption Cross Sections, Saturation Intensities, and Other Characteristics of Transitions Due to Impurities Doped in Semiconductors

Dopant	Host semiconductor	Transition[a]	λ (μm)	σ (10^{-16} cm^2)	$I_{sat}\tau \propto 1/\lambda\sigma$ (10^{-9} J/cm^2)	τ (nsec)	I_{sat} (W/cm^2)	T (K)	Ref.
Au	Ge	VB→AC	5	1.5	1.3×10^5	$\gtrsim 1$	$\lesssim 1.3 \times 10^5$	—	b
Hg	Ge	VB→AC	11	3.9	2.3×10^4	$\gtrsim 1$	$\lesssim 2.3 \times 10^4$	30	c,d
Cu	Ge	VB→AC	22	10	4.5×10^3	$\gtrsim 0.5$	$\lesssim 9 \times 10^3$	5	c,d
In	Si	VB→AC	4.8	0.33	6.3×10^5	—	—	80	c,d
Ga	Si	VB→AC	15	5	1.3×10^4	—	—	10	c,d
Al	Si	VB→AC	15	8	8.3×10^3	—	—	10	c,d
B	Si	VB→AC	23	14	3.1×10^3	—	—	5	c,d
P	Si	DO→CB	27	17	2.2×10^3	—	—	5	c,d
Cu	ZnSe	VB→AC	1.38	0.35	2×10^6	—	—	300	e,f
Cu	ZnS	VB→AC	0.73	1.6	8.5×10^5	—	—	80	g,h

[a] VB = valence band, CB = conduction band, AC = acceptor, DO = donor.
[b] Johnson and Levenstein (1960).
[c] Bube (1960).
[d] Bratt (1977).
[e] Stringfellow and Bube (1968).
[f] Grimmeiss *et al.* (1977).
[g] Bowers and Melamed (1955).
[h] Broser and Franke (1965).

GaAs are shown in Fig. 10. Note that although exciton absorption is described by transitions from states in the valence band to a set of discrete states, the resulting absorption spectrum consists of *discrete lines*, unlike the case of the valence-band-to-impurity absorption, which results in bandlike absorption spectra. This is because, in direct-gap semiconductors, momentum conservation requires that $\mathbf{k}_e = \mathbf{k}_h$, are nondissociative interaction between the electron and the hole requires that their group velocities be identical, i.e., $\partial E_c/\partial \mathbf{k}_e = \partial E_v/\partial \mathbf{k}_h$, a condition satisfied near the band edge only if $\mathbf{k} \simeq 0$. Thus, only states near the top of the valence band contribute to exciton absorption, which explains not only the narrow-line spectrum but also the facility with which the exciton absorption may saturate. Note that when phonon absorption or emission is necessary for momentum conservation, as in indirect-gap semiconductors, the exciton spectra are in the form of steps, and not sharp lines. Note also that symmetry considerations forbid the $n = 1$ transition in semiconductors (such as Cu_2O and Sn_2O) in which the direct transition at $\mathbf{k} = 0$ is forbidden. The reader is referred to a standard textbook (such as Smith, 1968) for a fuller discussion.

Returning to the hydrogen-atom analog of the free exciton, we may estimate the effective radius R_{x1} of the lowest exciton state by considering the radius of the first Bohr orbit of the dielectric-modified atom. Thus,

$$R_{x1} = \frac{\hbar^2 \epsilon}{e^2 m_{eh}^*} = \frac{\epsilon}{m_{eh}^*/m} a_0, \tag{65}$$

where $a_0 = 0.53$ Å is the first Bohr orbit of the hydrogen atom. From Eqs.

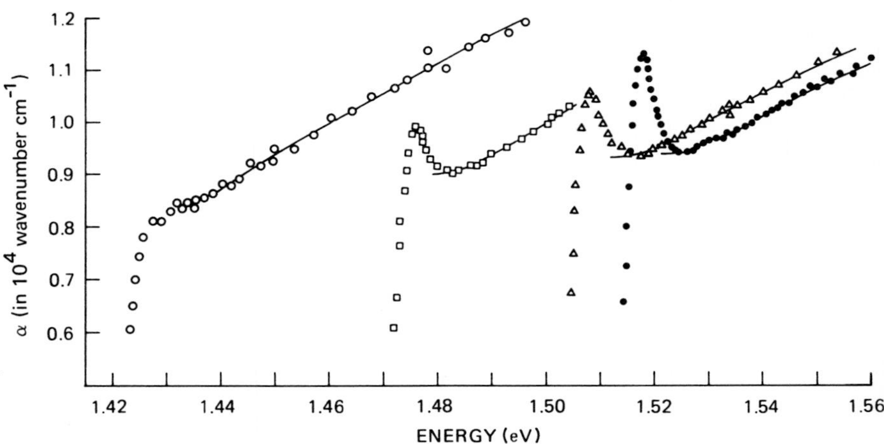

Fig. 10. Exciton absorption in GaAs at 294 K (○), 186 K (□), 90 K (△), and 21 K (●). (After Sturge, 1962, used with permission.)

(64) and (65) we see that in semiconductors with small effective electron–hole masses and large dielectric constants, the excitons (called Wannier excitons) are weakly bound (Wannier, 1937), and their wave functions cover many lattice spacings. Such excitons have been studied very intensively, partly because they may associate relatively easily to yield large molecular complexes (such as biexcitons and higher-order molecules) as well as the condensed electron–hole liquid (Knox, 1963; Hanamura and Haug, 1977). For example, two excitons may interact attractively if their electronic spin wave functions result in a combined singlet wave function (Hanamura and Haug, 1977). This results in a molecularlike biexciton complex, whose lowest excited state is of singlet-like character and has an energy lower than that of two free excitons by an amount equal to the binding energy Δ of the biexciton. Biexcitons have been observed in both indirect- and direct-gap semiconductors. In the direct gap semiconductors CuCl, CuBr, and CdS, such biexciton states may be addressed only via two-photon absorption. For example, in CuCl, the lowest state of the excitonic molecule, Γ_1^m (see Fig. 11), is described by a linear combination of two optically active Γ_5 (longitudinal and transverse) excitons and two optically inactive Γ_2 excitons. The two-photon excitation thus occurs principally via the optically active intermediate states. A "giant" oscillator strength has been observed, resulting in a two-photon absorption coefficient (at $\sim$10 MW/cm^2) that is comparable to the single exciton (one-photon) absorption coefficient. This giant oscillator strength stems principally from the large molecular radius of the biexciton; this may be understood from the fact that the second photon (in the two-photon absorption process leading to the biexciton state) can interact with any valence band electron within a large volume (determined by the large molecular radius) around the virtual exciton that represents the intermediate state. Large two-photon polarizabilities for DFWM have been observed in such biexcitons, as will be elaborated in Section IV.B.

Returning to the case of Wannier-like excitons of large radii, we note that significant modification of their character may be achieved via carrier confinement in ultrathin semiconductor structures. When the carriers are confined in planar layer-type structures such as isolated quantum wells, or multiquantum well "superlattices," the carrier masses become highly anisotropic (becoming extremely large in the direction normal to the layer). The intrinsic excitons are then essentially 2-dimensional with pancake-like orbitals, in contrast to the near-spherical orbitals that are found in bulk semiconductors and characterized by Eqs. (63) and (64). Similar to the case of the 2-dimensional hydrogenic atom and related 2-dimensional molecular complexes, such 2-dimensional excitons and excitonic complexes exhibit binding energies that are much larger than their 3-dimensional counterparts (Ralph, 1965; Kuramoto and Kamimura,

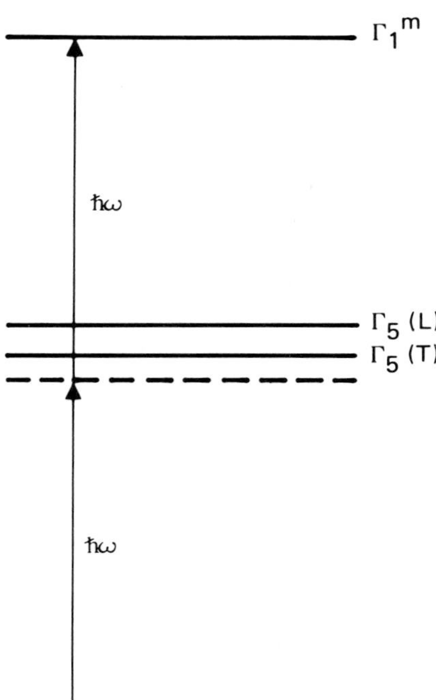

Fig. 11. Typical energy-level spacings of exciton and biexciton states in direct-gap compound semiconductors such as CuCl, CuBr, and CdS. The large two-photon polarizability of the biexciton level Γ_1^m is due to the resonant enhancement via the intermediate exciton states as well as to the "giant" oscillator strength of the exciton–biexciton transition. (After Hanamura and Haug, 1977, used with permission.)

1974). One consequence of the large binding energies of these 2-dimensional excitons in multiquantum well structures is that these excitons may be stably bound even at room temperatures (for instance, see Dingle, 1975; Miller *et al.*, 1976).

Saturation of the absorption in the vicinity of the transition energies of Wannier-like excitons has been studied by several workers, in bulk semiconductors as well as in multilayered structures. Dite *et al.* (1970) reported a saturation intensity of ~1 MW/cm² in bulk CdSe with the use of ~15 nsec pulses from a *Q*-switched ruby laser. However, since these laser pulse durations were smaller than the recombination time of the exciton, the measured saturation energy density (~15 mJ/cm²) is a more meaningful datum point for the description of this interaction (see Section III.C.3). With the use of quasi-cw radiation, Shah *et al.* (1977) and Leheny *et al.* (1978) reported much lower saturation intensities for the $n = 1$ A-excitons in GaAs and CdS, respectively; however these authors did not fully explore the nature (homogeneous or inhomogeneous) of the broadening mechanism. Subsequently, Gibbs *et al.* (1980) have studied saturation broadening for the $n = 1$ A-exciton in GaAs, and have fit their data to a homogeneous model with saturation intensities of only $\sim$150 W/cm^2 for their $\sim$500-nsec-long pulses ($\tau_L \gg \tau$) at 820 nm.

Saturation of the absorption in two-dimensional excitons was first studied by Miller *et al.* (1976) in GaAs–$Ga_{1-x}Al_xAs$ multiquantum well structures at room temperature and lowered temperatures. More recently, Chemla *et al.* (1982) and Gibbs *et al.* (1982) have measured saturation intensities of the order of 1 KW/cm^2 in their room-temperature experiments on multilayered GaAs-GaAlAs samples. The saturation behavior in neither the multilayered GaAs-GaAlAs structures nor the GaAs samples is yet totally clear; however in contrast to the single Lorentzian equations required to fit the GaAs saturation data (Gibbs, 1980; Chemla *et al.*, 1982), two Lorentzian expressions were required to fit the room temperature multiquantum well data (Chemla *et al.*, 1982), with saturation intensities of $\sim$500 W/cm^2 and $\sim$4.4 KW/cm^2, respectively.

For solids with large effective electron–hole masses and small dielectric constants, the exciton may be highly bound, with R_{xn} corresponding to an electron–hole pair (called a Frenkel exciton) that is localized to a few lattice spacings. Frenkel excitons have been studied in solid noble gases (Seitz, 1940), particularly solid argon, and in alkali halides, but are only seldom observed in semiconductors (Thomas and Timofeev, 1980).

Excitons and excitonic complexes may also occur tightly bound to lattice sites such as impurities. For instance, a free hole can combine with the electron bound to a donor to form a positively charged exciton–ion complex that includes the nucleus of the donor. Such a "bound exciton" was first observed in silicon (Haynes, 1960). Similarly, bound excitons with a spectrum of many sharp lines (see Fig. 12) have been observed in GaP doped with nitrogen. In this case the isoelectronic center (substitution of one atom of the crystal by another of the same valence) of nitrogen acts like a localized potential well trapping an electron, which in turn traps a hole; excitons bound to the nitrogen isoelectronic trap thus result. When the isoelectronic centers are close together, they may result in new potential fields that can bind their own excitons. This explains the absorption spectrum in Fig. 12. Each line corresponds to excitons bound to different pairs of isoelectronic centers, with the binding energy varying inversely with the distance between the two atoms of a pair. Thus the lowest-energy absorption peak (NN_1) corresponds to the most strongly bound exciton of a nearest-neighbor (NN) pair; other absorptions peaks (NN_i) correspond to more distant pairings. The intensity of the absorption peak depends on the number of pairings that can form about one nitrogen atom in a given crystallographic direction, and is not a direct measure of the absorption cross section or saturation intensity. Nevertheless, relatively low saturation intensities are estimated for several of these lines, including the distinct NN_5 line at 2.288 eV. Coherent

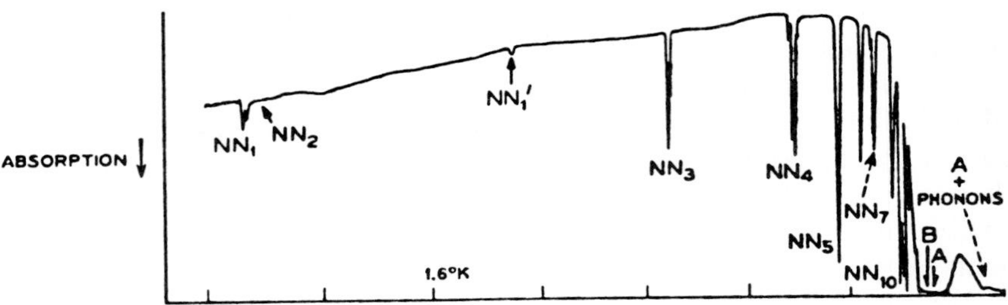

Fig. 12. Absorption lines due to excitons bound to pairs of isoelectronic nitrogen traps in GaP. The binding energy of the excitons to the isoelectronic centers depends on the distance between the two nitrogen atoms forming a pair, and is strongest for the nearest pairs (NN_1). The intensity of the absorption lines depends on the number of pairings in a given crystallographic direction. (Used with permission from Thomas *et al.*, 1965.)

optical phenomena have been studied recently in the vicinity of bound excitonic transitions in GaP: N by Gibbs and Hu (1978), and saturation intensities of $\sim 330\ W/cm^2$ have been very recently measured by Dagenois and Winful (1982) for the I_2 line of bound excitons in CdS (Henry and Nassau, 1970).

Because of their low saturation intensities, and by comparison with the other saturation-related nonlinearities, exciton resonances of various kinds are clearly important as potential candidates for degenerate four-wave mixing. A number of bulk semiconductor exciton transitions are listed in Table VIII: none of these has yet been used for backward DFWM. We note however that 2-dimensional excitons in multilayered GaAs/GaAlAs structures have been used recently for DFWM by Hegarty *et al.* (1982), and that experiments on nonlinear index changes and transient holography have been performed near the excitonic resonance of CdSe (Jarasiunas and Vaitkus, 1974). Moreover, the potential of excitonic resonances has also been elucidated by the experiments of Gibbs *et al.* (1979, 1980, 1981, 1982) on optical bistability in GaAs, and by the related theoretical work of Goll and Haken (1980) and Koch and Haug (1981). Note also that such exciton effects are easily destroyed at high intensities because of the screening of the electron–hole pairs by the large number of carriers, and care must be exercised not to confuse the observed nonlinear refraction due to the plasma with that due to exciton saturation (Koch *et al.*, 1981).

5. *Contributions from Other Absorption Processes*

A variety of other transitions and related processes may result in optical nonlinearities related to DFWM. We will discuss a few of them briefly: intraband and hot carrier absorption, absorption away from the

TABLE VIII

Wavelengths, Saturation Intensities, and Related Data for Exciton Transitions in Bulk Semiconductors

Material	Type of exciton	T (°K)	$\hbar\omega_{ex}$ [a] (eV)	λ_{ex} [a] (nm)	$\Delta\nu_{FWHM}$ (cm^{-1})	f [a] (10^4)	σ_{ex} [a] (10^{-14} cm^2)	$I_{sat}\tau$ (10^{-9} J/cm^2)	τ (10^{-9} sec)	I_{sat} (W/cm^2)	Ref.
GaAs	Intrinsic (A, n = 1)	21	1.515[b]	816	—	—	—	—	—	150[c]	[d]
		90	1.510[b]	821	~50[b]	—	—	—	—	—	—
		186	1.476[b]	840	~60[b]	—	—	—	—	—	—
InP	Intrinsic (A, n = 1)	6	1.416[e]	876	~25[e]	—	—	—	—	—	—
		77	1.409[e]	880	~40[e]	—	—	—	—	—	—
CdS, hex	Intrinsic (A, n = 1)	~10	2.554[f]	486	—	29[g]	—	—	—	—	[h,i,j]
CdS, hex	Intrinsic (A, n = 1)	77	—	487	39.5[k]	—	5.4	3.8×10^3 [m]	1.8	2.1×10^3	[i,j,k]
CdSe, hex	Intrinsic (A, n = 1)	~10	1.826[l]	679	—	9[g]	—	—	—	—	[j]
CdSe, hex	Intrinsic (A, n = 1)	77	1.817[m]	682	—	—	—	—	11[m]	10^6 [m]	—
CdSe, hex	Intrinsic (A, n = 1)	110	1.802[m]	688	—	—	—	—	9[m]	10^6 [m]	—
ZnS, cubic	Intrinsic (A, n = 1)	~10	3.871[n]	—	—	43[g]	—	—	—	—	[j]
ZnSe, cubic	Intrinsic (A, n = 1)	~10	2.860[o]	—	—	14[g]	—	—	—	—	[j]
Cu_2O	Intrinsic (A, n = 2)	77	2.140[p]	579	~30	—	—	—	—	—	[p,q,r]
	Intrinsic (A, n = 3)	77	2.154[p]	476	~20	—	—	—	—	—	[q,r]
ZnO	Intrinsic (A, n = 2)	4	3.397	365	17.1	—	19	1.4×10^3	0.32	4.4×10^3	[k,s,t]
GaP : N	Bound	77	2.31	537	18	—	0.1	1.8×10^5	1	1.8×10^5	[k,u,v]

[a] $\hbar\omega_{ex}$ ($= E_g - E_{xn}$) is the photon energy of the exciton transition, λ_{ex} the corresponding free-space wavelength, f the oscillator strength, and σ_{ex} the peak absorption cross section per exciton.

[b] Sturge (1962).
[c] Gibbs *et al.* (1980).
[d] Shah *et al.* (1977).
[e] Turner *et al.* (1964).
[f] Thomas and Hopfield (1961).
[g] Segall and Marple (1967).
[h] Leheny *et al.* (1978).
[i] Hopfield and Thomas (1961).
[j] Gutsche and Voight (1967).
[k] Chang (1981).
[l] Wheeler and Dimmock (1962).
[m] Dite *et al.* (1970).
[n] Wheeler and Mickloaz.
[o] Gross *et al.* (1962).
[p] Baumeister (1961).
[q] Nikitine *et al.* (1961).
[r] Green *et al.* (1961).
[s] Park *et al.* (1966).
[t] Reynolds (1969).
[u] Spitzer *et al.* (1959).
[v] Leheny and Shah (1975).

band edge (related to band-edge modification by various mechanisms), and photorefractive effects induced by space-charge fields created by carrier migration subsequent to their photogeneration by fundamental or impurity-assisted absorption.

Intraband and hot carrier absorption are phonon-assisted processes, which result in nonlinear absorption that increases with intensity. Such nonlinear absorption will result in an *absorption* grating due to the spatial interference of the fields, resulting in scattered DFWM signals. It can be easily shown, however, that the scattering efficiencies (or DFWM signals) from these processes are generally very weak.

Absorption may also occur away from the usual direct band edge. In the presence of high-intensity illumination, this may be due to band-gap renormalization (Koch *et al.*, 1981; Haug *et al.*, 1981) by many-body effects in the presence of a high-density plasma. This absorption results in further plasma generation, and the resulting nonlinearity is essentially due to refraction by an electron–hole plasma [Eq. (43)], but with revised expressions for the form of the absorption coefficient and for the density of states and transition matrix elements near the band edge. Similarly, band-edge modification (Urbach, 1953; Halperin and Lax, 1966) can also occur owing to impurity states, phonon-assisted absorption, local-field perturbations by impurities, lattice defects, etc. In the light of such possibilities, interpretation of the nonlinearity due to plasma generation and saturation of interband absorption [Eqs. (50) and (51)] should be treated with caution; this is particularly important when the band-tail absorption [Eq. (31)] is compared with a model based on the dephasing rates of Bloch states near the band edges (Miller *et al.*, 1980b).

Finally, an important nonlinearity used for degenerate four-wave mixing is the photorefractive effect, in which (a) photogenerated carriers are generated at locations corresponding to maxima in the interfering fields; (b) these carriers migrate and recombine at traps, thereby resulting in a space-charge field with the same period as the interfering fields; (c) the field changes result in an index grating via the electro-optic effect (the grating is of the same period but may be shifted in phase with respect to the interfering fields); and (d) diffraction of light by this index grating results in a signal similar to that commonly observed in DFWM. Note, however, that the nonlocal characteristics of this effect (Kukhtarev *et al.*, 1979) precludes description of this nonlinearity in the usual framework of $\chi^{(3)}$ or n_2. As will be discussed in Chapter 11, large-band-gap "semiconductors" such as $KNbO_3$, $BaTiO_3$ and $Sr_{1-x}Ba_xNb_2O_6$(SBN) have been used for efficient phase conjugation using such a nonlinearity (Gunter and Krumins, 1980; Feinberg and Hellwarth, 1980; White *et al.*, 1981; Fischer, *et al.*, 1982).

D. Other Nonlinearities

1. *Thermal Contributions to the Nonlinear Index*

The energy absorbed by a semiconductor as a consequence of transitions to excited states may decay radiatively or nonradiatively. If a significant fraction of this excited-state energy decays nonradiatively, the local temperature in the semiconductor may rise, causing local changes in the band gap and thus in the index of refraction. This intensity-dependent change in the index of refraction may be significant, and may result in sizable self-interaction effects and consequently in large DFWM signals. Assuming that the dominant contribution to the index change occurs via a change in the band gap with temperature (Tsay *et al.*, 1973), the size of the thermal nonlinearity can easily be estimated from

$$\Delta n \simeq \Delta T \frac{dn}{dT} \simeq \Delta T \frac{\partial n}{\partial E_g} \frac{\partial E_g}{\partial T}. \tag{66}$$

For DFWM, ΔT is described by a sinusoidal modulation that depends on the local light intensities. It can easily be shown (Eichler *et al.*, 1973; Jain and Steel, 1982) that for $\alpha L < 1$ the amplitude of the temperature modulation is given by

$$\Delta T \simeq \beta_{NR} I_m \alpha \, \Lambda^2 / 4\pi^2 K, \tag{67}$$

where I_m is the amplitude of intensity modulation due to the interfering fields, β_{NR} is the fraction of the absorbed energy that is lost nonradiatively ($\beta_{NR} = 1 - \eta_R$, where η_R is the quantum efficiency of radiation), and K is the thermal conductivity of the sample. Thus, from Eqs. (66) and (67), the thermal nonlinear index is given by

$$(n_I)_{th} = \frac{\alpha \beta_{NR} (dn/dT) \tau_{th}}{C_p}, \tag{68}$$

where τ_{th} is the thermal decay time of the grating ($\tau_{th} = \Lambda^2 / 4\pi^2 D_{th}$, where $D_{th} = K/\rho C_p$ is the thermal diffusion coefficient), ρ is the density, and C_p is the specific heat of the semiconductor.

The magnitude of the thermal nonlinearity is often large. However, it is usually characterized by a very slow response time τ_{th}, which can often be used to identify this nonlinear mechanism. As an example, for cw mixing of 10.6-μm radiation in HgCdTe (Jain and Steel, 1982), for a 1° pump–probe angle, $\Lambda \simeq 624$ μm. With $\rho = 7.6$ g/cm^3, $C_p = 0.15$ J/g K, and $K = 0.1$ W/cm K, a grating diffusion time $\tau_{th} = 1.1$ msec is estimated. A steady-state temperature modulation of $\Delta T = 1.7 \times 10^{-2}$ °C and a consequent index modulation of 2.7×10^{-5} (using $dn/dT = 1.6 \times$

$10^{-3}/°C$) are thereby obtained. This is a significant modulation, and its contribution to the DFWM signal must be carefully considered. Note also that the index change due to the thermal contribution is usually positive in most of the semiconductors of interest (Tsay *et al.*, 1973), in contrast to the negative index change due to most of the electronic nonlinearities considered above. Such differences in sign are significant when several nonlinearities of comparable magnitude are present (see Section III.E).

2. *Franz–Keldysh Contributions to the Third-Order Susceptibility*

The Franz–Keldysh effect describes the smearing of the band edge and the consequent reduction in the effective value of the band gap of a semiconductor in the presence of an external dc electric field (Keldysh, 1958; Franz, 1958). The reduction in the band gap is related to an increase in the tunneling probability of electrons from the valence to the conduction band; using simple uncertainty-principle arguments, Vavilov (1962) showed that the amount of reduction ΔE_g in the band gap in the presence of an external electric field $\vec{\mathscr{E}}_{dc}$ is given approximately by

$$\Delta E_g = \tfrac{3}{2}(m_{eh}^*)^{1/3}(e\hbar\vec{\mathscr{E}}_{dc})^{2/3}. \tag{69}$$

The change in the dielectric function, $\Delta\epsilon_1(\omega, \vec{\mathscr{E}}_{dc})$, due to an external dc electric field $\vec{\mathscr{E}}_{dc}$ was subsequently described by Aspnes (1967) and Aspnes *et al.* (1968) for the case of the one-electron approximation in a semiconductor with a direct band gap. By extending the analysis to an ac electric field, in the approximation of a two-band model and a single M_0-type critical point (see Kittel, 1971), Van Vechten and Aspnes (1969) were able to derive a *field-induced* third-order susceptibility

$$\begin{aligned}\chi^{(3)}(\omega) &= \Delta\epsilon_1(\omega, \vec{\mathscr{E}})/4\pi\epsilon^2 \\ &\simeq \tfrac{1}{32}(e^4C_0^2\hbar/m^2)(m_{eh}^*)^{1/2}E_g^{-9/2},\end{aligned} \tag{70}$$

where C_0 is the momentum matrix element at the M_0 critical point. Third-order susceptibilities based on this expression have been calculated by Van Vechten and Aspnes (1969) and found to be of the order of 10^{-10} esu in Ge, Si, and GaAs. These contributions may thus be important for DFWM at photon energies far from the band gap, especially because the Franz–Keldysh susceptibilities (positive) are comparable in magnitude and opposite in sign to the negative interband contributions.

E. Summary

In this section we have described a large variety of physical mechanisms in semiconductors that lead to optical self-interaction effects, such as nonlinear refraction and nonlinear absorption. Simplified expressions have been derived for each of these nonlinearities to help estimate the typical magnitudes and signs of the nonlinear susceptibilities and polarizations. To this end, relatively simple treatments have been given for each nonlinearity, with the goal of providing the reader with a good understanding of the overall nature of the nonlinearity and its dependence on easily quantifiable parameters in the semiconductor.

The physical origins of these nonlinearities have been examined by several authors, most recently by Wherrett (1983). By comparing groups of terms in a virtual transition treatment for a generalized expression of $\chi^{(3)}$ (following Butcher and Maclean, 1963), Wherrett provides additional insight into the role of interband transitions in each of these nonlinearities, including those (such as nonlinear motion of conduction band electrons) that essentially involve only the response of apparently free carriers. From a relatively simple viewpoint, the role of interband transitions in free carrier theories (such as that of conduction-band nonparabolicity) can be understood by noting the common influence of the lattice potential on electronic motion as well as on interband transitions, illustrated very graphically by the intricate relation between the effective mass m_{eh}^* and the momentum matrix element ($1/m_{\text{eh}}^* = 2|p_{\text{vc}}|^2\omega_{\text{g}}/m^2$, Kane, 1957). From a slightly different viewpoint, Wherrett notes that the source of nonparabolicity, a fourth-order $\mathbf{k} \cdot \mathbf{p}$ perturbation, is directly related to the $\mathbf{A} \cdot \mathbf{p}$ perturbation terms that are featured in all of these nonlinearities.

Despite such common physical origins, the external manifestation of the various physical mechanisms is generally different; nevertheless, they appear to share another common feature, e.g., the relationship between their magnitudes and speeds. We have seen that the magnitude of the third-order susceptibilities ranges from $\sim 10^{-12}$ esu all the way to $\sim 10^{-1}$ esu; however, most of the larger values of $\chi^{(3)}$ appear to occur at the expense of speed. To establish this more clearly, we have tabulated the magnitudes and speeds for representative cases of some of the principal nonlinearities for one specific wavelength in Table IX. This trend is clear for all the nonlinearities in the table; and despite the vast differences in the physical mechanisms and in the final expressions for the magnitude of each of these nonlinearities, the values of $\chi^{(3)}/\tau$ for all of these nonlinearities lie approximately between 10^4 and 10^5 esu/sec; i.e., they vary over less than two orders of magnitude, in spite of the fact that $\chi^{(3)}$ varies

TABLE IX

Estimated Speeds and Magnitudes of Various Third-Order Nonlinearities in Semiconductors at a Wavelength $\lambda = 10.6\ \mu m$

Nonlinear Mechanism	Example	Typical magnitude of $\chi^{(3)}$ (esu)	Typical speed τ (sec)
Bound-electron anharmonicity	Ge	$\lesssim 10^{-11}$	$\lesssim 10^{-15}$
Conduction-band nonparabolicity	HgCdTe	$\lesssim 10^{-8}$	$\lesssim 10^{-13}$
Inter-valence-band absorption	Ge	$\sim 10^{-7}$	$\sim 10^{-12}$
Valence-to-conduction-band absorption	HgCdTe	$\lesssim 10^{-2}$	$\sim 10^{-7}$

over approximately ten orders of magnitude. In this table we only wish to illustrate the *relative insensitivity* in the value of $\chi^{(3)}/\tau$ despite large variations in the value $\chi^{(3)}$, and we definitely are not suggesting any constancy in the value of $\chi^{(3)}/\tau$. The "typical" values of $\chi^{(3)}$ and τ tabulated here are only crude estimates; the estimates of τ for nonparabolicity and inter-valence-band absorption are based on incomplete information on scattering and relaxation time in bands, and values of $\chi^{(3)}$ are based on reported measured values. Constancy in the value of $\chi^{(3)}/\tau$ is also precluded by the different dependences of the various nonlinearities on parameters such as wavelength, effective mass, and resonant enhancement.

An important conclusion to be made, however, is that if the speed and magnitude of the nonlinearity are both important considerations, one must make an appropriate compromise. On a general basis, $\chi^{(3)}/\tau$ may be a better figure of merit than $\chi^{(3)}$ alone. Note also that the absorption coefficients encountered with many of the nonlinearities that yield large values of $\chi^{(3)}$ might also pose problems for a given experimental situation (e.g., in DFWM or in optical bistability applications), and $\chi^{(3)}/\alpha\tau$ is probably an even more important figure of merit for many experimental situations, as elaborated in Section III.C.1.

More than one nonlinear mechanism may be present for a given choice of material, wavelength, and pulse duration, and interference effects may occur between these multiple nonlinearities. For nonlinearities that are independent, the response may be described by simple bookkeeping in which the magnitudes and signs of the various nonlinearities must be considered. This summation must be considered as a function of time, because the temporal response of the nonlinearities and their buildup and decay properties may be different. This is exemplified by considering DFWM in HgCdTe at 10.6 μm, where conduction-band nonparabolicity, plasma generation, saturation of valence-to-conduction-band absorption, saturation of impurity absorption, and thermally induced index changes may all be simultaneously present. If plasma generation and thermally induced index changes are the dominant nonlinearities, the opposing signs of the index changes and the vastly different speeds of the nonlinearities may lead to unusual effects, which must be interpreted carefully. In particular, the inevitable presence of an external resonator (provided by the DFWM experimental configuration itself) around the nonlinear sample may result in regenerative pulsations (McCall, 1978; Jain and Steel, 1981, 1982).

IV. DFWM Experiments with Various Semiconductors

A. General Considerations

The experimental arrangement for forward DFWM is straightforward and will not be discussed here. The experimental arrangement for backward DFWM is a little more subtle because of the presence of three input waves as well as the constraints on temporal delays and the precise counterpropagation of the pumps. The basic optical arrangement, depicted in earlier chapters, allows matching of the path length traversed by the two pumps E_f and E_b and the probe E_p, so that all three waves arrive at the sample simultaneously, or at relative time delays shorter than the temporal coherence of the waves. Calibration of the DFWM signal for measurement of the DFWM reflectivity is usually done by removing the nonlinear medium and inserting a 100% retroreflector in the path of the probe beam. Variations of this basic experimental arrangement have often been used, particularly when the sample is highly absorbing; significantly different arrangements are noted in the discussion of specific experiments in Section IV. B.

The choice of the nonlinearity used for a DFWM experiment is determined generally by the laser sources used, including wavelength, pulse duration, and peak intensity, and by requirements of speed and magnitude of the DFWM signal, sample availability, etc. In backward DFWM experiments, $\chi^{(3)}$ is estimated by measuring the signal-to-probe reflectivity R as a function of $(I_f I_b)^{1/2}$, where I_f and I_b are the intensities of the forward and backward pumps, respectively. Then, for low reflectivities and low absorption coefficients, $\chi^{(3)}$ can be estimated from the expression (Yariv and Pepper, 1977)

$$R = |\kappa|^2 L^2, \tag{71}$$

where $\kappa = (2\pi\omega/nc)\chi^{(3)}\mathscr{E}_f\mathscr{E}_b$. For absorbing samples one must consider the attenuation of the beams as they propagate through the sample. As mentioned in Section II.A, Kukhtarev and Kovalenko (1980) have derived expressions for the reflectivity of the probe as a function of the intensities of the two pumps for the case where absorption accompanies the nonlinearity due to plasma generation. A simpler approach was taken by AuYeung *et al.* (1979), who included a "passive" absorption coefficient (no pump depletion effects) in the coupled propagation equations, by which relatively simple expressions were obtained for the reflectivity as a function of the linear absorption coefficient and the mean pump intensity $(I_f I_b)^{1/2}$. For most of the experiments described here, we will use the reflectivity R vs $(I_f I_b)^{1/2}$ equations of Yariv and Pepper (1977) or that of AuYeung *et al.* (1978) to estimate the magnitude of $\chi^{(3)}$ from the linear portion of the $\ln R$ vs $\ln (I_f I_b)^{1/2}$ plot. In typical experiments this plot saturates, owing either to the emerging influence of higher-order competing terms in the nonlinearity itself (as occurs during absorption saturation) or to the introduction of other higher-order processes (such as multiphoton absorption) that may be detrimental to the DFWM interaction (see, e.g., Jain *et al.*, 1979b).

A secondary consequence of the experimental arrangement typically employed in DFWM relates to the fact that it results quite naturally in a configuration where a nonlinear medium is located within an optical resonator one end mirror of which is the output coupler of the laser used for the DFWM experiments. Thus, when the nonlinearities are sufficiently high, nonlinear Fabry–Perot effects are easily observed (Jain and Steel, 1980, 1981). Also, the simultaneous interaction of DFWM and optical bistability can often be observed, leading to instabilities and regenerative pulsations (Jain and Steel, 1981, 1982; Agrawal, *et al.*, 1981).

B. Experiments in Specific Materials

In this section, we briefly describe several recent DFWM experiments in which semiconductors have been used. In most cases, the choice of the materials (and nonlinear mechanisms) has been determined largely by the laser wavelengths available, and only occasionally have the lasers been explicitly designed to study a specific nonlinearity (e.g., biexcitons in CuCl, Chemla *et al.*, 1979). The DFWM experiments that we discuss could be grouped either according to the nonlinear optical mechanism (i.e., the nature of the transition, if any) or according to the material used. We have chosen to group the experiments by material, in the hope that when the various experiments are viewed in the context of the physical parameters of the semiconductor (such as band structure, effective masses, band gap, and other band separation energies) vis-à-vis the characteristic of the incoming optical radiation (such as photon energy, light intensity, and pulse duration), the reader will have a better overall perspective on the various nonlinear interactions that may occur simultaneously within a chosen material, and on the dependence of the relative strengths of these interactions on experimental parameters such as photon energy.

Some of the more important material properties of several semiconductors of interest are shown in Table X. The band-gap energies are shown for 300 K, and we note that for a situation where the photon energy is close to the band gap, the band-gap energy (and the transition energies) can be tuned with temperature. For all the semiconductors shown (and for most semiconductors except the lead salts PbS, PbSe, and PbTe), the band gap *increases* with decreasing temperature as expected from simple lattice contraction; typically a change of ~ 0.1 eV is obtained with a 300-K change in temperature, with a room-temperature $\partial E_g/\partial T$ coefficient ranging from approximately 2.3×10^{-4} eV/K in Si to approximately 9.5×10^{-4} eV/K in ZnO (Pankove, 1971; Tsay *et al.*, 1973). Likewise, the band gap may also be tuned with pressure, although temperature tuning is generally more practical.

Another important parameter of particular relevance to our discussion is the recombination time τ_R of the electron–hole pairs, which not only varies significantly with carrier density and thus light intensity and temperature, but also may vary significantly from sample to sample, with a strong dependence on factors such as impurity type and concentration, trap density, and lattice imperfections. Such considerations preclude a reasonable tabulation of the recombination time in Table X.

Our discussion of the DFWM experiments in various materials will follow the same general order as Table X. We will start with the elemental group-IV semiconductors and then proceed to the III–V and II–VI compounds; within each group we will cover semiconductors in the order of increasing band-gap energy.

1. Group-IV Elements

The two semiconductors considered here, Ge and Si, are the most extensively studied semiconductors, and are perhaps the semiconductors best characterized in terms of their electrical and optical properties and band structure. They also appear to be the semiconductors used most extensively to date for degenerate four-wave mixing.

Both Ge and Si (and other group-IV solids) crystallize in the cubic "dia-

TABLE X

Band-Gap Energies, Carrier Masses, and Other Properties of Semiconductors

Semiconductor	Compound group	Crystalline symmetry		Minimum band gap at 300 K		Effective masses of carriers (in units of free-electron mass)			Ambipolar diffusion coefficient D_a at 300 K (cm²/sec)
		Point group	Structure[a]	E_g^b (eV)	λ_g (μm)	Electron (m_e^*)	Light hole (m_{LH}^*)	Heavy hole (m_{HH}^*)	
Ge	IV	m3m	Cubic, D	0.67[c] (I)	1.85	m_L, 1.58[c]; m_T, 0.08[c]	0.043[d]	0.34[d]	64
Si	IV	m3m	Cubic, D	1.11[c] (I)	1.12	m_L, 0.93[c]; m_T, 0.19[c]	0.16[d]	0.52[d]	15
InSb	III–V	$\bar{4}3m$	Cubic, Z	0.17[c] (D)	7.3	0.012[e]	0.015[e]	0.39[d]	37
InAs	III–V	$\bar{4}3m$	Cubic, Z	0.36[c] (D)	3.4	0.026[d]	0.025[d]	0.41[d]	23
InP	III–V	$\bar{4}3m$	Cubic, Z	1.35[f] (D)	0.92	0.073[d]	0.078[d]	0.4[d]	7.3
GaAs	III–V	$\bar{4}3m$	Cubic, Z	1.43[c] (D)	0.87	0.07[d]	0.12[d]	0.68[d]	19
AlAs	III–V	$\bar{4}3m$	Cubic, Z	2.16[c] (I)	0.57	0.5[c]	0.22[g]	0.65[h]	45
GaP	III–V	$\bar{4}3m$	Cubic, Z	2.25[c] (I)	0.55	0.35[g]	0.14[g]	0.86[g]	3.0
GaN	III–V	6mm	Hex, W	3.4[c] (D)	0.36	0.2[c]	—	—	—
$Hg_{0.8}Cd_{0.2}Te$	II–VI	$\bar{4}3m$	Cubic, Z	0.14–0.17[i] (D)	7.3–8.9	0.007[i]	0.01[i]	0.4[i]	—

CdTe	II–VI	$\bar{4}3m$	Cubic, Z	1.45[f] (D)	0.86	0.11[c]	0.10[h]	0.35[c]	3.5
CdSe	II–VI	6mm	Hex, W	1.74[c] (D)	0.71	0.13[c]	—	2.5[d] $\parallel c$; 0.42[d] $\perp c$	—
CdS	II–VI	6mm	Hex, W	2.42[f] (D)	0.51	0.20[c]	—	5.0[d] $\parallel c$; 0.7[d] $\perp c$	—
ZnSe	II–VI	$\bar{4}3m$	Cubic, Z	2.58[c] (D)	0.48	0.17[c]	0.15[h]	—	—
	II–VI	6mm	Hex, W	2.79[d] (D)	0.45	—	—	—	—
ZnO	II–VI	6mm	Hex, W	3.2[c] (D)	0.39	0.32[c]	—	—	—
ZnS	II–VI	6mm	Hex, W	3.8[c] (D)	0.33	0.28[c]	—	—	—
	II–VI	$\bar{4}3m$	Cubic, Z	3.6[c] (D)	0.34	0.39[c]	—	—	—
CuBr	I–VII	$\bar{4}3m$	Cubic	2.94[d] (D)	0.42	0.23[h]	—	—	—
CuCl	I–VII	$\bar{4}3m$	Cubic	3.1[h] (D)	0.40	0.42[h]	—	—	—
Cu_2O	II–VI	m3m	Cubic	2.17[d] (D, 0 K)	0.57	—	—	—	—

[a] D = diamond, Z = zinc blende, W = wurtzite.
[b] I = indirect band gap, D = direct band gap.
[c] Pankove (1971, p. 412).
[d] American Institute of Physics Handbook (1969).
[e] Moss *et al.* (1973, pp. 334–335).
[f] Kittel (1971, p. 210).
[g] Neuberger (1971).
[h] Thomas and Timofeev (1980).
[i] Dornhaus and Nimtz (1976).

mond'' (m3m) point-group structure, which is an fcc Bravais lattice (see Kittel, 1971) with a basis of two atoms at [0, 0, 0] and $[\frac{1}{4}, \frac{1}{4}, \frac{1}{4}]$. The band structures of these semiconductors are quite different, as seen in Fig. 13. We will not discuss details of the band structures here [see Long (1968) or Geballe (1960)], but only point out a few essentials relating to transition

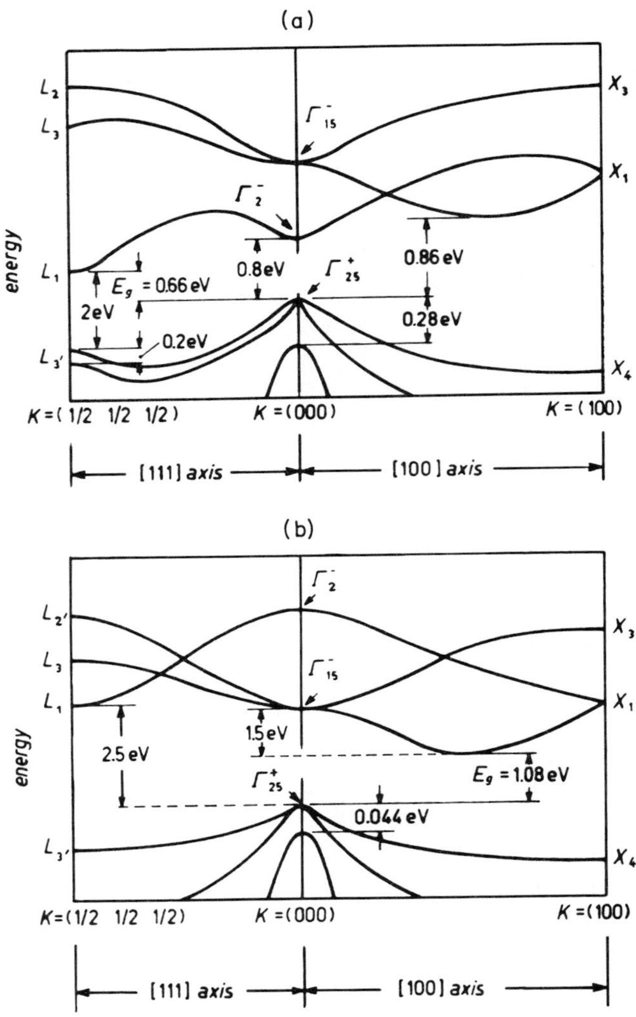

Fig. 13. Band structures of (a) germanium and (b) silicon. (Used with permission from Lax, 1963.)

energies and carrier effective masses. Both Ge and Si have indirect gaps that are smaller than the direct band gap, with conduction-band minima in the [1, 1, 1] and [1, 0, 0] directions, respectively; they are the only indirect-band-gap semiconductors considered in this chapter (GaP and AlAs also have indirect gaps, but are discussed only briefly). In Ge the constant energy surfaces near the bottom of the conduction band consist of four symmetrically equivalent ellipsoids whose lengths are along the [1, 1, 1] direction; in Si these consist of three equivalent ellipsoids whose lengths are along the [1, 0, 0] axes. Thus, the conduction-band electrons are characterized by longitudinal and transverse effective masses m_L and m_T, respectively, defined by the energy. Energy–wave-number equations for these ellipsoids are

$$E_c = \frac{\hbar^2}{2}\left(\frac{k_x^2 + k_y^2}{m_T} + \frac{k_z^2}{m_L}\right). \tag{72}$$

It can then be easily shown (Spitzer and Fan, 1957) that the optical effective mass of the conduction-band electron is given by

$$(m_e^*)^{-1} = \tfrac{1}{3}(2m_T^{-1} + m_L^{-1}) \tag{73}$$

as will be used below in our discussion of plasma generation (and as pointed out in Section III). Note that the valence-band maxima consist of three nearly spherical surfaces at $\mathbf{k} = 0$. Two of the surfaces meet (i.e., are degenerate) at $\mathbf{k} = 0$, and only a small spin–orbit interaction separates the third energy surface ($\Delta_{so} \approx 0.28$ and 0.044 eV for Ge and Si, respectively). If we assume that only a negligible number of carriers are thermally excited into the split-off band (which is quite valid at room temperature, especially for Ge), it can easily be shown that the optical effective mass of the holes is

$$1/m_h^* = (m_{LH}^{*1/2} + m_{HH}^{*1/2})/(m_{LH}^{*3/2} + m_{HH}^{*3/2}). \tag{74}$$

It follows from Table X that, for Ge, $m_e^* = 0.12m$ and $m_h^* = 0.34m$ and, for Si, $m_e^* = 0.27m$ and $m_h^* = 0.37m$. Note that in Ge the effective mass of conduction-band electrons at the center of the Brillouin zone (i.e., $\mathbf{k} = 0$) is only $\simeq 0.034m$, but that such electrons (created say, by transitions across the direct band gap) scatter very rapidly via phonon interactions into the valley at L_1. Note also that in Ge the indirect gap ($E_{gI} \sim 0.66$ eV) is only slightly smaller than the direct gap ($E_{gD} \sim 0.80$ eV), but in Si the indirect gap ($E_{gI} \sim 1.08$ eV) is *much* smaller than the direct gap ($E_{gD} \sim 2.5$ eV). Because of the proximity of the direct- and indirect-gap minima, the direct absorption easily dominates ($\alpha_D \gg \alpha_I$) in Ge at photon energies only slightly above the indirect band gap. (The absorption coefficient is proportional to the product of the matrix element and the joint density of states. The matrix elements for a phonon-assisted transition are much smaller than for a direct transition.)

a. Germanium Ge was the first semiconductor in which the phase conjugation and aberration correction properties of DFWM were explicitly studied. Early in the development of DFWM for phase conjugation, it was recognized that materials for the 10-μm region would be of particular interest, because of the need to correct aberrations in high-power CO_2 laser systems. Because of its ready availability and known third-order susceptibility at 10 μm, transparent (undoped or n-type) Ge was one of the first materials to be studied by several groups for phase conjugation of CO_2 laser radiation. However, because the only optical nonlinear mechanism in transparent Ge is the anharmonic motion of bound electrons, the third-order susceptibilities are relatively low ($\sim 10^{-10}$ esu), and high optical intensities are needed for reasonable DFWM reflectivities [see Eq. (71)]. Bergmann *et al.* (1978) were the first to demonstrate DFWM in Ge, by simply utilizing the high peak intensities within the cavity of a CO_2 TEA laser. The 5-mm-thick sample of polycrystalline intrinsic Ge was coated for high reflectivity ($R = 95\%$) on one side and AR coated on the other, and arranged to be both the DFWM medium and the output coupler of the TEA laser, as shown in Fig. 14. This intracavity mounting of the germanium sample ensured temporal overlap of the 50-nsec counterpropagating pump pulses, besides providing high pump intensities without focusing. One disadvantage of such an experimental arrangement is that it is difficult to discriminate against simple reflective backscatter of the probe pulse into the signal direction; in the traditional experimental arrangement the magnitude of this noise source is easily ascertained by blocking one or both of the pump beams. With a mean pump intensity of $(I_f I_b)^{1/2} \sim 40$ MW/cm^2 and a 5-mm interaction length, Bergmann *et al.* measured a probe-to-signal reflectivity of 2%, corresponding to an experimental third-order susceptibility of $\chi^{(3)} \simeq 1.5 \times 10^{-10}$ esu. They also demonstrated aberration correction and pulse narrowing consistent with expectations of electric field amplitude products in nonlinear media with extremely fast nonlinearities ($\tau \ll \tau_L$).

The light intensities used in this experiment were very close to the threshold values for optical breakdown. To obtain larger values of DFWM reflectivity, higher intensities and thus shorter laser pulses are necessary. Watkins *et al.* (1981) used a CO_2 laser with a 1.5-nsec pulse width, which allowed pump intensities as high as 120 MW/cm^2 to be used before damage occurred. With the use of such large intensities and a variety of Ge samples (particularly *p*-type), these authors were able to perform a comprehensive study of the three different nonlinear optical mechanisms that may be invoked in Ge at CO_2 laser wavelengths.

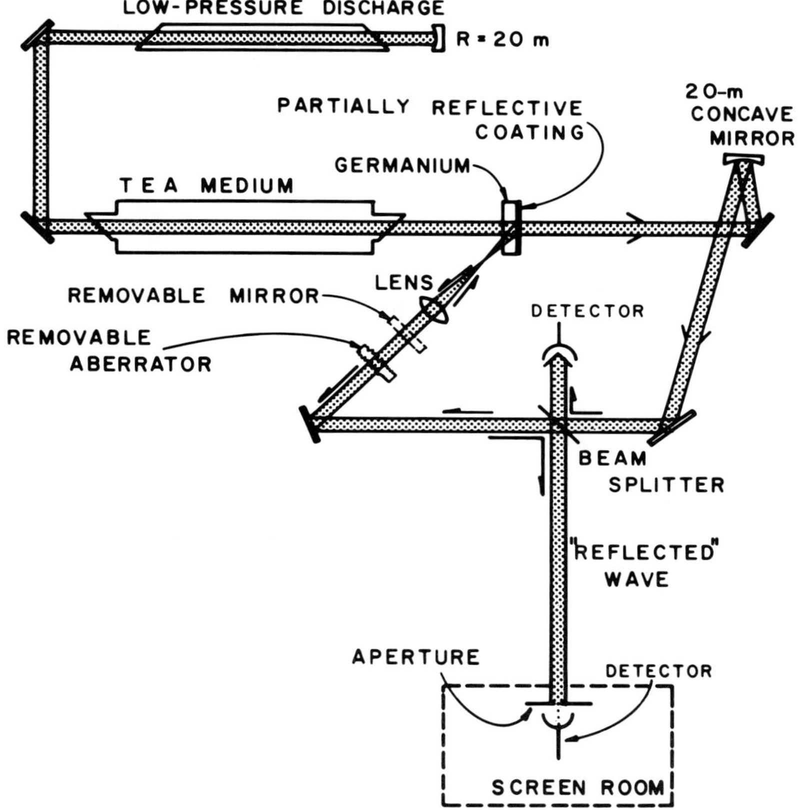

Fig. 14. Experimental arrangement for intracavity four-wave mixing in Ge at 10.6 μm. (Used with permission from Bergmann *et al.*, 1978.)

Besides the usual bound-electron nonlinearity (studied at low intensities in optical-grade samples), the two other mechanisms that were studied by Watkins *et al.* are: (1) saturation of inter-valence-band absorption in the p-type samples and (2) nonlinear plasma generation at very high pump intensities ($\gtrsim 80$ MW/cm^2). At pump intensities below and close to the saturation intensity, the experimental results for a 6×10^{15}-cm^{-3} doped p-type sample were found to be consistent with the nonlinearity due to inter-valence-band saturation, both in magnitude and in functional form. In Fig. 15 is shown a log–log plot of the DFWM reflectivity versus pump intensity. The experimental results (circles) were found to be in reasonable agreement with a simple model based on the two-level saturable absorber analysis of Abrams and Lind (1978).

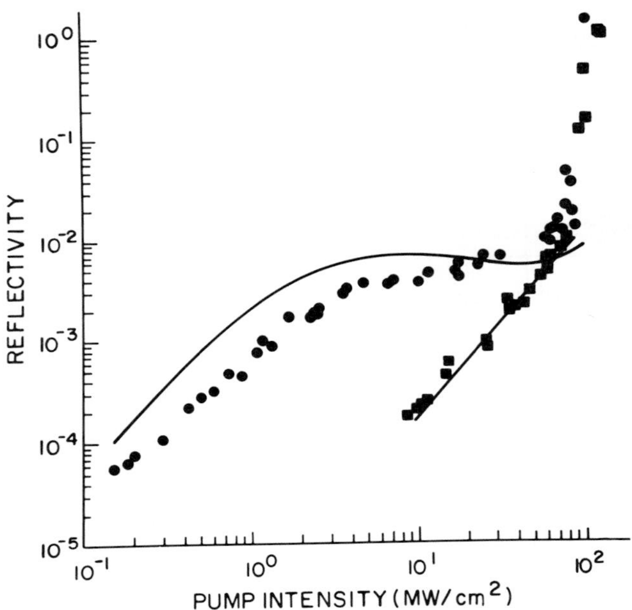

Fig. 15. Phase-conjugate reflectivity in 3-mm polycrystalline samples of p-type (●) and optical-grade (■) germanium. (After Watkins *et al.*, 1981, used with permission.)

At pump intensities above 80 MW/cm^2, Watkins *et al.* found that the DFWM reflectivity increased rapidly for both the optical-grade and the p-type samples, with an assumed 11th-power variation with the pump intensity (up to the damage threshold limit of 120 MW/cm^2). Note also that DFWM amplification of over a factor of 8 (i.e., a DFWM reflectivity of over 800%) was observed at intensities just below the damage threshold. The sharp increase in the DFWM signal was found to be related to the nonlinear generation of a plasma, whose presence was confirmed by photoconductivity measurements. The plasma itself leads to a large nonlinear index change similar to that described by Eq. (36); the rapid increase in the DFWM signal with intensity was confirmed to be caused by a rapid increase in the density of the photogenerated plasma.

The order of the nonlinearity involved for DFWM in this nonlinear plasma generation regime depends on the precise mechanism for plasma generation. The two most likely mechanisms are multiphoton absorption (Yuen *et al.*, 1980) and avalanche ionization (James, 1982). Watkins *et al.* (1981) initially attributed their data on nonlinear plasma generation, and the consequent data on DFWM reflectivities, to multiphoton absorption. Assuming a minimum band gap of 0.67 eV for the indirect transition in Ge with 10.6 μm (0.117 eV) radiation one requires at least a six-photon (phonon-assisted) process for multiphoton absorption in Ge. However, as shown by Yuen *et al.* (1980), the seven-photon direct absorption at $\mathbf{k} = 0$ is expected to dominate. Watkins *et al.* have attempted to explain their high-intensity DFWM data by assuming the onset of a six-photon transition. Under such an assumption, the *lowest-order* nonlinear polarization that may be invoked for the backward DFWM interactions is of 13th-order in the electric field (see Section II.C)[3]. Such a 13th-order polarization density will include a large number of terms representing all the possible combinations of the pump fields that satisfy both phase matching and energy conservation, as elucidated in Section II.C with examples for two-photon transitions. For the six-photon transition, two examples of terms[4] that are phase matched for backward DFWM are represented in $\mathscr{P}^{13}(\omega)$ as follows:

$$\begin{aligned} \mathscr{P}^{(13)}(-\omega;\, 6\omega,\, -6\omega,\, \omega) = {} & a[(\vec{\mathscr{E}}_f)^3(\vec{\mathscr{E}}_b)^3(\vec{\mathscr{E}}_f^*)^2(\vec{\mathscr{E}}_b^*)^3\vec{\mathscr{E}}_p^*]\vec{\mathscr{E}}_b \\ & + b[(\vec{\mathscr{E}}_f)^3(\vec{\mathscr{E}}_b)^3(\vec{\mathscr{E}}_b^*)^2(\vec{\mathscr{E}}_f^*)^3\vec{\mathscr{E}}_p^*]\vec{\mathscr{E}}_f \\ & + \cdots . \end{aligned} \tag{75}$$

Regardless of which term(s) in the overall polarization density dominate, for a six-photon absorption process the DFWM reflectivity should vary as the 12th order (I^{12}) of the total pump intensity (or as the 14th order if the dominant absorption is indeed due to a 7-photon transition), and not as the 11th order, as assumed by Watkins, *et al.* Considering the small range of pump intensities over which a DFWM signal could be obtained with this mechanism, the exponent in the intensity dependence of the experimental data is difficult to establish conclusively, and it is not possible to discriminate effectively between an I^{11} and an I^{14} variation.

James (1982) has suggested that avalanche ionization is the dominant mechanism for the nonlinear plasma generation in these experiments by Watkins *et al.* For such a mechanism, the functional form of the intensity dependence of both the plasma density as well as the DFWM reflectivity is expected to be exponential. Unfortunately, once again the small inten-

[3] This statement assumes the absence of any saturable intermediate levels, such as might be caused by the presence of a high concentration of specific impurities.

[4] The physical processes leading to the terms in Eq. (75) may be described as follows: Six photons are absorbed from the pump beams to create an optical coherence at 6ω (with no spatial variation). This coherence then mixes with five more photons from the pumps and then with one photon from the probe to form a free-carrier index grating. A photon from one of the pumps (the choice being determined by phase-matching considerations) then scatters from this grating to form the backward-traveling signal beam.

sity range of the DFWM data of Watkins *et al.* precludes a clear-cut determination of this hypothesis.

Backward-wave DFWM has also been studied at 3.8 μm in germanium (Depatie and Haueisen, 1980) by using a DF laser and the nonresonant bound-electron anharmonicity. In this work a 0.14% conjugate reflectivity was measured at a pump intensity of 12 MW/cm^2, corresponding to the anticipated third-order susceptibility of 4×10^{-11} esu. Multiline phase conjugation using six lines in the 3.7–3.9-μm range was also demonstrated.

Using photon energies much larger than those discussed above (1.06 μm $\equiv$ 1.17 eV), Smirl *et al.* (1980) have observed forward DFWM signals via a single-photon valence-to-conduction-band transition in germanium. Forward DFWM is a natural configuration in this situation, because the samples were thin ($\sim$6 μm) and highly absorbing. Using high-intensity picosecond pulses ($\sim$30 psec), Smirl *et al.* were able to perform a preliminary study of the carrier dynamics via these transient gratings. Two key differences between this experiment and the earlier parametric scattering observations (also using transient carrier gratings) of Shank and Auston (1975) and Kennedy *et al.* (1974) are (1) the study of the conjugate signal in the forward DFWM configuration by Smirl *et al.*, as opposed to the "parametrically coupled" signal in the probe direction, and (2) the use of a second "pump" pulse, at fixed temporal delay, to probe the decay characteristics of the grating. In the experimental arrangement, shown in Fig. 16, a single pulse from the mode-locked YAG laser is split into two temporally separated (delay t_p) copropagating pump pulses, and a third probe

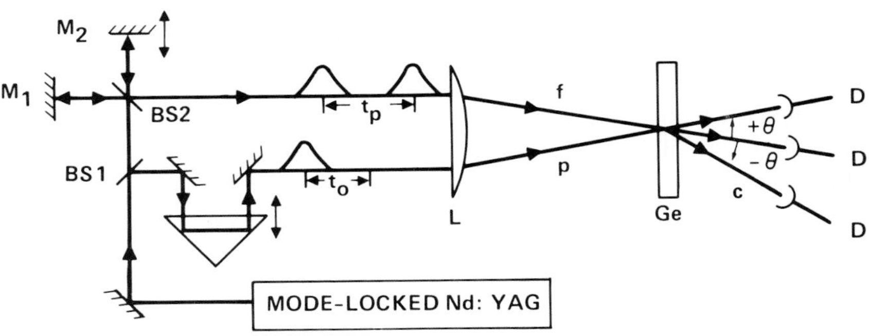

Fig. 16. Experimental arrangement for picosecond forward-traveling DFWM in Ge at 1.06 μm. BS1 and BS2 are beamsplitters, and the movable mirror M_2 and prism allow adjustment of the delays t_0 and t_p. Lens L focuses both beams into overlapping spots, and D represents detectors that measure the energy in the pumps, probe, and signal pulses. (After Smirl *et al.*, 1980, used with permission.)

pulse arrives at an angle of 3–5° at a delay t_0 with respect to the center of the two pump pulses. This highly unconventional technique for measurement and presentation of carrier dynamics results in data similar to those shown in Fig. 17, in which the energy in the conjugate wave is measured as a function of the probe delay t_p for two cases of fixed pump-probe separation ($t_p = 0$ and 105 psec). Corresponding theoretical curves are easily obtained by using a one-dimensional plasma diffusion equation similar to Eq. (37). Auger recombination effects are neglected, and the "two delay points" data of Fig. 17 are observed to fit a theoretical curve corresponding to $\tau = 250$ psec. Simply stated, Fig. 17a corresponds to the energy in the signal wave when the two pumps are coincident in time, and Fig. 17b shows the decay in the amplitude of the grating (caused by the first pump pulse and probe) to $\lesssim 50\%$ of its original value in a time duration of 105 psec, resulting in an estimated decay time of ~ 250 psec. These authors also studied the conjugate-wave properties of the signal wave via imaging experiments.

In separate experiments at comparable intensity levels, Smirl *et al.* (1980) have observed saturation of the direct absorption, which was attributed to the filling of the valence bands by holes up to and including the states from which absorption normally occurs. The filling of the hole states results from the lower density-of-states mass m^*_{LH} and the consequent lower density of states in the valence band. Smirl *et al.* initially

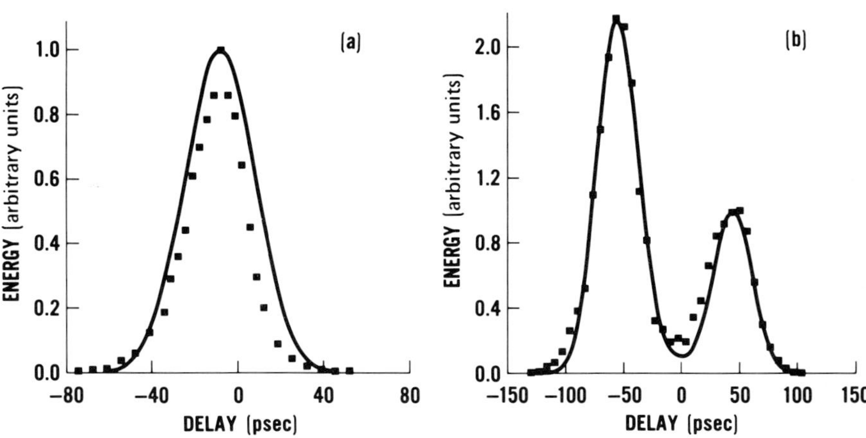

Fig. 17. Energy in the forward DFWM signal as a function of probe delay. Squares indicate experimental data for (a) $t_p = 0$ psec and (b) $t_p = 105$ psec. The solid lines are the corresponding theoretical curves based on a grating time $\tau = \tau_D\tau_R/(\tau_D + \tau_R) = 250$ psec. (Used with permission from Smirl *et al.*, 1980.)

attributed their DFWM signals to gratings that are dominantly absorptive, as may be expected from absorption saturation considerations. However, rough calculations on the index terms show that the refractive gratings still dominate, and that the reported results are essentially indistinguishable from a simple plasma generation model in which variations in $\Delta\epsilon_2(\omega)$ are also included. Either way, the equations of p. 337 provide good approximations for the description of this nonlinearity. In more recent and comprehensive studies of these transient gratings, Smirl *et al.* (1982) have confirmed that the index effects apparently dominant those of absorption.

In related picosecond transient-grating work, based on the technique of Phillion *et al.* (1975) and Aoyagi *et al.* (1980), Moss *et al.* (1981) have used the forward DFWM geometry to study carrier diffusion in germanium. Values of 53 and 142 cm^2/sec were estimated for the ambipolar diffusion coefficient D_a at room temperature and 135 K, in good agreement with the value of D_a measured by other techniques (see Table X). We note that a similar study might be performed with nanosecond pulses with the use of a smaller pump–probe angle. In addition to the possibility of improved temporal resolution and thus of improved accuracy, an important advantage of the picosecond pulse technique is the possibility of using much higher light intensities, by which the dynamics of high-density plasmas may be studied. Such studies have been performed by Smirl *et al.* (1982); the effect of nonlinear diffusion mechanisms was noted. Such transient grating studies involving high density plasmas should lead to a clearer understanding of the role of plasmas in laser annealing, which has been a topic of much theoretical speculation (Yoffa, 1980; Van Vechten and Wautelet, 1981; Wautelet and Van Vechten, 1981).

b. Silicon. The only nonlinear mechanism used for DFWM in silicon is that of plasma generation. One convenient optical feature of silicon is that its room-temperature band gap (see Fig. 13b) of ~1.10 eV is well matched to the photon energy (1.17 eV) of the Nd : YAG and Nd : glass lasers. Moreover, the indirect nature of the band gap results in a small enough value of the absorption coefficient (~10 cm^{-1} at low light intensity) that conveniently thick (>200 μm) samples may be used effectively for DFWM experiments. Another advantage of silicon, stemming from its widespread use in technology, is the ready availability of well-characterized high-purity samples.

The first DFWM-type experiments in silicon (Woerdman and Bolger, 1969; Woerdman, 1970, 1971; Jarasiunas and Vaitkus, 1977) were in the context of transient holography; these transient grating studies were designed to measure the transport properties of carriers in silicon. Woerdman (1971) developed a model based on the theory of volume gratings to calculate the amplitudes of the diffracted beams for both forward- and backward-wave scattering geometries. His calculated values were in good agreement with measured values. Imaging properties of the free-carrier gratings were also studied by Woerdman (1971). Jarasiunas and Vaitkus (1977) extended these measurements to a detailed study of semiconductor carrier parameters in silicon (and other semiconductors), and demonstrated that the forward-wave diffracted signals varied as the third power of laser intensity (up to 10 MW/cm^2), as expected from their volume diffraction grating model.

Backward-wave DFWM in silicon at 1.06 μm was studied by Jain and Klein (1979). Using the model for plasma generation via interband transitions and ignoring the resonant factor due to the distant (2.5 eV) direct gap, these authors considered the carrier grating problem in the formulation of nonlinear optics, leading to expressions for the steady-state and transient third-order susceptibility (see Section III.C.1). Experiments were performed using pulses of 15 nsec duration from a Q-switched Nd : YAG laser. In initial experiments using 0.2-mm-thick samples (so that $\alpha L \ll 1$), a third-order susceptibility of 1.1×10^{-7} esu was measured by Jain and Klein (1979). For this experimental estimate, consideration was given to the linear attenuation experienced by all the input waves (AuYeung *et al.*, 1979). This measured value corresponds to the linear portion of the log–log plot in Fig. 18. The deviation from linear behavior at high intensities is attributed to free-carrier absorption, which is well borne out by the more drastic deviations from linearity of these logarithmic (reflectivity versus intensity) plots for the thicker samples (Jain *et al.*, 1979b). Note that peak reflectivities of over 150% were observed in these experiments at relatively modest pump intensities (compared with those of Watkins *et al.* in Ge).

Comparison of the experimentally measured values of the third-order susceptibility with values estimated from the plasma model is complicated slightly by the fact that several decay times are involved, and, as indicated in Section III.C.1, it is the smallest of these that determines the size of the effective third-order susceptibility. It is thus necessary to estimate the grating decay times τ_D for each of the two gratings and to compare these with the recombination time τ_R and laser pulse duration τ_L. The relevant decay times for the given experimental conditions are tabulated in Table XI. The value of the effective susceptibility for the small-period grating is determined by the rapid diffusional decay of these carriers, whereas the effective susceptibility of the large-period grating is determined by the laser pulse duration. The small-period grating thus contrib-

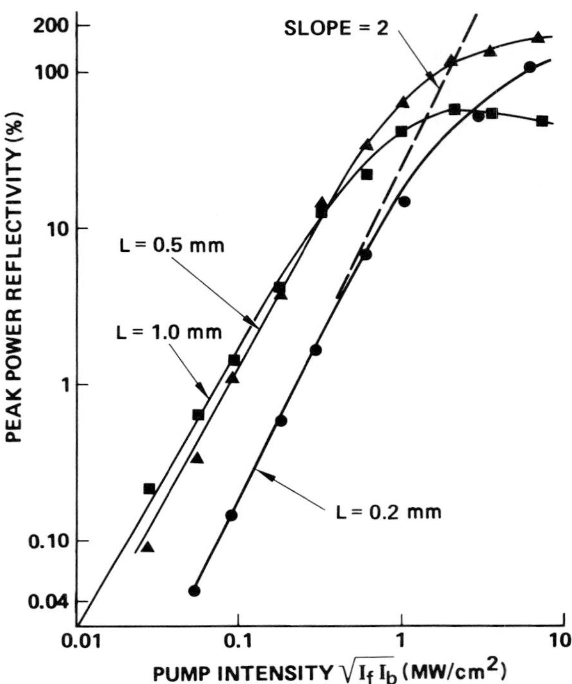

Fig. 18. DFWM peak power reflectivity (using ~15-nsec Nd : YAG pulses) versus mean pump intensity for various silicon sample thicknesses L. The sample "slices" were oriented with an angle of incidence $\psi = 60°$ at $\theta = 2°$, i.e., near the Brewster angle with $\alpha = 10\ \mathrm{cm}^{-1}$ (low intensity) for the p-polarized radiation. Resistivities of $\gtrsim 150\ \Omega$ cm were measured for these moderate-purity near-intrinsic samples. (Jain *et al.*, 1979b.)

utes negligibly to the DFWM signal, whereas the large-period grating results in an effective third-order susceptibility of $\sim 8 \times 10^{-8}$ esu, in good agreement with the value of 1.1×10^{-7} esu measured experimentally (from the data of Fig. 18).

Other experiments were also performed by Jain *et al.* (1979b), including a study of the DFWM signal as a function of the pump–probe angular separation θ for various combinations of the pump and probe polarizations. No signals were observed above the background noise level when the forward pump wave and probe wave were orthogonally polarized, whereas large and distinct signals were observed when these waves were copolarized. This again indicates the dominance of the large-period grating and suggests that the contribution of anisotropic nonlinearities (such as those due to bound-electron anharmonicity) is neg-

TABLE XI

Grating Period, Ambipolar Decay Time, and Susceptibility for f–p and f–b Gratings in Silicon[a]

Parameter	Forward-pump–probe grating	Backward-pump–probe grating
Grating period Λ	61 μm	0.15 μm
Ambipolar decay time τ_D	63 nsec	0.4 psec
Regime of operation	Transient	Steady state
Relevant temporal duration[b] (for calculating $\chi^{(3)}$)	$\tau = \tau_L = 15$ nsec	$\tau = \tau_D = 0.4$ psec
Effective susceptibility $\chi^{(3)}$	8×10^{-8} esu	4×10^{-12} esu

[a] Jain and Klein (1979).
[b] Laser pulse duration $\tau_L = 15$ nsec. Recombination time $\tau_R = 1$ μsec $\gg \tau_D, \tau_L$.

ligible. The signals increase monotonically with decreasing angle θ because of the increased grating period Λ and correspondingly larger diffusional decay times τ_D at the smaller angles. The latter result is consistent with similar measurements in transient holography studies, and DFWM may easily be used as a technique for studying ambipolar diffusion in silicon as a function of sample preparation techniques.

One detrimental effect of the large values of the self-induced nonlinear index changes is the lenslike action induced by transverse variations in the intensity of the beams. The large negative value of the nonlinear index change due to a free-carrier plasma in a semiconductor results in a large amount of self-defocusing, which inevitably affects the quality of the phase conjugation obtained when beams with large variations in the transverse intensity profile are used as DFWM pumps. Independent of the specific nonlinearities used, self-focusing is a basic problem encountered with high-efficiency DFWM. Recently, Hopf *et al.* (1981) have studied the quality of phase conjugation in forward DFWM by an interferometric technique, and noted that phase distortions as large as $\lambda/2$ occur at moderate reflectivities ($\gtrsim 50\%$). Also, owing to the lack of anticipated "weak-wave retardation" in their data, Hopf *et al.* (1981) have estimated a rapid ($\gtrsim 4$ nsec) washout of their gratings (formed with $\theta \simeq 5°$), implying an ambipolar diffusion coefficient over 16 times that reported in most of the earlier literature. However, this *indirect* measurement of the ambipolar diffusion coefficient should be treated with great caution, particularly because such high values of D_a are not observed in the more direct diffraction efficiency measurements obtained from transient grating experiments using similar optical configurations (Woerdman, 1971; Jarasiunas and Vaitkus, 1977; Jain and Klein, 1979).

2. *III–V Compounds*

This class of compounds comprises covalently bonded solids that are structurally similar to the group-IV elements, except that the two basis atoms in this cubic "diamondlike" crystalline structure are different, and belong to two different groups (III and V) in the periodic table. For example, in InSb the In ions are at [0, 0, 0] sites and the Sb ions at [$\frac{1}{4}$, $\frac{1}{4}$, $\frac{1}{4}$] sites. In such a crystalline structure, called the zinc-blende structure, the center of inversion symmetry between the two sublattices is destroyed; these compounds thus belong to a different symmetry group ($\bar{4}3m$).

The most common III–V compounds are the nine combinations of the three elements In, Ga, and Al from group III and the three elements Sb, As, and P from group V. Thus, while being structurally similar to the group-IV elements and relatively simple (compared with the other classes of compounds), III–V compounds represent a very wide selection of the basic semiconductor parameters (such as effective masses, energy gaps, and carrier mobilities), as can easily be seen from the seven representative compounds included in Table X. (GaN represents one of the more complex III–V compounds and behaves very differently from the nine compounds mentioned above.) An even larger dimension of flexibility is obtained in the control and choice of basic semiconductor parameters with the III–V compounds (and other compound semiconductors) when one realizes that they can easily be prepared in the form of ternaries (and even quaternaries) such as $InAs_{1-x}P_x$ and $GaAs_{1-x}P_x$. In such compounds, the band structure (and other semiconductor properties) is a smooth function of x, and thus one has an infinity of choices in the semiconductor parameters. (See Casey and Panish, 1978, Chapter 5). For example, the energy gap of $GaAs_{1-x}P_x$, one of the best-studied ternaries, is smoothly variable from a 1.43-eV direct gap (GaAs) to a 2.25-eV indirect gap (see Fig. 6 for the band structure of III–V's), with a smooth crossover at $x = 0.4$ from a direct to an indirect gap.

A final point about III–V compounds is that they are the simplest and the best-understood class of semiconductor compounds. The most common III–V compounds (e.g., InSb, GaAs, and InP) have direct gaps; as a consequence, the theoretical understanding of the optical properties of such semiconductors, particularly near $\mathbf{k} = 0$, is by and large even more complete than that of the elemental semiconductors Si and Ge.

Degenerate four-wave mixing may be performed efficiently in any semiconductor in this compound group, especially via interband transitions near the band edges and via exciton transitions in the larger-band-gap ($\gtrsim$1 eV) compounds (the latter constraint is determined by binding energy, exciton radius, and electrical screening considerations). Thus, exciton transitions in InP and GaAs represent important candidates for DFWM applications at the appropriate (exciton transition) wavelengths, and n-doped GaP is expected to result in nonlinearities of over 10^{-7} esu near 3 μm via inter-conduction-band transitions (see Section III.C). However, for valence-to-conduction-band transitions, the III–V compound with the largest nonlinearity is clearly InSb, as expected from the strong inverse dependence of this nonlinearity on the photon energy $\hbar\omega$ ($\simeq E_g$) and reduced effective mass, as seen in Section III.C.1.

a. Indium Antimonide. Until a few years ago, when GaAs and InP started to draw attention because of the technological potential of their high electron mobilities, InSb was the most extensively studied III–V compound, for three reasons. First, it is relatively easy to prepare InSb with moderately high purities; second, its small band gap and the extremely high mobilities and small effective masses of its conduction-band electrons lead to unusual electrical and optical properties; and third, it is still one of the semiconductors best characterized (for values of $\mathbf{k} \simeq 0$) in terms of the theory of its optical properties. The band structure of InSb is typical of III-V compounds, and is similar to that shown in Fig. 6, except that the conduction-band valley at $\mathbf{k} = 0$ is much lower, resulting in a direct band gap of $\simeq$0.17 eV at $T = 300$ K; the corresponding spin–orbit splitting Δ_{so} is $\simeq$0.75 eV. Note also that the band gap of InSb is also in excellent coincidence with the discretely tunable 5–6-μm emission from CO lasers, allowing large nonlinearities at these wavelengths.

Interestingly, the large nonlinearities in InSb were first observed quite accidentally as beam defocusing effects (Miller *et al.*, 1978) in the transmission of a Gaussian CO laser beam through a cooled InSb sample. The large defocusing effects were attributed to an electronic nonlinearity, which was subsequently exploited for optical bistability (Miller *et al.*, 1979) and DFWM (Miller *et al.*, 1980a). This valence-to-conduction-band nonlinearity, which is of the order of 10^{-2} esu, was discussed in Section III.C.

The experimental arrangement used by Miller *et al.* (1980a) for their DFWM studies is shown in Fig. 19. The low-power cw CO laser was tightly focused ($\sim$200-μm spot size) into the sample. With the InSb sample at 5 K ($E_g \sim 1899$ cm^{-1}) and the CO laser tuned to 1886 cm^{-1}, DFWM reflectivites of $\sim$1% were observed at pump intensities of $\sim$40 W/cm^2 in a 7.5-mm-long sample. These large returns imply a third-order susceptibility of much less than the anticipated 10^{-2} esu, possibly due to

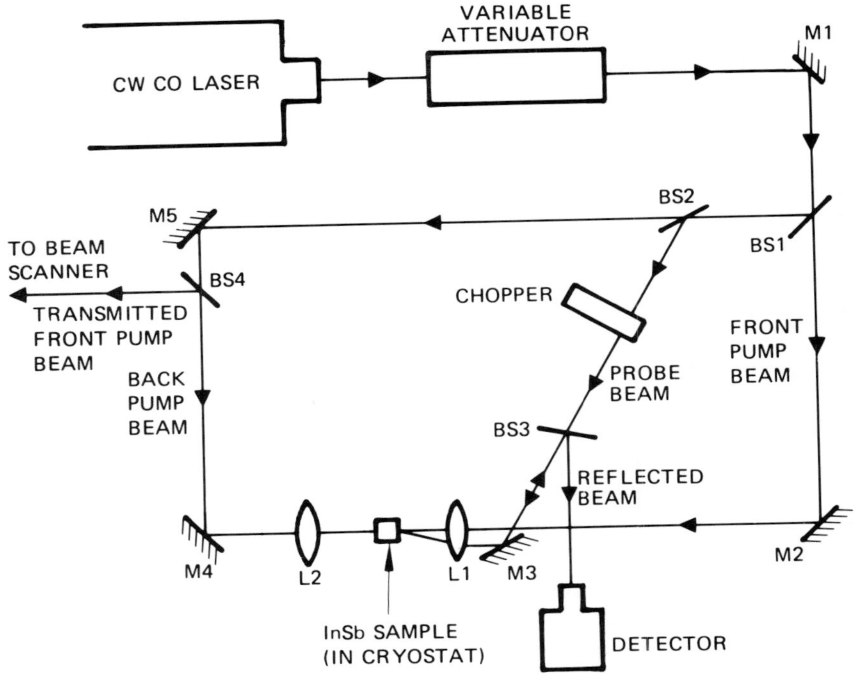

Fig. 19. Experimental arrangement for cw DFWM in InSb. (After Miller *et al.*, 1980a, used with permission.)

the alignment problems inevitable in the tight beam focusing arrangement used for these experiments.

InSb has also been used frequently for nearly degenerate four-wave mixing; in these difference frequency generation ($\omega_4 = 2\omega_1 - \omega_2$) experiments the much faster (but smaller) nonlinearity due to nonlinear carrier motion has been used. Of particular interest are the experiments of Yuen and Wolff (1982), using very small frequency differences ($\omega_1 - \omega_2 = 1.4\ \text{cm}^{-1}$, with $\omega_1 \approx \omega_2 \approx 1000\ \text{cm}^{-1}$), in which third-order susceptibilities on the order of 10^{-7} esu were measured. Of particular interest in these experiments is the data on the large increase in the third-order susceptibility as $\omega_1 - \omega_2 \rightarrow 0$, and the estimation of an energy relaxation time ($\tau_{\text{th}} \approx 4$ ps) for the optically-excited carriers. The latter was obtained by curve fitting the data on the dependence of $\chi^{(3)}$ on the difference frequency with an expression [Eq. (25)] based on a simple model that includes the effect of inelastic collisions. An independent and more direct measurement of τ_{th} under similar conditions of excitation would illustrate the validity of both this expression and the experimental technique for measuring energy relaxation times in semiconductors.

b. Gallium Arsenide. Although no DFWM experiments have as yet been performed with bulk GaAs samples, the experiments on absorption saturation and optical bistability by Gibbs *et al.* (1980, 1981) clearly indicate its promise at wavelengths in the near infrared. Very recently, Hegarty *et al.* (1982) have used backward DFWM for a preliminary study of the nonlinear dispersion and diffusion dynamics of the 2-dimensional excitons in GaAs–GaAlAs quantum-well structures.

The samples used by Hegarty *et al.* consisted of identical (80 or less) layers of GaAs, ~50–100 Å thick, each layer being separated by ~20 Å of $Ga_{0.75}$ $Al_{0.25}$ As. Low energy wavelength-tunable picosecond pulses from a synchronously-pumped (80MHz) cw mode-locked dye laser, tunable from 7500 to 8200 Å, were used for DFWM studies near the lowest energy exciton resonances of this structure. The pump energy densities used were typically between $\sim 10^{-8}$ and $\sim 2 \times 10^{-7}$ J/cm²/pulse, corresponding to average exciton densities of between 10^8 and 10^9/cm²/layer. The studies were performed at sample temperatures below 90 K, so that the exciton resonances remained within the tuning range of the dye laser. Most of the detailed data reported was for a sample temperature of 2 K.

Besides demonstrating the possibility of observing DFWM signals in such structures with such low excitation levels, Hegarty *et al.* noted that the nonlinearity was most strongly enhanced at the heavy exciton line, which appeared to shift to higher energies with increased exciton density. More interesting were their preliminary studies of the grating decay for excitation on the low energy side of the heavy exciton line, obtained by delaying the backward pump pulse with respect to the other two pulses. These studies revealed two distinct decay components. The slower decay component corresponded simply to the diffusional time of the heavy excitons, which is simply related to the grating period (see Eqs. (38a,b)) by the effective diffusion coefficient, D_e, for the heavy exciton; from this data, an effective diffusion coefficient of ~1–10 cm²/sec was estimated for the various samples. Although no data was presented, a strong dependence of D_e on wavelength was noted.

The faster decay component showed much more unusual behavior. It was found to be independent of the grating period and varied with GaAs layer thickness, ranging from 25–70 psec for the 51 Å and 200 Å thicknesses, respectively. Although it was not seen on the high energy side of the heavy exciton line, it had a much weaker wavelength dependence than the slow decay component. The authors tentatively attributed this poorly understood decay component to "local spectral diffusion" involving excitons that move over distances that are small compared to the grating period.

3. *II–VI Compounds*

The elements Cd, Zn, and Hg from group II of the periodic table combine with the elements Te, Se, S, and O from group VI to result in the most commonly used II–VI compound semiconductors. Among these, the most extensively studied are CdTe, CdSe, CdS, ZnSe, ZnS, and ZnTe, all of which (with the exception of CdTe) may crystallize in both the cubic zinc-blende structure and the hexagonal wurtzite structure (6mm point-group symmetry). In Table X, semiconductor data are tabulated only for the more commonly available structures, most of which have been used for DFWM and DFWM-type experiments. One consequence of the 6 mm hexagonal symmetry is that the overall band features are different from those in cubic semiconductors, although these differences are not major, because of the similar atomic arrangements of the neighboring atoms. Another consequence of particular interest here is the presence of a larger number of nonzero terms in the $\boldsymbol{\chi}^{(3)}$ and $\boldsymbol{\sigma}$ tensors, which can be significant for some of the nonlinear mechanisms of interest. Because the c-axis-directed asymmetry in wurtzite crystal structures is small (wurtzite structures are equivalent to zinc-blende crystals that are slightly stressed in the [111] direction), we will neglect the secondary consequences of this c-axis-oriented asymmetry in our discussion.

One of the II–VI compounds, HgTe, has an "inverted" band order, and is actually a semimetal (just like α-Sn in group IV). The band structures of CdTe and HgTe are shown in Fig. 20. Whereas the band structure of CdTe is that of a conventional semiconductor with a rather wide energy gap at Γ ($\mathbf{k} = 0$), in HgTe the Γ_6 state (ordinarily the conduction-band minimum) happens to be lower than the valence-band maximum, resulting in a "negative" value of the $\Gamma_6 - \Gamma_8$ energy gap, and thus in the semimetallic nature of HgTe. This property of HgTe leads to very interesting and unusual properties in the technologically important II–VI alloy $Hg_{1-x}Cd_xTe$.

a. Mercury Cadmium Telluride Because the band structures of CdTe and HgTe are basically of the same form (except that HgTe exhibits an inverted band order), an admixture of CdTe with HgTe results in gradual shifting of the Γ_6 band, and in a smooth variation of the $\Gamma_6 - \Gamma_8$ gap from negative to positive, via a smooth semimetal–semiconductor transition. This is illustrated in Fig. 21, where the band structure of the $Hg_{1-x}Cd_xTe$ alloy system near $\mathbf{k} = 0$ for various values of E_g (i.e., of x) is depicted. A

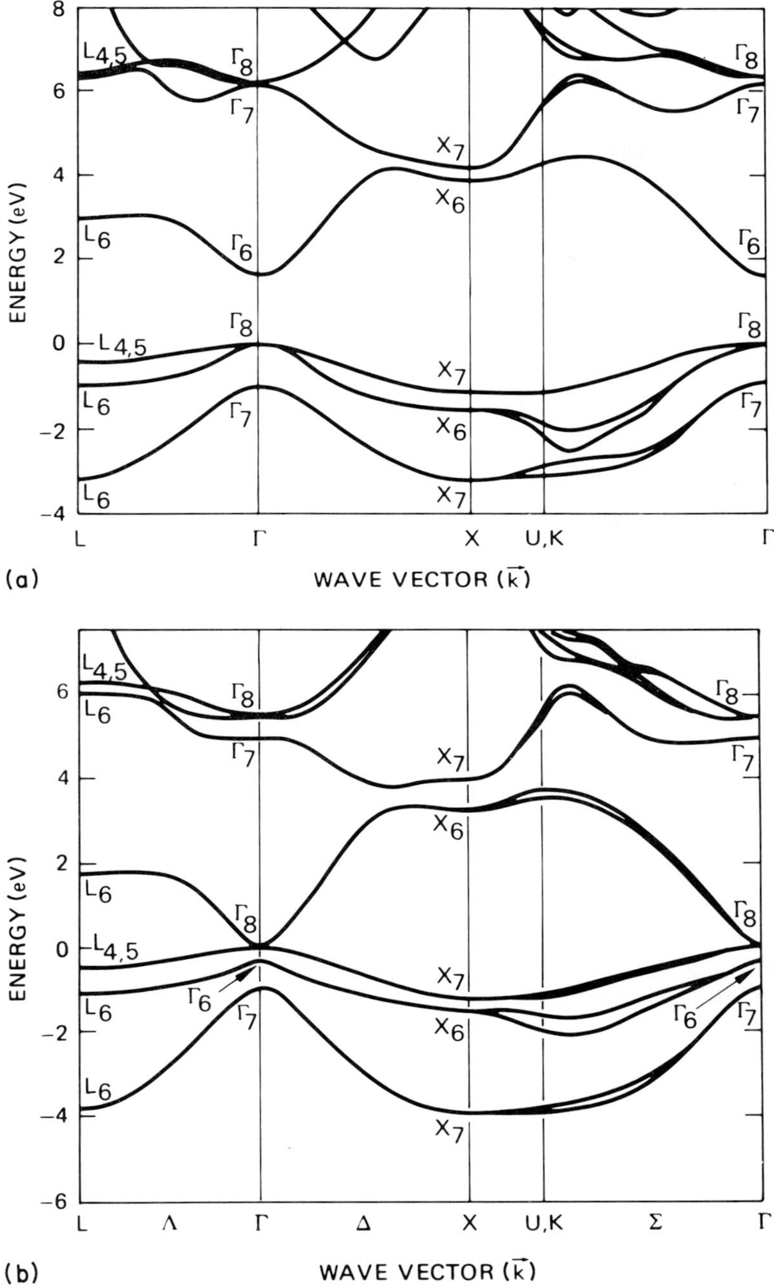

Fig. 20. Band structures of (a) CdTe and (b) HgTe. (Used with permission from Dornhaus and Nimtz, 1976; after Chadi *et al.*, 1972.)

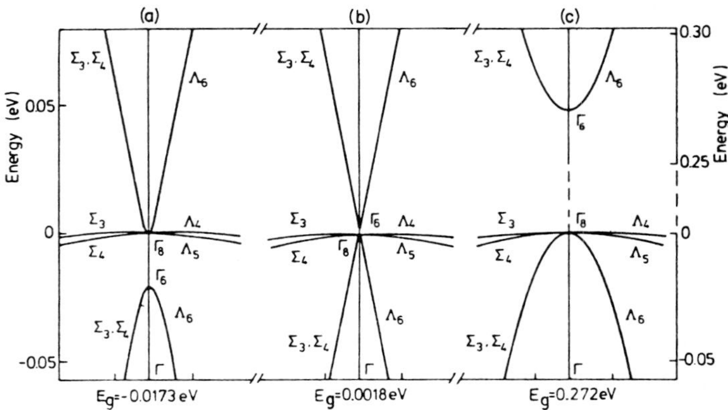

Fig. 21. Energy-band structure variation in $Hg_{1-x}Cd_xTe$ with x. (a) $E_g = -0.0173$ eV, $x \approx 0.15$ (4 K). (b) $E_g = +0.0018$ eV, $x \approx 0.17$ (4 K). (c) $E_g = +0.272$ eV, $x \approx 0.32$ (4 K). (Used with permission from Dornhaus and Nimtz, 1976; after Overhof, 1971.)

very important aspect of this behavior is the smooth, controllable manner in which the band gap passes through zero at a certain alloy composition ($x \simeq 0.15$ at $T \simeq 0$ K). As expected from Kane's $\mathbf{k} \cdot \mathbf{p}$ theory, the effective masses at the edges of the light-hole valence band and the conduction band become infinitesimally small for these zero-band-gap semiconductors, as has been confirmed by numerous experimental measurements. We thus see [from Eq. (41)] that with suitable choice of alloy compositions ($x \simeq 0.15$) and temperatures ($T \lesssim 4$ K), at wavelengths close to these small bandgaps ($\lambda \gtrsim 100\ \mu$m), *third-order susceptibilities of over* 10^3 esu *should be easily observable in such an alloy system.* (The alloy PbSnTe shows similar potential.)

The longest wavelengths at which DFWM experiments have been performed in HgCdTe are the 10-μm CO_2 laser wavelengths. The usefulness of HgCdTe for DFWM at long wavelengths was first predicted by Jain and Klein (1979), and preliminary results on 10-μm DFWM in HgCdTe were reported shortly thereafter by Jain *et al.* (1979a) in collaboration with Steel. A more detailed description of these results was subsequently published by Jain and Steel (1980), who used 180-nsec pulses from a 10.6-μm CO_2 TEA laser, and carefully selected their HgCdTe sample to have a room-temperature absorption coefficient given by $\alpha L \approx 1$. With an $x = 0.2$ n-type sample, a third-order susceptibility of 5.4×10^{-6} esu was measured, corresponding to a conjugate reflectivity of over 10% at a pump intensity of $(I_f I_b)^{1/2} \sim 160$ kW/cm^2. At higher intensities a dropoff in signal occurred owing to a sharp increase in sample absorption, presumably via avalanche generation of carriers (Jamison and Nurmikko, 1978).

Several sources of nonlinearity were considered by Jain and Steel (1980), and the magnitude of the nonlinearity was found to be in order-of-magnitude agreement with the generation of an electron–hole plasma, as targeted in the experiment. The small factor represented by the resonant term in Eq. (41) was neglected in the theoretical estimate, since larger uncertainties occur in the estimates of m_{eh}^* and τ, owing to a large scatter in the published values of these parameters (ideally, these parameters should be measured directly for the specific sample). However, a significant observation made by these authors was that, at the 300 K sample temperature and with the carrier densities used, the size of the nonlinearity was limited by Auger recombination. They predicted that with the use of lower temperatures ($T \lesssim 180$ K) and sample compositions (e.g., $x = 0.23$) with an appropriate match of the low-temperature band gaps to the laser photon energy, third-order susceptibilities of over 10^{-2} esu would be possible. In subsequent work (Steel *et al.*, 1981; Jain *et al.*, 1981; Jain and Steel, 1981, 1982), these authors reported demonstration of third-order susceptibilities of this magnitude, consistent with these predictions, as described below.

Using an unfocused 1-W cw CO_2 laser with a power density of a few watts per cm^2 and an $x = 0.22$ n-type sample, Jain and Steel observed a strong increase in the DFWM signal as the sample temperature was lowered (Fig. 22). A change in temperature affects both τ and ω_g, the latter resulting in a change in the absorption coefficient α. The observed increase in DFWM signals at the lowered temperatures is simply due to the combined effect of an increase in τ and α simultaneous with a decrease in the resonant denominator (due to the tuning of the band gap with temperature). The rapid dropoff in the DFWM signal at temperatures below -143 °C for this sample corresponds to the extremely rapid increase in the absorption coefficient, resulting in $\alpha L \gg 1$ and thus in significant reduction in the coupling of the counter-opposed waves in the DFWM interaction. At the optimum temperature (-143 °C), a reflectivity of ~2.2% was observed at a mean pump intensity of only 10 W/cm^2 (see Fig. 23), implying a third-order susceptibility of over 5×10^{-2} esu, consistent with the magnitudes predicted earlier. Comparable third-order susceptibilities were measured in several n- and p-type samples. Note that a close comparison of these results with those of a comparable DFWM experiment in InSb at 5.3 μm by Miller *et al.* (1980a) indicates that the observed nonlinearity in HgCdTe is larger than that in InSb.

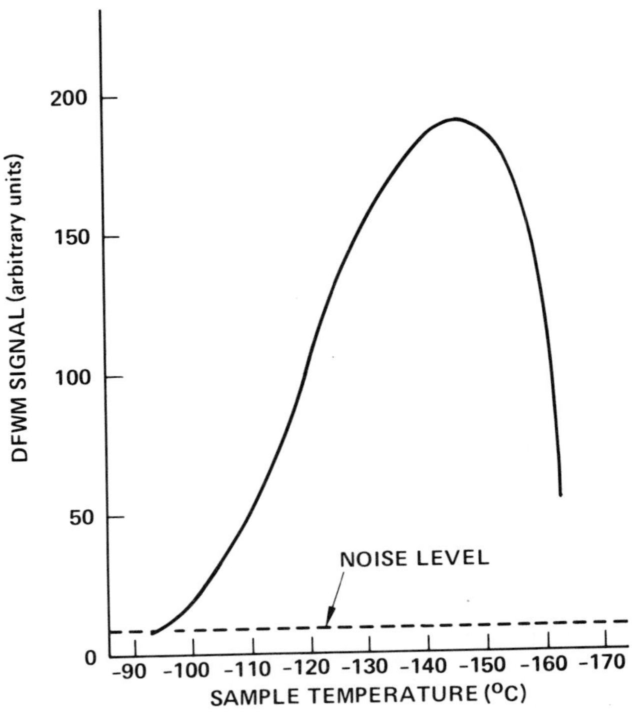

Fig. 22. DFWM signal versus sample temperature in n-type $Hg_{1-x}Cd_xTe$, where $x = 0.22$ and $L = 0.36$ mm. (Used with permission from Jain and Steel, 1981.)

Another nonlinear optical mechanism explored for DFWM in HgCdTe is that of conduction-band nonparabolicity. Khan *et al.* (1980) used several n-type samples with different composition values at temperatures of 12, 77, and 295 K for pulsed DFWM of CO_2 radiation. For the range of experimental conditions, the absorption edge of the samples varied discretely from 6.1 to 12.4 μm. The DFWM reflectivity was measured as a function of pump intensity; at low power levels the variation was quadratic, and evidence of saturation was observed at higher levels. The largest observed value of DFWM reflectivity was 9% in $Hg_{0.77}Cd_{0.23}Te$ at a temperature of 295 K and a pump intensity of 20 MW/cm^2. The measured values of susceptibility were in the range 5×10^{-8}–4×10^{-7} esu. The authors attributed these results to conduction-band nonparabolicity. This explanation is straightforward in samples for which the band gap is much larger than the photon energy, such that interband transitions are unim-

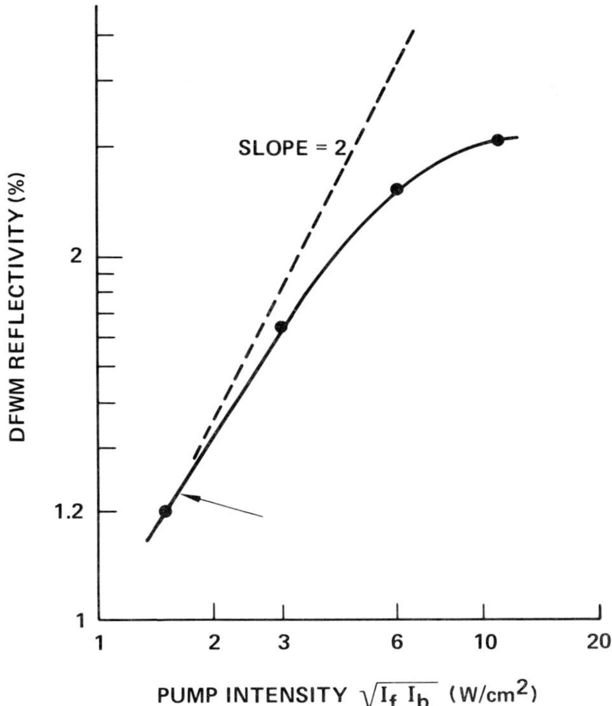

Fig. 23. DFWM reflectivity versus mean pump intensity for an $x = 0.22$ HgCdTe sample at $T = -143°C$ with $\chi^{(3)}_{meas} \gtrsim 5 \times 10^{-2}$ esu and $\chi^{(3)}_{calc} \approx 9 \times 10^{-2}$ esu. (Used with permission from Jain and Steel, 1982.)

portant. For operation near the band gap, plasma generation can also contribute to the nonlinear susceptibility. Details of the susceptibility calculation and comments on the resonant factor are given in Section III.B.

Subsequent to the prediction by Jain and Steel (1980) of large third-order susceptibilities at lowered temperatures with appropriately chosen band-gap resonant samples, several attempts were made by Khan *et al.* to measure such susceptibilities (Kruse, 1981). However, their use of nanosecond TEA laser pulses resulted in small effective susceptibilities and precluded such observations. With clarification of this point (Jain *et al.*, 1981) and the subsequent use of cw radiation, Khan *et al.* (1981) were able to confirm the results of Jain and Steel (1981), illustrating dramatically that the large third-order nonlinearities are indeed relatively slow "integrating" nonlinearities.

b. Cadmium Telluride Four-wave mixing due to plasma generation in CdTe has been studied in both the forward direction (Kremenitskii *et al.*, 1979) and the backward direction (Kremenitskii *et al.*, 1980). The pump laser in both cases was a Nd : YAG laser at 1.06 μm, whose photon energy (1.17 eV) was much less than the band gap of CdTe (1.605 eV). The generation of free carriers was thus attributed to two-step absorption via intermediate (impurity) states. These intermediate states apparently saturate at relatively low intensities, so that the material behaves almost like a linear absorber at the MW/cm² levels of the experimental intensities. The dependence of the measured DFWM reflectivity on the pump intensity was quadratic, which is consistent with the above model. At the highest pump intensity (10 MW/cm²), the DFWM reflectivity reached 200% and showed no sign of saturation.

c. Cadmium Selenide CdSe is a direct-band-gap semiconductor with a 300 K band edge at 0.71 μm. Transient holography has been used to study grating formation and decay mechanisms in CdSe at 1.06 and 0.694 μm. With the use of 1.06-μm radiation (Jarasiunas and Vaitkus, 1977), free carriers could be created either by two-photon interband absorption or by excitation from deep impurity levels. Photoconductivity measurements (Baltramiejunas *et al.*, 1973) indicate that two-photon absorption was the dominant mechanism at the pump intensity levels used (0.1–10 MW/cm²). This was confirmed by measurements of the forward-scattered signal as a function of pump intensity in the transient experiments. The scattered signal varied as $(I_f I_b)^2$, consistent with the behavior expected for a nonlinearity based on two-photon absorption.

The energy of the ruby laser (1.786 eV) is very close to the band gap of CdSe (1.825 eV at 13 K), so that nonlinearities due to the saturation of exciton absorption as well as to plasma generation can be observed. The exciton nonlinearity was studied by Baltrameyunas *et al.* (1976), who adjusted the sample temperature so that the sum of the laser photon energy and that of an optical LO phonon (E_p = 2.7 meV) was coincident with the energy of A (n = 1) excitons. Using the forward-beam geometry, they observed large diffraction efficiencies at pump intensities in the range 0.1–1 MW/cm². At the low end of this range the signal intensity varied as the cube of the laser intensity (see Fig. 24). At higher pump intensities, nearly complete saturation of the exciton transition was observed, leading to saturation in the signal intensity. This behavior is in qualitative agreement with the theory of Abrams and Lind for a saturable two-level system.

Jarasiunas and Gerritsen (1978) have also studied forward scattering in CdSe at 6943 Å. The sample temperature was not tuned to an exciton resonance, and the observed signal was attributed to scattering from a

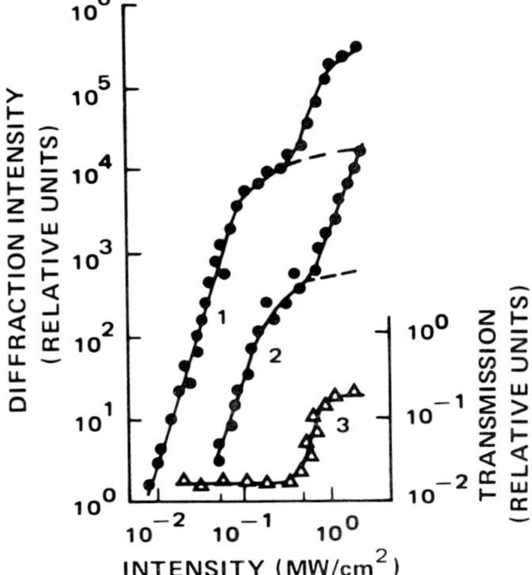

Fig. 24. Signal intensity for grating periods of 30 μm (curve 1) and 6.2 μm (curve 2) and transmission (curve 3) in CdSe versus input intensity. The dashed lines represent the variation of the observed signal if correction is made for the variation in the sample absorption with intensity, due to saturation of the exciton transition. (After Baltrameyunas *et al.*, 1976, used with permission.)

free-carrier grating excited by two-photon absorption. The ambipolar diffusion coefficient was deduced from the variation of the forward-scattered signal as a function of the angle between the two beams.

d. Cadmium Sulfide CdS is a direct-band-gap semiconductor with a room-temperature band gap of 2.42 eV. The large band gap and the prevalence of significant trap densities lead to strong photoconductivity in the visible spectral region, but also complicate estimation of intrinsic effects such as carrier recombination. DFWM resulting from a variety of nonlinear mechanisms has been studied in CdS. In experiments at 0.694 μm, forward signals have been attributed to scattering from two-photon-induced free-carrier gratings (Borshch *et al.*, 1973; Jarasiunas and Gerritsen, 1978). Jarasiunas and Gerritsen observed that the output signal varied as the fifth power of the overall input intensity, the expected variation for a fifth-order nonlinear process. The ambipolar diffusion coefficient was also deduced from the variation of the signal as a function of the angle between the two beams.

The output wavelength of the doubled Nd : YAG laser (0.53 μm) is very close to the room-temperature band edge in CdS, thus allowing the study of nonlinearities due to direct interband absorption. Jain and Lind (1983) used backward-wave DFWM with 15-nsec laser pulses at 0.53 μm to measure the DFWM reflectivity as a function of the pump intensity in the range 0.1–20 MW/cm^2. They estimated an experimental value of 3.4×10^{-9} esu for the effective third-order susceptibility. Assuming that the nonlinearity is due to interband transitions, and noting that the 15-nsec laser pulses are shorter than the ~1-μsec lifetime of the electron–hole plasma (Smith, 1957), Jain and Lind found that the experimental value was in good agreement with the effective value of 4×10^{-9} esu calculated from data on the semiconductor parameters of CdS.

e. CdS_xSe_{1-x}*-Doped Glasses* The advantage of using ternaries or semiconductor "alloys" to tune the band gap and other semiconductor parameters has already been discussed, as has the advantage of using semiconductors near their band edges to obtain moderate to high nonlinearities. One convenient mixed semiconductor system whose bandgap spans a broad range in the visible spectral region is the CdS_xSe_{1-x} system, samples of which can be obtained conveniently and inexpensively in the form of sharp cutoff color glass filters. Each CdS_xSe_{1-x} filter glass consists of independent microcrystals (typically 100–1000 Å in size) of fixed composition x suspended in a glass matrix. The individual microcrystals are of uniform composition throughout the sample and oriented randomly. The bulk material is thus isotropic and insulating.

DFWM due to plasma generation in the CdS_xSe_{1-x} glasses has been studied by Jain and Lind (1983), using a variety of pulsed ($\lesssim$15 nsec) laser sources with wavelengths in the range 0.532–0.694 μm. Susceptibilities of the order of 10^{-8} esu and reflectivities of up to 10% (at pump powers of ~1 MW/cm^2) were measured.

One interesting consequence of the microscopic nature of the CdS_xSe_{1-x} glasses is that the DFWM signal due to the small-period grating was found to be comparable with (actually somewhat larger than) the signal due to the large-period grating. The relative contribution of each grating (and of the two-photon term) to the DFWM signal was determined by a series of polarization experiments, as indicated in Table XII (Jain and Lind, 1983). Also included are data on single-crystal CdS, where diffusion leads to dominance of the large-period grating term. The contributions of the gratings in the CdS_xSe_{1-x} glasses are comparable because the semiconductor crystallites are generally much smaller (100–1000 Å) than even the 1030-Å period of the small-period grating. Thus unlike the situation in bulk semiconductors, the free-carrier plasma is effectively confined to these isolated pockets, and plasma diffusion does not reduce the diffraction efficiency of the small-period grating. To our knowledge, this is the only "diffusion-free" DFWM or transient grating experiment reported.

f. Zinc Selenide and Zinc Oxide In the context of transient holography, backward-traveling DFWM via two-photon generation of a

TABLE XII

Relative Intensities of DFWM Signals in $CdS_{0.9}Se_{0.1}$ Glass and CdS single Crystal for Four Polarization Conditions

POLARIZATION CONDITION	DFWM SIGNAL RELATIVE UNITS	
	$CdS_{0.9}Se_{0.1}$ IN GLASS	CdS CRYSTAL
f, b, p, θ	1	1
f, b, p — LARGE PERIOD (f–p) GRATING	0.35 (±0.07)	1
f, b, p — SMALL PERIOD (b–p) GRATING	0.45 (±0.08)	0
f, b, p — 2ω TEMPORAL COHERENCE	0	0

free-carrier plasma has been studied in ZnO and ZnSe with a ruby laser ($\lambda = 0.694\ \mu m$) by Dean and Collins (1973) and Jarasiunas and Vaitkus (1977). More recently, Borshch *et al.* (1980) have repeated similar experiments in ZnSe and modeled this interaction in the formulation of nonlinear optics, in contrast to the earlier transient grating models. Describing the interaction as a fifth-order nonlinear process, and using coupled-wave equations to model the six-photon interaction, Borshch *et al.* specifically consider the effect of phase distortion by the "self-action," i.e., self-induced index change, of each beam. Their theoretical model shows that optimum phase matching at high intensities occurs for a slight misalignment (approximately 5 min) of one of the pump beams, consistent with their experimental observations. With 15-nsec pulses and a mean pump intensity of $\sim 50\ MW/cm^2$, a peak reflectivity of $\sim 200\%$ was obtained.

4. *Other Semiconductors*

DFWM and transient gratings may be used to study the physics and phase-conjugation potential of numerous other semiconductors. Attractive choices include the use of PbSnTe at 10.6 μm and exciton and biexciton transitions in Cu_2O, CuBr, and CuCl. In particular, DFWM provides a very powerful technique for the spectroscopy of biexcitons in these materials.

Biexciton energy states lie well within the conduction bands, so that they cannot easily be studied via linear absorption spectroscopy. One obvious alternative is to use two-photon absorption; as discussed in Section III, the two-photon absorption coefficient is enhanced by the single-photon resonances with the excitonic states, which in turn are connected to the biexcitonic states by very large oscillator strengths. However, two-photon absorption spectra of biexcitons are still influenced by the rapidly varying structure of the linear exciton absorption. Other diagnostic techniques of biexcitons require the actual creation of biexcitonic molecules, whose properties are difficult to study by optical or electrical means. As an alternative, the use of DFWM in the forward-traveling geometry (Maruani *et al.*, 1978; Chemla *et al.*, 1979; Chemla, 1980; Aoyagi *et al.*, 1980) allows the study of the biexciton states by producing only virtual excitation through the "two-photon" coherence term in the nonlinear polarization [see Eq. (2)]. Maruani *et al.* have observed as many as five orders in the forward-scattering experiments with their thin, highly absorbing samples of CuCl. The lowest-order scattered beam results from a two-photon resonant third-order nonlinear interaction involving the scattering of the probe beam from a temporal coherence at 2ω. The spectral dependence of the signal in this first-order scattered beam is shown in Fig. 25, and is explained by the competition between the nonlinear gain and two-photon absorption. The asymmetry originates in the two-photon absorption line profile, and has been attributed to Fano interferences in the autoionizing character of the biexcitons.

Maruani and Chemla (1981) attributed the higher scattered orders to multiple third-order scattering, or higher-order single scattering. We note that similar results have been observed by numerous other workers using thin samples (Woerdman, 1971; Jain and Klein, unpublished), and that the high order multiple scatterings may be equivalently interpreted as higher-order Raman–Nath-type scattering from a thin grating. In the language of nonlinear optics, the smaller interaction length L corresponds to a larger spread in $\mathbf{k}$ values, and the multiple orders in scattering may then be simply related to the angular profile of the $[(\sin\theta)/\theta]^2$ phase-matching function.

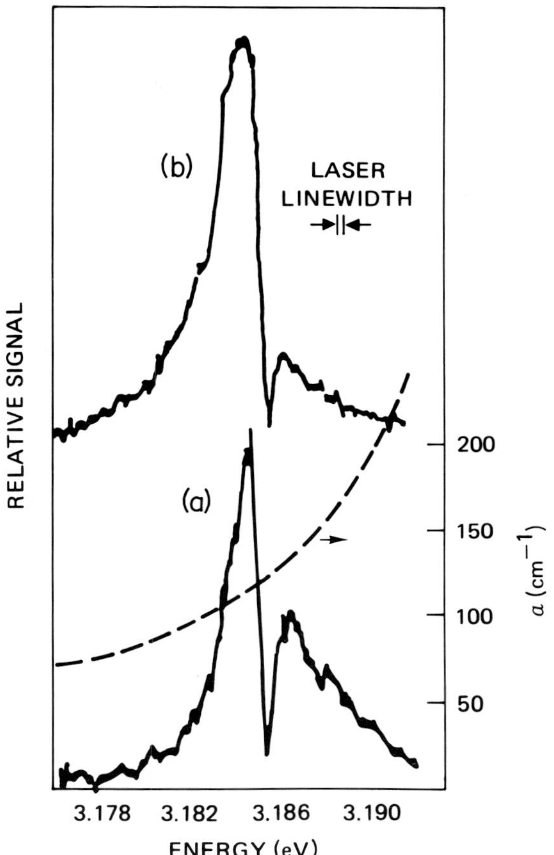

Fig. 25. First-order forward-scattered signal near 3900 Å in CuCl for sample thicknesses of (a) 80 and (b) 200 μm. The dashed line is the absorption coefficient. In both spectra the dip is located at half the formation energy of the biexciton. (After Maruani and Chemla, 1981, used with permission.)

V. Conclusions

Semiconductors clearly allow a large variety of choices of nonlinear mechanisms and speeds for DFWM and other self-interaction-type applications (such as intrinsic optical bistability) at any wavelength from the middle infrared to the near ultraviolet. Although the magnitudes of the nonlinearities are generally not as large as those obtained near the resonance lines of atomic vapors, semiconductors are less demanding on the wavelength and linewidth constraints of the laser sources. Moreover, large values of third-order susceptibilities may be attained in semiconductors at higher speeds than in most other classes of nonlinear materials.

It is hoped that this chapter will not only be of tutorial benefit, but will also enable researchers to select appropriate semiconductors for related nonlinear optics applications, and to identify the various nonlinearities that might be simultaneously present. Other choices, beyond the classes of semiconductors discussed here, are also available. These include amorphous semiconductors and crystalline semiconductors under the influence of large magnetic fields. The extension of the ideas discussed here to such cases is relatively straightforward. For instance, the nonlinear index change via plasma generation in an amorphous semiconductor can be modeled in the same manner as that of a nonresonant band-edge transition, with the mobility gap replacing the usual band gap; also, for the "plasma nonlinearity," the lowered values of the mobilities and recombination times in amorphous semiconductors will clearly yield lower values of the third-order susceptibilities, albeit at higher speeds. Likewise, in the magnetic field case, nonlinearities using resonant transitions invoke sharper resonances with large densities of states and well-defined selection rules. Large resonant nonlinear susceptibilities should thus be possible, as exemplified by the recent observation of two-photon transitions in InSb with low-power cw radiation by Seiler *et al.* (1981).

In addition, it is clear that DFWM itself can be used as a powerful technique for the study of semiconductor properties in both the time and the frequency domains, with a sophistication unrivaled by other techniques. Such studies might include the nondestructive testing of carrier transport properties such as mobility and diffusion, particularly at high carrier densities, and the study of the coherence properties of biexciton transitions.

Acknowledgments

We are grateful to J. F. Lam for many discussions regarding Sections II, III.A, and III.B, and to D. H. Auston for his numerous comments and a critical reading of the manuscript. We also acknowledge several useful comments by R. A. Fisher, A. L. Smirl, B. S. Wherrett, A. V. Nurmikko, and D. M. Pepper. We are grateful to Cheryl Krieger and Joanne Ordean for their cheerful and diligent typing of the manuscript.

References and Supplementary Bibliography

Abrams R. L. and Lind, R. C. (1978). Degenerate four-wave mixing in absorbing media, *Opt. Lett.* **2,** 94; Erratum (1978). *Opt. Lett.* **3,** 205.

Agrawal, G. P., Flytzanis, C., Frey, R., and Pradere, F. (1981). Bistable reflectivity of phase-conjugated signal through intracavity degenerate four-wave mixing, *Appl. Phys. Lett.* **38,** 492.

Allen, J. W., and Hodby, J. W. (1962). Infra-red absorption in gallium phosphide–gallium arsenide alloys, I. absorption in n-type materials, *Proc. Phys. Soc., London* **82,** 315.

"American Institute of Physics Handbook," 3rd ed., pp. 9-59–9-61. McGraw-Hill, New York, 1969.

Aoyagi, Y., Segawa, Y., Baba, T., and Namba, S. (1980). Determination of dispersion curves of excitonic polaritons by picosecond time-of-flight method, *Top. Meet. Picosecond Phenom., North Falmouth, Mass.* Pap. FA-5.

Arthur, J. B., Baynham, A. C., Fawcett, W., and Paige, E. G. S. (1966). Optical absorption due to free holes in germanium: a comparison of theory and experiment, *Phys. Rev.* **152,** 740.

Ashcroft, N. W., and Mermin, N. D. (1976). "Solid State Physics." Holt, New York.

Aspnes, D. E. (1967). Electric field effects on the dielectric constant of solids, *Phys. Rev.* **153,** 972.

Aspnes, D. E., Handler, P., and Blossey, D. F. (1968). Interband dielectric properties of solids in an electric field, *Phys. Rev.* **166,** 921.

Auston, D. H., McAfee, S., Shank, C. V., Ippen, E. P., and Teschke, O. (1978). Picosecond spectroscopy of semiconductors, *Solid-State Electron.* **21,** 147.

AuYeung, J., Fekete, D., Pepper, D. M., Yariv, A., and Jain, R. K. (1979). Continuous backward-wave generation by degenerate four-wave mixing in optical fibers, *Opt. Lett.* **4,** 42.

Aven, M., and Prener, J. S. (1967). "Physics and Chemistry of II–VI Compounds." Wiley, New York.

Balkanski, M. ed., (1980). "Handbook on semiconductors," Vol. 2. *Optical Properties of Solids* North-Holland Publ., Amsterdam.

Baltramiejunas, R., Grivickas, V., Storasta, J., and Vaitkus, J. (1973). The influence of mobility of photoconductivity in CdSe single crystals at high excitation density, *Phys. Status Solidi A* **19,** K115.

Baltrameyunas, R., Vaitkus, Y., Yarashyunas, K. (1976). Dynamic holographic gratings based on excitons in CdSe, *Sov. Phys.—Semicond.* **10,** 572.

Baranovskii, I. V., Borshch, A. A., Brodin, M. S., and Kamuz, A. M. (1971). Self-focusing of ruby laser radiation in CdS crystals, *Sov. Phys.—JETP* **33,** 861.

Bassani, F. (1966). Band structure and interband transitions, *in* "Optical Properties of Solids" (J. Tauc, ed.). Academic Press, New York.

Bassani, F. (1969). Interband transitions and optical properties, *in* "Optical Properties of Solids" (E. D. Haidemenakis, ed.). Gordon & Breach, New York.

Baumeister, P. W. (1961). Optical absorption of cuprous oxide, *Phys. Rev.* **121,** 359.

Bebb, H. and Williams, E. W. (1972). Photoluminescence I: theory, *in* "Semiconductors and Semimetals" (R. K. Willardson and A. C. Beer, eds.), Vol. 8. Academic Press, New York.

Bechtel, J. H. and Smith, W. L. (1976). Two-photon absorption in semiconductors with picosecond laser pulses, *Phys. Rev. B* **13,** 3515.

Bergmann, E. E., Bigio, I. J., Feldman, B. J., and Fisher, R. A. (1978). High-efficiency pulsed 10.6-μm phase-conjugate reflection via degenerate four-wave mixing, *Opt. Lett.* **3,** 82.

Bivas, A., Levy, R., Phach, V. D., and Grun, J. B. (1979). Biexciton two-photon absorption in the nanosecond and picosecond range in copper halides, *in* "Physics of Semiconductors, 1978" (B. L. H. Wilson, ed.). Institute of Physics, London.

Blakemore, J. S. (1962). "Semiconductor Statistics." Pergamon, New York.

Bloembergen, N. (1965). "Nonlinear Optics," Chap. 1. Benjamin, New York.

Bonch-Bruevich, N. L. (1966). Fundamentals of solid-state optics *in* "The Optical Properties of Solids" (J. Tauc, ed.). Academic Press, New York.

Borshch, A. A., Brodin, M. S., Ovchar, V. V., Odoulov, S. G., and Soskin, M. S. (1973). Dynamic holographic gratings in cadmium sulfide, *JETP Lett.* **18,** 397.

Borshch, A. A., Brodin, M. S., and Krupa, N. N. (1976). Nature of the nonlinearity that leads to self-action of laser radiation in semiconductors of the A_2B_6 group, *Sov. Phys.—JETP* **43,** 940.

Borshch, A. A., Brodin, M. S., Krupa, N. N., Lukomskii, V. P., Pisarenko, V. G., Petropavlovskii, A. I., and Chernyl, U. V. (1978). Determination of the coefficients of the nonlinear refractive index of a CdS crystal by the nonlinear refraction method, *Sov. Phys.—JETP* **48,** 41 (1978).

Borshch, A., Brodin, M., Volkov, V., and Kukhtarev, N. (1980). Phase conjugation by the degenerate six-photon mixing in semiconductors, *Opt. Commun.* **35,** 287.

Bowers, R. and Melamed, N. T. (1955). Luminescent centers in ZnS:Cu:Cl phosphors, *Phys. Rev.* **99,** 1781.

Bratt, P. R. (1977). Impurity germanium and silicon infrared detectors *in* "Semiconductors and Semimetals" (R. K. Willardson and A. C. Beer, eds.), Vol. 12. Academic Press, New York.

Braunstein, R. and Kane, E. D. (1962). The valence-band structure of the III-V compounds, *J. Phys. Chem. Solids* **23,** 1423.

Briggs, H. B. and Fletcher, R. C. (1953). Absorption of infrared light by free carriers in germanium, *Phys. Rev.* **91,** 1342.

Broser, I. and Franke, K. H. (1965). Erzeugung "Utraroter" Kupferzentren in ZnS durch kernzerfall von ^{65}Zn. *J. Phys. Chem. Solids* **26,** 1013.

Bube, R. H. (1960). "Photoconductivity of Solids." Wiley, New York.

Bube, R. H. (1974). "Electronic Properties of Crystalline Solids." Academic Press, New York.

Burstein, E., Picus, G. S., and Sclar, N. (1956). Optical and photoconductive properties of silicon and germanium, *in* "Proc. Conf. on Photoconductivity," p. 353. Wiley, New York.

Butcher, P. N. (1965). "Nonlinear Optical Phenomena." Ohio State Univ., Columbus, Ohio.

Butcher, P. N. and McLean, T. P. (1963). The nonlinear constitutive relation in solids at optical frequencies, *Proc. Phys. Soc., London* **81,** 219.

Butcher, P. N., Loudon, R., and McLean, T. P. (1965). Parametric amplification near resonance in nonlinear dispersive media, *Proc. Phys. Soc., London* **85,** 565.

Cardona, M. Shaklee, K. L., and Pollak, F. H. (1967). Electroreflectance at a semiconductor electrolyte interface, *Phys. Rev.* **154,** 696.

Carlson, R. L., Montgomery, M. D., Ladish, J. S., and Lockhart, C. M. (1977). Simultaneous multiline saturation of gallium doped germanium and SF_6 on the 10.6 μm band, *IEEE J. Quant. Electron.* **QE-13,** 35D.

Casey, H. C. Jr., and Panish, M. B. (1978). "Heterostructure Lasers." Academic Press, New York.

Casey, H. C., Jr., and Stern, F. (1976). Concentration-dependent absorption and spontaneous emission of heavily doped GaAs, *J. Appl. Phys.* **47,** 631.

Chadi, D. J., Walter, J. P., Cohen, M. L., Petroff, Y., and Balkanski, M. (1972). Reflectivities and electronic band structures of CdTe and HgTe, *Phys. Rev. B* **5,** 3058.

Chang, T. Y. (1981). Fast self-induced refractive index changes in optical media: a survey, *Opt. Eng.* **20,** 220.

Chemla, D. S. (1980). Nonlinear optical properties of condensed matter, *Rep. Prog. Phys.* **43,** 1191.

Chemla, D. S. and Jerphagnon, J. (1980). Nonlinear optical properties, *in* "Handbook on semiconductors," Vol. 2 "Optical Properties of Solids" (M. Balkanski, ed.). North-Holland Publ., Amsterdam.

Chemla, D. S., Maruani, A., and Batifol, E. (1979). Evidence of the autoionizing character of biexcitons, *Phys. Rev. Lett.* **42,** 1075.

Chemla, D. S., Miller, D. A. B., Smith, P. W., Gossard, A. C., and Tsang, W. T. (1982). Large optical nonlinearities in room-temperature GaAs-GaAlAs multiple quantum well structures, *Conf. Lasers Electro-Opt., Phoenix, Arizona,* Paper ThU5.

Cohen, M. H. (1966). Introductory lectures *in* "The Optical Properties of Solids" (J. Tauc, ed.). Academic Press, New York.

Cohen, M. L. and Bergstrasser, T. K. (1966). Band structures and pseudopotential form factors for fourteen semiconductors of the diamond and zinc-blende structures, *Phys. Rev.* **141,** 789.

Conwell, E. M. (1967). "High Field Transport in Semiconductors." Academic Press, New York.

Dagenais, M. and Winful, H. (1982). Low-power saturation of bound excitons in cadmium sulfide platelets, *Ann. Meet. Opt. Soc. Am., Tucson, Arizona,* Paper ThE3.

Dean, D. R. and Collins, R. J. (1973). Transient phase gratings in ZnO induced by two-photon absorption, *J. Appl. Phys.* **44,** 5455.

Dennis, R. B., Pidgeon, C. R., Smith, S. D., Wherrett, B. S., and Wood, R. A. (1972). Stimulated spin-flip scattering in InSb, *Proc. R. Soc. London, Ser. A* **331,** 203.

Depatie, D. and Haueisen, D. (1980). Multiline phase conjugation at 4 μm in germanium, *Opt. Lett.* **5,** 252.

Dimmock, J. O. (1967). Introduction to the theory of exciton states in semiconductors, *in* "Semiconductors and Semimetals" (R. K. Willardson and A. C. Beer, eds.), Vol. 3. Academic Press, New York.

Dingle, R. (1975). Confined carrier quantum states in ultrathin semiconductor heterostructures, in "Festkörperprobleme" ("Advances in Solid State Physics"), (H. J. Queisser, ed.). Vol. 15, p. 21. Pergamon/Vieweg, Braunschweig.

Dite, A. F., Timofeev, V. B., Fain, V. M., and Yashchin, E. G. (1970). Absorption saturation in the exciton-phonon spectrum of CdSe, *Sov. Phys.—JETP* **31,** 245.

Dornhaus, R. and Nimtz, G. (1976). "The properties and applications of the HgCdTe system," Springer-Verlag Series on Solid State Physics. Springer-Verlag, Berlin and New York.

Dresselhaus, G. and Dresselhaus, M. S. (1966). Magneto-optical effects in solids, *in* "Optical Properties of Solids" (J. Tauc, ed.). Academic Press, New York.

Ehrenreich, H. (1966). Electro-magnetic transport in solids: optical properties and plasma effects, *in* "The Optical Properties of Solids" (J. Tauc, ed.), p. 106. Academic Press, New York.

Eichler, H. J. (1977). Laser-induced grating phenomena, *Opt. Acta* **24,** 631.

Eichler, H. J. (1978). Forced light scattering at laser-induced gratings—a method for investigation of optically excited solids, *Festkoerperprobleme* **18,** 241.

Eichler, H. J., Salje, G., and Stahl, H. (1973). Thermal diffusion measurements using spatially periodic temperature distributions induced by laser light, *J. Appl. Phys.* **44,** 5383.

Eichler, H. J., Hartig, C., and Knof, J. (1978). Laser induced gratings in CdS, *Phys. Status Solidi A* **45,** 433.

Elci, A. and Rogovin, D. (1981). Four-wave mixing and phase conjugation near the band edge, *Phys. Rev. B* **24,** 5796.

Ermanis, J., and Wolfstirn, K. (1966). Hall effect and resistivity of Zn-doped GaAs, *J. Appl. Phys.* **37,** 1963.

Fan, H. Y. (1967). Effects of free carriers on optical properties, *in* "Semiconductors and Semimetals" (R. K. Willardson and A. C. Beer, eds.), Vol. 3. Academic Press, New York.

Feinberg, J. and Hellwarth, R. W. (1980). Phase-conjugate mirror with continuous-wave gain, *Opt. Lett.* **5,** 519.

Fischer, T., Cronin-Golomb, M., White, J. D., Yariv, A., and Neurgaonkar, R. (1970). Amplifying continuous wave phase conjugate mirror with strontium barium niobate, *Appl. Phys. Lett.* **40,** 863.

Flytzanis, C. (1970). Third-order optical susceptibilities in II-VI and III-V semiconductors, *Phys. Lett.* **31A,** 273.

Flytzanis, C. (1975). Theory of nonlinear optical susceptibilities, *in* "Quantum Electronics. Vol. 1: Nonlinear Optics," Part A, (H. Rabin and C. L. Tang, eds.). Academic Press, New York.

Franz, W. (1958). Einfluss eines elektrischen feldes auf eine optische absorptionskante, *Z. Naturforsch.*, **13A,** 484.

Frenkel, J. (1931). On the transformation of light into heat in solids, *Phys. Rev.* **37,** 17.

Geballe, T. H. (1960). Group IV semiconductors, *in* "Semiconductors" (N. B. Hannay, ed.). Reinhold, New York.

Gibbs, H. M. and Hu, P. (1978). Picosecond self-induced transparency and photon echoes in sodium vapor, *in* "Proc. First Int. Conf. on Picosecond Phenomena" (C. V. Shank, E P. Ippen, and S. L. Shapiro, eds.). Springer-Verlag, Berlin and New York.

Gibbs, H. M., McCall, S. L., Venkatesan, T. N. C., Gossard, A. C., Passner, A., and Wiegmann, W. (1979). Optical bistability in semiconductors, *Appl. Phys. Lett.* **35,** 451.

Gibbs, H. M., McCall, S. L., Venkatesan, T. N. C., Passner, A., Gossard, A. C., and Wiegmann, W. (1980). Saturation of the free exciton resonance in GaAs, *Solid State Commun.* **30,** 271.

Gibbs, H. M., McCall, S. L., Venkatesan, T. N. C., Passner, A., Gossard, A. C., and Wiegmann, W. (1981). Optical bistability and optical nonlinearities in GaAs, *in* "Proc. Int. Conf. on Excited States and Multiresonant Nonlinear Optical Processes in Solids, Aussois, France." Editions Physique, Orsay, France.

Gibbs, H. M., Tarng, S. S. Jewell, J. L., Weinburger, D. A., Tai, K., Gossard, A. C., McCall, S. L., Passner, A., and Weigmann, W. (1982). Room temperature excitonic optical bistability in GaAs etalons, *Conf. Lasers Electro-Opt., Phoenix, Arizona,* Paper FL6-1.

Gibson, A. F., Rosito, C. A., Raffo, C. A., and Kimmitt, M. F. (1972). Absorption saturation in germanium, silicon, and gallium arsenide at 10.6 μm, *Appl. Phys. Lett.* **21,** 356.

Gibson, A. J. (1963). Transport of excess carriers in semiconductors, *in* "Semiconductors" (R. A. Smith, ed.). Academic Press, New York.

Goll, J. and Haken, H. (1980). Theory of optical bistability of excitons, *Phys. Status Solidi B* **101,** 489.

Grimmeiss, H. G., Dvren, C., Ludwig, W., and Mack, R. (1977). Identification of deep centers in ZnSe, *J. Appl. Phys.* **48,** 5122.

Gross, E. F., Suslina, L. G., and Kon'kov, P. A. (1962). Exciton spectrum of hexagonal ZnSe single crystals, *Sov. Phys. Solid State* **4,** 287.

Grun, J. B., Levy, R. Ostertag, E. Phach, H. V. D., and Port, H. (1976). Biexciton luminescence in CuCl and CuBr, *in* "Physics of Highly Excited States in Solids" (M. Ueta and Y. Nishina, eds.). Springer-Verlag, Berlin and New York.

Grun, J. B., Sieskind, M., and Nikitine, S. (1961). Détermination de l'Intensité d'Oscillateur des Raies de la Série Verte de Cu_2O aux Basses Températures, *J. Phys. Radium* **22,** 176.

Gunter, P. and Krumins, A. (1980). High sensitivity read–write volume holographic storage in reduced $KNbO_3$ crystals, *Appl. Phys.* **23,** 199.

Gutsche, E. and Voigt, J. (1967). Electron-phonon interaction in CdS, in *II-VI Semiconducting compounds,* D. G. Thomas, ed., p. 337. Benjamin, New York.

Halperin, B. I. and Lax, M. (1966). Impurity band tails in the high-density limit. I. Minimum counting methods, *Phys. Rev.* **148,** 722.

Hanamura, E. and Haug, H. (1977). Condensation effects of excitons, *Phys. Rep.* **33C,** 209.

Harbeke, G. (1972). Optical properties of semiconductors, *in* "Optical Properties of Solids" (J. Abeles, ed.). North-Holland Publ., Amsterdam.

Harrison, W. A. (1970). "Solid State Theory." McGraw-Hill, New York.

Haug, H., Koch, S. W., Marz, R., and Schmitt-Rink, S. (1981). Optical nonlinearity and instability in semiconductors due to biexciton formation, *in* "Proc. Int. Conf. on Excited States and Multiresonant Nonlinear Optical Processes in Solids, Aussois, France." Editions Physique, Orsay, France.

Haynes, J. R. (1960). Experimental proof of the existence of a new electronic complex in silicon, *Phys. Rev. Lett.* **4,** 361.

Hegarty, J., Sturge, M. D., Gossard, A. C., and Wiegmann, W. (1982). Resonant degenerate four-wave mixing in GaAs multiquantum well structures, *Appl. Phys. Lett.* **40,** 132.

Hellwarth, R. W. (1977). Third-order optical susceptibilities of liquids and solids, *in* "Progress in Quantum Electronics" (J. H. Sanders and S. Stenholm, eds.), Vol. 5, Part I. Pergamon, Oxford.

Henry, C. H. and Nassau, K. (1970). Lifetimes of bound excitons in CdS, *Phys. Rev.* **B1,** 1628.

Herman, F. (1954). Speculations on the energy band structures of Ge-Si alloys, *Phys. Rev.* **95,** 847.

Hilsum, C. (1966). Some key features of III-V compounds, *in* "Semiconductors and Semimetals" (R. K. Willardson and A. C. Beer, eds.), Vol. 1. Academic Press, New York.

Hilsum, C. and Rose-Innes, A. L. (1961). "Semiconducting III-V compounds" Pergamon, Oxford.

Hoffman, C. A., Jarasiunas, K., Gerritsen, H. J., and Nurmikko, A. V. (1978). Measurement of surface recombination velocity in semiconductors by diffraction from picosecond transient free-carrier gratings, *Appl. Phys. Lett.* **33,** 536.

Hopf, F. A., Tomita, A., and Liepmann, T. (1981). Quality of phase conjugation in silicon, *Opt. Commun.* **37,** 72.

Hopfield, J. J. and Thomas, D. G. (1961). Fine structure and magneto-optic effects in the exciton spectrum of cadmium sulfide, *Phys. Rev.* **122,** 35.

Jain, R. K. and Klein, M. B. (1979). Degenerate four-wave mixing near the band gap of semiconductors, *Appl. Phys. Lett.* **35,** 454.

Jain, R. K. and Lind, R. C. (1983). Degenerate four-wave mixing in semiconductor-doped glasses, *J. Opt. Soc. Am.,* to be published, Special Issue on Phase Conjugation.

Jain, R. K. and Steel, D. G. (1980). Degenerate four-wave mixing of 10.6 μm radiation in $Hg_{1-x}Cd_xTe$, *Appl. Phys. Lett.* **37,** 1.

Jain, R. K. and Steel, D. G. (1981). Large Optical nonlinearities and cw degenerate four-wave mixing in HgCdTe, *Conf. Laser Eng. Electro-Opt., Washington, D. C.* Paper FL3.

Jain, R. K. and Steel, D. G. (1982). Large optical nonlinearities and cw degenerate four-wave mixing in HgCdTe, *Opt. Commun.* **43,** 72.

Jain, R. K., Klein, M. B., and Lind, R. C. (1979a). High-efficiency degenerate four-wave mixing in silicon and other semiconductors, *Conf. Laser Eng. Appl., Washington, D. C.* Pap. 11.8.

Jain, R. K., Klein, M. B., and Lind, R. C. (1979b). High-efficiency degenerate four-wave mixing of 1.06 μm radiation in silicon, *Opt. Lett.* **4,** 328.

Jain, R. K., Giuliano, C. R., Klein, M. B., Lind, R. C., and Steel, D. G. (1981). Nonlinear optical mechanisms for degenerate four-wave mixing in semiconductors, *in* "Proc. Int. Conf. on Excited States and Multiresonant Nonlinear Optical Processes in Solids, Aussois, France." Editions Physique, Orsay, France.

James, R. B. (1982). Personal communication.

James, R. B. and Smith, D. L. (1979). Theory of nonlinear infrared absorption in p-type germanium, *Phys. Rev. Lett.* **42,** 1495.

James, R. B. and Smith, D. L. (1980a). Saturation of intervalence-band transitions in p-type semiconductors, *Phys. Rev. B* **21,** 3502.

James, R. B. and Smith, D. L. (1980b). Saturation characteristics of p-type semiconductors over the CO_2 laser spectrum, *J. Appl. Phys.* **51,** 2836.

James, R. B. and Smith, D. L. (1980c). Dependence of the saturation intensity of p-type Ge on impurity concentration and residual absorption at 10.59 μm, *Solid State Commun.* **33,** 395.

James, R. B. and Smith, D. L. (1981a). Laser-induced changes in the dispersive properties of p-Ge due to intervalence-band absorption, *Phys. Rev. B* **23,** 4044.

James, R. B. and Smith, D. L. (1981b). Theoretical description of intervalence band photoconductivity of p-Ge at 10.6 μm, *Phys. Rev. B* **23,** 4049.

James, R. B., Smith, D. L., and McGill, T. C. (1981). Nonlinear optical properties of semiconductors induced by high intensity infrared light, *in* "Proc. Int. Conf. on Excited States and Multiresonant Nonlinear Optical Properties in Solids, Aussois, France." Editions Physique, Orsay, France.

Jamison, S. A. and Nurmikko, A. V. (1978). High-intensity infrared transmission limit in $Hg_{1-x}Cd_xTe$, *Appl. Phys. Lett.* **33,** 182.

Jamroz, W. (1980). The third order electric susceptibilities of NaCl and KCl monocrystals, *Opt. Quantum Electron.* **12,** 443.

Jarasiunas, K. and Gerritsen, H. J. (1978). Ambipolar diffusion measurements in semiconductors using nonlinear transient gratings, *Appl. Phys. Lett.* **33,** 190.

Jarasiunas, K. and Vaitkus, J. (1974). Properties of a laser induced phase grating in CdSe, *Phys. Status Solidi A* **23,** K19.

Jarasiunas, K. and Vaitkus, J. (1977). Investigation of non-equilibrium processes in semiconductors by the method of transient holograms, *Phys. Status Solidi A* **44,** 793.

Jha, S. S. and Bloembergen, N. (1968a). Nonlinear optical susceptibilities in group IV and III-V semiconductors, *Phys. Rev.* **171,** 891.

Jha, S. S. and Bloembergen, N. (1968b). Nonlinear optical coefficients in group IV and III-V semiconductors, *IEEE J. Quantum Electron.* **QE-4,** 670.

Johnson, E. J. (1967). Absorption near the fundamental edge, *in* "Semiconductors and Semimetals" (R. K. Willardson and A. C. Beer, eds.), Vol. 3. Academic Press, New York.

Johnson, E. J. and Fan, H. Y. (1965). Impurity and exciton effects on the infrared absorption edges of III-V compounds, *Phys. Rev.* **139A,** 1991.

Johnson, L. F. and Levinstein, H. (1960). Infrared properties of gold in germanium, *Phys. Rev.* **117,** 1191.

Kaiser, W., Collins, R. J., and Fan, H. Y. (1953). Infrared absorption in p-type germanium, *Phys. Rev.* **91,** 1380.

Kane, E. O. (1957). Band structure of indium antimonide, *J. Phys. Chem. Solids* **1,** 249.

Kane, E. O. (1966). The **k · p** Method, *in* "Semiconductors and Semimetals" (R. K. Willardson and A. C. Beer, eds.), Vol. 1. Academic Press, New York.

Kaw, P. (1968). Optical mixing by mobile carriers in semiconductors, *Phys. Rev. Lett.* **21,** 539.

Keilmann, F. (1976). Infrared saturation spectroscopy in p-type germanium, *IEEE J. Quantum Electron.* **12,** 592.

Keilmann, F. and Kuhl, J. (1978). Broadband modulation of 3 μm light by 10 μm light in p-germanium, *IEEE J. Quantum Electron.* **QE-14,** 203.

Keldysh, L. V. (1958). The effect of a strong electric field on the optical properties of insulating crystals, *Sov. Phys.—JETP* **7,** 788.

Kennedy, C. J., Matter, J. C., Smirl, A. L., Weiche, H., Hopf, F. A., and Pappu, S. V. (1974).. Nonlinear absorption and ultrashort carrier relaxation times in germanium under irradiation by picosecond pulses, *Phys. Rev. Lett.* **32,** 419.

Khan, M. A., Kruse, P. W., and Ready, J. F. (1980). Optical phase conjugation in $Hg_{1-x}Cd_xTe$, *Opt. Lett.* **5,** 261.

Khan, M. A., Bennet, R. L. H., and Kruse, P. W. (1981). Bandgap-resonant optical phase conjugation in n-type $Hg_{1-x}Cd_xTe$ at 10 μm, *Opt. Lett.* **6,** 560.

Kittel, C. (1971). "Introduction to Solid State Physics," 4th ed. Wiley, New York.

Klingshirn, C. (1981). Excitation dependent variations of the dielectric function of direct gap semiconductors. Experimental aspects, *in* "Proc. Int. Conf. on Excited States and Multiresonant Nonlinear Optical Processes in Solids, Aussois, France." Editions Physique, Orsay, France.

Knox, R. S. (1963). "Theory of excitons," Solid State Physics, Suppl. 5. Academic Press, New York.

Koch, S. W. and Haug, H. (1981). Two-photon generation of excitonic molecules and optical bistability, *Phys. Rev. Lett.* **46,** 450.

Koch, S. W., Schmitt-Rink, S., and Haug, H. (1981). Calculation of the intensity-dependent changes of the index of refraction in GaAs, *Solid State Commun.* **38,** 1023.

Kohn, W. (1957). Shallow impurity states in silicon and germanium, in "Solid State Physics" (F. Seitz and D. Turbull, eds.), Vol. 5, p. 257. Academic Press, New York.

Kremenitskii, V., Odoulov, S., Soskin, M. (1979). Dynamic gratings in cadmium telluride, *Phys. Status Solidi A* **51,** K63.

Kremenitskii, V., Odoulov, S., and Soskin, M. (1980). Backward degenerate four-wave mixing in cadmium telluride, *Phys. Status Solidi A* **57,** K71.

Krishnamurity, B. S. and Paranjape, V. V. (1969).. Note on optical mixing by mobile carriers in semiconductors, *Phys. Rev.* **181,** 1153.

Kruse, P. W. (1981). Personal communication with R. K. Jain at Int. Conf. on Excited States and Multiresonant Nonlinear Optical Processes in Solids, Aussois, France. March 1981.

Kruse, P. W., Ready, J. F., and Khan, M. A. (1979). Nonlinear optical effects in $Hg_{1-x}Cd_xTe$, *Infrared Phys.* **19,** 497.

Kukhtarev, N. V., Markov, V. B., Odoulov, S. G., Soskin, M. S., and Vinetskii, V. L. (1979). Holographic storage in electrooptic crystals. I. Steady State, *Ferroelectrics* **22,** 949.

Kukhtarev, N. V. and Kovalenko, G. V. (1980). Wavefront reversal via interband absorption in semiconductors, *Sov. J. Quantum Electron.* **10,** 446.

Kuramoto, Y. and Kamimura, H. (1974). Theory of two-dimensional electron-hole liquids—application to layer-type semiconductors, *J. Phys. Soc. Japan* **37,** 716.

Lax, B. (1963). Cyclotron resonance and magneto-optic effects in semiconductors, *in* "Semiconductors" (R. A. Smith, ed.). Academic Press, New York.

Leheny, R. F. and Shah, J. (1975). Optical pumping in nitrogen-doped GaP, *Phys. Rev. B* **12,** 3268.

Leheny, R. F., Shah, J., and Chiang, G. C. (1978). Exciton-contribution to the reflection spectrum at high excitation intensities in CdS, *Solid State Commun.* **25,** 621.

Long, D. (1968). "Energy Bands in Semiconductors." Wiley (Interscience), New York.

Long, D. and Schmit, J. L. (1970). Mercury–Cadmium telluride and closely related alloys, *in* "Semiconductors and Semimetals" (R. K. Willardson and A. C. Beer, eds.), Vol. 5. Academic Press, New York.

McCall, S. L. (1978). Instability and regenerative pulsation phenomena in Fabry–Perot nonlinear optic media devices, *Appl. Phys. Lett.* **32,** 284.

McKelvey, J. P. (1966). "Solid State and Semiconductor Physics." Harper, New York.

MacKenzie, H. A., Dennis, R. B., Smith, S. D., Voge, D., and Wang, W. (1981). Intensity dependent faraday rotation and saturable interband magneto-absorption in InSb, *in* "Proc. Int. Conf. on Excited States and Multiresonant Nonlinear Optical Processes in Solids, Aussois, France." Editions Physique, Orsay, France.

McLean, T. P. (1960). The absorption edge spectrum of semiconductors, *Prog. Semicond.* **5,** 87.

McLean, T. P. (1963). The effective-mass approximation *in* "Semiconductors" (R. A. Smith, ed.). Academic Press, New York.

Maker, P. O. and Terhune, R. W. (1965). Study of optical effects due to an induced polarization third order in the electric field strength, *Phys. Rev.* **137,** A801.

Marburger, J. H. (1975). Self focusing: theory, *Prog. Quantum Electron.* **4,** p. 35.

Maruani, A. (1980). Propagation analysis of forward degenerate four-wave mixing, *IEEE J. Quantum Electron.* **QE-16,** 558.

Maruani, A. and Chemla, D. S. (1981). Active nonlinear spectroscopy of biexcitons in semiconductors: propagation effects and Fano interferences, *Phys. Rev. B* **23,** 841.

Maruani, A., Oudar, J. L., Batifol, E., and Chemla, D. S. (1978). Nonlinear spectroscopy of biexcitons in CuCl by resonant coherent scattering, *Phys. Rev. Lett.* **41,** 1372.

Maruani, A., Chemla, D. S., and Bonnouvrier, F. (1981). High order scatterings and multiple resonant susceptibilities associated to biexcitons in semiconductors, *in* "Proc. Int. Conf. on Excited States and Multriesonant Nonlinear Optical Processes in Solids, Aussois, France." Editions Physique, Orsay, France.

Miller, A. (1979). Nonlinear effects induced by lasers, *in* "Physics of Semiconductors, 1978" (B. L. H. Wilson, ed.), p. 63. Institute of Physics, London.

Miller, D. A. B., Mozolowski, M. H., Miller, A., and Smith, S. D. (1978). Non-linear optical effects in InSb with a C.W. CO laser, *Opt. Commun.* **27,** 133.

Miller, D. A. B., Smith, S. D., and Johnston, A. (1979). Optical bistability and signal amplification in a semiconductor crystal: applications of new low-power nonlinear effects in InSb, *Appl. Phys. Lett.* **35,** 658.

Miller, D. A. B., Harrison, R. G., Johnston, A. M., Seaton, C. T., and Smith, S. D. (1980a). Degenerate four-wave mixing in InSb at 5K, *Opt. Commun.* **32,** 478.

Miller, D. A. B., Smith, S. D., and Wherrett, B. S. (1980b). The microscopic mechanism of third-order optical nonlinearity in InSb, *Opt. Commun.* **35,** 221.

Miller, D. A. B., Seaton, C. T., Prise, M. E., and Smith, S. D. (1981). Band-gap-resonant nonlinear refraction in III-V sermiconductors, *Phys. Rev. Lett.* **47,** 197.

Miller, R. C. (1964). Optical second harmonic generation in piezoelectric crystals, *Appl. Phys. Lett.* **5,** 17.

Miller, R. C., Dingle, R., Gossard, A. C., Logan, R. A., Nordland, W. A., Jr., and Wiegmann, W. (1976). Laser oscillation with optically pumped very thin GaAs-$Al_xGa_{1-x}As$ multilayer structures and conventional double heterostructures, *J. Appl. Phys.* **47,** 4509.

Mita, T. and Nagasawa, H. (1978). Light mixing induced by the giant two-photon absorption of excitonic molecules in CuCl, *Opt. Commun.* **24,** 345.

Morgan, T. N. (1968). Symmetry of electron states in GaP, *Phys. Rev. Lett.* **21,** 819.

Moss, S. C., Lindle, J. R., Mackey, H. J., and Smirl, A. L. (1981). Measurement of the diffusion coefficient and recombination effects in Ge by picosecond transient gratings, *Appl. Phys. Lett.* **39,** 227.

Moss, T. S. (1980). Theory of intensity dependence of refractive index, *Phys. Status Solidi B* **101,** 555.

Moss, T. S., Burrell, G. J., and Ellis, B. (1973). "Semiconductor Opto-Electronics." Wiley, New York.

Mysyrowicz, A. (1976). Biexcitons in CuCl, *in* "Physics of Highly Excited States in Solids" (M. Ueta and Y. Nishima, eds.). Springer-Verlag, Berlin and New York.

Narducci, L. M., Mitra, S. S., Shates, R. A., Pfeiffer, P. A., and Vaidyanthan, A. (1976). One-photon Keldysh absorption in direct-gap semiconductors, *Phys. Rev. B* **14,** 2508.

Nee, T. W., Cantrell, C. D., Scott, J. F., and Scully, M. O. (1978). Nonlinear optical properties of InSb: Hot-electron effects, *Phys. Rev. B* **17,** 3936.

Neuberger, M. (1971). "Handbook of Electronic Materials," Vol. 2. Plenum, New York.

Nikitine, S., Grun, J. B., and Sieskind, M. (1961). Étude spectophotometrique de la série jaune de Cu_2O aux basses températures, *J. Phys. Chem. Solids* **17,** 292.

Nurmikko, A. V. and Jamison, S. A. (1979). High intensity infrared transmission limit in narrow band gap semiconductors, *in* "Physics of Semiconductors, 1978" (B. L. H. Wilson, ed.). Institute of Physics London.

Nurmikko, A. V. (1982). Personal communication.

Olsson, A. and Tang, C. L. (1981). Injected-carrier induced refractive index changes in semiconductor lasers, *Appl. Phys. Lett.* **39,** 24.

Onton, A. (1970). AlAs and AlP: Raman scattering and electric field modulated reflectance, in *Proc. Int. Conf. Phys. Semicond., 10th, Cambridge, Mass.* p. 107.

Overhof, H. (1971). A model calculation for the energy bands in the $Hg_{1-x}Cd_xTe$ mixed crystal system, *Phys. Status Solidi B* **45,** 315.

Pankove, J. I. (1971). "Optical Processes in Semiconductors." Dover, New York.

Park, Y. S., Litton, C. W., Collins, T. C., and Reynolds, D. C. (1966). Exciton spectrum of ZnO. *Phys. Rev.* **143,** 512.

Patel, C. K. N., Slusher, R. E., and Fleury, P. A. (1966). Optical nonlinearities due to mobile carriers in semiconductors, *Phys. Rev. Lett.* **17,** 1011.

Paul, W., ed., (1980). "Handbook on Semiconductors," Vol. 1. (North-Holland Publ., Amsterdam.

Pearson, G. L. and Bardeen, J. (1949). Electrical properties of pure silicon and silicon alloys containing boron and phosphorous, *Phys. Rev.* **75,** 865.

Pershan, P. S. (1966). Nonlinear optics of III-V compounds, *in* "Semiconductors and Semimetals" Vol. 2, (R. K. Willardson and A. C. Beer, eds.), Vol. 2. Academic Press, New York.

Phillion, D. W., Kuizenga, D. J., and Siegman, A. E. (1975). Subnanosecond relaxation time measurements using a transient induced grating method, *Appl. Phys. Lett.* **27,** 85.

Phillips, J. C. (1966). Excitons, *in* "Optical Properties of Solids" (J. Tauc, ed.). Academic Press, New York.

Phipps, C. R. Jr. and Thomas, S. J. (1977). Saturation behaviour of *p*-type germanium at CO_2 laser wavelengths, *Opt. Lett.* **1,** 93.

Picus, G., Burstein, E., Henvis, B. W., and Hass, M. (1959). Infrared lattice vibration studies of polar character in compound semiconductors, *J. Phys. Chem. Solids* **8,** 282.

Pidgeon, C. R., Groves, S. H., and Feinleib, J. (1967). Electroreflectance study of interband magneto-optical transitions in InAs and InSb at 1.5° K, *Solid State Commun.* **5,** 677.

Pincherle, L. (1963). Band structure of semiconductors, *in* "Semiconductors" (R. A. Smith, ed.). Academic Press, New York.

Ralph, H. I. (1965). The electronic absorption edge in layer type crystals, *Solid State Comm.* **3,** 303.

Reynolds, D. C. (1969). "Optical Properties of Solids." Plenum, New York.

Reynolds, D. C. (1969). Excitons in II-VI compounds, in "Optical Properties of Solids" (S. Nudelman and S. S. Mitra, eds.). Plenum, New York.

Schmitt-Rink, S., Tran Thoal, D. B., and Haug, H. (1980). Calculation of the dielectric function for highly excited polar semiconductors, *Z. Phys. B* **39,** 25.

Scott, J. F. (1975). Spin-flip light scattering and spin-flip lasers, *in* "Laser Applications to Optics and Spectroscopy" (S. F. Jacobs, M. Sargent, J. F. Scott, and M. O. Sbully, eds.), Vol. 2. Addison-Wesley, Reading, Massachusetts.

Segall, B. and Marple, D. T. F. (1967). Intrinsic exciton absorption, *in* "Physics and Chemistry of II-VI Compounds" (M. Aven and J. S. Prener, eds.). Wiley, New York.

Seiler, D. G., Goodwin, M. W., and Weiler, M. H. (1981). High resolution two-photon magnetospectroscopy in InSb at milliwatt CO_2 cw powers, *in* "Proc. Int. Conf. on Excited States and Multiresonant Nonlinear Optical Processes in Solids, Aussois, France." Editions Physique, Orsay, France.

Seitz, F. (1940). "Modern Theory of Solids." McGraw-Hill, New York.

Shah, J., Leheny, R. F., and Wiegmann, W. (1977). Low temperature absorption spectrum in GaAs in the presence of optical pumping, *Phys. Rev. B* **16,** 1577.

Shank, C. V. and Auston, D. H. (1975). Parametric coupling in an optically excited plasma in Ge, *Phys. Rev. Lett.* **34,** 479.

Shen, Y. R. (1975). Self-focusing: experimental, *Prog. Quantum Electron.* **4,** 1.

Smirl, A. L., Boggess, T. F., and Hopf, F. A. (1980). Generation of a forward-traveling phase conjugate wave in germanium, *Opt. Commun.* **34,** 463.

Smirl, A. L., Moss, S. C., and Lindle, J. R. (1982). Picosecond dynamics of high density laser-induced transient plasma gratings in Ge, *Phys. Rev. B* **25,** 2645.

Smith, R. A. (1968). "Semiconductors." Cambridge Univ. Press, London and New York.

Smith, R. W. (1957). Low-field electroluminescence in insulating crystals of cadmium sulfide, *Phys. Rev.* **105,** 900.

Smith, S. D. (1967). Magneto-optics in crystals, *in* "Encyclopedia of Physics" (L. Genzel, ed.), Vol. XXV/2a. Springer-Verlag, Berlin and New York.

Spitzer, W. G. and Fan, H. Y. (1957). Determination of optical constants and carrier effective masses of semiconductors, *Phys. Rev.* **106,** 882.

Spitzer, W. G., Gershenzon, M., Frosh, C. J., and Gibbs, D. F. (1959). Optical absorption in n-type gallium phosphide, *J. Phys. Chem. Solids* **11,** 339.

Steel, D. G., Jain, R. K., Lind, R. C., Lam, J. F., and Dunning, G. J. "Phase Conjugation of CO_2 Laser Wavelengths," Final Rep., Dep. Energy/Los Alamos Sci. Lab., Contract No. 4-L90-7269M-1. January 1981.

Stern, F. (1963). Elementary theory of the optical properties of solids, *in* "Solid State Physics" (F. Seitz and D. Turnbull, eds.), Vol. 15, p. 163. Academic Press, New York.

Stern, F. (1964). Dispersion of the index of refraction near the absorption edge of semiconductors, *Phys. Rev.* **133,** A1653.

Streetman, B. G. (1972). "Solid State Electronic Devices." Prentice-Hall, Englewood Cliffs, New Jersey.

Stringfellow, G. B. and Bube, R. H. (1968). Photoelectronic properties of ZnSe crystals, *Phys. Rev.* **171,** 903.

Sturge, M. D. (1962). Optical absorption of gallium arsenide between 0.6 eV and 2.75 eV, *Phys. Rev.* **127,** 768.

Sze, S. (1969). "Physics of Semiconductor Devices." Wiley (Interscience), New York.

Sze, S. and Irwin, J. C. (1968). Resistivity, mobility and impurity levels in GaAs, Ge and Si at 300° K, *Solid State-Electron.* **11,** 599.

Tauc, J. (1966). Optical properties of semiconductors, *in* "Optical Properties of Solids" (J. Tauc, ed.). Academic Press, New York.

Thomas, D. G. and Hopfield, J. J. (1961). A magneto-stark effect and exciton motion in CdS, *Phys. Rev.* **124,** 657.

Thomas, G. A. and Timofeev, V. B. (1980). A review of $N = 1$ to ∞ particle complexes in semiconductors, *in* "Optical Properties of Solids" (M. Balkanski, ed.). North-Holland Publ., Amsterdam, New York, Oxford.

Thomas, D. G., Hopfield, J. J., and Frosch, C. J. (1965). Isoelectronic traps due to nitrogen in gallium phosphide, *Phys. Rev. Lett.* **15,** 857.

Tsay, Y., Bendow, B., and Mitra, S. S. (1973). Theory of the temperature derivative of the refractive index in transparent crystals, *Phys. Rev. B* **8,** 2688.

Turner, W. J., Reese, W. E., and Pettit, G. D. (1964). Exciton absorption and emission in InP, *Phys. Rev.* **136,** A1467.

Ueta, M. and Nagasawa, N. (1976). Two photon generation of excitonic molecules in CuCl and CuBr, *in* "Physics of Highly Excited States in Solids" (M. Ueta and Y. Nishina, eds.). Springer-Verlag, Berlin and New York.

Urbach, F. (1953). The long-wavelength edge of photographic sensitivity and of the electronic absorption of solids, *Phys. Rev.* **92,** 1324.

Vaitkus, Y., Gaubas, E., and Yarashyunas, K. (1978). Investigation of the mechanism of self-diffraction of light on the surface of germanium, *Sov. Phys.—Solid State* (*Engl. Transl.*) **20,** 1824.

Van Vechten, J. A. and Aspnes, D. E. (1969). Franz–Keldysh contributions to third-order optical susceptibilities, *Phys. Lett. A* **30A,** 346.

Van Vechten, J. A. and Wautelet, M. (1981). Variation of semiconductor band gaps with lattice temperature and with carrier temperature when these are not equal, *Phys. Rev. B* **23,** 5543.

Van Vechten, J. A., Cardona, M., Aspnes, D. E., and Martin, R. M. (1970). Theory of the third-order susceptibility of semiconductors, *in* "Proc. Tenth Int. Conf. on the Physics of Semiconductors" (S. P. Keller, J. C. Hensel, and F. Stern, eds.). U. S. At. Energy Comm., Oak Ridge, Tennessee.

Vavilov, V. S. (1962). Radiation ionization processes in germanium and silicon crystals, *Sov. Phys.—Usp.* **4,** 761.

Vinetskii, V. L., Kukhtarev, N. V., Sal'kova, E. N., and Sukhoverkhova, L. G. (1980). Mechanisms of dynamic conversion of coherent optical beams in CdS, *Sov. J. Quantum Electron.* **10,** 684.

Wang, C. C. (1970). Empirical relation between the linear and the third-order nonlinear optical susceptibilities, *Phys. Rev. B* **2,** 2045.

Wang, C. C. and Ressler, N. W. (1969). Nonlinear optical effects of conduction electrons in semiconductors, *Phys. Rev.* **188,** 1291.

Wang, C. C. and Ressler, N. W. (1970). Observation of optical mixing due to conduction electrons in n-type germanium, *Phys. Rev. B* **2,** 1827.

Wannier, G. H. (1937). The structure of electronic excitation levels in insulating crystals, *Phys. Rev.* **52,** 191.

Watanabe, N. (1966). Infrared absorption in p-type semiconductors with zincblende structure-application to zinc telluride, *J. Phys. Soc. Jpn.* **21,** 713.

Watanabe, N. and Usui, S. (1966). Near-infrared absorption in phosphorus-doped p-type ZnTe, *Jpn. J. Appl. Phys.* **5,** 569.

Watkins, D. E., Figueira, J. F., and Thomas, S. J. (1980a). Observation of resonantly enhanced DFWM in doped alkali halides, *Opt. Lett.* **5,** 169.

Watkins, D. E., Phipps, C. R. Jr., C. R., and Thomas, S. J. (1980b). Determination of the third-order nonlinear optical coefficients of germanium through ellipse rotation, *Opt. Lett.* **5,** 248.

Watkins, D. E., Phipps, Jr., C. R., and Thomas, S. J. (1981). Observation of amplified reflection through degenerate four-wave mixing at CO_2 laser wavelengths in germanium, *Opt. Lett.* **6,** 76.

Wautelet, M. and Van Vechten, J. A. (1981). Carrier diffusion in semiconductors subject to large gradients of excited carrier density, *Phys. Rev. B* **23,** 5551.

Wheeler, R. G. and Dimmock, J. O. (1962). Exciton structure and Zeeman effects in cadmium selenide, *Phys. Rev.* **125,** 1805.

Wheeler, R. G. and Mickloaz, J. C. (1964). Exciton structures and Zeeman effects in ZnS, in *Proc. Seventh Int. Conf. Phys. Semicond.,* Paris, p. 873.

Whelan, J. M. (1960). Properties of some covalent semiconductors, *in* "Semiconductors" (N. B. Hannay, ed.). Reinhold, New York.

Wherrett, B. S. (1983). A comparison of theories of resonant nonlinear refraction in semiconductors, *Proc. Roy. Soc.,* to be published.

Wherrett, B. S. and Higgins, N. A. (1982). Theory of nonlinear refraction near the band edge of a semiconductor, *Proc. R. Soc. London, A* **379,** 67.

White, J. O., Cronin-Golomb, M., Fischer, B., and Yariv, A. (1982). Coherent oscillation by self-induced gratings in photorefractive crystals, *Appl. Phys. Lett.* **40,** 450.

Wiggins, T. A. and Carrieri, A. H. (1979). Reflectivity changes of Ge due to illumination by beams of unequal frequency, *Appl. Opt.* **18,** 1921.

Wiggins, T. A., Bellay, J. A., and Carrierri, A. H. (1978). Refractive index changes in germanium due to intense radiation, *Appl. Opt.* **17,** 526.

Wight, D. R., Blenkinsop, I. D., Harding, W., and Hamilton, B. (1981). Diffusion-limited lifetime in semiconductors, *Phys. Rev. B* **23,** 5495.

Woerdman, J. P. (1970). Formation of a transient free carrier hologram in silicon, *Opt. Commun.* **2,** 212.

Woerdman, J. P. (1971). Some optical and electrical properties of a laser-generated free-carrier plasma in Si, *Philips Res. Rep., Suppl.* No. 7.

Woerdman, J. P. and Bolger, B. (1969). Diffraction of light by a laser induced grating in silicon, *Phys. Lett. A* **30A,** 164.

Wolff, P. A. (1966a). Thomson and Raman scattering by mobile electrons in crystals, *Phys. Rev. Lett.* **16,** 225.

Wolff, P. A. (1966b). Giant density fluctuations in semiconductors?, *J. Chem. Phys. Solids* **27,** 685.

Wolff, P. A. and Pearson, G. A. (1966). Theory of optical mixing by mobile carriers in semiconductors, *Phys. Rev. Lett.* **17,** 1015.

Wynne, J. J. (1969). Optical third-order mixing in GaAs, Ge, Si, and InAs, *Phys. Rev.* **178,** 1295.

Wynne, J. J. and Boyd, G. D. (1968). Study of optical difference mixing in Ge and Si using a CO_2 laser, *Appl. Phys. Lett.* **12,** 191.

Yariv, A. and Pepper, D. M. (1977). Amplified reflection, phase conjugation, and oscillation in degenerate four-wave mixing, *Opt. Lett.* **1,** 16.

Yoffa, E. (1980). Dynamics of dense laser-induced plasmas, *Phys. Rev. B* **21,** 2415.

Yuen, S. Y., Aggarwal, R. L., Lee, N., and Lax, B. (1979). Nonlinear absorption of CO_2 laser radiation by nonequilibrium carriers in germanium, *Opt. Commun.* **28,** 237.

Yuen, S. Y., Aggarwal, R. L., and Lax, B. (1980). Saturation of transmitted intensity of CO_2 laser pulses in germanium, *J. Appl. Phys.* **51,** 1146.

Yuen, S. Y. and Wolff, P. A. (1982). ''Difference-frequency variation of the free-carrier-induced, third order nonlinear susceptibility in n-InSb,'' Appl. Phys. Lett. **40,** 457.

Ziman, J. M. (1964). ''Principles of the Theory of Solids.'' Cambridge Univ. Press, London and New York.

11 Optical Phase Conjugation in Photorefractive Materials

Jack Feinberg†

University of Southern California
Los Angeles, California

I. Introduction

Four-wave mixing was proposed only a few years ago as a means of generating a phase-conjugate replica of an optical wave (Hellwarth, 1977). Since then, there has been an ongoing search for materials in which to perform four-wave mixing, especially for materials that will respond to extremely weak optical beams. This search has explored many different physical mechanisms in a wide variety of materials. The goal of this search is to find a physical mechanism that will enable light, especially weak light, to efficiently alter the index of refraction of some material.

Photorefractive materials are some of the most sensitive materials to date for performing optical phase conjugation. These materials have a light sensitivity comparable with that of optical film, and include $BaTiO_3$,

† This work was supported, in part, by a Joseph H. DeFrees grant of the Research Corporation.

Optical Phase Conjugation

ISBN 0-12-257740-X

$Bi_{12}SiO_{20}$ (BSO), and $KTa_{1-x}Nb_xO_3$ (KTN). The light sensitivity of photorefractive materials cannot be directly compared with that of the other materials mentioned in this book because, as we will show, the photorefractive index change depends on the optical energy and not on the optical intensity. Because comparisons will nevertheless be made, let it be noted that optical phase conjugation has been performed in photorefractive materials with the use of only milliwatt (Huignard *et al.*, 1979; Feinberg *et al.*, 1980; Feinberg and Hellwarth, 1980) and even microwatt beams (White *et al.*, 1982). These recent four-wave mixing experiments using the photorefractive effect in a crystal of $BaTiO_3$ succeeded in producing a continuous-wave phase-conjugate replica of a wave with an intensity 100 times greater than that of the incident beam.

Because the photorefractive effect may be unfamiliar, some care will be taken in this chapter to describe the effect in detail. It is wonderfully complicated, and can be summarized as follows.

1. Light causes charge to migrate and separate in a crystalline material.
2. The separation of charge produces a strong electrostatic field, on the order of 10^5 V/m.
3. The electrostatic field causes a change in the refractive index of the crystal by the linear electro-optic effect (the Pockels effect).

The photorefractive effect has some peculiarities that set it apart from other physical mechanisms responsible for light-induced refractive index changes. For example, for most mechanisms the refractive index change increases with any increase in the intensity of the incident light. In contrast, the steady-state photorefractive index change is independent of the total intensity of the incident beams, and instead depends on their relative intensity. However, the speed of the photorefractive effect increases as the light intensity is increased. An optical beam of almost arbitrarily small intensity can eventually produce a large refractive index change in a photorefractive material by interacting with the material for a long enough time. This integration time is limited by the dark storage time of the material, which ranges from a few seconds to a few years. However, the finite coherence time of the writing beams often sets a maximum useful writing time and therefore a minimum useful intensity. This intensity can still be quite small, on the order of microwatts per square centimeter.

From a theoretical viewpoint, the photorefractive effect is curious because it is not spatially local; the maximum refractive index change does not necessarily occur where the intensity of the incident beams is the greatest.

In this chapter we will discuss these and other features of the photorefractive effect, and describe recent applications of the photorefractive effect to phase-conjugate wave generation. Particular care will be taken to explain the key features of the photorefractive effect in terms of simple physical models. In Section II we present a brief history and qualitative description of the photorefractive effect. In Section III we discuss the size of the electrostatic field $\mathbf{E}(\mathbf{x})$ caused by light-induced charge migration. In Section IV we discuss the peculiar intensity dependence of $\mathbf{E}(\mathbf{x})$, and in Section V the dependence of $\mathbf{E}(\mathbf{x})$ on incident beam angles. In Sections VI and VII we give expressions for the diffraction efficiency and mirror reflectivity of a photorefractive index grating. We digress a bit in Section VIII to discuss the nonlocal nature of the photorefractive effect and its consequences. In Section IX we discuss the speed of photorefractive devices, and in Section X we describe some of the recent applications of the photorefractive effect to optical phase conjugation.

II. The Photorefractive Effect

The photorefractive effect was first noticed because it was a nuisance. During early attempts at second harmonic generation using pulsed lasers, it was noticed that the efficiency of second harmonic generation in $LiNbO_3$ crystals became seriously degraded after only a couple of laser shots. The incident laser pulse was found to cause a local, semi-permanent change in the refractive index of $LiNbO_3$ and KTN crystals, which ruined phase matching for second harmonic generation (Ashkin *et al.*, 1966; Chen, 1967). Consequently, the crystal had to be moved around after every few laser shots to avoid "optical damage." A few years later, Chen and co-workers (Chen *et al.*, 1968) used optical damage to their advantage when they demonstrated that the same light-induced refractive index change could be used to store high-quality holographic images in $LiNbO_3$ and $LiTaO_3$. The possible computer storage application led to a flurry of theoretical and experimental papers on the "photorefractive effect," which have been summarized in review papers (von der Linde and Glass, 1975; Kim *et al.*, 1977; Glass, 1978).

The explanation of the photorefractive effect given here assumes the existence of charges in a crystalline material as first proposed by Chen (1969). The origin of these charges is uncertain, but it will be assumed that the charges inhabit low-lying traps formed by impurity or defect sites in the crystal. [In $LiNbO_3$, for example, trapping sites have been caused by trace amounts of iron doped into the sample (Peterson *et al.*, 1971; Staebler and Phillips, 1974).] In most photorefractive materials the origin of the traps is unknown.

In the presence of light, the trapped charges can migrate between trapping sites. In the absence of light, the charges are "frozen" in place by the small dark conductivity of the crystal. As shown below, the light-induced migration and separation of charge in the crystal gives rise to an electrostatic field in the crystal that eventually prevents further charge separation. However, this electrostatic field also produces a refractive index change in the crystal by the linear electro-optic (Pockels') effect, provided the crystal lacks inversion symmetry. (If the crystal has inversion symmetry, a first-order electro-optic effect is forbidden. However, a refractive index change linear with the light-induced electric field can still be obtained by applying an additional dc electric field across the crystal.) The light-induced electrostatic field can be quite large, on the order of 10^5 V/m, and can cause a large change (10^{-3}) in the refractive index of materials with large linear electro-optic coefficients, such as $BaTiO_3$ or KTN. The theory presented here for the formation of the light-induced electrostatic field and for its effects agrees quite well with the experimental data, and has led to the successful prediction and explanation of new effects.

In the following sections we will present equations for the electrostatic field caused by the light-induced separation of charge in a crystal. We will discuss how this electric field depends on the charge density, the incident light intensity, and the geometry of the optical beams relative to the crystal. We will then compute the refractive index change caused by the electrostatic field, and give expressions for the reflectivity of a photorefractive, phase-conjugating mirror.

III. Light-Induced Electric Field E(x)

There exist detailed descriptions of the photorefractive effect using both hopping and diffusion models (Chen, 1969; Townsend and LaMacchia, 1970; Amodei, 1971; Young *et al.*, 1974; Alphonse *et al.*, 1975; Kim *et al.*, 1976; Peltier and Micheron, 1977; Kukhtarev *et al.*, 1979; Feinberg *et al.*, 1980). We will give the results but not the details of these models here, and will most closely follow the hopping treatment of Feinberg *et al.* (1980). For further details, see the above references.

We now describe the steady-state electric field that builds up owing to light-induced charge migration in a crystal. We only consider the case where the writing beams have the same optical frequency, so that the intensity pattern does not move through the crystal. In particular, let two uniform optical beams with nonorthogonal unit polarizations $\hat{\mathbf{e}}_1$ and $\hat{\mathbf{e}}_2$ and slowly varying electric field amplitudes $\vec{\mathscr{E}}_1$ and $\vec{\mathscr{E}}_2$ intersect at an angle 2θ in the crystal. They will create a spatially periodic intensity pattern $I(\mathbf{x})$ given by

$$I(\mathbf{x}) = I_0(1 + m \cos \mathbf{k} \cdot \mathbf{x}), \tag{1}$$

where $I_j = |\vec{\mathscr{E}}_j|^2$, $I_0 = I_1 + I_2$, $\mathbf{k} = \mathbf{k}_1 - \mathbf{k}_2$, and $\mathbf{k}_1$ and $\mathbf{k}_2$ are the wave vectors of the incident beams, as shown in Fig. 1. The vector $\mathbf{k}$ is the "grating" wave vector, and has a magnitude $k = |\mathbf{k}| = 2|\mathbf{k}_1| \sin\theta$. The dimensionless modulation index m is given by

$$m = 2\vec{\mathscr{E}}_1 \cdot \vec{\mathscr{E}}_2^*/I_0. \tag{2}$$

Now assume that there exists an initially uniform density N of charges per unit volume that can migrate in the presence of light, and an equal density of immobile charges of opposite sign to make the crystal electrically neutral. In the presence of light, these charges migrate from the high-intensity regions towards the low-intensity regions. If the light intensity is spatially periodic with wave vector $\mathbf{k}$, as above, the resulting steady-state distribution of charges will have a Fourier component with periodicity $\mathbf{k}$. These charges will produce a static, electric field $\mathbf{E}(\mathbf{x}) = \mathrm{Re}[E\hat{\mathbf{k}} \exp(i\mathbf{k} \cdot \mathbf{x})]$, as shown in Fig. 2.

The direction of this electrostatic field $\mathbf{E}(\mathbf{x})$ is parallel to the direction of

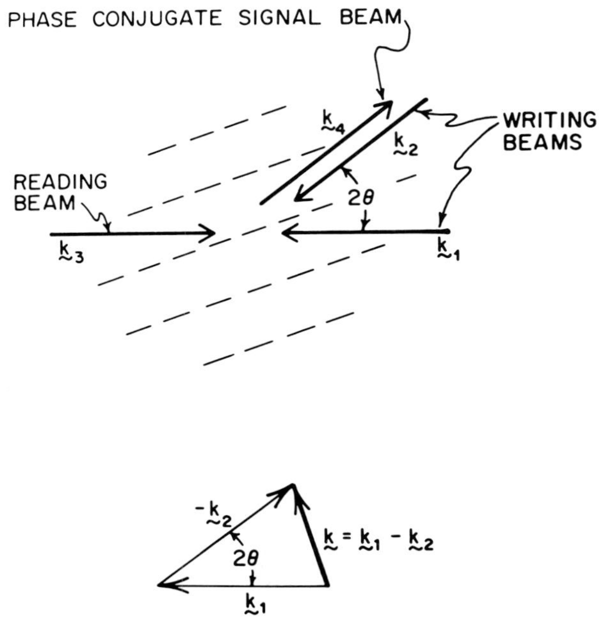

Fig. 1. Geometry for optical phase conjugation. Two beams with wave vectors $\mathbf{k}_1$ and $\mathbf{k}_2$ write a refractive index grating with wave vector $\mathbf{k} = \mathbf{k}_1 - \mathbf{k}_2$. A reading beam with wave vector $\mathbf{k}_3 = -\mathbf{k}_1$ Bragg-scatters off of this grating to form the phase-conjugate signal beam with wave vector $\mathbf{k}_4 = -\mathbf{k}_2$. The two writing beams cross at an angle 2θ in the material.

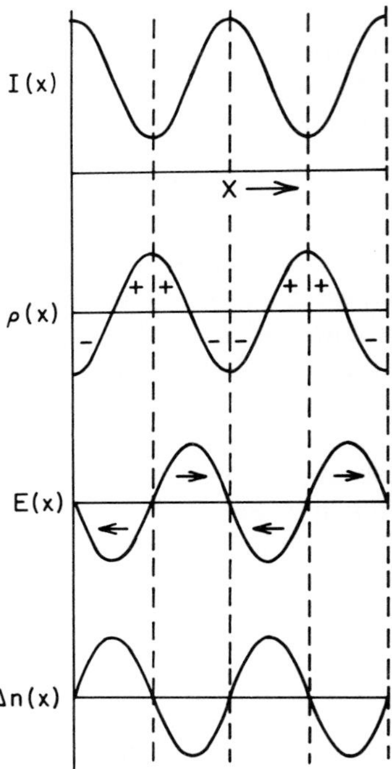

Fig. 2. Formation of a photorefractive index grating. From top to bottom: Light with spatially periodic intensity $I(\mathbf{x})$ rearranges the charge density $\rho(\mathbf{x})$ in the material. The mobile charges, here with positive charge, tend to accumulate in the dark regions of the intensity pattern. The resulting periodic charge distribution $\rho(\mathbf{x})$ causes a periodic electrostatic field $\mathbf{E}(\mathbf{x})$ by Poisson's equation. This electric field then causes a change in the refractive index $\Delta n(\mathbf{x})$ of the crystal by the linear electro-optic (Pockels') effect. The photorefractive effect is nonlocal; the maximum refractive index change does not occur at the intensity peak. In this figure the spatial shift between $\Delta n(\mathbf{x})$ and $I(\mathbf{x})$ is $\frac{1}{4}$ of the grating period.

the unit grating wave vector $\hat{\mathbf{k}}$. The complex amplitude E of the steady-state electric field is, for $m \ll 1$,

$$E = i\frac{k_B T}{q} k_0 m \frac{\alpha + i\mathbf{f} \cdot \hat{\mathbf{k}}}{1 + \alpha^2 + i\boldsymbol{\alpha} \cdot \mathbf{f}} \tag{3}$$

where $k_B T$ is the thermal energy of the crystal lattice, q the charge of the mobile charge carriers, and k_0 a constant of the material that depends on the number density N of charge carriers available for charge migration, according to

$$k_0 = (Nq^2/\epsilon\epsilon_0 k_B T)^{1/2}. \tag{4}$$

Also, $\boldsymbol{\alpha} = \mathbf{k}/k_0$, and $\mathbf{f} = \mathbf{E}_0 q/k_B T k_0$, where $\mathbf{E}_0$ is any constant and uniform electric field in the crystal, either applied or intrinsic. Equation (3) does not include the effects of a photoinduced current, such as that proposed by Glass *et al.* (1974) to account for data in $LiNbO_3$. In the absence of an intrinsic or applied field $\mathbf{E}_0$, this result simplifies considerably, to

$$\mathbf{E}(\mathbf{x}) = -\frac{k_B T}{q} m \frac{\mathbf{k}}{1 + (k/k_0)^2} \sin \mathbf{k} \cdot \mathbf{x}. \tag{5}$$

In the above expressions for the light-induced electric field, it has been assumed that the hopping length between trapping sites (or the drift length in the conduction band) is much smaller than the spacing $2\pi/k$ of the intensity pattern, and that there is a plethora of empty sites for the charges to occupy. Comparison of Eqs. (5) and (1) shows that $\mathbf{E}(\mathbf{x})$ and $I(\mathbf{x})$ are shifted in phase, as shown in Fig. 2 and discussed in Section VIII.

IV. Intensity Dependence

Suppose that there are only two coherent optical beams, here called the writing beams, incident on the crystal. For most physical mechanisms, increasing the intensity of these beams increases any refractive index change caused by these beams. In contrast, the steady-state photorefractive index change is independent of the total intensity of the writing beams, and instead depends only on their relative intensity. This can be understood as follows. The steady-state distribution of charges in the crystal is the result of a battle between the periodic part of the intensity pattern, which tries to force the charges into a periodic charge distribution, and the uniform component I_0, which tends to randomize the positions of the charges and erase any periodic charge distribution. The ratio of the periodic intensity component to the uniform intensity component is $|m|$, given by Eq. (2).

The modulation index m is largest ($|m| = 1$) when the two writing beams are equal in intensity. Because E is proportional to m, the resulting light-induced electric field is also largest when the writing beam intensities are equal. Note that m is unchanged if the intensities of the two writing beams are both increased by the same multiplicative factor, so that the strength of the light-induced electric field depends only on the ratio of I_1 and I_2. If one of the writing beams is more intense than the other ($|m| < 1$), the uniform component of the intensity dominates over the periodic component, and tends to diminish the light-induced electric field. Note that m is a very nonlinear function of the intensities of the incident beams, as shown in Fig. 3. This nonlinearity can result in dramatic and useful distortions of a holographic image, as shown later.

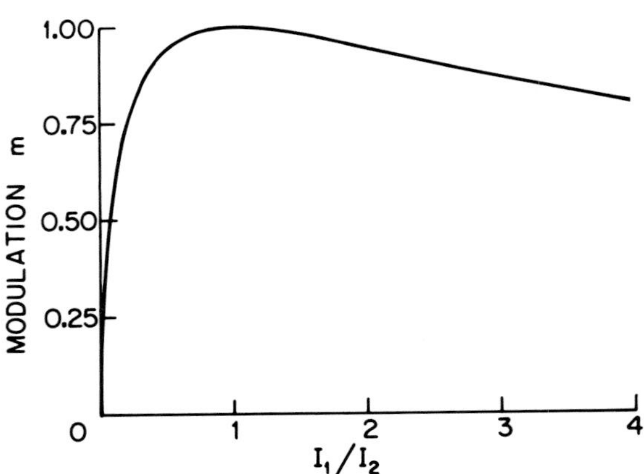

Fig. 3. Variation of the intensity modulation ratio $|m|$ versus the relative intensity I_1/I_2 of the writing beams. The maximum value of $|m|$ occurs when the intensities of the writing beams are equal. See Eq. (2).

V. Grating Wave Vector k Dependence and Effect of an Applied Field

The size and direction of the grating wave vector **k** is determined by the wavelength and the crossing angle 2θ of the two incident writing beams in the crystal. As shown in Fig. 1, the direction of **k** is given by $\mathbf{k} = \mathbf{k}_1 - \mathbf{k}_2$, and the magnitude of **k** is $k = |\mathbf{k}| = 2k_1 \sin \theta$. According to Eq. (5), the strength of the light-induced electric field increases linearly with grating wave vector k if $k \ll k_0$ and $\mathbf{E}_0 = 0$, because the charges diffuse down the optical intensity gradient until they create a restoring force equal to the diffusional force mk_BTk. At equilibrium the amplitude of this restoring force per unit charge is

$$E = -mk_BTk/q, \tag{6}$$

which agrees with Eq. (5) in the limit of small k.

As k increases, however, the electric field cannot become arbitrarily large. From Poisson's equation, a periodic electric field of magnitude E requires a charge density $\rho = Ek\epsilon\epsilon_0$. Because there is a maximum number density N of charges available for charge migration, for large enough k the product Ek becomes constant, and the light-induced electric field amplitude decreases as $1/k$. From Eq. (5) the maximum electric field is produced when $k = k_0$, as shown in Fig. 4.

As an example, consider the case of $BaTiO_3$. With $\mathbf{E}_0 = 0$ and a number

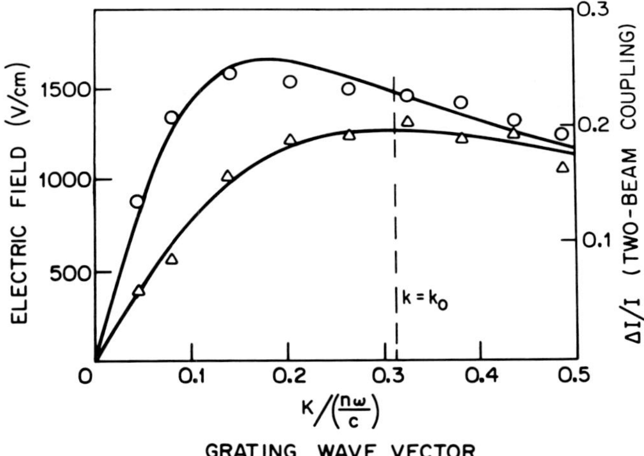

Fig. 4. Variation of the imaginary part of the electrostatic field amplitude E versus grating wave vector k. [The imaginary part of E is the component of $\mathbf{E}(\mathbf{x})$ shifted by 90° with respect to $I(\mathbf{x})$; see Fig. 2.] The actual measured quantity (○) is the fractional change $\Delta I/I$ in the intensity of a writing beam due to two-beam coupling with the other writing beam, shown on the right axis. In the absence of an applied voltage (△), the two-beam coupling efficiency and light-induced electric field amplitude E reach a maximum when $k = k_0$. The solid curves are the best theoretical fits to the data; in this crystal of $BaTiO_3$ they correspond to $k_0 = 0.31 \pm 0.02(n\omega/c)$, where n and ω are the refractive index and optical circular frequency of the writing beams. The density N of mobile charges can then be computed from k_0, and is $N \sim 1.3 \times 10^{16}$ cm^{-3} in this sample. The sign of the charges is positive in this crystal. (From Feinberg *et al.*, 1980.)

density of $N = 10^{16}$ cm^{-3} of singly charged carriers in this crystal at room temperature, the maximum light-induced electric field is $E = k_B T k_0 / 2q =$ 1200 V/cm. Note that, for a fixed density of charge, an increase in the static dielectric constant ϵ (taken along the direction of k) will decrease the maximum value of the light-induced electric field. In $BaTiO_3$ at room temperature, ϵ is highly anisotropic and varies from $\epsilon_{\parallel} = 106$ for an electric field along the c axis of the crystal to $\epsilon_{\perp} = 4900$ for an electric field perpendicular to the c axis. Consequently, the light-induced electric field will vary according to the orientation of the c axis with respect to the grating wave vector k by a factor of $(\epsilon_{\perp}/\epsilon_{\parallel})^{1/2}$ (≈ 7 in $BaTiO_3$). To produce a large refractive index grating with no applied field ($\mathbf{E}_0 = 0$), it is necessary to choose a grating wave vector k comparable with k_0.

If there is a uniform electric field in the crystal during grating formation ($\mathbf{E}_0 \neq 0$), a large refractive index grating can be produced even for small values of the grating wave vector k. The uniform electric field $\mathbf{E}_0$ produces a force $q\mathbf{E}_0$ on the charge carriers in addition to the diffusion force. Because this additional force is independent of k, the resulting light-induced electric field will no longer go to zero for small k as before for $\mathbf{E}_0 = 0$. Instead, the limiting value of the light-induced electric field for small k is now $m\mathbf{E}_0 \cdot \hat{\mathbf{k}}$, which can be made as large as desired simply by applying a large electric field $\mathbf{E}_0$. Inspection of Eq. (3) reveals that E can be made almost independent of k for small k by choosing the right value for the applied electric field $\mathbf{E}_0$. This feature is particularly important for imaging applications, where it may be desirable to have an optical transfer function with a flat response over a wide range of spatial frequencies k.

VI. Grating Reflectivity

As we have shown, when two optical writing beams create a periodic intensity interference pattern in a photorefractive material, a periodic electrostatic field $\mathbf{E}(\mathbf{x}) = \mathrm{Re}[E\hat{\mathbf{k}} \exp(i\mathbf{k} \cdot \mathbf{x})]$ is produced. This electric field makes a phase grating, which can be described by a change in the optical susceptibility of the material. The amplitude $\overleftrightarrow{\chi}$ of the change in the optical susceptibility $\mathrm{Re}[\overleftrightarrow{\chi} \exp(i\mathbf{k} \cdot \mathbf{x})]$ caused by $\mathbf{E}(\mathbf{x})$ is

$$\overleftrightarrow{\chi} = -\overleftrightarrow{\epsilon}_\omega \cdot (\vec{\mathscr{R}} \cdot E\mathbf{k}) \cdot \overleftrightarrow{\epsilon}_\omega . \tag{7}$$

The optical dielectric tensor $\overleftrightarrow{\epsilon}_\omega$ at frequency ω is diagonal for Cartesian coordinates coinciding with the principal axes of the crystal. Here $\vec{\mathscr{R}}$ is the third-rank electro-optic tensor, whose nonzero elements are determined by the symmetry of the crystal.

This periodic change in the optical susceptibility can be thought of as a phase grating. The grating can scatter a "reading" beam (incident at the Bragg angle if the grating is thick). If this beam has an intensity I_3 and a polarization $\hat{\mathbf{e}}_3$, the efficiency $\eta = I_4/I_3$ of scattering into polarization $\hat{\mathbf{e}}_4$ is given by

$$\eta = |(\omega L/4n_3c)\hat{\mathbf{e}}_4^* \cdot \overleftrightarrow{\chi} \cdot \hat{e}_3|^2, \tag{8}$$

where L is the interaction length of the beams. If there is a background nonsaturable optical absorption coefficient γ, the interaction length is $L = \ell \exp(-\gamma\ell/2)$ for a crystal length ℓ. Also, ω is the circular frequency, and n_3 the refractive index for the reading beam. In Eq. (8), it is assumed that $\eta \ll 1$, so that the scattered beam is much weaker than the reading beam. The tensor notation is necessary in Eq. (8) to keep track of the various polarizations of the reading and scattered beams, because, in general, the refractive index grating can be birefringent.

As an example of a birefringent grating, consider the case of $Bi_{12}SiO_{20}$

(BSO). This crystal has cubic symmetry, so $(\epsilon_\omega)_{ij} = n^2\delta_{ij}$. (We will ignore the optical activity of BSO.) The nonzero components of the electro-optic tensor of BSO are $\mathcal{R}_{xyz} = r_{63}$, $\mathcal{R}_{yzx} = r_{41}$, and $\mathcal{R}_{zxy} = r_{52}$ (the electro-optic tensor elements always obey $\mathcal{R}_{ijk} = \mathcal{R}_{jik}$, and we have given the conventional contracted notation of these elements). In BSO these tensor components all have the same value of 5×10^{-12} m/V. If there is a light-induced electrostatic field E along the (001) direction of the crystal, it will produce a susceptibility change given by $\overleftrightarrow{\chi} = -n^4 r_{63} E(\hat{\mathbf{x}}\hat{\mathbf{y}} + \hat{\mathbf{y}}\hat{\mathbf{x}})$. The cross dyadic will cause the grating to be birefringent. From Eq. (8), a reading beam with $\hat{\mathbf{x}}$ polarization will emerge with $\hat{\mathbf{y}}$ polarization. Similarly, a reading beam with $\hat{\mathbf{y}}$ polarization will emerge with $\hat{\mathbf{x}}$ polarization.

The grating reflectivity in the above example will be

$$\eta = |(\omega L/4c)n^3 r_{63} E|^2. \tag{9}$$

In principle, the grating efficiency can be increased by increasing the crossing angle of the two writing beams, until one eventually reaches the optimum angle determined by the finite charge density in the material, as discussed in Section V. In general, the charge density available for migration will depend on the optical frequency of the writing beams. However, the following rather strange fact is noted: for many materials, studied in different laboratories, the diffraction efficiency always seems to peak for the same full crossing angle of about 40° in air. According to the model used here, this implies a similar number density of charges (10^{16} cm^{-3}) for all of these quite different materials, a rather suspect coincidence. This point must be pursued further.

As another example of the richness of Eqs. (7) and (8), consider the case of $BaTiO_3$. At room temperature this material has tetragonal symmetry and is uniaxial. Let the c axis (001) be taken as the z direction. The optical dielectric tensor then has components $\epsilon_{\omega_{xx}} = n_o^2$, $\epsilon_{\omega_{yy}} = n_o^2$, and $\epsilon_{\omega_{zz}} = n_e^2$, where n_o and n_e are the ordinary and extraordinary indices of refraction. The nonzero components of the electro-optic tensor for $BaTiO_3$ are $\mathcal{R}_{xxz} = r_{13} = 8$, $\mathcal{R}_{zzz} = r_{33} = 28$ and $\mathcal{R}_{zyy} = \mathcal{R}_{zxx} = r_{42} = r_{51} = 820$, all in units of 10^{-12} m/V. If the crystal is oriented so that the light-induced electric field is along the (001) direction, as in the previous example, there will be no contribution to $\overleftrightarrow{\chi}$ from the unusually large r_{42} coefficient of $BaTiO_3$.

To make a grating in $BaTiO_3$ with a large scattering efficiency η, it is therefore necessary to align the crystal so that the light-induced electric field makes some nonzero angle β with the c axis. For a reading beam polarized ordinary, the scattering efficiency is

$$\eta_o = |(\omega L/4c)n_o^3 r_3 E(\beta)\cos\beta|^2, \tag{10}$$

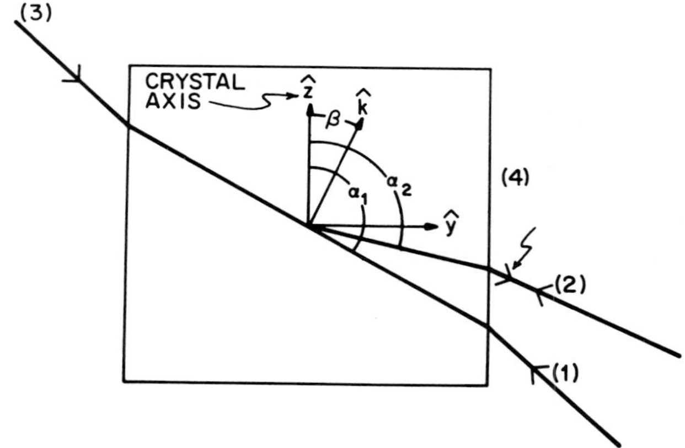

Fig. 5. Geometry for obtaining mirror reflectivity R greater than unity in a crystal with a large Pockels coefficient r_{42}. (1) Reference writing beam; (2) image writing beam; (3) reading beam; and (4) phase conjugate beam. The direction $\hat{\mathbf{k}}$ of the grating wave vector and the direction of the c axis of the crystal make a *nonzero* angle β. (From Feinberg and Hellwarth, 1980.)

and for a reading beam polarized extraordinary,

$$\eta_e = |(\omega L/4cn_3)E(\beta) \cos \beta \, (n_e^4 r_{33} \sin \alpha_1 \sin \alpha_2 + 2n_e^2 n_0^2 r_{42} \sin^2 \beta + n_0^4 r_{13} \cos \alpha_1 \cos \alpha_2)|^2, \tag{11}$$

where the angles α_1, α_2 are shown in Fig. 5. The magnitude $E(\beta)$ of the light-induced electric field depends on the angle β through the change in the static dielectric constant ϵ with crystal orientation.

Comparison of Eqs. (10) and (11) shows that in $BaTiO_3$ only the extraordinary reading beam feels the effects of the electro-optic tensor element r_{42}; it is possible to produce large scattering efficiencies ($\eta \sim 50\%$) by using an extraordinary reading beam and a $BaTiO_3$ crystal only a few millimeters in length. The effect of an applied electric field $\mathbf{E}_0$ on the grating reflectivity is shown in Fig. 6.

VII. Mirror Reflectivity

Up to now, we have discussed only the scattering efficiency $\eta = I_4/I_3$ of a grating (I_4 is the signal intensity and I_3 the reading beam intensity). It is also useful to define the phase-conjugate mirror reflectivity R of the grating,

$$R = I_4/I_2. \tag{12}$$

Unlike the scattering efficiency η, which is always less than unity, the mirror reflectivity R can exceed unity. When this occurs, one has a

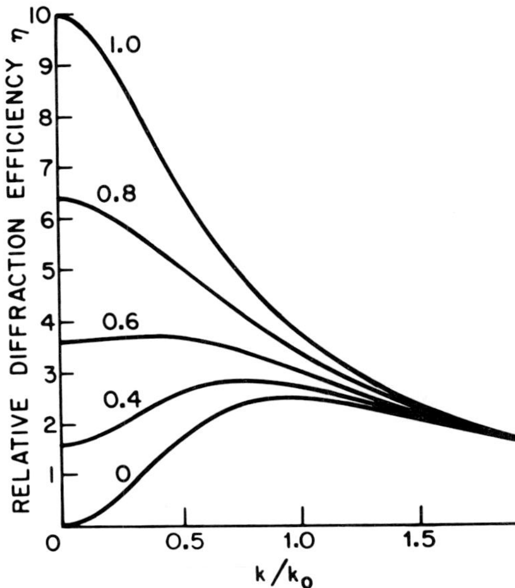

Fig. 6. Theoretical plots of grating diffraction efficiency η versus grating wave vector k, for various uniform applied electric fields E_0. The curves are labeled by their value of the applied electric field normalized by the factor $k_B T k_0/q$. Note that the spatial frequency response can be made relatively flat. (From Feinberg *et al.*, 1980.)

phase-conjugating mirror with gain. Because $R = \eta I_3/I_2$, one can use the results of the previous section, in all their tensorial glory, to derive the following expression for the mirror reflectivity of a phase-conjugating mirror:

$$R = (I_3/I_2)|(\omega LE/4n_3c)\overleftrightarrow{\epsilon}_\omega \cdot (\vec{\mathscr{R}} \cdot \hat{\mathbf{k}}) \cdot \overleftrightarrow{\epsilon}_\omega|^2. \tag{13}$$

Now, because $E \propto m = 2(I_1I_2)^{1/2}/(I_1 + I_2 + I_3)$, the intensity dependence of R is

$$R \propto I_1I_3/(I_1 + I_2 + I_3)^2, \tag{14}$$

which for $I_2 \ll I_1$ or $I_2 \ll I_3$ is independent of I_2. Thus for small I_2 the system acts like a mirror with a constant reflectivity fixed by the relative intensity of the counterpropagating beams. According to Eq. (14), the highest reflectivity occurs when $I_1 = I_3$, that is, when the counterpropagating pumping beams have equal intensities. This does not include the two-beam coupling effects discussed earlier, however, and in practice the

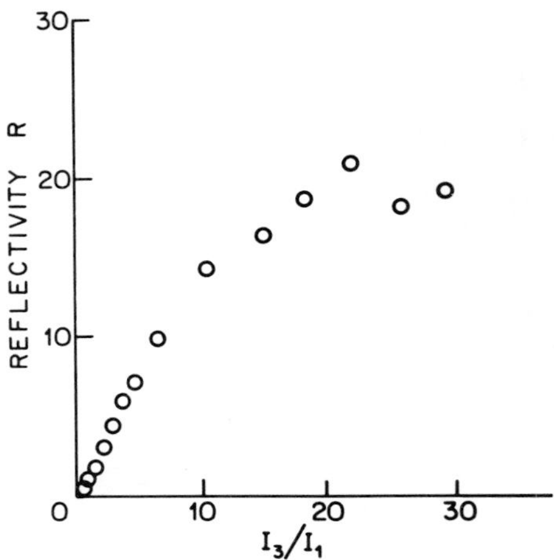

Fig. 7. Variation of the mirror reflectivity R versus reading beam intensity I_3 in $BaTiO_3$. The writing beam intensities are fixed at $I_2 = \frac{1}{4}I_1$. The reading beam intensity is normalized by the fixed intensity I_1 of the reference writing beam. The optimum reflectivity occurs when $I_3 \gg I_1$. (From Feinberg and Hellwarth, 1980.)

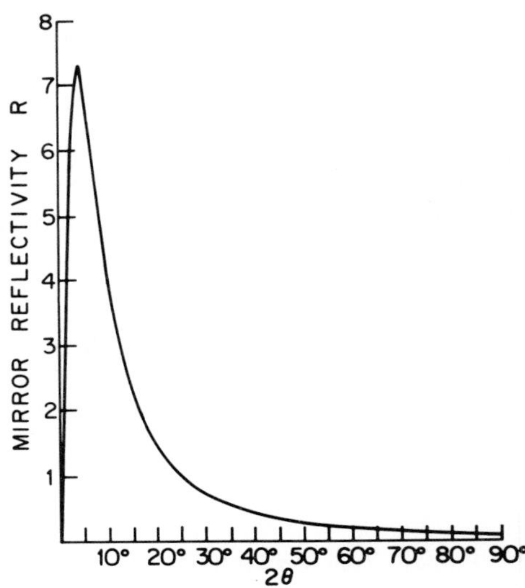

Fig. 8. Theoretical variation of the mirror reflectivity R versus the full crossing angle 2θ of the writing beams for a 4-mm-long crystal of $BaTiO_3$ with a number density of charges of 2×10^{16} cm^{-3}. For each crossing angle 2θ, the crystal was oriented (i.e., the angle β varied) to maximize the reflectivity.

optimum reflectivity will occur for unequal pumping beams (Feinberg and Hellwarth, 1980; Fischer *et al.*, 1981), as shown in Fig. 7. For $BaTiO_3$ with extraordinary polarized pumping beams, Eq. (13) gives a mirror reflectivity of

$$R = (I_3/I_2)|[\omega LE(\beta)/4cn_3] \cos \beta(n_e^4 r_{33} \sin \alpha_1 \sin \alpha_2 + 2r_{42}n_e^2 n_o^2 \sin^2 \beta + n_o^4 r_{13} \cos \alpha_1 \cos \alpha_2)|^2. \quad (15)$$

A plot of R versus various beam angles is shown in Fig. 8. This plot includes the parametric dependence of E on β, but not the effects of two-beam energy coupling. It is clear that the mirror reflectivity can exceed unity over a wide range of beam angles and pumping beam intensities in this crystal.

VIII. Spatial Shift between I(x) and n(x): The Origin of Two-Beam Coupling

The photorefractive effect is not spatially local; i.e., the maximum refractive index change does not necessarily occur where the intensity is largest. This is in marked contrast to most of the physical mechanisms responsible for light-induced refractive index changes discussed in this book. This spatial shift between the peak refractive index change and the region of peak light intensity leads to the following effect: when two optical beams of the same frequency intersect in a photorefractive crystal, one of the beams can experience gain and the other loss as they propagate through the crystal (Staebler and Amodei, 1972). This two-beam coupling can be quite large. In $BaTiO_3$, for example, one beam can deplete essentially all of the power of the other beam after traveling through only a few millimeters of the crystal (Feinberg *et al.*, 1980). The direction of the c axis of the crystal determines which beam will experience gain or loss, as shown in Fig. 9.

The nonlocal nature of the photorefractive effect is a consequence of Poisson's equation $\nabla \cdot \mathbf{E} = \rho/\epsilon\epsilon_0$. A region of large charge density produces not a large electric field but a large electric field gradient. Suppose that an intensity pattern has a spatially periodic component $I(\mathbf{x}) \propto \cos \mathbf{k} \cdot \mathbf{x}$. In the absence of any uniform field ($\mathbf{E}_0 = 0$), the light will cause charges to migrate and diffuse in the crystal until they eventually collect in the low-intensity regions, giving a charge density $\rho(\mathbf{x}) \propto -\cos \mathbf{k} \cdot \mathbf{x}$. From Poisson's equation, this charge density will produce an electrostatic field $\mathbf{E}(\mathbf{x}) \propto -\sin \mathbf{k} \cdot \mathbf{x}$, shifted 90° relative to $I(\mathbf{x})$. From Eq. (7) $\Delta n(\mathbf{x}) \propto -\mathbf{E}(\mathbf{x})$, so the refractive index change is also shifted 90° relative to the intensity pattern, as shown in Fig. 2. This phase shift between $\Delta n(\mathbf{x})$ and

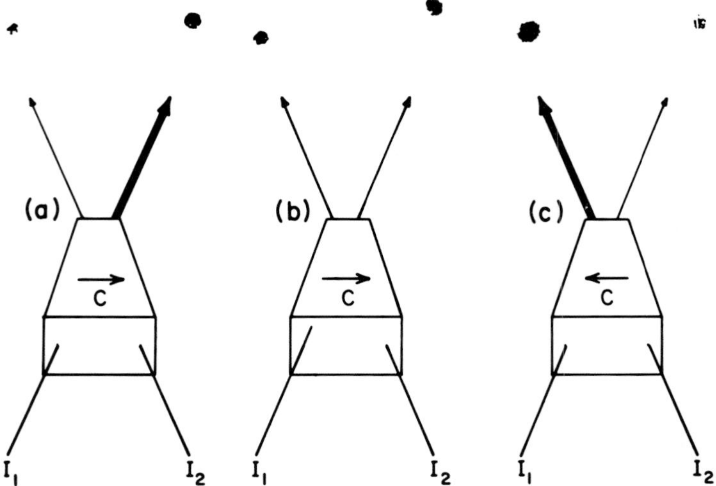

Fig. 9. Two-beam coupling in a photorefractive crystal. The direction of the positive *c* axis is indicated. In (a) two beams with equal intensities emerge with different intensities. In (b) the beams are misaligned so as not to intersect in the crystal, and no coupling occurs. In (c) the crystal is rotated 180° to reverse the direction of the *c* axis. The direction of the coupling also reverses. The spots were produced by exposing a strip of photographic paper to the beams after they had passed through the crystal. No dc field was applied (except prior to the experiment to pole the crystal). (From Feinberg *et al.*, 1980.)

$I(\mathbf{x})$ causes the writing beams to couple as they propagate through the crystal. Let the incident optical beams have slowly varying electric field amplitudes $\vec{\mathscr{E}}_1$ and $\vec{\mathscr{E}}_2$, and propagation vectors $\mathbf{k}_1$ and $\mathbf{k}_2$ at small angles $+\theta$ and $-\theta$ with respect to what is arbitrarily called the x direction. The coupled-wave equations for two beams are

$$\begin{aligned} \frac{\partial}{\partial x}\vec{\mathscr{E}}_1 &= \frac{(\vec{\mathscr{E}}_1 \cdot \vec{\mathscr{E}}_2^*)\overleftrightarrow{\kappa}\vec{\mathscr{E}}_2}{I_0}, \\ \frac{\partial}{\partial x}\vec{\mathscr{E}}_2 &= \frac{-(\vec{\mathscr{E}}_2 \cdot \vec{\mathscr{E}}_1^*)\overleftrightarrow{\kappa}^*\vec{\mathscr{E}}_1}{I_0}. \end{aligned} \tag{16}$$

Here, $I_0 \equiv |\vec{\mathscr{E}}_1|^2 + |\vec{\mathscr{E}}_2|^2$, as before, and $\overleftrightarrow{\kappa}$ is a second-rank tensor

$$\overleftrightarrow{\kappa} \equiv \frac{i\omega}{2nc\cos\theta}\frac{\overleftrightarrow{\chi}}{m}. \tag{17}$$

Note that $\overleftrightarrow{\kappa}$ has no dependence on $\vec{\mathscr{E}}_1$ or $\vec{\mathscr{E}}_2$.

Consider the case when $\vec{\mathscr{E}}_1$ and $\vec{\mathscr{E}}_2$ have polarizations $\hat{\mathbf{e}}_1$ and $\hat{\mathbf{e}}_2$, respectively. The coupled-wave equations for the intensities are (Vahey, 1975)

$$\frac{\partial}{\partial x} I_1 = \frac{2(\mathrm{Re}\ \kappa_{1,2})I_1 I_2}{I_0},$$
$$\frac{\partial}{\partial x} I_2 = \frac{-2(\mathrm{Re}\ \kappa_{1,2})I_1 I_2}{I_0}, \tag{18}$$

where $\kappa_{1,2} = (e_1^* \cdot \overleftrightarrow{\kappa} \cdot \hat{e}_2)(\hat{e}_1 \cdot \hat{e}_2^*)$.

If Re $\kappa_{1,2} > 0$, beam 1 will experience gain (and beam 2 loss) as it propagates through the crystal. Note that only the real part of $\kappa_{1,2}$, which corresponds to the part of the refractive index grating shifted $\pi/2$ with respect to the intensity pattern, contributes to the two-beam coupling. However, both the real and the imaginary part of $\kappa_{1,2}$ contribute to the diffraction efficiency.

Two-beam coupling is a convenient way to measure both the sign and the density of the mobile charges in the material. The sign of the mobile charges can be determined by the direction of the two-beam coupling with respect to the c axis of the crystal. The density of mobile charges can be determined by the variation of two-beam coupling strength with the crossing angle of the two beams, because the coupling is at a maximum when $k = k_0$, as shown in Fig. 4, and k_0 depends on the charge density.

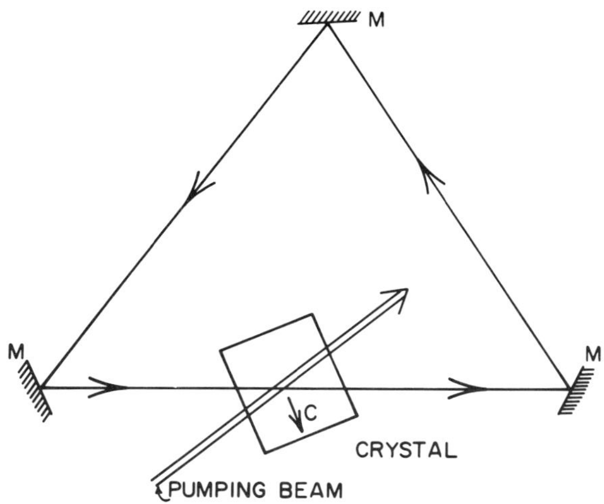

Fig. 10. Unidirectional ring resonator using two-beam coupling in a photorefractive crystal.

The charge density can also be determined experimentally by measuring the speed of grating formation or grating decay, as discussed below.

Practical applications of two-beam coupling include a self-pumped phase conjugator and a ring resonator cavity in which light circulates in only one direction (White *et al.*, 1982), shown in Fig. 10. If the crystal is aligned with its c axis as shown, a beam traveling around the ring clockwise interferes with and experiences gain from an incident pumping beam. A beam traveling counterclockwise around the ring experiences no gain, because it and the pumping beam make an interference pattern aligned parallel to the c axis of the crystal, so that symmetry precludes coupling. Two-beam coupling can also play a major role in four-wave mixing, as discussed below.

IX. Speed

In this section we briefly discuss the speed of photorefractive phase conjugators. The rate of formation or erasure of a photorefractive index grating increases with the intensity of the incident optical beams and depends on the charge-transport properties of the material. For example, a uniform beam of light will tend to erase a grating with a decay rate

$$A_{\text{decay}} = 2I_0Dd^2(k_0^2 + k^2), \tag{19}$$

as shown in Fig. 11. The light-assisted diffusion constant D and the average hop length d of the mobile charges are usually unknown. However, the speed can also be predicted from a measurement of the photoconductivity σ, because

$$\sigma = \epsilon\epsilon_0 I_0 Dd^2k_0^2. \tag{20}$$

The density of mobile charges can be inferred from measurements of the speed of grating formation and decay for different grating wave vectors $\mathbf{k}$. According to Eq. (19), the rates are proportional to $k_0^2 + k^2$, and so a graph of speed versus k will yield the unknown parameter k_0. Equations (19) and (20) assume nearest neighbor hopping and $kd \ll 1$. If these conditions are not met, or if charge transport through the conduction band is considered, the resulting expressions for the grating erasure and formation rate are more complex (Kukhtarev *et al.*, 1979).

When illuminated by low-intensity optical beams (mW/cm^2), photorefractive crystals are amazingly slow, with time constants ranging from milliseconds to minutes. There have been attempts to increase the writing speed in various photorefractive materials by using intense optical beams (Gaylord *et al.*, 1973; Shah *et al.*, 1974; Berg *et al.*, 1977; Chen *et al.*,

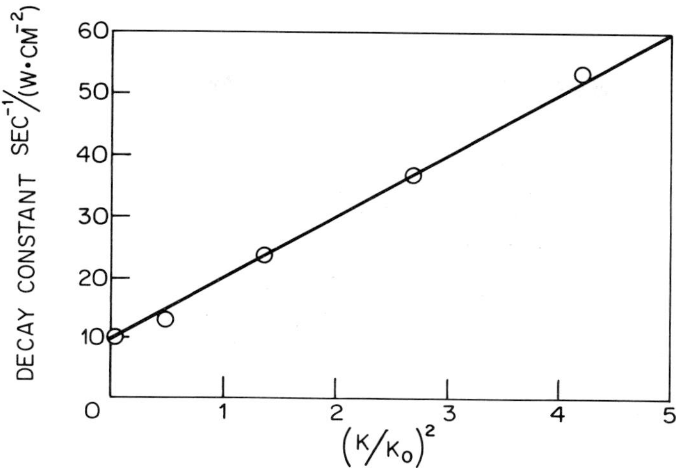

Fig. 11. Variation of the decay rate of a photorefractive index grating in $BaTiO_3$ versus grating wave vector **k**. The charge density can be determined by the slope of this graph, and the photoconductivity by the intercept. (From Feinberg *et al.*, 1980.)

1980; Lam *et al.*, 1981), and response times of less than a nanosecond have been reported.

X. Applications

One of the more dramatic applications of the photorefractive effect has been the demonstration of a photorefractive phase-conjugating mirror having continuous-wave gain (Feinberg and Hellwarth, 1980; White *et al.*, 1982). This device has a mirror reflectivity greater than unity when pumped by milliwatt optical beams in any part of the visible spectrum. This "mirror" uses four-wave mixing in a crystal of $BaTiO_3$. The large cw gain is made possible, in part, by the large electro-optic tensor coefficient ($r_{42} = 820 \times 10^{-12}$ m/V) of this crystal.

The phase-conjugating mirror and an ordinary mirror can form an optical cavity that self-oscillates, owing to the gain of the phase-conjugating mirror. The phase-conjugating mirror tracks the ordinary mirror. See Fig. 12, in which is shown a phase-conjugating mirror tracking a kitchen spatula. If the ordinary mirror is misaligned in a time shorter than the response time of the phase-conjugating mirror (for instance, by giving the ordinary mirror a sharp blow), the power in the resonator cavity will momentarily drop, but then be completely restored.

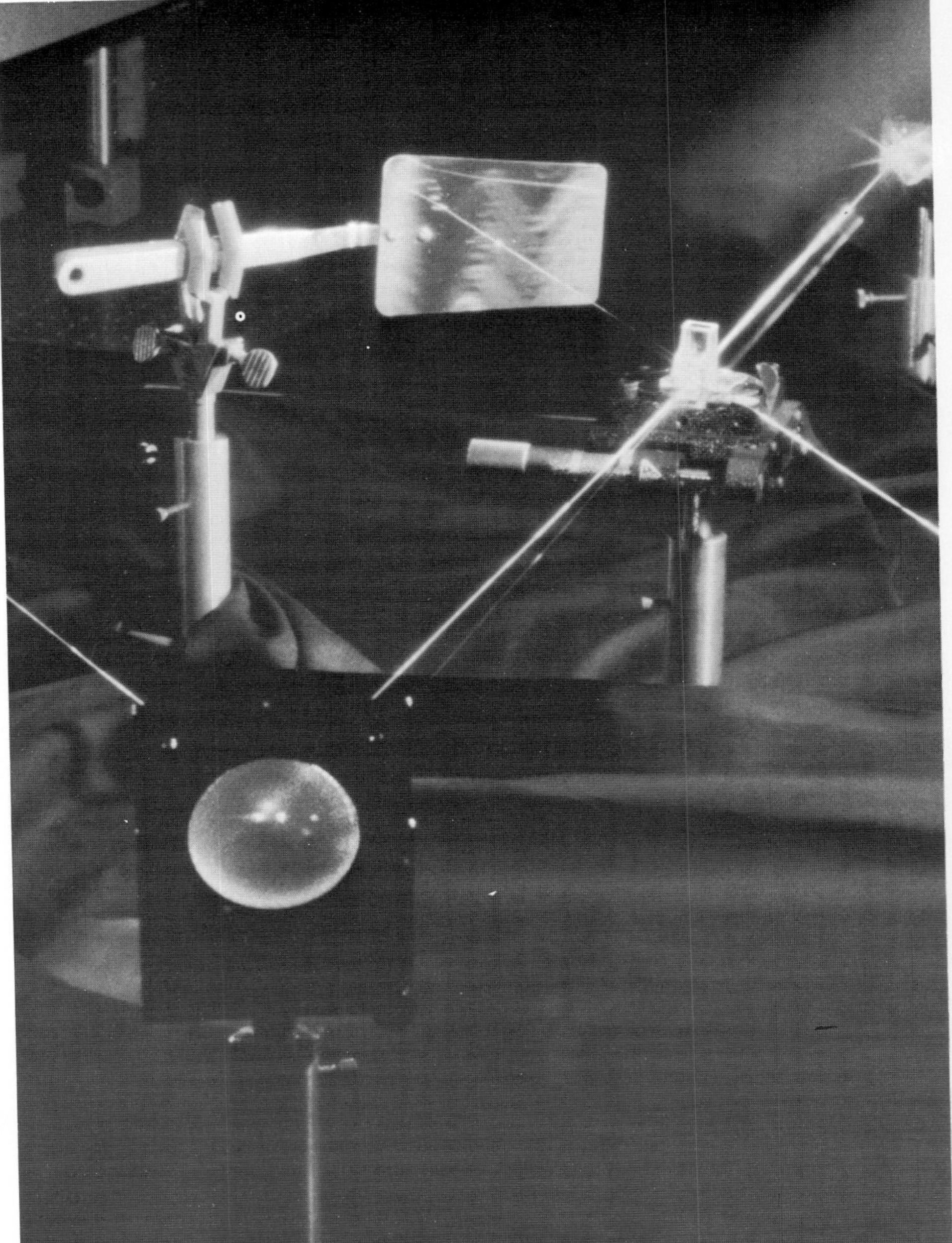

Such a cavity is not only self-aligning; it can also correct for phase aberrations inside the optical resonator (AuYeung *et al.*, 1979). The cavity has stable modes for any cavity length and for any (even negative) curvature of the ordinary mirror. (For a theoretical treatment of this aberration correction, see Chapter 13).

If the beam that goes through the crystal is taken as the output beam, then, depending on the orientation of the *c* axis of the crystal, the power of the output beam can be either greater or less than the power inside the optical resonator cavity, because the output beam experiences gain or loss as it traverses the crystal owing to two-beam coupling between the output beam and the nearest pumping beam. (Two-beam coupling between the output beam and the *other* pumping beam is usually greatly reduced because of the orientation of the resulting grating with respect to the *c* axis of the crystal.) In a conventional laser resonator, the power inside the resonator is greater that the power of the output beam. However, in a resonator cavity with a photorefractive phase-conjugating mirror, the power of the output beam can exceed the power in the resonator, as shown in Fig. 13.

White *et al.* (1982) have also demonstrated a "self-pumped" phase-conjugating mirror that requires no external pumping beams. It relies on two-beam coupling to generate the pumping beams from the incident beam itself. The pumping beams are then confined between two mirrors.[1]

The slow time response of a photorefractive phase-conjugating mirror using milliwatt beams makes it ideal for filming. To date, at least two movies have been made showing a $BaTiO_3$ phase conjugator in action (Feinberg, 1981; Cronin-Golomb *et al.*, 1981).

Increasing the intensity of both pumping beams does not alter the gain but does increase the speed of the photorefractive mirror. The variable time response of a photorefractive mirror has been used to enhance the stationary parts of a vibrating image (Marrakchi *et al.*, 1980). In Fig. 14 is shown the image of a loudspeaker after four-wave mixing in a crystal of

[1] Two recent "self-pumped" phase conjugators have been demonstrated in a photorefractive crystal of $BaTiO_3$. The device of Cronin-Golomb *et al.* (1982) requires only *one* external mirror; the device of Feinberg (1982) requires *no* external photorefractive mirror.

Fig. 12. Photograph of a phase-conjugating mirror with gain "finding" a kitchen spatula. The phase-conjugating mirror is formed by a crystal of $BaTiO_3$ (in the glass cuvette) and two milliwatt cw pumping beams, incident from the top right and bottom left of the photo. A self-oscillating beam is visible between the spatula and the crystal. This beam experiences additional gain owing to two-beam coupling as it passes through the back of the crystal and exits off to the right side of the photo. (Cover photo of December 1981, *Laser Focus*, used with permission.)

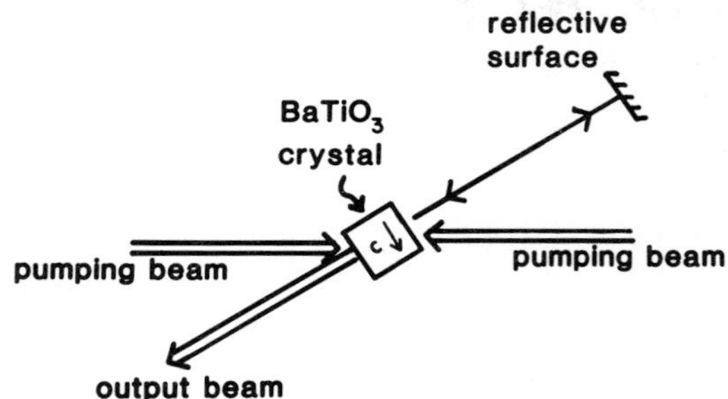

Fig. 13. Self-oscillating resonator cavity formed by a reflective surface and a phase-conjugating mirror with gain. Owing to two-beam coupling between the self-oscillating beam and the pumping beam incident from the right, the power of the output beam is stronger than the power oscillating in the resonator cavity, as seen in Fig. 12.

BSO. The moving regions of the loudspeaker move faster than the time response of the BSO crystal (as set by the pumping beam intensity), and therefore light from these regions cannot write a grating in the crystal. Light reflected off the vibrational nodes of the loudspeaker, however, will write a hologram in the crystal, so the nodes will be visible in the reconstructed image. The remarkable light sensitivity of BSO permitted this image to be formed without any lens between the crystal and the loudspeaker.

Another application of the photorefractive effect is in optical image processing. In four-wave mixing, the return image is the phase conjugate of the incident image only if the two *pumping* beams are phase conjugates of each other. If this condition is violated, the return image will not be the phase conjugate of the original image, but instead some function of the spatial intensity pattern of all three optical beams (Pepper *et al.*, 1978; White and Yariv, 1980). For example, the return beam can be the spatial convolution or correlation of two of the other beams, as described in Chapter 14.

Another example of image processing in photorefractive materials is edge enhancement of an optical image. For faithful imaging in photorefractive materials, the intensity of the image writing beam must be less than the intensity of the reference writing beam. (This condition is also necessary for production of faithful holograms in film.) If this condition is violated, a distinctive distortion of the final image results: all of its edges appear enhanced (Huignard and Herriau, 1978). See Fig. 15.

The explanation for this edge enhancement (Feinberg, 1980) relies on the peculiar intensity dependence of the photorefractive grating efficiency. As shown in Section VI, the photorefractive grating efficiency η is

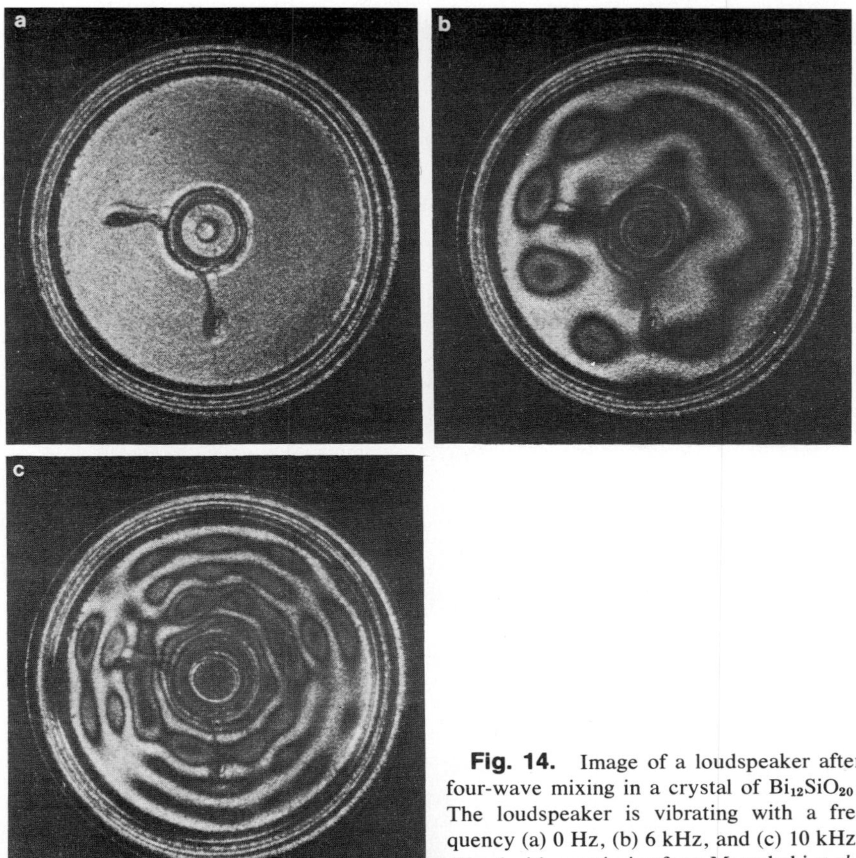

Fig. 14. Image of a loudspeaker after four-wave mixing in a crystal of $Bi_{12}SiO_{20}$. The loudspeaker is vibrating with a frequency (a) 0 Hz, (b) 6 kHz, and (c) 10 kHz. (Used with permission from Marrakchi *et al.*, 1980.)

largest when the two writing beams have the same intensity. Consider the case where a lens is used to focus a real image of the object into the photorefractive crystal, so that bright regions of the object correspond to bright regions in the crystal. Now make the average intensity $I_2(\mathbf{x})$ of the object writing beam greater than the intensity of the reference writing beam I_1, for $\mathbf{x}$ in the crystal, so that bright regions of the crystal have $I_2(\mathbf{x}) \gg I_1$. Because the intensities of the two writing beams are unequal, the resulting grating written in these regions will have a poor diffraction efficiency. Similarly, dark regions of the object correspond to $I_2(\mathbf{x}) \ll I_1$ in the crystal, and the resulting grating efficiency is again small. However, in transition regions between light and dark regions of

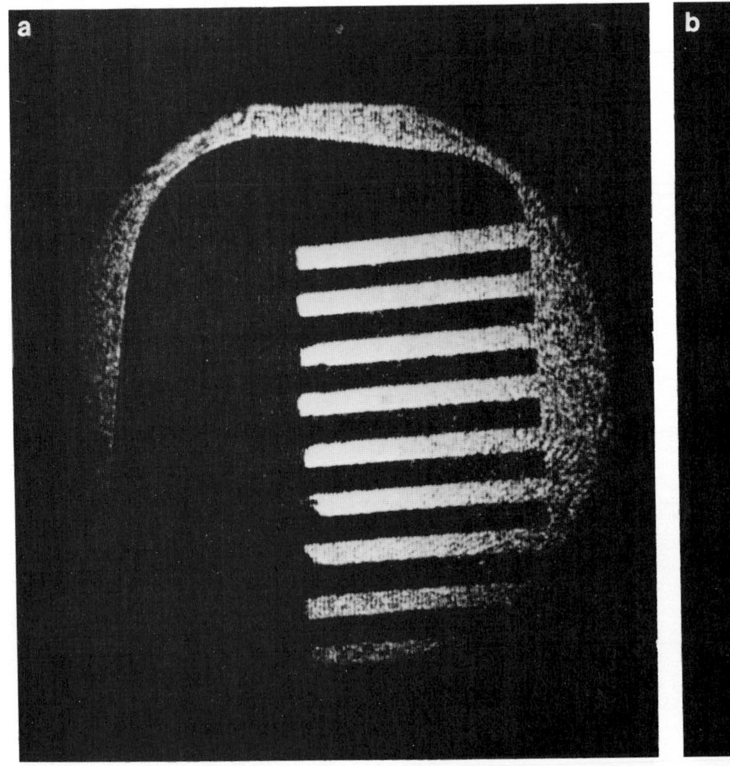

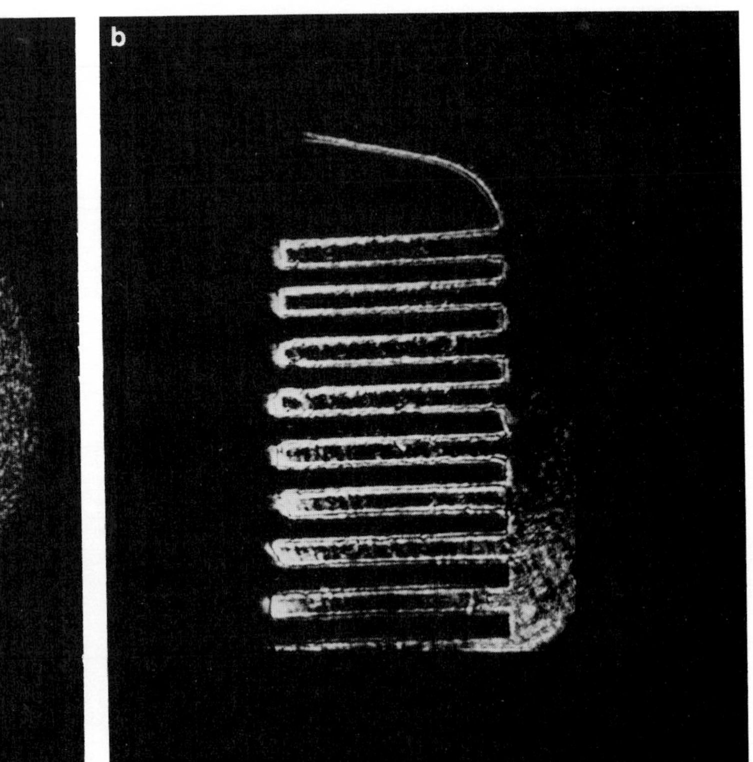

Fig. 15. Edge enhancement of the image of a moustache comb produced by four-wave mixing in $BaTiO_3$. (a) Normal image, (b) an edge-enhanced image. The amount of edge enhancement is controlled simply by adjusting the relative intensity of the two writing beams. (From Feinberg, 1980.)

the object, i.e., at the intensity edges, the intensity of the object beam equals the intensity of the reference beam, making $I_2(\mathbf{x}) = I_1$ and writing a grating with optimal diffraction efficiency. The very bright and very dark regions of the object are relatively ineffective in producing a grating; only at the intensity edges does the intensity of the object beam match the intensity of the reference beam to write an efficient grating.

XI. Conclusion

We have given a brief theory of the photorefractive effect and some of its applications, including both two- and four-wave mixing. We have shown how the photorefractive effect has been used to make the first phase-conjugating mirror with continuous-wave gain, and how this mirror has been used to form a self-oscillating resonator cavity that corrects its own aberrations. Further research on photorefractive materials is needed to determine the identity of the charge carriers, and to develop a material which has both a high optical sensitivity and a large beam-coupling efficiency.

References

Alphonse, G. A., Alig, R. C., Staebler, D. L. and Phillips, W. (1975). Time-dependent characteristics of photo induced space charge field and phase holograms in lithium niobate and other photo-refractive media, *RCA Rev.* **36,** 213.

Amodei, J. J. (1971). Analysis of transport processes during holographic recording in insulators, *RCA Rev.* **32,** 185.

Ashkin, A., Boyde, G. D., Dziedzic, J. M., Smith, R. G., Ballman, A. A., Levenstein, H. J., and Nassau, K. (1966). Optically-induced refractive index inhomogeneities in $LiNbO_3$ and $LiTaO_3$, *Appl. Phys. Lett.* **9,** 72.

AuYeung, J. Fekete, D., Pepper, D. M., and Yariv, A. (1979). A theoretical and experimental investigation of the modes of optical resonators with phase-conjugate mirrors, *IEEE J. Quantum Electron.* **QE-15,** 1180.

Berg, N. J., Udelson, B. J., and Lee, J. N. (1977). A new acousto-photorefractive effect in lithium niobate, *Appl. Phys. Lett.* **31,** 555.

Chen, C. T., Kim, D. M., and von der Linde, D. (1980). Efficient pulsed photorefractive process in $LiNbO_3$:Fe for optical storage and deflection, *IEEE J. Quantum Electron.* **QE-16,** 126.

Chen, F. S. (1967). A laser-induced inhomogeneity of refractive indices in KTN, *J. Appl. Phys.* **38,** 3418.

Chen, F. S. (1969). Optically induced change of refractive indices in $LiNbO_3$ and $LiTaO_3$, *J. Appl. Phys.* **40,** 3389.

Chen, F. S., LaMacchia, J. T. and Fraser, D. B. (1968). Holographic storage in lithium niobate, *Appl. Phys. Lett.* **13,** 223.

Cronin-Golomb, M., Fischer, B., White, J. O. and Yariv, A. (1981). Theory and experiments of four-wave mixing and optical phase conjugation in photorefractive materials, *J. Opt. Soc. Am.* **12,** 1564.

Cronin-Golomb, M., Fischer, B., White, J. O., and Yariv, A. (1982). A passive (self-pumped) phase conjugate mirror: a theoretical and experimental investigation, *Appl. Phys. Lett.* **41,** 689.

Feinberg, J. (1980). Real-time edge enhancement using the photorefractive effect, *Opt. Lett.* **5,** 330.

Feinberg, J. (1981). Laser with a phase-conjugating mirror, 16 mm film. J. Feinberg, Manhattan Beach, California.

Feinberg, J. (1982). Self-pumped, continuous-wave phase-conjugator using internal reflection, *Opt. Lett.* **7,** 486.

Feinberg, J. and Hellwarth, R. W. (1980). Phase-conjugating mirror with continuous-wave gain, *Opt. Lett.* **5,** 519; Erratum, *Opt. Lett.* **6,** 257 (1981).

Feinberg, J., Heiman, D., Tanguay, Jr., A. R. and Hellwarth, R. W. (1980). Photorefractive effects and light induced charge migration in barium titanate, *J. Appl. Phys.* **51,** 1297; Erratum, *J. Appl. Phys.* **52,** 537 (1981).

Fischer, B., Cronin-Golomb, M., White, J. O. and Yariv, A. (1981). Amplified reflection, transmission, and self oscillation in real-time holography, *Opt. Lett.* **6,** 519.

Gaylord, T. K., Rabson, T. A., Tittel, F. K. and Quick, C. R. (1973). Pulsed writing of solid state holograms, *Appl. Opt.* **12,** 414.

Glass, A. M. (1978). The photorefractive effect, *Opt. Eng.* **17,** 470.

Glass, A. M., von der Linde, D. and Negran, T. J. (1974). High-voltage bulk photovoltaic effect and the photorefractive process in $LiNbO_3$, *Appl. Phys. Lett.* **25,** 233.

Hellwarth, R. W. (1977). Generation of time-reversed wave fronts by nonlinear refraction, *J. Opt. Soc. Am.* **67,** 1.

Huignard, J. P. and Herriau, J. P. (1978). Real-time coherent object edge reconstruction with $Bi_{12}SiO_{20}$ crystals, *Appl. Opt.* **17,** 2671.

Huignard, J. P., Herriau, J. P., Auborg, P., and Spitz, E. (1979). Phase-conjugate wavefront generation via real-time holography in $Bi_{12}SiO_{20}$, *Opt. Lett.* **4,** 21.

Kim, D. M., Shah, R. R., Rabson, T. A., and Tittel, F. K. (1976). Nonlinear dynamic theory for photorefractive phase hologram formation, *Appl. Phys. Lett.* **28,** 338.

Kim, D. M., Rabson, T. A., Shah, R. R., and Tittel, F. K. (1977). Photorefractive materials for optical storage and display, *Opt. Eng.* **16,** 189.

Kukhtarev, N., Markov, V., Odulov, S., Soskin, M. and Vinetskii, V. (1979). Holographic storage in electro-optic crystals. I. Steady state, *Ferroelectrics* **22,** 949.

Lam, L. K., Chang, T. Y., Feinberg, J. and Hellwarth, R. W. (1981). Photorefractive-index gratings formed by nanosecond optical pulses in $BaTiO_3$, *Opt. Lett.* **6,** 475.

Marrakchi, J. M., Huignard, J. P. and Herriau, J. P. (1980). Application of phase conjugation in $Bi_{12}SiO_{20}$ crystals to mode pattern visualization of diffuse vibrating structures, *Opt. Commun.* **34,** 15.

Peltier, M. and Micheron, F. (1977). Volume hologram recording and charge transfer process in $Bi_{12}SiO_{20}$ and $Bi_{12}GeO_{20}$, *J. Appl. Phys.* **48,** 3683.

Pepper, D. M., AuYeung, J., Fekete, D. and Yariv, A. (1978). Spatial convolution and correlation of optical fields via degenerate four-wave mixing, *Opt. Lett.* **3,** 7.

Peterson, G. E., Glass, A. M. and Negran, T. J. (1971).Control of the susceptibility of lithium niobate to laser-induced refractive index changes, *Appl. Phys. Lett.* **19,** 130.

Shah, P., Rabson, T. A., Tittel, F. K. and Gaylord, T. K. (1974). Volume holographic recording and storage in Fe-doped $LiNbO_3$ using optical pulses, *Appl. Phys. Lett.* **24,** 130.

Staebler, D. L. and Amodei, J. J. (1972). Coupled-wave analysis of holographic storage in $LiNbO_3$, *J. Appl. Phys.* **43,** 1042.

Staebler, D. L. and Phillips, W. (1974). Fe-doped $LiNbO_3$ for read–write applications, *Appl. Opt.* **13,** 788.

Townsend, R. L. and LaMacchia, J. T. (1970). Optically induced refractive index changes in $BaTiO_3$, *J. Appl. Phys.* **41,** 5188.

Vahey, D. W. (1975). A nonlinear coupled-wave theory of holographic storage in ferroelectric materials, *J. Appl. Phys.* **46,** 3510.

von der Linde, D. and Glass, A. M. (1975). Photorefractive effects for reversible holographic storage of information, *Appl. Phys.* **8,** 85.

White, J. and Yariv, A. (1980). Real-time image processing via four-wave mixing in a photorefractive medium, *Appl. Phys. Lett.* **37,** 5.

White, J., Cronin-Golomb, M., Fischer, B. and Yariv, A. (1982). Coherent oscillation by self-induced gratings in photorefractive crystals, *Appl. Phys. Lett.* **40,** 450.

Young, L., Wong, W. K. Y., Thewalt, M. L. W. and Cornish, W. D. (1974). Theory of formation of phase holograms in lithium niobate, *Appl. Phys. Lett.* **24,** 264.

12 *Wave-Front Reversal by a Reflecting Surface*

N. F. Pilipetskii
V. V. Shkunov
A. N. Sudarkin
B. Ya. Zel'dovich
Institute for Problems in Mechanics
USSR Academy of Sciences
Moscow, USSR

I. Introduction

As discussed in previous chapters, the process of optical phase conjugation is most commonly realized either by conventional techniques of stimulated scattering (Zel'dovich *et al.*, 1972), or by holography and parametric interactions (Denisyuk, 1962; Stepanov *et al.*, 1971; Yariv, 1976; Hellwarth, 1977) in the bulk of a transparent medium. In this chapter we discuss a novel method using the reflecting interface between two media. This has recently been proposed by Zel'dovich *et al.* (1980) and realized by Kulikov *et al.* (1980), Golubtsov *et al.* (1981a), and Vasil'ev *et al.* (1982).

The method of phase conjugation, or wave-front reversal (WFR), by a

Optical Phase Conjugation

ISBN 0-12-257740-X

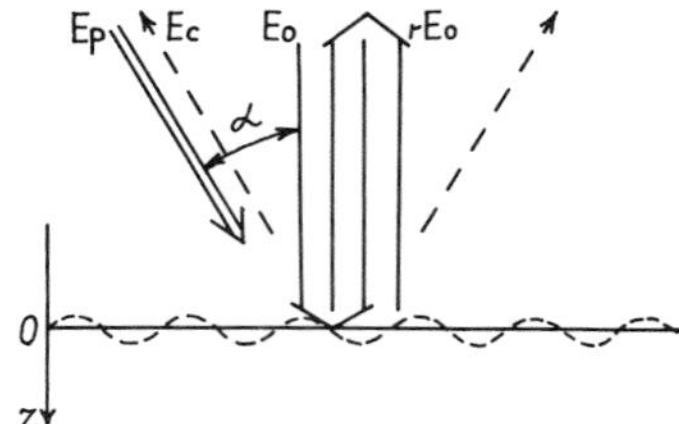

Fig. 1. General geometry of WFR by a reflecting surface.

surface is based on the fact that optical properties of the surface can change under the influence of laser radiation. If a wave incident on such a surface is spatially nonuniform, the intensity profile of the incident radiation is imprinted as a surface reflectivity distribution; i.e., a plane reflective hologram is recorded. The surface with the hologram in turn generates the wave front or amplitude distribution of the reflected radiation directly during the recording process. A nonlinearly responding reflecting surface thus makes it possible to realize dynamic holograms that are read out concurrently with their recording.

II. Phenomenological Calculation of Reflectivity into Reversed Wave

Reversal by a surface can be understood most simply in an example where the local reflectivity r depends on the instantaneous intensity I of the incident field:

$$r(I) = r(I_0) + \beta \, \Delta I, \tag{1}$$

where I_0 is the incident intensity averaged over the surface, ΔI are small-intensity variations about the mean value, and $\beta = \delta r/\delta I|_{I=I_0}$. In the scheme of WFR by a surface (Fig. 1), two waves interfere on the reflecting surface: the plane reference wave $E_0 e^{-i\omega t + ikz}$ normal to the surface and the probe wave to be reversed $E_\text{p} e^{-i\omega t + i\mathbf{k}_\text{p}\cdot\mathbf{R}}$, propagating at an angle with respect to the reference wave. Without loss of generality, we assume the probe to be a plane wave. The intensity distribution of the laser field on the surface is modulated by the function $\sin(\mathbf{q} \cdot \mathbf{r}_\perp)$, where $\mathbf{q}$ is the projection of the wave vector $\mathbf{k}_\text{p}$ onto the nonlinear surface, and $\mathbf{r}_\perp$ is the two-dimensional coordinate in the surface plane. If the light is sufficiently intense, the initially uniform reflecting surface also acquires a periodic structure

$$r = r_0 + \beta(E_0 E_\text{p}^* \, e^{-i\mathbf{q}\cdot\mathbf{r}_\perp} + E_0^* E_\text{p} e^{i\mathbf{q}\cdot\mathbf{r}_\perp}), \tag{2}$$

where $r_0 = r(|E_0|^2 + |E_\text{p}|^2)$.

Equation (2) is sufficient for the calculation of the wave E_{ref} reflected from the surface. Except for a common phase, we have

$$E_{\mathrm{ref}} = r_0 E_0 e^{-ikz} + \beta |E_0|^2 E_p e^{-i\mathbf{q}\cdot\mathbf{r} - ik_z z} + \beta E_0^2 E_p^* e^{-i\mathbf{k}_p \cdot \mathbf{R}}, \tag{3}$$

Where the first term on the right-hand side corresponds to the reflection of the reference wave exactly backward from the undisturbed surface, and the last terms correspond to the symmetric maxima of diffraction from the periodic sinusoidal lattice (2). The second term corresponds to the wave diffracted in the direction of the signal beam reflected by the undisturbed surface (Fig. 1). The third term corresponds to the wave E_c, the conjugate of the signal E_p. The reflection coefficient η, defined through $|E_c|^2 = \eta |E_p|^2$, depends only on the reference wave intensity I_0 and nonlinearity coefficient β: $\eta = |\beta|^2 I_0^2$.

The exact conjugation of a signal is achieved by this method only if the reference wave propagates precisely normal to the reflecting surface. Otherwise WFR occurs with the rotation (Zel'dovich and Shkunov, 1979; Blashchuk *et al.*, 1979a,b) to the double angle of detuning of the reference wave from the normal. Note that although the absolute diffraction efficiency of the plane surface hologram may not necessarily be high, reversal may occur with an effective gain because of energy transfer from the intense reference wave, if the dependence $r(I)$ is rather sharp.

III. Mechanism of the Change in Optical Properties of Surfaces

Let us now consider possible mechanisms of reflectivity variation. Intense laser radiation may result in variation of both the modulus of reflectivity r and its phase.

In the first case, for the surface of a metal or semiconductor, variation of r can arise from free-electron concentration changes in the surface layer (Wiggins, 1980). Although such a mechanism is reversible, the amplitude of the reflectivity modulation is rather low, and the diffraction efficiency of corresponding surface holograms appears to be small. Another possible mechanism is surface destruction (Kulikov *et al.*, 1980). Specular reflectivity changes in this case vary greatly, so the conjugation efficiency is high. An experiment on WFR due to surface destruction is described in detail in the next section. Phase reflectance surface holograms are recorded, for example, when the surface shape changes. The incident radiation exerts pressure on the surface, inducing mechanical movements, or heating the surface layer of the absorbing substance, deforming the surface through spatially nonuniform thermal expansion. The thermal mechanism is investigated in detail in Section V. Here we consider the radiation pressure mechanism of deformation for a liquid surface.

If the pressure on the surface $p(\mathbf{r}_\perp)$, is distributed according to harmonic law, $p(\mathbf{r}_\perp) = p_0 \sin(\mathbf{q} \cdot \mathbf{r}_\perp)$, then for the surface modulation of the form h $\sin(\mathbf{q} \cdot \mathbf{r}_\perp)$ we can write the capillary wave equation

$$\ddot{h} + 2\Gamma\dot{h} + \Omega^2 h = qp_0/\rho, \tag{4}$$

where $\Gamma = 2\nu q^2/\rho$, $\Omega^2 = \sigma q^3/\rho$, where ν is the liquid viscosity, σ is the surface tension coefficient, and ρ is the liquid density. For the lattice spatial frequencies of interest ($q \approx 10^3$–$10^4\ \mathrm{cm}^{-1}$), the oscillation frequency Ω and damping coefficient Γ of capillary waves are about 10^4–$10^7\ \mathrm{sec}^{-1}$ for ordinary liquids. If the light pulse duration does not exceed the characteristic times Γ^{-1} and Ω^{-1}, the process is transient. The response is $h(t) = qp_0t^2/2\rho$ if the recording pulse is instantaneously switched on to a constant level. After time $t \gtrsim \Gamma^{-1}$ the response becomes steady, and exactly reproduces the pressure-distribution profile $h_{\mathrm{st}} = (qp_0/\rho\Omega^2)\sin(\mathbf{q} \cdot \mathbf{r}_\perp)$. A much greater modulation depth of the surface profile for nonviscous liquids can be achieved if the radiation disturbing the surface is temporally modulated at frequency Ω. The amplitude of synchronous surface vibrations in this case is $h_{\mathrm{res}} = \Omega h_{\mathrm{st}}/2\Gamma$. The nature of light-induced pressure forces can be different, but their magnitude is proportional to the local intensity of the incident light, $p_0 = 2\gamma\sqrt{I_0 I_{\mathrm{p}}}$. Electrostrictive forces draw the dielectric toward the incident radiation [$\gamma_{\mathrm{es}} = -(2/c)(n-1)/(n+1)$] (Zel'dovich *et al.*, 1980). The forces are directed oppositely in the case of reflection from a metallic surface [$\gamma_{\mathrm{m}} = (1+R)/c$, where $R = |r_0|^2$ is the surface reflectivity]. Pressure may also arise under laser evaporation of a substance from the surface, in which case $\gamma_{\mathrm{ev}} \approx 1/v$, where v is the velocity of particles leaving the surface.

If the surface modulation amplitude is small, the diffraction efficiency of the relief hologram is equal to $R(kh)^2$; hence we obtain for the WFR efficiency in the transient case

$$\eta = R(\gamma I_0 kq/\rho)^2\, t^4 \tag{5}$$

Equation (5) is valid for the steady-state regime if Ω^{-1} is substituted for t. Let us consider acetone, for example. With a power density of the reference wave of 30 MW/cm^2 and an angle between the waves of 10^{-2} rad, the WFR efficiency due to ponderomotive forces reaches unity in approximately 5 μsec.

Certain photosensitive layers, recently developed for various purposes, are of great interest from the standpoint of recording dynamic surface holograms and using them to reverse the wave front of low-power radiation. These include thermoplastics (Glenn, 1959), liquid-crystal-based controlled light-valves (Bleha *et al.*, 1978), and FTIROS-type[1] materials (Bugaev *et al.*, 1975). All these materials have very high photosensitivity and potentially may reverse the wave front of radiation with intensities of less than 1 $\mu W/cm^2$. The results of a WFR experiment using liquid-crystal-based controlled light valves are given in Section VI.

The sluggish response of hologram recording is a shortcoming of such materials and of the surface deformation mechanism. The recording characteristic times change from microseconds (Zel'dovich *et al.*, 1980; Bugaev *et al.*, 1975) up to several seconds (Golubtsov *et al.*, 1981a; Glenn, 1959; Bleha *et al.*, 1978).

The search for effective surface WFR is now extremely urgent. In the next section we describe some recent results.

IV. WFR under Metallic Mirror Destruction

The scheme of an experiment performed by Kulikov *et al.* (1980) is shown in Fig. 2. The amplified radiation from a single-mode *Q*-switched master laser on neodymium glass ($\lambda = 1.06\ \mu m$) passed through the diaphragm $A_\perp$. The reference beam E_0 was directed onto the operating mirror M through the beam splitter M_1 and the prism Pr. The signal beam, reflected from the mirror M_1, passed through the diaphragm A_2, the glass plates P_1 and P_2 of the recording system, and the aberrating phase plate PP, the image of which was formed by the lens L_1 ($f = 6$ cm) with the unit magnification of the mirror. The divergence of the signal beam after passage through the phase plate increased up to 5×10^{-3} rad. The diaphragm A_2, 4 mm in diameter, was used for the signal beam to illuminate on the operating surface an area where the reference wave intensity was uniform.

The operating surface was a $\sim$1-μm-thick aluminum layer evaporated onto the glass substrate. Such a mirror is destroyed by laser pulses if the light energy density on the surface exceeds $\simeq 0.3\ J/cm^2$. When the surface is destroyed, its specular reflectivity decreases from unity to zero.

The reversed beam compensated for its aberrations upon retraversing PP and L_1, and was directed into the recording system. The energies of the signal and reference beams were measured by photocells PD_1 and PD_2, and the angular distributions of the signal and reversed beams were

[1] FTIROS is the abbreviation for the name of a device invented at the Ioffe Physical-Technical Institute in Leningrad. It is a surface coated by a thin semiconductor layer, with varying reflectivity due to thermal effects induced by incident light absorption. Such devices are widely described (see Bugaev *et al.*, 1975).

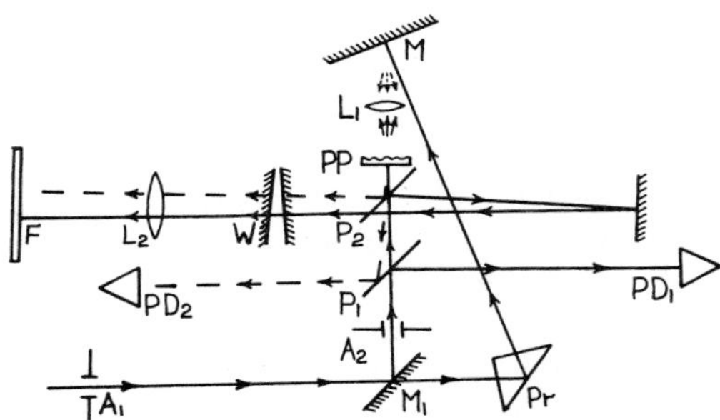

Fig. 2. Experimental scheme for WFR due to profiled destruction of a metallic mirror.

compared with the use of beam-splitting wedge W and photoplate F in the focal plane of lens L_2 according to a conventional scheme (Zel'dovich *et al.*, 1972; Kulikov *et al.*, 1980; Blashchuk *et al.*, 1979a).

A. Experimental Results

The undisturbed signal beam produced on the photoplate F a set of successively reduced spots with dimensions corresponding to the diffraction-limited divergence. The angular structure of the reversed beam after correction by the phase plate was more complicated. The main portion of energy ($\gtrsim 60\%$) corresponded to high-directivity radiation with a divergence about 1.3–1.5 times greater than the diffraction limit. About half of the energy of the high-directivity component was contained in the diffraction-limited spot corresponding to exactly reversed radiation. The broadening of the divergence of the high-directivity component of the reversed beam is caused by the effective aperturing of the radiation area of the reversed wave due to some nonuniformity of the reference wave. The holographic grating is stronger in the central part of the operating area, where the intensity of the Gaussian reference beam is greatest.

In Fig. 3 are shown the angular spectra of the signal beam in front of the phase plate (a) and behind the phase plate (b) and of the reversed wave corrected by the phase plate (c).

In addition to the highly directional component, the reversed radiation contained the background not recorded by the photoplate. Its source is the high contrast (in time and space) of the dynamic hologram recording by destruction. For quantitative determination of the energy properties of the WFR process (the signal reflection coefficient into the reversed wave and the reversal fraction), photocells PD_1 and PD_2 were used.

The total energy of the reversed wave, propagating within a solid angle

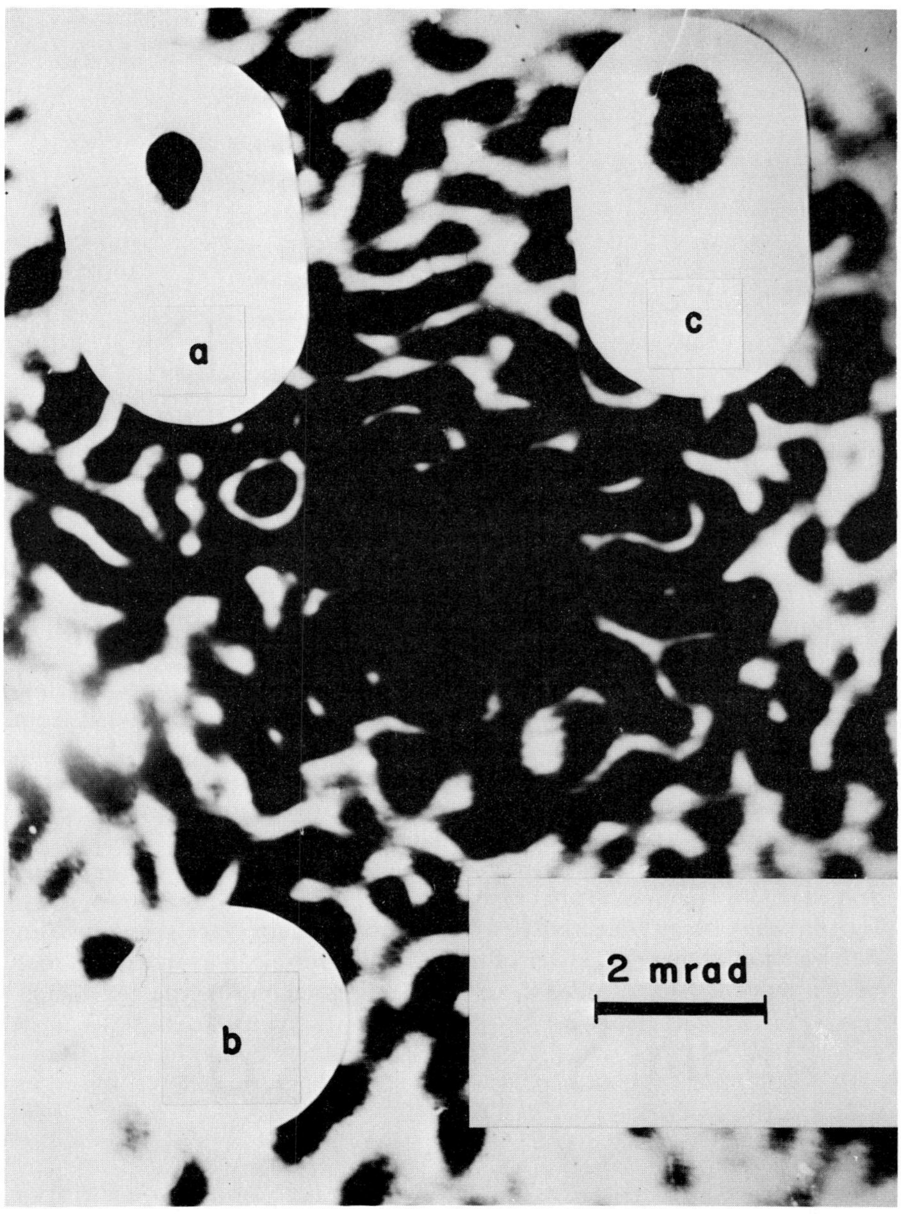

Fig. 3. Angular spectra obtained in an experiment using mirror destruction, for the signal beam in front of the phase plate (a) and behind it (b) and for the conjugate beam after retraversing the same plate (c).

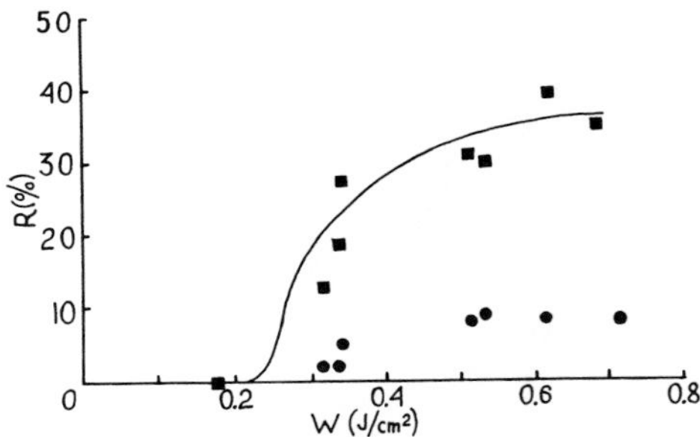

Fig. 4. WFR energy efficiency versus the reference wave energy density for an experiment using mirror destruction. ■, experimental points for the total energy reflected in the conjugate wave direction; ●, experimental points for the energy in the exactly conjugated fraction. The solid curve is the theoretical dependence. [Used with permission from Kulikov *et al.* (1980) and Golubstov *et al.* (1981a).]

of ~0.15 sr, was compared with the total energy of the signal wave; the brightness of the far-field patterns of both waves were also compared. In Fig. 4 is shown the reflection coefficient with respect to the total energy (squares) and to the energy in the exactly reversed component (circles) for different densities of the reference beam. It may be concluded from the experimental dependences of Fig. 4 that the process is of threshold character, and can be observed in this case with reference wave energy density higher than 0.2 J/cm^2. Reversal occurs with an energy efficiency of up to 40%, the reversal fraction reaching 25–30%.

In the experiment, a comparison of the time dependences of the signal and reversed beams was also made. The oscillograms of these processes for the case of high reflectivity (with respect to energy) are shown in Fig. 5. It was found that the power of the reversed pulse can even exceed that of the incident pulse. In this case, its duration decreases; the half-height

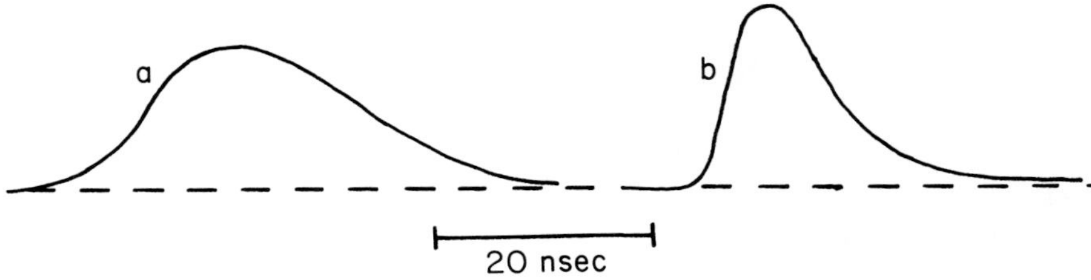

Fig. 5. Oscillograms for the signal a and reversed pulses b.

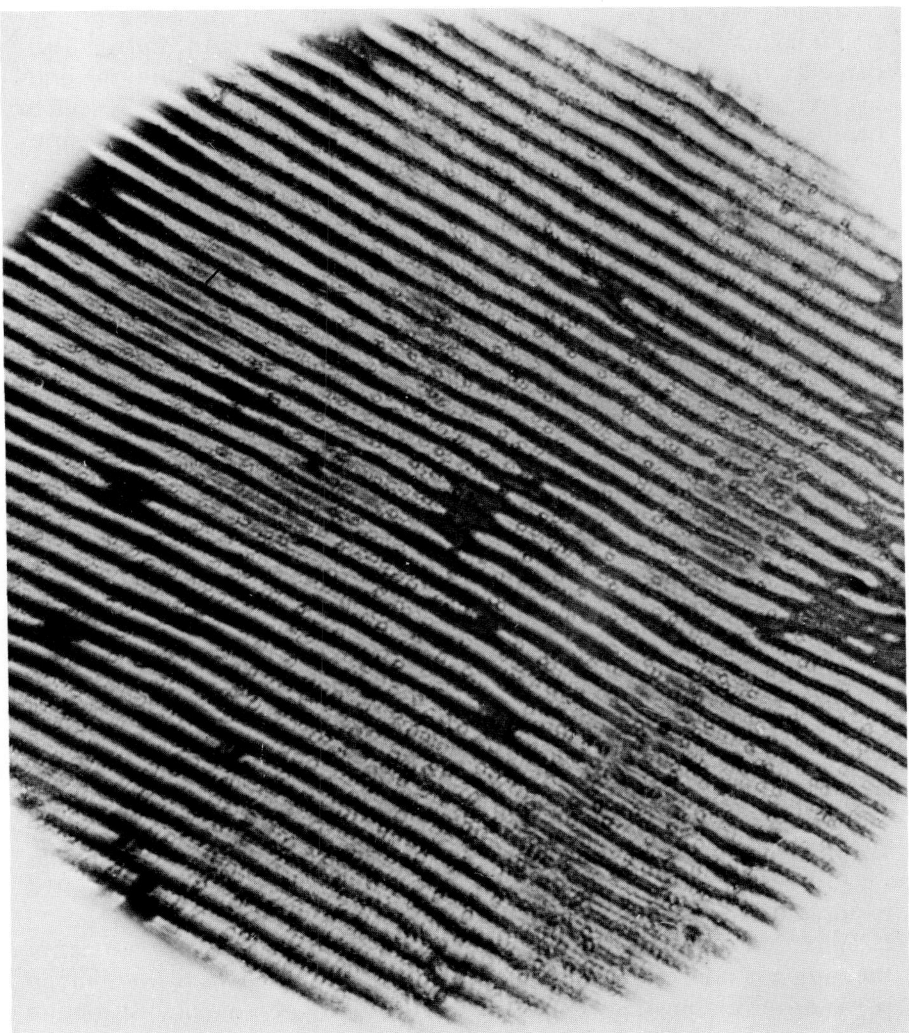

Fig. 6. Magnified photograph of the hologram remaining on the surface.

pulse duration was reduced from 21 to 10 nsec, and the rise time of the leading edge (10–90%) was reduced from 11 to 4 nsec.

An enlarged photograph of the mirror surface after the reference and signal waves have interacted upon it is shown in Fig. 6. If the pulse energies are not too high and the mirror is not destroyed even in the minima of the interference pattern, as in the case recorded in Fig. 6, the diffraction holographic grating with period $\sim 2\pi/q$ remains on the mirror. Incidentally, this hologram can be reconstructed by laser radiation, the grating giving several diffraction orders simultaneously. The grating is seen to be nonuniform both in phase and in amplitude. This nonuniformity caries information about the distorted signal wave E_p.

B. Theory

Let us describe the dynamics of the formation of this grating (Kulikov *et al.*, 1980), in the framework of the following simple model. We assume that as soon as the incident energy density in some area of the film exceeds the threshold value W_{th}, the specular reflectivity of the surface changes irreversibly in this area from unity to zero. (This has been confirmed by additional experiments.) Let us introduce the instantaneous mean energy density

$$W(t) = (I_0 + I_p) \int_0^t f(t')\, dt'$$

and the coefficient $\psi = (I_0 + I_p)/2\sqrt{I_0 I_p}$, characterizing the correlation between the intensities of the signal beam, I_p, and the reference beam, I_0 [$f(t)$ is the time envelope of the pulses]. The reflectivity grating forms on the mirror at moment t_1, when the energy density in the maxima of the interference pattern achieves the threshold value $W(t_1) = \psi W_{th}/(1 + \psi)$; it reaches maximum diffraction efficiency at moment t_2, when the widths of the destroyed and undestroyed regions become equal, and where $W(t_2) = W_{th}$; and, finally, it disappears at moment t_3, when the specular reflecting fringes vanish in interference pattern minima. At that time, the energy density is $W(t_3) = \psi W_{th}/(\psi - 1)$.

The efficiency of scattering the field E_0 into the first diffraction order of the lattice is nonzero in the interval $t_1 < t < t_3$, and depends on time in terms of the grating linewidth $\Lambda(t)$. For the exactly harmonic distribution of the intensity it is easy to obtain

$$\cos(q\Lambda) = \psi[W_{th}/W(t) - 1], \tag{6}$$

whence, for the instantaneous WFR efficiency $\eta(t)$ at $t_1 < t < t_3$, we have

$$\eta(t) = \frac{1}{\pi^2}\frac{I_0}{I_p}\sin^2\left(\frac{q\Lambda}{2}\right) = \frac{1}{\pi^2}\frac{I_0}{I_p}\left[1 - \psi\left(\frac{W_{th}}{W(t)} - 1\right)^2\right]. \tag{7}$$

The time dependence of the reversed pulse is determined by $I_c(t) =$

$I_p f(t)\eta(t)$. The pulse appears at moment t_1, and lasts either until the end of the reference pulse t_p or until moment t_3, if $t_p > t_3$ (Fig. 6). The duration of the reversed wave pulse is always less than that of the exciting light pulse.

The integral (over time) efficiency of WFR, η_Σ, can easily be obtained by the integration over time

$$\eta_\Sigma \equiv \frac{\int_0^{t_p} \eta(t) f(t)\, dt}{\int_0^{t_p} f(t)\, dt} = \frac{1}{W_p} \int_{W_1}^{W_2} \eta(W)\, dW, \tag{8}$$

where W_p is the total energy density of the radiation on the surface, $W_1 = W(t_1)$, and $W_2 = W(t_3)$ [or $W_2 = W_p = W(t_p)$ if $t_p < t_3$]. The solid curve in Fig. 4 represents the dependence of the coefficient of reflection into the reversed wave (with respect to the total energy) as calculated from Eq. (8), for $\psi = 1.48$, $I_0 = 6.7 I_p$, and $W_{th} \simeq 0.3\ \mathrm{J/cm^2}$, corresponding to the experimental situation.

The calculations obtained with this model agree satisfactorily with the energy measurements (Fig. 4) and show the twofold reduction in the pulse duration of the reversed wave for $W_p \simeq W_{th}$.

The value of the instantaneous reversal coefficient obtained experimentally was 150% (Fig. 5), i.e., twice the theoretical value, indicating a greater contrast holographic grating in the real situation.

The limit case of weak signals ($\psi \gg 1$, $W_p \gtrsim W_{th}$) is of interest. In this situation, when the reference wave intensity is rather uniform over the interaction region, the reversed wave pulse appears at the moment the energy density of the reference wave achieves the threshold value $t \simeq t_p W_{th}/W_p$, and is much shorter than the signal pulse $\Delta t \simeq t_p (W_{th}/W_p)(2/\psi)$. In this case, reversal may take place with simultaneous signal amplification (with respect to intensity) by a factor of $\sim 0.1 I_0/I_p$ and (with respect to energy) by $\sim 0.3\sqrt{I_0/I_p}\, W_{th}/W_p$.

V. WFR under a Change in the Surface Shape

In this section we consider the results of studying the mechanism of recording dynamic holograms using the variation of the surface shape of a highly absorbing medium due to thermal expansion under nonuniform absorption of the laser interference pattern. For substances with high coefficient of bulk expansion, this mechanism appears to be very efficient (Golubtsov *et al.*, 1981a,b), and permits us to reverse the wave front of He–Ne laser radiation using milliwatt beams.

A. Theory

We assume here the model of the local isotropic thermal expansion, valid for liquids if hydrodynamic effects are not significant, i.e., for pulses of short duration and liquids of high viscosity. In elastic solids the anisotropy of the thermal expansion in the surface layer is rather small, and the results of this model differ from actual experimental results by no more than a factor of 1.5 (Golubtsov *et al.*, 1981b). The system of equations for small variations of temperature, δT, and density, $\delta\rho$, due to laser heating and for the surface velocity $\mathbf{v}$ includes the equation for the thermal conductivity with the boundary condition of surface heat release,

$$\rho c_T \frac{\partial\, \delta T}{\partial t} = H\, \nabla^2\, \delta T, \tag{9a}$$

$$\left.\frac{\partial\, \delta T}{\partial z}\right|_{z=0} = -\frac{1}{H}(1 - R)I(\mathbf{r}_\perp, t), \tag{9b}$$

the continuity equation, the definition of the bulk expansion coefficient μ,

$$\operatorname{div} \mathbf{v} = -\frac{1}{\rho}\frac{\partial\, \delta\rho}{\partial t} = \mu \frac{\partial\, \delta T}{\partial t}, \tag{10}$$

and the condition of the isotropy of the thermal expansion,

$$\frac{\partial\, v_z}{\partial z} = \tfrac{1}{3} \operatorname{div} \mathbf{v} = \frac{\partial v_x}{\partial x} = \frac{\partial v_y}{\partial y}. \tag{11}$$

ρ, c_T, and H are the density, specific heat capacity, and thermal conductivity of the heated substance.

The equation for the height variation,

$$h(\mathbf{r}_\perp, t) = -\int_0^t v_z\,(\mathbf{r}_\perp, t')|_{z=0}\, dt'$$

($t = 0$ is the moment at which the laser pulse appears), can easily be rewritten by direct integration of the thermal conductivity equation (9a) with respect to dz in the limits $(0, +\infty)$ with the substitution of the boundary condition (9b) and taking into account the relation

$$h(\mathbf{r}_\perp, t) = \tfrac{1}{3}\mu \int_0^\infty \delta T\, dz,$$

which follows directly from Eqs. (10) and (11). We obtain

$$\rho c_T \frac{\partial h}{\partial t} = H\, \nabla_\perp^2\, h - \frac{1}{3}\mu(1 - R)I(\mathbf{r}_\perp, t). \tag{12}$$

For the intensity profile of the form of interest, $I(\mathbf{r}_\perp) = 2\sqrt{I_0 I_p} \cos(\mathbf{q} \cdot \mathbf{r}_\perp)$, we have

$$h(\mathbf{r}_\perp, t) = h_0(1 - e^{-t/\tau}) \cos(\mathbf{q} \cdot \mathbf{r}_\perp),$$
$$h_0 = 2\mu(1 - R)\sqrt{y_0 y_p}/3Hq^2, \qquad \tau = \rho c_T/Hq^2. \tag{13}$$

For the absorbing substances in the absence of evaporation from the surface, the thermal mechanism of surface reshaping dominates over the mechanism connected with the light pressure. Let us estimate the corresponding quantities for pulsed lasers interacting at an air-mercury interface. To reverse the wave front of a signal beam with a WFR efficiency of ~ 1 in the transient regime, the energy density of the reference wave, $W \sim 1$ J/cm², is required. In this case the condition of the absence of boiling, $W \lesssim \Delta T_{\mathrm{Boil}} \sqrt{H \rho c_T t_p}$, limits the pulse duration to $t_p \gtrsim 5 \times 10^{-6}$ sec. If the angle between the reference and signal waves exceeds the value $\alpha_0 \simeq 2 \times 10^{-2}$ rad, the surface shape has time to achieve steady state, and the reversal efficiency falls as $(\alpha_0/\alpha)^4$ for the same power density of the reference wave, ~ 0.2 MW/cm².

B. Experiment[2]

An epoxy resin ($R \simeq 0.05$) was mixed with highly absorbing green dye[3] *Viridis ninentis* as the nonlinear medium. The radiation of a He–Ne laser (LG-38, $\lambda = 0.63$ μm) was split into two beams, which were focused by the optical system to the same spot (~ 0.8 mm diameter) on the surface investigated. The radiation reflected from the surface was displayed on the screen, where the first diffraction orders (in addition to the zeroth order) were observed from the thermal grating. The temporal history of the reversed wave was recorded by a photocell and an oscilloscope. The angle between the beams and the intensities of the beams were varied.

To eliminate thermal defocusing of the reflected radiation, which can reduce the WFR quality (Golubtsov *et al.*, 1981b), a mechanical light modulator was used that produced a train of pulses with a duration of $\sim 3 \times 10^{-2}$ sec separated by intervals of ~ 0.3 sec. The process of the formation of the thermal lens, corresponding to a larger spatial scale and hence to greater inertia, could not develop significantly for 30 msec; but the surface returned to its smooth condition before the beginning of the next pulse.

[2] Golubtsov *et al.* (1981a).

[3] *Viridis ninentis* is the Latin name of the green dye we used. Its alcoholic dilution (like iodine) is popular in our country as a medium for disinfecting small scratches on the body.

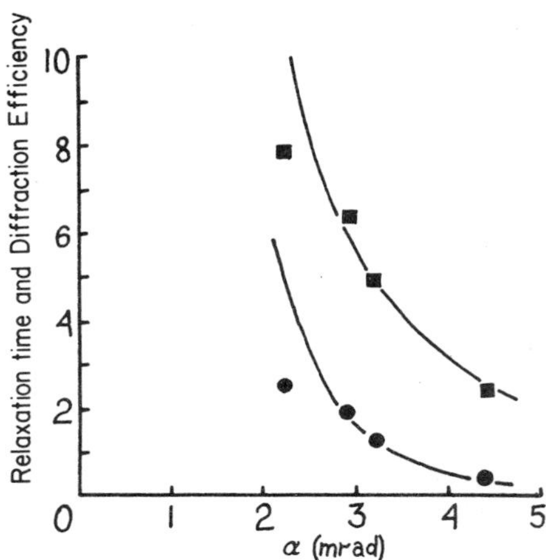

Fig. 7. Relaxation time $\tau_{0.5}$ (■) and diffraction efficiency $\eta_{gr} \times 10^2$ of the grating (●), versus angle α between the waves in an epoxy resin experiment.

The divergence of the reversed wave, determined by the spot diameter in the far-field zone on the screen, coincided with the signal wave divergence of ~1 mrad. The time dependence of the reversed wave intensity corresponded to the dependence of Eq. (13) i.e., a quadratic increase at the initial stage, followed by a steady-state value. The buildup time, measured as the 50% level of the steady-state efficiency at different angles between the waves, is represented by squares in Fig. 7; the measured values of the grating efficiency in the steady-state regime, $\eta_{gr} = (kh_0)^2$, are represented by circles. The corresponding theoretical dependences, calculated from Eq. (13) for the epoxy resin with $c_T = 0.7$ J/g $H =$ 1.6 W/cm °C, and $\rho = 1$ g/cm³, are also shown. In this set of measurements, the power of the recording waves was constant (4×10^{-3} W).

The WFR efficiency at a constant angle between the waves and in the steady-state regime increased as the square of the reference wave intensity, in accordance with theory of Golubtsov *et al.* (1981a), and did not depend on the signal beam intensity.

VI. WFR of Low-Intensity Radiation with the Aid of a Liquid-Crystal-Based Controlled Light Valve

The operating principle of the liquid-crystal-based controlled light valve (see, e.g., Bleha *et al.*, 1978), discussed by Vasil'ev *et al.* (1983), and

Zel'dovich and Tabiryan (1981) is rather simple. If a layer of the planar (along the surface) oriented nematic liquid crystal (DeGennes, 1974) is placed in a static electric field between the plates of a capacitor, the direction of the nematic director may change. If the field strength B exceeds some critical value B_{Fr}—the Frederix threshold (DeGennes, 1974)—the nematic is reoriented in the direction normal to the faces, and for fields just slightly greater than B_{Fr}, the angle of the director inclination is $\theta \simeq \sqrt{(B - B_{Fr})/B_{Fr}}$. As a result, the effective permittivity of the liquid crystal changes for the radiation with polarization parallel to the initial director, $\Delta\epsilon \approx \frac{1}{2}\epsilon_0\theta^2$, and this leads to an additional, field-induced phase shift $\Delta\Phi_{NL} \approx \frac{1}{4}k\epsilon_0\theta^2 l$ of the radiation passing through the layer of thickness L.

In our light valve (Zel'dovich and Tabiryan, 1981) a photoconductor layer was used as one of the capacitor plates. The light incident onto the light valve modulates the charge distribution in this layer, which in turn modulates the electric field applied to the liquid-crystal layer. The interference pattern of the incident light is recorded as a profile of the phase shift $\Delta\Phi_{NL}$ of the radiation passed through the light valve. The dependence of $\Delta\Phi_{NL}$ on the incident light intensity is shown in Fig. 8. The liquid-crystal layer with a thickness $L = 5\ \mu m$ was adjacent to a mirror with reflectivity $R \simeq 5\%$; therefore the dynamic holograms recorded were reflecting.

The scheme of the experiment (Vasil'ev *et al.*, 1983) is shown in the Fig. 9. The radiation of a single-mode He–Cd laser ($\lambda = 0.44\ \mu m$) was expanded in the telescope to a spot diameter of 2.5 cm, and was then split by a semitransparent mirror SM into the reference and signal beams, which were brought together on the light valve LC at an angle $\alpha = 10^{-2}$ rad. The

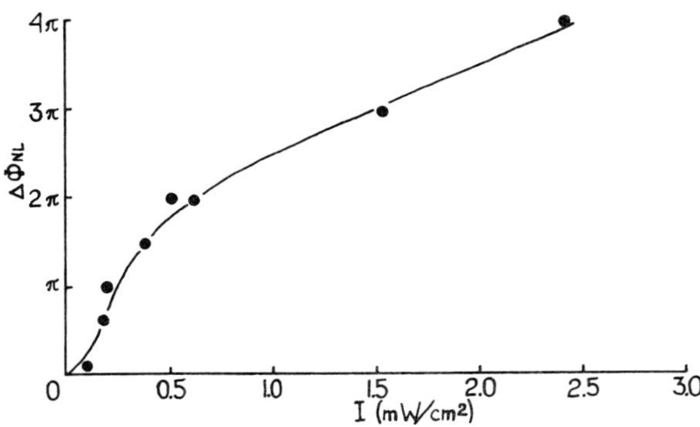

Fig. 8. Nonlinear phase shift $\Delta\Phi_{NL}$ for the wave, reflected from the liquid-crystal modulator, versus wave intensity.

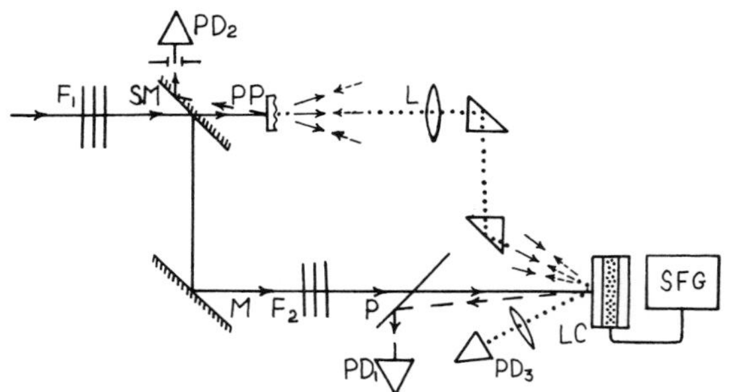

Fig. 9. Experimental scheme for reversal by a liquid-crystal light modulator.

radiation on the light valve was apertured to avoid the large-scale phase nonuniformity induced by the reference wave in the layer LC. The beam intensities were changed by the light filters F_1 and F_2. The angular characteristics (in the far-field zone) of the reference, signal, and reversed waves, as well as the energies of these waves and the reversal quality, were recorded by the photocells PD_1, PD_2, and PD_3. The phase plate PP, imaged on the transparent surface, was placed in the path of the signal beam. After retraversing the phase plate PP, the conjugate wave produced in the far-field zone a bright spot with a divergence only ~1.5 times greater than the initial divergence of the signal beam. A halo with a divergence of ~2 mrad was observed around the spot. This halo was caused by nonexact reversal of the signal wave on the hologram recorded by the light valve. In addition, it contained a component that did not vanish even when the voltage acrross the light valve capacitor was turned off. This caused by scattering of the reference wave from optical irregularities in the liquid-crystal layer and to imperfect optical elements in the light valve. The energy ratio in these three parts of the angular spectrum of the reversed wave did not change with reference wave intensity, and was approximately 10:9:4.

The signal wave intensity was $\frac{1}{3}$ of the reference beam intensity in all the measurements. When the power of the reference beam was varied within the range 10–90 μW, the 1% reflection coefficient in the reversed wave was constant within the 20% experimental accuracy. The minimum power of the signal beam, the reversal of which could be recorded (3 μW), was limited by the performance characteristics of our devices. The buildup times of the WFR process were several seconds for our light valve. The constancy of the reflectivity is explained by the fact that the nonlinearity coefficient β, determined by the slope of the dependence $\Delta\Phi_{BL}(I)$ (Fig. 8), decreases with increasing reference wave intensity.

Thus, by using special photosensitive surfaces, it is possible to reverse the wave front of extremely-low-power radiation.

VII. Advantages of WFR by a Reflecting Surface

In summary, the new method of WFR by a reflecting surface is distinguished by high reversal efficiency. In principle, up to $\frac{1}{3}$ of the reference wave energy can be transferred to the reversed wave. In the first experiment to use this technique (Kulikov *et al.*, 1980), amplified signal reversal was obtained by simultaneous amplification.

In bulk nonlinear processes in the medium, the relative variations of the optical properties of substances under the action of light are usually small, and high efficiency may be achieved by the coherent superposition of weak sources in a large interaction volume. When the interaction in thin layers is nonlinear, the relative variations of the optical properties may be of the order of unity. However, distortions of the participating fields, inevitable under bulk interaction in such a situation, are much less of a problem in thin layers.

It is easy to eliminate the distortions caused by the self-action of the reference wave when using a nonlinear surface, by making the reference wave uniform over the interaction region.

An important advantage of the surface method is the independence of the wavelength to be reversed; this wavelength determines only the period of the grating recorded on the surface, $\sim\lambda/\sin\alpha$, where α is the angle between the signal and reference waves. If this angle is chosen appropriately, the method is equally effective for all modern lasers. The materials needed for the operating surface are extensively used in optics and readily available.

In WFR via destruction of the reflecting coating, the pulse duration is reduced and the steepness of the leading edge of the pulse increases. These conditions are very important for laser-induced fusion.

The advantages of reversal by a surface have been demonstrated in an experiment with a controlled optical light value, in which a beam of extremely low power ($\sim 3\ \mu W$) was reversed.

The difficulties in WFR for broadband radiation when the reference and signal beams are not coherent must be regarded as disadvantages of this method, because the surfaces known to us are not capable of recording traveling interference gratings. However, WFR of broadband light is not trivial for any other reversal method. At present, the search for a nonlinear surface with a quick reversible response is a very urgent problem.

In conclusion, the method described here is very promising for applications of wave-front reversal.

Acknowledgments

We are greatly indebted to V. V. Ragul'skiy for discussions of the problems considered in this chapter, and to A. A. Talishev and A. I. Ioffe, who helped us with the translation.

References

Blashchuk, V. N., Mamaev, A. V., Pilipetskii, N. F., Shkunov, V. V., and Zel'dovich, B. Ya, (1979a). Wave front reversal with angular tilting theory and experiment for the four wave mixing, *Opt. Commun.* **31,** 383.

Blashchuk, V. N., Zel'dovich, B. Ya, Mamaev, A. V., Pilipetskii, N. F., and Shkunov, V. V. *In* "Trudi VI Vavilovskoi Konferenzii," Part 2, p. 197. Novosibirsk, 1979b.

Bleha, W. P., Lipton, L. T., Wiener-Avnear, E., Grinberg, J., Reif, P. G., Casasent, D., Brown, H. B., and Markevitch, B. (1978). Application of the Liquid Crystal Light Valve to Real-Time Optical Data Processing, *Opt. Eng.* **17,** 371.

Bugaev, A. A., Zakharchenya, B. P., Meshkovskii, I. K., Ovchinnikov, V. M., and Chudnovskii, F. A. (1975). Holography with Reversing Light Reflector, *Pis'ma Zh. Tekh. Fiz.* **1,** 209; *Sov. Tech. Phys. Lett.* (*Engl. Transl.*) **1,** 99 (1975).

de Gennes, P. G. (1974). "The Physics of Liquid Crystals." Oxford Univ. Press (Clarendon), London and New York.

Denisyuk, Y. N. (1962). On the Reproduction of the Optical Properties of an Object by the Wave Field of Its Scattered Radiation, *Opt. Spektrosk.* **15,** 522; *Opt. Spectrosc.* (*Engl. Transl.*) **15,** 279 (1962).

Glenn, W. E. (1959). Thermoplastic Recording, *J. Appl. Phys.* **30,** 1870.

Golubtsov, A. A., Pilipetskii, N. F., Sudarkin, A. N., and Shkunov, V. V. (1981a). Wavefront Reversal by Light-Induced Shaping of the Surface of an Absorbing Material, *Kvantovaya Elektron.* **8,** 663; *Sov. J. Quantum Electron.* (*Engl. Transl.*) **11,** 402 (1981).

Golubtsov, A. A., Pilipetskii, N. D., Sudarkin, A. N., and Shkunov, V. V. (1981b). Self-Defocusing of He–Ne Laser Radiation as a Result of Thermoelastic Deformations of a Reflecting Surface, *Kvantovaya Elektron.* **8,** 370; *Sov. J. Quantum Electron.* (*Engl. Transl.*) **11,** 218 (1981).

Hellwarth, R. W. (1977). Generation of Time-Reversed Wave Fronts by Nonlinear Refraction, *J. Opt. Soc. Am.* **67,** 1.

Kulikov, O. L., Pilipetskii, N. F., Sudarkin, A. N., and Shkunov, V. V. (1980). Inversion of a Wave Front by a Surface, *Pis'ma Zh. Eksp. Teor. Fiz.* **31,** 377; *JETP Lett.* (*Engl. Transl.* **31,** 345 (1980).

Stepanov, B. I., Ivakin, E. V., and Rubanov, A. S. (1971). Recording Two-Dimensional and Three-Dimensional Dynamic Holograms in Bleachable Substances, *Dokl. Akad. Nauk SSSR* **196,** 567; *Sov. Phys.—Dokl.* (*Engl. Transl.*) **16,** 46 (1971).

Vasil'ev, A. A., Garibjan, O. V., Zel'dovich, B. Ya, Kompanets, J. N., Parfenov, A. V., Pilipetskii, N. F., Sudarkin, A. N., Suhov, A. V., Tabiryan, N. V., and Shkunov, V. V. (1983). Wave-Front Reversal in the Presence of mkW Power of the Reference Wave by the Aid of the Optical Light Spatial Modulator, to be published.

Wiggins, T. A. (1980). Reflectivity Changes of Ge and the Efficiency of Moving Diffraction Gratings, *Appl. Opt.* **19,** 521.

Yariv, A. (1976). On Transmission and Recovery of Three-Dimensional Image Information in Optical Waveguides, *J. Opt. Soc. Am.* **66,** 301.

Zel'dovich, B. Ya, and Shkunov, V. V. (1979). Wave Front Reversal in Nonlinear Media, *Gorkii* p. 23.

Zel'dovich, B. Ya, and Tabiryan, N. V. (1981). Possibility of Optical Wavefront Reversal with the Aid of Liquid–Crystal Transparencies, *Kvantovaya Elektron.* **8,** 421; *Sov. J. Quantum Electron.* (*Engl. Transl.*) **11,** 257 (1981).

Zel'dovich, B. Ya, Popovichev, V. I., Ragul'skii, V. V., and Faizullov, F. S. (1972). Connection Between the Wave Fronts of the Reflected and Exciting Light in Stimulated Mandel'shtam–Brillouin Scattering, *Zh. Eksp. Teor. Fiz., Pis'ma Red.* **15,** 160; *Sov. Phys. JETP* (*Engl. Transl.*) **15,** 190 (1972).

Zel'dovich, B. Ya, Pilipetskii, N. F., Sudarkin, A. N., and Shkunov, V. V. (1980). Wave Front Reversal by an Interface, *Dokl. Akad. Nauk SSSR* **252,** 92; *Sov. Phys.—Dokl.* (*Engl. Transl.*) **25,** 377 (1980).

Zel'dovich, B. Ya, and Tabiryan, N. V. (1981). "Possibility of optical wavefront reversal with the aid of liquid–crystal transparencies" *Kvantovaya Elektron.* **8,** 421; *Sov. J. Quantum Electron.* (*Engl. Transl.*) **11,** 257 (1981).

13 Optical Resonators Using Phase-Conjugate Mirrors†

A. E. Siegman
Stanford University
Stanford, California

Pierre A. Belanger
Laval University
Quebec, Canada

Amos Hardy
Weizmann Institute of Science
Rehovot, Israel

† This research was carried out at Stanford University under support from the U.S. Air Force Office of Scientific Research.

Optical Phase Conjugation

ISBN 0-12-257740-X

I. Introduction

A. Phase-Conjugate Resonators

In earlier chapters, the unique reflective properties associated with optical phase conjugation were discussed, for conjugation accomplished by three- or four-wave mixing, stimulated scattering processes, or pseudo-conjugators. In this chapter, we describe the transverse and longitudinal modes and other properties of optical resonators, or optical cavities, in which one end mirror is a more or less ideal optical conjugator or "phase-conjugate mirror" (PCM). A general resonator of this type is shown in Fig. 1. Such a resonator may contain an internal laser gain medium or rely on a phase-conjugate reflectivity greater than unity to achieve oscillation inside the cavity.

A primary reason for being interested in phase-conjugate resonators is the expectation that the phase-conjugate mirror will be able to cancel or at least partially correct for imperfections and phase aberrations in the laser medium and other optical elements inside the resonator. This turns out to be true, at least in part. Such resonators also have a variety of other interesting properties and potential applications, which we discuss briefly later on.

The reults presented in this chapter apply primarily to phase-conjugate mirrors produced by four-wave mixing, or by related kinds of dynamic holography, such as photorefractive effects. Some of the results, especially concerning transverse modes, should also apply to some extent to stimulated scattering processes, but in general not to pseudoconjugators or passive retroreflective mirrors, which are treated elsewhere.

The formal analysis in this chapter summarizes and substantially extends analyses of phase-conjugate resonators that have appeared earlier in the literature, including those of AuYeung *et al.* (1979), Bel'dyugin *et al.* (1979), Bel'yugin and Zemskov (1979), Belanger *et al.* (1980a,b), Lam and Brown (1979, 1980), and Reznikov and Khizhnyak (1980). Modes in resonators with passive retroreflective mirrors have been analyzed by Zhou and Casperson (1981). Phase-conjugate resonator concepts have also been patented by Wang and Yariv (1980).

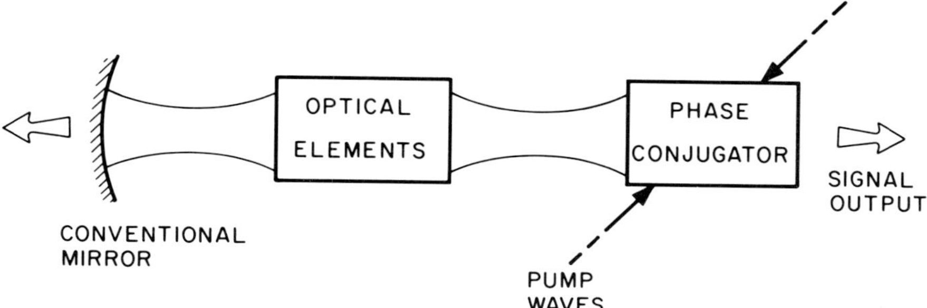

Fig. 1. A general phase-conjugate optical resonator using four-wave mixing. The optical elements inside the resonator may include lenses, aberrating elements, apertures, and/or a laser gain medium.

B. Phase-Conjugate Resonator Experiments and Applications

When the analyses presented in this chapter were first undertaken, useful phase conjugation with large reflectivity seemed to be limited to large pump powers, short pulses, or narrow bandwidths, or all three of these. Hence the analysis of phase-conjugate mirrors applied to optical resonators seemed a largely academic exercise. The dramatic demonstration of low-power cw PCM resonator oscillation recently accomplished by Feinberg and Hellwarth (1980) makes it clear that this view was very pessimistic. Not only can phase-conjugate resonators be made to oscillate, but they also have very interesting and significant differences from conventional optical resonators (for reviews of conventional resonators, see, e.g., Arnaud, 1976; Kogelnik, 1965; Kogelnik and Li, 1966; Siegman, 1971, 1974). Other recent experimental results on phase-conjugate resonators include those of AuYeung, *et al.* (1979), Lind and Steel (1981), Vanherzeele and Van Eck (1981), and Vanherzeele *et al.* (1981).

Phase-conjugate resonators, with or without internal gain media, may be of interest for correcting aberrations in laser optical elements or active laser media; for shaping novel optical wave fronts; for generating ultrashort mode-locked pulses; and for other novel applications yet to be invented. Novel properties that would result from using a phase-conjugate mirror in a ring laser gyroscope have, for example, been suggested by Diels and McMichael (1981). Applications of phase-conjugate mirrors to Mach–Zender and Michelson interferometers have been suggested by Hopf (1980) and Fainman *et al.* (1981) and experimentally demonstrated by Bar Joseph *et al.* (1981). Holographic mirrors have been used for coupling the complex fields radiated by the open end of one optical fiber into another fiber. This suggests that phase conjugation might be very useful in optical waveguide resonators, for reflecting the equally complex fields radiated by the open end of an optical waveguide back down into that waveguide with high efficiency.

C. Outline of This Chapter

The modal properties of an optical resonator, whether conventional or phase conjugate, can almost always be separated into *transverse eigenmodes,* which describe the amplitude and phase variations of the field across planes perpendicular to the resonator axis, and *axial* or *longitudinal eigenmodes* and *resonant frequencies,* which describe the essentially independent variation of the field amplitude and phase along the resonator axis. This separation applies in the present discussion, with unique properties appearing in both the transverse and the longitudinal behavior of phase-conjugate resonators. We therefore first devote several sections to examining the transverse-mode properties of PCM resonators, considering first some general properties of PCM resonators (Section II), then Gaussian resonators with Hermite–Gaussian transverse modes, (Sections III and IV), and finally hard-edged resonators with more complex diffractive effects (Section V). The correction of phase aberrations and other disturbances comes to the fore in this section.

We then consider the longitudinal-mode properties and resonant frequencies of phase-conjugate resonators (Section VI). Phase-conjugate resonators turn out, on the one hand, to have a central resonant frequency that is given by the pumping frequency and is entirely independent of the cavity length L, and, on the other, to have a sequence of paired "half-axial" modes centered at this frequency, whose frequency spacing is equal to $c/4L$, or half the usual axial-mode spacing in conventional laser cavities.

In the remaining sections we summarize the experimental results to date (Section VII), and discuss some of the unique mode-locking and self-Q-switching properties in phase-conjugate resonators (Section VIII). Effects that are similar or closely related to phase conjugation also play a significant role in conventional laser cavities, and we summarize some of these effects in the final section of the chapter (Section IX).

II. Transverse Eigenmodes: Basic Properties

A general model for a phase-conjugate resonator is illustrated schematically in Fig. 2a. The transverse eigenmodes of an optical resonator are conventionally defined as those transverse field patterns (i.e., transverse amplitude and phase distributions) which exactly reproduce themselves in transverse form after one round trip around the resonator, including aperturing and diffraction effects. The overall wave amplitude after one round trip is usually multiplied, however, by a complex eigenvalue with magnitude less than unity that represents the round-trip phase shift and diffraction losses. We wish to find both the eigenmodes and the eigenvalues for PCM resonators.

Mirror reflection coefficients are usually assumed to have unity magnitude in transverse-mode analyses. Phase-conjugate reflectors produced by four-wave mixing with sufficiently intense pump fields may, of course, have a reflection coefficient greater than unity. However, for simplicity we will treat both the conventional and the phase-conjugate mirrors in our

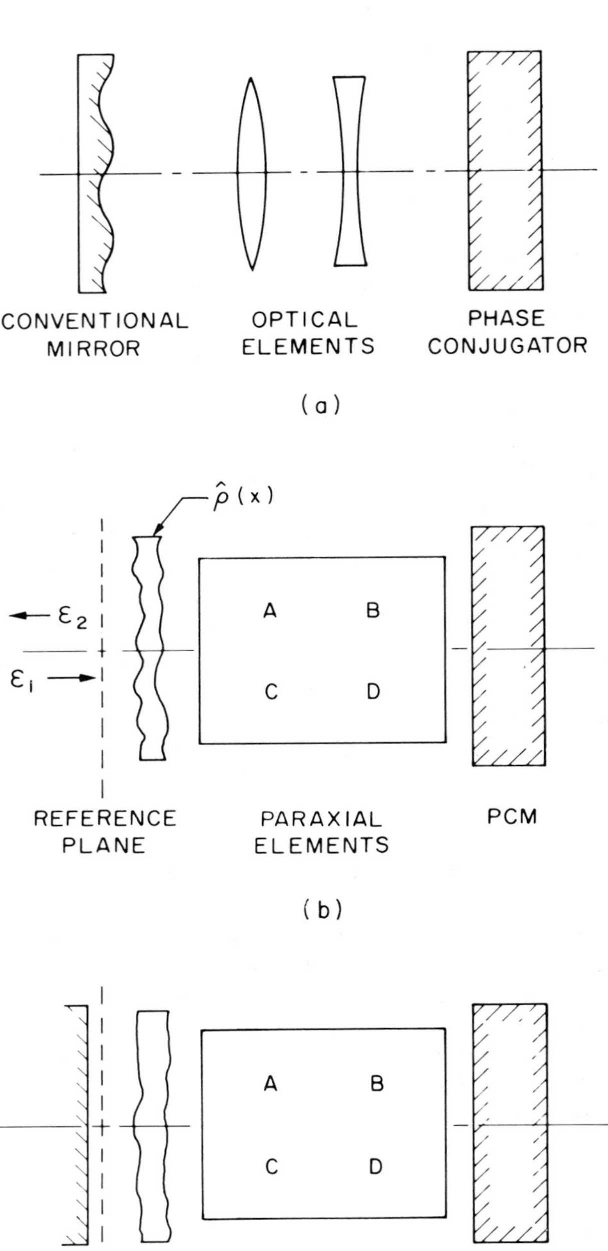
CONVENTIONAL MIRROR
OPTICAL ELEMENTS
PHASE CONJUGATOR
(a)
$\hat{\rho}(x)$
ε_2
ε_1
A
B
C
D
REFERENCE PLANE
PARAXIAL ELEMENTS
PCM
(b)
A
B
C
D
CM
$\hat{\rho}(x)$
PARAXIAL ELEMENTS
PCM
(c)

resonators as having unity magnitude for their reflection coefficients. We will also neglect any gain or loss mechanisms that may be present inside the cavity so long as these are transversely uniform and thus do not change the mode shape.

Just as in conventional laser resonators, the net round-trip amplitude gain or loss produced by the combination of lossy or finite-reflectivity optical elements, diffraction losses, laser gain media, and the amplitude reflectivity of the phase-conjugate mirror will determine the oscillation threshold level, or the net growth or decay with time of the resonant fields in the cavity. However, only the diffraction losses associated with hard or soft apertures inside the cavity need be included in the loss calculations for the transverse eigenmodes.

A. Reflection Properties of a General Phase-Conjugate Mirror

We first establish some useful results for the reflection from an ideal phase-conjugate reflector when it is viewed through various combinations of apertures, arbitrary wave-front-perturbing screens, and general paraxial optical systems, as illustrated in Fig. 2b. Suppose an arbitrary field distribution $\mathscr{E}_1(x_1)$ traveling to the right at a reference plane as in Fig. 2b passes first through a general phase- or amplitude-perturbing screen with a complex amplitude transmission $\hat{\rho}(x)$, and then through an arbitrary series of cascaded paraxial optical elements (lenses, interfaces, ducts, etc.), before reaching an ideal phase conjugator and being reflected back out through the same system. We can develop simple formulas for the overall propagation through this series of optical components.

The set of cascaded conventional paraxial optical elements Fig. 2a can be described by an overall ray matrix or *ABCD* matrix. The elements of this matrix may in general be complex if Gaussian transverse gain or loss

---

Fig. 2. (a) Schematic diagram of a general phase-conjugate resonator with a conventional mirror (CM) at one end, a phase-conjugate mirror (PCM) at the other end, and arbitrary optical elements in between. (b) A transverse field pattern $\mathscr{E}_1(x)$, after passing through a series of optical elements, reflecting from an ideal phase conjugator, and passing back out through the same elements again, produces a field $\mathscr{E}_2(x)$ upon returning to the same reference plane. (c) The distorted conventional mirror shown in (a) can be replaced with a planar conventional mirror plus a purely phase-distorting element $\hat{\rho}(x)$, and the paraxial optical elements in (a) can be replaced with a (possibly complex) *ABCD* matrix. The field distribution immediately in front of the plane mirror in (c) will then be the same as the field pattern on the distorted mirror surface in (a).

variations are present, but will otherwise be real. Huygens' integral for the propagation of a wave front through such a paraxial system by itself can then be written within the Fresnel or paraxial approximation in the general form (Collins, 1970)

$$\mathscr{E}(x) = e^{-jkL} \sqrt{\frac{jk}{2\pi B}} \int_{-w_0}^{w_0} \mathscr{E}_0(x_0) \exp\left[-\frac{jk}{2B} (Ax_0^2 - 2xx_0 + Dx^2) \right] dx_0, \quad (1)$$

where $2w_0$ is the width of the aperture (if any) at the input plane, and $\mathscr{E}_0(x_0)$ is the input wave. We assume the form $\exp[j(\omega t - kz)]$ for the underlying variation of all fields in this section.

In most practical optical resonators, field variations in the x and y transverse coordinates can be separated, assuming that the optical system may have astigmatism but not image rotation. For simplicity, therefore, we have written Eq. (1) in one transverse dimension only; and because we are primarily interested in the transverse wave-front variations only, we will also ignore the on-axis or plane-wave phase-shift term $\exp(-jkL)$ from now on. We now consider the overall reflection from $\mathscr{E}_1$ to $\mathscr{E}_2$ in Fig. 2b.

1. *General Case*

Suppose the phase conjugator in Fig. 2 is pumped at frequency ω_0 and the incident field $\mathscr{E}_1(x_1)$ is at frequency $\omega_1 \equiv \omega_0 + \Delta\omega$. The reflected signal from the phase-conjugate mirror will then be at frequency $\omega_2 = \omega_0 - \Delta\omega$. Let A_1, B_1, C_1, D_1 be the paraxial *ABCD* matrix elements for a signal at frequency ω_1 traveling to the right from just beyond the perturbing screen to the phase-conjugate mirror in Fig. 2b. (We will use capital letters A_i, B_i, C_i, D_i to refer to conventional paraxial *ABCD* matrices, real or complex, for a wave traveling to the right through a conventional optical system, with the subscript indicating the frequency at which the matrix elements are evaluated; the symbols $\tilde{\mathscr{A}}$, $\tilde{\mathscr{B}}$, $\tilde{\mathscr{C}}$, $\tilde{\mathscr{D}}$ will be used later to refer to a special kind of *ABCD* matrix arising in phase-conjugate systems.) The paraxial matrix for a wave traveling from the PCM back to the reference plane through the same elements at frequency ω_2 will then be D_2, B_2, C_2, A_2. Any small differences between A_1, B_1, C_1, D_1 and A_2, B_2, C_2, D_2 will arise only from the small frequency difference between ω_1 and ω_2. In particular, the two matrices become identical for the degenerate case $\omega_1 = \omega_2$.

The reflected field $\mathscr{E}_2(x_2)$ coming out of the reference plane after traveling through the perturbing screen and into the PCM, reflecting off the PCM, and traveling back out through the screen again will be given by the double integral

$$\mathscr{E}_2(x_2) = \hat{\rho}(x_2)\sqrt{\frac{k_1k_2}{4\pi^2B_1^*B_2}}\int_{-w_{\text{pcm}}}^{w_{\text{pcm}}} dx_{\text{pc}} \int_{-w_{\text{cm}}}^{w_{\text{cm}}} dx_1\, \hat{\rho}^*(x_1)\mathscr{E}_1^*(x_1)$$
$$\times \exp\left[\frac{jk_1}{2B_1^*}(A_1^*x_1^2 - 2x_{\text{pc}}x_1 + D_1^*x_{\text{pc}}^2)\right.$$
$$\left. - \frac{jk_2}{2B_2}(D_2x_{\text{pc}}^2 - 2x_2x_{\text{pc}} + A_2x_2^2)\right]. \tag{2}$$

Here $2w_{\text{cm}}$ is the width of any aperture at the conventional mirror or reference plane, $2w_{\text{pcm}}$ is the width of any aperture at the phase-conjugate mirror end, and the wave front $\mathscr{E}_2(x_2)$ is now at frequency ω_2. The aperture at the conventional mirror end can of course be absorbed into the screen function $\hat{\rho}(x)$ by setting $\hat{\rho}(x) = 0$ for $|x| > w_{\text{cm}}$, after which the corresponding limits of integration can be extended to infinity.

2. *Unbounded Phase-Conjugate Mirror*

The ideal case is an unbounded phase-conjugate mirror ($w_{\text{pcm}} \to \infty$). Reversing the order of integration and evaluating the integral over dx_{pc} then reduces Eq. (2) to the form

$$\mathscr{E}_2(x_2) = \hat{\rho}(x_2)\sqrt{\frac{jk_2}{2\pi\tilde{\mathscr{B}}}}\int_{-\infty}^{\infty} dx_1\, \hat{\rho}^*(x_1)\mathscr{E}_1^*(x_1)$$
$$\times \exp\left[-\frac{jk_2}{2\pi\tilde{\mathscr{B}}}(\tilde{\mathscr{A}}x_1^2 - 2x_1x_2 + \tilde{\mathscr{D}}x_2^2)\right]. \tag{3}$$

The complex quantities $\tilde{\mathscr{A}}$, $\tilde{\mathscr{B}}$, $\tilde{\mathscr{C}}$, $\tilde{\mathscr{D}}$ appearing in this integral are the elements of an "equivalent phase-conjugate ray matrix," or an equivalent *ABCD* matrix for the double-passed system, given by

$$\begin{bmatrix} \tilde{\mathscr{A}} & \tilde{\mathscr{B}} \\ \tilde{\mathscr{C}} & \tilde{\mathscr{D}} \end{bmatrix} = \begin{bmatrix} D_2 & B_2 \\ C_2 & A_2 \end{bmatrix}\begin{bmatrix} A_1^* & -(k_2/k_1)B_1^* \\ -(k_1/k_2)C_1^* & D_1^* \end{bmatrix}$$
$$= \begin{bmatrix} A_1^*D_2 - (k_1/k_2)B_2C_1^* & B_2D_1^* - (k_2/k_1)B_1^*D_2 \\ A_1^*C_2 - (k_1/k_2)A_2C_1^* & A_2D_1^* - (k_2/k_1)B_1^*C_2 \end{bmatrix} \tag{4}$$

Note that for the degenerate case where $k_1 = k_2$ the off-diagonal elements $\tilde{\mathscr{B}}$ and $\tilde{\mathscr{C}}$ of this matrix become purely imaginary and $\tilde{\mathscr{D}} = \tilde{\mathscr{A}}^*$. The two-way trip into the PCM and back out is evidently equivalent to taking the complex conjugate of the input field $\mathscr{E}_1(x_1)$, and then propagating this conjugated field at frequency ω_2 by applying the usual Huygens integral of Eq. (1), but using the phase-conjugate equivalent $\tilde{\mathscr{A}}\tilde{\mathscr{B}}\tilde{\mathscr{C}}\tilde{\mathscr{D}}$ matrix given by Eq. (4).

3. *Finite Phase-Conjugate Mirror, Purely Real Paraxial Elements*

Suppose the paraxial optical elements remain degenerate ($\omega_1 = \omega_2 = \omega_0$) and also purely real (no transverse loss or gain variations), but the phase conjugator has only a finite width extending from $-\omega_{\mathrm{pcm}}$ to ω_{pcm}. Then the finite integral of Eq. (2) over dx_{pc} can still be carried out, leading to the equation given by Lam and Brown (1980),

$$\mathscr{E}_2(x_2) = \hat{\rho}(x_2) \int_{-\infty}^{\infty} \hat{\rho}^*(x_1)\mathscr{E}_1^*(x_1) \exp\left[-\frac{jkA}{2B}(x_2^2 - x_1^2)\right] \times \frac{\sin[(kw_{\mathrm{pcm}}/B)(x_2 - x_1)]}{\pi(x_2 - x_1)}\, dx_1. \tag{5}$$

The kernel in this equation differs in an interesting way from the usual Huygens kernel. The equation says in effect that the diffractive effects of a sharp aperture immediately in front of a phase conjugator appear as a kind of filtering of the usual Huygens kernel with a filter of the form $\mathrm{sinc}[(kw_{\mathrm{pcm}}/B)(x_2 - x_1)]$, where $\mathrm{sinc}\, x \equiv (\sin x)/x$.

4. *Unbounded Phase Conjugator, Phase Perturbations Only*

When the width of the phase conjugator becomes very large and $w_{\mathrm{pcm}} \to \infty$, the sinc function becomes effectively a Dirac δ function, and Eq. (5) reduces to

$$\mathscr{E}_2(x) = |\hat{\rho}(x)|^2 \mathscr{E}_1^*(x). \tag{6}$$

Suppose in addition the perturbing screen contains only phase and not amplitude perturbations, so that $|\hat{\rho}(x)|^2 \equiv 1$. Then this result becomes simply $\mathscr{E}_2(x) = \mathscr{E}_1^*(x)$. This verifies mathematically the semiobvious conclusion that *an ideal phase-conjugate mirror seen through any optical system containing only arbitrary* phase *perturbations is still an ideal phase-conjugate mirror*. Any transverse *amplitude* variations, however, whether they occur in the *ABCD* elements or in the perturbing screen $\rho(x)$, will reduce or destroy the ideal phase conjugation.

B. Unbounded Phase-Conjugate Resonators

Suppose now an unbounded 100% reflecting conventional mirror with an arbitrary surface contour as in Fig. 2a is set up facing an unbounded phase-conjugate system as just described. The arbitrarily distorted end mirror can be replaced with an ideal plane mirror plus a suitable phase-perturbing screen with a phase perturbation proportional to the surface deviation of the mirror, as illustrated in Fig. 2c. The field variation $\mathscr{E}(x)$ across the plane mirror in Fig. 2c will then correspond to the field profile on the surface of the distorted end mirror in Fig. 2a. Any phase distortion in the end mirror can thus be absorbed into the total perturbing element $\hat{\rho}(x)$.

For the limiting case of degenerate signals, real optical elements, and pure phase perturbations only, the resonator eigenvalue equation then becomes simply

$$\mathscr{E}_2(x) = \mathscr{E}_1^*(x) = \gamma\mathscr{E}_1(x), \tag{7}$$

where γ is the resonator eigenvalue and $\mathscr{E}_1(x)$ is the field on the surface of the conventional end mirror. Equation (7) is evidently satisfied by any field distribution that has constant phase but arbitrary amplitude variations, i.e., $\mathscr{E}_1(x) = |\mathscr{E}_1(x)| \exp(j\theta_1)$, with $\theta_1 = \text{const}$. The associated eigenvalue is $\gamma = \exp(-2j\theta_1)$, so that $|\gamma| = 1$. We conclude, therefore, that in an ideal degenerate phase-conjugate resonator (i.e., one with an unbounded phase conjugator, arbitrary phase perturbations, but no transverse amplitude variations) *any wave front with a phase surface matching the conventional mirror surface, but with any arbitrary amplitude profile, is self-reproducing after one round trip*. Such an ideal PCM resonator thus has, depending on one's viewpoint, either an infinite variety of transverse modes or no unique eigenmodes at all.

A second conclusion is that *the distinction between geometrically stable and unstable periodic focusing systems, so fundamental to conventional optical resonators, completely disappears in phase-conjugate resonators*. In Fig. 3 we show, for example, how in a phase-conjugate resonator, even with a strongly divergent conventional end mirror, the diverging wave front of a resonant mode will simply be focused back onto the mirror, rather than being diverged and magnified on successive round

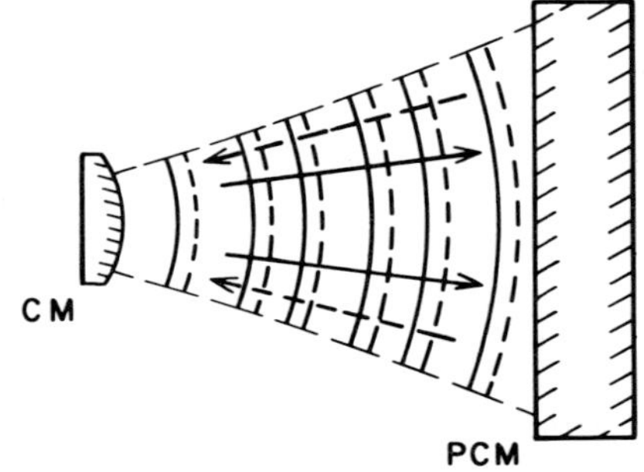

Fig. 3. A phase-conjugate mirror can reflect the diverging beam from a divergent conventional mirror back to the same mirror surface again, with no net magnification per round trip. If both the conventional mirror and the phase conjugator are unbounded transversely, any field distribution whose phase front matches the conventional mirror surface, with any arbitrary transverse amplitude variation, will be self-reproducing after a single round trip.

trips as would be the case in a conventional unstable resonator. There is essentially no such thing as an "unstable" phase-conjugate resonator. This observation may point the way to obtaining large-mode-volume but still low-loss phase-conjugate resonators.

A third conclusion can be derived by cascading Eq. (6) through two complete round trips to obtain $\mathscr{E}_3(x) = |\hat{\rho}|^4\mathscr{E}_1(x)$, where $\mathscr{E}_3$ is the output after two round trips. If $\hat{\rho}$ has only phase variations, we can say that *in an ideal phase-conjugate resonator (i.e., no apertures) any amplitude and phase pattern, whether the phase front matches the mirror surface or not, is self-reproducing after two complete round trips*. This behavior is illustrated for a simple example in Fig. 4, in which the second round trip is shown as if it were folded over onto the opposite side of the conjugate reflector. Note that this result does not mean that the double-passed wave front is therefore a transverse mode. Rather, this situation should be viewed as a mixture of transverse modes beating with each other on alternate round trips.

These general properties of ideal phase-conjugate resonators have the following implications for active PCM resonators, whether they employ intracavity laser gain or a phase-conjugate mirror with net gain, i.e., with a reflection coefficient greater than unity. First, the optical resonator shown in Fig. 1 may have serious optical phase aberrations either in the intracavity elements or in the conventional end mirror. If the intracavity distortions are purely phase distortions, no matter how "thick" these may be, then an ideal phase-conjugate mirror (without any aperturing) will produce a uniphase wave front on the conventional mirror surface. This means an essentially diffraction-limited output beam from the left-

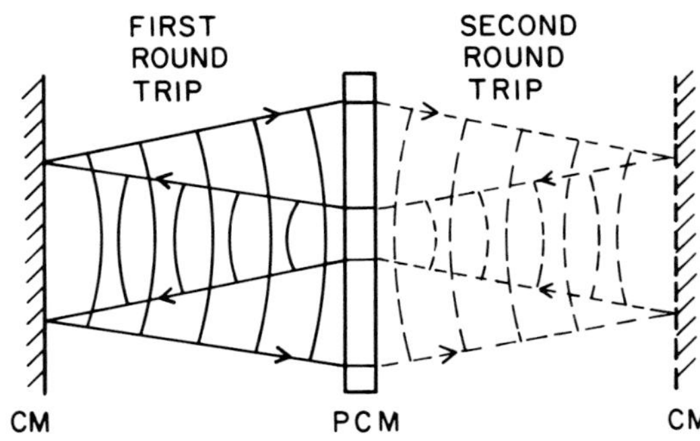

Fig. 4. An ideal phase conjugator causes any wave front, whether it matches the conventional mirror surface or not, to be self-reproducing after *two* complete round trips. The phase conjugator acts like a convergent mirror or positive lens for diverging waves, but like a negative lens for converging waves.

hand or conventional mirror end of this resonator (given that the diffraction spreading of a laser beam is much more sensitive to its phase profile than to its amplitude profile). Severe phase distortion in the optical elements can be effectively canceled. The cavity will not experience either the severe reduction in beam quality or the severe power losses that intracavity phase perturbations commonly produce (Siegman, 1977a).

Of course, if the conventional mirror itself has a badly wrinkled surface, the resonator will still oscillate with low losses; but the output beam through the conventional mirror will reproduce the mirror's wrinkles exactly, and thus have bad beam quality. Some other solution must then be found.

If either the intracavity elements or the end mirrors have *amplitude* perturbations, however, the phase conjugator will not in general cancel these perturbations. It is not entirely clear as yet exactly what the conjugator will then accomplish, although it is likely to produce some improvement, certainly in reduced cavity losses and probably in improved beam quality. Further numerical calculations addressed to the effects of amplitude perturbations in PCM resonators are needed.

C. Phase-Conjugate Resonators with Finite Apertures

The infinite variety of transverse amplitude patterns possible in an unbounded phase-conjugate resonator is destroyed as soon as any finite aperture, or any other transverse gain or loss variation, is added to the resonator. Transverse gain variations or apertures produce diffraction effects for the waves going in both directions inside a phase-conjugate resonator. These diffraction effects cannot be fully canceled by the phase-conjugate mirror, because the conjugator cannot in general intercept all the light. Sharp-edged apertures in particular will introduce losses and edge diffraction effects that cannot be compensated for by the PCM.

We will explore this behavior in subsequent sections, both for soft (i.e., Gaussian) transverse apertures and for hard-edged apertures. As a general rule, however, we can say that *adding either soft or hard apertures to a phase-conjugate resonator converts the continuous distribution of possible transverse amplitude patterns into a discrete set of lowest-order and higher-order transverse eigenmodes.* These modes generally continue to have phase fronts that match closely, but no longer exactly, the surface of the conventional end mirror. They also have amplitude patterns that generally minimize the diffraction losses produced by the aperture on the transverse eigenmodes. Optical resonators in general have a marvelous ability to adapt their mode patterns so as to minimize their diffraction losses, while still remaining within the constraints of the wave equation. Phase-conjugate resonators with finite apertures by and large simply take advantage of the properties of phase conjugation to do an even more effective job of this. All of these points will be illustrated by examples in the following sections.

D. An Important Limitation

Finally, we must note that phase-conjugating devices act in general to cancel the phase perturbations in an optical system by conjugating the wave *reflected* from the conjugator. However, most phase-conjugating devices—at least all those based on degenerate four-wave mixing—also have a net power gain, *with no conjugation effects,* in the *forward* direction. The forward gain of a four-wave mixing device is in many cases larger than the backward reflectivity. Thus the output from the right-hand end of the cavity in Fig. 1 will retain any distortions caused by the intracavity perturbations, and will also be amplified by the four-wave mixing process. In this situation most of the output power may come from the wrong end of the PCM cavity. A double-ended PCM cavity, with optical elements and conventional mirrors folded about both sides of the phase-conjugate element, may be a solution to this difficulty.

III. Hermite–Gaussian Transverse Eigenmodes

The transverse eigenmodes of a conventional optical resonator or lens waveguide can be closely approximated by Hermite–Gaussian functions so long as (a) the resonator is geometrically stable, (b) diffraction effects due to hard apertures are negligible, and (b) no optical elements with transverse gain or phase-shift variations of order higher than quadratic are present (Kogelnik and Li, 1966; Siegman, 1971). The transverse eigenmodes for phase-conjugate resonators will similarly turn out to be of Hermite–Gaussian form if (a) hard-edge diffraction effects are small and (b) the resonator contains only conventional (though possibly complex) quadratic paraxial elements in addition to the phase-conjugate mirror. Geometrical stability in the conventional resonator sense is not, however, required (or even meaningful) in the PCM case.

We will solve for these Hermite–Gaussian PCM modes in this section. These solutions will not tell us about the behavior of PCM resonators in the presence of aberrations, but they will give us the fundamental mode shapes in simple paraxial PCM resonators.

A. Complex Paraxial System Elements

We first remind the reader that paraxial optical elements whose amplitude transmission varies transversely in the form $\exp(-\alpha_2 x^2)$ [or possibly in the reverse form $\exp(+\alpha_2 x^2)$] can be described analytically by complex elements in the *ABCD* matrices of the optical system. We will refer to such transversely varying elements in general as Gaussian apertures, Gaussian ducts, or Gaussian variable-reflectivity mirrors (Zucker, 1970; Yariv and Yeh, 1975).

We also note again that the surface of a curved conventional mirror can be replaced with a planar mirror behind an appropriate thin lens. This lens can then be absorbed into the overall single-pass *ABCD* matrix from the mirror surface to the PCM. We shall assume that this has been done, so that the field patterns we describe are in all cases the fields on the curved surface of a real end mirror, or on the planar surface of an equivalent planar end mirror.

In two earlier papers (Belanger *et al.*, 1980a,b) we analyzed the Hermite–Gaussian modes of PCM resonators using as our reference plane the surface immediately in front of the phase conjugator. However, experience shows that using a reference plane located directly on the surface of the conventional mirror, as done by AuYeung *et al.* (1979), is generally a better choice—the analytical results become somewhat simpler, and the results are more physically useful. Therefore in this section we derive the Hermite–Gaussian transverse modes characteristic of this class of phase-conjugate resonators, starting from the conventional mirror surface as the reference plane.

B. Hermite–Gaussian Mode Propagation

We first examine the propagation properties of the most general complex Hermite–Gaussian modes, in the most general paraxial case. A general Hermite–Gaussian transverse mode at frequency ω with complex radius of curvature $\hat{q}$ and with a "complex spot size" $\hat{v}$ may be written as

$$\mathscr{E}(x) = \alpha_m \hat{v}^m H_m \left(\frac{\sqrt{2}x}{\hat{v}}\right) \exp\left(-j\frac{kx^2}{2\hat{q}}\right), \tag{8}$$

where α_m is a complex amplitude coefficient and H_m the Hermite polynomial of order m. The complex radius of curvature $\hat{q}$ is defined as usual by

$$\frac{1}{\hat{q}} \equiv \frac{1}{R} - j\frac{\lambda}{\pi w^2}, \tag{9}$$

with R being the real radius of curvature and w the real Gaussian spot size, so that the fields have a transverse amplitude variation like $\exp(-x^2/w^2)$. The parameter $\hat{v}$, which appears only in the higher-order Hermite functions, is a complex generalization of the real spot size w. This parameter is normally real and equal to w in elementary Gaussian beam theory. However, it can become an independent complex quantity in general, and especially so when complex $ABCD$ matrices are involved.

Let a Hermite–Gaussian mode in the general form of Eq. (8) with parameters α_{m1}, $\hat{v}_1$, k_1, and $\hat{q}_1$ be substituted into the round-trip propagation integral of Eq. (3). (In other words, this mode will be the input wave propagating to the right, just inside the planar mirror surface in Fig. 2c.) The reflected field at this same plane at frequency ω_2 after one round trip will then be another Hermite–Gaussian function of the same order and the same form as Eq. (8), but with transformed values α_{m2}, $\hat{v}_2$, and $\hat{q}_2$ that are given in terms of the phase-conjugate $\tilde{\mathscr{A}}$, $\tilde{\mathscr{B}}$, $\tilde{\mathscr{C}}$, $\tilde{\mathscr{D}}$ matrix elements of Eq. (4) by

$$\begin{aligned} \alpha_{m2} &= \frac{\alpha_{m1}^*}{[\tilde{\mathscr{A}} - (k_1/k_2)\tilde{\mathscr{B}}/\hat{q}_1^*]^{m+1/2}}, \\ \hat{v}_2^2 &= [\tilde{\mathscr{A}} - (k_1/k_2)\tilde{\mathscr{B}}/\hat{q}_1^*]^2\hat{v}_1^{*2} + \frac{4j\tilde{\mathscr{B}}}{k_2}[\tilde{\mathscr{A}} - (k_1/k_2)\tilde{\mathscr{B}}/\hat{q}_1^*], \\ \frac{1}{\hat{q}^2} &= \frac{\tilde{\mathscr{D}}/\hat{q}_1^* - (k_2/k_1)\tilde{\mathscr{C}}}{\tilde{\mathscr{B}}/\hat{q}_1^* - (k_2/k_1)\tilde{\mathscr{A}}}. \end{aligned} \tag{10}$$

These formulas summarize what happens to a Hermite–Gaussian beam in one round trip around a phase-conjugate resonator. The tranformation rule for α_m ignores the round-trip axial phase shift $\exp[-j(k_2 - k_1)L]$, as well as any constant phase shift due to reflection at the PCM. Equations (10) are very similar to the propagation rules for Gaussian beams in conventional optical systems, but differ in the minus signs, the complex conjugation of $\hat{q}$, and the use of the special phase-conjugate $\tilde{\mathscr{A}}\tilde{\mathscr{B}}\tilde{\mathscr{C}}\tilde{\mathscr{D}}$ matrix.

C. Hermite–Gaussian Resonator Eigenmodes

The self-consistent Hermite–Gaussian eigenmodes of a phase-conjugate resonator are those values of $\hat{q}_1 = \hat{q}_2 = \hat{q}_{\mathrm{cm}}$ and $\hat{v}_1 = \hat{v}_2 = \hat{v}_{\mathrm{cm}}$ at the conventional mirror which obey Eqs. (10), and which are self-reproducing after one round trip for the degenerate case $\omega_1 = \omega_2$, or after two round trips for the nondegenerate case. We will examine these Hermite–Gaussian eigenmodes for some typical PCM resonators shortly. The resulting self-consistent $\hat{q}$ values can be conveniently represented as points or loci in a complex $1/\hat{q}$ plane (actually the $1/\hat{q}^*$ plane) with x and y

axes defined by

$$\frac{1}{\hat{q}} \equiv x - jy = \frac{1}{R} - j\frac{\lambda}{\pi w^2}. \tag{11}$$

Points in the upper half plane ($y > 0$) then correspond to *confined* Hermite–Gaussian modes, i.e., modes for which the transverse field variation $\exp(-x^2/w^2)$ decays rather than grows at infinite radius. To be physically significant, Hermite–Gaussian resonator eigenmodes must be not only confined ($w_{\mathrm{cm}}^2 > 0$) but also *perturbation stable* (Casperson, 1974). Perturbation stability requires that small perturbations $\delta\hat{q}$ or $\delta\hat{v}$ about the self-consistent eigensolutions $\hat{q}_{\mathrm{cm}}$ and $\hat{v}_{\mathrm{cm}}$ should decay rather than growing exponentially on successive round trips. For complex *ABCD* matrices, not all confined eigensolutions are necessarily perturbation stable, and vice versa.

D. Degenerate Eigensolutions: Real Paraxial Elements

The Hermite–Gaussian eigensolutions of a PCM resonator for the degenerate case $\omega_1 = \omega_2 = \omega_0$ are found as follows. The $\tilde{\mathscr{A}}$, $\tilde{\mathscr{B}}$, $\tilde{\mathscr{C}}$, $\tilde{\mathscr{D}}$ matrix elements for the PCM cavity as given in Eq. (4) may be written in general in the convenient forms

$$\begin{aligned} \tilde{\mathscr{A}} &= \mathscr{A}_m e^{-j\theta_m} = \mathscr{A}_1 - j\mathscr{A}_2, \\ \tilde{\mathscr{B}} &= j\mathscr{B}_m, \\ \tilde{\mathscr{C}} &= -j\mathscr{C}_m, \\ \tilde{\mathscr{D}} &= \tilde{\mathscr{A}}^*, \end{aligned} \tag{12}$$

where $\mathscr{A}_m^2 - \mathscr{B}_m\mathscr{C}_m = 1$. (We will use $\mathscr{A}_m$, etc., for the magnitudes of these complex matrix elements, and $\mathscr{A}_1$ and $\mathscr{A}_2$ for the real and imaginary parts of $\tilde{\mathscr{A}}$.) The transformation of the $1/\hat{q}$ value in one round trip is then given by

$$\frac{1}{\hat{q}_2} = \frac{\mathscr{A}_m e^{j\theta_m}/\hat{q}_1^* + j\mathscr{C}_m}{j\mathscr{B}_m/\hat{q}_1^* - \mathscr{A}_m e^{-j\theta_m}}, \tag{13}$$

where $\hat{q}_1$ is the value before and $\hat{q}_2$ the value after one round trip. With purely real *ABCD* elements, which means $\mathscr{A}_m = 1$ and $\theta_m = \mathscr{B}_m = \mathscr{C}_m = 0$, this becomes

$$1/\hat{q}_2 = -1/\hat{q}_1^*, \tag{14}$$

which has the self-consistent eigensolutions

$$\operatorname{Re}\left(\frac{1}{\hat{q}_{\mathrm{cm}}}\right) \equiv \frac{1}{R_{\mathrm{cm}}} = 0, \qquad \operatorname{Im}\left(\frac{1}{\hat{q}_{\mathrm{cm}}}\right) \equiv \frac{-j\lambda}{\pi w_{\mathrm{cm}}^2} = \text{arbitrary}. \tag{15}$$

This family of solutions for real *ABCD* elements maps into a locus of self-reproducing solutions located anywhere on the *y* axis in the complex $1/\hat{q}$ plane, as illustrated in Fig. 5a. These solutions correspond physically to a uniphase wave front on the end mirror (i.e., $R_{\mathrm{cm}} = \infty$) and an arbitrary Gaussian spot size w_{cm}. This simply reiterates our earlier conclusion that in an ideal PCM resonator any field pattern with phase front matching the end mirror profile and with any amplitude profile will be self-reproducing.

Equation (13) also says that a Gaussian beam launched from any initial point $1/\hat{q}_1$ in such a system, with finite initial curvature R_1 and any spot size w_1, is simply reflected in the imaginary axis in the $1/\hat{q}$ plane after each round trip, as illustrated in Fig. 5a. The beam returns to the mirror after one round trip with the same spot size w_1 and reversed curvature $R_2 = -R_1$. We conclude that any Gaussian beam in a real-matrix PCM resonator is marginally perturbation stable, and moreover is automatically self-consistent after *two* round trips. This behavior has been illustrated schematically in Fig. 4.

The self-consistent locus for $1/\hat{q}_{\mathrm{cm}}$ as measured at the conventional mirror end of the resonator can be transformed to any other plane within the phase-conjugate resonator by using the appropriate *ABCD* matrix and the rules for Gaussian beam propagation. For example, the equivalent locus at the phase-conjugate mirror end is given by

$$\hat{q}_{\mathrm{pcm}} = (A\hat{q}_{\mathrm{cm}} + B)/(C\hat{q}_{\mathrm{cm}} + D). \tag{16}$$

This represents a conformal transformation in the $1/\hat{q}$ plane. For purely real *ABCD* elements, Eq. (16) transforms the straight-line *y*-axis locus at the conventional mirror end into a circle at the PCM end, as shown in earlier analyses (Belanger *et al.*, 1980a,b) and illustrated in Fig. 5b. Even for complex paraxial systems with transversely varying losses, the imaginary parts of the *ABCD* elements are usually small, and the same interpretation remains approximately correct.

E. Complex Paraxial Elements

For *complex-valued ABCD* elements (i.e., systems with transversely varying loss or gain) the self-consistent solutions to Eq. (13) are given by

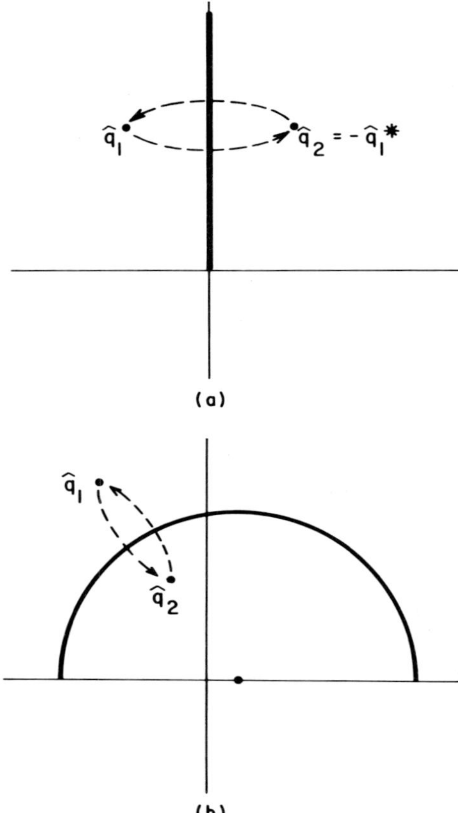

Fig. 5. (a) For an unbounded degenerate phase-conjugate resonator with $\omega_1 = \omega_2$ and with purely real paraxial elements (i.e., no transversely varying gain or loss), the Hermite–Gaussian eigensolutions on the conventional mirror surface can be located anywhere along the y axis in the $1/\hat{q}$ plane ($x \equiv 1/R_{\text{cm}}$ and $y \equiv \lambda/\pi w_{\text{cm}}^2$). This means physically that the phase front matches the mirror curvature, with $R_{\text{cm}} = \infty$, while the spot size w_{cm} is arbitrary. Any Hermite–Gaussian beam launched from the conventional mirror surface with an arbitrary initial parameter $\hat{q}_1$ will simply reflect back and forth on successive round trips between $\hat{q}_1$ and a mirror image value $\hat{q}_2 = -\hat{q}_1^*$. (b) The locus for the degenerate eigenvalues at the phase-conjugator end becomes a semicircle in the $1/\hat{q}$ plane with $x \equiv 1/R_{\text{pcm}}$ and $y \equiv \lambda/\pi w_{\text{pcm}}^2$. Round-trip propagation from an arbitrary starting point then transforms into inversion with respect to this circle. (From Belanger *et al.*, 1980a.)

the roots of

$$\frac{1}{\hat{q}\hat{q}^*} + \frac{j}{\mathcal{B}_m}\left(\frac{\tilde{\mathcal{A}}}{\hat{q}} + \frac{\tilde{\mathcal{A}}^*}{\hat{q}^*}\right) - \frac{\mathcal{C}_m}{\mathcal{B}_m} = 0. \tag{17}$$

This equation may be separated into the two parts

$$\left|\frac{1}{\hat{q}}\right|^2 = \frac{\mathscr{C}_m}{\mathscr{B}_m} \quad \text{and} \quad \mathrm{Re}\left(\frac{\tilde{\mathscr{A}}}{\hat{q}}\right) = 0. \tag{18}$$

Provided that $\mathscr{C}_m$ and $\mathscr{B}_m$ have the same sign, there are then two unique self-consistent points in the $1/\hat{q}$ plane given by

$$\begin{aligned} x_{\mathrm{cm}} &\equiv \frac{1}{R_{\mathrm{cm}}} = \pm\sqrt{\frac{\mathscr{C}_m}{\mathscr{B}_m}\frac{\mathscr{A}_2}{\mathscr{A}_m}} = \pm\sqrt{\frac{\mathscr{C}_m}{\mathscr{B}_m}}\sin\theta_m, \\ y_{\mathrm{cm}} &\equiv \frac{\lambda}{\pi w_{\mathrm{cm}}^2} = \sqrt{\frac{\mathscr{C}_m}{\mathscr{B}_m}\frac{\mathscr{A}_1}{\mathscr{A}_m}} = \pm\sqrt{\frac{\mathscr{C}_m}{\mathscr{B}_m}}\cos\theta_m, \end{aligned} \tag{19}$$

or, in condensed form,

$$\frac{1}{\hat{q}_{\mathrm{cm}}} = \frac{1}{R_{\mathrm{cm}}} - j\frac{\lambda}{\pi w_{\mathrm{cm}}^2} = \mp j\sqrt{\mathscr{C}_m/\mathscr{B}_m}\,\exp(j\theta_m). \tag{20}$$

Of these two roots, of course, only the confined solution with $y_{\mathrm{cm}} > 0$ and hence with $w_{\mathrm{cm}}^2 > 0$ can be physically meaningful.

For purely real paraxial systems, the angle θ_m will become vanishingly small, and the magnitude ratio $\mathscr{C}_m/\mathscr{B}_m$ will become indeterminate. In this limit Eqs. (19) convert to the continuous y-axis locus of Eq. (15). For complex systems with a transmission maximum on axis and with not too rapid transverse variation, which is the usual situation, the angle θ_m will normally be small. The two discrete eigensolutions given by Eqs. (19) or (20) will then lie close to the y axis, at equal distances above and below the x axis, as we will illustrate shortly. The point below the x axis of course represents a nonphysical solution whose fields increase rather than decrease with distance away from the resonator axis.

F. Mode Stability

Suppose a Gaussian beam is in some fashion launched into a phase-conjugate resonator just inside the conventional mirror end with an initial Gaussian beam parameter that differs from the exact self-consistent value $1/\hat{q}_{\mathrm{cm}} \equiv \hat{z}_{\mathrm{cm}}$ by a small perturbation $\delta\hat{z} \equiv \delta(1/\hat{q}) \ll 1/\hat{q}$. We can then write the Gaussian curvature parameter for this beam one round trip later, which we will call $\hat{z}' \equiv 1/\hat{q}'$, in the following form:

$$\hat{z}' \equiv \hat{z}_{\mathrm{cm}} + \delta\hat{z}' = \frac{\tilde{\mathscr{D}}(\hat{z}_{\mathrm{cm}} + \delta\hat{z})^* - \tilde{\mathscr{C}}}{\tilde{\mathscr{B}}(\hat{z}_{\mathrm{cm}} + \delta\hat{z})^* - \tilde{\mathscr{A}}}. \tag{21}$$

Expanding this to first order in the small perturbation $\delta\hat{z}$ and making use of $\tilde{\mathscr{A}}\tilde{\mathscr{D}} - \tilde{\mathscr{B}}\tilde{\mathscr{C}} = 1$ gives the growth ratio for the perturbation after one round trip:

$$\frac{\delta\hat{z}'}{\delta\hat{z}^*} = \frac{-1}{(\tilde{\mathscr{A}} - \tilde{\mathscr{B}}/\hat{q}^*)^2}, \tag{22}$$

where $\hat{q}$ is one or the other of the self-consistent solutions from Eqs. (19). Each of these self-consistent solutions will or will not be perturbation stable depending on whether the small perturbation $\delta\hat{z} \equiv \delta(1/\hat{q})$ decays or enlarges on successive round trips. We must ask, therefore, whether $|\delta\hat{z}'/\delta\hat{z}^*|$ has magnitude grater or less than unity. Putting the degenerate self-consistent results from Eqs. (20) into Eq. (22) while keeping proper track of the plus and minus signs yields

$$\left|\frac{\delta\hat{z}'}{\delta\hat{z}^*}\right| = \frac{1}{(\mathscr{A}_m \pm \mathscr{B}_m\sqrt{\mathscr{C}_m/\mathscr{B}_m})^2}. \tag{23}$$

Noting that $\mathscr{A}_m^2 - \mathscr{B}_m\mathscr{C}_m \equiv (\mathscr{A}_m + \sqrt{\mathscr{B}_m\mathscr{C}_m})(\mathscr{A}_m - \sqrt{\mathscr{B}_m\mathscr{C}_m}) = 1$ leads us to define a magnification M given by

$$M \equiv \mathscr{A}_m + \sqrt{\mathscr{B}_m\mathscr{C}_m}. \tag{24}$$

This parameter is vaguely similar though not identical to the magnification M commonly defined for conventional unstable resonators. Evidently M is greater than unity, at least when $\mathscr{B}_m$ and $\mathscr{C}_m$ have the same sign, as they must if any self-consistent solutions are to exist at all. The growth ratio for the perturbations then becomes one or the other of

$$\left|\frac{\delta\hat{z}'}{\delta\hat{z}^*}\right| = \frac{1}{M^2} \quad \text{or} \quad \left|\frac{\delta\hat{z}'}{\delta\hat{z}^*}\right| = M^2. \tag{25}$$

The first case applies if $\mathscr{B}_m > 0$ and the upper sign in Eqs. (19) and (20) is chosen, or if both of these conditions are reversed. Otherwise, the second case applies, in which case the perturbations grow on successive round trips.

To be physically useful a transverse eigenmode most be both confined ($y > 0$) and perturbation stable ($|\delta\hat{z}'/\delta\hat{z}^*| = 1/M^2$). This evidently can occur either if $\mathscr{A}_1$ and $\mathscr{B}_m$ [in the notation of Eq. (12)] are both >0, in which case the upper sign in Eqs. (19) and (20) must be chosen, or if $\mathscr{A}_1$ and $\mathscr{B}_m$ are both <0, in which case the lower sign must be chosen. A condensed way of expressing this is to require that

$$\text{Im}(\tilde{\mathscr{A}}/\tilde{\mathscr{B}}) < 0. \tag{26}$$

This happens to be formally the same as in an earlier analysis (Belanger *et al.*, 1980a,b), but the definitions of $\tilde{\mathscr{A}}$ and $\tilde{\mathscr{B}}$ are here quite different. For all cases where this is not satisfied, the confined Gaussian solution will be perturbation unstable, and the perturbation-stable solution will diverge radially and hence be unphysical.

G. Resonator Eigenvalues and Mode Losses

The complex amplitude coefficient for a degenerate Hermite–Gaussian wave of order m and complex curvature $\hat{q}_1$ decreases after each round trip by the ratio

$$\frac{\alpha_{m2}}{\alpha^*_{m1}} = \left(\frac{1}{\tilde{\mathscr{A}} - \tilde{\mathscr{B}}/\hat{q}^*_1}\right)^{m+1/2}. \tag{27}$$

If $\hat{q}_1$ corresponds to the confined and perturbation-stable eigensolution of Eqs. (19) or (20), this ratio becomes simply

$$\frac{\alpha_{m2}}{\alpha^*_{m1}} = \frac{1}{M^{m+1/2}}\, e^{j(m+1/2)\theta_m}. \tag{28}$$

The parameter M thus defines not only the perturbation stability but also the mode losses per pass experienced by the Hermite–Gaussian modes. These losses are of course produced by the presence of the quadratic transverse loss variations or ''soft apertures'' in a complex paraxial phase-conjugate resonator.

An interesting observation is that the mode discrimination between lowest-order and higher-order modes depends only on the magnification M, and hence only on the loss value of the lowest-order mode, but is otherwise entirely independent of the resonator parameters and of the mode diameter of the transverse eigenmodes. This property is also true of conventional complex paraxial resonators (Casperson and Lunnam, 1975; Ganiel and Hardy, 1976).

H. Higher-Order Modes: The Complex $\hat{v}$ Parameter

The complex spot size $\hat{v}$ in a degenerate phase-conjugate resonator transforms in one complete round trip according to

$$\hat{v}^2_2 = (\tilde{\mathscr{A}} - \tilde{\mathscr{B}}/\hat{q}^*_1)^2 \hat{v}^{*2}_1 + j\frac{2\tilde{\mathscr{B}}\lambda}{\pi}(\tilde{\mathscr{A}} - \tilde{\mathscr{B}}/\hat{q}^*_1). \tag{29}$$

If we let $\hat{q}_1 = \hat{q}_{\mathrm{cm}}$ be the confined perturbation-stable eigensolution at the conventional mirror end, then $\tilde{\mathscr{A}} - \tilde{\mathscr{B}}/\hat{q}^*_{\mathrm{cm}} = Me^{-j\theta_m}$, and Eq. (35) becomes

$$v^2_2 e^{j\theta_m} = M^2 \hat{v}^{*2}_1 e^{-j\theta_m} - \frac{2M\mathscr{B}_m\lambda}{\pi}. \tag{30}$$

The self-reproducing value $\hat{v}_1 = \hat{v}_2 = \hat{v}_{\text{cm}}$ for the complex spot size at the conventional mirror end is thus given by

$$\hat{v}_{\text{cm}}^2 = \frac{2M}{M^2 - 1} \frac{\mathcal{B}_m \lambda}{\pi} e^{-j\theta_m}. \tag{31}$$

For complex paraxial systems the $\hat{v}$ parameter evidently has a complex value at the conventional mirror end. This indicates a complex argument in the Hermite polynomials. Among other things this will produce slightly nonspherical phase fronts at the conventional mirror end for all the Hermite–Gaussian modes of order $m \geq 2$.

By using the relation $M = \mathcal{A}_m + \sqrt{\mathcal{B}_m \mathcal{C}_m}$, Eq. (31) can also be converted into

$$\frac{1}{\hat{v}_{\text{cm}}^2} = \frac{1}{w_{\text{cm}}^2} + j \frac{\pi}{\lambda R_{\text{cm}}}, \tag{32}$$

where w_{cm} and R_{cm} are the Gaussian eigensolutions given earlier. The complex spot size $\hat{v}_{\text{cm}}$ at the conventional mirror end thus differs from the real spot size w_{cm} at the same end essentially to the same (small) extent that $1/R_{\text{cm}}$ differs from zero, or to the same extent that the phase-front curvature at the conventional end mirror differs from exact coincidence with the mirror surface itself.

It is also possible, after some rather tangled algebra, to show that the complex spot size $\hat{v}_{\text{cm}}$ when transformed to the phase-conjugate mirror end of the resonator becomes real and identically equal to the real spot size w_{pcm} at that end, i.e., $\hat{v}_{\text{pcm}} \equiv w_{\text{pcm}}$. This agrees with an earlier analysis of the $\hat{v}$ parameter carried out directly at the PCM end (Belanger *et al.*, 1980b).

I. Nondegenerate Hermite–Gaussian Mode Solutions

We will show in section VI that whenever an off-resonance signal at a frequency $\omega_1 = \omega_0 + \Delta\omega$ is present in a phase-conjugate resonator, a mirror-image signal at frequency $\omega_2 = \omega_1 - \Delta\omega$ must be present also. In essence any energy in the cavity at frequency ω_1 is continuously conjugated and converted into energy at frequency ω_2, and vice versa. It is not possible for a phase-conjugate resonator to oscillate at a single off-resonance frequency ω_1; the conjugate frequency ω_2 must be present also.

The most meaningful way of defining transverse eigenmodes in this case is r lear. Should we require that the parameters $\hat{q}_1$ and $\hat{q}_2$ corresponding to the two frequency components ω_1 and ω_2 each be separately self-consistent, though not necessarily equal to each other, after two round trips, in which each is converted into the other and then back? If so, the instantaneous output field resulting from the sum of these two spectral components will be a more complicated "double-Gaussian" total field, whose wave front may in fact appear to change shape periodically (along with the complex wave amplitude) at the beat frequency $2\Delta\omega$ between the two spectral components. Or, should we require not only self-consistency but also $k_2/\hat{q}_2 = k_1/\hat{q}_1$, where the k factors are included to allow for the different wavelengths in the definition of the $\hat{q}$ parameters? If so, the output beam will then have constant wave-front parameters R and w in time, though the instantaneous field amplitude and phase will still fluctuate at the beat frequency.

Whichever choice is made, the algebra describing the double-round-trip transformation in the nondegenerate case becomes extremely lengthy (though straightforward); and we have not been able to develop any simple or meaningful formulas for the general case. We note, however, that the fractional frequency offsets $\Delta\omega/\omega_0$ involved in practical resonators will be extremely small ($\sim 10^{-6}$); and the $ABCD$ matrices in many cases will have no explicit frequency dependence at all. This argues that the practical difference between the degenerate and slightly nondegenerate cases will be extremely small.

The case of complex and nondegenerate $ABCD$ matrices becomes even more difficult. However, we again appeal to continuity to argue that the self-consistent solutions in the complex nondegenerate case will probably be unique, and very close to the unique degenerate results given by Eqs. (20) or (21).

IV. Examples of Hermite–Gaussian Eigenmodes

The analysis in the preceding section is rather abstract. It can best be illustrated, and considerable insight into the properties of PCM resonators can be obtained as well, by considering a few simple examples.

A. Phase-Conjugate Mirror plus Conventional Curved Mirror

The simplest phase-conjugate resonator would seem to be a conventional end mirror with arbitrary radius of curvature R_0 spaced a distance L from a phase-conjugate mirror having a Gaussian aperture (GA) located just in front of it, as in Fig. 6a. Note that the actual location of the phase conjugator behind the Gaussian aperture is irrelevant, because the phase

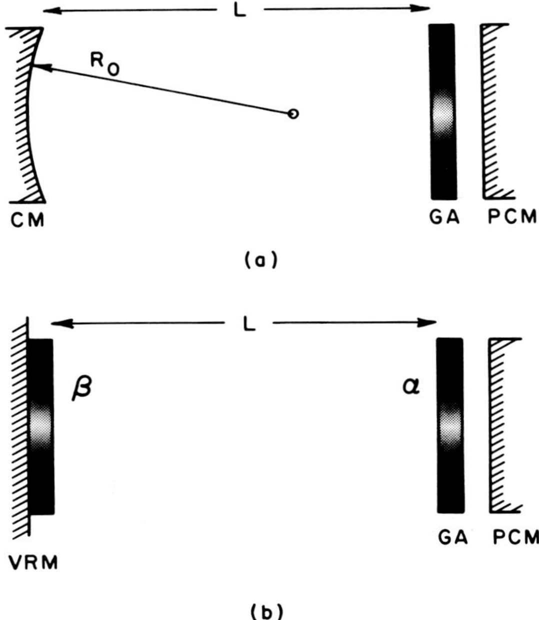

Fig. 6. (a) Example of an elementary PCM resonator consisting of an unbounded conventional mirror (CM) with radius of curvature R_0 spaced a distance L from a phase-conjugate mirror (PCM) that has a Gaussian soft aperture (GA) immediately in front of it. (b) Another elementary PCM resonator consisting of a phase conjugator with a Gaussian aperture of strength α at one end and a Gaussian variable-reflectivity mirror (VRM) of strength β at the other.

conjugator will be effectively transformed forward to immediately behind the Gaussian aperture in any case. In practical cases this Gaussian aperture may be an inherent characteristic of the phase conjugator itself, produced by finite-width Gaussian pump beams used in the four-wave mixing process (Trebino and Siegman, 1980).

The curved end mirror with radius of curvature R_0 can be replaced with a plane mirror plus a thin lens with focal length $f \equiv 1/R_0$. Traversing this lens twice gives the mirror the usual focal length of $2/R_0$ for reflection from a curved mirror. (Note that we use sign conventions in which $R > 0$ referring to a *wave* indicates a diverging spherical wave front, whereas $R > 0$ referring to a *mirror* indicates a concave or converging-type mirror.) The single-pass voltage transmission of the Gaussian aperture may be written

$$t(x) \equiv \mathscr{E}(x)/\mathscr{E}_0 = \exp(-x^2/w_\alpha^2), \tag{33}$$

so that w_α is the $1/e$ radius for the amplitude transmission through the aperture. It is then convenient to define a Fresnel number

$$N_\alpha \equiv w_\alpha^2/2L\lambda, \tag{34}$$

which characterizes the Gaussian aperture size relative to the length of the PCM cavity. It is also convenient to define a g parameter

$$g \equiv 1 - L/R_0, \tag{35}$$

which is the same as the usual g parameter for conventional optical resonators.

The conventional one-way $ABCD$ elements for this cavity are then given by

$$\begin{bmatrix} A & B \\ C & D \end{bmatrix} = \begin{bmatrix} 1 & 0 \\ -j\alpha & 1 \end{bmatrix}\begin{bmatrix} 1 & L \\ 0 & 1 \end{bmatrix}\begin{bmatrix} 1 & 0 \\ -1/R_0 & 1 \end{bmatrix}, \tag{36}$$

where $\alpha \equiv 1/2\pi N_\alpha L$. Note the the individual matrices are cascaded in reverse order to the order in which they are encountered by the beam, and that the Gaussian aperture is represented by a matrix with a complex C element. The phase-conjugate $\tilde{\mathscr{A}},\tilde{\mathscr{B}},\tilde{\mathscr{C}},\tilde{\mathscr{D}}$ matrix defined by Eq. (4) then has the elements

$$\begin{aligned} \tilde{\mathscr{A}} &= 1 - 2j\alpha L(1 - L/R_0) = 1 - 2jg\alpha L, \qquad \tilde{\mathscr{B}} = 2j\alpha L^2, \\ \tilde{\mathscr{C}} &= -2j\alpha(1 - L/R_0)^2 = -2jg^2\alpha, \qquad \tilde{\mathscr{D}} = \tilde{\mathscr{A}}^*. \end{aligned} \tag{37}$$

The self-consistent solutions for the $\hat{q}$ parameter on the conventional mirror surface are

$$\frac{1}{\hat{q}_{\mathrm{cm}}} = \mp j\frac{|g|}{L}\exp(j\theta_m), \tag{38}$$

where

$$\theta_m = \tan^{-1}(g/\pi N_\alpha). \tag{39}$$

Perhaps we should emphasize once again that this is the Gaussian $\hat{q}$ parameter for the wave *on the conventional mirror surface, traveling inward to the right*. Note that only the upper sign in Eq. (38) corresponds to a physically meaningful confined and perturbation-stable solution.

The self-consistent radius and spot size on the mirror surface at the conventional mirror end are given by

$$R_{\mathrm{cm}} = \frac{L}{|g|\sin\theta_m} \approx \frac{\pi N_\alpha L}{g|g|} \tag{40}$$

and

$$w_{\rm cm}^2 = \frac{L\lambda}{\pi} \frac{1}{|g| \cos \theta_m} \approx \frac{L\lambda}{\pi} \frac{1}{|g|}, \tag{41}$$

where the approximations are valid for $N_\alpha \gg |1 - L/R_0|$. This approximation basically means that the Gaussian aperture is large compared with the spot size of the mode at that aperture. The Gaussian mode loss factor—that is, the reduction in mode amplitude after one round trip—contains essentially the same ratio, in the form

$$\frac{\alpha_2}{\alpha_1} = \frac{1}{1 + |g/\pi N_\alpha|}, \tag{42}$$

where α_1 and α_2 are the mode amplitudes before and after one round trip. Note that the wave-front radius $R_{\rm cm}$ at the conventional mirror end is nearly always $\gg L$; i.e., the wave front very nearly matches the mirror

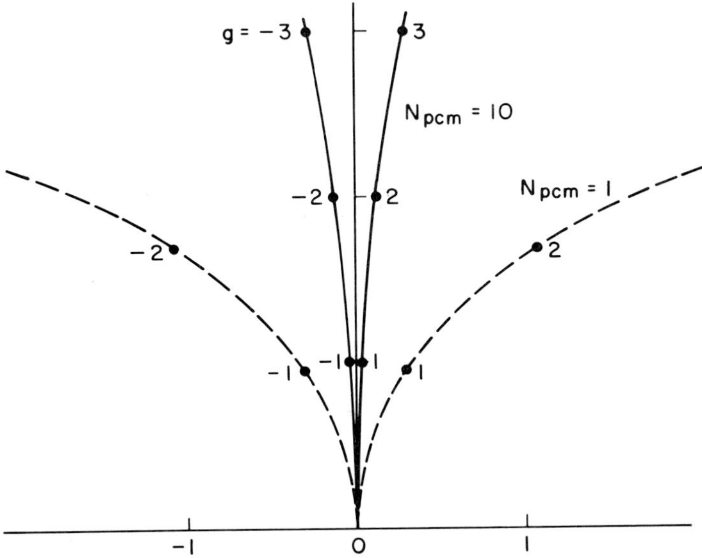

Fig. 7. Loci of the self-consistent Hermite–Gaussian $\hat{q}_{\rm cm}$ parameters for varying values of the parameter $g \equiv 1 - L/R_0$ in the elementary resonator of Fig. 6a, shown for two different values of the Gaussian Fresnel number N_α with $x \equiv 1/R_{\rm cm}$ and $y \equiv \lambda/\pi w_{\rm cm}^2$. A smaller Fresnel number (i.e., a smaller spot size for the Gaussian aperture) causes the locus to deviate more rapidly from the y axis; the phase front of the Hermite–Gaussian eigenmode then deviates more significantly from the conventional mirror surface profile.

surface, except for very highly divergent or convergent end mirrors, which have $|R_0| \to 0$.

In Fig. 7 we show the locus of the self-consistent $\hat{q}$ parameter for this resonator in the $1/\hat{q}$ plane as a function of the parameters $g \equiv 1 - L/R_0$ and the Gaussian Fresnel number N_α; and in Fig. 8 we show how, for a typical pair of values of g and N_α, a Gaussian beam launched with an arbitrary initial $\hat{q}_0$ value will converge toward the appropriate self-consistent point on the locus on subsequent round trips. With a finite soft aperture the $1/\hat{q}$ value no longer oscillates back and forth between two points on opposite sides of a continuous locus, but converges slowly in to a final discrete eigensolution.

The Gaussian beam parameters at the conventional mirror end can be transformed to the phase-conjugate mirror end, just inside the Gaussian aperture, by using the conventional matrix parameters $A = 1 - L/R_0$, $B = L$, $C = -1/R_0$, and $D = 1$. The beam parameters just before hitting the Gaussian aperture at the PCM end are then

$$R_{\mathrm{pcm}} = 2gL/(2g - 1) \tag{43}$$

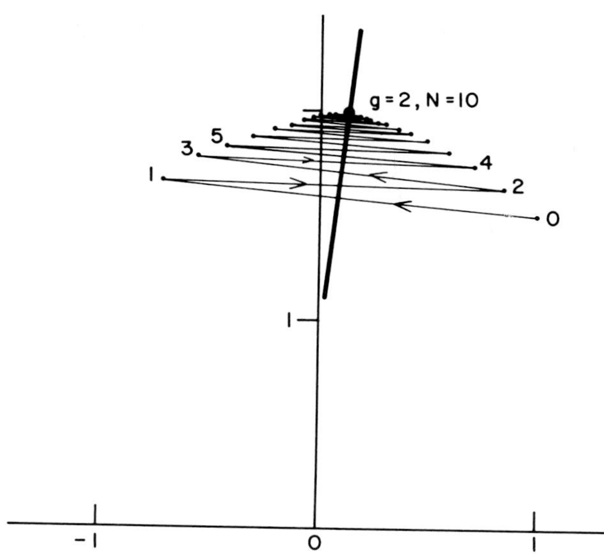

Fig. 8. When a PCM resonator contains a positive Gaussian aperture, as in Fig. 6a, a Gaussian beam starting from an arbitrary initial point converges rapidly on successive bounces to a unique self-consistent eigenvalue, as in this figure, rather than oscillating between two points as in Fig. 5. The self-consistent point in this example is the value from Fig. 7 appropriate to $g = 2$ and $N_\alpha = 10$, with $x \equiv 1/R$ and $y \equiv \lambda/\pi w^2$.

and

$$w_{\text{pcm}}^2 = \frac{2L\lambda}{\pi} \frac{|g| + g \sin \theta_m}{\cos \theta_m} \approx \frac{2|g|\lambda L}{\pi}. \tag{44}$$

Note the universal rule that $w_{\text{cm}} w_{\text{pcm}} \approx \sqrt{2} L\lambda/\pi$.

The Gaussian mode in this type of PCM resonator will also have associated with it a waist located in most cases very close to the center of curvature of the conventional end mirror. In fact the location z_0 of this waist, with z_0 measured positive to the right of the mirror, will be given by

$$\frac{z_0 - R_0}{R_0} \approx \frac{g^2}{g^2 + (1 - g)^2}, \tag{45}$$

and the spot size of the mode waist will be

$$w_0^2 \approx \frac{L\lambda}{\pi} \frac{|g|}{g^2 + (1 - g)^2}. \tag{46}$$

In Fig. 9 we show mode spot sizes at the conventional mirror CM and PCM ends versus the g parameter for this elementary example of a PCM

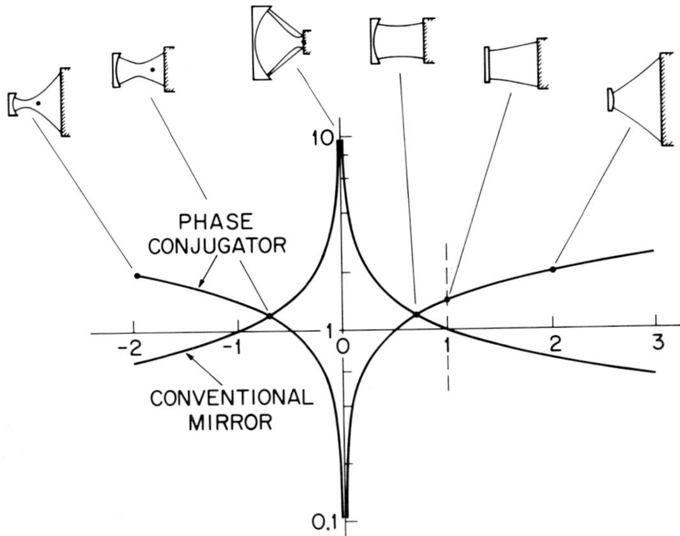

Fig. 9. Gaussian spot sizes (w) at the conventional mirror and phase-conjugator ends of the elementary PCM resonator of Fig. 6a versus the parameter $g \equiv 1 - L/R_0$. The sketches indicate mode profiles at significant points, plotted to a common transverse scale. Only the region between $g = 0$ and $g = 1$ would have stable Gaussian modes in a conventional (non-PCM) resonator.

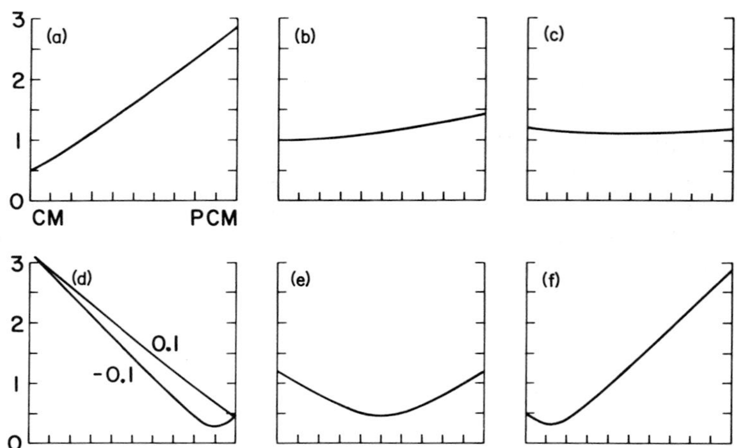

Fig. 10. Gaussian mode profiles similar to those in Fig. 9, showing the spot size $w(z)$ versus z inside the elementary PCM resonator of Fig. 6a. (a) divergent end mirror, $g = 2$, with $R_0 = -L/3$; (b) planar end mirror, $g = 1$, with $R_0 = \infty$; (c) symmetric case 1, $g = 1/\sqrt{2}$, with $R_0 = (2 + \sqrt{2})L$; (d) near-hemispherical end mirror, $R_0 \approx L$; (e) symmetric case 2, $g = -1/\sqrt{2}$, with $R_0 = \sqrt{2}L/(1 + \sqrt{2})$; (f) and over-convergent end mirror, $g = -4$, with $R_0 = L/5$. In each case the conventional mirror is at the left-hand end, the phase conjugator and weak Gaussian aperture are at the right-hand end, and the spot sizes are normalized to the confocal value $\sqrt{L\lambda/\pi}$.

resonator, and in Fig. 10 we show mode profiles for some of the same cases.

B. Properties of an Elementary PCM Resonator

A conventional resonator as in Fig. 6a, but with a flat conventional mirror in place of the PCM, would have confined Gaussian modes only over the stable region $0 < g < 1$, with unstable modes outside. The Gaussian mode behavior of the PCM resonator differs significantly from this. Adding an arbitrarily weak Gaussian-aperturing effect to an unbounded PCM (i.e., $N_\alpha \to \infty$) leads to discrete, confined, perturbation-stable Gaussian eigenmodes for all values of g. These spot sizes and mode profiles are, moreover, essentially independent of the aperture strength so long as $N_\alpha \gg 1$. The PCM resonator mode behavior then includes the following significant points, as illustrated in Figs. 9 and 10.

(a) For the "divergent" regime $g > 1$, or $R_0 < 0$ (the positive-branch unstable regime in a conventional resonator), the PCM resonator exhibits behavior similar to that of Fig. 3, with diverging spot size at the PCM end and small spot size at the CM end. This behavior will eventually be limited at large g, when the PCM spot size expands to where it begins to be influenced by the weak Gaussian aperture.

(b) At the planar point $g = 1$ or $R_0 \to \infty$, the PCM mode is half confocal in form, with a waist of spot size $\sqrt{L\lambda/\pi}$ at the planar mirror end and a spot size $\sqrt{2}$ larger at the PCM end.

(c) In the region between $g = 1$ and $g = 0$, at the point $g = 1/\sqrt{2}$ or $R_0 = (2 + \sqrt{2})L$, is a "symmetric point 1," where $w_{\mathrm{cm}} = w_{\mathrm{pcm}}$, and where there is a shallow waist located in the exact center of the cavity.

(d) As the mirror curvature is increased, the resonator approaches the hemispherical point $g = 0$ or $R_0 = L$. The spot size at the conventional mirror end blows up here (or in practice expands until it encounters a finite aperture) while the PCM spot size goes toward zero.

(e) Between $g = 0$ and $g = -1(R_0 = L$ to $R_0 = L/2)$ the resonator is "overconvergent" and nominally unstable (in the negative-branch sense.) However, the mode behavior here is a mirror image of the region between $g = 0$ and $g = +1$, except with deeper waists inside the resonator. In particular, the mode behavior comes to "symmetric point 2," located at $g = -1/\sqrt{2}$ or $R_0 = [\sqrt{2}/(1 + \sqrt{2})]L$. Here the mode again has $w_{\mathrm{cm}} = w_{\mathrm{pcm}}$ as at symmetric point 1, but with a considerably deeper waist at the center of the resonator.

(f) Finally, for $g < -1$ or $R_0 < L/2$ the overconvergent mode becomes much like the divergent mode for $g > +1$, with a large spot size at the PCM end, except that in this case there is a very tight internal focus just in front of the conventional mirror (rather than just behind it as occurs for $g \to +\infty$).

A distinguishing feature of all these cases is that the resonator always selects the waist location (from among the infinite family available to it) that *minimizes the Gaussian spot size at the Gaussian aperture while keeping the mode curvature matched to the mirror surface at the CM end*. Moving the Gaussian aperture to any other plane in the resonator will change the mode to minimize the spot size at the new aperture location.

C. Mode Spot-Size Sensitivity

One fundamental property of phase-conjugate resonators is greatly reduced sensitivity of the mode parameters to perturbations in the cavity design parameters. The spot size w of the lowest-order Gaussian mode at any reference plane in a conventional real and stable optical resonator may be written in terms of the conventional round-trip $ABCD$ matrix of the same resonator, in the form

$$w^2 = \frac{B\lambda}{\pi} \frac{1}{\sqrt{1 - [(A + D)/2]^2}} = \frac{w_0^2}{(1 - g^2)^{1/2}}, \tag{47}$$

where $g \equiv (A + D)/2$, $w_0^2 \equiv B\lambda/\pi$, and the $ABCD$ elements are the conventional elements for one complete round trip starting from the same reference plane. The parameter B in a conventional resonator is often comparable with the physical length of the laser cavity. The "confocal spot size" $w_0 = \sqrt{B\lambda/\pi}$ is then usually much smaller than is desirable to obtain efficient energy extraction from a large-diameter laser medium.

Obtaining a large spot size w and hence a large mode volume in a laser cavity thus usually requires a cavity design that operates close to (or beyond) the geometrical stability boundary $g^2 \to 1$. In this limit, however, the sensitivity of the mode spot size to small fluctuations δg in the resonator geometrical parameters becomes very large. One can write this sensitivity for a conventional stable resonator, for example, by differentiating Eq. (47) in the form

$$\frac{\delta w}{w} = \frac{g}{2}\left(\frac{w}{w_0}\right)^4 \delta g \approx \frac{1}{2}\left(\frac{w}{w_0}\right)^4 \frac{\delta g}{g}, \qquad w \gg w_0. \tag{48}$$

The fluctuations in g are evidently multiplied by the very large ratio $(w/w_0)^4$. Conventional stable resonators with large mode diameters are thus inordinately sensitive to very small perturbations in their physical dimensions.

The results obtained above, for example, in Eqs. (41) and (44), show that in phase-conjugate resonators we have instead

$$\frac{\delta w}{w} \approx \frac{1}{2}\frac{\delta g}{g}. \tag{49}$$

The spot sizes in PCM resonators are thus in general much less sensitive to small changes in the resonator parameters. Larger mode volumes can be achieved without operating on the boundary between geometrically stable and unstable systems; yet good transverse-mode discrimination can still be obtained, for example, by using an appropriate diverging end mirror plus a weak Gaussian aperture in the PCM cavity.

D. Phase-Conjugate Mirror plus Gaussian–Apertured Mirror

As a second example of Gaussian apertures and phase-conjugate resonators, we consider the somewhat artificial but instructive example of a cavity consisting of a Gaussian-apertured phase-conjugate mirror plus a

Gaussian-apertured planar mirror spaced by distance L, as in Fig. 6b. This example is conceptually similar to the first example of Fig. 6a, except that the conventional mirror now has a variable "imaginary curvature" rather than a variable real curvature.

Let the strengths of the two Gaussian apertures be denoted by $\alpha L \equiv 1/2\pi N_\alpha$ and $\beta L \equiv 1/2\pi N_\beta$, in the same notation as above. The matrix elements of Eq. (4) are then given by

$$\begin{aligned} \tilde{\mathscr{A}} &= \mathscr{A}_1 - j\mathscr{A}_2 = 1 + 2\alpha\beta L^2 - 2j\alpha L, \\ \tilde{\mathscr{B}} &= j\mathscr{B}_m = 2j\alpha L^2, \\ \tilde{\mathscr{C}} &= -j\mathscr{C}_m = -2j[\alpha + \beta + \alpha(\beta L)^2]. \end{aligned} \tag{50}$$

If we allow the possibility that either α or β may become negative, meaning that one or the other of the Gaussian apertures may have a transmission that *increases* with radius, the mode properties of this resonator become fairly complicated.

It may not seem particularly useful to analyze the negative aperture case, because clearly no real aperture can have a transmission function that increases indefinitely with increasing radius. There can be laser gain media, however, in which the gain does increase with radial distance from the axis, at least for some finite distance about the axis. One can also synthesize Gaussian variable-reflectivity mirrors with an inverted inflection function versus radius, at least over the finite diameter of the mirror. An understanding of idealized negative Gaussian apertures will allow us to assess the possible behavior of such systems, at least up to the point where the Gaussian mode itself spreads beyond the finite edges of such a system.

We will only summarize briefly here the rather complex results for this example. It is convenient to diagram the properties of this resonator in different regions of an α,β plane, as shown in Fig. 11. From Eqs. (19) or (20), physically meaningful and self-consistent solutions can exist only for $\mathscr{C}_m/\mathscr{B}_m > 0$. Hence from Eqs. (50) one can determine that there is a "forbidden region," with $\mathscr{C}_m/\mathscr{B}_m < 0$, located between the $\alpha = 0$ axis and the curve $\alpha = -\beta/[1 + (\beta L)^2]$, as indicated by the shaded area in Fig. 11. In addition, the quantity $\mathscr{A}_1 \equiv \mathscr{A}_m \cos\theta_m$, which compels the choice of upper or lower signs in Eqs. (19), (20), and (23), changes from positive to negative on the boundary curves defined by $\alpha L = -1/2\beta L$. Note also that the sign of $\mathscr{B}_m$, which combines with $\mathscr{A}_1$ to determine perturbation stability, is the same as the sign of α.

The various combinations that can result are indicated in Fig. 11. Confined and perturbation-stable solutions at the conventional mirror end can exist, according to the analytical criteria, everywhere in the first quadrant

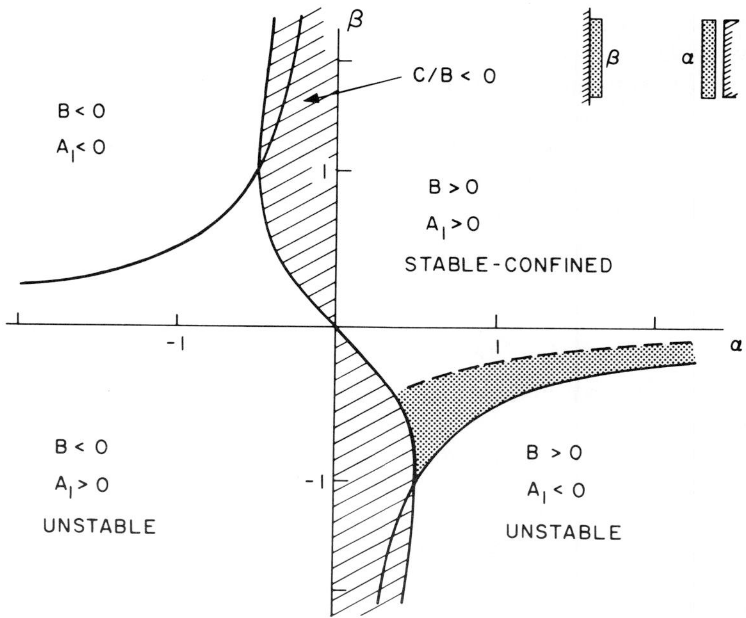

Fig. 11. Significant regions and boundary curves in the αL, βL plane for the double-Gaussian-aperture PCM resonator of Fig. 6b, including possibly negative values for the transverse variation of either one of the apertures (i.e., transmission increasing with radius). Modes that are both perturbation stable and transversely confined at all cross sections within the resonator exist only for values of αL and βL located in the upper right region of the plane, above the shaded area bounded by the heavy dashed line. Negative values of β for the conventional mirror extending as far down as this dashed line are allowed.

($\alpha > 0$, $\beta > 0$), as would be expected. Confined and stable solutions can also exist in a limited portion of the lower right quadrant, where $\alpha > 0$ but $\beta < 0$; and confined solutions can apparently also exist in the outer portion of the upper left quadrant, where both $\alpha < 0$ and $\beta < 0$.

Closer examination, however, reveals additional complications. In the nominally allowed regions of the upper left quadrant, and also in the portion of the allowed region in the lower right quadrant below the dashed curve in Fig. 11, the solutions for $\hat{q}$ are indeed perturbation stable and confined at the conventional mirror end. However, for values of α and β falling in these regions, the modes turn out not to remain confined at all planes elsewhere in the resonator. To illustrate this, we show in Fig. 12 a series of trajectories for $\hat{q}(z)$ in the $1/\hat{q}^*$ plane for resonators with different values of α and β. The first three plots show the trajectories for

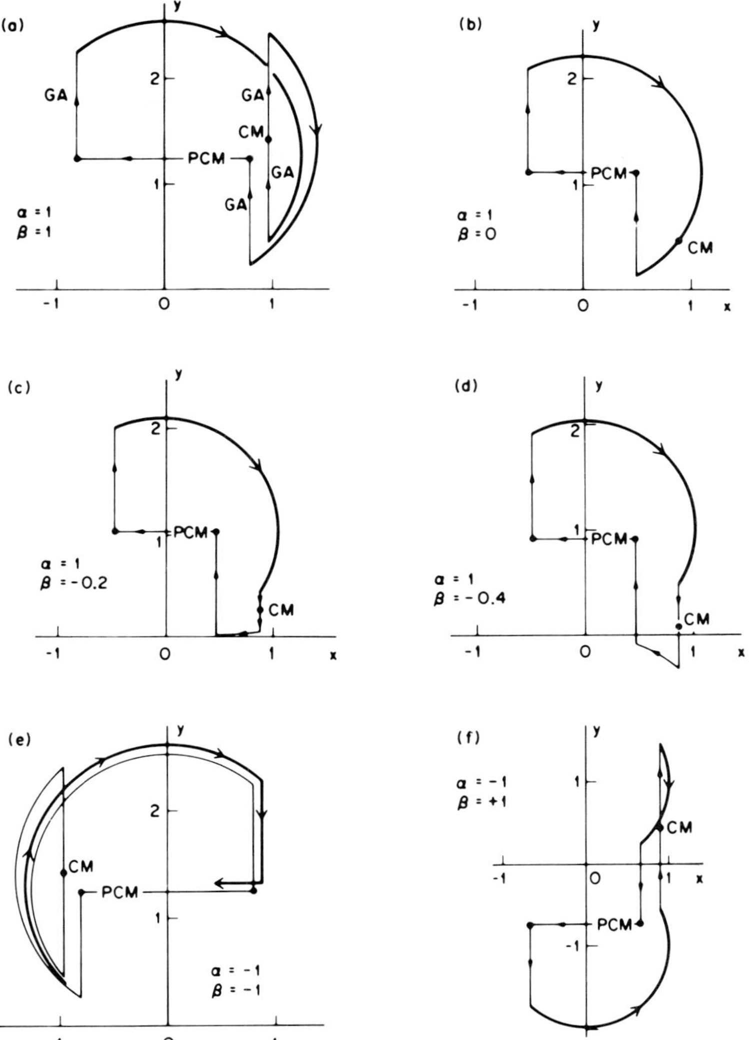

Fig. 12. Trajectories of $\hat{q}(z)$ in the $1/\hat{q}$ plane during successive round trips for different choices of the Gaussian aperture parameters of αL and βL falling in different regions of the $1/\hat{q}$ plane shown in Fig. 11, where $x \equiv 1/R$ and $y \equiv \lambda/\pi w^2$. Gaussian apertures produce vertical translations in this plane (upward for positive apertures, downward for negative), whereas free-space propagation between the mirrors translates into arcs of circles that pass through the origin. Only the top three cases, corresponding generally to the upper right region in Fig. 11, are both perturbation stable and transversely confined at all planes within the resonator. Cases (d) and (f) are perturbation stable but not everywhere transversely confined (i.e., $y \equiv \lambda/\pi w^2$ goes negative at certain planes). In case (e), corresponding to the lower left quadrant of Fig. 11, there is a closed and transversely confined orbit in the $1/\hat{q}$ plane, but this orbit is unstable against small perturbations.

fully satisfactory modes corresponding to two points located in the upper right quadrant, plus one point slightly below the positive-α axis, namely, $(\alpha L, \beta L) = (1, 1), (1, 0)$, and $(1, -0.2)$. It is evident in the third example that the combination of the positive Gaussian aperture and phase-conjugate mirror is barely able to override the destabilizing effects of the negative Gaussian aperture at the conventional mirror end.

The fourth trajectory in Fig. 12 corresponds to the case $\alpha L = 1$, $\beta L = -0.4$, which should still be an allowed solution within the analytical criteria of Eqs. (19) and (23). The transverse profile of the Gaussian beam, however, though confined at both the conventional mirror and PCM ends, passes into the $y < 0$ region and thus beomes unconfined during the propagation path from the conventional mirror end to the PCM. We might emphasize that this trajectory, though clearly nonphysical, is still entirely perturbation stable over the entire trajectory. The dashed line in Fig. 11, defined by the analytical criterion that $y_{cm}(\alpha, \beta) = -\beta$, marks the contour in the α, β plane below which this behavior occurs.

The remaining two trajectories in Fig. 12 show typical $1/\hat{q}$ trajectories for resonator parameters in the lower left and upper left regions of the αL, βL plane. The trajectory for $\alpha = -1$, $\beta = -1$ is fully confined [with the appropriate choice of sign in Eqs. (19)], but is perturbation unstable. Any small deviation from this trajectory grows rapidly on successive round trips, and the trajectory rapidly diverges away and then reconverges to a stable but unconfined trajectory that is the same trajectory inverted in the x and y axes. The trajectory for $\alpha = -1$, $\beta = +1$ is stable and confined at the conventional mirror end, but is unconfined at the PCM end and in the portion of the resonator coming from the PCM to the CM.

A general lesson from these results is that when an optical system contains a combination of positive and negative Gaussian apertures, *the ABCD matrix and the analytical stability criteria evaluated at any one plane are necessary but not sufficient to determine completely the physically allowable modes of the system.* When negative Gaussian apertures are present, one must apparently examine the mode profile at every point in the system (or at least after every negative aperture in the system), to make sure the mode does not become unconfined anywhere in the resonator.

V. Transverse Modes with Hard-Edged Apertures

Hard-edged apertures are of course easier to achieve in practice than soft or Gaussian apertures (see, however, Giuliani *et al.*, 1980; Eggleston *et al.*, 1981). Hence we must next consider the effects of finite, hard-edged mode-controlling apertures on phase-conjugate resonators.

Phase-conjugate resonators with hard-edged apertures will exhibit Fresnel diffraction effects due to these edges, much as in conventional optical resonators. These effects, in PCM as in conventional resonators, will generally be complicated and analytically intractable. As a result, exact mode properties with hard-edged apertures can generally be determined only by numerical calculation methods, such as the well-known Fox–Li and Prony methods (Fox and Li, 1961; Siegman and Miller, 1970). In this section we review briefly the small amount of work carried out thus far on finite-aperture PCM resonators, and attempt to deduce some of the more general aspects of the behavior of these resonators.

A. Numerical Mode Calculations

The integral equation appropriate for round-trip propagation in a simple PCM resonator is Eq. (5), assuming separable rectangular transverse coordinates and a finite aperture of width $2w_{\text{pcm}}$ at the conjugator end. A similar aperture of width $2w_{\text{cm}}$ at the conventional mirror end can be incorporated into the perturbation function $\hat{\rho}(x)$, or can be accounted for by truncating the integration at $\pm w_{\text{cm}}$. Extension of this integral equation to accommodate apertures elsewhere in the cavity, or to express it in cylindrical coordinates, is a straightforward exercise.

Eigensolutions of this type of integral equation are routinely calculated in conventional resonator theory, most effectively by the use of fast Fourier or fast Hankel transform methods (Sziklas and Siegman, 1974, 1975; Siegman, 1977b; Sheng and Siegman, 1980). These methods can be coupled with Prony methods if higher-order eigenmodes are to be extracted (Siegman and Miller, 1970; Murphy and Bernabe, 1978). Note that the kernel in Eq. (5) differs in an interesting way from the usual kernel in conventional resonators. However, the integral still retains the form of a convolution integral, namely, $\mathscr{E}_2(x_2) \sim \int \mathscr{E}_1^*(x_1)K(x_2 - x_1)\,dx_1$, so that fast transform methods can still be employed.

For the simple case of a finite mirror plus a finite PCM, the significant parameters are the Fresnel numbers, which can be defined as

$$N_{\text{cm}} \equiv w_{\text{cm}}^2/B\lambda, \qquad N_{\text{pcm}} \equiv w_{\text{pcm}}^2/B\lambda. \tag{51}$$

The parameter B is simply the cavity length L if no lenses or other elements intervene between the mirror surface and the PCM. If both of these Fresnel numbers are $\gg 1$, the Hermite–Gaussian analysis of the previous section can be expected to apply. If one or both of the Fresnel

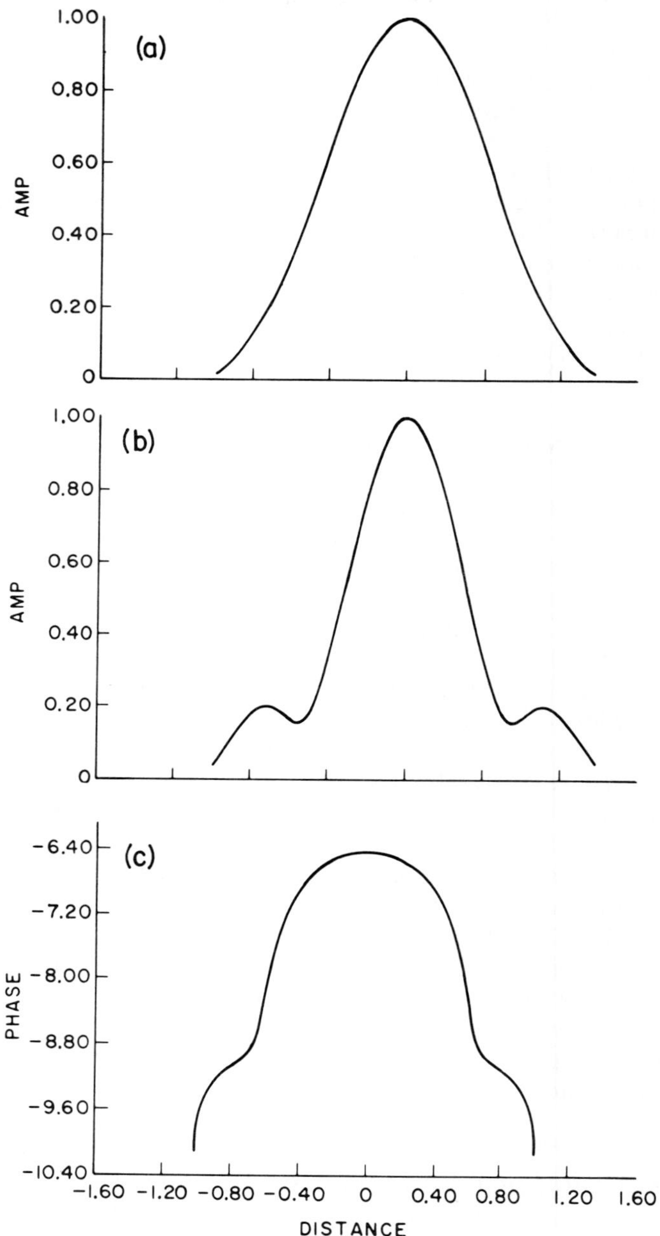
(a)
AMP
1.00
0.80
0.60
0.40
0.20
0
(b)
AMP
1.00
0.80
0.60
0.40
0.20
0
(c)
PHASE
-6.40
-7.20
-8.00
-8.80
-9.60
-10.40
-1.60 -1.20 -0.80 -0.40 0 0.40 0.80 1.20 1.60
DISTANCE

numbers becomes small enough that the aperture impinges on the Hermite–Gaussian solution, significant diffraction effects can be expected to occur. As a general rule, the transverse modes in resonators with hard apertures will display considerable ingenuity in distorting their amplitude patterns and phase profiles to minimize their diffraction losses at the apertures, while continuing to obey the optical wave equation.

Numerical calculations of transverse modes in a phase-conjugate resonator with a finite aperture have been carried out by Lam and Brown (1980), (1980b), considering only one transverse coordinate (strip resonator) and using fast Fourier transform methods. Their results are shown in Figs. 13 and 14. For example, in Fig. 13 are shown amplitude and phase patterns at the conventional mirror end for two elementary phase-conjugate resonators ($g = 0$ and $g = 2$) with moderately strong aperturing ($N_{cm} = 1$). The mode patterns show the expected Fresnel ripples and diffractive perturbations to the underlying Gaussian approximations.

Figure 13a corresponds to the hemispherical case $R_0 = L$ or $g = 0$. The mode in this case will be apertured most heavily at the conventional mirror end, and the PCM appears to do a good job of removing most diffraction ripples. Figures 13b and 13c correspond to the divergently unstable case $R_0 = -L$ or $g = 2$ (similar to Fig. 3). The mode in this case is most sharply apertured at the PCM end, and this shows up as fairly strong diffraction ripples at the conventional mirror end. The phase distribution plotted in Fig. 13c is the phase profile on a transverse plane perpendicular to the resonator axis, not on the conventional mirror surface. Note that it has a roughly quadratic shape, with superimposed ripples, corresponding to a phase front converging back down onto the mirror surfaces, as in Fig. 3.

The effects of end mirror distortion on a hemispherical PCM resonator are illustrated in Fig. 14. A deliberate phase distortion pattern imposed across the conventional mirror surface is shown in Fig. 14a. The resulting

---

Fig. 13. Results of exact numerical calculations by Lam and Brown (1980) for PCM resonators with discrete apertures ($N = 1$): (a) Numerically calculated amplitude distribution across the conventional mirror of a finite-aperture hemispherical PCM resonator ($L = R_0$) with an effective Fresnel number equal to unity. The calculated mode pattern is very close to the results of the Hermite–Gaussian analysis. (b) Amplitude distribution across the conventional mirror for a positive-branch unstable case ($R_0 = -L$) with Fresnel number of unity. The mode pattern is still roughly Gaussian but has acquired significant diffraction ripples. (c) Phase distribution across a perpendicular transverse plane at the conventional mirror end for the same case. The phase front still roughly matches the divergent curvature of the conventional mirror surface, but has also acquired significant ripples due to hard-edged diffraction effects. (Used with permission from Lam and Brown, 1980.)

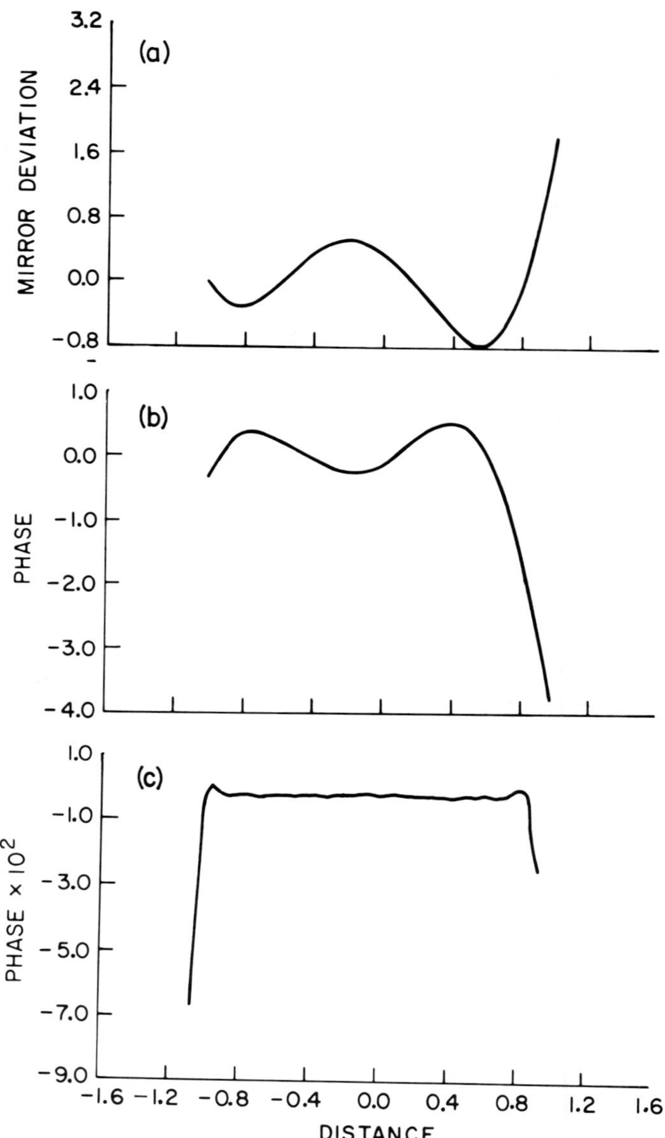

Fig. 14. Effects of mirror distortion on a PCM resonator. (a) Deliberate phase distortion (in radians) introduced into the surface of the conventional end mirror in a finite-aperture hemispherical phase-conjugate resonator. (b) and (c) Resulting phase profiles of the output wave on the conventional end mirror surface for small and large Fresnel numbers ($N_{\rm cm}$ = (a) $N_{\rm pcm}$, (b) $2/\pi$, and (c) $15/\pi$). (Used with permission from Lam and Brown, 1980.)

phase profile on the mirror surface is then shown in Figs. 14b and 14c for Fresnel numbers of $N_{cm} = N_{pcm} = 2/\pi$ and $15/\pi$, respectively. For the lower Fresnel number the wave at the mirror surface remains significantly distorted by the mirror distortion, but for the larger Fresnel number the phase conjugator can resolve and almost totally correct for the phase aberrations.

Additional numerical calculations on more complex PCM resonators have been carried out by Hardy (1981). They consider strip resonators of length $2L$ with an unbounded planar conventional mirror at the left end, an aperture of width $2a$ at the resonator midplane, and either a PCM or another conventional plane mirror with an aperture of width $2b$ at the right end. In Table I we tabulate the resonator eigenvalues squared for some of the cases considered, using the definition $N_\alpha \equiv a^2/L\lambda$. The phase-conjugate reflector is obviously able to greatly reduce the resonator losses by canceling nearly all the diffraction effects from the midplane aperture.

For the PCM cases with $N_\alpha = 1.875$ the calculated mode patterns agree very closely with the Gaussian profiles predicted by the Hermite–Gaussian analysis. Even for the case $N_\alpha = 0.3$, where the analytically predicted Hermite–Gaussian mode is heavily clipped by the midplane aperture, the mode between the aperture and the conventional mirror remains very close to Gaussian, with minor side lobes. The conjugate reflector captures, and largely cancels the diffractive scattering from the midplane aperture. By contrast, the conventional mirror results show the usual diffractive ripples as well as much larger mode losses.

The effects of an intracavity phase perturbation on this type of resonator are explored in Fig. 15 (Hardy, 1981). The lowest-order mode was calculated for a resonator with a phase grating inserted at the resonator mid-

TABLE I

Eigenvalues for One-dimensional Strip Resonators of Overall Length $2L$[a]

$N_a = a^2/L\lambda$	b/a	γ^2(cm)	γ^2(pcm)
1.875	1.2	0.849	0.990
1.875	4.0	0.835	0.999
0.300	10.0	0.260	0.530

[a] Each resonator has an infinite planar conventional mirror at one end, a slit aperture of width $2a$ at the resonator midplane, and either a planar conventional mirror [γ^2(cm)] or a phase-conjugate mirror [(γ^2(pcm)] of width $2b$ at the other end.

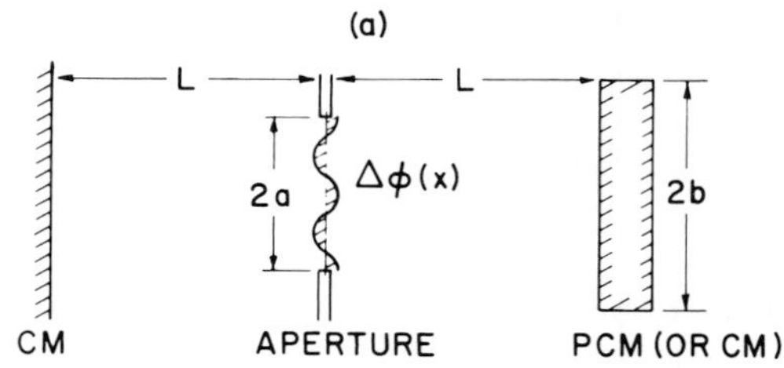

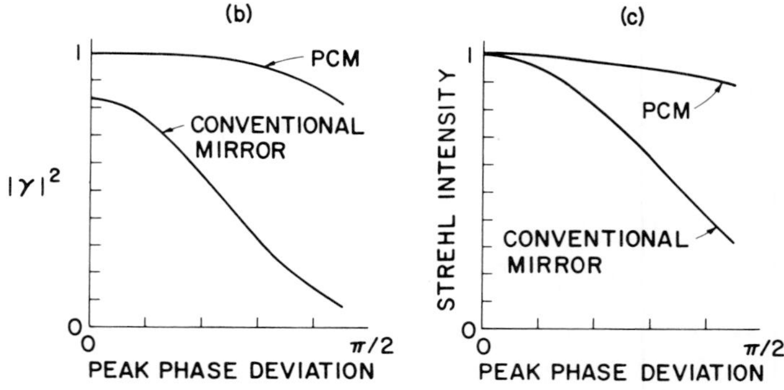

Fig. 15. (a) Combined effects of a midplane aperture and a midplane phase grating on the diffraction losses and beam quality of a resonator with an apertured phase-conjugate mirror (PCM) at the right-hand end and a planar conventional mirror at the left-hand end, compared with the same effects with the PCM replaced with a planar conventional mirror (used with permission from Hardy, 1981). (b) The lowest-mode eigenvalue squared versus peak phase deviation of the phase grating, with $N_\alpha \equiv a^2/L\lambda = 1.875$, $b/a = 4$, and $2a/T = 10$ (i.e., ten full cycles of the phase grating across the aperture); (c) the Strehl intensity ratio for the beam emerging from the conventional mirror (left-hand) end in each case. The Strehl ratio is defined here as the ratio of far-field peak intensity to total power transmitted, with and without the grating present.

plane (i.e., in the aperture plane) so that the circulating wave passes through the grating going in both directions. The grating has a phase deviation $\Delta\phi(x) = \Delta\phi_d \cos(2\pi x/T)$ across the aperture.

The lower left plot in Fig. 15 shows the resonator eigenvalue (magnitude squared) versus peak phase deviation $\Delta\phi_d$ for both phase-conjugate and conventional mirror cases with $2a/T = 10$ (i.e., with ten fringes across the aperture), and with parameters $N_\alpha = 1.875$ and $b/a = 4$. The lower right plot shows the Strehl ratio of the beam emerging from the conventional mirror end versus $\Delta\phi_d$ for the same two cases. Evidently both the diffraction loss and the beam quality are much less affected by an intracavity phase grating in the PCM case than in the conventional mirror case. Indeed, the PCM resonator manages to compensate quite effectively for the scattering effects that the grating has on the intracavity waves going both toward the PCM and toward the conventional mirror in the cavity.

The failure of the PCM to correct for all of the perturbation introduced by the phase grating is largely due to the finite apeture of the PCM. An identical set of calculations was also carried out with the width of the phase conjugator reduced to $b = a$, or $N_\alpha = N_{\text{pcm}} = 1.875$. As might be expected, the phase conjugate mirror then provides little or no improvement (except for smaller diffraction losses to start with) over the conventional plane mirror.

B. Transverse-Mode Orthogonality Properties

Phase-conjugate resonators that have hard-edged apertures, or that depart in other ways from the Hermite–Gaussian approximations, will almost certainly possess higher-order transverse modes also. None of these higher-order mode patterns appears to have been published as yet for other than the Hermite–Gaussian case, although such calculations are under way (Hardy, personal communication). The orthogonality properties of these higher-order transverse modes can, however, be established in a fairly general fashion, similar to the orthogonality properties of the transverse modes in conventional optical resonators (Siegman, 1979). Only the final result will be stated here, with the proof left to a separate publication (Hardy *et al.*, 1982).

We consider a general PCM resonator in which the reflection from the phase conjugator may have an arbitrary transverse variation in magnitude but not in phase, i.e.,

$$\mathscr{E}_{\text{refl}}(x) = |\hat{\rho}_{\text{pcm}}(x)| e^{j\psi} \mathscr{E}^*_{\text{inc}}(x). \tag{52}$$

Here $\mathscr{E}_{\text{inc}}(x)$ is the incident field striking the PCM, $\mathscr{E}_{\text{refl}}(x)$ is the phase-conjugated reflected field, and ψ is not a function of the transverse coordinates. (That is, the phase conjugator may be apertured but is otherwise ideal.) The conventional optics in the remainder of the PCM resonator may be quite arbitrary, including paraxial elements, perturbing screens, and finite apertures. Let an additional index m or n label the different-order transverse eigenmodes in the PCM resonator. Then one can show under very general conditions (with proper normalization) that

$$\int_{-\infty}^{\infty} |\hat{\rho}_{\mathrm{pcm}}(x)| \mathscr{E}_{\mathrm{inc},m}(x) \vec{\mathscr{E}}^{\,*}_{\mathrm{inc},n}(x)\, dx \quad = \int_{-\infty}^{\infty} \mathscr{E}_{\mathrm{inc},m}(x) \mathscr{E}_{\mathrm{refl},n}(x)\, dx = \delta_{nm}. \tag{53}$$

This calculation is made initially at the input plane to the phase conjugator. However, it can also be shown under very general conditions (Arnaud, 1976) that the second integral will continue to hold at any other plane within the resonator, if we write it in the form

$$\int_{-\infty}^{\infty} \mathscr{E}_{\mathrm{for},m}(x) \mathscr{E}_{\mathrm{back},n}(x)\, dx = \delta_{nm}, \tag{54}$$

where $\mathscr{E}_{\mathrm{for}}$ now represents the forward-traveling and $\mathscr{E}_{\mathrm{back}}$ the backward- or reverse-traveling wave at that plane.

Under the most general conditions in the PCM resonator, as in conventional optical resonators, neither the transverse eigenmodes going in the forward direction, $\mathscr{E}_{\mathrm{for},n}(x)$, nor the same modes going in the backward direction at the same plane, $\mathscr{E}_{\mathrm{back},n}(x)$, form a separately orthogonal set, with or without complex conjugation (unless the PCM is ideal, with no transverse variation). Rather, each transverse eigenmode going in one direction at a given transverse plane is orthogonal without complex conjugation to all the eigenmodes in the family going in the opposite direction at the same plane. This is essentially the same as the orthogonality relation for conventional optical resonators (Siegman, 1979). It represents again the fact that Huygens' integral written in a phasor formulation has a complex symmetric but not a Hermitian kernel. Alternatively, it can be viewed as arising from the fact that the optical resonator problem obeys a fundamentally Hermitian operator (the wave equation), but in general does not have adjoint boundary conditions whenever diffraction losses are present.

VI. Resonance Frequencies and Longitudinal Modes

We next derive the *resonance frequencies* and *axial-mode properties* of an idealized phase-conjugate resonator consisting of a PCM reflector produced by four-wave mixing at one end and a conventional mirror at the other. The special properties derived in this section will arise primarily from the fact that phase conjugation produced by four-wave mixing is an *active* nonlinear process, produced by pumping beams with definite frequency and phase. The resonance frequency properties associated with resonators using stimulated scattering processes or passive pseudoconjugators would be quite different, and are not treated in this chapter.

A. Frequency Characteristics of Phase-Conjugate Reflectors

The resonance properties of a phase-conjugate resonator depend in general not only on the cavity length and related parameters but also on the phase-versus-frequency characteristic of the PCM reflector. Hence we need to understand how the reflectivity of the phase-conjugate mirror changes for signals tuned off resonance. Detailed derivations of the complex reflection coefficient versus frequency for a phase conjugator using four-wave mixing are given in Section II.E of Chapter 2. We will present, however, another brief derivation that may give additional insight into this process.

The physical reasoning behind this derivation is illustrated in Fig. 16. Suppose a four-wave mixing cell is continuously pumped at frequency ω_0 and has total interaction length L_c, so that the transit time through the conjugating cell is $T_c = L_c/c$ along the direction of the signal waves. Let a

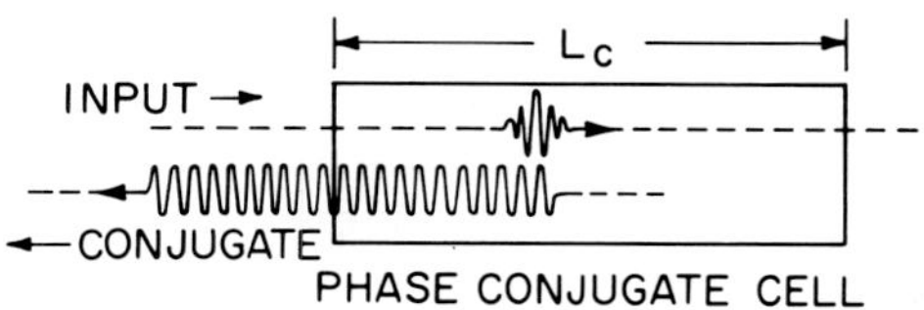

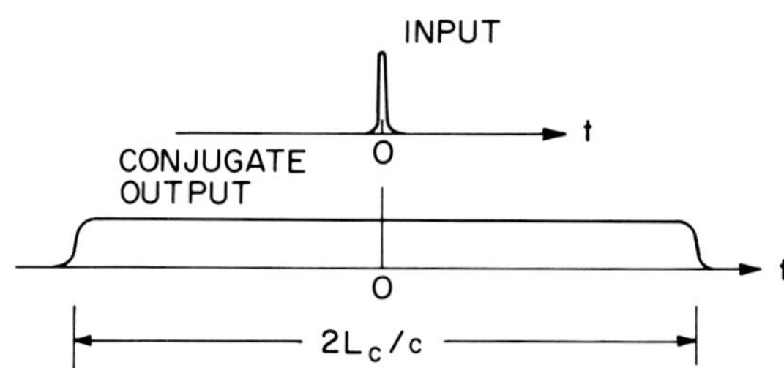

Fig. 16. A δ-function pulse with carrier frequency ω_0 coming into a four-wave mixing cell of length L_c produces an extended square conjugate pulse of length $2L_c/C$ at the conjugate frequency ω_0 (assuming small reflectivity and no saturation effects). This pulse is centered in time on the instant when the signal pulse reaches the *center* of the phase-conjugate cell. The Fourier transform of this reflected pulse gives the frequency response of the conjugate reflector.

very short signal pulse with carrier frequency equal to the pump frequency ω_0 and a pulse duration much shorter than the transit time T_c be sent into the cell traveling in the $+z$ direction. The signal pulse, as it travels through the cell and interacts with the cw pump waves, will shed a continuous "tail" of reflected waves, or idler waves, at frequency ω_0. These phase-conjugated waves will radiate backward through the cell, emerging in the $-z$ direction. This conjugated or idler-frequency radiation begins to appear at the instant the signal pulse arrives at the front face of the cell, and continues to emerge back through the front face of the cell for one full transit time after the input signal impulse leaves the cell at the opposite end.

For the idealized case of small absolute reflectivity and small pump depletion, the impulse response of a four-wave conjugator will thus be essentially a square, frequency-shifted pulse having constant amplitude and a time duration $2T_c = 2L_c/c$. [For expanded discussions of transient responses in four-wave mixing, see Bobroff and Haus (1967), Fisher *et al.* (1981), and Suydam (1980, 1982).] If the reference plane for the phase-conjugate mirror is chosen at the center rather than at the front face of the four-wave cell, this impulse response will be symmetric from $-T_c$ to $+T_c$ about $t = 0$.

The frequency response of the cell, if it were a conventional linear system, would then be the Fourier transform of this rectangular time response. The phase-conjugate case is slightly more complicated, because the four-wave mixing cell also inverts the frequency of each input signal component with respect to the pumping frequency ω_0. In the example we have chosen, however, the input δ-function pulse has a flat spectrum over an arbitrarily large range about ω_0. and so this frequency flipping has no directly visible effects. The power reflectivity $R(\omega)$ of the cell, with this frequency-inverting property kept in mind, is still the magnitude of the rectangle's Fourier transform squared, or

$$R(\omega) = R(\omega_0)\left|\frac{\sin T_c(\omega - \omega_0)}{T_c(\omega - \omega_0)}\right|^2. \tag{55}$$

This agrees with the results derived by Pepper and Abrams (1978) or by Marburger (1978) in the weak coupling limit. The conjugate reflector thus has a frequency-dependent reflectivity by itself, of the form illustrated in Fig. 17. The power reflectivity has maximum value $R(\omega_0)$ on resonance and falls off with the familiar $[(\sin x)/x]^2$ dependence on frequency on either side.

There are then two significant limiting cases, short cells and long cells, to be considered. If the PCM cell length L_c is small compared with the length L of the remainder of the optical cavity, the central lobe of the fre-

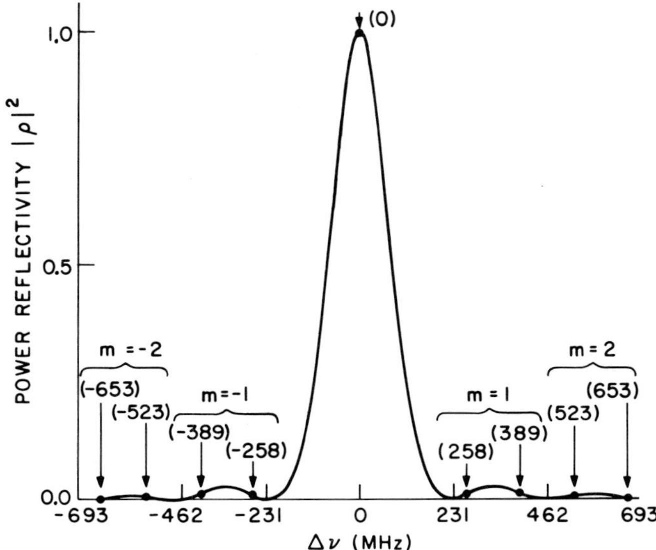

Fig. 17. Power reflectivity versus frequency for a typical finite-length four-wave mixing cell, together with axial-mode resonances for a typical PCM resonator using this cell. An additional $m = 0$ resonance at ~ 123 MHz was overlooked in the original reference. Physical dimensions are: ordinary cavity length $L = 25$ cm; four-wave-cell optical length $L_c = 1.62 \times 40$ cm; and pumping parameter $(\kappa c)^2 = \pi/4$. (Used with permission from AuYeung *et al.*, 1979. Copyright 1979, IEEE.)

quency response of the conjugator will be broad compared with the axial-mode spacing of the optical cavity. Also, with the PCM reference plane taken at the center of the four-wave mixing cell, the reflectivity for the weak coupling case will have constant phase across the main frequency lobe. We can therefore evaluate the resonance properties of a PCM cavity in this short-cell limit by assuming simply a constant amplitude and constant phase reflectivity for the PCM reflector.

The short-cell, broad-response condition requires a strong nonlinearity or strong pumping to produce significant reflectivity, and thus is more difficult to achieve in realistic experimental situations, though this condition has been experimentally achieved by Lind and Steel (1981). In the opposite limit, however, where the cell length L_c is comparable with or longer than the cavity length L, the frequency response of the PCM reflector itself will be narrow compared with the axial-mode spacing of the cavity. The off-resonance axial modes of the cavity will then in general be far down on the side lobes of the PCM reflection profile. Hence these side modes will have little chance of oscillating in competition with the central mode in any case. The off-resonance axial modes in this long-cell or narrow-response case thus appear to be of much less interest. The latter situation has been analyzed in more detail, however, by AuYeung *et al.* (1979).

The round-trip phase shift versus frequency in the conventional part of an optical cavity will also depend slightly, in second order, on the transverse-mode proerties of the cavity, because of the well-known Guoy phase shifts. These effects cause slight frequency offsets for higher-order transverse modes in conventional as well as in PCM cavities (Siegman, 1971). However, these effects are basically equivalent to a minuscule change in the apparent length L of the cavity. We will therefore ignore these effects and use an essentially transmission-line or plane-wave approach for the main part of the optical cavity.

B. Frequency Analysis of a Short-Cell PCM Resonator

We therefore consider a phase-conjugate resonator consisting of an ideal (i.e., very short) PCM or four-wave mixing cell located at $z = 0$ plus a normal mirror located at $z = -L$. Suppose that incident on the PCM from inside the cavity is a signal wave of the general form

$$E(t) = \mathrm{Re}[\mathscr{E}_{\mathrm{inc}}(t)e^{j\omega_0 t}]. \tag{56}$$

We ignore transverse variations for simplicity, but assume that both the phase and the amplitude of the complex phasor $\mathscr{E}_{\mathrm{inc}}(t)$ may be time varying in general. The absolute phases of all waves are measured at the reference plane $z = 0$ in the center of the four-wave mixing cell. For small incident fields, the reflected wave from a PCM is given by

$$E_{\mathrm{refl}}(t) = \mathrm{Re}[\hat{\kappa}\mathscr{E}_{\mathrm{inc}}^{*}(t)e^{j\omega_0 t}]. \tag{57}$$

The reflection coefficient $\hat{\kappa}$ due to four-wave mixing may in general be complex, with a fixed phase angle and magnitude that depend on the cell geometry, the magnitude and phase of the nonlinear susceptibility, and the pumping-wave phasor amplitudes.

Let the round-trip transit time around the cavity from the PCM to the normal mirror and back be $T \equiv 2L/c$. Then the waves $\mathscr{E}_{\mathrm{inc}}(t)$ and $\mathscr{E}_{\mathrm{refl}}(t)$ will also be related by

$$\mathscr{E}_{\mathrm{inc}}(t)e^{j\omega t} = \sqrt{G}\mathscr{E}_{\mathrm{refl}}(t - T)e^{j\omega_0(t-T)}, \tag{58}$$

where G is the net round-trip power gain due to any laser material (or losses) inside the cavity. For simplicity we assume that G has no dispersion. Because we are interested primarily in the resonant frequencies of the cavity, we will assume that this round-trip power gain and the PCM power reflectivity combine to give unity net gain, i.e., $|G\kappa^2| \equiv 1$. Let the net phase angle of the quantity $\hat{\kappa}$ be ψ, i.e., $\hat{\kappa} \equiv \kappa e^{j\psi}$. Then from Eqs. (56)–(58) the round-trip self-consistency condition for the cavity becomes

$$\mathscr{E}_{\text{inc}}(t + T)e^{j\omega_0 T} = \mathscr{E}_{\text{refl}}(t) = e^{j\psi}\mathscr{E}^*_{\text{inc}}(t). \tag{59}$$

Solutions to this equation determine the resonant frequencies of the cavity.

1. Central Resonant Frequency

Consider first the case when $\mathscr{E}_{\text{inc}}(t)$ is independent of t, i.e., when all signals are exactly at ω_0. If we write the phasor amplitude as

$$\mathscr{E}_{\text{inc}} \equiv |\mathscr{E}_{\text{inc}}|e^{j\theta_0}, \tag{60}$$

the self-consistency condition becomes

$$\exp[j(\theta_0 + \omega_0 T)] = \exp[j(\psi - \theta_0)]. \tag{61}$$

This result says the cavity can *always* have a resonance at ω_0 independent of the cavity transit time T or the cavity length L. However, the signal at ω_0 must have a definite *time* phase θ_0 (measured at $z = 0$) given by

$$2\theta_0 = \psi - \omega_0 T (\text{mod } 2\pi). \tag{62}$$

The cavity standing waves are in essence not only frequency synchronized but also phase synchronized (in time, not space) to the phases of the pumping waves as contained in the phase angle of $\hat{\kappa}$.

2. Nondegenerate Axial Modes

To find the nondegenerate or off-resonance "axial modes" of the ideal PCM resonator, we next suppose that $\mathscr{E}(t)$ contains an upshifted signal component at frequency $\omega_0 + \omega_m$, where ω_m is some as yet unspecified modulation frequency. The phasor amplitude $\mathscr{E}_{\text{inc}}(t)$ then varies in time as

$$\mathscr{E}_{\text{inc}}(t) \sim e^{j\omega_m t}. \tag{63}$$

This will produce a reflected wave from the PCM given by

$$\mathscr{E}_{\text{refl}}(t) = \hat{\kappa}\mathscr{E}^*_{\text{inc}}(t) \sim e^{-j\omega_m t}. \tag{64}$$

An upshifted component of $\mathscr{E}_{\text{inc}}(t)$ at $\omega_0 + \omega_m$ thus produces a reflected and downshifted component of $\mathscr{E}_{\text{refl}}(t)$ at $\omega_0 - \omega_m$. But this wave $\mathscr{E}_{\text{refl}}(t)$ at $\omega_0 - \omega_m$ will travel around the cavity eventually to become $\mathscr{E}_{\text{inc}}(t)$ at $\omega_0 - \omega_m$ rather than $\omega_0 + \omega_m$. We must therefore write $\mathscr{E}_{\text{inc}}(t)$ more generally as a combination of both of these frequencies, i.e.,

$$\mathscr{E}_{\text{inc}}(t) = e^{j\theta_0}(\mathscr{E}_+ e^{j\omega_m t} + \mathscr{E}_- e^{-j\omega_m t}), \tag{65}$$

where $\mathscr{E}_+$ and $\mathscr{E}_-$ are both complex phasor amplitudes. Putting Eq. (65) into the frequency self-consistency relation (59) and separating like-frequency terms gives

$$\mathscr{E}_+ = e^{-jT\omega_m}\mathscr{E}_-^*, \qquad \mathscr{E}_- = e^{jT\omega_m}\mathscr{E}_+^*. \tag{66}$$

Combining the first of these relations with the complex conjugate of the second then produces the condition

$$\exp(j2T\omega_m) = \exp(jn2\pi) \tag{67}$$

where n is any integer. The modulation or side-band frequency ω_m can thus have any of the discrete values given by

$$\omega_m = n(\pi/T) = 2\pi n(c/4L). \tag{68}$$

These modes or cavity resonances are spaced by $c/4L$ Hz, *half* the usual axial-mode spacing for a cavity of length L.

The instantaneous field $E(t)$ in the cavity may thus be expanded in resonant modes in the form

$$\begin{aligned} E(t) &= \text{Re}\{\mathscr{E}(t)\exp[j(\omega_0 t + \theta_0)]\} \\ &= \text{Re}\left(e^{j\theta_0}\sum_{n=-\infty}^{\infty}\mathscr{E}_n \exp\{j[\omega_0 t + n(\pi/T)t]\}\right), \end{aligned} \tag{69}$$

with the condition that

$$\mathscr{E}_{-n} = (-1)^n\mathscr{E}_n^*. \tag{70}$$

This is equivalent to writing

$$\mathscr{E}(t) = \sum_{n=0,2,4,\ldots}\mathscr{E}_n\cos(n\pi t/T + \theta_n) + \sum_{n=1,3,\ldots}\mathscr{E}_n\sin(n\pi t/T + \theta_n). \tag{71}$$

The PCM cavity thus has a central resonance frequency determined by the pumping frequency ω_0, not the cavity length L. This central frequency is surrounded by a set of paired "half-axial-mode" resonances spaced by $c/4L$ Hz, or *half* the usual spacing for a cavity of length L, as shown in Fig. 18. The existence of these half-axial modes has recently been confirmed experimentally by Lind and Steel (1981).

This behavior may be understood physically by the following picture. In Fig. 19 are shown the circulating waves traveling in both directions inside the cavity at an initial instant (top sketch) when the cavity is assumed to be entirely filled with radiation at an upshifted frequency $\omega_0 + \omega_m$. This

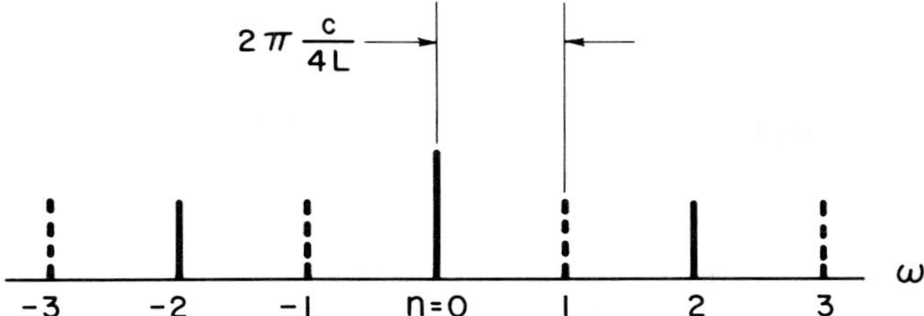

Fig. 18. The resonance frequencies of an ideal PCM resonator occur in symmetric pairs about the pump frequency ω_0 (corresponding to $n = 0$), and are spaced by *half* the axial-mode interval for a conventional cavity of the same length L.

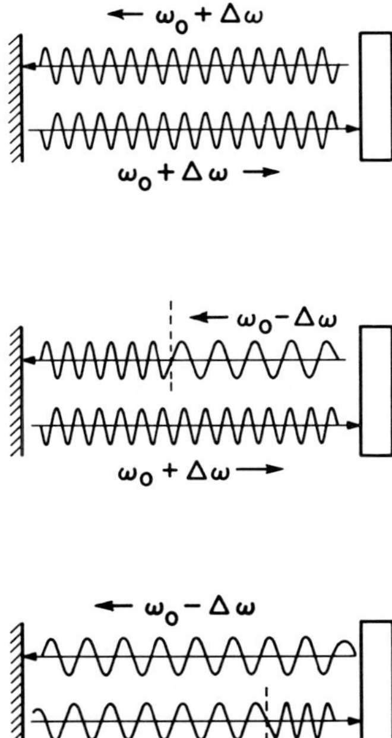

Fig. 19. A phase-conjugate resonator initially filled with radiation at an upshifted frequency $\omega_0 + \Delta\omega$ (top sketch) gradually converts all this radiation to the downshifted frequency $\omega_0 - \Delta\omega$ (middle and bottom sketches) during one complete round trip. This process then reverses and converts the resonator back to its initial condition during the next round trip. Hence it takes *two* full round trips for the initial conditions to be fully reproduced.

radiation, as it impinges on the PCM, will be reflected at the downshifted frequency $\omega_0 - \omega_m$. Hence after a short time the signal inside the cavity will appear as in the middle sketch, with lower-frequency radiation emerging from and moving away from the PCM. The PCM will continue to take in the higher frequency $\omega_0 + \omega_m$ and emit the lower frequency $\omega_0 - \omega_m$ for one complete round-trip time T, after which the cavity will be entirely filled with lower-frequency radiation (bottom sketch). But this process will then reverse, and the incident signal at $\omega_0 - \omega_m$ will again be converted back to the original value $\omega_0 + \omega_m$. This cycle will be complete after a time $2T$ or *two* round trips. It is this doubling of the time interval before the signal returns to its original state that leads to the halving of the effective axial-mode spacing.

C. Resonant Frequencies with Finite Cell Length

The resonant frequencies of a PCM resonator in the more general case when the phase-conjugate cell is not short compared with the remainder of the cavity have been discussed by AuYeung *et al.* (1979), who develop an expression for the phase shift on reflection from the front face of a four-wave mixing cell for the upper and lower sidebands at frequencies $\omega_0 \pm \omega_m$. Their expression for a cell of optical length L_c (i.e., with index of refraction included in the definition of length), with constant-phase terms and small transverse-mode factors omitted, becomes

$$\phi_\pm = \tan^{-1}\left[\frac{\omega_m}{c\beta_m}\tan(\beta_m L_c)\right], \tag{72}$$

where to a good approximation $\beta_m = \sqrt{\kappa^2 + (\omega_m/c)^2}$. The power reflectivity magnitude versus frequency is then

$$R(\omega) = \frac{\kappa^2 \sin^2(\beta_m L_c)}{\kappa^2 \cos^2(\beta_m L_c) + (\omega_m/c)^2}, \tag{73}$$

which agrees with Eq. (55) in the small-β limit. A round-trip self-consistency analysis similar to the one we have just presented then leads to a resonance expression that in our notation becomes

$$\frac{2\omega_m L}{c} + \tan^{-1}\left[\frac{\omega_m}{c\beta_m}\tan(\beta_m L_c)\right] = m\pi, \tag{74}$$

or

$$\tan\left(\frac{2\omega_m L}{c}\right) + \frac{\omega_m L_c}{c}\operatorname{tanc}(\beta_m L_c) = 0, \tag{75}$$

where by analogy to the function sinc $x \equiv (\sin x)/x$ we have defined tanc $x \equiv (\tan x)/x$. For simplicity we have absorbed all index of refraction factors into the optical lengths of the four-wave cell L_c and of the remainder of the cavity L. The roots of Eq. (75) give the pairs of axial-mode frequencies $+\omega_m$ and $-\omega_m$ for the cavity, taking into account the frequency dependence of the phase conjugator.

Note first that for a short cavity and not too strong coupling we can re-

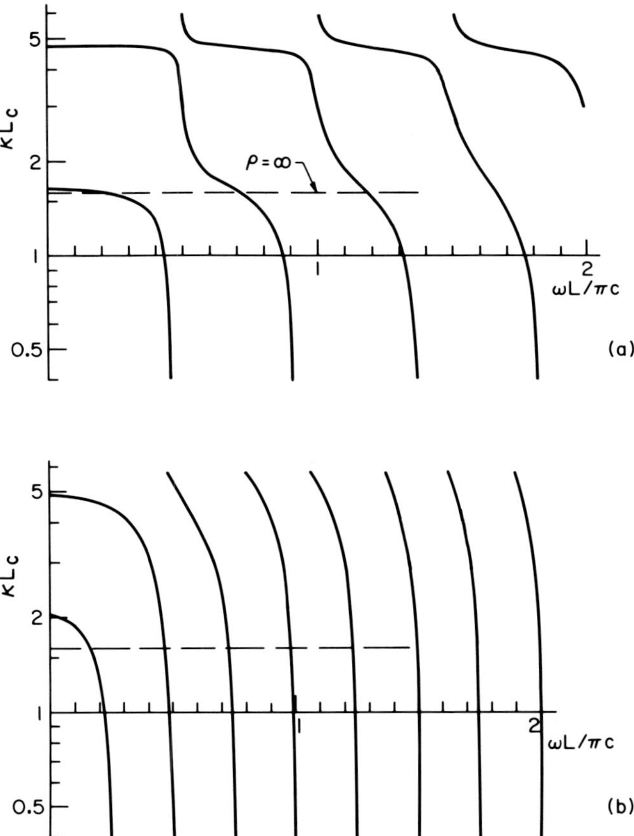

Fig. 20. Axial-mode frequencies versus pumping strength κL_c in a PCM resonator for two different ratios of four-wave-cell length L_c to ordinary cavity length L (the distance from the conventional mirror to the front face of the cell): (a) $L_c = 0.2L$, and (b) $L_c = 2L$. The axial-mode spacing under nearly all conditions remains very close to $\Delta\omega_{ax} = 2\pi c/(L + L_c/2)$, indicating that the constant-phase reference plane for the four-wave cell is at the midplane, not the front face, of the cell.

place the tangents with their arguments and obtain

$$\omega_m \approx 2m\pi c/(4L + 2L_c). \tag{76}$$

The effective length of the cavity is evidently not just L, but rather $L + L_c/2$. This agrees with our earlier physical argument that for weak reflectivities the effective constant-phase reference plane in a four-wave mixing cell is located at the *midplane* of the cell rather than at the front face.

In Fig. 17 are shown calculated mode frequency results from AuYeung *et al.* for the particular case of a strongly reflecting phase-conjugate cell considerably longer than the remainder of the cavity, namely, $L = 25$ cm and $L_c = 1.62 \times 40 = 64.8$ cm, with $\kappa L_c = \pi/4$. Even in this rather strong coupling case with $L_c > L$ the axial resonances are still spaced quite close to an even 131 MHz apart, with only slight irregular deviations from regular spacing. (AuYeung *et al.* appear to have overlooked a second $m = 0$ resonance at ~123 MHz.) The 131-MHz interval matches very closely with $c/4L_{\text{eff}}$ for their cavity, assuming an effective cavity length of $L_{\text{eff}} \approx L + L_c/2 = 57.4$ cm.

In Fig. 20 are shown further examples of mode spectra versus pumping strength κL_c for different ratios of $L_c/2L$ that we have calculated from Eq. (74). Even for very large values of κL_c, and for large or small values of the ratio $L_c/2L$, the axial modes are spaced by very close to $c/(4L + 2L_c)$ Hz. All the modes shift over in frequency, however, and the lowest off-resonance mode moves in and disappears, each time κL_c approaches one of the roots of the function $\kappa L_c + (L_c/2L)\tan(\kappa L_c) - 0$. The dashed line in each figure marks the value $\kappa L_c = \pi/2$ at which the power reflectivity $|\rho|^2 \to \infty$. Above this value the four-wave cell can oscillate by itself, without the assistance of any external cavity. The presence of an external cavity would, however, presumably modify both the threshold and the off-resonance frequencies of this oscillation, in a manner that could be determined only by a more detailed nonlinear analysis.

D. Double-Ended PCM Resonators

What if one has phase conjugators at *both* ends of a PCM resonator? Consider a cavity with phase conjugators at both ends, pumped at two possibly different frequencies ω_a and ω_b, with absolute phases ψ_a and ψ_b for the phase-conjugate reflectivities. Suppose a signal of the general form

$$E_1(t) = \mathscr{E}_1 \exp[j(\omega_1 t + \theta_1)] \tag{77}$$

strikes the first of these conjugators. The reflected signal then has time dependence $E_2(t) = \hat{\kappa}_a \mathscr{E}_1^* \exp[j(2\omega_a - \omega_1)t - j\theta_1]$, and the signal striking the second conjugator is $E_3(t) = E_2(t - T)$. This signal is similarly phase conjugated and inverted about the second pumping frequency ω_b to produce a signal $E_4(t)$. Finally, the signal returning to the original starting point will be $E_5(t) = E_4(t - T)$, which may be written

$$E_5(t) = \kappa_a \kappa_b \mathscr{E} \exp\{j[(\omega_1 + 2\omega_b - 2\omega_a)t + (4\omega_a - 2\omega_b - 2\omega_1)T + \theta_1 + \psi_b - \psi_a]\}. \tag{78}$$

The original signal at ω_1 acquires an additional frequency offset of $2\omega_b - 2\omega_a$ on each round trip, unless $\omega_b - \omega_a \equiv 0$. Evidently *no steady-state resonant modes are possible in a cavity having two conjugators separately pumped at different frequencies*. Of course, one unrealistic possibility might be to cancel this frequency offset against a Doppler shift obtained by moving one of the conjugate mirrors. Another equally unrealistic possibility would be to develop a ring-type PCM resonator with three phase conjugators pumped with appropriately matched frequencies such that the total frequency offset equaled zero.

Steady-state resonant solutions with $E_5(t) = E_1(t)$ can be obtained for degenerate pumping of both conjugators by the same frequency, namely, $\omega_a = \omega_b = \omega_0$. The resonant frequencies will be given by

$$\omega_1 = \omega_0 + \frac{\psi_b - \psi_a + 2n\pi}{2T} = \omega_0 + \frac{\psi_b - \psi_a}{2T} + 2n\pi \frac{c}{2L}, \tag{79}$$

where n is any integer. Clearly these modes have the usual $c/2L$ axial-mode spacing. The frequencies ω_1, however, apply only to waves going in one direction (from phase conjugator b to phase conjugator a) inside the cavity. The corresponding waves going in the opposite direction will be symmetrically located on the opposite side of ω_0.

These descriptions of single- and double-ended phase-conjugate resonator modes can be unified by noting that a double-ended PCM resonator is essentially equivalent to a folded single-ended resonator of half the length.

VII. Experimental Results

There have been only a few published experimental results on PCM resonators up to this point. Au Yeung *et al.* (1979) were the first to demonstrate experimentally a four-wave phase-conjugate resonator, using CS_2 as the nonlinear medium and a ruby laser for both the four-wave pump beams and the PCM cavity gain medium. They used a collinear pump–signal arrangement as illustrated in Fig. 21a to obtain maximum interac-

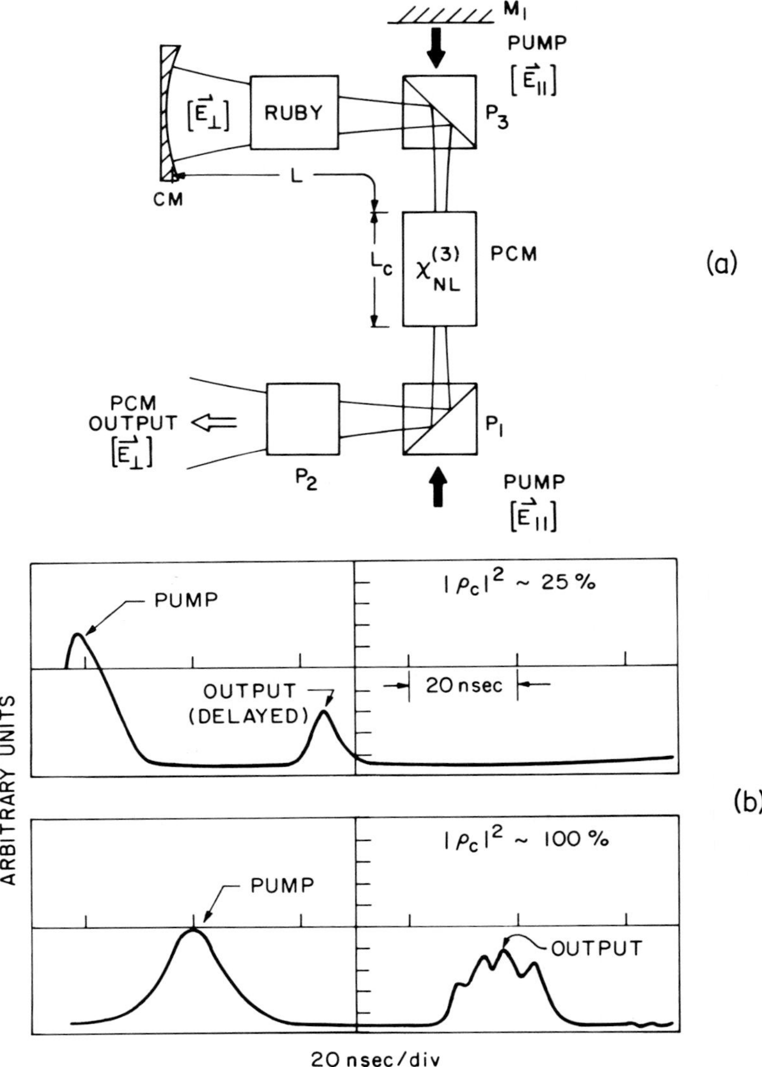

Fig. 21. (a) Phase-conjugate oscillator experiment carried out by AuYeung *et al.* (1979). The counterpropagating pump waves (incident upon and reflected from the upper mirror M_1 are collinear with the phase-conjugate resonator axis to obtain maximum interaction length, and are distinguished from the phase-conjugated signals by separate polarizations. (b) Temporal form of the pumping pulse and the phase-conjugated oscillation pulse for two different pumping strengths. (The phase-conjugate oscillation pulse is delayed in each trace for clarity.) The upper signals represent a PCM power reflectivity of ~25% and the lower signals an estimated PCM reflectivity of ~100%. Note that for the larger pumping case the output is on for a longer time interval and exhibits an oscillatory character that may represent PCM resonator mode beating. (Used with permission from AuYeung *et al.*, 1979. Copyright 1979, *IEEE*.)

tion length and beam overlap, while using polarization discrimination to discriminate between the pumping laser beam and the PCM signal waves.

When the flash-lamp excitation of the ruby gain medium within the PCM resonator was properly synchronized with the pumping laser beam coming into the four-wave mixing cell in these experiments, an intense s-polarized signal pulse was detected that presumably arose from phase-conjugate oscillation in this polarization. This type of output was observed with either a convex (stable) or concave (unstable) conventional mirror at the other end of the resonator. No measurements of the transverse-mode shape were made, however, and thus no detailed comparison with the theory is possible here. There can be some difficulty in this type of setup in distinguishing between regenerative amplification of depolarized pump beams and true phase-conjugate oscillation, especially because the number of passes in the PCM resonator during the pump pulse is limited. Their results at the higher pump level do show an interesting periodic ripple at ~250 MHz. This could possibly be due to axial-mode beating, because this would be the appropriate axial (not half-axial) value for their cavity.

Russian workers have reported laser action with a PCM resonator using stimulated Brillouin scattering (Bel'dyugin and Zemskov, 1980, Lesnik *et al.*, 1979, Bezrodnyi *et al.*, 1980). Recently, Mathieu and Belanger (1980) also published experimental results on CO_2 TEA and visible dye lasers using retroreflective pseudo-conjugate-reflectors in combination with planar conventional mirrors. In these experiments they were able to correct aberrations obtained by placing cylindrical lenses just in front of the PCM. This is a confirmation of the model presented in Section II, where we have shown that a lens located in front of the PCM does not affect the beam waists either on the PCM or on the mirror. In their experiments one can imagine that a cylindrical lens of focal length F_x was located along the x axis and a cylindical lens of focal length F_y was located along the y axis. According to our theoretical results, the beam waists should still be the same along the x and y axes, resulting in a circular beam, which was observed experimentally.

In one of the most recent and most striking experiments as of the time of writing this chapter, Feinberg and Hellwarth (1980) have shown that strong phase-conjugation effects and, more dramatically, strong PCM resonator oscillation can be obtained on a cw basis, using the photorefractive effect in $BaTiO_3$ as the nonlinear medium and a cw argon laser or even a He–Ne laser as the pump source.

The phase-conjugation effect in this material arises from a slowly developing and reversible optically induced electronic damage effect peculiar to at least certain samples of $BaTiO_3$. Though this damage mechanism is not yet fully understood, it appears that mobile electrons are optically induced in the illuminated regions of the crystal, notably at the bright fringes of any interference patterns present between the pump and signal beams. These electrons drift to the dark fringes in the interference pattern and become trapped there, producing an electrostatic field grating in the crystal. These electrostatic fields then produce a phase grating, or index of refraction grating, through a straightforward Pockels effect mechanism. Because of its slow response—the grating can take several seconds to develop, or to decay—the phase conjugation will be extremely narrow band; but its effect is also very strong. Reflectivities substantially greater than unity can be obtained in a few millimeters of crystal, with a few milliwatts of pumping power.

No systematic studies of mode properties have been published as yet using this system. However, the phase conjugator exhibits the expected abilities to oscillate with a badly distorted conventional end mirror (e.g., a common kitchen utensil) placed nearby; to oscillate with good beam quality and low loss even with severely aberrated phase elements inside the cavity; and to oscillate even with a highly divergent mirror (specifically a small polished metal sphere) as the conventional end mirror.

It is important to note that, because of the peculiar nature of the phase-conjugation mechanism in this material, the index gratings formed inside the crystal are displaced transversely by one-quarter of a spatial cycle from the intensity fringes formed by the pumping and signal beams. This is a very different situation from the usual case, e.g., the optical Kerr effect, where the index changes are directly proportional to and in spatial phase with the local light intensity. This leads to unusual and somewhat unexpected properties for this particular phase conjugator.

Another important demonstration of continuous phase-conjugate resonator oscillation has been reported by Lind and Steel (1981) at Hughes Research Laboratories using four-wave mixing in sodium vapor as the nonlinear effect. By pumping a short sodium vapor cell (1 cm long) with two counterpropagating pump beams from a monochromatic cw dye laser tuned close to the sodium D_2 resonance line, phase-conjugate power reflectivities greater than 80% (and in some cases greater than unity) could be obtained. This cell was then used as one end mirror for a second modified dye laser cavity containing a dye laser gain medium at the same wavelength. The product of phase-conjugate cell reflectivity and dye laser amplification was sufficient to bring this second cavity above threshold on a continuous basis.

The output beam from the conventional mirror end of this second cavity appeared to be a smooth TEM_{00} mode as expected, and the mode quality of this beam appeared to remain unchanged when an aberrating element was placed inside the cavity. The beam from the opposite end of the cavity, coming through the sodium cell, on the other hand, was seriously distorted as expected.

Of more immediate significance, this device also provided the first experimental demonstration of the half-axial-mode spectrum predicted by the analysis in this chapter. When pumped exactly on the atomic resonance frequency, the nonlinearity in the sodium cell had a narrow bandwidth and response time ($\sim$50 MHz and $\sim$16 nsec) determined by the radiative lifetime for the upper-level sodium atoms. By detuning the pumping frequency a small amount from the resonant frequency of the sodium atoms, however, it was possible to broaden the frequency response of the phase-conjugate mirror into a frequency profile with several subsidiary amplitude peaks and a bandwidth up to $\sim$200 MHz. Under these conditions the phase-conjugate cavity clearly oscillated in three to five modes separated by the expected half-axial-mode spacing of $c/4L \approx$ 40 MHz.

Another novel and promising technique for laser mode locking and Q switching in a conventional laser cavity with a "self-pumped" phase-conjugate mirror has also very recently been demonstrated (Vanherzeele *et al.*, 1981), as described more fully in the following section.

VIII. Q-Switching and Mode Locking in PCM Resonators

The strong dependence of phase-conjugate reflectivity on instantaneous pumping power can lead to very interesting possibilities for Q-switched and/or mode-locked operation of phase-conjugate resonators, especially if a laser medium is included inside the PCM cavity, or if part of the oscillation power of the PCM cavity is used to provide the pumping power for the phase conjugator (which we will refer to as self-pumping). In this section we will briefly explore a few of these possibilities.

A. Mode Locking of Phase-Conjugate Resonators

It should be possible at least in theory to mode-lock a PCM resonator and to obtain short pulses recurring at the usual round-trip transit time T, either by using an intracavity amplitude modulator operating at the usual mode-locking frequency $\omega_{ax} = 2\pi(c/2L)$ or by synchronously pumping the phase conjugator with short pump pulses at the same repetition frequency. Note, however, that the circulating mode-locked pulse inside the cavity may have a carrier frequency on one round trip given by $\omega_0 + \Delta\omega$, where $\Delta\omega$ is any arbitrary frequency offset. Then, by the same agrument as illustrated in Fig. 19, this same pulse will have carrier frequency $\omega_0 - \Delta\omega$ on the next round trip, and so forth. The train pulses emerging from the resonator will thus seem to have the usual interpulse interval $2L/c$; but the fundamental repetition period for the signal spectrum, including phase and frequency considerations, will be $4L/c$. The fundamental sideband spacing of the mode-locked pulse spectrum will thus be $2\pi(c/4L)$ as discussed in an earlier section, even though the pulse repetition rate is $2\pi(c/2L)$.

Active mode locking using an intracavity modulator, though entirely possible in principle, does not appear promising with presently available phase-conjugating materials. The usual tradeoff between bandwidth and required pump power in most phase conjugators means that any conjugator with sufficiently low pumping requirements to be operated cw, so that mode-locked pulses can develop, is unlikely to have sufficient bandwidth to produce pulses short enough to be of interest. In addition, the finite bandwidth of the PCM reflectivity, and the axial-mode pulling effects arising from finite conjugator length, as analyzed in a previous section, add further difficulties.

Synchronous pumping of the phase-conjugate mirror using two pulses that collide from opposite directions in the wave mixing cell seems a much more promising method for PCM mode locking. The pump pulses should, of course, arrive at a repetition rate exactly equal to the round-trip time for pulses circulating inside the PCM resonator. The PCM reflectivity is then modulated in time even more sharply than the pumping pulses. Hence, as discussed elsewhere (Bloom *et al.*, 1978), the reflected pulses should be even shorter than the pumping pulses themselves. An important side advantage is, of course, that much higher peak reflectivities can be obtained for the same average power from the pumping-pulse source.

With sufficiently short pumping pulses, synchronous pumping should be applicable for compressing pulses down to something like the interaction length $T_c = 2L_c/c$ of the pulses in the four-wave cell. Suydam (1980, 1982) has shown, in fact, how one can use a specially tailored input pulse, something like a string of zero-pi pulses, to produce a phase-conjugated output pulse significantly shorter than the impulse response function of the four-wave cell. Further research would appear to be desirable to explore the oscillation, pulse compression, and pulse stability properties of this approach.

B. Self-Pumped Phase-Conjugate Resonators

There are also a variety of very promising new methods for producing self-Q-switching effects, pulse regeneration effects, and unusually short

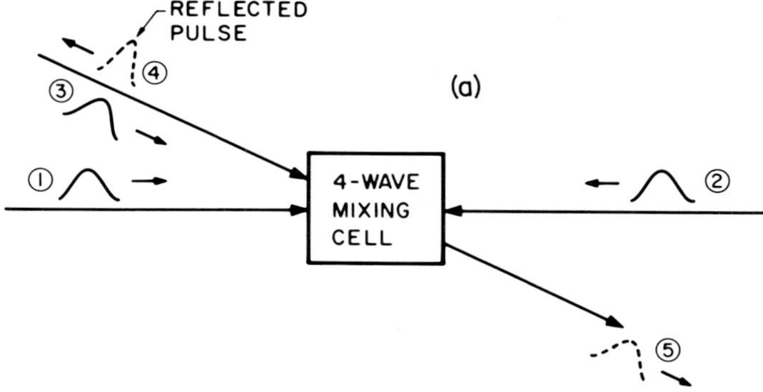

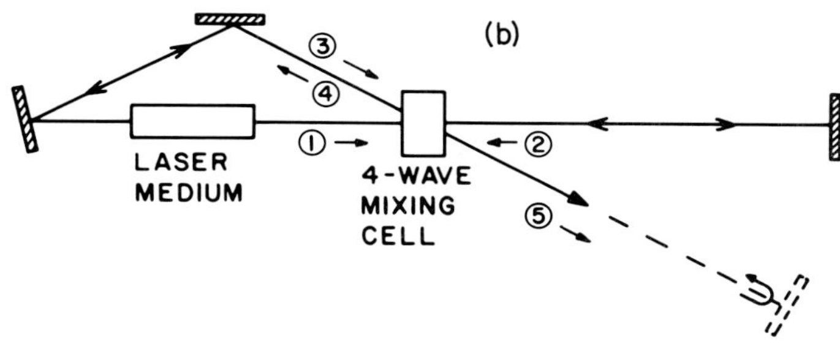

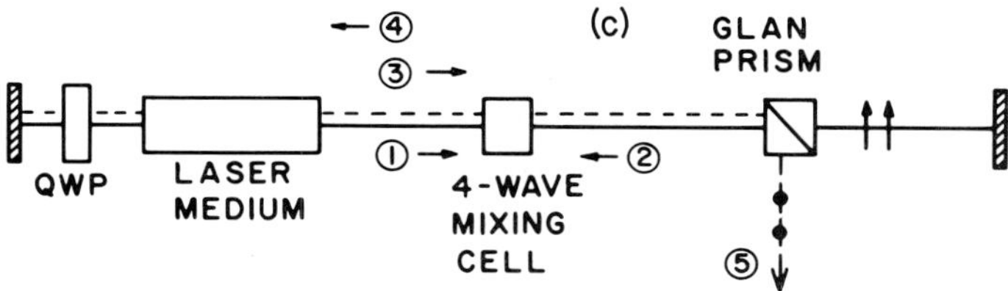

Fig. 22. Method for generating mode-locked oscillation in a conventional laser oscillator using "self-pumped" phase conjugation. (a) Three pulses colliding in a four-wave mixing cell produce a fourth phase-conjugated and temporally shortened pulse. (b) Pulses recirculated so as to produce self-Q-switching and mode locking in an otherwise open-ended laser cavity. (c) An in-line version of the same cavity, using polarization rather than angular discrimination between the waves.

mode-locked pulses in a conventional laser, using phase conjugation as both the switching and the pulse compression mechanism.

One such method is outlined in Fig. 22. Suppose three pulses 1, 2, and 3 collide in a four-wave mixing cell as illustrated in Fig. 22a. If depletion effects are neglected, and if the interaction length in the cell is short compared with the optical pulse lengths, the reflected pulse envelope will have the time dependence $\mathscr{E}_4(t) = \kappa\mathscr{E}_1(t)\mathscr{E}_2(t)\mathscr{E}_3^*(t)$. This will produce a reflected pulse 4 that in general will be shorter than any of the three incident pulses (Bloom *et al.*, 1978). The cell thus serves as a nonlinear pulse-compressing mirror from pulse 3 to pulse 4. (This is of course also the basis of the synchronous mode-locking technique described above.) For the case of three incident Gaussian pulses of equal pulse width τ_p, the reflected pulse width will be shortened to $\tau_p/\sqrt{3}$.

Suppose, however, that this cell is located inside a cavity including a laser gain medium, as in Fig. 22b. The conjugate-reflected pulse 4 then travels back around (and is amplified) to become pump pulse 1; pulse 1 is transmitted and reflected back to become pump pulse 2 of the next collision; and pump pulse 2 travels around to become signal pulse 3. The cavity lengths are assumed to be adjusted to give equal transit times for each of these paths.

This cavity can then apparently oscillate as a "self-pumped" mode-locked laser cavity, using the phase-conjugate reflection from pulse 3 to pulse 4 to accomplish both feedback and intracavity pulse compression. In contrast to other common active or passive mode-locking methods, this should produce continued pulse shortening by a factor of ~ 1.36 on each complete round trip, independent of pulse width or pulse intensity, at least until depletion or transit-time effects take over. In Fig. 22c is shown a collinear version of this same scheme, using orthogonal polarizations and quarter-wave plates rather than angular separation to distinguish between the various beams.

A fundamental difficulty with either of these proposed schemes is evidently that the laser cavity is open ended and has no net feedback at low intensities. Hence, laser action must be initiated either by external pulse injection (requiring three properly spaced pulses), or by providing an externally switched mirror to reflect beam 5 back into the cavity until laser action starts, or possible by self-Q-switching at very high laser gains. In any of these cases, because of the strongly power-dependent reflectivity from the pulse-compressing conjugate mirror, one can expect a very strong self-Q-switching behavior once threshold has been reached.

Preliminary results demonstrating self-sustained mode-locked oscillation in the in-line version of Fig. 22c were first obtained by Vanherzeele and Van Eck (1981). The pulses in this case were initiated by external in-

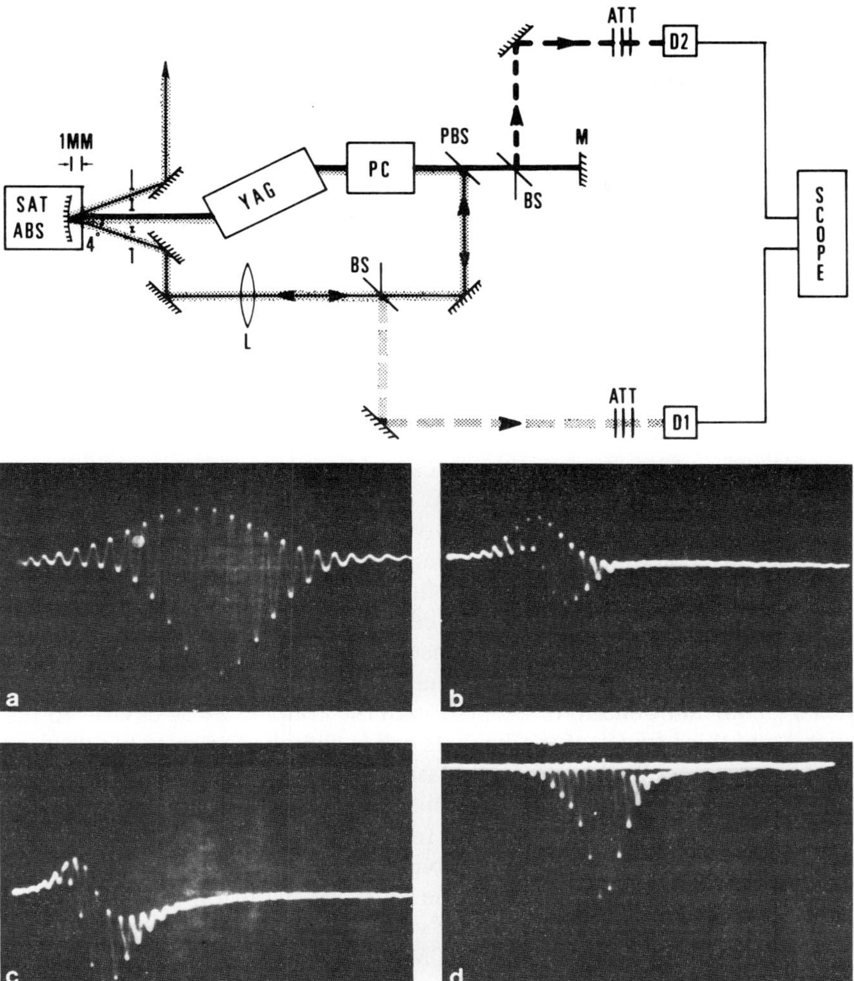

Fig. 23. Experimental results demonstrating self-sustained oscillation and mode locking in a PCM resonator using "self-pumped" phase-conjugate reflection. (From Vanherzele *et al.*, 1981.) In the schematic diagram, the solid line indicates a conventional Nd:YAG laser cavity that can be Q switched and mode locked by the saturable absorber cell at one end; the dashed line indicates a phase-conjugate resonator closed by phase-conjugate reflection of the beam incident at a small angle into the same cell. Adjustable coupling between the two cavities is provided by a Pockels cell (PC) and polarizing beam splitter (PBS). Experimental curves at the bottom show how mode-locked operation in the conventional cavity (upward deflection) can be switched over into self-sustained and mode-locked oscillation in the phase-conjugate cavity (downward deflection) at successively higher excitation.

jection from a separate mode-locked laser. More extensive experimental results clearly demonstrating this self-pumped mode of operation have recently been obtained by Vanherzeele *et al.* (1981) as shown in Fig. 23.

In the schematic diagram of Fig. 23, the solid line represents a conventional Nd:YAG laser cavity, Q switched and mode locked by the saturable absorber cell at one end. The dashed line evidently represents a PCM resonator in which feedback is supplied by phase-conjugate reflection of the beam projected at a small angle into the same saturable absorber cell. A Pockels cell (PC) and polarizing beam splitter (PBS) provide for variable coupling between these two resonators. If the Pockels cell is switched so that all of the energy passing through the YAG rod is switched into the phase-conjugate ring (dashed line), this cavity evidently corresponds to a folded version of the left-hand half of Fig. 22b. By operating the Pockels cell so that mode-locked oscillation initially builds up in the conventional laser cavity, and then switching this oscillation over into the phase-conjugate ring cavity, one can obtain self-pumped phase-conjugate resonator operation.

The oscilloscope traces at the bottom of Fig. 23 illustrate both the conventional laser action (upward deflection) and the self-pumped phase-conjugate oscillation (downward deflection) observed in this cavity for different switching of the Pockels cell. In the first trace the coupling between the two cavities is held constant at $\sim 50\%$. Mode-locked operation builds up first in the conventional cavity, and then evidently stimulates a significant buildup of oscillation in the coupled PCM cavity. In the second and third traces of the Pockels cell is switched so as to block more or less completely the conventional laser cavity approximately at the peak of the Q-switched burst. Self-sustained and mode-locked operation evidently continues in the PCM cavity. In the third and especially the fourth traces a second, time-synchronized Nd:YAG rod is added to provide additional gain in the shaded part of the PCM resonator only, and the conventional laser cavity is switched off very early in the burst. Nearly all of the laser action then occurs in the PCM cavity only. Further investigation of this mode of operation appears very desirable.

IX. Phase-Conjugation Effects in Conventional Lasers

Phase-conjugate interactions, in addition to their potential for correcting aberrations and generating new types of resonators, can also play an unintended and often unrecognized role in ordinary laser resonators. As noted elsewhere in this book, phase conjugation can readily be demonstrated in the saturable gain medium of an oscillating laser, for example, by sending a probe wave through the gain medium at a small angle to the resonator axis (see, e.g., Tomita, 1979). The recirculating waves inside the laser cavity then function as the pump waves (Fisher and Feldman, 1979). Other types of nonlinear phase-conjugating materials can be similarly deployed inside the laser cavity to produce phase conjugation of a probe wave coming in at a small angle (e.g., Bergmann *et al.*, 1978; Vanherzeele and Van Eck, 1981; Vanherzeele *et al.*, 1981). In this section we briefly explore some additional ways in which phase conjugation and related effects can appear in conventional laser cavities.

A. Saturable Grating Effects

Wave-mixing or induced-grating effects will always occur between two waves intersecting collinearly from opposite directions in any kind of nonlinear material, including, for example, a saturable laser gain medium. Such effects are thus likely to occur between the left- and right-traveling waves in any conventional standing-wave laser cavity, as illustrated in Fig. 24. These wave coupling or grating effects can play a significant and not always fully recognized role in normal lasers.

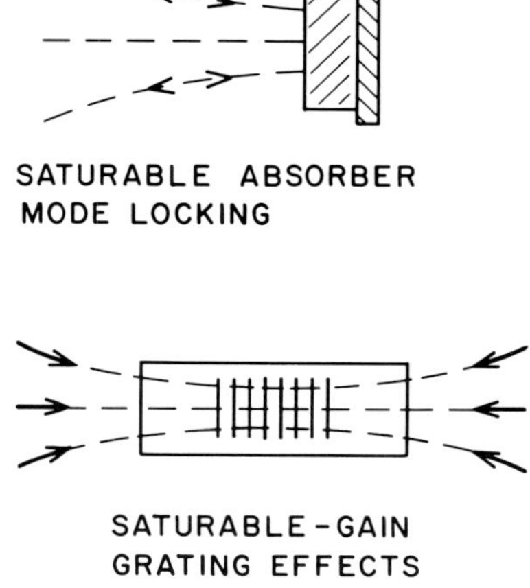

Fig. 24. Induced-grating or wave-mixing effects that are similar if not identical to phase conjugation occur between the forward- and reverse-traveling waves in conventional laser cavities, in either (a) saturable absorbers, or (b) saturable gain media. These interactions can have a significant effect on the operation of conventional laser cavities.

It should first be noted that true phase conjugation (that is, production of a complex-conjugated reflection) can occur for collinear or in-line waves in a saturable absorber or amplifier only between two different eigenmodes of polarization. That is, there can be conjugate reflection in a saturable absorber from a forward-going E_x wave only to a reverse-traveling E_y wave, and vice versa; or only from a right-handed circular to a left-handed circular wave, and vice versa. Induced-grating or wave-mixing effects will also produce wave coupling and grating reflection effects between forward- and reverse-traveling waves of the same sense of polarization; but the reflected wave amplitude E_{refl} in this case will be proportional only to the incident wave amplitude E_{inc}, not to its complex conjugate. Hence, while these may be called wave-mixing or distributed-feedbck effects, they will not be true phase-conjugate effects.

Mode coupling effects between forward and backard waves in conventional laser cavities are nonetheless very similar and sometimes identical to phase-conjugation effects, though they are not commonly described in phase-conjugation terms. Such effects are important, for example, in the gain saturation behavior of standing-wave laser oscillators, in the mode competition of ring lasers, and in the mode-locking behavior of passively mode-locked lasers. The impact of these effects on a conventional laser may be complicated, depending among other things on whether one is dealing with saturable gain, saturable absorption, or a nonlinear refractive effect (saturable index); on whether the nonlinear process involved follows the optical fields instantaneously or in an integrating fashion; on whether the laser transition involved is homogeneous or inhomogeneous; on whether the laser is operating cw or pulsed; on whether the laser cavity is a standing-wave or ring laser cavity; and on the simultaneous mode coupling, backscattering, and saturation effects caused by other elements elsewhere in the laser cavity.

It has been pointed out, for example, that the cw energy extraction versus coupling in a high-gain standing-wave laser cavity can depend significantly on standing-wave grating effects in the laser medium (Agrawal and Lax, 1979, 1981; Bambini *et al.,* 1979). Inclusion of these effects slightly but noticeably modifies the well-known laser power output analysis by Rigrod (1965). The coupling and competition between oppositely traveling waves in a ring laser cavity depends even more strongly on the grating effects that will be induced by saturable gain or absorption (Sargent, 1976; Kuhlke and Horak, 1979). The resulting mode coupling effects can be very complex, especially in Doppler-broadened systems. Effects of this type have been discussed in an inordinately large number of papers, especially with respect to ring laser gyroscopes.

Distributed-feedback laser action (Kogelnik and Shank, 1972) is another example of collinear wave coupling where the backscattering of each wave into the opposite direction is produced either by an externally generated grating structure in the laser medium or by a self-induced grating arising from a two-or four-wave type of mixing process. The case of self-induced distributed feedback, where the lasing waves produce their own induced grating, is a form of phase conjugation with particularly interesting dynamic stability properties (Idiatulin, 1979a,b).

B. Colliding-Pulse Mode Locking

A particularly important example in which wave-mixing effects play a vital role in a conventional laser occurs with the recently developed "colliding-pulse" form of mode locking. In passively mode-locked lasers, best performance in the past has usually been obtained with the saturable absorber placed against one end mirror of the laser cavity. Tocho *et al.* (1980a,b) have pointed out that phase conjugation clearly plays a significant role in the saturable absorber behavior when the saturable absorber is against one end mirror, as illustrated in the upper sketch in Fig. 24.

In one of the most important recent developments in passive mode locking, a Bell Laboratories group (Fork *et al.,* 1980) have shown that passive saturable absorber mode locking appears to work very much better (i.e., produces much shorter pulses with much greater stability) if two counterpropagating mode-locked pulses collide in a suitably thin absorber cell. They obtained mode-locked pulses shorter than 80 fsec from a cw passively mode-locked dye laser in this fashion. The colliding-pulse condition was produced using either a standing-wave laser cavity with the

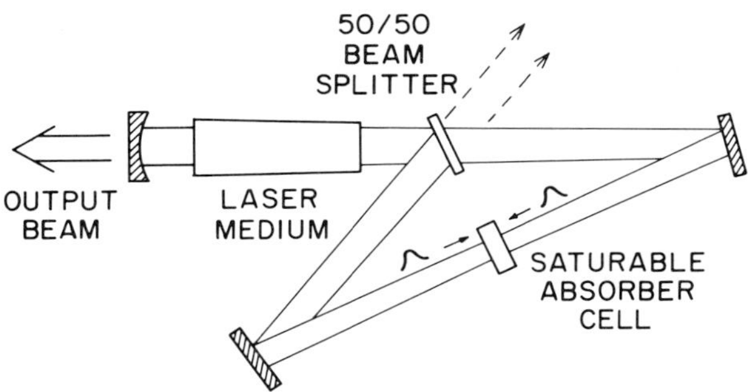

Fig. 25. Colliding-pulse mode locking produces greatly improved pulse shortening partly due to phase-conjugation effects, when two mode-locked pulses collide in a saturable absorber cell, for example, in the "antiresonant ring" structure shown here.

cell located exactly at the cavity center and with two pulses per cycle circulating in the cavity, or a ring laser cavity with two pulses circulating in opposite directions.

An alternative resonator that can have significant advantages for achieving the same kind of colliding-pulse mode locking is the "antiresonant ring" configuration illustrated in Fig. 25 (Siegman, 1973; Trutna and Siegman, 1977). The application of this configuration to colliding-pulse mode locking has recently been proposed by one of us (Siegman, 1981) and subsequently demonstrated using a Nd:YAG laser (Vanherzeele *et al.*, 1981). Mode-locked pulses as short as 12 psec or less were obtained from a Nd:YAG laser, in which the normal pulse width with standing-wave mode locking ranges from 30 to 40 psec.

REFERENCES

Agrawal, G. P. and Lax, M. (1979). Effects of interference on gain saturation in laser resonators, *J. Opt. Soc. Am.* **69,** 1717.

Agrawal, G. P. and Lax, M. (1981). Analytic evaluation of interference effects on laser output in a Fabry–Perot interferometer, *J. Opt. Soc. Am.* **71,** 515.

Arnaud, J. A. (1976). "Beam and Fiber Optics." Academic Press, New York.

Au Yeung, J., Fekete, D., Pepper, D. M., and Yariv, A. (1979) A theoretical and experimental investigation of the modes of optical resonators with phase-conjugate mirrors, *IEEE J. Quantum Electron.* **QE-15,** 1180.

Bambini, A., Vallauri, R., and Karamaliev, R. (1979). Saturation effects and stimulated scattering in two-standing mode operation of a lasing system with stationary atoms, *Phys. Rev. A* **19,** 1673.

Bar, Joseph I., Hardy, A., Katzir, Y., and Silberberg, Y. (1981). Low power phase conjugate interferometry, personal communication.

Belanger, P. A., Hardy, A., and Siegman, A. E. (1980a). Resonant modes of optical cavities with phase-conjugate mirrors, *Appl. Opt.* **19,** 602.

Belanger, P. A., Hardy, A., and Siegman, A. E. (1980b). Resonant modes of optical cavities with phase-conjugate mirrors; higher-order modes, *Appl. Opt.* **19,** 479.

Bel'dyugin, I. M. and Zemskov, E. M., (1979). Theory of resonators with wavefront-reversing mirrors. *Sov. J. Quantum Electron.* **9,** 1198.

Bel'dyugin, I. M. and Zemskov, E. M. (1980). Calculation of the field in a laser resonator with a wavefront-reversing mirror, *Kvantovaya Elektron.* **7,** 1334; *Sov. J. Quantum Electron.* (*Engl. Transl.*) **10,** 764 (1980).

Bel'dyugin, I. M., Galushkin, M. G. and Zemskov, E. M. (1979). Properties of resonators with wavefront-reversing mirrors, *Kvantovaya Elektron.* **6,** 38; *Sov. J. Quantum Electron.* (*Engl. Transl.*) **9,** 20 (1979).

Bergmann, E. E., Bigio, I. J., Feldman, B. J. and Fisher, R. A. (1978). High-efficiency pulsed 10.6 μm phase-conjugate reflection via degenerate four-wave mixing, *Opt. Lett.* **3,** 82.

Bezrodnyi, V. I., Ibragimov, F. I., Kislenko, V. I. K., Petrenko, R. A. Strizhevskii, V. L. and Tikhonov, E. A. (1980). Mechanism of laser Q switching by intracavity stimulated scattering, *Kvantovaya Elektron.* **7,** 664; *Sov. J. Quantum Electron.* (*Engl. Transl.*) **10,** 382 (1980).

Bloom, D. M., Shank, C. V., Fork, R. L. and Teschke, O. (1978). Sub-picosecond optical grating and wavefront conjugation by four wave mixing, *in* "Picosecond Phenomena" (C. V. Shank, E. P. Ippen, and S. L. Shapiro, eds.), p. 96. Springer-Verlag, Berlin and New York.

Bobroff, D. L. and Haus, H. A. (1967). Impulse response of active coupled wave systems, *J. Appl. Phys.* **38,** 390.

Casperson, L. W. (1974). Mode stability of lasers and periodic optical systems, *IEEE J. Quantum Electron.* **QE-10,** 629.

Casperson, L. W. and Lunnam, S. D. (1975). Gaussian modes in high loss laser resonators, *Appl. Opt.* **14,** 1193.

Chiao, R. Y., Kelley, P. L., Garmire, E. (1966). Stimulated four-photon interaction and its influence on stimulated Rayleigh-wing scattering, *Phys. Rev. Lett.* **17,** 1158.

Collins, S. A. (1970). Lens-system diffraction integral written in terms of matrix optics, *J. Opt. Soc. Am.* **60,** 1168.

Diels, J.-C. and McMichael, I. C. (1981). Influence of wave-front-conjugated coupling on the operation of a laser gyro, *Opt. Lett.* **6,** 219.

Eggleston, J. M., Giuliani, G. and Byer, R. L. (1981) Radial intensity filters using radial birefringent elements, *J. Opt. Soc. Am.* **71,** 1264.

Fainman, Y., Leng, E. and Shamir, J. (1981). Contouring by phase conjugation, *Appl. Opt.* **20,** 158.

Feinberg, J. and Hellwarth, R. W. (1980). Phase-conjugating mirror with continuous-wave gain, *Opt. Lett.* **5,** 519.

Fisher, R. A. and Feldman, B. J. (1979) On-resonant phase-conjugate reflection and amplification at 10.6 μm in inverted CO_2, *Opt. Lett.* **4,** 140.

Fisher, R. A., Suydam, B. R., and Feldman, B. J. (1981). Transient analysis of Kerr-like phase conjugators using frequency-domain techniques, *Phys. Rev. A* **23,** 3071.

Fork, R. L., Greene, B. I. and Shank, C. V. (1980). Generation of optical pulses shorter than 0.1 picoseconds by colliding pulse modelocking, *Appl. Phys. Lett.* **38,** 671.

Fox, A. G. and Li, T. (1961). Resonant modes in a maser interferometer, *Bell Syst. Tech. J.* **40,** 453.

Ganiel, U. and Hardy, A. (1976). Eigenmodes of optical resonators with mirrors having Gaussian reflectivity profiles, *Appl. Opt.* **15,** 2145.

Giuliani, G., Park, Y. K. and Byer, R. L. (1980). Radial birefringent element and its application to a laser resonator design, *Appl. Opt.* **5,** 491.

Hardy, A. (1981). Sensitivity of phase-conjugate resonators to intracavity phase perturbations, *IEEE J. Quantum Electron.* **QE-17,** 1581.

Hardy, A., Belanger, P. A., and Siegman, A. E. (1982). Orthogonality properties of phase conjugate optical resonators, *Appl. Opt.* **21,** 1122.

Hopf, F. A. (1980). Interferometry using conjugate-wave generation, *J. Opt. Soc. Am.* **70,** 1320.

Idiatulin, V. S. (1979a). On the interaction of counterrunning radiation in a mirrorless device, *Opt. Commun.* **30,** 419.

Idiatulin, V. S. (1979b). Stability of the self-induced distributed feedback in a system without a resonator, *Zh. Tekh. Fiz.* **49,** 2268. *Sov. Phys.–Tech. Phys.* (*Engl. Transl.*) **24,** 1257 (1979).

Kogelnik, H. (1965). Modes in optical resonators, *in* "Lasers: A Series of Advances" (A. K. Dekker, ed.), Vol. 1, p. 295. Dekker, New York.

Kogelnik, H. and Li, T. (1966). Lasers beams and resonators, *Proc. IEEE* **54,** 1312; *Appl. Opt.* **5,** 1550.

Kogelnik, H., and Shank, C. V. (1972). Coupled wave theory of distributed feedback lasers, *J. Appl. Phys.* **43,** 2327.

Kuhlke, D. and Horak, R. (1979). Intensity fluctuations in a ring laser taking into consideration a spatial population-inversion grating, *Opt. Quantum Electron.* **11,** 485.

Lam, J. F. (1979). Optical resonators with phase conjugate mirrors, *1979 IEEE/OSA Conf. Laser Eng. Appl., Washington, D.C.* Pap. 11.1 (1979); Abstr., *IEEE J. Quantum Electron.* **QE-15,** 69D.

Lam, J. F. and Brown, W. P. (1980). Optical resonators with phase-conjugate mirrors, *Opt. Lett,* **5,** 61.

Lesnik, S. A., Soskin, M. S. and Khiznyak, A. I. (1979). Laser with a stimulated-Brillouin-scattering complex-conjugate mirror, *Zh. Tekh. Fiz.* **49,** 2257; *Sov. Phys.–Tech. Phys.* (*Engl. Transl.*) **24,** 1249 (1979).

Lind, R. C. and Steel, D. G. (1981). Demonstration of the longitudinal mode and aberration–correction properties of a continuous-wave dye laser with a phase-conjugate mirror, *Opt. Lett.* **6,** 554.

Marburger, J. H. (1978). Optical pulse integration and chirp reversal in degenerate four-wave mixing, *Appl. Phys. Lett.* **32,** 372.

Mathieu, P and Belanger, P. A. (1980). Retroreflective array as resonator mirror, *Appl. Opt.* **19,** 2262.

Murphy, W. D. and Bernabe, M. L. (1978). Numerical procedures for solving nonsymmetric eigenvalue problems associated with optical resonators, *Appl. Opt.* **17,** 2358.

Pepper, D. M. and Abrams, R. (1978). Narrow optical bandpass filter via nearly degenerate four-wave mixing, *Opt. Lett.* **3,** 212.

Reznikov, M. G. and Khizhnyak, A. I. (1980). Properties of a resonator with a wavefront-reversing mirror, *Kvantovaya Elektron.* **7,** 1105; *Sov. J. Quantum Electron.* (*Engl. Transl.*) **10,** 633 (1980).

Rigrod, W. W. (1965). Saturation effects in high-gain lasers, *J. Appl. Phys.* **36,** 2487.

Sargent, M. III (1976). Laser saturation grating phenomena, *Appl. Phys.* **9,** 127.

Sheng, S.-C. and Siegman, A. E. (1980). Nonlinear optical calculations using fast transform methods: Second harmonic generation with depletion and diffraction, *Phys. Rev. A.* **21,** 599.

Siegman, A. E. (1971). "Lasers," p. 293. McGraw-Hill, New York.

Siegman, A. E. (1973). An antiresonant ring interferometer for coupled laser cavities, laser output coupling, mode locking, and cavity dumping, *IEEE J. Quantum Electron.* **QE-9,** 247.

Siegman, A. E. (1974). Unstable optical resonators, *Appl. Opt.* **13,** 353.

Siegman, A. E. (1977a). Effects of small-scale phase perturbations on laser oscillator beam quality, *IEEE J. Quantum Electron.* **QE-13,** 334.

Siegman, A. E., (1977b). Quasi fast Hankel transform, *Opt. Lett.* **1,** 13.

Siegman, A. E. (1979). Orthogonality properties of optical resonator eigenmodes, *Opt. Commun.* **31,** 369.

Siegman, A. E. (1981). Passive mode locking using an antiresonant-ring laser cavity, *Opt. Lett.* **6,** 334.

Siegman, A. E. and Miller, H. Y. (1970). Unstable resonator loss calculations using the Prony method, *Appl. Opt.* **9,** 2729.

Suydam, B. R. (1980). Can optical phase conjugators produce a very short pulse? *Trans. Int. Conf. Lasers, Orlando, Florida.*

Suydam, B. R. (1982). Generation of very short conjugate pulses in a four wave optical conjugator, *IEEE J. Quantum Electron.* (submitted).

Sziklas, E. A. and Siegman, A. E. (1974). Diffraction calculations using fast Fourier transform methods, *Proc. IEEE* **62,** 410.

Sziklas, E. A. and Siegman, A. E. (1975). Mode calculations in unstable resonators with flowing saturable gain. II. Fast Fourier transform method, *Appl. Opt.* **14,** 1874.

Tocho, J. O., Sibbett, W., and Bradley, D. J. (1980a). Picosecond phase-conjugation reflection and gain by degenerate four-wave mixing, *Top. Meet. Picosecond Phenom., North Falmouth, Mass.* Pap. WA9-A.

Tocho, J. O., Sibbett, W. and Bradley, D. J. (1980b). Picosecond phase-conjugate reflection from organic dye saturable absorbers, *Opt. Commun.* **34,** 122.

Tomita, A. (1979). Phase conjugation using gain saturation of a Nd:YAG laser, *Appl. Phys. Lett.* **34,** 463.

Trebino, R. and Siegman, A. E. (1980). Phase-conjugate reflection at arbitrary angles using TEM_{00} pump beams, *Opt. Commun.* **32,** 1.

Trutna, R. and Siegman, A. E. (1977). Laser cavity dumping using an antiresonant ring, *IEEE J. Quantum Electron.* **QE-13,** 955.

Vanherzeele, H. and Van Eck, J. L. (1981). Pulse compression by intracavity degenerate four-wave mixing, *Appl. Opt.* **20,** 524.

Vanherzeele, H., Van Eck, J. L. and Siegman, A. E. (1981). Colliding pulse mode-locking of a Nd:YAG laser with an antiresonant ring structure, *Appl. Opt.* **20,** 3484.

Wang, V. and Yariv, A. (1980). Laser having a nonlinear phase conjugating reflector, U.S. Patent 4,233,571.

Yariv, A. and Yeh, P. (1975). Confinement and stability in optical resonators employing mirrors with Gaussian reflectivity tapers, *Opt. Commun.* **13,** 370.

Zhou, G.-S. and Casperson, L. W. (1981). Modes of a laser resonator with a retroreflective mirror, *Appl. Opt.* **20,** 1621.

Zucker, H. (1970). Optical resonators with variable reflectivity mirrors, *Bell Syst. Tech. J.* **49,** 2349.

14 *Applications of Nonlinear Optical Phase Conjugation*†

Thomas R. O'Meara
David M. Pepper
Hughes Research Laboratories
Malibu, California

Jeffrey O. White
California Institute of Technology
Pasadena, California

I. Introduction

In this chapter we describe many of the potential applications of nonlinear optical phase conjugation (NOPC), all of which incorporate the real-time processing of electromagnetic fields. With the successful demonstrations of NOPC, potential applications of nonlinear optical (NLO) interactions have been substantially increased. Many of the classical NLO functions of frequency conversion and mixing, modulation, switching, image up conversion, and amplification can now be approached in new

† The work of two of the authors (T.R.O. and D.M.P.) was supported by Hughes Research Laboratories.

Optical Phase Conjugation

ISBN 0-12-257740-X

ways. However, the major thrust of NOPC was, and perhaps still is, in the real-time generation of conjugated (or spatially inverted) wave fronts. It was soon appreciated that the NLO processes used to realize NOPC could be generalized to yield new classes of "all-optical" information processors. For example, in four-wave mixing (FWM), one can replace the canonical cw and/or planar probe and pump waves with spatially and/or temporally encoded fields. By controlling the properties[1] of the various interacting fields (propagation direction, sequencing, polarization, or wavelength), as well as the properties of the NLO medium that couples them (geometry, response time, the NLO mechanism, or wavelength dependence), one can construct many interesting and potentially useful optical devices (for previous reviews, see, e.g., Giuliano, 1981; Pepper, 1982). These processors can offer many advantages over their conventional (e.g., linear optical, electronic, or mechanical) counterparts. Among these are enhanced spatial and temporal bandwidths, reduced size, cost, weight, and power consumption, and improved environmental resistance to RFI, vibration, and temperature.

This chapter is organized as follows. In Section II we discuss applications that exploit the *spatial* properties of the various NLO interactions. The use of NOPC wave-front aberration correction is treated in the areas of the focusing, imaging, transmission, and generation of monochromatic (or nearly monochromatic) light. Here, applications such as optical train aberration compensation, pointing and tracking, and compensated imaging are treated. This is followed by discussions of imaging through multimode optical fibers and lensless photolithography. Next, the processing of two-dimensional functions and images via NOPC is reviewed. Spatial convolution (and correlation) and edge enhancement provide examples under this category. Applications to iterative optical algorithms and interferometry are also described.

In Section III we discuss applications that exploit the *frequency* or *temporal* properties of NLO interactions. Examples include temporal signal processing schemes (e.g., convolution and correlation of temporally encoded pulses, and envelope reversal and encoding), synthesis of time domain optical filters, pulse compression, and optical computing. Then, in Section IV, we present a brief review of laser resonators employing phase-conjugate mirrors (PCMs), which use many of the spatial and frequency domain properties of NOPC. Section V contains potential applications that employ the quantum optical properties of the so-called two-photon coherent state (TCS) of the radiation field, with emphasis on possible ultralow-noise detection schemes.

[1] These same properties (as well as the reversed propagation nature of the conjugate wave) can be useful to separate (or discriminate) the desired output field(s) from the various input waves.

We conclude by discussing, in Section VI, the use of a particular NOPC interaction, FWM, in the field of nonlinear laser spectroscopy. By probing various parametric dependences of a PCM, one can gain insight into the properties of the conjugator medium on an atomic or molecular level.

In studying this chapter, the reader should keep in mind that these applications need not be confined to the optical spectrum. Indeed, given suitable nonlinearities, sources, guided structures, detectors, etc., one may extend these ideas to other portions of the electromagnetic spectrum (e.g., rf, microwaves, millimeter waves, IR, and UV) and to other classes of propagation, e.g., acoustic waves.

II. Spatial Domain Applications

A. Aberration Compensation

As a consequence of turbulence, vibration, thermal heating, and imperfect optical elements, aberrations commonly appear in the generation, transmission, processing, and imaging of coherent light. Most phase aberrations can be compensated by letting the wave front retrace its path through the aberrating medium, following wave-front reversal via a PCM, as illustrated in Fig. 1. The complex amplitudes of such forward- and backward-going waves are complex conjugates of each other, and the utility of the PCM in Fig. 1 lies in its ability to generate a conjugate replica of

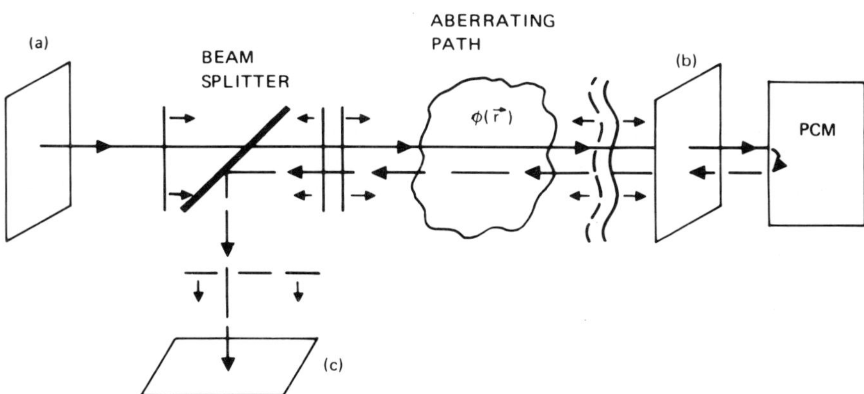

Fig. 1. A basic phase-conjugation compensation system. An input field at plane (a) becomes aberrated at plane (b) after propagation through a distortion path $\phi(\mathbf{r})$. As a result of the conjugation process and subsequent traversal over the same path, the conjugate of the initial field is recovered at plane (a). The beam splitter allows one to view the same compensated (or recovered) field at the plane (c). PCM, phase-conjugate mirror.

a general incident field, having an arbitrary spatial phase and amplitude distribution.

Preceding the development of NOPC concepts, the art of optical phase conjugation using "conventional" adaptive optical techniques has made much progress over the past ten years (see, e.g., Fried, 1977). These systems have achieved compensation for aberrated wave fronts by the use of electro-optic devices, acoustic devices, and deformable mirrors, all systems being driven with electronic signals generated by wave-front error sensors, in iterative, closed-loop feedback systems. The earliest of these were also called phase-conjugate systems. As mentioned in the Introduction, NLΘ techniques have several advantages, which stem primarily from their "all-optical" nature.

Most of the distortion compensation schemes to be discussed here—especially those involving four-wave mixing (FWM) and degenerate four-wave mixing (DFWM)—can be viewed as real-time holographic analogs of earlier static holographic compensation schemes (Kogelnik, 1965; Lukosz, 1968). The examples of compensation schemes employing double-pass geometries that we will discuss can be operationally related to the real-time generation of a pseudoscopic image. The wave-front reversal properties of such systems using conventional (static) holograms were treated some time ago (Kogelnik, 1965; Lukosz, 1968). Another example of a compensation scheme to be discussed involves a single-pass geometry; this can be viewed as a dynamic real-time extension of a classic (static) holographic approach (Goodman *et al.*, 1966). The holographic analogs of FWM are discussed in greater detail in Chapter 2.

Before discussing certain classes of aberrations along with various NOPC compensation techniques, we describe two examples of canonical demonstrations in which a PCM was shown to correct for phase aberrations encountered as a result of the *focusing* and *imaging* of coherent radiation through a fixed aberrating medium. These basic demonstrations provide a background and motivation for the use of more elaborate NOPC schemes for the real-time compensation of many classes of fixed and dynamic aberrations encountered in the processing of coherent radiation.

The optimal focusing of a laser beam calls for diffraction-limited output from a given optical system. There have been many demonstrations of this basic compensation concept using the geometry of Fig. 1 (Jain and Lind, 1982). In one experiment, the input field was that of a focused, pulsed ruby laser beam operating in a TEM_{00} mode. The phase aberrator consisted of an etched sheet of glass, and the PCM was realized via DFWM (using a semiconductor-doped glass as the NLO medium, pumped by a ruby laser). In Fig. 2 we show photographs taken in the far

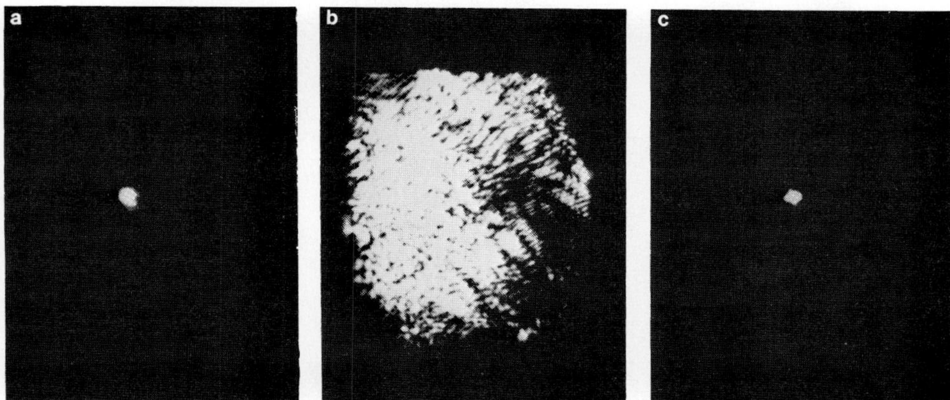

Fig. 2. Results of an aberration correction demonstration with the configuration of Fig. 1. Photograph (a) shows an unperturbed focused laser beam; (b) shows the same beam after it has passed through a phase aberrator (an etched glass plate); and (c) shows the beam after a phase-conjugate reflection and a second pass through the same aberrator. The photographs are taken in the far field of the planes (a)–(c) of Fig. 1. (After Jain and Lind, 1982 as appeared in Giuliano, 1981).

field of the initial beam [photograph (a), corresponding to plane (a) in Fig. 1], the now distorted beam [(b), corresponding to plane (b)] after passage through the etched glass plate, and the exiting beam (from the beam splitter) [(c), corresponding to plane (c)] after conjugation and subsequent traversal back through the distorter. The near-perfect recovery of a diffraction-limited focal spot using NOPC techniques is thus dramatically demonstrated.

As a second demonstration of the canonical compensation geometry of Fig. 1, we show in Fig. 3 photographic results of an experiment by Bloom and Bjorklund (1977). A resolution chart, illuminated by a planar probe wave, is placed at plane (a) of Fig. 1. The PCM is realized using a DFWM process, with carbon disulfide as the nonlinear medium and a doubled Nd:YAG laser as the source. This scheme demonstrates both the compensation ability of the system and its lensless imaging capabilities. Both photographs in Fig. 3 were taken at plane (c), with a phase aberrator (a piece of etched glass) in the system. In photograph (a), the counterpropagating pump waves were properly aligned, resulting in a near-perfect two-dimensional (2D) image reconstruction. By contrast, photograph (b) was taken with the counterpropagating beam misaligned by only 0.25 mrad. The conjugate wave was therefore offset from the incident wave and traversed through a different part of the phase aberrator compared

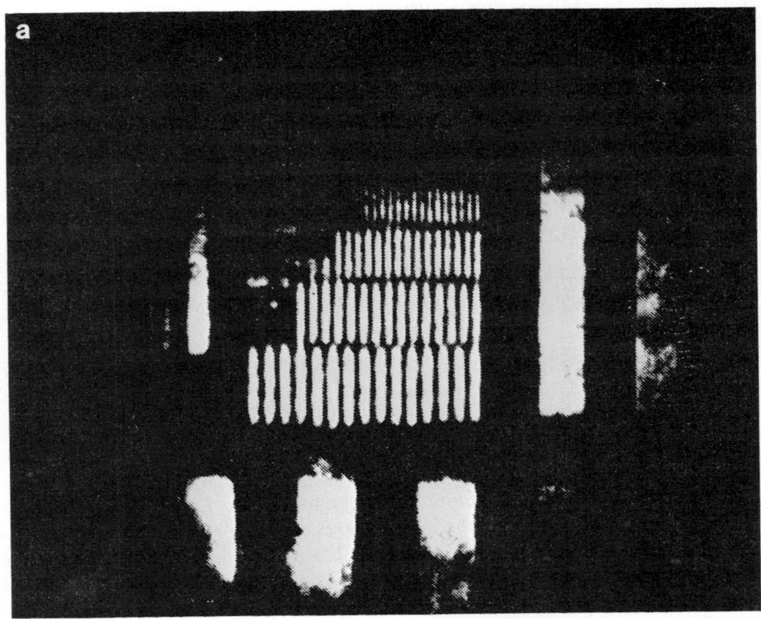

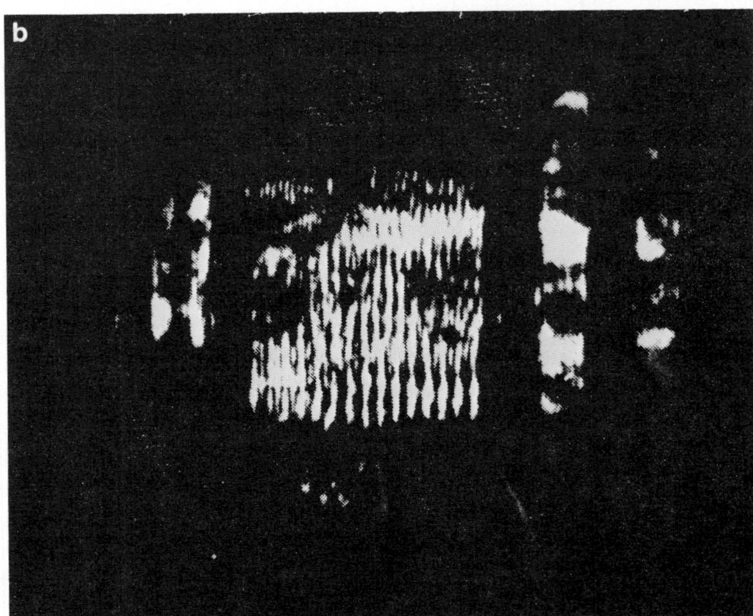

Fig. 3. Photomicrographs of reconstructed images of a resolution chart using a DFWM phase conjugator and the double-pass geometry shown in Fig. 1 [plane (c)]. Part (a) shows the quality of the reconstructed image; (b) shows the degradation experienced as a result of a pump-wave misalignment of 0.25 mrad, which caused the conjugated wave to traverse different portions of the phase aberrator. (After Bloom and Bjorklund, 1977.)

with its forward-going counterpart, causing poor compensation and leading to a badly degraded image. This points out the importance of generating a true phase-conjugate replica, both in terms of its wave-front contours and in terms of the propagation direction. In this context, one can appreciate the usefulness of this scheme in compensating for phase aberrations, in both lensless and conventional imaging applications.

1. Compensation for Aberrations Encountered in the Focusing and Transmission of Coherent Light

The usual goal of a laser focusing system is to deliver maximum energy from the laser to a target in the face of aberrations in the propagating path or in the laser amplifiers themselves (Kruzhilin, 1978; Ilyukhin *et al.*, 1979; Peregudov *et al.*, 1979). The compensation sequence shown in Fig. 4 begins with illumination of the target, which may be a fusion pellet, for example, with an unfocused pulsed auxiliary laser. Some of the light scattered from the pellet is captured in the aperture of the focusing element and provides a reference. Upon passing through the system, this reference wave front may be aberrated by imperfect elements in the amplifier train. After conjugation, the wave undergoes further amplification, and the aberrations suffered during the first pass are (in principle) canceled in the reverse sequence. On the second pass the majority of the inverted population energy stored in the amplifiers is extracted.

The basic compensation scheme gives substantial improvement even in the face of systems which include nonlinear amplifier response (for linear distortions), provided that the high spatial frequency components of the distorted wave are ultimately collected by the PCM (O'Meara, 1982b). Since reference wave tilts convey information about target position, these systems have an automatic target tracking capability exactly like that of their adaptive optics forerunners. Although these systems offer many potential advantages over conventional adaptive optical systems, there are many practical problems in implementing such NOPC systems (O'Meara, 1982b).

When employing references at long ranges (e.g., transmitting a laser signal to a communications satellite in synchronous orbit), the round-trip propagation time may be sufficiently large that one must establish a lead or point-ahead angle for the outgoing signal if it is to hit the reference (presumed to be located on the satellite). Four-wave mixers may be "programmed" into providing such offsets in several ways. Small angular offsets, of the order of a few milliradians, have been demonstrated by simply misaligning the pumps (recall Fig. 3b). Beyond this amount, one typically encounters serious degradations in conjugator efficiency and precision because of phase mismatch. More complex, three-dimensional pump geometries have been demonstrated that yield still larger angular offsets

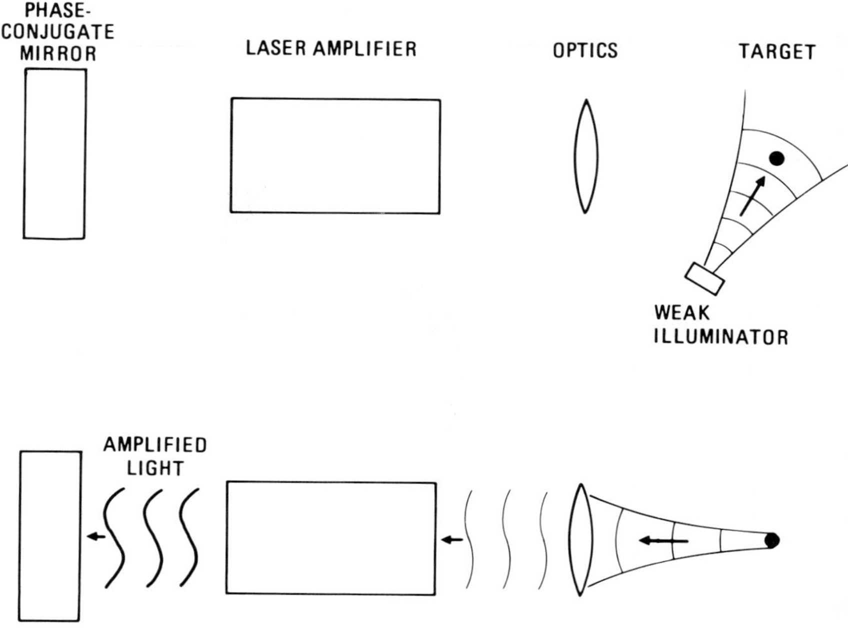

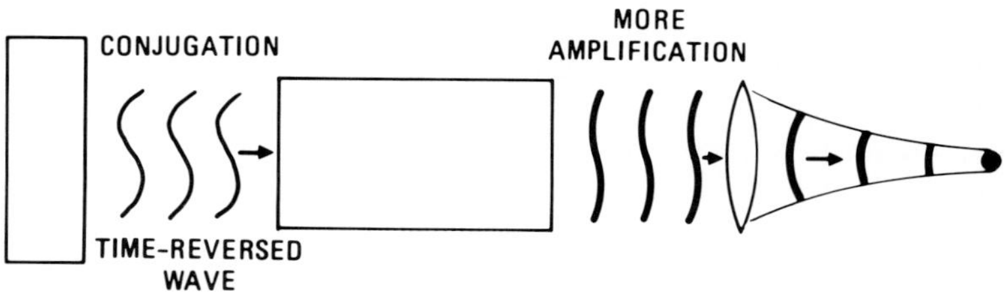

Fig. 4. Potential application of optical phase conjugation to laser fusion. (a) The illuminator irradiates the target, with some of the scattered light being collected by the focusing elements of the optical train. (b) Light passing through the system is amplified and acquires phase and/or polarization aberrations as a consequence of imperfect optics, heating in the amplifier (e.g., index variations or stress-induced birefringence), etc. (c) The phase-conjugate mirror wave-front-reverses the propagation of the aberrated beam. After passage back through the same amplifier chain and optical train, an amplified, nearly diffraction-limited intense beam of light is focused onto the target. (After Guiliano, 1981.)

without phase mismatch. The differences between the reference and return paths can lead to severe "anisoplanatic" degradations even though the conjugation is perfect, as was discussed in conjunction with Fig. 3b.

If the gain of the amplifier and the reflectivity of the target are sufficiently high, it is possible to establish an oscillation mode in which the target serves as one reflector while the conjugate reflector completes the resonator. In this mode of oscillation, the beam autotracks the moving target, and the auxiliary illuminating laser is not needed (AuYeung and Yariv, 1979; Pepper, 1980).

We next discuss how a PCM can compensate for undesirable linear and/or nonlinear phase aberrations resulting from transmission through bulk media (e.g., turbulent atmospheres). Schemes involving guided-wave structures (e.g., optical fibers) will be discussed in Section II A 3.

A laser beam transmitted from point A to point B will spread by diffraction due to the finite size of the beam at A. Ideally, given an unaberrated laser source focused at point B and a propagation path in free space, the size of the beam at B would be diffraction limited (and could be made smaller only by use of a waveguide). However, in realistic systems, this diffraction limit may be difficult to achieve. For example, the beam can encounter major spreading from propagation through atmospheric turbulence. Further, if the beam is sufficiently intense, it can also generate nonlinear phase aberrations such as thermal blooming.

To compensate for the turbulence spreading, one may first transmit a weak probe beam from B to A, so that it acquires the total path transverse phase aberration in transit. Next, the phase conjugate is generated at A. On reversing its path, the phase-conjugate wave should arrive at B with a close approximation to the phase and amplitude distribution of the originally transmitted wave (Kogelnik, 1965; Yariv, 1977). One implementation of this scheme (after Wang, 1978) is illustrated in Fig. 5. Here, as in conventional adaptive optical schemes, the severity of the aberrations that can be compensated will be limited, because not all of the light diffracted by the aberrator may be collected by the PCM.

The system shown in Fig. 5 can also be used in iterative focusing applications, by allowing the process to repeat. That is, the conjugate wave (after propagation back to the target) acts as a new probe beam and scatters off the target. By enabling this sequence to continue, the conjugate wave can ultimately converge to a focus on a dominate glint or highlight on the target. An application of this scheme as part of a compensated active imaging system is discussed in the next section.

We conclude this section with a brief discussion of the application of NOPC techniques to another class of phase aberrations: *nonlinear* (e.g., intensity-dependent) phase distortions. A detailed understanding of the

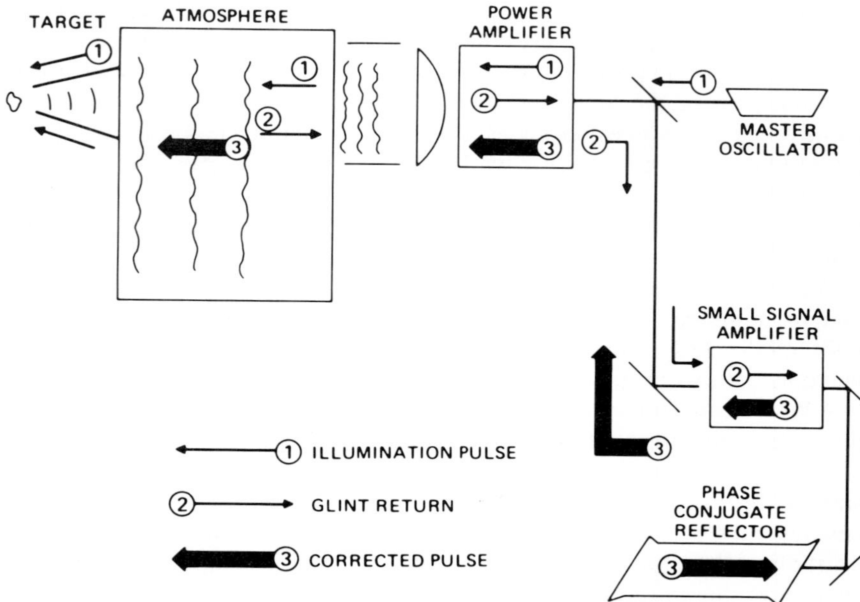

Fig. 5. Conceptual pulsed laser system using a phase conjugator for the real-time correction of atmospheric and laser-induced aberrations. The illumination pulse (1) strikes the target, reflecting off a glint, of spatial extent less than a diffraction-limited spot. The glint return (2), appearing as if it emanated as a point source from the target, samples the atmospheric and/or power amplifier aberrations. After conjugation, the corrected pulse (3) is amplified, resulting in a nearly diffraction-limited beam focused at the glint site. (After Wang, 1978.)

ability of phase-conjugate systems to compensate for nonlinear phase aberrations is not yet available. One special case has been investigated by Pepper and Yariv (1980) and Bol'shov *et al.* (1980). They show that if the probe and conjugate intensities are equal at all points (i.e., identical spatial profiles) and the PCM has unity reflectivity, perfect compensation is achieved. Their steady-state analysis assumed a *local,* intensity-dependent index change for the aberrator. They also assumed the standard double-pass geometry, with no attenuation on the path. For other classes of nonlinear path distortions, one can also obtain excellent compensation under the more practical operating conditions of highly unbalanced intensities for the probe and conjugate waves (O'Meara, 1982b). Two important examples are near-field thermal blooming and irradiance mapping distortions of reflective surfaces.

2. *Compensation for Aberrations Encountered in Imaging with Coherent Light*

The imaging of coherent light through turbulent atmospheres or via conventional imperfect optical imaging devices is subject to various classes of aberrations that can degrade their performance. The use of one or more properly positioned PCMs can compensate for these degradations.

Three basic classes of active imaging problems have been considered in which phase conjugation may play an important compensation role.

(1) An object wave is transmitted through an aberrating path, conjugated, and retransmitted back through the path, emerging as a compensated image on the same side of the aberration as the original object (after the basic scheme of Fig. 1).

(2) A (cooperative or, with suitable techniques, noncooperative) reference and object are located on one side of an aberrating path and a compensated image is to be transmitted through the path.

(3) A reference and object are located on opposite sides of an aberrating path and, by using the "information" from a transmitted reference wave, a compensated image of the object is to be transmitted back through the path and formed on the reference side.

The first class of schemes involves a *double pass* of image information through an aberrating path; the other two classes involve a *one-way* transmission of image information through the path, with a reference wave sampling the aberration.

The major utility of the first class of imaging systems appears to be in the field of photolithography, which will be discussed separately in Section II.B.

The second class of systems appears at first sight to have limited utility, because it is difficult to establish a reference in close proximity to an object in practice. Nevertheless, in cooperative situations it is possible to transmit a compensated image through an aberrating path, as was first demonstrated by Goodman *et al.* (1966) using conventional holography as illustrated in Fig. 6. This technique has since been extended to *real-time* compensation by substituting a FWM conjugator for the hologram (Ivakhnik *et al.,* 1980). Moreover, the basic concept can be further extended to the imaging of *noncooperative* objects (under active illumination) by employing the convergence action of phase conjugators to form a reference on an extended object: an artificial reference. This concept leads to many classes of active imaging systems, one of which is illus-

(a)

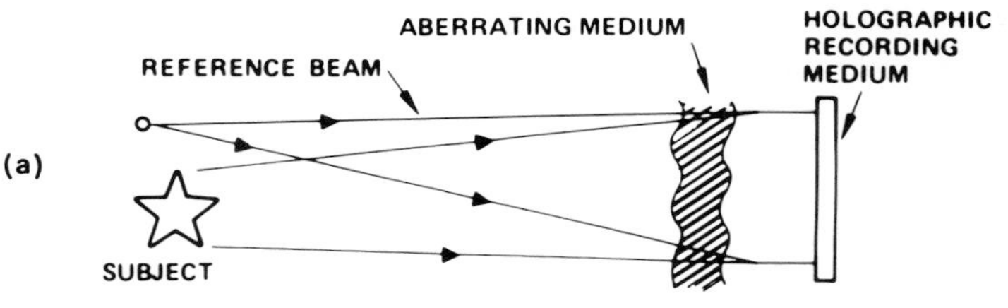

(b)

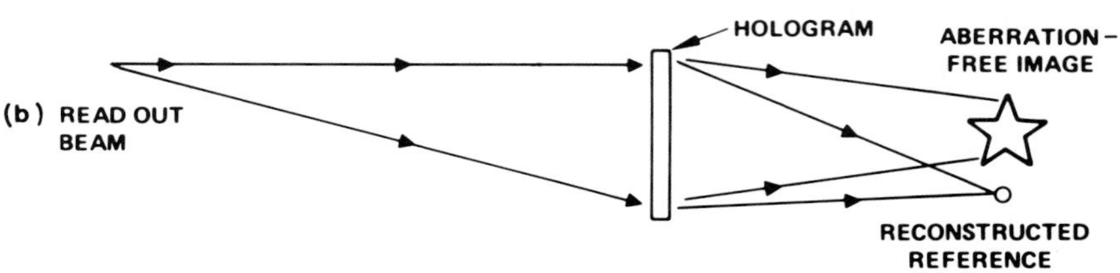

Fig. 6. Technique for imaging through a phase-aberrating medium with equal phase distortion of reference and subject waves. (After Goodman *et al.*, 1966.) (a) Recording. A point-source reference beam interferes with the subject wave to form a hologram that is recorded in the medium. Because these waves transit essentially the same aberrating path, the resulting wave distortions cancel in the recording process. (b) Reconstruction. Upon illumination, the hologram generates an aberration-free image as well as a reconstructed offset reference.

trated in Fig. 7. This scheme employs two four-wave mixers: one four-wave mixer (at frequency $\omega_p - \Delta$) is used to establish a single-glint reference on the object; another four-wave mixer is used to perform real-time compensated imaging of the object through the aberrating path, in the spirit of Ivakhnik *et al.* (1980). One can appreciate the need for the first four-wave mixer by recalling the discussion of Section II.A.1, in which the focusing of a diffraction-limited laser beam on an extended object was treated. The focusing property mentioned in Section II.A.1 can be used to form a single diffraction-limited focal spot on the object. This spot provides an artificial reference for situations where there is no access to the object side of the aberrating path. There is a further advantage to this approach. Recall that the classical compensation scheme of Fig. 6 employs a point-source reference, which is *offset* from the object. It is therefore subject to anisoplanatic problems (Goodman *et al.*, 1966). These problems can be alleviated by bringing the reference onto the object, that is, by forming an artificial reference. The artificial reference wave can be generated by the reflection of an illumination beam (via laser oscillator 1 at frequency $\omega_p - \Delta$) from a single small glint or highlight on the object.

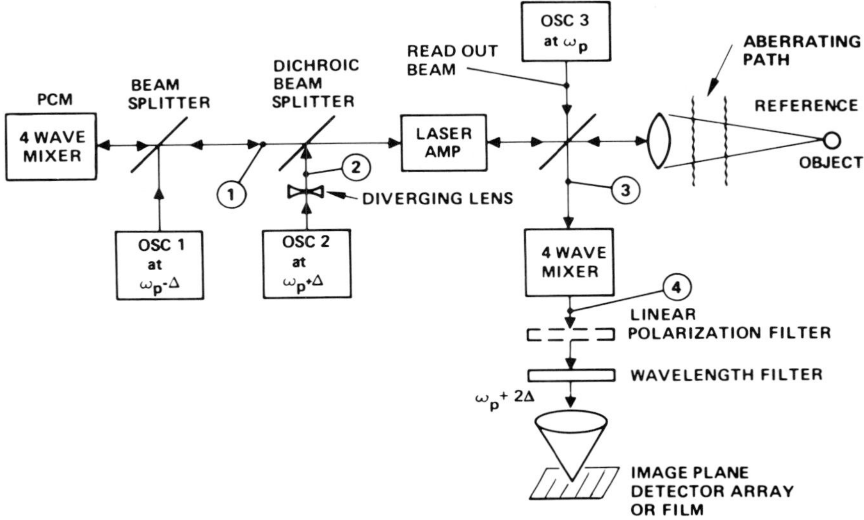

Fig. 7. Compensated imaging concept using a phase conjugator and a four-wave mixer. (1) Laser oscillator 1, with the aid of the phase conjugator, iterates to form a diffraction-limited spot (or reference) on the highest reflectivity area of the object to be imaged. (2) Laser oscillator 2 simultaneously floodlights the object at a wavelength slightly offset from oscillator 1. (3) The return "image" fields from the object beat together to form a real-time hologram in a four-wave mixer (after Fig. 6, but without angular offset), which is unaberrated by the propagation path. (4) Laser oscillator 3 reads out the desired on-axis "holographic" image, which can be separated from the unwanted probe and pump waves by frequency and polarization offsets in the system. (After O'Meara, 1982a.)

However, because most objects include multiple glints or highlights, one must resort to a more sophisticated approach to realize a single-glint reference. This can be achieved by using a NOPC system as illustrated on the far left-hand side of Fig. 7. The fraction of the light scattered from the glint that is collected by the PCM will be conjugated. This conjugated wave will retrace the path of the illuminating beam, and converge onto the object in a manner similar to that shown in Fig. 5. By allowing this process to repeat, the self-iterative action of the conjugator–object system will establish a beam that comes to a focus on the dominant glint, providing the necessary (artificial) point-source reference (O'Meara, 1978, 1982a). Having established the desired (diffraction-limited) reference on the object, the second four-wave mixer is then used to generate the compensated image. The reference glint and the object wave via laser oscillator 2 (at frequency $\omega_p + \Delta$) are used to form a real-time hologram, which is read out by laser oscillator 3 (at frequency ω_p). The desired compensated image at frequency $\omega_p + 2\Delta$ is then formed at the image plane (polarization and/or frequency discrimination techniques can be used to separate out the desired field). An alternative active image compensation system can be realized by raster-scanning the artificial glint spot across an object (O'Meara, 1978, 1982a).

We now discuss one compensation technique for the third class of imaging problems (i.e., an object and a cooperative reference on opposite sides of an aberrator). A scheme utilizing NOPC for the one-way transmission of the intensity and phase of a picture field through a thin phase aberrator was recently proposed (Yariv and Koch, 1982) and demonstrated (Fischer *et al.*, 1982). As shown in Fig. 8, the geometry involves the use of a plane-wave reference located on one side of a given phase aberrator, which mixes in a NLO medium with a second reference field and a picture field, both located on the opposite side of the aberration. The scheme can be viewed as a FWM process using a point source as one pump wave, with pictorial information encoded on the other pump wave. A planar probe wave samples the aberrating path. This system appears to be limited to thin NLO media (owing to phase-matching constraints) and thin phase aberrators located in the near field.

Most (perhaps all) of the classes of imaging systems discussed here can be approached in principle by using either conventional adaptive op-

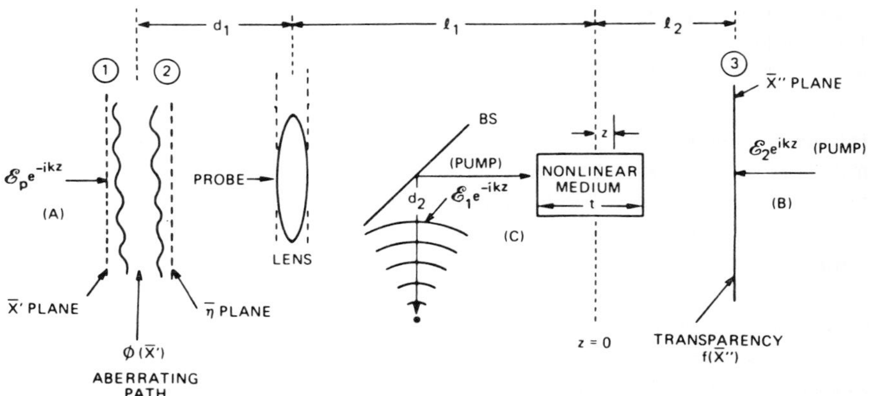

Fig. 8. Configuration for projecting an image from plane (3) through an aberrating medium to plane (1) without distortion. This scheme can be viewed as a FWM process with a planar probe wave (A) sampling an aberrating path ϕ and imaged onto the conjugator. The conjugator is pumped by a spatially encoded picture field (B) and a spherical wave (C). (After Yariv and Koch, 1982.)

tical schemes (Fried, 1977) or NOPC techniques. Each approach has certain advantages in terms of source requirements and bandwidth. For example, conventional adaptive optics systems can function well with white light sources, whereas the NOPC systems demonstrated to date have employed coherent sources. The two types of systems can also be compared with regard to the available spatial bandwidth of the aberrations that can be compensated. The experimental demonstrations to date suggest a strong advantage of the NOPC systems. Finally, one must consider the relative temporal bandwidths of the two types of system. In conventional adaptive optical compensated imaging systems (Fried, 1977), the system response time is limited by either the speed of the deformable mirror or the integration time required to capture sufficient photons for wave-front error estimation. In NOPC systems, the speed of response is usually much faster, being limited by the response time of the nonlinear medium. However, the compensation speed in both cases may be limited by the round-trip propagation time between the error source and the compensation device.

3. *Compensation of Aberrations Encountered in the Transmission of Images through Multimode Optical Waveguides*

Phase aberration problems can arise even when transmitting a coherent (2D or 3D) image through an *ideal* multimode waveguide, as depicted in Fig. 9a. In this case, the degradation is due to modal dispersion. A theoretical treatment of compensation for this class of phase aberration by the use of optical phase conjugation has been given (Yariv, 1976a,b, 1978, 1979). An experimental demonstration of such a compensation scheme using a two-dimensional image in a double-pass geometry through the same fiber (with a configuration similar to that in Fig. 1) has recently been carried out (Dunning and Lind, 1982). The photographs of a resolution chart shown in Fig. 9b were taken at the indicated positions along a forward and reverse pass through the same section of fiber.

To analyze the image transmission through fibers, consider the geometry and fields as defined in Fig. 9a. A picture field f_0, illuminated at frequency ω and coupled into a fiber of length $2L$, will, at the entrance of the fiber, excite many (real) eigenmodes A_{mn}, such that the picture field at the input plane can be represented by the superposition

$$f_0(x, y, z = 0, t) = \sum_{mn} \mathscr{E}_{mn} A_{mn}(x, y) e^{i\omega t}, \tag{1}$$

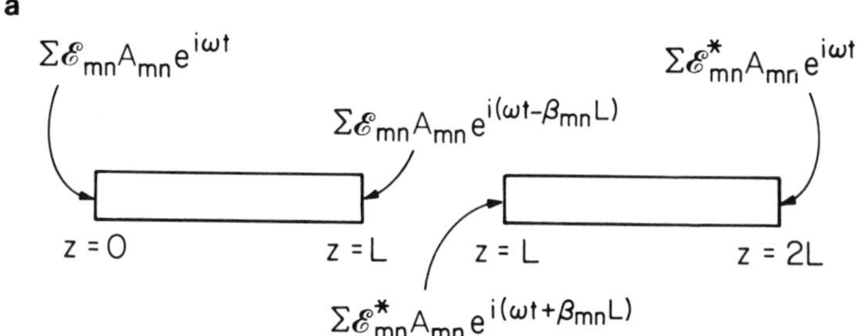

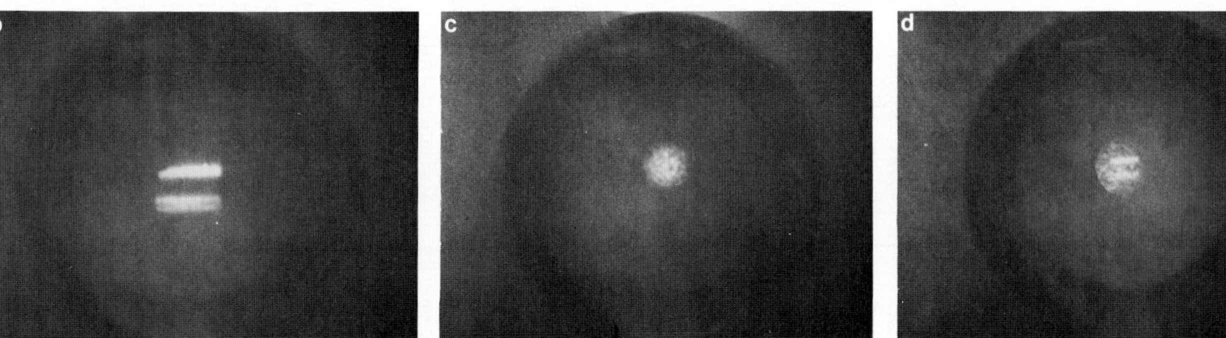

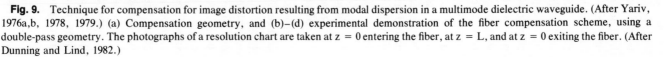

Fig. 9. Technique for compensation for image distortion resulting from modal dispersion in a multimode dielectric waveguide. (After Yariv, 1976a,b, 1978, 1979.) (a) Compensation geometry, and (b)–(d) experimental demonstration of the fiber compensation scheme, using a double-pass geometry. The photographs of a resolution chart are taken at $z = 0$ entering the fiber, at $z = L$, and at $z = 0$ exiting the fiber. (After Dunning and Lind, 1982.)

where $\mathscr{E}_{mn}$ are the complex amplitudes[2] of the eigenmodes A_{mn}. The dependence of the phase velocity, β_{mn} on the mode number is responsible for an accumulation (over a given fiber length) of phase differences between the modes. This results in a new (aberrated) image field f_1 as it exits the midpoint of the fiber link, given by

$$f_1(x, y, z = \mathrm{L}, t) = \sum_{mn} \mathscr{E}_{mn} A_{mn}(x, y) e^{i(\omega t - \beta_{mn} L)}. \tag{2}$$

To recover an undistorted image field, one possibility is to conjugate the field f_1 midway down the fiber path, to obtain

$$f_2(x, y, z = L, t) \propto \sum_{mn} \mathscr{E}^*_{mn} A_{mn}(x, y) e^{i(\omega t + \beta_{mn} L)}. \tag{3}$$

The conjugated field f_2 now traverses the second half of the fiber, presumed to have an identical modal phase delay ($\beta_{mn}L$), or, alternatively, f_2 retraverses the initial fiber (see photographs in Fig. 9). Passage through the second half of the fiber produces the same phase changes encountered in the first half. Thus, the exiting field f_3 becomes

$$f_3(x, y, z = 0 \quad \text{or} \quad 2L, t) \propto \sum_{mn} \mathscr{E}^*_{mn} A_{mn}(x, y) e^{i\omega t}. \tag{4}$$

The *intensity* distribution of the image can thus be recovered by recording $|f_3|^2$ on a film or a vidicon. However, if the original 3D object *field* distribution is required, a second stage of conjugation is necessary. This results in an output field proportional to the original 3D object field f_0 given by Eq. (1).

4. *Compensation of Aberrations Encountered in the Generation of Coherent Light*

The angular divergence of a laser beam, as it exits from a laser, can far exceed the diffraction limit as a consequence of poor-quality or misaligned optical components, thermal distortions, acoustic waves, or turbulence in the lasing media (e.g., in gas laser discharges or chemical lasers). As discussed both in Chapter 13 and in Section IV in the present chapter, a phase-conjugate resonator consisting of a conventional mirror and a phase-conjugate mirror (PCM), as shown in Fig. 10, has been shown to emit an

[2] In this chapter, we use the same field description as defined in Chapter 2, Eq. (7): For simplicity we assume the scalar version of these fields. **E** is the total field; $\mathscr{E}$ is the corresponding, slowly varying complex amplitude envelope.

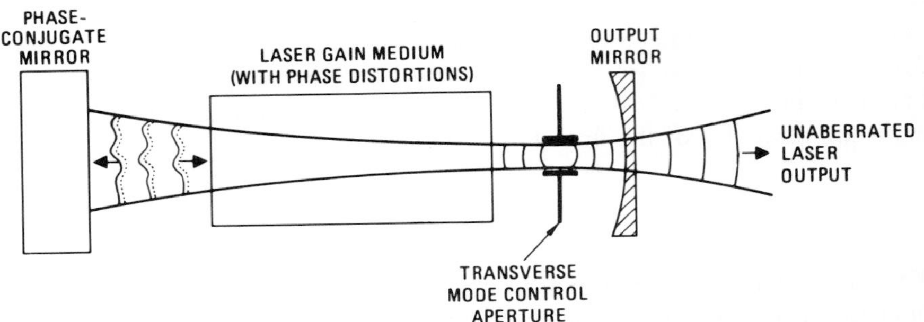

Fig. 10. One basic phase-conjugate resonator (PCR) structure. A conventional resonator is modified by replacing one or both of the cavity mirrors with a PCM. An optional aperture may be situated within the cavity for transverse-mode control. An unaberrated output beam is coupled out through the conventional mirror, if this mirror is well figured. (After Pepper, 1982).

output beam with reduced aberration, even in the presence of intracavity distorting elements (Feinberg and Hellwarth, 1980; Lind and Steel, 1981).

One can also employ a PCM to improve the output beam quality of a master oscillator–power amplifier (MOPA) configuration. A weak plane-wave reference is sent through the amplifier chain in the reverse direction so as to acquire all linear phase aberrations. The phase-conjugate replica then propagates back through the system, emerging as an amplified plane wave of enhanced optical quality (Anan'ev, 1975; Wang, 1978; Basov, 1978, 1979; Ragozin and Plotkin, 1980; Hon, 1980a; O'Meara, 1982c).

Finally, NOPC techniques can be used to improve the performance of optical parametric devices, such as in second harmonic generation (SHG). For example, it is well known that inhomogeneities in SHG crystals can increase the angular divergence of the second harmonic output. In one experiment (Andreev *et al.*, 1980), a weak probe beam was passed through an imperfect nonlinear doubling crystal, thereby sampling the distortion. This distorted beam was then reflected and conjugated by a stimulated Mandelstam–Brillouin PCM, amplified, and returned through the crystal as the pump wave. The angular divergence of the second harmonic output was observed to decrease by a factor of 5 to nearly the diffraction limit.

5. *Compensation for Polarization Aberrations*

Thus far, we have considered aberrations that distort the equiphase surfaces of a wave front while conserving its polarization state. *Polarization* aberrations can also degrade the performance of optical systems. Such aberrations can result from dynamic birefringence effects induced by nonuniform thermal or mechanical perturbations of optical elements, gain media, etc. Aside from static polarization distortions, which in some cases can be compensated by using inverted matching media, NOPC offers an approach to the compensation of *dynamic* polarization distortions (Zel'dovich and Shkunov, 1979). Such PCMs may find use in high-power laser systems and imaging devices.

The experimental demonstrations performed (Blashchuk *et al.*, 1980a; Martin *et al.*, 1980) typically involve the use of a DFWM conjugator in a double-pass geometry similar to that illustrated in Fig. 1. The phase aberration in this case is a sheet of randomly oriented birefringent elements; polarization beam splitters and wave plates are used to evaluate the conjugated wave. In these systems, the polarization of the pump waves, the tensorial aspects of the nonlinear medium, and the propagation directions of the interacting fields must be considered to obtain the desired output field. In addition to unscrambling polarization distortions, the wave front of the incident field is also inverted, enabling one to use this scheme to compensate simultaneously for the aberrations discussed earlier.

B. Lensless Imaging: Application to High-Resolution Photolithography Using Four-Wave Mixing

Resolution requirements place stringent limitations on the aberrations that can be tolerated in a conventional projection photolithographic system. Although a phase-conjugate imaging system may compensate defects in optical elements (such as lenses) through which an image makes both a forward and a return path, a major advantage of NOPC appears to be in the elimination of the lenses altogether, thereby circumventing a major aberration source. Further, this lensless approach permits a major reduction in the effective F number, thereby improving the spatial resolution of the system.

A typical photolithography requirement is diffraction-limited resolution over a wafer as large as 3–4 in. in diameter. For the reasons cited, lensless photolithography by wave-front conjugation, using the system illustrated in Fig. 11, has demonstrated an attractive solution to noncontacting photolithography (Levenson, 1980; Levenson *et al.*, 1981). A resolution of 800 lines/mm over a 6.8-mm^2 field and a feature

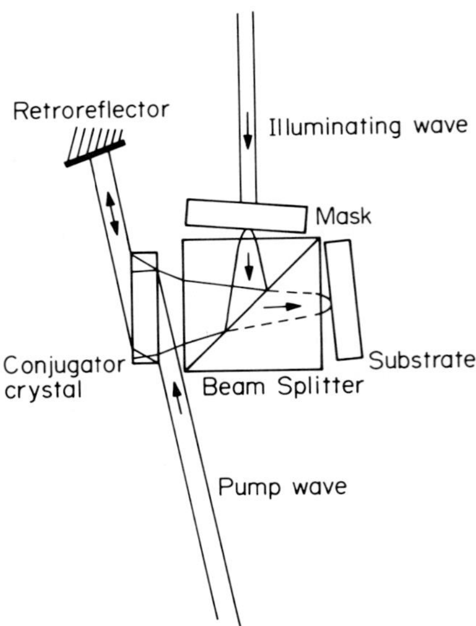

Fig. 11. Schematic diagram of a lensless imaging system for photolithography using a DFWM phase conjugator. The illumination beam, after passage through the mask, reflects from the beam splitter, striking the conjugator crystal. After conjugation, the mask pattern is imaged onto the substrate surface, free of physical contacting and common-path aberration problems. (After Levenson *et al.*, 1981.)

size of 0.75 μm were obtained; this is equivalent to a numerical aperture (N.A.) of 0.48, which far exceeds the N.A. of conventional imaging systems. The FWM phase-conjugate mirror employed a crystal of $LiNbO_3$, pumped by a 413-nm krypton ion laser. The resolving ability of this scheme is demonstrated in the photograph of Fig. 12. As in all phase conjugators, the accuracy of the conjugated wave produced by the PCM is critically dependent upon the quality of the nonlinear medium and of the pump waves.

Speckle was not present in the image plane of the conjugator used above, although it is a typical problem in laser imaging systems. The elimination of speckle is said to be a consequence of using the PCM in conjunction with plane-wave illumination (Levenson, 1980). Speckle-free phase-conjugate images can also be obtained, even with diffuse illumination of a film, by the usual speckle averaging techniques (Huignard *et al.*, 1980). In one demonstration (Huignard *et al.*, 1980), film in the image plane was used to time-integrate ~100 conjugate images with independent speckle patterns.

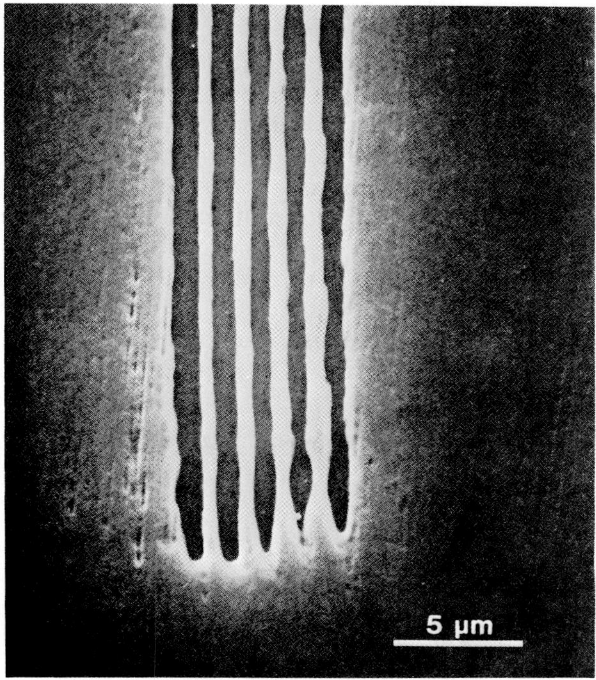

Fig. 12. Demonstration of the resolution capability of the lensless photolithography system shown in Fig. 11. The image pattern is formed by five 0.75-μm lines with 0.50-μm gaps in Shipley AZ1350B photoresist. Note the high-quality, speckle-free resolution of the system. (After Levenson *et al.*, 1981.)

C. Spatial Information Processing

A major area of potential applications of NOPC is the real-time spatial information processing of multidimensional phase and/or amplitude images. In this section, we discuss examples of arrangements for coherent image amplification, product integration (including convolution and correlation), edge enhancement, logical processing of 2D fields, and iterative optical processors.

The first four topics can be categorized as single-pass optical processors: the various input fields form a real-time grating in the nonlinear medium, which is read out on a one-time basis. The last topic involves a multiple-pass geometry whereby the nonlinearly generated output field is reinserted back into the nonlinear medium, thus successively forming new gratings. Hence, iterative algorithms can be realized optically in a short time span.

Perhaps the simplest example of information "processing" is the (coherent) amplification of an image without distortion. In a recent experiment using a DFWM scheme, a two-dimensional image has been amplified by a PCM with gain (Feinberg and Hellwarth, 1980). In addition to the image amplification property, the experiment demonstrated that, owing to the freedom from phase-matching constraints, fields with large spatial frequency bandwidths may be processed using DFWM. Further, the versatility of DFWM allows for the concurrent incidence of several fields, so that multiple image processing operations (e.g., correlation and convolution) can be performed simultaneously. Accordingly, we limit the present treatment to systems that employ FWM processes.

1. Addition, Subtraction, Differentiation, and Integration of Images

The use of gratings (in photographic film and nonlinear media) for optical processing has been established (Lee, 1974). For example, a grating placed in the filter plane of a coherent optical processor can translate, add, or subtract input pattern functions, depending on the period and phase of the grating. Multiple gratings can also be used to obtain first- or higher-order derivatives of intensity distributions, restore aberrated images, and multiply exposed images. By forming a grating in a nonlinear optical medium, it is possible to change the grating rapidly *in situ*.

These concepts can be generalized beyond gratings in the filter plane to the production of more complicated Van der Lugt filters (Goodman, 1968). In the conventional Van der Lugt technique, a filter or mask is first synthesized by exposing a film to the interference pattern of a plane reference wave and the Fourier transform $\tilde{H}$ of the desired impulse response H. The developed film is then reinserted into the filter plane, and the object field $\mathscr{E}$ to be processed is inserted into the input plane.[3] Part of the field leaving the transform plane is proportional to the product $\tilde{H}\tilde{\mathscr{E}}$. Propagating through a second lens gives $H*\mathscr{E}$ at the output plane, where $*$ denotes convolution.

As before, a nonlinear medium can be substituted for the film, allowing for rapid modification of the filter. Some caution must be exercised when making such a substitution. Depending on the confocal parameter of the system, the accuracy of the transform may degrade because of the

[3] $\mathbf{E}$ is the total field; $\vec{\mathscr{E}}$ is the corresponding, slowly varying complex amplitude envelope.

thickness of the nonlinear medium (Goodman, 1968). This places a limit on the thickness t of both volume filters and film emulsions,

$$t < 2f^2\lambda/r_{max}^2 , \tag{5}$$

where r_{max} is the spatial extent of the input field with the largest dimensions, and f is the focal length of the lens.

For the real-time operation of a processor, the fields H and $\mathscr{E}$ must be present concurrently. Because they can no longer both be on axis, the filter cannot be read from the same angle at which it was written (Petrov *et al.*, 1979). For thin nonlinear filters this is not a problem, because the amplitude transmittance does not depend strongly upon the angle of illumination. However, the diffraction efficiency will be limited owing to the thinness of the filter. In the next section we will see that the DFWM process can be used to overcome this and other problems.

There are other differences between conventional and nonlinear optical processors. Whereas filters made from photographic film are unaffected by light after the developing process, nonlinear media remain sensitive to light in the absence of a fixing process (Burke *et al.*, 1978). Furthermore, many nonlinear optical materials are birefringent and/or optically active. In many cases, these complexities can turn out to be useful for improving performance (Herriau *et al.*, 1978; Miridinov *et al.*, 1978.)

2. *Spatial Convolution and Correlation*

Nonlinear processing with DFWM offers an attractive solution to the above problems (Lee, 1974). Specifically, spatial convolution and correlation has been proposed and analyzed (Pepper *et al.*, 1978b) for the configuration illustrated in Fig. 13. These systems combine the multiplicative properties of DFWM with the Fourier-transforming properties of lenses. As shown in Fig. 13, all three input fields contain arbitrary amplitude and/or phase information. The input complex amplitudes $\mathscr{E}_1(x, y)$, $\mathscr{E}_2(x, y)$, and $\mathscr{E}_p(x, y)$ at their respective outer focal planes are spatially Fourier transformed by lenses L upon propagation to a common focal plane. The transformed fields $\tilde{\mathscr{E}}_1$, $\tilde{\mathscr{E}}_2$, and $\tilde{\mathscr{E}}_p$ induce nonlinear polarizations in the DFWM medium. The resulting third-order nonlinear polarization

$$P = \chi^{(3)}\tilde{\mathscr{E}}_1\tilde{\mathscr{E}}_2\tilde{\mathscr{E}}_p^* \tag{6}$$

radiates an output field that propagates essentially backward relative to the input field $\mathscr{E}_p$, returning through lens L. It can be shown (Pepper *et al.*, 1978b) that the output field, when evaluated at the plane located a dis-

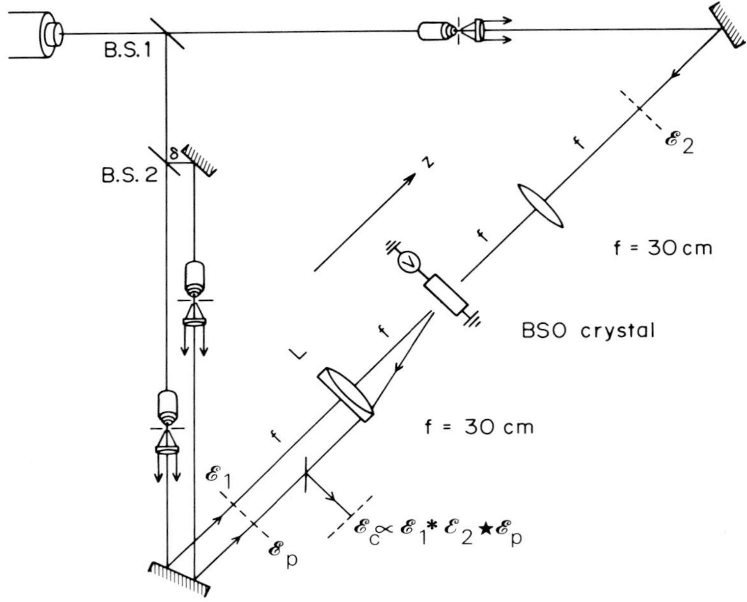

Fig. 13. Experimental apparatus for performing spatial convolution and correlation. Input and output planes are shown by dashed lines. (After White and Yariv, 1980.)

tance f in front of lens L, is of the form

$$\mathscr{E}_c \propto \mathscr{E}_1 * \mathscr{E}_2 \star \mathscr{E}_p, \tag{7}$$

where ★ denotes correlation and $*$ convolution (for definitions, see, e.g., Goodman, 1968).

Spatial convolution and correlation of two-dimensional phase and amplitude images was demonstrated (White and Yariv, 1980; Odoulov *et al.*, 1980b) using $Bi_{12}SiO_{20}$ as the nonlinear medium, with the results shown in Fig. 14. The first three columns show the input fields $\mathscr{E}_1$, $\mathscr{E}_2$, and $\mathscr{E}_p$, and the fourth column shows photographs taken at the output plane. Rows (a)–(c) illustrate correlation and row (d) illustrates convolution. The output in row (c) clearly demonstrates real-time pattern recognition.

3. *Intensity Filtering: Application to Edge Enhancement of Two-Dimensional Images*

As discussed in greater detail in Chapter 11, the special type of nonlinearity generated by the photorefractive effect (Glass, 1979) can be used to

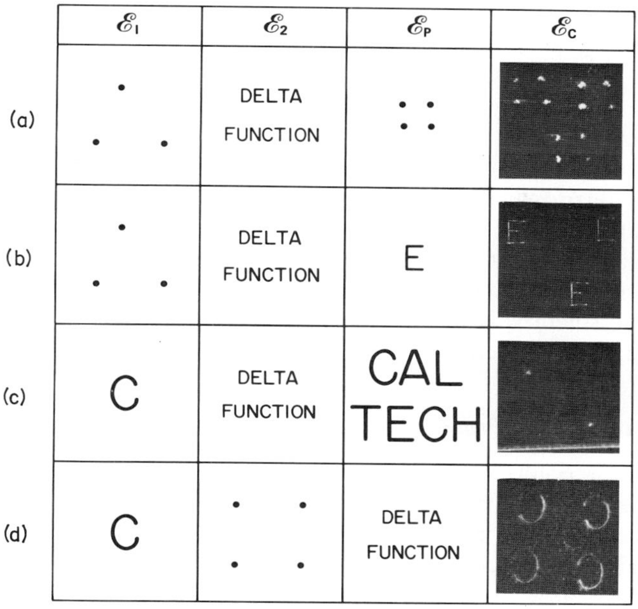

	$\mathscr{E}_1$	$\mathscr{E}_2$	$\mathscr{E}_P$	$\mathscr{E}_C$
(a)	• • •	DELTA FUNCTION	• • • •	
(b)	• • •	DELTA FUNCTION	E	
(c)	C	DELTA FUNCTION	CAL TECH	
(d)	C	• • • •	DELTA FUNCTION	

Fig. 14. Results demonstrating real-time spatial convolution and correlation of two-dimensional images using the geometry of Fig. 13. (a)–(c) the input objects $\mathscr{E}_1$, $\mathscr{E}_2$, and $\mathscr{E}_p$; (d) the output field $\mathscr{E}_c$. (After White and Yariv, 1980.)

advantage in another type of image processing: edge enhancement (Huignard and Herriau, 1978; Feinberg, 1980). By properly selecting the field amplitudes of the various beams the dark-to-light transitions of a two-dimensional image can be enhanced.

When a photorefractive material (e.g., $BaTiO_3$) is exposed to a sinusoidally varying intensity distribution, resulting, for example, from the interference of two plane waves, the steady-state amplitude of the nonlinearity-induced index modulation is proportional to the fringe modulation (Huignard *et al.*, 1979; Feinberg *et al.*, 1980). Unlike the optical Kerr effect, an increase in the intensity of one of the optical fields will not necessarily yield a larger index change. The largest index modulation (resulting in a hologram with the greatest diffraction efficiency) will be produced when two optical fields of the *same* intensity interfere. Thus, the nonlinearity can be used as an *intensity* filter.

One can employ this property to enhance the boundary of an object field containing a light and a dark region, if one chooses a reference beam with an intermediate intensity. The largest interference fringe modulation and index modulation can be made to occur at the boundary between the light and dark regions of the image. The entire medium is next illuminated with a *weak*, third plane wave at the Bragg angle, counterpropagating with respect to the reference beam. The boundary of the image will then diffract the most light. A different type of filtering can also be performed by imaging the Fourier transform of an object onto the nonlinear medium (Feinberg, 1980). Horizontal or vertical edges can be selectively enhanced by a slight misalignment of the beams. Some typical experimental results are shown in Fig. 15.

4. *Logic and Digital Processing of Binary Signals*

By using the polarization dependence of various four-wave mixing processes, one can realize new optical logic processors (O'Meara, 1982c). Such processing can occur in a *parallel* fashion (via spatially dependent polarization encoding of two-dimensional arrays or images) or *serial* fashion (via temporal sequencing of the various polarization-encoded input fields), or both. One can also employ on/off states (via the existence or nonexistence of the various interacting fields) of nonlinear elements for digital signal processing. A brief description of on/off processing via four-wave mixers is given in Section III.B.2; a description of processing via on/off states in nonlinear bistable elements is given in Section III B 5.

One can employ several different representations of optical binary logic states (true/false or ones/zeroes). Among these are (a) on/off states, (b) 0°/180° phase states, and (c) orthogonal polarization states. It has been suggested (O'Meara, 1982c) that orthogonal linear polarization states are particularly appropriate to logical processing via nonlinear optics—in particular, by four-wave mixers. In these systems, one associates a one/zero or true/false state with a given orthogonal polarization state. In what follows, we assume the existence of a NLO medium with the proper third-order susceptibility tensor element(s) to achieve the desired output polarization state, given a set of input polarization-encoded optical fields.

For example, a logical inverter or negator may be realized by FWM in which the pumps are oppositely polarized. Whichever polarization state enters the mixer as a probe induces an oppositely polarized output (in the form of a conjugate signal). Of course, a half-wave plate achieves the same objective more simply; however, the mixer offers the advantage of level control and/or gain.

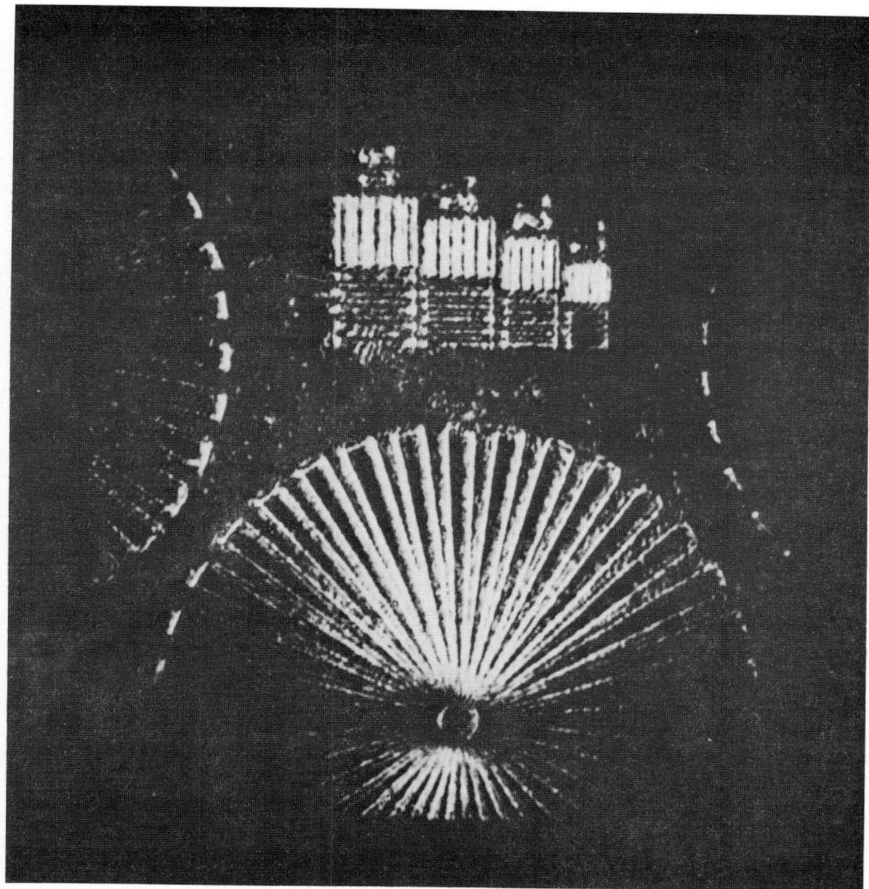

Fig. 15. Experimental demonstration of edge enhancement of two-dimensional images using four-wave mixing in a photorefractive medium. As a result of the intentional misalignment of the object beam in the vertical direction, only the horizontal edges are enhanced. (After Feinberg, 1980.)

One approach to a logical optical circuit is illustrated in Table I. BS denotes a conventional beam splitter and PSBS a polarization-selecting beam splitter (e.g., a Glan prism surrounded by half-wave plates). This truth table also illustrates the inputs and outputs along with the corresponding optical paths for this particular circuit. Note that in states 1 and 2 the output signal is generated by four-wave mixing action, whereas the four-wave mixer is essentially unactivated in states 3 and 4; for this pair of states, pump A is coupled out via splitter PSBS, thereby

TABLE I

Optical Logic Elements Using Polarization States in a FWM Interaction[a]

Optical processing operation	State identification number	A	B	S(OUT)
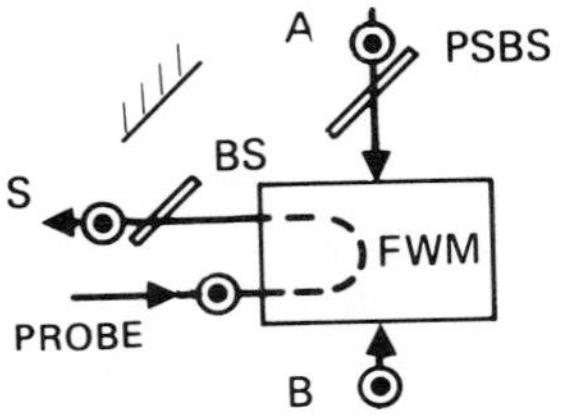	(1)	T	T	T
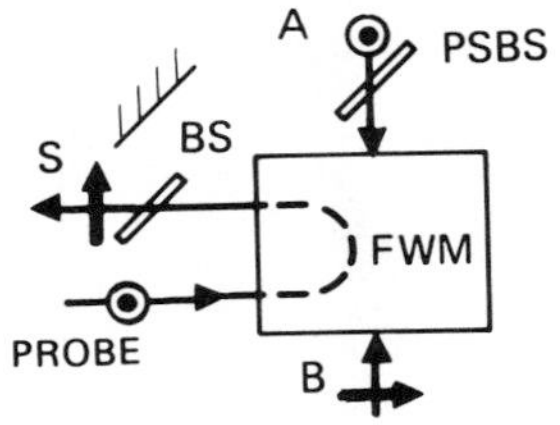	(2)	T	F	F
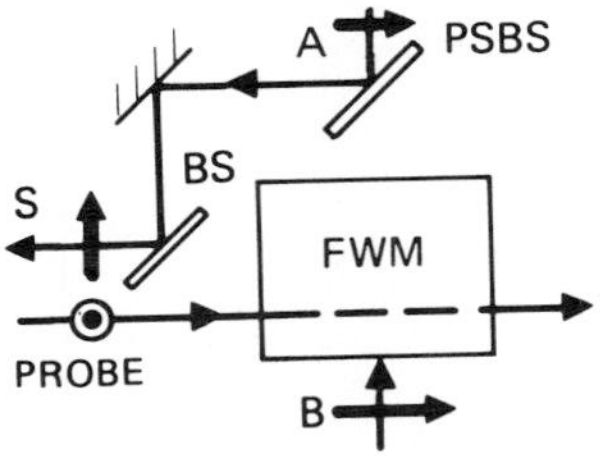	(3)	F	F	F
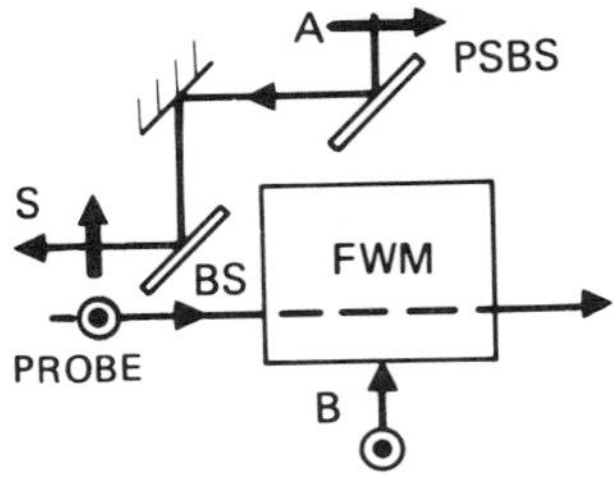	(4)	F	T	F

[a] The various input and output fields are encoded in an orthogonal, linearly polarized fashion (linear optical polarization states (·) perpendicular to the page and (→) parallel to the page). The proper third-order nonlinear susceptibility tensor elements are assumed to be nonzero. The (small) angle offset between the pump fields (A and B) and the probe field is exaggerated for illustrative purposes. BS, beam splitter; PSBS, polarization selective beam splitter. Logic states T and F have been recorded. After O'Meara (1982c).

never entering the NLO medium. By extending this concept one can achieve an exclusive "or," or any other two-state logic function.

Using these concepts, one can realize *parallel* logical processing by encoding (polarization) information onto two-dimensional *arrays* and using a collinear mixing process. In this case, angle, wavelength, or temporal offsets offer techniques to separate out the desired output array.

5. *Coherent Optical Processing Using Iterative Algorithms*

Iterative techniques have been proposed for restoring linearly degraded images (Kawata and Ichioka, 1980) and for restoring finite-energy optical objects from limited spatial and spectral information (Stark *et al.*, 1981). Optical implementations of Gerschberg-type algorithms for reconstructing an object from the modulus of its Fourier transform (Fienup, 1978) and image extrapolation (Marks, 1980) have also been proposed. Coherent optical processors for solving the above restoration problems, and processors designed to solve differential and integral equations (Cederquist, 1981; Cederquist and Lee, 1981), typically rely on mirrors to send a wave front on multiple passes through a specified filter. PCMs should find use here because of their wave-front reversal properties and, owing to the possibility of gain, may breathe new life into systems that in the past were deemed impractical because of reflection losses (Szu, 1980).

D. Interferometry

PCMs have been proposed for use in several conventional and unconventional interferometers. In the Mach–Zender type, as shown in Fig. 16a, the object wave front interferes with its phase conjugate instead of a plane-wave reference (Hopf, 1980). The input signal $\mathscr{E}(x, y)$ evaluated at plane (x) of Fig. 16a is given by

$$\mathscr{E}(x, y) = A(x, y) \exp[i\phi(x, y)]. \tag{8}$$

This field is imaged to the observation plane with an arbitrary transmission T, yielding an amplitude $\mathscr{E}_T = T\,\mathscr{E}(x, y)$. The conjugate of the input is generated by a FWM process in reflection from a PCM and arrives at the observation plane with an amplitude

$$\mathscr{E}_R = R\,A(x, y) \exp[-i\phi(x, y)], \tag{9}$$

where R and T depend on the reflectivity and transmittance of the PCM and other components within the interferometer. The time-averaged in-

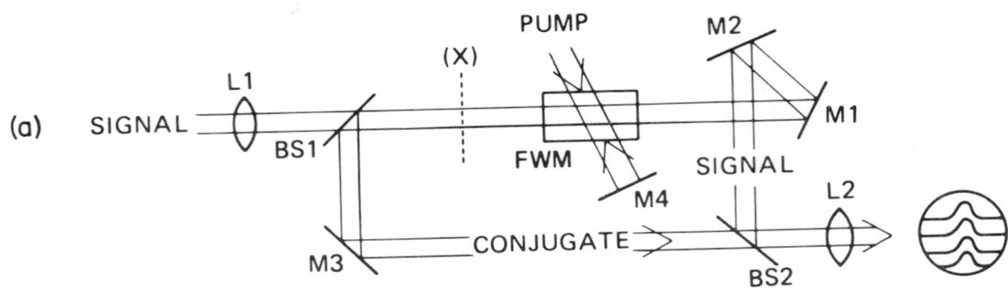

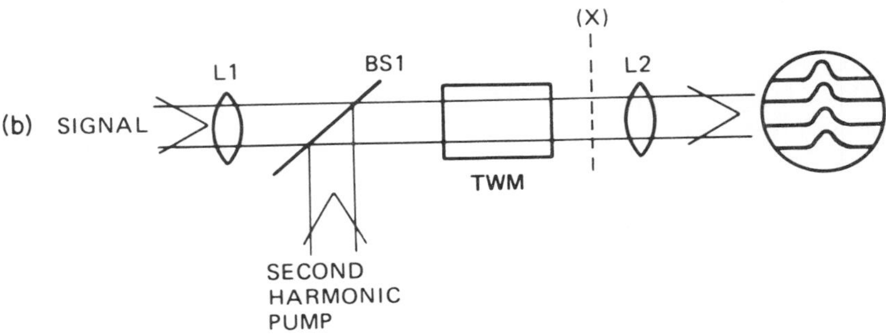

Fig. 16. Schematic diagram of a conjugating interferometer based on (a) FWM and (b) TWM. Lens L1 images the object onto beam splitter BS1; lens L2 images BS1 onto the observation plane. Mirrors M1 and M2 are necessary for proper orientation of the forward-going field. See text regarding the field evaluation plane (x). (After Hopf, 1980.)

tensity at the observation plane is proportional to

$$I = |\mathscr{E}_T + \mathscr{E}_R|^2 = [A(x, y)]^2\{T^2 + R^2 + 2RT\cos[2\phi(x, y)]\}. \tag{10}$$

One important advantage here is that the reflectivity of the components, including the PCM, can be adjusted so that $T = R$, yielding a fringe visibility of unity, independent of intensity variations in or across the sample beam. These predictions were verified in both pulsed and cw modes of operation (Bar-Joseph *et al.*, 1981). The DFWM phase conjugator was realized by using a nonlinear medium consisting of the organic dye–saturable absorber eosin, dissolved in thin (30–80 μm) gelatin films, pumped by an argon ion laser, with intensities of the order of watts per cm^2.

Another type of phase-conjugate interferometer uses a forward-going phase-conjugate beam generated by three-wave mixing (TWM) inside a crystal possessing a second-order nonlinear susceptibility (Hopf, 1980), as sketched in Fig. 16b. The phase-conjugate beam has the same frequency as the signal, so they interfere within the crystal, with the output pattern being imaged to the observation plane.

Rapid changes in an irregular phase object could be observed using the interferometer shown in Fig. 17 (Siegman, 1979). A stationary object would introduce fixed aberrations onto a planar wave front that would be exactly canceled on the return trip through the object if the PCM is ideal. The stationary aberration compensation could be verified by interfering with a planar reference beam. A changing object, however, would introduce uncompensated aberrations as a consequence of the time delays in the compensation path. The resultant phase difference may be displayed by the interference with a plane-wave reference at the observation plane. The time between passes may range from a fraction of a nanosecond up to several milliseconds by using optical delay lines. A pulsed illumination sequence could therefore sample the phase changes, and cw illumination would yield an output signal approximating the first time derivative of the phase changes.

Dynamic interferometry has been demonstrated using the nonlinear

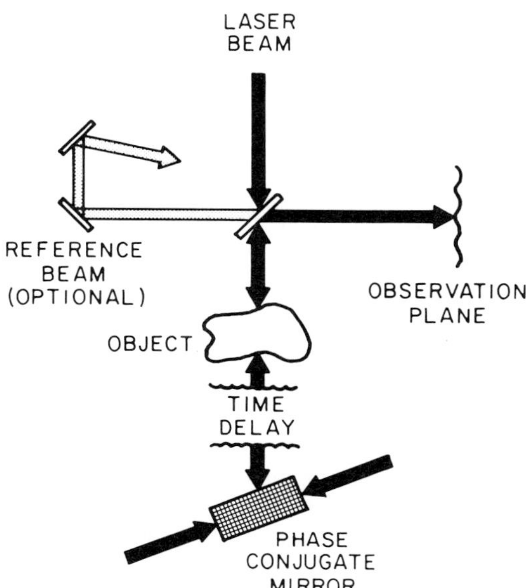

Fig. 17. Experimental arrangement for dynamic interferometry and/or differential holography using phase conjugation. (After Siegman, 1979.)

medium $Bi_{12}SiO_{20}$ in DFWM configurations (Huignard *et al.*, 1977). Also, the response time of the crystal is such that mode patterns of a diffuse vibrating structure can be visualized using a similar system (Marrakchi *et al.*, 1980).

A two-beam wave-front-reversing interferometer in which both beams are reflected simultaneously from the same Stimulated Brillouin Scattering (SBS) mirror was found to have a lowered sensitivity to path-length changes in the two beams (Basov *et al.*, 1980); a phase difference appears at the output because the frequency of the conjugate wave differs from that of the signal (i.e., the probe wave) by the Stokes frequency. The period of the interference pattern at the output was measured to be 8 cm, indicating a Brillouin frequency shift of 0.125 cm^{-1} for stimulation by 1.06-μm laser light.

We note, finally, that is has been proposed that phase conjugation may be applied to the phase problem in optical coherence theory (Agrawal, 1978). By generating a phase-conjugate wave and superimposing it on the original wave field, the phase of the second-order coherence function of the source may be determined.

III. Temporal and Frequency Domain Applications

A. Frequency Filtering

In addition to the wave-front reversal properties of DFWM processes (and other NLO interactions), one can exploit the frequency dependence of certain NLO interactions to realize new classes of optical devices. In the next section, we discuss the application of *nearly degenerate* FWM to optical filtering. Nearly degenerate means the frequency of the probe wave is *slightly detuned* from that of the counterpropagating pump waves. The frequency of the resultant conjugate wave is essentially the same as (but, in general, not equal to) those of the incident fields. The frequency dependence of this process and other NLO processes is basic to many other time domain applications to be discussed in subsequent sections.

Several effects can give rise to narrow-bandpass filtering: phase-matching, gain, and resonance. Phase mismatch occurs when a small frequency shift between pump and signal destroys the synchronization between contributions to the output wave generated throughout a *macroscopic* interaction region. Nonresonant gain can amplify the difference between being in and out of synchronization. In resonant media, it is the *microscopic* properties of individual atoms and molecules that are strongly wavelength dependent. Depending on the specific NLO material and its dimensions, the filter character may be dominated by phase-matching (e.g., in carbon disulfide, using visible wavelengths) or resonance effects (e.g., in sodium vapor, using radiation near a resonance, such as the famous D line). For more theoretical background, the reader is referred to Chapters 2 and 8, respectively. In what follows, we provide a basic review of the consequences of these processes, with emphasis on various areas of application.

Nearly degenerate FWM in a nonresonant medium has been shown to be capable of yielding an active, narrow-bandpass optical filter based on phase-matching constraints (Pepper and Abrams, 1978). In principle, any nonlinear optical interaction can be used in this way, but in general the field of view is restricted because, to avoid phase mismatch, the wave to be filtered must be incident at a certain angle. Moreover, the input and output wavelengths may differ appreciably. The nearly degenerate FWM approach circumvents both of these limitations, providing for a large field of view for the input wave, and a (potentially amplified) filtered output of essentially the same wavelength. In Fig. 18 we show a calculation of the reflectivity of such a filter as a function of the normalized wavelength detuning Ψ, where Ψ is the phase mismatch ΔkL divided by 2π. Finally, a filter that produces a reflected wave with a controllable frequency offset can be realized by having the pump waves offset in frequency relative to each other (Blashchuk *et al.*, 1980b). The theory for nearly degenerate four-wave mixing in nonresonant media is treated in Chapter 2.

Resonantly enhanced nonlinear media[4] can also be useful for constructing filters (Fu and Sargent, 1979; Nilsen and Yariv, 1979, 1981; Nilsen *et al.*, 1981; Raj *et al.*, 1980a,b; Woerdman and Schuurmans, 1981; Saikan and Wakata, 1981). Such a filter may have an ultranarrow bandpass determined by the linewidth of the atomic resonance. The bandwidth of the filter is narrowest when the two pump beams are exactly at line center of the atomic resonance, and increases as the pumps are detuned. Note, however, that the pump frequency yielding the optimum (i.e., greatest) reflectivity may be off line center because of the effects of resonant absorption. Bandwidths approaching $1/T_2$ should be possible, where T_2 is the inhomogeneous lifetime of the transition. A filter using the sodium D line with a fractional bandpass of $\sim 10^{-8}$ has been demonstrated (Nilsen *et al.*, 1981). In gaseous media, the filter field of view may be limited by the effects of atomic motion (Wandzura, 1979; Nilsen and Yariv,

[4] In addition to narrow-band filtering, another advantage of constructing a device that uses a resonantly enhanced nonlinear medium is the potential realization of large nonlinear coupling coefficients. In this and many other NOPC applications, the resonant nature of the nonlinear medium can result in high-efficiency mixers, thereby reducing the necessary pump-wave power requirements.

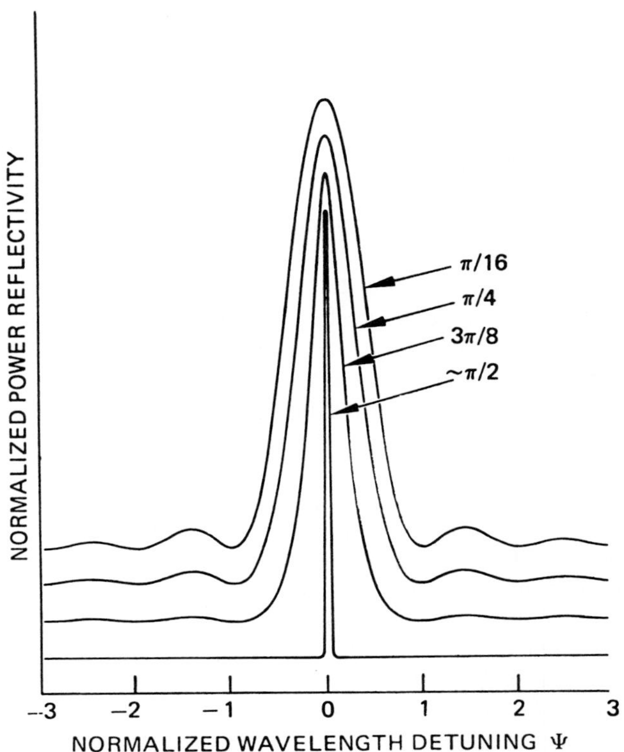

Fig. 18. Power reflectivity of a FWM filter versus normalized wavelength detuning Ψ for several values of the nonlinear gain κL. All curves are normalized to unity reflectivity to emphasize the frequency bandpass of the interaction. (After Pepper and Abrams, 1978.)

1981). Finally, owing to atomic structure and/or dynamic effects (e.g., hyperfine structure, optical pumping, intermediate resonances, Doppler effects, and ac Stark shifts), the filter bandpass may have a narrow, yet detailed bandpass characteristic. The theory and experimental results for nearly degenerate four-wave mixing in resonant media are treated in Chapter 8.

In addition to the filtering action, the nearly degenerate FWM interaction causes the spectrum of the conjugate wave to be inverted about the pump frequency ω_0. That is, a probe wave at frequency $\omega_0 + \delta$ will induce a conjugate output having a frequency $\omega_0 - \delta$. This frequency inverting or "flipping" property of the filter is relevant to other time domain applications such as group-velocity dispersion compensation schemes, and to phase-conjugate resonator mode analysis, as will be discussed separately in subsection B. 6 and in Section IV, respectively.

The filtering ability of a PCM also plays an important role in the simultaneous precision conjugation of multifrequency fields, because narrowband operation can eliminate certain (frequency) cross-talk problems (Depatie and Haueisen, 1980; Steel and Lam, 1980). Such multifrequency conjugation has been demonstrated (Depatie and Haueisen, 1980; Steel and Lam, 1980) using multiline lasers (DF and CO_2). A potential application of this property may be the incorporation of PCMs in high-power multiline lasers and MOPA systems.

We remind the reader that, aside from the (small) frequency shifts in these nearly degenerate processes, the wave-front reversal properties of the PCM are still operative, and may thus be exploited. The phase-conjugate resonator (Section IV) is an example of a device that uses *both* the spatial and the frequency properties of the PCM simultaneously to advantage.

B. Temporal Signal Processing

Signal processing in the temporal domain can also be performed using NOPC. Structures have been proposed for the functions of wave-train chirping, optical gating, time delay control, logic gating, convolution and correlation, envelope reversal, and space-to-time modulation/demodulation. Applications of these devices include pulse compression for communications as well as optical and microwave radars, data processing, encoding and decoding, and high-bandwidth filtering in communication systems.

1. Pulse Envelope Shaping

Perhaps the most fundamental time domain operation is pulse shaping. An important special case, because of its relevance to laser fusion, is pulse narrowing. In several experiments, the ability of a PCM to temporally *shorten* pulses was demonstrated. In the first experiments, all three inputs to a four-wave mixer were pulsed (Pepper *et al.,* 1978a; Bloom *et al.,* 1978b). The output, being proportional to the product of the three inputs, was a narrower pulse. In another set of experiments using DFWM (Vanherzeele and Van Eck, 1981; Vanherzeele *et al.,* 1981), the conjugate pulse was reinserted into a system consisting of a laser oscillator–amplifier in addition to the conjugator, thereby generating subsequent pulses of even shorter duration. The ultimate pulse-width narrowing was limited by the response time of the nonlinear medium and the oscillator–amplifier system.

Pulse compression was also demonstrated via stimulated Brillouin scattering (SBS) inside a *tapered* waveguide (Hon, 1980b), consisting of either solid glass or a hollow tube filled with methane. The tapering action of the waveguide, along with the threshold nature of the SBS interaction, was responsible for the compression effect. To understand the scheme, assume that the waveguide is sufficiently long that the input pulse can "spatially fill" the lightpipe longitudinally. As the input field intensity (assumed to be below the threshold intensity for the onset of SBS at the input face of the guide) propagates down the *tapered* waveguide, the intensity increases with the distance traversed owing to the decreasing guide diameter. As the intensity along the waveguide exceeds the SBS threshold, a backscattered Stokes pulse begins to form. This short pulse builds up in amplitude (but not in length) as it is amplified by the trailing part of the input pulse. In the cited experiment, the input was a 20-nsec pulse provided by a Q-switched Nd:YAG laser, and the output was a 1-nsec conjugate pulse. Iterative approaches using this scheme have also been demonstrated to further sharpen the output pulse (this topic is reviewed in a special journal issue on NOPC; see Pepper, 1982).

Pulse *expansion* with nonlinear optics can also be realized. For example, the generation of an arbitrary impulse response function by spatially encoding the pumps in a DFWM system has been theoretically proposed (O'Meara and Yariv, 1982). That is, an optical time domain filter can be synthesized that, when excited by a short optical input pulse, produces an output pulse whose complex envelope can have a controllable amplitude and/or phase distribution in time. The synthesis is subject to the constraint that the maximum rate of change in the output envelope is limited either by the width of the excitation pulse or by the bandwidth of the nonlinear medium, whereas the pulse duration is limited by the length of the mixer.

2. *Optical Gating, Time Delay Control, and Logic Functions*

A subpicosecond optical gate can be realized using a PCM. In an experiment by Bloom *et al.* (1978a), an optical gate was demonstrated using 0.5-psec pulses from a mode-locked dye laser. The amplified output was split into parallel pump and object beams, which were then focused by a single lens into a CS_2 cell. An internal mirror retroreflected the pump to form the counterpropagating pump wave. A backward-propagating pulse was observed only when the object pulse was incident on the PCM in time

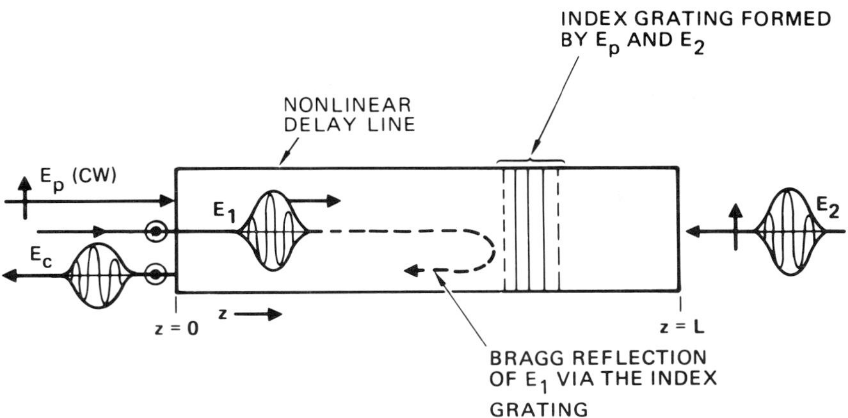

Fig. 19. Time delay control of one optical pulse (E_1) by another (E_2) using a nonlinear delay line with a collinear four-wave mixer. The inputs E_p (cw) and E_2 (pulsed) are copolarized; the signal pulse to be controlled, E_1, is orthogonally polarized. The same geometry can be used for pulse reshaping. (After O'Meara and Yariv, 1982.)

coincidence with the pump pulse. The bandwidth of the signal that can be gated is limited by the response time of the medium, but the gate time is not. This geometry is conceptually similar to measuring the autocorrelation width of a pulse by second harmonic generation (see, e.g., Ippen and Shank, 1977).

This optical gating scheme can be conceptually extended to give rise to new all-optical devices, such as controllable time delay lines, and optical logic elements. These devices can be realized by the proper choice of system geometry (NLO medium configuration and tensor elements, type of nonlinearity, etc.) and interacting field parameters (field propagation directions, temporal sequencing, field polarizations, etc.).

For example, a controllable delay line can be constructed with three collinear input fields E_1, E_2, E_p incident on a slab or fiber of material possessing a nonlinear refractive index (O'Meara and Yariv, 1982). A typical geometry for the scheme is sketched in Fig. 19. The input fields E_p and E_2 are counterpropagating and copolarized, E_p being cw and E_2 being a short pulse. As E_2 propagates through the medium, it beats with E_p, forming an interference pattern that modulates the refractive index of the medium. If the medium response is sufficiently fast, the envelope of the index grating will replicate the envelope of the pulse E_2 as it moves down the rod. The fine structure of the interference pattern remains fixed in space. A short-duration cross-polarized pulse E_p enters the fiber from the opposite

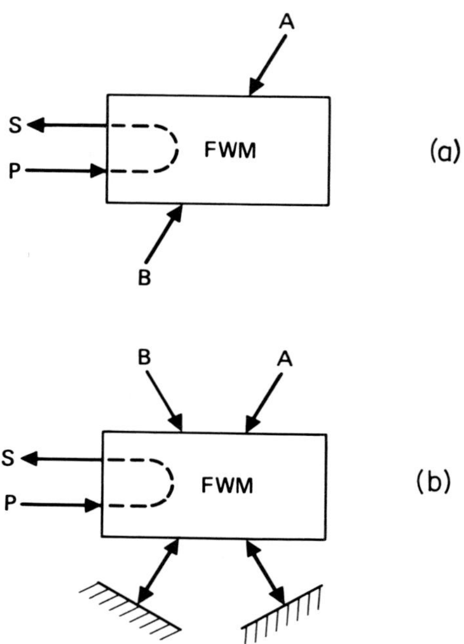

Fig. 20. Optical logic elements formed via FWM using On/Off states (i.e., presence or absence of the various interacting fields). (a) "And" gate (A · B). (b) "Inclusive Or" gate (A ∨ B). (After O'Meara, 1982c.)

side and scatters off an essentially localized grating, giving rise to the conjugate field E_c. The (complex) output field envelope $\mathscr{E}_c$ can be shown to be related to the corresponding input (complex) field envelopes by

$$\mathscr{E}_c(0, t) = -i\alpha\mathscr{E}_2(t) \int_0^L \mathscr{E}_1(t - 2z/r)\mathscr{E}_p^*(t - 2z/r)\, dz. \tag{11}$$

If the entrance time of E_2 is controllable, the exit time of E_c at $z = 0$ can thereby be electronically controlled.

Other possible applications are in pulse reshaping and in the synthesis of logic functions (O'Meara, 1982c) such as And or Or gates. For example, using the canonical FWM geometry shown in Fig. 20a, one sees that unless fields A *and* B are simultaneously present, no output is produced. Similarly, using the double-pump configuration of Fig. 20b, one can realize an Inclusive Or operating on the two pump fields A and B. Alternative logical processing systems are discussed in Sections II.C.4 and III.B.5.

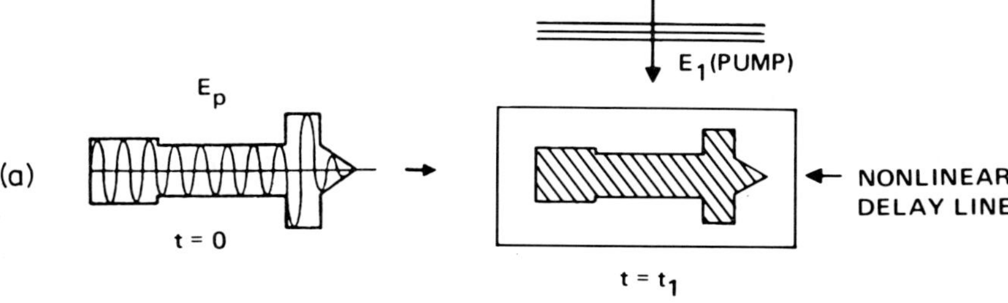

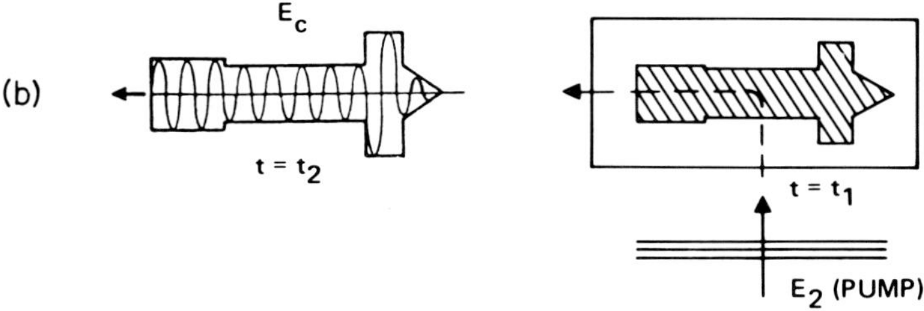

Fig. 21. Envelope time reversal using a nonlinear delay line via FWM. The interaction length is chosen to be long enough for the input wave train E_p to fit inside the delay line. At the moment the entire wave train is situated within the nonlinear slab, the (short) pump pulses $E_{1,2}$ irradiate the device, resulting in the wave train E_c exiting the delay line in a temporally reversed sequence. (Concept after Miller, 1980.) (a) Holographic recording process. The input wave-train amplitude E_p beats E_1. The resultant interference with one pump wave modulates the strength of the holographic fringe pattern that fills the nonlinear delay line. (b) Holographic playback process. The wave-train envelope is retained in the playback process, wherein the opposing pump wave E_2 reflects off of the hologram, resulting in the output wave train E_c. (After O'Meara and Yariv, 1982.)

3. Envelope Time Reversal

As discussed in Chapter 3, another application of signal processing with nonlinear optics is the time reversal of pulse envelopes (Miller, 1980) using a long, thin FWM conjugator. As illustrated in Fig. 21, short pulses $E_{1,2}$ are arranged to pump the conjugator at the moment when the signal to be envelope reversed, E_p, is within the conjugator. This generates a backward-propagating output signal pulse E_c whose envelope shape replicates that of the input pulse, but is time reversed.

4. *Temporal Convolution and Correlation*

The envelopes of two counterpropagating fields E_1 and E_2 can be convolved or correlated (O'Meara and Yariv, 1982) using the orthogonal pumping geometry shown in Fig. 22. A third input field E_p, uniform in z and essentially cw, enters through the side of the delay line normal to the propagation direction of $E_{1,2}$. Where the three fields overlap, a backward-going wave E_c is generated. If E_c is collected with a lens, the amplitude $\mathscr{E}_c$ at the focus would have the basic form of a convolution integral,

$$\mathscr{E}_c(0, t) \propto \int \mathscr{E}_1(z - vt)\mathscr{E}_2(z + vt)\, dz. \tag{12}$$

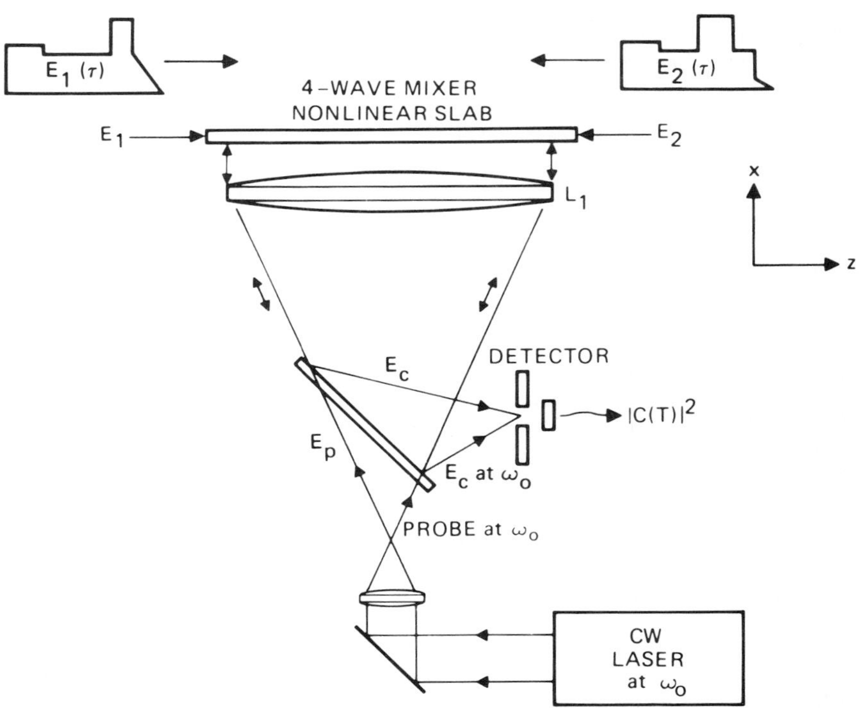

Fig. 22. The four-wave mixer as a time domain correlator. The modulation envelopes $\mathscr{E}_1(z)$ and $\mathscr{E}_2(z)$ are cross correlated in the nonlinear slab as they pass each other. The detector output gives the correlation function as a function of time. (After O'Meara and Yariv, 1982.)

5. *Optical Bistability*

Bistable optical devices (BODs) are of interest for optical computing. It has been shown that, by placing a nonlinear medium within a Fabry–Perot cavity, bistable operation may be realized (Smith, 1981). This stems from the intensity-dependent phase shifts acquired by the circulating Fabry–Perot fields. As the input optical field (the driving field) to the cavity is varied, the output optical field can exhibit hysteresis in the transmitted or reflected output wave. An all-optical logic device can be founded on this principle. One variation is to consider the counter-propagating Fabry–Perot fields as pump waves and the intracavity element as a $\chi^{(3)}$ medium, thereby forming a PCM, as shown in Fig. 23a. The logic state can be read out with a weak beam by probing the PCM reflection coefficient as a function of the driving field. The predicted PCM reflectivity (Abrams and Lind, 1977, 1978) within the Fabry–Perot resonator has been analyzed, and predicted to exhibit a bistable behavior (Agrawal and Flytzanis, 1981), as shown in Fig. 23b. That the probe can be incident at some nonzero angle with respect to the optic axis of the Fabry–Perot allows for simultaneous probing, remote access, and isolation from the driving fields. We conclude by noting that a recent analysis has shown that an isolated nonlinear medium also has the potential to exhibit bistable behavior (Flytzanis and Tang, 1980), as do distributed-feedback geometries (Winful and Marburger, 1980).

6. *Compensation for Pulse Spreading Effects*

One can use the frequency inversion and conjugation properties of optical parametric interactions (as occur, e.g., in TWM and FWM) to compensate for the undesirable spreading of pulses resulting from either linear dispersion or self-phase-modulation effects along a transmission medium. This *temporal* compensation scheme can be useful for increasing the bandwidth capability (or data rate) of a *serial* communications link.[5]

As an example of the former effect, consider the spreading of a pulse when propagating through a dispersive channel, e.g., a single-mode fiber. A PCM placed at the midpoint of the channel, as shown in Fig. 24, has been proposed as a means of compensating this temporal broadening ef-

[5] We note an interesting parallel here: the wave-front reversal properties of NOPC interactions in the *spatial* domain, which can compensate spatial phase aberrations, can be viewed as a scheme useful for increasing the bandwidth capability of a *parallel* (i.e., 2D or 3D image) communications link.

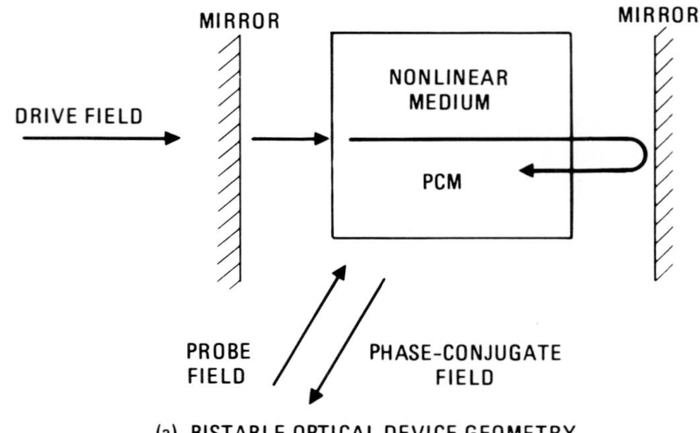

(a) BISTABLE OPTICAL DEVICE GEOMETRY

TRACE NO. 1
PCM REFLECTIVITY →
"X" "Y"
DRIVE FIELD →

TRACE NO. 2
PCM REFLECTIVITY →
"X" "Y"
DRIVE FIELD →

(b) CALCULATED OPTICAL HYSTERESIS

Fig. 23. A bistable optical device (BOD) using four-wave mixing. (a) The interaction of the counterpropagating Fabray–Perot fields with the intracavity nonlinear element can be viewed as forming a PCM via DFWM. (b) Calculated PCM reflectivity as a function of the input driving field. The critical points (X and Y) correspond to the BOD switching points. In trace No. 1 both points are realized. In trace No. 2 the input drive field is raised to a level that does not reach that of the second critical point. (After Agrawal and Flytzanis, 1981 as depicted in Pepper, 1982.)

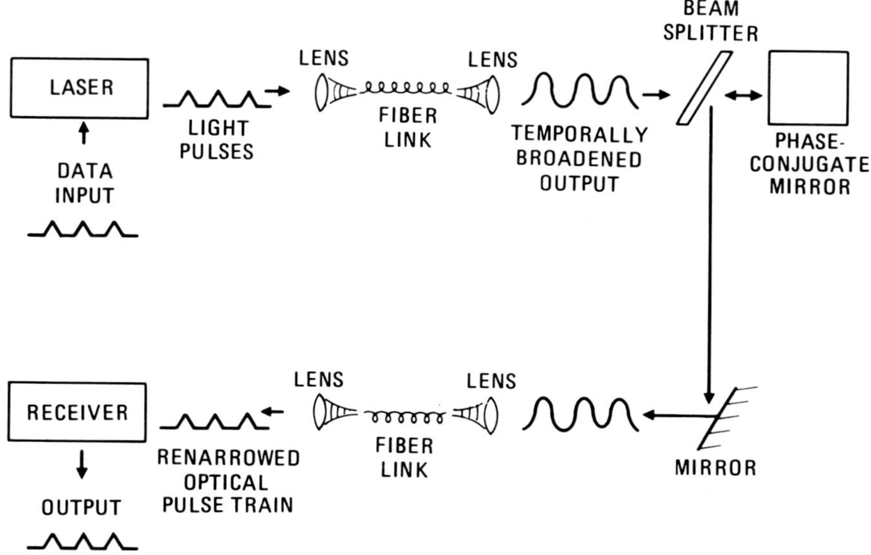

Fig. 24. Potential approach for compensating for the temporal spreading of pulses, encountered in propagation over dispersive networks. After the pulse train is broadened by the first part of the link, the PCM conjugates and inverts its frequency spectrum. Thus, upon traversal through a matching link, the broadened pulses are renarrowed, thereby increasing the bandwidth of the network. (After Yariv *et al.*, 1979.)

fect. The broadened pulse is reflected and conjugated by the PCM, with the reflected frequency spectrum inverted about the pump frequency. This frequency reordering results in a conjugated pulse whose initial frequency component propagates more slowly, thereby allowing its faster counterpart to "catch up" during propagation through the second half of the fiber link (Yariv *et al.*, 1979).

Ideally, two conditions for the realization of this scheme are (1) that the bandwidth of the PCM is at least as large as that of the input pulse and (2) that the group-velocity spreads in both halves of the fiber are identical.

Compensation schemes with forward-going conjugators can yield much higher data rates because their bandwidths are limited only by the linear dispersion of the conjugator (Pepper, 1980). These forward-going schemes may also offer geometrical advantages.

A similar compensation geometry has been proposed for reversal of the chirp generated by the self-phase-modulation of optical pulses, which occurs at high optical intensity levels (Marburger, 1978).

IV. Phase-Conjugate Resonators

In this section, we briefly discuss some key properties of optical resonators in which one of the conventional mirrors is replaced with a PCM, forming the so-called phase-conjugate resonator (PCR) as shown in Figs. 10 and 25. We limit the discussion to PCMs realized by four-wave mixing. A detailed description of this novel device is presented in Chapter 13; we mention the PCR here because, from an applications viewpoint, it is a device that exploits many of the spatial and temporal (or frequency) properties of the PCM discussed thus far. The properties of PCRs to be discussed here include aberration compensation, spatial field characteristics, transverse and longitudinal modes, and cavity stability. Where applicable, we first treat the ideal PCR (i.e., neglecting diffractive losses from end mirrors and internal stops, saturation, and temporal response) and then the nonideal PCRs. We conclude with a summary of recent experiments that have demonstrated many of the expected properties of PCRs.

Perhaps the most compelling reason for considering the properties of PCRs is the capability of removing, or at least alleviating, phase and polarization aberrations that may accumulate within the resonator. Even time-varying (i.e., dynamic) aberrations, resulting, for example, from thermal, mechanical, or nonlinear optical effects, can be compensated. (Odoulov, *et al.*, 1980a demonstrated the ability of a PCM to compensate for time-dependent aberrations using the canonical geometry). In ideal PCRs, this can result in a diffraction-limited output from the conventional mirror end of the resonator, subject to the precision fabrication of the mirror. For example, if the normal output-coupling–feedback mirror is

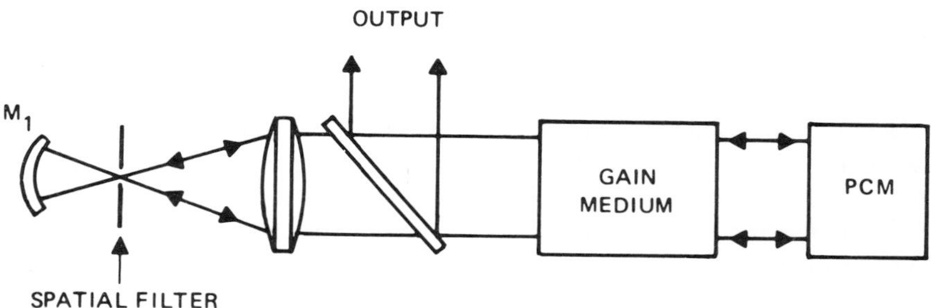

Fig. 25. One form of a phase-conjugate resonator. If the spatial filter is diffraction limited, distortions in mirror M_1 do not degrade the beam output quality. However, computer simulations indicate that it is sometimes preferable to employ a high-quality mirror (for M_1) and to use the spatial filter for transverse-mode control.

well figured, the requirement of satisfying the boundary conditions at this mirror (after one round trip within the cavity) ensures that the feedback wave is unaberrated; that is, it matches the contour of the output mirror (AuYeung *et al.*, 1979; Belanger *et al.*, 1980a,b; Lam and Brown, 1980). One may accommodate imperfectly figured feedback mirrors by incorporating a spatial filter into the system, after Fig. 25, with a separate beam splitter or diffraction grating for outcoupling. In either case, if the *transmission* properties of the output mirror (or output coupler) are also aberration-free, the outcoupled wave is unaberrated.

These arguments are predicted on the assumption of a self-consistent field formed after a single round trip. However, a two-round-trip self-consistent field is possible (AuYeung *et al.*, 1979). This follows from the property that any general field can replicate itself after two round trips in a phase-conjugate resonator. Replication can occur even though the wave front of the field may not match the contour of the output mirror. Because this argument can apply to any transverse field distribution, an aberrated PCR output is possible. Although computer simulations suggest that such fields may sometimes exist, diffraction losses and intracavity apertures tend to suppress the high spatial frequency components of such aberrated PCR fields.

Recent computer simulations of certain phase-conjugate resonator configurations (Valley and Fink, 1983), an example of which is sketched in Fig. 25, have shown that the substitution of a phase-conjugate mirror for a conventional mirror can appreciably improve the output field quality and far-field Strehl ratio (for definitions, see, e.g., Born and Wolf, 1975), as illustrated in Fig. 26. This improvement can occur even when gain-region diffraction effects are present and the laser gain medium is well saturated. In fact, the PCR continues to oscillate satisfactorily in the face of aberrations sufficiently severe to quench the oscillation of a matching conventional resonator. Nevertheless, the combined effects of diffraction and saturation in the laser gain medium produce a significant limitation in the strength of the phase errors that can be compensated for, as evidenced in Fig. 26.

Assuming that aberrations are not present or that the aberration compensation is perfect, what are the shapes of the modes formed within the resonator? The *transverse*-mode structures of *ideal, large-mirror* phase-conjugate resonators may be described by Hermite–Gaussian transverse eigenmodes. The properties of such modes have been analyzed by using Gaussian beam matrix techniques (AuYeung *et al.*, 1979; Belanger *et al.*, 1980a,b; Hardy, 1981).

In the case of *nonideal* phase-conjugate resonators (cavities employing mirrors and PCMs of finite dimensions and possibly other apertures, etc.),

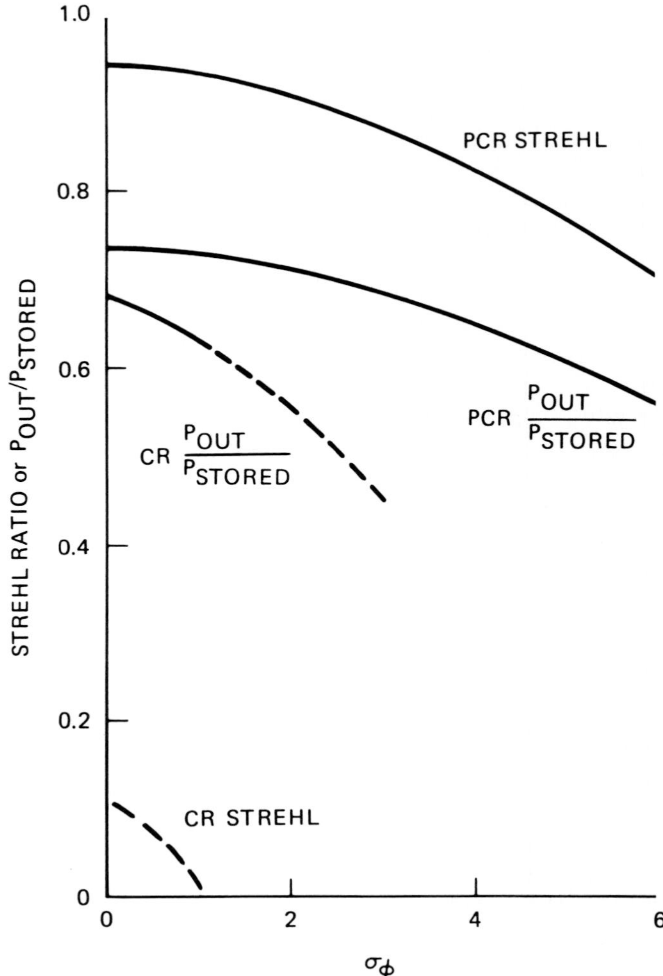

Fig. 26. Comparison of the performance of a phase-conjugate resonator (PCR) (using the geometry of Fig. 25) and a conventional resonator (CR). The output power and Strehl ratio are plotted as a function of the phase standard deviation at the input to the gain medium. (After Valley and Fink, 1983.)

the study of transverse modes can be carried out by using the Huygens principle in the Huygens–Fresnel approximation. For certain special cases (e.g., confocal configurations), the modal solutions may be written analytically in terms of prolate spheroidal wave functions (Lam and Brown, 1980). More general resonators can be analyzed numerically by using Fox–Li diffraction techniques (Bel'dyugin *et al.*, 1979; Lam and Brown, 1980; Reznikov and Khizhnyah, 1980; Valley and Fink, 1983). Computer simulations have also been employed (Valley and Fink, 1983) to examine mode filling and aberration compensation properties of the PCRs, including the effects of diffraction losses at the mirrors and internal stops (e.g., spatial filters).

The mechanisms that produce *longitudinal* modes in a PCR are rather different from the mechanisms that produce longitudinal modes in conventional resonators. Because the total phase shift accumulated in a single round trip within the PCR cavity is zero (independent of the cavity length), the normal $c/2\ell$ modes do not exist. Further, as discussed in Section III. A, any modal frequency offset (relative to the pumps) of a wave impinging on a PCM using four-wave mixing is, upon reflection from the PCM, frequency offset in the opposite direction. Thus, a necessary condition for the field to replicate in a single round trip is that the frequency of the mode matches that of the pumps (i.e., the offset is zero). Such a mode has been designated as a "degenerate mode" of the PCR (AuYeung *et al.*, 1979). Note, however, that a two-round-trip longitudinal mode can also exist, in accordance with the frequency-flipping properties of the four-wave PCM. That is, a field can exist that satisfies a self-consistency analysis after two round trips within the resonator. In this case, the effective length of the cavity that establishes a stable longitudinal mode is doubled, resulting in a mode spacing of $\pm c/4\ell$, forming the so-called paired half-axial modes (AuYeung *et al.*, 1979; Belanger *et al.*, 1980a,b). The spectrum of these higher-order modes is centered at the frequency of the pump waves that drive the PCM. (The length of the PCM, as well as its frequency-dependent phase shift, modifies this mode spacing slightly.) This is in contrast to a conventional resonator, wherein the modes are spaced by $c/2\ell$, without a "central" frequency in general. Whether or not the $\pm c/4\ell$ modes are observed has been shown to depend on the round-trip gain and on the bandwidth of the phase-conjugate mirror. For example, it has been experimentally verified that when a sodium four-wave mirror was operated in a narrow-bandwidth condition, the higher-order PCR longitudinal modes did not occur, but when the PCM bandwidth was increased, these higher-order modes were observed (Lind and Steel, 1981). Hence, the output frequency spectrum of the PCR is expected to be equal to or less than that of the pump waves. Further, if the resonance bandwidth of the PCM is less than that of the pump laser spectral bandwidth, the spectral bandwidth (or monochromaticity) of the PCR output may actually be less than that of the pump laser (at the cost of driving the pump power requirements higher). In addition to the frequency locking of the PCR to the PCM, the *coherent* coupling action of the interacting fields within the four-wave mixer can be viewed as an effective frequency and phase injection locking[6] of the PCR fields to the pump fields.

Several stability analyses of PCRs have been performed, to determine whether (stable resonators) or not (unstable resonators) a ray near the center of the resonator remains confined within the cavity. A paraxial ray analysis has been performed (AuYeung *et al.*, 1979; Belanger *et al.*, 1980a,b; Lam and Brown, 1980) with the conclusion that the PCR is always stable, independently of the curvature of the conventional mirror or the length of the cavity. This behavior is also in sharp contrast to that of conventional resonators.

For the above reasons, one can conclude that, in principle, the overall result of forming a resonator with a four-wave mixer as the PCM is a stable laser possessing a high-quality output, both in terms of its wavefront character and in terms of its improved monochromaticity, even in the face of dynamic external perturbations (mirror misalignments, cavity length fluctuations, phase and/or polarization aberrations, etc.).

There have been a number of experimental demonstrations of PCRs, first without and then with intracavity gain media (e.g., loaded resonators). The combination of a pulsed PCM with gain (i.e., a nonlinear reflectivity greater than unity) and a normal mirror has been demonstrated to oscillate with no additional intracavity gain media required, i.e., an unloaded cavity (Pepper *et al.*, 1978a; Bloom *et al.*, 1978b).

The operation of a loaded PCR was first studied by replacing one of the mirrors of a pulsed ruby laser with a PCM; the PCM was realized via a DFWM process in a CS_2 cell, pumped by another *Q*-switched ruby laser (AuYeung *et al.*, 1979). The PCR output was $\sim 1\%$ of the pump power. It was possible to operate the PCR with a convex normal mirror which otherwise would have prohibited lasing.

A cw (unloaded) PCR was constructed with a PCM having a gain up to ~ 100. The PCM was realized via DFWM in a photorefractive crystal ($BaTiO_3$), pumped by an argon ion laser (Feinberg and Hellwarth, 1980). By using an intracavity spatial filter to define the mode, oscillation was observed with no loss in output power or beam quality when an aberrating plate was inserted. Several other oscillators based on self-induced gratings in $BaTiO_3$ have been demonstrated, including a unidirectional ring laser, a self-pumped PCM (White *et al.*, 1982). A self-pumped PCM has been used as an end mirror on an argon–ion laser at 488 nm. The PCM was observed to compensate for distortions produced by a piece of etched glass placed within the cavity (Cronin–Golomb *et al.*, 1982).

In another experiment, the output coupler of a standard jet-stream cw

[6] It may be more precise to say that the system functions more like a master oscillator–power amplifier than an injection-locked oscillator system.

dye laser was replaced with a sodium vapor PCM using a FWM process, pumped by a second cw dye laser in near resonance with the Na D_2 line (Lind and Steel, 1981). The cw output had a spectrum with the anticipated (AuYeung *et al.*, 1979) "paired half-axial mode" spacing, and with the central (degenerate) mode locked onto the frequency of the pump waves. The dynamic aberration compensation property of the PCR was established when a piece of etched glass was placed and rotated within the resonator; an unaberrated output beam still emerged from the conventional mirror. This same aberrator caused oscillation to cease in a matching, conventional dye laser.

Oscillation has also been observed, and theoretical analyses presented, using ring lasers containing PCMs (Diels and McMichael, 1981; Ivakhnik *et al.*, 1981; White *et al.*, 1982) and lasers with one mirror replaced with a pseudoconjugator retroreflector array (Orlov *et al.*, 1978; Mathieu and Belanger, 1980; Jacobs, 1982) or a SBS PCM (this topic is reviewed in Chapter 13, and in the special journal issue on NOPC; see Pepper, 1982).

V. Applications of Two-Photon Coherent States

Two-photon coherent states (TCSs) have been proposed for use in low-noise quantum-limited detection and amplification schemes (Yuen, 1976). It has also been proposed that optical parametric interactions such as DFWM may give rise to a TCS (Yuen and Shapiro, 1979; Schubert and Vogel, 1981). The state of the radiation field inside a single-photon transition laser operating well above threshold is an example of a (single-photon) coherent state (Loudon, 1978). In such a state, the quantum-mechanical uncertainties in the quadrature components of the radiation field are equal. The state of the radiation field inside a two-photon transition laser may (Yuen, 1976) or may not (Schubert and Vogel, 1981) be an example of a TCS. In such a state, the quadrature components may have unequal uncertainties. If the TCS is generated by DFWM, the uncertainty distribution may be controlled by varying the nonlinear gain.

A TCS can be detected by its effect on photon statistics and noise fluctuation properties. Although the TCS remains to be observed experimentally, it has been proposed for use in several detection schemes.[7]

The performance of a communications system can be improved if the transmitter generates information-carrying radiation on a TCS instead of a single-photon coherent state (Yuen, 1976; Yuen and Shapiro, 1980). The

[7] There is active debate in the optics community regarding the current analyses of the TCS. Specifically, issues such as pump-wave-induced quantum fluctuations and the quantum nature of the nonlinear medium may require further attention.

TABLE II
Laser Spectroscopy Using Four-Wave Mixing

Phase-conjugate mirror reflectivity versus	Physical mechanism	Reference
Angle	Atomic-motional effects	*a*
Buffer gas pressure	Collisional excitation	*b*
	Pressure-broadening mechanisms	*c*
Pump–probe polarization	Coherent-state phenomena	*d*
	Multiphoton transitions	*e*
	Quadrupole optical transitions	*f*
	Electronic and nuclear contributions to nonlinear optical susceptibility	*g*
Magnetic fields	Optical pumping	*h*
	Zeeman state coupling	*i*
	Liquid-crystal phase transitions	*j*
Electric fields	Stark effects	*f*
	Liquid-crystal phase transitions	*k*
rf and microwave fields	Hyperfine state coupling	*f*
Pump–probe detuning	Atomic population relaxation rates	*l*
	Atomic linewidth effects	*l*
	Optical pumping effects	*l*
	Doppler-free one- and two-photon spectroscopy	*l*
Frequency ($\omega_{pump} = \omega_{probe}$)	Natural linewidth measurements	*m*
	Atomic-motionally-induced nonlinear coherences	*m*
Pump frequency scanning	Laser-induced cooling of vapors	*n*
	Sub-Doppler spectroscopy	*f*
Transient regime (temporal effects)	Atomic coherence times	*o*
	Population relaxation rates	*p*
	Inter- and intramolecular relaxation	*q*
	Carrier diffusion coefficients	*r*
Pump-wave intensity	Saturation effects	*s*
	Inter- and intramolecular population coupling	*t*
	Optically induced level shifts and splittings	*u*

[a] Wandzura (1979); Steel *et al.* (1979); Nilsen *et al.* (1981); Nilsen and Yariv (1979, 1981); Humphrey *et al.* (1980); Fu and Sargent (1980); Saiken and Wakata (1981); Bloch *et al.* (1981).

[b] Liao *et al.* (1977); Fujita *et al.* (1979); Bogdan *et al.* (1981).

[c] Raj *et al.* (1980a,b); Woerdman and Schuurmans (1981); Bloch *et al.* (1981).

information capacity of the transmitted radiation can be increased by transferring the uncertainty from the quadrature component carrying the signal to the other quadrature component. An increase in the signal-to-noise ratio can be realized even if no information is to be carried by the noisy component. A proper reception scheme should be able to observe the signal quadrature component without degradation from the noisy quadrature component.

An interesting potential application using a TCS has been proposed by Shapiro (1980): an efficient waveguide tap. Through the use of TCS, a waveguide tap can be configured in such a manner so as to minimize the insertion loss. Hence, a communication link can be realized possessing a large number of tapping networks, all of which have an appreciable signal-to-noise figure.

Gravitational wave detectors based on interferometry (Forward, 1978) may be optimized by the use of a TCS (Caves, 1981). The two sources of quantum-mechanical noise in such detectors are: fluctuations in the number of output photons (photon counting error) and fluctuations in the radiation pressure on the interferometer mirrors. The first source is the limiting factor in the current state-of-the-art interferometers that employ low-power lasers. If the radiation state in the interferometer is a TCS, the ratio of the two sources of noise can be controlled by varying the nonlinear gain, thereby optimizing the interferometer sensitivity (Caves, 1981). The conservation of photon momentum in an ideal loss-free

---

[d] Lam *et al.* (1981).

[e] Steel and Lam (1979); Bloch *et al.* (1981).

[f] No evidence regarding this mechanism has been reported to date.

[g] Hellwarth (1977).

[h] Economou and Liao (1978); Steel *et al.* (1981a).

[i] Economou and Liao (1978); Yamada *et al.* (1981); Steel *et al.* (1981a).

[j] Khoo (1981).

[k] Jain and Lind (1982).

[l] Bloch (1980); Bloch *et al.* (1981); Steel *et al.* (1981a).

[m] Bloom *et al.* (1978b); Lam *et al.* (1981).

[n] Palmer (1979).

[o] Liao *et al.* (1977).

[p] Fujita *et al.* (1979); Steel *et al.* (1981b).

[q] Steel and Lam (1979); Steel *et al.* (1981b).

[r] Eichler (1977); Hamilton *et al.* (1979); Moss *et al.* (1981).

[s] Fu and Sargent (1979); Harter and Boyd (1980); Raj *et al.* (1980a,b); Woerdman and Schuurmans (1981); Steel and Lind (1981).

[t] Dunning and Lam (1981).

[u] Bloom *et al.* (1978b); Nilsen *et al.* (1981); Nilsen and Yariv (1979, 1981); Fu and Sargent (1980); Saikan and Wakata (1981); Steel and Lind (1981).

DFWM interaction may make it possible to minimize the radiation pressure errors by replacing the interferometer mirrors with PCMs (Pepper, 1981), subject to the Heisenberg uncertainty principle.

VI. Nonlinear Laser Spectroscopy

By probing the properties of the PCM itself, one can gain insight into the physics of the atomic or molecular species that constitute the nonlinear medium. Measurements of the FWM frequency response, angular sensitivity, or polarization dependence can yield information about linewidths, atomic motion, excitation diffusion, etc. Some of these spectroscopic studies to date are listed in Table II.

Finally, the spatial structure of the nonlinear susceptibilities can be studied using nonlinear microscopy by TWM (Hellwarth and Christensen, 1974, 1975) or FWM (Pepper *et al.*, 1978b).

VII. Conclusion

The emergence of nonlinear optical phase conjugation has added a new dimension to the field of nonlinear optics, which, up to this point, has been primarily to generate new wavelength sources, modulation devices, and image up conversion. Nonlinear optical phase conjugation appears to offer much promise in extending the field of quantum electronics applications. From the beginning, it held potential as a key element in optical aberration compensation systems. More recently, its applications to optical computing, communications, lensless imaging, laser fusion, image processing, temporal signal processing, low-noise detection schemes, and nonlinear laser spectroscopy appear quite promising.

The extension of the techniques discussed here to other regions of the electromagnetic spectrum may provide new classes of quantum electronic processors. Applications in the fields of radar, millimeter imaging, high-resolution photolithography, and molecular spectroscopy may benefit from real-time nonlinear field processing. Indeed, such real-time, all-optical interactions, which are free of electromechanical components, electronic feedback networks, and servo loops, can provide for large spatial and temporal bandwidths. Still in its infancy (the field is less than ten years old), nonlinear optical phase conjugation may open the door to a new and exciting generation of coherent optical information processors.

Acknowledgments

We wish to thank A. Yariv, G. C. Valley, R. C. Lind, and J. F. Lam for numerous fruitful technical discussions, and are grateful to Mrs. Carol O'Meara and Mrs. Eugenie Z. Anderson for their help with preparation of the manuscript.

References

Abrams, R. L. and Lind, R. C. (1977). Degenerate four-wave mixing in absorbing media, *Opt. Lett.* **2,** 94; and (1978) *Opt. Lett.* **3,** 205.

Agrawal, G. P. (1978). Phase determination by conjugate wave-front generation, *J. Opt. Soc. Am.* **68,** 1135.

Agrawal, G. P. and Flytzanis, C. (1981). Bistability and Hysteresis in Phase-Conjugated Reflectivity, *IEEE J. Quantum Electron.* **QE-17,** 374.

Anan'ev, Y. A. (1975). Possibility of Dynamic Correction of Wave Fronts, *Sov. J. Quantum Electron.* **4,** 929.

Andreev, N. F., Bespalov, V. I., Kiselev, A. M., Matveev, A. Z., and Pasmanik, G. A. (1980). Coherent Frequency Doubling in Nonlinear Nonhomogeneous Elements, *JETP Lett.* **30,** 286.

AuYeung, J. and Yariv, A. (1979). Phase Conjugate Optics, *Opt. News* Spring, p. 13.

AuYeung, J., Fekete, D., Pepper, D. M., and Yariv, A. (1979). A Theoretical and Experimental Investigation of the Modes of Optical Resonators with Phase-Conjugate Mirrors, *IEEE J. Quantum Electron.* **QE-15,** 1180.

Bar-Joseph, I., Hardy, A., Katzir, Y., and Silverberg, Y. (1981). Low-Power Phase-Conjugate Interferometry, *Opt. Lett.* **6,** 414.

Basov, N. (1978). "Principles of High-Power Laser Construction for Controlled Thermonuclear Fusion," Reprint No. 153. P. N. Lebedev Phys. Inst., Moscow.

Basov, N. and Zubarev, I. (1979). Powerful Laser Systems with Phase Conjugation by SMBS Mirror, *Appl. Phys.* **20,** 261.

Basov, N. G., Zubarev, I. G., Mironov, A. B., Mikhailov, S. I., and Okulov, A. Y. (1980). Laser Interferometer with Wavefront-Reversing Mirrors, *Sov. Phys.—JETP* **52,** 847.

Belanger, P. A., Hardy, A., and Siegman, A. E. (1980a). Resonant Modes of Optical Cavities with Phase Conjugate Mirrors: Higher-Order Modes, *Appl. Opt.* **19,** 479.

Belanger, P. A., Hardy, A., and Siegman, A. E. (1980b). Resonant Modes of Optical Cavities with Phase Conjugate Mirrors, *Appl. Opt.* **19,** 602.

Bel'dyugin, I. B., Galushkin, M. G., and Zemskov, E. M. (1979). Properties of Resonators with Wave-Front Reversing Mirrors, *Sov. J. Quantum Electron.* **9,** 20.

Blashchuk, V. N., Zel'dovich, B. Y., Mamaev, A. V., Pilipetsky, N. F., and Shkunov, V. V. (1980a). Full Wave Front Reversal of Depolarized Radiation in Degenerate Four Wave Mixing (Theory and Experiment), *Sov. J. Quantum Electron.* **10,** 356.

Blashchuk, V. N., Pilipetskii, N. F., and Shkunov, V. V. (1980b). Four-Wave Interaction as a Controllable Frequency Filter, *Sov. Phys.—Dokl.* **25,** 185.

Bloch, D. (1980). Ph.D. Thesis.

Bloch, D., Giacobino, E., and Ducloy, M. (1981). *Int. Conf. Laser Spectrosc., 5th, Jasper, Alberta.*

Bloom, D. M. and Bjorklund, G. C. (1977). Conjugate Wavefront Generation and Image Reconstruction by Four-Wave Mixing, *Appl. Phys. Lett.* **31,** 592.

Bloom, D. M., Shank, C. V., Fork, R. L., and Teschke, O. (1978a). Sub-Picosecond Optical Gating by Optical Wavefront Conjugation, *in* "Picosecond Phenomena" (C. V. Shank, E. P. Ippen, and S. L. Shapiro, eds.), p. 96. Springer-Verlag, Berlin and New York.

Bloom, D. M., Liao, P. F., and Economou, N. P. (1978b). Observation of Amplified Reflection by Degenerate Four-Wave Mixing in Atomic Sodium Vapor, *Opt. Lett.* **2,** 58.

Bogdan, A. R., Prior, Y., and Bloembergen, N. (1981). Pressure-Induced Degenerate Frequency Resonance in Four-Wave Light Mixing, *Opt. Lett.* **6,** 82.

Bol'shov, L. A., Vlasov, D. V., Dykhne, M. A., Korobkin, V. V., Saidov, K. S., and Starostin, A. N. (1980). On the Possibility of a Complete Compensation for Nonlinear Distortions of a Light Beam by Means of an Inversion of its Wavefront, *JETP Lett.* **31,** 286.

Born, M. and Wolf, E. (1975). "Principles of Optics," 5th ed. Pergamon, Oxford.

Burke, W. D., Staebler, D. L., Phillips, W., and Alphonse, G. A. (1978). Volume Phase Holographic Storage in Ferroelectrics, *Opt. Eng.* **17,** 308.

Caves, C. V. (1981). Quantum-Mechanical Noise in an Interferometer, *Phys. Rev. D* **23,** 1693.

Cederquist, J. (1981). Integral-Equation Solution Using Coherent Optical Feedback, *J. Opt. Soc. Am.* **71,** 651.

Cederquist, J. and Lee, S. H. (1981). Confocal Feedback Systems with Space Invariance, Time Sampling, and Secondary Feedback Loops, *J. Opt. Soc. Am.* **71,** 643.

Cronin-Golomb, M., Fischer, B., Nilsen, J., White, J. O., and Yariv, A. (1982). A Laser with Dynamic Holographic Intracavity Distortion Correction Capability, *Appl. Phys. Lett.* **41,** 219.

Depatie, D. and Haueisen, D. (1980). Multiline Phase Conjugation at 4 μm in Germanium, *Opt. Lett.* **5,** 252.

Diels, J. C. and McMichael, I. C. (1981). Influence of Wave-Front-Conjugated Coupling on the Operation of a Laser Gyro, *Opt. Lett.* **6,** 219.

Dunning, G. J. and Lind, R. C. (1982). Demonstration of Image Transmission through Fibers by Optical Phase Conjugation, *Opt. Lett.*, 558.

Dunning, G. J. and Lam, J. F. (1981). Optical phase conjugation in saturable absorbers, *CLEO* Pap. #FP2.

Economou, N. P. and Liao, P. F. (1978). Magnetic-Field Quantum Beats in Two-Photon Free-Induction Decay, *Opt. Lett.* **3,** 172.

Eichler, H. J. (1977). Laser-Induced Grating Phenomena, *Opt. Acta* **24,** 631.

Feinberg, J. (1980). Real-Time Edge Enhancement Using the Photorefractive Effect, *Opt. Lett.* **5,** 330.

Feinberg, J. and Hellwarth, R. W. (1980). Phase-Conjugating Mirror with Continuous Wave Gain, *Opt. Lett.* **5,** 519.

Feinberg, J., Heiman, D., Tanguay, A. R., Jr., and Hellwarth, R. W. (1980). Photorefractive Effects and Light-Induced Charge Migration in Barium Titanate, *J. Appl. Phys.* **51,** 1297.

Fienup, J. R. (1978). Reconstruction of an Object from the Modulus of its Fourier Transform, *Opt. Lett.* **3,** 27.

Fischer, B., Cronin-Golomb, M., White, J. O., and Yariv, A. (1982). Real-Time Phase Conjugate Window for One-Way Optical Imaging through a Distortion, *Appl. Phys. Lett.* **41,** 141.

Flytzanis, C. and Tang, C. L. (1980). Light-Induced Critical Behavior in the Four-Wave Interaction in Nonlinear Systems, *Phys. Rev. Lett.* **45,** 441.

Forward, R. L. (1978). Wideband Laser-Interferometer gravitational-radiation Experiment, *Phys. Rev. D* **17,** 379.

Fried, D. L. guest ed., (1977). Special Issue on Adaptive Optics, *J. Opt. Soc. Am.* **67.**

Fu, T. Y. and Sargent, M. III, (1979). Effects of Signal Detuning on Phase Conjugation, *Opt. Lett.* **4,** 366.

Fu, T. Y. and Sargent, M. III, (1980). Theory of Two-Photon Phase Conjugation, *Opt. Lett.* **5,** 433.

Fujita, M., Nakatsuka, H., Nakanishi, H., and Matsuoka, M. (1979). Backward Echo in Two-Level Systems, *Phys. Rev. Lett.* **42,** 974.

Giuliano, C. R. (1981). Applications of Optical Phase Conjugation, *Phys. Today* **34,** 27.

Glass, A. M. (1978) The photorefractive effect, *Opt. Eng.* **17,** 470.

Goodman, J. W. (1968). "Introduction to Fourier Optics," p. 87 and Chap. 7, p. 141. McGraw-Hill, New York.

Goodman, J. W., Huntley, W. H., Jr., Jackson, D. W., and Lehman, M. (1966). Wavefront Reconstruction Imaging through Random Media, *Appl. Phys. Lett.* **8,** 311.

Hamilton, D. S., Heiman, D., Feinberg, J., and Hellwarth, R. W. (1979). Spatial Diffusion Measurements in Impurity-Doped Solids by Degenerate Four-Wave Mixing, *Opt. Lett.* **4,** 124.

Hardy, A. (1981). Sensitivity of Phase-Conjugate Resonators to Intracavity Phase Perturbations, *IEEE J. Quantum Electron.* **QE-17,** 1581.

Harter, D. J. and Boyd, R. W. (1980). Nearly degenerate four-wave mixing enhanced by the ac stark effect, *IEEE J. Quantum Electron.* **QE-16,** 1126.

Hellwarth, R. W. (1977). Third Order Optical Susceptibilities of Liquids and Solids, *Prog. Quantum Electron.* **5,** 1.

Hellwarth, R. W. and Christensen, P. (1974). Nonlinear Optical Microscopic Examination of Structure in Polycrystalline ZnSe, *Opt. Commun.* **12,** 318.

Hellwarth, R. W. and Christensen, P. (1975). Nonlinear Optical Microscope Using SHG, *Appl. Opt.* **14,** 247.

Herriau, J. P., Huignard, J. P., and Auborg, P. (1978). Some Polarization Properties of Volume Holograms in $Bi_{12}SiO_{20}$ Crystals and Applications, *Appl. Opt.* **17,** 1851.

Hon, D. T. (1980a). High-Brightness Nd:YAG Laser Using SBS Phase Conjugation, *J. Opt. Soc. Am.* **70,** 635.

Hon, D. T. (1980b). Pulse Compression by Stimulated Brillouin Scattering, *Opt. Lett.* **5,** 516.

Hopf, F. A. (1980). Interferometry Using Conjugate-Wave Generation, *J. Opt. Soc. Am.* **70,** 1320.

Huignard, J. P. and Herriau, J. P. (1978). Real-Time Coherent Object Edge Reconstruction with $Bi_{12}SiO_{20}$ Crystals, *Appl. Opt.* **17,** 2671.

Huignard, J. P., Herriau, J. P., and Valentin, T. (1977). Time Average Holographic Interferometry with Photoconductive Electro-Optic $Bi_{12}SiO_{20}$ Crystals, *Appl. Opt.* **16,** 2796.

Huignard, J. P., Herriau, J. P., Auborg, P., and Spitz, E. (1979). Phase-Conjugate Wavefront Generation via Real-Time Holography in $Bi_{12}SiO_{20}$ Crystals, *Opt. Lett.* **4,** 21.

Huignard, J. P., Herriau, J. P., Pichon, L., and Marrakchi, A. (1980). Speckle-Free Imaging in Four-Wave Mixing Experiments with $Bi_{12}SiO_{20}$ Crystals, *Opt. Lett.* **5,** 436.

Humphrey, L. M., Gordon, J. P., and Liao, P. F. (1980). Angular Dependence of Line Shape and Strength of Degenerate Four-Wave Mixing in a Doppler-Broadened System with Optical Pumping, *Opt. Lett.* **5,** 56.

Ilyukhin, A. A., Peregudov, G. V., Plotkin, M. E., Ragozin, E. N., and Chirkov, V. A. (1979). Focusing of a Laser Beam on a Target Using the Effect of Wavefront Inversion

(WFI) Produced as a Result of Stimulated Mandel'shtam–Brillouin Scattering (SMBS), *JETP Lett.* **29,** 328.

Ippen, E. P. and Shank, C. V. (1977). Techniques of Measurement, *in* "Ultrashort Light Pulses" (S. L. Shapiro, ed.), p. 83. Springer-Verlag, Berlin and New York.

Ivakhnik, V. V., Petnikova, V. M., Solomatin, V. S., Kharchenko, M. A., and Shuvalov, V. V. (1980). Single-Pass Systems for Compensation of Phase Distortions, *Sov. J. Quantum Electron.* **10,** 514.

Ivakhnik, V. V., Petnikova, V. M., and Shuvalov, V. V. (1981). Enhancement of the Efficiency of Wavefront Reversal Systems Using Ring Resonators, *Sov. J. Quantum Electron.* **11,** 275.

Jacobs, S. F. (1982). Experiments with Retrodirective Arrays, *Opt. Eng.* **21,** 281.

Jain, R. K. and Lind, R. C. (1983). Degenerate four-wave mixing in semiconductor-doped glass, *J. Opt. Soc. Am.*, to be published, Special Issue on Phase Conjugation.

Kawata, S. and Ichioka, Y. (1980). Iterative Image Restoration for Linearly Degraded Images. I. Basis, *J. Opt. Soc. Am.* **70,** 762.

Khoo, I. C. (1981). Degenerate Four-Wave Mixing in the Nematic Phase of a Liquid Crystal, *Appl. Phys. Lett.* **38,** 123.

Kogelnik, H. (1965). Holographic Image Projection Through Inhomogeneous Media, *Bell Syst. Tech. J.* **44,** 2451.

Kruzhilin, Y. I. (1978). Self-Adjusting Laser-Target System for Laser Fusion, *Sov. J. Quantum Electron.* **8,** 359.

Lam, J. F. and Brown, W. P. (1980). Optical Resonators with Phase Conjugate Mirrors, *Opt. Lett.* **5,** 61.

Lam, J. F., Steel, D. G., McFarlane, R. A., and Lind, R. C. (1981). Atomic Coherence Effects in Resonant Degenerate Four-Wave Mixing, *Appl. Phys. Lett.* **38,** 977.

Lee, S. H. (1974). Mathematical Operations by Optical Pumping, *Opt. Eng.* **13,** 196.

Levenson, M. D. (1980). High-Resolution Imaging by Wave-Front Conjugation, *Opt. Lett.* **5,** 182.

Levenson, M. D., Johnson, K. M., Hanchett, V. C., and Chaing, K. (1981). Projection Photolithography by Wave-Front Conjugation, *J. Opt. Soc. Am.* **71,** 737.

Liao, P. F., Economou, N. P., and Freeman, R. R. (1977). Two-Photon Coherent Transient Measurements of Doppler-Free Linewidths with Broadband Excitation, *Phys. Rev. Lett.* **39,** 1473.

Lind, R. C. and Steel, D. G. (1981). Demonstration of the Longitudinal Modes and Aberration-Correction Properties of a Continuous-Wave Dye Laser with a Phase-Conjugate Mirror, *Opt. Lett.* **6,** 554.

Loudon, R. (1978). "Quantum Theory of Light," Chap. 7. Oxford Univ. Press (Clarendon), London and New York.

Lukosz, W. (1968). Equivalent-Lens Theory of Holographic Imaging, *J. Opt. Soc. Am.* **58,** 1084.

Marburger, J. H. (1978). Optical Pulse Integration and Chirp Reversal in Degenerate Four-Wave Mixing, *Appl. Phys. Lett.* **32,** 372.

Marks, R. J., II, (1980). Coherent Optical Extrapolation of 2D Band-Limited Signals: Processor Theory, *Appl. Opt.* **19,** 1670.

Marrakchi, A., Huignard, J. P., and Herriau, J. P. (1980). Application of Phase Conjugation in $Bi_{12}SiO_{20}$ Crystals to Mode Pattern Visualization of Diffuse Vibrating Structures, *Opt. Commun.* **34,** 15.

Martin, G., Lam, L. K., and Hellwarth, R. W. (1980). Generation of a Time-Reversed Replica of a Nonuniformly Polarized Image-Bearing Optical Beam, *Opt. Lett.* **5,** 185.

Mathieu, P. and Belanger, P. A. (1980). Retroreflective Array as a Resonator Mirror, *Appl. Opt.* **19,** 2262.

Miller, D. A. B. (1980). Time Reversal of Optical Pulses by Four-Wave Mixing, *Opt. Lett.* **5,** 300.

Miridinov, S. V., Petrov, M. P., and Stepanov, S. I. (1978). Light Diffraction by Volume Holograms in Optically Active Photorefractive Crystals, *Sov. Tech. Phys. Lett.* **4,** 393.

Moss, S. C., Lindle, J. R., Mackey, H. J., and Smirl, A. L. (1981). Measurement of the Diffusion Coefficient and Recombination Effects in Germanium by Diffraction from Optically-Induced Picosecond Transient Gratings, *Appl. Phys. Lett.* **39,** 227.

Nilsen, J. and Yariv, A. (1979). Nearly Degenerate Four-Wave Mixing Applied to Optical Filters, *Appl. Opt.* **18,** 143.

Nilsen, J. and Yariv, A. (1981). Nondegenerate Four-Wave Mixing in a Doppler-Broadened Resonant Medium, *J. Opt. Soc. Am.* **71,** 180.

Nilsen, J. Gluck, N. S., and Yariv, A. (1981). Narrow-Band Optical Filter Through Phase Conjugation by Nondegenerate Four-Wave Mixing in Sodium Vapor, *Opt. Lett.* **6,** 380.

Odoulov, S., Soskin, M., and Vasnetsov, M. (1980a). Compensation for Time Dependent Phase Inhomogeneity via Degenerate Four-Wave Mixing in $LiTaO_3$, *Opt. Commun.* **32,** 355.

Odoulov, S. G., Soskin, M. S., and Vasnetsov, M. (1980b). Correlation Analysis of Images Under Degenerate Four-Wave Mixing in Colliding Beams, *Sov. Phys.—Dokl.* **25,** 380.

O'Meara, T. R. (1982a). Applications of nonlinear phase conjugation in compensated active imaging, *Opt. Eng.* **21,** 231.

O'Meara, T. R. (1982b). Compensation of Laser Amplifier Trains with Nonlinear Conjugation Techniques, *Opt. Eng.* **21,** 243.

O'Meara, T. R. (1982c). Digital Logic Processing with Nonlinear Optics Operating on Polarization States, to be published.

O'Meara, T. R. and Yariv, A. (1982). Time-Domain Signal Processing via Four-Wave Mixing in Nonlinear Delay Lines, *Opt. Eng.* **21,** 237.

Orlov, V. K., Virnik, Y. Z., Vorotilin, S. P., Gerasinov, V. B., Kalimin, Y. A., and Sagalovish, A. Y. (1978). Retroreflecting Mirror for Dynamic Compensation of Optical Inhomogeneities, *Sov. J. Quantum Electron.* **8,** 799.

Palmer, A. J. (1979). Nonlinear Optics in Radiatively Cooled Vapors, *Opt. Commun.* **30,** 104.

Pepper, D. M. Phase Conjugate Optics: On the Theory, Observation, and Utilization of Temporally-Reversed Wavefronts as Generated via Nonlinear Optical Parametric Interactions, Ph.D. Thesis, California Inst. Technol. Pasadena, 1980 (available through Univ. Microfilms No. 80-14 305).

Pepper, D. M. (1981). Unpublished observations.

Pepper, D. M. (1982) guest ed., Special Issue on Nonlinear Optical Phase Conjugation, *Opt. Eng.* **21,** No. 2.

Pepper, D. M. and Abrams, R. L. (1978). Narrow Optical Bandpass Filter via Nearly Degenerate Four-Wave Mixing, *Opt. Lett.* **3,** 212.

Pepper, D. M. and Yariv, A. (1980). Compensation for Phase Distortions in Nonlinear Media by Phase Conjugation, *Opt. Lett.* **5,** 59.

Pepper, D. M., Fekete, D., and Yariv, A. (1978a). Observation of Amplified Phase-Conjugate Reflection and Optical Parametric Oscillation by Degenerate Four-Wave Mixing in a Transparent Medium, *Appl. Phys. Lett.* **33,** 41.

Pepper, D. M., AuYeung, J., Fekete, D., and Yariv, A. (1978b). Spatial Convolution and Correlation of Optical Fields via Degenerate Four-Wave Mixing, *Opt. Lett.* **3,** 7.

Peregudov, C. V., Plotkin, M. E., and Ragozin, E. N. (1979). Use of the Wave-Front Reversal Effect to Investigate a Jet Formed by Focusing Laser Radiation on a Plane Target, *Sov. J. Quantum Electron.* **9,** 1413.

Petrov, M. P., Miridonov, S. V., Stepanov, S. I., and Kulikov, V. V. (1979). Light Diffraction and Nonlinear Image Processing in Electro-Optic $Bi_{12}SiO_{20}$ Crystals, *Opt. Commun.* **31,** 301.

Ragozin, E. N. and Plotkin, M. E. (1980). Multiple Use of Wavefront Reversal in Laser Systems, *Sov. J. Quantum Electron.* **10,** 915.

Raj, R. K., Bloch, D., Snyder, J. J., Carney, G., and Ducloy, M. (1980a). High-Sensitivity Nonlinear Spectroscopy Using a Frequency-Offset Pump, *Opt. Lett.* **5,** 163, 326.

Raj, R. K., Bloch, D., Snyder, J. J., Carney, G., and Ducloy, M. (1980b). High-Frequency Optically Heterodyned Saturation Spectroscopy via Resonant Degenerate Four-Wave Mixing, *Phys. Rev. Lett.* **44,** 1251.

Reznikov, M. G. and Khizhnyak, A. I. (1980). Properties of a Resonator with a Wavefront-Reversing Mirror, *Sov. J. Quantum Electron.* **10,** 633.

Saikan, S. and Wakata, H. (1981). Configuration Dependence of Optical Filtering Characteristics in Backward Nearly Degenerate Four-Wave Mixing, *Opt. Lett.* **6,** 281.

Schubert, N. and Vogel, W. (1981). Quantum Statistical Properties of the Radiation Field in the Degenerate Two-Photon Emission Process, *Opt. Commun.* **36,** 164.

Shapiro, J. H. (1980). Optical Waveguide Tap with Minimal Insertion Loss, *Opt. Lett.* **5,** 351 and references therein.

Siegman, A. E. (1979). Dynamic Interferometry and Differential Holography of Irregular Phase Objects Using Phase Conjugate Reflection, *Opt. Commun.* **31,** 257.

Smith, P. W. (1981) guest ed., Special Issue on Optical Bistability, *IEEE J. Quantum Electron.* **QE-17**.

Stark, H., Cahana, D., and Webb, H. (1981). Restoration of Arbitrary Finite-Energy Optical Objects from Limited Spatial and Spectral Information, *J. Opt. Soc. Am.* **71,** 635.

Steel, D. G. and Lam, J. F. (1979). Two-Photon Coherent-Transient Measurement of the Nonradiative Collisionless Dephasing Rate in SF_6 via Doppler-Free Degenerate Four-Wave Mixing, *Phys. Rev. Lett.* **43,** 1588.

Steel, D. G. and Lam, J. F. (1980). Multiline Phase Conjugation in Resonant Materials, *Opt. Lett.* **5,** 297.

Steel, D. G. and Lind, R. C. (1981). Multiresonant Behavior in Nearly Degenerate Four-Wave Mixing: The AC Stark Effect, *Opt. Lett.* **6,** 587.

Steel, D. G., Lind, R. C., Lam, J. F., and Giuliano, C. R. (1979). Polarization–Rotation and Thermal-Motion Studies via Resonant Degenerate Four-Wave Mixing, *Appl. Phys. Lett.* **35,** 376.

Steel, D. G., Lam, J. F., and McFarlane, R. A. (1981a). *Int. Conf. Laser Spectrosc., 5th, Jasper, Alberta.*

Steel, D. G., Lind, R. C., and Lam, J. F. (1981b). Degenerate Four-Wave Mixing in a Resonant Homogeneously Broadened System, *Phys. Rev. A* **23,** 2513.

Szu, H. H. (1980). Foundation of Single Frame Image Processing, *ICO Conf., Opt. Four Dimens., Ensenada, Mex.*

Valley, G. C. and Fink, D. (1983). Three dimensional phase conjugate resonator performance I; Ideal conjugator *J. Opt. Soc. Am.* **73,** to be published.

Vanherzeele, H. and Van Eck, J. L. (1981). Pulse Compression by Intracavity Degenerate Four-Wave Mixing, *Appl. Opt.* **20,** 524.

Vanherzeele, H., Van Eck, J. L., and Siegman, A. E. (1981). Mode-Locked Laser Oscillation Using Self-Pumped Phase-Conjugate Reflection, *Opt. Lett.* **6,** 467.

Wandzura, S. M. (1979). Effects of Atomic Motion on Wavefront Conjugation by Resonantly Enhanced Degenerate Four-Wave Mixing, *Opt. Lett.* **4,** 208.

Wang, V. (1978). Nonlinear Optical Phase Conjugation for Laser Systems, *Opt. Eng.* **17,** 267.

White, J. O. and Yariv, A. (1980). Real-Time Image Processing via Four-Wave Mixing in a Photorefractive Medium, *Appl. Phys. Lett.* **37,** 5.

White, J. O., Cronin-Golomb, M., Fischer, B., and Yariv, A. (1982). Coherent Oscillation by Self-Induced Gratings in the Photorefractive Crystal $BaTiO_3$, *Appl. Phys. Lett.* **40,** 450.

Winful, H. G. and Marburger, J. H. (1980). Hysteresis and Optical Bistability in Degenerate Four-Wave Mixing, *Appl. Phys. Lett.* **36,** 613.

Woerdman J. P. and Schuurmans, M. F. H. (1981). Effect of Saturation on the Spectrum of Degenerate Four-Wave Mixing in Atomic Sodium Vapor, *Opt. Lett.* **6,** 239.

Yamada, K., Fukuda, Y., and Haski, T. (1981). Time-Development of the Population Grating in Zeeman Sublevels in Sodium Vapor—Detection of Zeeman Quantum Beats, *J. Phys. Soc. Jpn.* **50,** 592.

Yariv, A. (1976a). Three-Dimensional Pictorial Transmission in Optical Fibers, *Appl. Phys. Lett.* **28,** 88.

Yariv, A. (1976b). On Transmission and Recovery of Three-Dimensional Image Information in Optical Waveguides, *J. Opt. Soc. Am.* **66,** 301.

Yariv, A. (1977). Compensation for Atmospheric Degradation of Optical Beam Transmission, *Opt. Commun.* **21,** 49.

Yariv, A. (1979). Phase Conjugate Optics and Real-Time Holography, *IEEE J. Quantum Electron.* **QE-14,** 650 (1978); **QE-15,** 256, 523.

Yariv, A. and Koch, T. L. (1982). One-Way Coherent Imaging through a Distorting Medium Using Four-Wave Mixing, *Opt. Lett.* **7,** 113.

Yariv, A., Fekete, D., and Pepper, D. M. (1979). Compensation for Channel Dispersion by Nonlinear Optical Phase Conjugation, *Opt. Lett.* **4,** 52.

Yuen, H. P. (1976). Two-Photon Coherent States of the Radiation Field, *Phys. Rev. A* **13,** 2226.

Yuen, H. P. and Shapiro, J. H. (1979). Generation and Detection of Two-Photon Coherent States in Degenerate Four-Wave Mixing, *Opt. Lett.* **4,** 334.

Yuen, H. P. and Shapiro, J. H. (1980). Optical Communication with Two-Photon Coherent States. Part III: Quantum Measurements Realizable with Photoemissive Detectors, *IEEE Trans. Inf. Theory* **IT-26,** 78.

Zel'dovich, B. Y. and Shkunov, V. V. (1979). Spatial-Polarization Wavefront Reversal in Four-Photon Interaction, *Sov. J. Quantum Electron.* **9,** 379.

Optical Phase Conjugation Bibliography

Compiled by
Robert A. Fisher
Los Alamos National Laboratory
Los Alamos, New Mexico

R. L. Abrams and R. C. Lind, Degenerate Four-Wave Mixing in Absorbing Media, *Opt. Lett.* **2,** 94 (1978); Erratum, *Opt. Lett.* **3,** 205 (1978).

R. L. Abrams, C. R. Guiliano, and J. F. Lam, On the Equality of Stimulated Brillouin Scattering Reflectivity to Conjugate Reflectivity of a Weak Probe Beam, *Opt. Lett.* **6,** 131 (1981).

G. P. Agrawal, Optical Bistability in a Phase-Conjugate Fabry–Perot Cavity, *Opt. Commun.* **37,** 366 (1981).

G. P. Agrawal, Phase Conjugation and Degenerate Four-Wave Mixing in Three-Level Systems, *IEEE J. Quantum Electron.* **QE-17,** 2335 (1981).

G. P. Agrawal, Phase Conjugation through Two-Photon Resonant Nondegenerate Four-Wave Mixing, *Opt. Commun.* **39,** 272 (1981).

G. P. Agrawal, Phase Conjugation in Biharmonically Pumped Two-Photon Resonant Systems, *Opt. Commun.* **42,** 366 (1982).

G. P. Agrawal, and J.-L. Boulnois, Waveguide Resonators with a Phase-Conjugate Mirror, *Opt. Lett.* **7,** 159 (1982).

G. P. Agrawal and C. Flytzanis, Bistability and Hysteresis in Phase-Conjugated Reflectivity, *IEEE J. Quantum Electron.* **QE-17,** 374 (1981).

G. P. Agrawal, C. Flytzanis, R. Frey, and F. Pradere, Bistable Reflectivity of Phase-Conjugated Signal through Intracavity Degenerate Four-Wave Mixing, *Appl. Phys. Lett.* **38,** 492 (1981).

G. S. Agrawal, Dipole Radiation in the Presence of a Phase Conjugate Mirror *Opt. Commun.* **42,** 205 (1982).

G. S. Agrawal and E. Wolf, Theory of Phase Conjugation with Weak Scatterers, *J. Opt. Soc. Am.* **72,** 321 (1982).

G. S. Agrawal, A. T. Friberg, and E. Wolf, Effect of Backscattering in Phase Conjugation with Weak Scatterers, *J. Opt. Soc. Am.* **72,** 861 (1982).

M. H. Ahmed, The Effect of Strong Wave Front Distribution in Optical Wave Conjugation by Three-Photon Mixing, *IEEE J. Quantum Electron.* **QE-15,** 870 (1979).

Y. A. Anan'ev, On the Possibility of Dynamic Correction of Wave Fronts, *Kvantovaya Elektron.* **1,** 1669 (1974); *Sov. J. Quantum Electron.* (*Engl. Transl.*) **4,** 929 (1975).

Y. A. Anan'ev, A. V. Gorlanov, N. I. Grishmanova, N. A. Sventsis-

Optical Phase Conjugation

ISBN 0-12-257740-X

kaya, and V. D. Solov'ev, Transient Self-Diffraction of Coherent Light Beams in an Absorbing Liquid, *Kvantovaya Elektron.* **6,** 1813 (1979); *Sov. J. Quantum Electron.* (*Engl. Transl.*) **9,** 1072 (1979).

N. F. Andreev, V. I. Bespalov, A. M. Kiselev, A. Z. Matveev, and G. A. Pasmanik, Coherent Frequency Doubling in Nonlinear Nonhmogeneous Elements, *Pis'ma Zh. Eksp. Teor. Fiz.* **30,** 308 (1979); *JETP Lett.* (*Engl. Transl.*) **30,** 286 (1979).

N. F. Andreev, V. I. Bespalov, A. M. Kiselev, A. Z. Matveev, G. A. Pasmanik, and A. A. Shilov, Wave-Front Reversal of Weak Optical Signals with Large Reflection Coefficients, *Pis'ma Zh. Eksp. Teor. Fiz.* **32,** 639 (1980).

J. R. Andrews and R. Hochstrasser, Transient Grating Effects in Resonant Four-Wave Mixing Experiments, *Chem. Phys. Lett.* **76,** 213 (1980).

P. A. Apanasevich, A. A. Afanas'ev, and A. I. Urbanovich, Mechanism of Diffraction of Light by Optically Induced Gratings in Absorbing Media, *Kvantovaya Electron.* **2,** 2423 (1975); *Sov. J. Quantum Electron.* (*Engl. Transl.*) **5,** 1320 (1976).

P. A. Apanasevich, A. A. Afanas'ev, and S. P. Zhvavyi, Efficiency of Wavefront Reversal of Optical Beams in Four-Wave Interactions in a Resonant Medium, *Kvantovaya Elektron.* **7,** 1572 (1980); *Sov. J. Quantum Electron.* (*Engl. Transl.*) **10,** 906 (1980).

P. Aubourg, J. P. Bettini, G. P. Agrawal, P. Cottin, D. Guerin, O. Meunier, and J. L. Boulnois, Doppler-Free Continuous Wave Phase-Conjugate Spectrum of SF by Resonant Degenerate Four-Wave Mixing at 10.6 μm, *Opt. Lett.* **6,** 383 (1981).

J. AuYeung and A. Yariv, Phase-Conjugate Optics, *Opt. News* Spring, p. 13 (1979).

J. AuYeung and A. Yariv, Phase Conjugate Optics, *in* "Laser Spectroscopy IV" (H. Walther and K. W. Rothe, eds.), p. 492. Springer-Verlag, Berlin and New York, 1979.

J. AuYeung, D. Fekete, D. M. Pepper, and A. Yariv, A Theoretical and Experimental Investigation of the Modes of Optical Resonators with Phase-Conjugate Mirrors, *IEEE J. Quantum Electron.* **QE-15,** 1180 (1979).

J. AuYeung, D. Fekete, D. M. Pepper, A. Yariv, and R. K. Jain, Continuous Backward-Wave Generation by Degenerate Four-Wave Mixing in Optical Fibers, *Opt. Lett.* **4,** 42 (1979).

P. Avizonis, F. A. Hopf, W. D. Bomberger, S. F. Jacobs, A. Tomita, and K. H. Womack, Optical Phase Conjugation in a Lithium Formate Crystal, *Appl. Phys. Lett.* **31,** 435 (1977).

N. B. Baranova and B. Y. Zel'dovich, Transverse Enhancement of Coherence of the Scattered Field in Wavefront Reversal, *Kvantovaya Elektron.* **7,** 299 (1980); *Sov. J. Quantum Electron.* (*Engl. Trans.*) **10,** 172 (1980).

N. B. Baranova and B. Y. Zel'dovich, Wavefront Reversal of Focused Beams (Theory of Stimulated Brillouin Backscattering, *Kvantovaya Elektron.* **7,** 973 (1980); *Sov. J. Quantum Electron.* (*Engl. Transl.*) **10,** 555 (1980).

N. B. Baranova, B. Y. Zel'dovich, and V. V. Shkunov, Wavefront Reversal in Stimulated Light Scattering in a Focused Spatially Inhomogeneous Pump Beam, *Kvantovaya Elektron.* **5,** 973 (1978); *Sov. J. Quantum Electron.* (*Engl. Transl.*) **8,** 559 (1978).

I. Bar-Joseph, A. Hardy, Y. Katzir, and Y. Silberberg, Low-Power Phase-Conjugate Interferometry, *Opt. Lett.* **6,** 414 (1981).

H. H. Barrett and S. F. Jacobs, Retroreflective Arrays as Approximate Phase Conjugators, *Opt. Lett.* **4,** 190 (1979).

N. G. Basov, "Principles of High-Power Laser Construction for Controlled Thermonuclear Fusion," Prepr. No. 153. P. N. Lebedev Phys. Inst., Moscow, 1978.

N. Basov and I. Zubarev, Powerful Laser Systems with Phase Conjugation by SMBS Mirror, *Dig. Rep. Natl. Conf. Coherent Nonlinear Opt., 9th, Leningrad* (1978).

N. G. Basov and I. G. Zubarev, Powerful Laser Systems with Phase Conjugation by SMBS Mirror, *Appl. Phys.* **20,** 261 (1979).

N. G. Basov, V. F. Efimkov, I. G. Zubarev, A. V. Kotov, S. I. Mikhailov, and M. G. Smirnov, Inversion of Wavefront of SMBS of Depoparized Pump, *Pis'ma Zh. Eksp. Teor. Fiz.* **28,** 215 (1978); *JETP Lett.* (*Engl. Lett.*) **28,** 197 (1978).

N. G. Basov, V. F. Efimkov, I. G. Zubarev, A. V. Kotov, A. B. Mironov, S. I. Mikhailov, and M. G. Smirnov, Influence of Certain Radiation Parameters on Wavefront Reversal in a Brillouin Mirror, *Kvantovaya Elektron.* **6,** 765 (1979); *Sov. J. Quantum Electron.* (*Engl. Transl.*) **9,** 455 (1979).

N. G. Basov, I. G. Zubarev, A. V. Kotov, S. I. Mikhailov, and M. G. Smirnov, Small-Signal Wavefront Reversal in Nonthreshold Reflection from a Brillouin Mirror, *Kvantovaya Elektron.* **6,** 394 (1979); *Sov. J. Quantum Electron.* (*Engl. Transl.*) **9,** 237 (1979).

N. G. Basov, I. G. Zubarev, A. B. Mironov, S. I. Mikhailov, and A. Y. Okulov, Phase Fluctuations of the Stokes Wave Produced as a Result of Stimulated Scattering of Light, *Pis'ma Zh. Eksp. Teor. Fiz.* **31,** 685 (1980); *JETP Lett.* (*Engl. Transl.*) **31,** 645 (1980).

N. G. Basov, I. G. Zubarev, A. B. Mironov, S. I. Mikhailov, and

A. Y. Okulov, Laser Interferometer with Wavefront-Reversing Mirrors, *Zh. Eksp. Teor. Fiz.* **79,** 1678 (1980); *Sov. Phys.–JETP (Engl. Transl.)* **52,** 847 (1980).

N. G. Basov, B. Ya. Zel'dovich, V. I. Kovalev, F. S. Faizullov, and V. B. Federov, Reflection of Multifrequency Signal in Four-Wave Interaction in Germanium at 10.6 μ, *Kvantovaya Elektron.* **8** 860 (1981); *Sov. J. Quantum Electron. (Engl. Transl.)* **11,** 514 (1981).

N. G. Basov, V. F. Efimkov, I. G. Zubarev, A. V. Kotov, and S. I. Mikhailov, Control of the Characteristics of Reversing Mirrors in the Amplification Regime, *Kvantovaya Elektron.* **8,** 2191 (1981); *Sov. J. Quantum Electron. (Engl. Transl.)* **11,** 1335 (1981).

P. A. Belanger, Phase Conjugation and Optical Resonators, *Opt. Eng.* **21,** 266 (1982).

P. L. Belanger, A. Hardy, and A. E. Siegman, Resonant Modes of Optical Cavities with Phase Conjugate Mirrors: Higher-Order Modes, *Appl. Opt.* **19,** 479 (1980).

P. A. Belanger, A. Hardy, and A. E. Siegman, Resonant Modes of Optical Cavities with Phase Conjugate Mirrors, *Appl. Opt.* **19,** 602 (1980).

I. M. Bel'dyugin and E. M. Zemskov, Influence of Changes in the Pump Field on the Form of the Field of a Signal Amplified under Stimulated Scattering Conditions, *Kvantovaya Elektron.* **5,** 2055 (1978); *Sov. J. Quantum Electron. (Engl. Transl.)* **8,** 1163 (1978).

I. M. Bel'dyugin and E. M. Zemskov, Theory of Resonators with Wavefront-Reversing Mirrors, *Kvantovaya Elektron.* **6,** 2036 (1979); *Sov. J. Quantum Electron. (Engl. Transl.)* **9,** 1198 (1979).

I. M. Bel'dyugin and E. M. Zemskov, Calculation of the Field in a Laser Resonator with a Wavefront-Reversing Mirror, *Kvantovaya Elektron.* **7,** 1334 (1980); *Sov. J. Quantum Electron. (Engl. Transl.)*, **10,** 764 (1980).

I. M. Bel'dyugin, M. G. Galushkin, E. M. Zemskov, and V. I. Mandrosov, Complex Conjugation of Fields in Stimulated Brillouin Scattering, *Kvantovaya Elektron.* **3,** 2467 (1976); *Sov. J. Quantum Electron. (Engl. Transl.)* **6,** 1349 (1976).

I. M. Bel'dyugin, E. M. Zemskov, and V. I. Chernen'kii, Theory of Amplification of the First Stokes Component in a Nonmonochromatic Pump Field under Stimulated Raman Scattering Conditions, *Kvantovaya Elektron.* **5,** 1349 (1978); *Sov. J. Quantum Electron. (Engl. Transl.)* **8,** 769 (1978).

I. M. Bel'dyugin, V. N. Seminogov, and E. M. Zemskov, Possible Wavefront Reversal of Fields Using Nonlinear-Optics Methods,

Kvantovaya Elektron. **6,** 638 (1979); *Sov. J. Quantum Electron.* (*Engl. Transl.*) **9,** 385 (1979).

I. M. Bel'dyugin, M. G. Galushkin, and E. M. Zemskov, Properties of Resonators with Wavefront-Reversing Mirrors, *Kvantovaya Elektron.* **6,** 38 (1979); *Sov. J. Quantum Electron.* (*Engl. Transl.*) **9,** 20 (1979).

I. M. Bel'dyugin, M. G. Galushkin, and E. M. Zemskov, Stimulated Scattering of Nonmonochromatic Spatially Inhomogeneous Radiation, *Kvantovaya Elektron.* **6,** 587 (1979); *Sov. J. Quantum Electron.* (*Engl. Transl.*) **9,** 348 (1979).

I. M. Bel'dyugin, E. M. Zemskov, and V. N. Klushin, Problem of Wavefront Reversal by Stimulated Brillouin Scattering, *Kvantovaya Elektron.* **6,** 2039 (1979); *Sov. J. Quantum Electron.* (*Engl. Transl.*) **9,** 1200 (1979).

I. M. Bel'dyugin, E. M. Zemskov, V. N. Klushin, and V. N. Seminogov, Wavefront Reversal Based on Anti-Stokes Stimulated Raman Scattering, *Kvantovaya Elektron.* **6,** 2481 (1979); *Sov. J. Quantum Electron.* (*Engl. Transl.*) **9,** 1469 (1979).

I. M. Bel'dyugin, I. G. Zubarev, and S. I. Mikhailov, Analysis of Conditions for Stimulated Raman Scattering of Multimode Pump Radiation in Dispersive Media, *Kvantovaya Elektron.* **7,** 1471 (1980); *Sov. J. Quantum Electron.* (*Engl. Transl.*) **10,** 847 (1980).

V. N. Belousov, L. A. Bolshov, N. G. Kovalskii, and Y. K. Nizienko, Experimental Study of the Properties of Mandelstam–Brillouin Stimulated Scattering and Stimulated Temperature Scattering Mirrors, *Zh. Eksp. Teor. Fiz.* **79,** 2119 (1980).

E. E. Bergmann, I. J. Bigio, B. J. Feldman, and R. A. Fisher, High-Efficiency Pulsed 10.6 μM Phase-Conjugate Reflection via Degenerate Four-Wave Mixing, *Opt. Lett.* **3,** 82 (1978).

V. I. Bespalov, A. A. Betin, and G. A. Pasmanik, Reconstruction Effects in Stimulated Scattering, *Pis'ma. Zh. Tekh. Fiz.* **3,** 215 (1977); *Sov. Tech. Phys. Lett.* (*Engl. Transl.*) **3,** 85 (1977).

V. I. Bespalov, A. A. Betin, and G. A. Pasmanik, Experimental Investigation of the Threshold of the Stimulated Scattering (SS) of Multimode Light Beams and the Degree of Regeneration of the Pumping in Scattered Radiation, *Izv. Vyssh. Uchebn. Zaved. Radiofiz.* **20,** 791 (1977); *Radiophys. Quantum Electron.* (*Engl. Transl.*) **20,** 544 (1977).

V. I. Bespalov, A. A. Betin, and G. A. Pasmanik, Reproduction of the Pumping Wave in Stimulated-Scattering Radiation, *Izv. Vyssh. Uchebn. Zaved. Radiofiz.* **21,** 961 (1978); *Radiophys. Quantum Electron.* (*Engl. Transl.*) **21,** 675 (1979).

V. I. Bespalov, A. A. Betin, G. A. Pasmanik, and A. A. Shilov, Wave-front Inversion in the Raman Conversion of the Stokes Wave in Oppositely Directed Pump Beams, *Pis'ma Zh. Tekh. Fiz.* **5,** 242 (1979); *Sov. Tech. Phys. Lett.* (*Engl. Transl.*) **5,** 97 (1979).

V. I. Bespalov, V. G. Manishin, and G. A. Pasmanik, Nonlinear Selection of Optical Radiation on Reflection from a Stimulated Mandel'shtam–Brillouin Scattering Mirror, *Zh. Eksp. Teor. Fiz.* **77,** 1756 (1979); *Sov. Phys.—JETP* (*Engl. Transl.*) **50,** 879 (1979).

V. I. Bespalov, A. A. Betin, A. I. Dyatlov, S. N. Kulagina, V. G. Manishin, G. A. Pasmanik, and A. A. Shilov, Reversal of Wave Front in Four-Photon Processes under Conditions of Two-Quantum Resonance, *Zh. Eksp. Teor. Fiz.* **79,** 378 (1980); *Sov. Phys.—JETP* (*Engl. Transl.*) **52,** 190 (1980).

V. I. Bespalov, A. A. Betin, G. A. Pasmanik, and A. A. Shilov, Observation of Transient Field Oscillations in the Radiation of Stimulated Mandel'shtam–Brillouin Scattering, *Pis'ma Zh. Eksp. Teor. Fiz.* **31,** 668 (1980); *JETP Lett.* (*Engl. Transl.*) **31,** 630 (1980).

V. I. Bezrodnyi, F. I. Ibragimov, V. I. Kislenko, R. A. Petrenko, V. L. Strizhevskii, and E. A. Tikhonov, Mechanism of Laser Q Switching by Intracavity Stimulated Scattering, *Kvantovaya Elektron.* **7,** 664 (1980); *Sov. J. Quantum Electron.* (*Engl. Transl.*) **10,** 382 (1980).

I. J. Bigio, B. J. Feldman, R. A. Fisher, and E. E. Bergmann, High-Efficiency Phase-Conjugate Reflection in Germanium and in Inverted CO_2, *Proc. Int. Conf. Lasers, 1978* (R. V. Corcoran, ed.), p. 531. STS Press, McLean, Virginia, 1979.

I. J. Bigio, B. J. Feldman, R. A. Fisher, and E. E. Bergmann, High-Efficiency Wavefront Reversal in Germanium and in Inverted CO_2 (Review), *Kvantovaya Elektron.* **6,** 2318 (1979); *Sov. J. Quantum Electron.* (*Engl. Transl.*) **9,** 1365 (1979).

V. N. Blashchuk, B. Y. Zel'dovich, N. A. Mel'nikov, N. F. Pipipetskii, V. I. Popovichev, and V. V. Ragul'skii, Wave-Front Inversion in Stimulated Scattering of Focused Light Beams, *Pis'ma Zh. Tekh. Fiz.* **3,** 211 (1977); *Sov. Tech. Phys. Lett.* (*Engl. Transl.*) **3,** 83 (1977).

V. N. Blashchuk, V. N. Krasheninnikov, N. A. Melnikov, N. F. Pilipetsky, V. V. Ragulsky, V. V. Shkunov, and B. Y. Zel'dovich, SBS Wave Front Reversal for the Depolarized Light-Theory and Experiment, *Opt. Commun.* **27,** 137 (1978).

V. N. Blashchuk, B. Y. Zel'dovich, V. N. Krasheninnikov, N. A. Mel'nikov, N. F. Pilipetskii, V. V. Ragul'skii, and V. V. Shkunov, Stimulated Scattering of Depolarized Radiation, *Dokl. Akad. Nauk SSR* **241,** 1322 (1978); *Sov. Phys. Dokl.* (*Engl. Transl.*) **23,** 588 (1978).

V. N. Blashchuk, A. V. Mamaev, N. F. Pilipetsky, V. V. Shkunov, and B. Y. Zel'dovich, Wave Front Reversal with Angular Tilting Theory and Experiment for the Four Wave Mixing, *Opt. Commun.* **31,** 383 (1979).

V. N. Blashchuk, N. F. Pilipetskii, and V. V. Shkunov, Four Wave Interaction as a Tuning Frequency Filter, *Dokl. Akad. Nauk SSSR* **251,** 70 (1980); *Sov. Phys. Dokl.* (*Engl. Transl.*) **25,** 185 (1980).

V. N. Blashchuk, B. Y. Zel'dovich, A. V. Mamaev, N. F. Pilipetskii, and V. V. Shkunov, Complete Wavefront Reversal of Depolarized Radiation under Degenerate Four-Photon Interaction Conditions (Theory and Experiment), *Kvantovaya Elektron.* **7,** 627 (1980); *Sov. J. Quantum Electron.* (*Engl. Transl.*) **10,** 356 (1980).

G. L. Blekhovskikh, A. D. Kudryavtseva, and A. I. Sokolovskaya, Reconstruction of the Wavefront of Light Beams in Stimulated Raman Scattering, *Kvantovaya Elektron.* **5,** 1812 (1978); *Sov. J. Quantum Electron.* (*Engl. Transl.*) **8,** 1028 (1978).

D. Bloch, R. K. Raj, J. J. Snyder, and M. Ducloy, Heterodyne Detection of Phase-Conjugate Emission in an Ar Discharge with a Low-Power CW Laser, *J. Phys., Lett.* (*Orsay, Fr.*) **42,** L31 (1981).

D. M. Bloom and G. C. Bjorklund, Conjugate Wave-Front Generation and Image Reconstruction by Four-Wave Mixing, *Appl. Phys. Lett.* **31,** 592 (1977).

D. M. Bloom, P. F. Liao, and N. P. Economou, Observation of Amplified Reflection by Degenerate Four-Wave Mixing in Atomic Sodium Vapor, *Opt. Lett.* **2,** 58 (1978).

D. M. Bloom, C. V. Shank, R. L. Fork, and O. Teschke, Sub-Picosecond Optical Gating by Optical Wavefront Conjugation, *in* "Picosecond Phenomena" (C. V. Shank, E. P. Ippen, and S. L. Shapiro, eds.), p. 96. Springer-Verlag, Berlin and New York, 1978.

D. L. Bobroff, Coupled Modes Analysis of the Phonon–Photon Parametric Backward-Wave Oscillator, *J. Appl. Phys.* **36,** 1760 (1965).

D. L. Bobroff and H. A. Haus, Impulse Response of Active Coupled Wave Systems, *J. Appl. Phys.* **38,** 390 (1967).

L. A. Bol'shov, F. V. Bunkin, and D. V. Vlasov, Compensation of Nonlinear Distortions of Beams with Arbitrary Polarization State, *Kvantovaya Elektron.* **7,** 2057 (1980); *Sov. J. Quantum Electron.* (*Engl. Transl.*) **10,** 1197 (1980).

L. A. Bol'shov, D. V. Vlasov, M. A. Dykhne, V. V. Korobkin, K. S. Saidov, and A. H. Starostin, On the Possibility of Full Compensation of Nonlinear Distortions of a Light Beam by Means of Its Wave Front Reversal, *Pis'ma Zh. Eksp. Teor. Fiz.* **31,** 311 (1980).

L. A. Bol'shov, D. V. Vlasov, A. M. Dykhne, and A. N. Starostin,

Theory of Compensation of Nonlinear Distortions of a Light Beam by Reversal of its Wavefront, *Dokl. Akad. Nauk SSR* **251,** 1371 (1980); *Sov. Phys.—Dokl.* (*Engl. Transl.*) **25,** 283 (1980).

B. N. Borisov, Ya. I. Krizhilin, S. A. Nashchekin, V. K. Orlov, and S. V. Shklyarik, Wavefront Inversion in Induced Mandel'shtam–Brillouin Scattering in a Glass without Failure, *Zh. Tekh. Fiz.* **50,** 1073 (1980); *Sov. Phys.–Tech. Phys.* (*Engl. Transl.*) **25,** 645 (1980).

A. A. Borshch, M. S. Brodin, V. I. Volkov, and N. V. Kukhtarev, Nonlinear Optics Effects in a Degenerate Four-Wave Interaction in ZnSe crystals, *Pis'ma V. Zh. Tekh. Fiz.* **5,** 1240 (1979); *Sov. Tech. Phys. Lett.* (*Engl. Transl.*) **5,** 520 (1979).

A. Borshch, M. Brodin, V. Volkov, and N. Kukhtarev, Phase Conjugation by the Degenerate Six-Photon Mixing in Semiconductors, *Opt. Commun.* **35,** 287 (1980).

A. A. Borshch, M. S. Brodin, V. I. Volkov, and N. V. Kukhtarev, Optical Bistability and Hysteresis in a Reversed Wave in the Course of Degenerate Six-Photon Interaction in Cadmium Sulfide, *Kvantovaya Elektron.* **8,** 1304 (1981); *Sov. J. Quantum Electron.* (*Engl. Transl.*) **11,** 777 (1981).

A. Borshch, M. Brodin, V. Volkov, and N. Kukhtarev, "Optical Bistability and Hysteresis in Phase Conjugation by Degenerate Six-Photon Mixing, *Opt. Commun.* **41,** 213 (1982).

J. A. Buck, A. Dienes, and J. R. Whinnery, Effects of Separated Absorption and Emission Spectra on Degenerate Four-Wave Mixing in Organic Dyes, *J. Opt. Soc. Am.* **71,** 1381 (1981).

A. A. Bugaev, B. P. Zakharchenya, I. K. Meshkovskii, V. M. Ovchinnikov, and F. A. Chudnovskii, Holography with Reversing Light Reflector, *Pis'ma Zh. Tekh. Fiz.* **1,** 209 (1975); *Sov. Tech. Phys. Lett.* (*Engl. Transl.*) **1,** 99 (1975).

F. V. Bunkin, V. V. Savranskii, and G. A. Shafeev, Resonant Wavefront Reversal in an Active Medium Containing Copper Vapor, *Kvantovaya Elektron.* **8,** 1346 (1981); *Sov. J. Quantum Electron.* (*Engl. Transl.*) **11,** 810 (1981).

F. V. Bunkin, D. V. Vlasov, and Yu. A. Kravtsov, Problem of Reversal of an Acoustic Wavefront and Amplification of the Reversed Wave, *Kvantovaya Elektron.* **8,** 1144 (1981); *Sov. J. Quantum Electron.* (*Engl. Transl.*) **11,** 687 (1981).

F. V. Bunkin, D. V. Vlasov, and Yu. A. Kravtsov, Problem of Reversal of an Acoustic Wavefront and Amplification of the Reversed Wave, *Kvantovaya Elektron.* **8,** 1144 (1981); *Sov. J. Quantum Electron.* (*Engl. Transl.*) **11,** 687 (1981).

F. V. Bunkin, D. V. Vlasov, E. S. Zabolotskaya, and Yu. A. Kravtsov,

Thermal and Bubble Mechanisms for Four-Phonon Acoustic Phase Conjugation, *Pis'ma Zh. Tekh. Fiz.* **7,** 560 (1981); *Sov. Tech. Phys. Lett.* (*Engl. Transl.*) **7,** 239 (1981).

M. V. Bunkina, M. V. Morozov, and K. N. Firsov, Possibility of Using Vibrational-Transational Relaxation in an Amplifying Medium for Wavefront Reversal, *Kvantovaya Elektron.* **7,** 2026 (1980); *Sov. J. Quantum Electron.* (*Engl. Transl.*) **10,** 1173 (1980).

R. G. Caro and M. C. Gower, Phase Conjugation of KrF Laser Radiation, *Opt. Lett.* **6,** 557 (1981).

R. G. Caro and M. C. Gower, Amplified Phase Conjugate Reflection of KrF Laser Radiation, *Appl. Phys. Lett.* **39,** 855 (1981).

R. G. Caro and M. C. Gower, Phase Conjugation by Degenerate Four-Wave Mixing in Absorbing Media, *IEEE J. Quantum Electron.* **QE-18,** 1376 (1982).

B. Carquille and C. Froehly, Real-Time High-Resolvance Image Correlation by Bragg Diffraction in Saturable Absorbers, *Appl. Opt.* **19,** 207 (1980).

W. T. Cathey, Holographic Simulation of Compensation for Atmospheric Wavefront Distortion, *Proc. IEEE* **56,** 340 (1968).

G.-Y. Chu, H.-N. Zhu, Z.-H. Yu, Z.-G. Zhang, Y.-W. Qiu, G.-S. Ji, P.-X. Ye, Y.-R. Shen, and H.-T. Zhou, The Four-Wave Mixing in Liquid Crystals and Its Relaxation Effect, *Wu Li Hsueh Pao* **28,** 887 (1979).

M. Cronin-Golomb, B. Fischer, J. O. White, and A. Yariv, Passive (Self-Pumped) Phase Conjugate Mirror: Theoretical and Experimental Investigation, *Appl. Phys. Lett.* **41,** 689 (1982).

M. Cronin-Golomb, J. O. White, B. Fischer, and A. Yariv, Exact Solution of a Nonlinear Model of Four-Wave Mixing and Phase Conjugation, *Opt. Lett.* **7,** 313 (1982).

B. Crosignani and P. Di Porto, Influence of Chromatic Dispersion on Degenerate Four-Wave Mixing, *Opt. Lett.* **7,** 499 (1982).

L. C. Davis, K. A. Marko, and L. Rimai, Diffraction Coupled Laser Beam. Configurations for High Spatial Resolution, *Appl. Opt.* **20,** 1685 (1981).

Y. N. Denisyuk, On the Reproduction of the Optical Properties of an Object by the Wave Field of Its Scattered Radiation, *Opt. Spektrosk.* **15,** 522 (1962); *Opt. Spectrosc.* (*Engl. Transl.*) **15,** 279 (1962).

Yu. N. Denisyuk, Optical Phase Conjugation by Doppler Dynamic Holograms, *Pis'ma Zh. Tekh. Fiz.* **7,** 641 (1981); *Sov. Tech. Phys. Lett.* (*Engl. Transl.*) **7,** 275 (1981).

D. Depatie and D. Haueisen, Multiline Phase Conjugation at 4 μm in Germanium, *Opt. Lett.* **5,** 252 (1980).

D. Depatie, D. Haueisen, A. Elci, and D. Rogovin, Observation of Infrared Phase Conjugation in Molecular Gases, *Proc. Int. Conf. Lasers, 1978* (V. J. Corcoran, ed.), p. 525. STS Press, McLean, Virginia, 1979.

J.-C. Diels and I. C. McMichael, Influence of Wave-Front-Conjugated Coupling on the Operation of a Laser Gyro *Opt. Lett.* **6,** 219 (1981).

Y. V. Dolgopolov, V. A. Komarevskii, S. B. Kormer, G. G. Kochemasov, S. M. Kulikov, V. M. Murugov, V. D. Nikolaev, and S. A. Sukharev, Experimental Investigation of the Feasibility of Application of the Wavefront Reversal Phenomenon in Stimulated Mandel'shtam–Brillouin Scattering, *Zh. Eksp. Teor. Fiz.* **76,** 908 (1979); *Sov. Phys.—JETP (Engl. Transl.)* **49,** 458 (1979).

M. Ducloy, Nonlinear Optical Phase Conjugation, Festkörper Probleme **XXII,** 35 (1982).

M. Ducloy and D. Bloch, Theory of Degenerate Four-Wave Mixing in Resonant Doppler-Broadened Systems. I. Angular Dependence of Intensity and Lineshape of Phase-Conjugate Emission, *J. Phys. (Orsay, Fr.)* **42,** 711 (1981).

M. Ducloy and D. Bloch, Theory of Degenerate Four-Wave Mixing in Resonant Doppler-Broadened Media. II. Doppler-Free Heterodyne Spectroscopy via Collinear Four-Wave Mixing in Two- and Three-level Systems, *J. Physique* **43,** 57 (1982).

M. Ducloy, R. K. Raj, and D. Bloch, Polarization Characteristics of Phase-Conjugate Mirrors Obtained by Resonant Degenerate Four-Wave Mixing, *Opt. Lett.* **7,** 60 (1982).

G. J. Dunning and D. G. Steel, Effects of Unequal Pump Intensity in Resonantly Enhanced Degenerate Four-Wave Mixing, *IEEE J. Quantum Electron.* **QE-18,** 3 (1982).

N. P. Economou and P. F. Liao, Magnetic-Field Quantum Beats in Two-Photon Free-Induction Decay, *Opt. Lett.* **3,** 172 (1978).

V. F. Efimikov, I. G. Zubarev, A. V. Kotov, A. B. Mironov, S. I. Mikhailov, and M. G. Smirnov, Generation of High-Power Short Pulses with Wavefront Reversal under Steady-State Stimulated Brillouin Scattering Conditions, *Kvantovaya Elektron.* **6,** 2031 (1979); *Sov. J. Quantum Electron. (Engl. Transl.)* **9,** 1194 (1979).

V. F. Efimkov, I. G. Zubarev, A. V. Kotov, A. B. Mironov, S. I. Mikhailov, and M. G. Smirnov, Investigation of Systems for Obtaining Short High-Power Pulses by Wavefront Reversal of the Radiation in a Stimulated Brillouin Scattering Mirror, *Kvantovaya Elektron.* **7,** 372 (1980); *Sov. J. Quantum Electron. (Engl. Transl.)* **10,** 211 (1980).

K. Eidmann and R. Sigel, Backscatter Experiments, *in* "Laser Interaction and Related Plasma Phenomena'' (H. J. Schwarz and H. Hora, eds.), Vol. 3B, p. 667. Plenum, New York, 1974.

A. Elci and D. Rovovin, Phase Conjugation in Nonlinear Molecular Gases, *Chem. Phys. Lett.* **61,** 407 (1979).

A. Elci and D. Rogovin, Phase Conjugation in an Inhomogeneously Broadened Medium, *Opt. Lett.* **5,** 255 (1980).

A. Elci, D. Rogovin, D. Depatie, and D. Haueisen, Phase Conjugation in Ammonia, *J. Opt. Soc. Am.* **70,** 990 (1980).

P. Ewart, A. I. Ferguson, and S. V. O'Leary, Degenerate Four-Wave Mixing in Excited States Formed by Collision Assisted Transitions, *Opt. Commun.* **40,** 147 (1981).

Y. Fainman, E. Lenz, and J. Shamir, Contouring by Phase Conjugation, *J. Appl. Opt.* **20,** 158 (1981).

J. Y. Fan, C. K. Wu, and Z. Y. Wang, The Theory of Resonantly Enhanced Four-Wave Mixing in Absorbing Media, *Wu Li Hsueh Po* **29,** 897 (1980); *Chin. J. Phys.* (*Peking*) (*Engl. Transl.*) **1,** 570 (1981).

J. Y. Fan, C. K. Wu, and Z. Y. Wang, Nature of the Phase Conjugation in Stimulated Raman Scattering Backward Wave, *Laser J.* **7**(3), 14 (1980).

J. Feinberg, Real-Time Edge Enhancement Using the Photorefractive Effect, *Opt. Lett.* **5,** 330 (1980).

J. Feinberg, Self-Pumped, Continuous-Wave Phase Conjugator Using Internal Reflection, *Opt. Lett.* **7,** 486 (1982).

J. Feinberg and R. W. Hellwarth, Phase-Conjugating Mirror with Continuous-Wave Gain, *Opt. Lett.* **5,** 519 (1980).

D. Fekete, J. AuYeung, and A. Yariv, Phase-Conjugate Reflection by Degenerate Four-Wave Mixing in a Nematic Liquid Crystal in the Isotropic Phase, *Opt. Lett.* **5,** 51 (1980).

B. J. Feldman, R. A. Fisher, E. E. Bergmann, R. G. Tercovich, F. C. Sena, and I. J. Bigio, Intracavity Techniques for High Reflectivity Phase Conjugation at 10 μm in Germanium and Inverted CO_2, *Proc. Los Alamos Conf. Opt.* p. 412 (1979).

B. J. Feldman, R. A. Fisher, and S. L. Shapiro, Ultraviolet Phase Conjugation, *Opt. Lett.* **6,** 84 (1981).

J. L. Ferrier, Z. Wu, J. Gazengel, N. Phu Xuan, and G. Rivoire, Backward Scatterings in the Picosecond Range: Generation and Geometrical Conditions for Wave Front Reconstruction, *Opt. Commun.* **41,** 35 (1982).

J. L. Ferrier, Z. Wu, X. Nguyen Phu, and G. Rivoire, Four Wave Mixing in the Picosecond Range: Intensities, Durations, Wave Front Reconstruction, *Opt. Commun.* **41,** 207 (1982).

B. Fischer, M. Cronin-Golomb, J. O. White, and A. Yariv, Amplified

Reflection, Transmission, and Self-Oscillation in Real-Time Holography, *Opt. Lett.* **6,** 519 (1981).

B. Fischer, M. Cronin-Golomb, J. O. White, and A. Yariv, Real-Time Phase-Conjugate Window for One-Way Optical Field Imaging Through a Distortion, *Appl. Phys. Lett.* **41,** 141 (1982).

R. A. Fisher and B. J. Feldman, On-Resonant Phase-Conjugate Reflection and Amplification at 10.6 μm in Inverted CO_2, *Opt. Lett.* **4,** 140 (1979).

R. A. Fisher and B. J. Feldman, Optical Phase Conjugation, *in* "McGraw-Hill (1980) Yearbook of Science and Technology," p. 300. McGraw-Hill, New York, 1980.

R. A. Fisher, B. R. Suydam, and B. J. Feldman, Transient Analysis Kerr-Like Phase Conjugators Using Frequency-Domain Techniques, *Phys. Rev. A* **23,** 3071 (1981).

C. Flytzanis and C. L. Tang, Light-Induced Critical Behavior in the Four-Wave Interaction in Nonlinear Systems, *Phys. Rev. Lett.* **45,** 441 (1980).

T.-Y. Fu and M. Sargent, III, Effects of Signal Detuning on Phase Conjugation, *Opt. Lett.* **4,** 366 (1979).

T.-Y. Fu and M. Sargent, III, Effects of Signal Detuning on Phase Conjugation, *Proc. Los Alamos Conf. Opt.* (D. H. Liebenberg, ed.), p. 419. SPIE, Washington, D.C., 1979.

T.-Y. Fu and M. Sargent, III, Theory of Two-Photon Phase Conjugation, *Opt. Lett.* **5,** 433 (1980).

M. Fujita, H. Nakatsuka, H. Nakanishi, and M. Matsuoka, Backward Echo in Two-Level Systems, *Phys. Rev. Lett.* **42,** 974 (1979).

L. Fu-Li, J. A. Hermann, and J. N. Elgin, Effects of Two-Photon Optical Bistability Upon Phase Conjugation, *Opt. Commun.* **40,** 446 (1982).

V. B. Gerasimov, Answer to the Letter of B. Ya. Zel'Dovich, I. G. Zubarev, G. A. Pasmanik, and V. G. Sidorovich, *Kvantovaya Elektron.* **6,** 592 (1979); *Sov. J. Quantum Electron.* (*Engl. Transl.*) **9,** 352 (1979).

V. B. Gerasimov and V. K. Orlov, Influence of Wavefront Reversal on the Operation of Brillouin Laser, *Kvantovaya Elektron.* **5,** 906 (1977); *Sov. J. Quantum Electron.* (*Engl. Transl.*) **8,** 517 (1978).

V. B. Gerasimov and V. K. Orlov, Reproduction of Wavefronts in Scattering of Light by Acoustic Waves and Its Relationship to Dynamic Holography, *Kvantovaya Elektron.* **5,** 436 (1978); *Sov. J. Quantum Electron.* (*Engl. Transl.*) **8,** 253 (1978).

V. B. Gerasimov, S. A. Gerasimova, and V. K. Orlov, Wavefront of the Stokes component in Stimulated Brillouin Backscattering, *Kvantovaya Elektron.* **4,** 930 (1977); *Sov. J. Quantum Electron.* (*Engl. Transl.*) **9,** 352 (1979).

V. B. Gerasimov, S. A. Gerasimova, and V. K. Orlov, Substantial Improvement in Selective Properties of a Stimulated Brillouin Scattering Mirror Resulting from Use of Wide-Band Pumping, *Kvantovaya Elektron.* **4,** 932 (1977); *Sov. J. Quantum Electron.* (*Engl. Transl.*) **7,** 529 (1977).

H. J. Gerritsen, Nonlinear Effects in Image Formation, *Appl. Phys. Lett.* **10,** 237 (1967).

A. A. Golubtsov, N. F. Pilipetskii, A. N. Sudarkin, and V. V. Shkunov, Self-Defocusing of He–Ne Laser Radiation as a Result of Thermoelastic Deformations of a Reflecting Surface, *Kvantovaya Elektron.* **8,** 370 (1981); *Sov. J. Quantum Electron.* (*Engl. Transl.*) **11,** 218 (1981).

A. A. Golubtsov, N. F. Pilipetskii, A. N. Sudarkin, and V. V. Shkunov, Wavefront Reversal by Light-Induced Shaping of the Surface of an Absorbing Material, *Kvantovaya Elektron.* **8,** 663 (1981); *Sov. J. Quantum Electron.* **11,** 402 (1981).

A. V. Gorlanov, N. I. Grishmanova, N. A. Sventsitskaya, and V. D. Solov'ev, Angular Characteristics of Neodymium Laser Radiation After Wavefront Reversal by Three-Wave Parametric Interaction, *Kvantovaya Elektron.* **9,** 415 (1982); *Sov. J. Quantum Electron.* (*Engl. Transl.*) **23,** 240 (1982).

M. C. Gower, KrF Laser Amplifier with Phase-Conjugate Brillouin Retroreflectors, *Opt. Lett.* **7,** 423 (1982).

M. C. Gower and R. G. Caro, KrF Laser with a Phase-Conjugate Brillouin Mirror, *Opt. Lett.* **7,** 162 (1982).

N. C. Griffen and C. V. Heer, Focusing and Phase Conjugation of Photon Echoes in Na Vapor, *Appl. Phys. Lett.* **33,** 865 (1978).

D. Grischkowsky, N. S. Shiren, and R. J. Bennett, Generation of Time-Reversed Wave Fronts Using a Resonantly Enhanced Electronic Nonlinearity, *Appl. Phys. Lett.* **33,** 805 (1978).

W. M. Grossman and D. M. Shemwell, Coherence Lengths and Phase Conjugation by Degenerate Four-Wave Mixing, *J. Appl. Phys.* **51,** 914 (1980).

C. R. Guiliano, Applications of Optical Phase Conjugation, *Phys. Today* Apr., p. 27 (1981).

P. N. Günter, Electric-Field Dependence of Phase-Conjugate Wave-Front Reflectivity in Reduced $KNbO_3$ and $Bi_{12}GeO_{20}$, *Opt. Lett.* **7,** 10 (1982).

A. L. Gyulamiryan, A. V. Mamaev, N. F. Pilipetskii, V. V. Ragulskiy, and V. V. Shkunov, Investigation of the Efficiency of Nondegenerate Four-Wave Interaction, *Kvantovaya Elektron.* **8,** 196 (1981); *Sov. J. Quantum Electron.* (*Engl. Transl.*) **11,** 115 (1981).

D. S. Hamilton, D. Heiman, J. Feinberg, and R. W. Hellwarth, Spatial-Diffusion Measurements in Impurity-Doped Solids by Degenerate Four-Wave Mixing, *Opt. Lett.* **4,** 124 (1979).

A. Hardy, Sensitivity of Phase-conjugate Resonators to Intracavity Phase Perturbations, *IEEE J. Quantum Electron.* **QE-17,** 1581 (1981).

A. Hardy and S. Hochhauser, Higher-Order Modes of Phase-Conjugate Resonators, *Appl. Opt.* **21,** 2330 (1982).

D. J. Harter and R. W. Boyd, Nearly Degenerate Four-Wave Mixing Enhanced by the AC Stark Effect, *IEEE J. Quantum Electron.* **QE-16,** 1126 (1980).

D. C. Haueisen, Doppler-Free Two-Photon Spectroscopy Using Degenerate Four-Wave Mixing, *Opt. Commun.* **28,** 183 (1979).

C. V. Heer and N. C. Griffin, Generation of a Phase-Conjugate Wave in the Forward Direction with Thin Na-Vapor Cells, *Opt. Lett.* **4,** 239 (1979).

C. V. Heer and N. C. Griffen, Phase Conjugation in the Forward Direction with Thin Na Vapor Cells, *Proc. Los Alamos Conf. Opt.* (D. H. Liebenberg, ed.), p. 413. SPIE, Washington, D.C., 1979.

C. V. Heer and P. F. McManamon, Wavefront Correction with Photon Echoes, *Opt. Commun.* **23,** 49 (1977).

C. V. Heer and R. L. Sutherland, Optical Properties of Photon Echoes Stimulated by Three Frequencies, *Phys. Rev. A* **19,** 2026 (1979).

J. Hegarty, M. D. Sturge, A. C. Gossard, and W. Wiegmann, Resonant Degenerate Four-Wave Mixing in GaAs Multiquantum Well Structures, *Appl. Phys. Lett.* **40,** 132 (1982).

E. J. Heilweil, R. M. Hochstrasser, and H. Souma, Applications of the Phase Conjugate Configuration in Nonlinear Spectroscopy, *Opt. Commun.* **35,** 227 (1980).

R. W. Hellwarth, Generation of Time-Reversed Wave Fronts by Nonlinear Refraction, *J. Opt. Soc. Am.* **67,** 1 (1977).

R. W. Hellwarth, Theory of Phase Conjugation by Stimulated Scattering in a Waveguide, *J. Opt. Soc. Am.* **68,** 1050 (1978).

R. W. Hellwarth, Theory of Phase-Conjugation by Four-Wave Mixing in a Waveguide, *IEEE J. Quantum Electron.* **QE-15,** 101 (1979).

R. W. Hellwarth, Optical Beam Phase Conjugation by Stimulated Backscattering, *Opt. Eng.* **21,** 257 (1982).

R. W. Hellwarth, Optical Beam Phase Conjugation by Four-Wave Mixing in a Waveguide, *Opt. Eng.* **21,** 263 (1982).

J. P. Herriau, J. P. Huignard, and P. Auborg, Some Polarization Properties of Volume Holograms in $B_{12}SiO_{20}$ Crystals and Applications, *Appl. Opt.* **17,** 1851 (1978).

R. M. Hochstrasser, G. R. Meredeeth, and H. P. Trommsdorff, Resonant Four Wave Mixing in Molecular Crystals, *J. Chem. Phys.* **73,** 1009 (1981).

D. T. Hon, Pulse Compression by Stimulated Brillouin Scattering, *Opt. Lett.* **5,** 516 (1980).

D. T. Hon, Applications of Wavefront Reversal by Stimulated Brillouin Scattering, *Opt. Eng.* **21,** 252 (1982).

F. A. Hopf, Interferometery Using Conjugate-Wave Generation, *J. Opt. Soc. Am.* **70,** 1320 (1980).

F. A. Hopf, Optical Distortion in Phase Conjugation in High-Gain Three-Wave Mixing, *J. Opt. Soc. Am.* **70,** 1379 (1980).

F. A. Hopf, A. Tomita, K. H. Womack, and J. L. Jewell, Optical Distortion in Nonlinear Phase Conjugation by Three-Wave Mixing, *Opt. Soc. Am.* **69,** 968 (1979).

F. A. Hopf, A. Tomita, and T. Liepmann, Quality of Phase Conjugation in Silicon, *Opt. Commun.* **37,** 72 (1981).

H. Hsu, Large-Signal Theory of Phase-Conjugate Backscatterings, *Appl. Phys. Lett.* **34,** 855 (1979).

H. Hsu, Large Signal Characteristics of Phase Conjugate Back Scatterings, *Proc. Int. Conf. Lasers, 1978* (V. J. Corcoran, ed.), p. 538. STS Press, McLean, Virginia; 1979.

H. Hsu and C. Yu, Parametric Amplification, Oscillation, and Mixing in Nonlinear Backward Scattering, *Proc. Conf.* (*Int.*) Solid State Devices, 5th, Tokyo, 1973; Oyo Butsuri, Suppl. **43,** 75 (1974).

J. Hubbard, Resolving Power of Time-Reversed Wave-Front Imaging Devices, *J. Opt. Soc. Am.* **71,** 1029 (1981).

J. P. Huignard and J. P. Herriau, Real-Time Double-Exposure Interferometry with $Bi_{12}SiO_{20}$ Crystals in Transverse Electrooptic Configuration, *Appl. Opt.* **16,** 1807 (1977).

J. P. Huignard and J. P. Herriau, Real-Time Coherent Object Edge Reconstruction with $B_{12}SiO_{20}$ Crystals, *Appl. Opt.* **17,** 2671 (1978).

J. P. Huignard and F. Micheron, High-Sensitivity Read–Write Volume Holographic Storage in $B_{12}SiO_{20}$ and $Bi_{12}GeO_{20}$ Crystals, *Appl. Phys. Lett.* **29,** 591 (1976).

J. P. Huignard, J. P. Herriau, and T. Valentin, Time-Average Holographic Interferometry with Photoconductive Electrooptic $Bi_{12}SiO_{20}$ Crystals, *Appl. Opt.* **16,** 2796 (1977).

J. P. Huignard, J. P. Herriau, P. Aubourg, and E. Spitz, Phase-Conjugate Wavefront Generation via Real-Time Holography in $Bi_{12}SiO_{20}$ Crystals, *Opt. Lett.* **4,** 21 (1979).

J. P. Huignard, J. P. Herriau, L. Pichon, and A. Marrakchi, Speckle-Free Imaging in Four-Wave Mixing Experiments with $Bi_{12}SiO_{20}$ Crystals, *Opt. Lett.* **5,** 436 (1980).

J. P. Huignard, J. P. Herriau, G. Rivet, and P. Gunter, Phase-conjugation and Spatial-Frequency Dependence of Wave-Front Reflectivity $Bi_{12}SiO_{20}$ Crystals, *Opt. Lett.* **5,** 102 (1980).

L. M. Humphrey, J. P. Gordon, and P. F. Liao, Angular Dependence of Line Shape and Strength of Degenerate Four-Wave Mixing in a Doppler-Broadened System with Optical Pumping, *Opt. Lett.* **5,** 56 (1980).

V. S. Idiatulin, On the Interaction of Counterrunning Radiation in a Mirrorless Device, *Opt. Commun.* **30,** 419 (1979).

V. S. Idiatulin, Stability of the Self-Induced Distributed Feedback in a System without a Resonator, *Zh. Tekh. Fiz.* **49,** 2268 (1979); *Sov. Phys.—Tech. Phys.* (*Engl. Transl.*) **24,** 1257 (1979).

A. A. Ilyukhin, G. V. Peregudov, M. E. Plotkin, E. N. Ragozin, and V. A. Chirkov, Focusing of a Laser Beam on Target Using the Effect of Wave Front Inversion (WFI) Produced as a Result of Stimulated Mandel'shtam–Brillouin Scattering (SMBS), *Pis'ma Zh. Eksp. Teor. Fiz.* **29,** 364 (1979); *JETP Lett.* (*Engl. Transl.*) **29,** 328 (1980).

V. V. Ivakhnik, V. M. Petnikova, V. S. Solomatin, and V. V. Shuvalov, Compensation of Wavefront Distortions in a Thick Inhomogeneous Medium, *Kvantovaya Elektron.* **7,** 652 (1980); *Sov. J. Quantum Electron.* (*Engl. Transl.*) **10,** 373 (1980).

V. V. Ivakhnik, V. M. Petnikova, V. S. Solomatin, M. A. Kharchenko, and V. V. Shuvalov, Single-Pass Systems for Compensation of Phase Distortions, *Kvantovaya Elektron.* **7,** 898 (1980); *Sov. J. Quantum Electron.* (*Engl. Transl.*) **10,** 514 (1980).

V. V. Ivakhnik, V. M. Petnikova, and V. V. Shuvalov, Enhancement of the Efficiency of Wavefront Reversal Systems Using Ring Resonators, *Kvantovaya Elektron.* **8,** 455 (1981); *Sov. J. Quantum Electron.* (*Engl. Transl.*) **11,** 275 (1981).

V. V. Ivakhnik, V. M. Petnikova, and V. V. Shuvalov, Compensation for Phase Distortions of Spatially Modulated Fields, *Kvantovaya Elektron.* **8,** 774 (1981); *Sov. J. Quantum Electron.* **11,** 467 (1981).

E. V. Ivakin, I. P. Petrovich, and A. S. Rubinov, Self-Diffraction of Radiation by Light-Induced Phase Gratings, *Kvantovaya Elektron.* **1,** 96 (1973); *Sov. J. Quantum Electron.* (*Engl. Transl.*) **3,** 52 (1973).

E. V. Ivakin, I. P. Petrovich, A. S. Rubinov, and B. I. Stepanov, Dynamic Holograms in an Amplifying Medium, *Kvantovaya Elektron.* **2,** 1556 (1975); *Sov. J. Quantum Electron.* (*Engl. Transl.*) **5,** 840 (1975).

E. V. Ivakin, I. P. Petrovich, and A. S. Rubinov, Dynamic Holograms and Some Problems in Real-Time Information Processing, *in* "Optical Methods of Information Processing." Nauk Tekn., Minsk. 1978.

E. V. Ivakin, V. G. Koptev, A. M. Lazaruk, I. P. Petrovich, and A. S.

Rubanov, Phase Conjugation of Light Fields as a Result of Nonlinear Interaction in Saturable Media, *Pis'ma Zh. Eksp. Teor. Fiz.* **30,** 648 (1979); *JETP Lett.* (*Engl. Transl.*) **30,** 613 (1979).

E. V. Ivakin, A. M. Lazaruk, A. S. Rubanov, and B. I. Stephanov, Comments on "Phase Conjugate Optics and Real-Time Holography," *IEEE J. Quantum Electron.* **QE-15,** 523 (1979).

V. M. Izgorodin, S. B. Kormer, G. G. Kochemasov, V. D. Nikolaev, and A. V. Pinegin, Wavefront Reversal by Four-Wave Mixing in a Medium Exhibiting Raman Nonlinearity, *Kvantovaya Elektron.* **9,** 229 (1982); *Sov. J. Quantum Electron.* (*Engl. Transl.*) **12,** 119 (1982).

Y. H. Ja, Phase-Conjugate Wavefront Generation via Four Wave Mixing in $Bi_{12}GeO_{20}$ Crystals-Reflection Hologram Type, *Opt. Commun.* **41,** 159 (1982).

Y. H. Ja, Real-Time Image Subtraction in Four-Wave Mixing with Photorefractive $Bi_{12}GeO_{20}$ Crystals, *Opt. Commun.* **42,** 377 (1982).

S. F. Jacobs, Experiments with Retrodirective Arrays, *Opt. Eng.* **21,** 281 (1982).

R. K. Jain, Degenerate Four-Wave Mixing in Semiconductors: Application to Phase Conjugation and to Picosecond Time-Resolved Studies of Transient Carrier Dynamics, *Opt. Eng.* **21,** 199 (1982).

R. K. Jain and G. J. Dunning, Spatial and Temporal Properties of a Continuous-Wave Phase-Conjugate Resonator Based on the Photorefractive Crystal $BaTiO_3$, *Opt. Lett.* **7,** 420 (1982).

R. K. Jain and M. B. Klein, Degenerate Four-Wave Mixing Near the Band Gap of Semiconductors, *Appl. Phys. Lett.* **35,** 454 (1979).

R. K. Jain and D. G. Steel, Degenerate Four-Wave Mixing of 10.6 μm Radiation in $Hg_{1-x}Cd_xTe$, *Appl. Phys. Lett.* **37,** 1 (1980).

R. K. Jain and D. K. Steel, Large Optical Nonlinearities and CW Degenerate Four-Wave Mixing in HgCdTe, *Opt. Commun.* **43,** 72 (1982).

R. K. Jain, M. B. Klein, and R. C. Lind, High-Efficiency Degenerate Four-Wave Mixing of 1.06 μM Radiation in Silicon, *Opt. Lett.* **4,** 328 (1979).

S. M. Jensen and R. W. Hellwarth, Observation of the Time-Reversed Replica of a Monochromatic Optical Wave, *Appl. Phys. Lett.* **32,** 166 (1978).

S. M. Jensen and R. W. Hellwarth, Generation of Time-Reversed Waves by Nonlinear refraction in a Waveguide, *Appl. Phys. Lett.* **33,** 404 (1978).

R. Kachru, T. W. Mossberg, and S. R. Hartmann, Stimulated Photon Echo Study of $Na(3^2S_{1/2})$-CO Velocity-Changing Collisions, *Opt. Commun.* **30,** 57 (1979).

A. A. Kalinina, V. V. Lyubimov, L. V. Nosova, and I. B. Orlova, Telescopic Small-Signal Amplifier with a Brillouin Mirror, *Kvantovaya Elektron.* **6,** 2269 (1979); *Sov. J. Quantum Electron.* (*Engl. Transl.*) **9,** 1336 (1979).

M. A. Khan, D. J. Muehlner, and P. A. Wolff, Resonant Four-Wave Mixing in n-type Silicon, *Opt. Commun.* **30,** 206 (1979).

M. A. Khan, P. W. Kruse, and J. F. Ready, Optical Phase Conjugation in $Hg_{1-x}Cd_xTe$, *Opt. Lett.* **5,** 261 (1980).

M. A. Khan, R. L. H. Bennet, and P. W. Kruse, Bandgap-Resonant Optical Phase Conjugation in N-Type $Hg_{1-x}Cd_xTe$ at 10.6 μm, *Opt. Lett.* **6,** 560 (1981).

I. C. Khoo, Degenerate Four-Wave Mixing in the Nematic Phase of a Liquid Crystal, *Appl. Phys. Lett.* **38,** 123 (1981).

I. C. Khoo, Optically Induced Molecular Reorientation and Third-Order Nonlinear Optical Processes in Nematic Liquid Crystals, *Phys. Rev. A* **23,** 2077 (1981).

I.-C. Khoo and S. L. Zhuang, Wavefront Conjugation in Nematic Liquid Crystal Films, *IEEE J. Quantum Electron.* **QE-18,** 246 (1982).

Y. F. Kir'yanov, G. G. Kochemasov, S. M. Martynova, and V. D. Nikolaev, Four-Wave Mixing in Resonantly Amplifying Media in Inversion Saturation Regime, *Kvantovaya Elektron.* **8,** 1734 (1981); *Sov. J. Quantum Electron.* (*Engl. Transl.*) **11,** 1047 (1981).

G. G. Kochemasov and V. D. Nikolaev, Reproduction of the Spatial Amplitude and Phase Distributions of a Pump Beam in Stimulated Brillouin Scattering, *Kvantovaya Elektron.* **4,** 115 (1977); *Sov. J. Quantum Electron.* (*Engl. Transl.*) **7,** 60 (1977).

G. G. Kochemasov and V. D. Nikolaev, Wavefront Reversal in Stimulated Scattering of Two-Frequency Pump Radiation, *Kvantovaya Elektron.* **5,** 1837 (1978); *Sov. J. Quantum Electron.* (*Engl. Transl.*) **8,** 1043 (1978).

G. G. Kochemasov and V. D. Nikolaev, Inaccuracy of Reproduction of the Spatial Structure of a Beam in a Laser Amplifying Medium with a Reversing Mirror, *Kvantovaya Elektron.* **6,** 864 (1979); *Sov. J. Quantum Electron.* (*Engl. Transl.*) **9,** 514 (1979).

G. G. Kochemasov and V. D. Nikolaev, Investigation of the Spatial Characteristics of Stokes Radiation in Stimulated Scattering under Saturation Conditions, *Kvantovaya Elektron.* **6,** 1960 (1979); *Sov. J. Quantum Electron.* **9,** 1155 (1979).

H. Kogelnik, Holographic Image Projection through Inhomogeneous Media, *Bell Sys. Tech. J.* **44,** 2451 (1965).

H. Kogelnik and K. S. Pennington, Holographic Imaging through a Random Medium, *J. Opt. Soc. Am.* **58,** 273 (1968).

V. G. Koptev, A. M. Lazaruk, I. P. Petrovich, and A. S. Rubanov,

Wavefront Inversion in Superradiance, *Pis'ma Zh. Eksp. Teor. Fiz.* **28,** 468 (1978); *JETP Lett. (Engl. Transl.)* **28,** 434 (1979).

V. Kremenitskii, S. Odulov, and M. Soskin, Dynamic Gratings in Cadmium Telluride, *Phys. Status Solidi A* **51,** K63 (1979).

V. Kremenitskii, S. Odulov, and M. Soskin, Backward Degenerate Four-Wave Mixing in Cadmium Telluride, *Phys. Status Solidi A* **57,** K71 (1980).

G. V. Krivoshchekov, S. G. Struts, and M. F. Stupak, Stimulated-Thermal-Scattering Spectrum with Wavefront Inversion, *Pis'ma Zh. Tekh. Fiz.* **6,** 428 (1980); *Sov. Tech. Phys. Lett. (Engl. Transl.)* **6,** 184 (1980).

G. V. Krivoschekov, M. F. Stupak, and T. T. Timofeev, Broadening of the Spectral Components of Stimulated Brillouin Scattering Accompanied by Optical Phase Conjugation, *Pis'ma Zh. Tekh. Fiz.* **7,** 506 (1981); *Sov. Tech. Phys. Lett. (Engl. Transl.)* **7,** 217 (1981).

M. Kroll, Saturation Spectroscopy and Resonant Degenerate Four-Wave Mixing in Hg at 546.1 nm, *Opt. Lett.* **7,** 151 (1982).

Y. I. Kruzhilin, Self-Adjusting Laser-Target System for Laser Fusion, *Kvantovaya Elektron. (Moscow)* **5,** 625 (1978); *Sov. J. Quantum Electron. (Engl. Transl.)* **8,** 359 (1978).

V. I. Kryzhanovskii, A. A. Mak, V. A. Serebryakov, and V. E. Yashin, Use of Phase Conjugation to Suppress Small-Scale Self-Focusing, *Pis'ma Zh. Tekh. Fiz.* **7,** 400 (1981); *Sov. Tech. Phys. Lett.* **7,** 170 (1981).

A. D. Kudriavtseva, A. I. Sokolovskaia, J. Gazengel, N. P. Xuan, and G. Rivoire, Reconstruction of the Laser Wave-Front by Stimulated Scatterings in the Pico-Second Range, *Opt. Commun.* **26,** 446 (1978).

N. V. Kukhtarev, Wavefront Reversal of Optical Beams in Anisotropic Media, *Kvantovaya Elektron.* **8,** 1451 (1981); *Sov. J. Quantum Electron. (Engl. Transl.)* **11,** 878 (1981).

N. V. Kukhtarev and G. V. Kovalenko, Wavefront Reversal in Interband Absorption in Semiconductors, *Kvantovaya Elektron.* **7,** 781 (1980); *Sov. J. Quantum Electron. (Engl. Transl.)* **10,** 446 (1980).

N. Kukhtarev and S. Odulov, Degenerate Four-Wave Mixing in $LiNbO_3$ and $LiTaO_3$, *Opt. Commun.* **32,** 183 (1980).

N. V. Kukhtarev and T. I. Semenets, Wavefront Reversal of Optical Beams in Resonant Media, *Kvantovaya Elektron.* **7,** 1721 (1980); *Sov. J. Quantum Electron. (Engl. Transl.)* **10,** 994 (1980).

O. L. Kulikov, N. F. Pilipetskii, A. N. Sudarkin, and V. V. Shkunov, Inversion of a Wave Front by a Surface, *Pis'ma Zh. Eksp. Teor. Fiz.* **31,** 377 (1980); *JETP Lett. (Engl. Transl.)* **31,** 345 (1980).

R. T. V. Kung and J. H. Hammond, Phase Front Reproduction in Raman Conversion, *IEEE J. Quantum Electron.* **QE-18,** 1306 (1982).

J. F. Lam, Doppler-Free Laser Spectroscopy via Degenerate Four-Wave Mixing, *Opt. Eng.* **21,** 219 (1982).

J. F. Lam and W. P. Brown, Optical Resonators with Phase-Conjugate Mirrors, *Opt. Lett.* **5,** 61 (1980).

J. F. Lam, D. G. Steel, R. A. McFarlane, and R. C. Lind, Atomic Coherence Effects in Resonant Degenerate Four-Wave Mixing, *Appl. Phys. Lett.* **38,** 977 (1981).

R. Landauer, Parametric Standing Wave Amplifiers, *Proc. IRE* **48,** 1328 (1960).

C. M. Lawson, R. C. Powell, and W. K. Zwicker, Measurement of Exciton Diffusion Lengths in $Nd_xLa_{1-x}P_5O_{14}$ by Four-Wave Mixing Techniques, *Phys. Rev. Lett.* **46,** 1020 (1981).

A. M. Lazaruk, Wavefront Reversal in Amplifying Dynamic Dye-solution Holograms, *Kvantovaya Elektron.* **6,** 1770 (1979); *Sov. J. Quantum Electron.* (*Engl. Transl.*) **9,** 1041 (1979).

A. M. Lazaruk, Compensation for Small Scale Nonlinear Distortions of Optical Beams Using Wavefront Reversal, *Kvantovaya Elektron.* **8,** 2461 (1981); *Sov. J. Quantum Electron.* (*Engl. Transl.*) **11,** 1502 (1981).

A. M. Lazaruk and A. S. Rubanov, Energy Efficiency of the Generation of a Reversed Wavefront in a Four-Wave Interaction of Light Fields, *Zh. Prikl. Spektrosk.* **31,** 1099 (1979); *J. Appl. Spectrosc.* (*Engl. Transl.*) **31,** 1576 (1979).

A. M. Lazaruk and A. S. Rubanov, Energy Efficiency of Wavefront Reversal in Stimulated Four-Wave Parametric Scattering, *Kvantovaya Elektron.* **7,** 1992 (1980); *Sov. J. Quantum Electron.* (*Engl. Transl.*) **10,** 1148 (1980).

R. H. Lehmberg, Theory of Optical Ray Retracing in Laser–Plasma Backscatter, *Phys. Rev. Lett.* **41,** 863 (1978).

R. H. Lehmberg and K. A. Holder, Numerical Study of Optical Ray Retracing in Laser–Plasma Backscatter, *Phys. Rev. A* **22,** 2156 (1980).

S. A. Lesnik, M. S. Soskin, and A. I. Khiznyak, Laser with a Stimulated-Brillouin-Scattering Complex-Conjugate Mirror, *Zh. Tekh. Fiz.* **49,** 2257 (1979); *Sov. Phys. Tech. Phys.* (*Engl. Transl.*) **24,** 1249 (1979).

M. D. Levenson, High-Resolution Imaging by Wave-Front Conjugation, *Opt. Lett.* **5,** 182 (1980).

M. D. Levenson, K. M. Johnson, V. C. Hanchett, and K. Chiang, Projection Photolithography by Wave-Front Conjugation, *J. Opt. Soc. Am.* **71,** 737 (1981).

P. F. Liao and D. M. Bloom, Continuous-Wave Backward-Wave Generation by Degenerate Four-Wave Mixing in Ruby, *Opt. Lett.* **3,** 4 (1978).

P. F. Liao, N. P. Economou, and R. R. Freeman, Two-Photon Coherent Transient Measurements of Doppler-Free Linewidths with Broadband Excitation, *Phys. Rev. Lett.* **39,** 1473 (1977).

P. F. Liao, D. M. Bloom, and N. P. Economou, CW Optical Wave-Front Conjugation by Saturated Absorption in Atomic Sodium Vapor, *Appl. Phys. Lett.* **32,** 813 (1978).

P. F. Liao, L. M. Humphrey, D. M. Bloom, and S. Geschwind, Determination of Upper Limits for Spatial Energy Diffusion in Ruby, *Phys. Rev. B* **20,** 4145 (1979).

I. Liberman, Application of Phase Conjugation to CO_2 Lasers, *Proc. Los Alamos Conf. Opt.* (D. H. Liebenberg, ed.), p. 426. SPIE, Washington, D.C., 1979.

R. C. Lind and D. G. Steel, Demonstration of the Longitudinal Modes and Aberration–Correction Properties of a Continuous-Wave Dye Laser with a Phase-Conjugate Mirror, *Opt. Lett.* **6,** 554 (1981).

R. C. Lind, D. G. Steel, M. B. Klein, R. L. Abrams, C. R. Giuliano, and R. K. Jain, Phase Conjugation at 10.6 μm by Resonantly Enhanced Degenerate Four-Wave Mixing, *Appl. Phys. Lett.* **34,** 457 (1979).

R. C. Lind, D. G. Steel, and G. J. Dunning, Phase Conjugation by Resonantly Enhanced Degenerate Four-Wave Mixing, *Opt. Eng.* **21,** 190 (1982).

H. I. Mandelberg, Phase-Modulated Conjugate-Wave Generation in Ruby, *Opt. Lett.* **5,** 258 (1980).

J. H. Marburger, Optical Pulse Integration and Chirp Reversal in Degenerate Four-Wave Mixing, *Appl. Phys. Lett.* **32,** 372 (1978).

J. H. Marburger and J. F. Lam, Nonlinear Theory of Degenerate Four-Wave Mixing, *Appl. Phys. Lett.* **34,** 389 (1979).

J. H. Marburger and J. F. Lam, Effect of Nonlinear Index Changes on Degenerate Four-Wave Mixing, *Appl. Phys. Lett.* **35,** 249 (1979).

A. Marrakchi, J. P. Huignard, and J. P. Herriau, Applications of Phase Conjugation in $Bi_{12}SiO_{20}$ Crystals to Mode Pattern Visualization of Diffuse Vibrating Structures, *Opt. Commun.* **34,** 15 (1980).

G. Martin and R. W. Hellwarth, Infrared-To-Optical Image Conversion by Bragg Reflection from Thermally-Induced Index Gratings, *Appl. Phys. Lett.* **34,** 371 (1979).

G. Martin, L. K. Lam, and R. W. Hellwarth, Generation of a Time-Reversed Replica of a Nonuniformly Polarized Image-Bearing Optical Beam, *Opt. Lett.* **5,** 185 (1980).

A. Maruani and D. S. Chemla, Active Nonlinear Spectroscopy of Biexcitons in Semiconductors' Propagation Effects and Fano Interferences, *Phys. Rev. B* **23,** 841 (1981).

P. Mathieu and P. A. Belanger, Retroreflective Array as Resonator Mirror, *Appl. Opt.* **19,** 2262 (1980).

R. Mays and R. J. Lysiak, Phase Conjugated Wavefronts by Stimulated Brillouin and Raman Scattering, *Opt. Commun.* **31,** 89 (1979).

R. L. Melcher, Backward-Wave Phonon Spectroscopy in Si:In *Phys. Rev. Lett.* **43,** 939 (1979).

D. A. B. Miller, Time Reversal of Optical Pulses by Four-Wave Mixing, *Opt. Lett.* **5,** 300 (1980).

D. A. B. Miller, R. G. Harrison, A. M. Johnston, C. T. Seaton, and S. D. Smith, Degenerate Four-Wave Mixing in InSb at 5K, *Opt. Commun.* **32,** 478 (1980).

D. A. B. Miller, S. D. Smith, and C. T. Seaton, Optical Bistability in Semiconductors, *IEEE J. Quantum Electron.* **QE-17,** 312 (1981).

E. I. Moses and F. Y. Wu, Amplification and Phase Conjugation by Degenerate Four-Wave Mixing in a Saturable Absorber, *Opt. Lett.* **5,** 64 (1980).

T. Mossberg, A. Flusberg, R. Kachru, and S. R. Hartmann, Total Scattering Cross-Section for Na on He Measured by Stimulated Photon Echoes, *Phys. Rev. Lett.* **42,** 1665 (1979).

T. W. Mossberg, R. Kachru, S. R. Hartmann, and A. M. Flusberg, Echoes in Gaseous Media: A Generalized Theory of Rephasing Phenomena, *Phys. Rev. A* **20,** 1976 (1979).

T. W. Mossberg, R. Kachru, E. Whittaker, and S. R. Hartmann, Temporally Recurrent Spatial Ordering of Atomic Population in Gases: Grating Echoes, *Phys. Rev. Lett.* **43,** 851 (1979).

J. Nilsen and A. Yariv, Nearly Degenerate Four-Wave Mixing Applied to Optical Filters, *Appl. Opt.* **18,** 143 (1979).

J. Nilsen and A. Yariv, Nondegenerate Four-Wave Mixing in a Doppler-Broadened Resonant Medium, *J. Opt. Soc. Am.* **71,** 180 (1981).

J. Nilsen, N. S. Gluck, and A. Yariv, Narrow-Band Optical Filter Through Phase Conjugation by Nondegenerate Four-Wave Mixing in Sodium Vapor, *Opt. Lett.* **6,** 380 (1981).

O. Y. Nosach, V. I. Popovichev, V. V. Ragul'skii, and F. S. Faizullov, Cancellation of Phase Distortions in an Amplifying Medium with a "Brillouin Mirror," *Pis'ma Zh. Eksp. Teor. Fiz.* **16,** 617 (1972); *JETP Lett. (Engl. Transl.)* **16,** 435 (1972).

S. G. Odulov and M. S. Soskin, Correlation Analysis of Images Under Degenerate Four-Wave Mixing in Colliding Beams, *Dokl. Akad. Nauk SSSR* **252,** 336 (1980); *Sov. Phys. Dokl. (Engl. Transl.)* **25,** 380 (1980).

S. G. Odulov, E. N. Sal'kova, M. S. Soskin, and L. G. Sukhoverkhova, Removal of Distortions Induced in Laser Beam Amplifiers by Methods of Dynamic Holography, *Ukr. Fiz. Zh.* **23,** 562 (1978).

S. Odulov, M. Soskin, and M. Vasnetsov, Compensation for Time-Dependent Phase Inhomogeneity via Degenerate Four-Wave Mixing in $LiTaO_3$, *Opt. Commun.* **32,** 355 (1980).

T. R. O'Meara, Applications of Nonlinear Phase Conjugation in Compensated Active Imaging, *Proc. Int. Conf. Lasers, 1978* (V. J. Corcoran, ed.), p. 542. STS Press, McLean, Virginia, 1979.

T. R. O'Meara, Applications of Nonlinear Phase Conjugation in Compensated Active Imaging, *Opt. Eng.* **21,** 231 (1982).

T. R. O'Meara, Compensation of Laser Amplifier Trains with Nonlinear Conjugation Techniques, *Opt. Eng.* **21,** 243 (1982).

T. R. O'Meara, Wavefront Compensation with Pseudoconjugation, *Opt. Eng.* **21,** 271 (1982).

T. R. O'Meara and A. Yariv, Time-Domain Signal Processing via Four-Wave Mixing in Nonlinear Delay Lines, *Opt. Eng.* **21,** 237 (1982).

A. N. Oraevskii, Possible Use of Resonantly Excited Media in Phase Conjugation, *Kvantovaya Elektron.* **6,** 218 (1979); *Sov. J. Quantum Electron. (Engl. Transl.)* **9,** 119 (1979).

V. K. Orlov, Y. Z. Virnik, S. P. Vorotilin, V. B. Gerasimov, Y. A. Kalinin, and A. Y. Sagalovich, Retroreflecting Mirror for Dynamic Compensation of Optical Inhomogeneities, *Kvantovaya Elektron.* **5,** 1389 (1978); *Sov. J. Quantum Electron. (Engl. Transl.)* **8,** 799 (1978).

Y. I. Ostrovskii, Wavefront Reconstruction in Stimulated Scattering of Light, *Pis'ma Zh. Tekh. Fiz.* **5,** 769 (1979); *Sov. Tech. Phys. Lett. (Engl. Transl.)* **5,** 315 (1979).

Y. I. Ostrovskii, V. G. Sidorovich, D. I. Stastl'ko, and L. V. Tanin, Dynamic Holograms in Sodium Vapor, *Pis'ma Zh. Tekh. Fiz.* **1,** 1030 (1975); *Sov. Tech. Phys. Lett. (Engl. Transl.)* **1,** 442 (1975).

A. J. Palmer, Nonlinear Optics of Radiatively Cooled Vapors, *Opt. Commun.* **30,** 104 (1979).

A. J. Palmer, Nonlinear Optics in Aerosols, *Opt. Lett.* **5,** 54 (1980).

G. A. Pasmanik and M. S. Sandler, Some Possibilities of Controlling the Phase of an Optical Signal by Wavefront Reversal and Parametric Mixing of Light Waves, *Kvantovaya Elektron.* **8,** 726 (1981); *Sov. J. Quantum Electron. (Engl. Transl.)* **11,** 439 (1981).

Y. Peixuan and Y. R. Shen, Four-Wave Mixing and Optical-Field-Induced Helical structure in Liquid Crystalline Materials, *Appl. Phys.* **25,** 49 (1981).

D. M. Pepper, Phase Conjugate Optics: On the Theory, Observation, and Utilization of Temporally-Reversed Wavefronts as Generated via

Nonlinear Optical Parametric Interactions, Ph.D. Thesis, California Inst. Technol., Pasadena, 1980.

D. M. Pepper, Nonlinear Optical Phase Conjugation, *Opt. Eng.* **21,** 156 (1982).

D. M. Pepper and R. L. Abrams, Narrow Optical Bandpass Filter via Nearly Degenerate Four-Wave Mixing, *Opt. Lett.* **3,** 212 (1978).

D. M. Pepper and A. Yariv, Compensation for Phase Distortions in Nonlinear Media by Phase Conjugation, *Opt. Lett.* **5,** 59 (1980).

D. M. Pepper, J. AuYeung, D. Fekete, and A. Yariv, Spatial Convolution and Correlation of Optical Fields via Degenerate Four-Wave Mixing, *Opt. Lett.* **3,** 7 (1978).

D. M. Pepper, D. Fekete, and A. Yariv, Observation of Amplified Phase-Conjugate Reflection and Optical Parametric Oscillation by Degenerate Four-Wave Mixing in a Transparent Medium, *Appl. Phys. Lett.* **33,** 41 (1978).

G. V. Peregudov, M. E. Plotkin, and E. N. Ragozin, Use of Wave-Front Reversal Effect to Investigate a Jet Formed by Focusing Laser Radiation on a Plane Target, *Kvantovaya Elektron.* **6,** 2401 (1979); *Sov. J. Quantum Electron.* **9,** 1413 (1979).

M. P. Petrov, S. V. Miridonov, S. I. Stepanov, and V. V. Kulikov, Light Diffraction and Nonlinear Image Processing in Electrooptic $Bi_{12}SiO_{20}$ Crystals, *Opt. Commun.* **31,** 301 (1979).

N. F. Pilipetsky and V. V. Shkunov, Narrowband Four-Wave Reflecting Filter with Frequency and Angular Tuning, *Opt. Commun.* **37,** 217 (1981).

N. F.Pilipetskii, V. I. Popovichev, and V. V. Ragul'skii, Concentration of Light by Inverting Its Wavefront, *Pis'ma Zh. Eksp. Teor. Fiz.* **27,** 619 (1978); *JETP Lett.* (*Engl. Transl.*) **27,** 595 (1978).

N. F. Pilipetskii, V. I. Popovichev, and V. V. Regul'skii, Accuracy of Reconstruction of a Light Field after Stimulated Scattering, *Dokl. Akad. Nauk SSSR* **248,** 1097 (1979); *Sov. Phys. Dokl.* (*Engl. Transl.*) **24,** 845 (1979).

N. F. Pilipetsky, V. I. Popovichev, and V. V. Ragulsky, The Reproduction of Weak Components of a Light Field at Stimulated Scattering, *Opt. Commun.* **31,** 97 (1979).

N. F. Pilipetskii, V. I. Popovichev, and V. V. Ragulskiy, Comparison of the Amplification Coefficients of Conjugate and Non-Conjugate Waves in Stimulated Light Scattering'', *Opt. Commun.* **40,** 73 (1981).

A. K. Popov and V. M. Shalaev, Doppler-Free Spectroscopy and Wave-Front Conjugation by Four-Wave Mixing of Nonmonochromatic Waves, *Appl. Phys.* **21,** 93 (1980).

V. I. Popovichev, V. V. Ragul'skii, and F. S. Faizullov, Stimulated

Mandel'shtam–Brillouin Scattering Excited by Radiation with a Broad Spectrum, *Pis'ma Zh. Eksp. Teor. Fiz.* **19,** 350 (1974); *JETP Lett.* (*Engl. Transl.*) **19,** 196 (1974).

E. N. Ragozin and M. E. Plotkin, Multiple Use of Wavefront Reversal in Laser Systems, *Kvantovaya Elektron.* **7,** 1583 (1980); *Sov. J. Quantum Electron.* (*Engl. Transl.*) **10,** 915 (1980).

V. V. Ragulskiy, Stimulated Mandel'shtam–Brillouin Scattering, *Lebedev Tr.* **85,** 1 (1976).

V. V. Ragulskiy, Wavefront Inversion of Weak Beams in Stimulated Scattering, *Pis'ma Zh. Tekh. Fiz.* **5,** 251 (1979); *Sov. Tech. Phys. Lett.* (*Engl. Transl.*) **5,** 100 (1979).

V. V. Ragulskiy, Detection of Slight Absorption of Light Through Wavefront Inversion, *Pis'ma Zh. Tekh. Phys.* **6,** 687 (1980); *Sov. Tech. Phys. Lett.* (*Engl. Transl.*) **6,** 297 (1980).

R. K. Raj, D. Bloch, J. J. Snyder, G. Camy, and M. Ducloy, High Frequency Optically Heterodyned Saturation Spectroscopy via Resonant Degenerate Four-wave Mixing, *Phys. Rev. Lett.* **44,** 1251 (1980).

M. G. Reznikov and A. I. Khizhnyak, Properties of a Resonator with a Wavefront-Reversing Mirror, *Kvantovaya Elektron.* **7,** 1105 (1980); *Sov. J. Quantum Electron.* (*Engl. Transl.*) **10,** 633 (1980).

W. W. Rigrod, R. A. Fisher, and B. J. Feldman, Transient Analysis of Nearly Degenerate Four-Wave Mixing, *Opt. Lett.* **5,** 105 (1980).

G. Rivoire and A. Sokolovakaya, Phase Conjugation and Image Reconstruction in Stimulated Scatterings, *Opt. Commun.* **42,** 138 (1982).

S. Saikan, Directional Anisotropy of the Backward Four-Wave Mixing in Gases, *J. Phys. Soc. Jpn.* **50,** 230 (1981).

S. Saikan and H. Wakata, Configuration Dependence of Optical Filtering Characteristics in Backward Nearly Degenerate Four-Wave Mixing, *Opt. Lett.* **6,** 281 (1981).

D. A. Sealer and H. Hsu, Stimulated Brillouin Scattering as a Parametric Interaction, *IEEE J. Quantum Electron.* **QE-1,** 116 (1965).

S. A. Shakir, Zero-Area Optical Pulse Processing by Degenerate Four-Wave Mixing, *Opt. Commun.* **40,** 151 (1981).

W. Shao-Min and H. Weber, A Spherical Resonator Equivalent to Arbitrary Phase-Conjugate Resonators, *Opt. Commun.* **41,** 360 (1982).

N. S. Shiren, Generation of Time-Reversed Optical Wave Fronts by Backward-Wave Photon Echoes, *Appl. Phys. Lett.* **33,** 299 (1978).

N. S. Shiren, W. Arnold, and T. G. Kazyaka, Backward-Wave Phonon Echoes in Glass, *Phys. Rev. Lett.* **39,** 239 (1977).

V. V. Shkunov and B. Y. Zel'dovich, Recording and Reconstruction of an Object's Wave State of Polarization by a Volume Hologram, *Appl. Opt.* **18,** 3633 (1979).

R. C. Shockley, Simplified Theory of the Impulse Response of an Optical Degenerate Four-Wave Mixing Cell, *Opt. Commun.* **38,** 221 (1981).

R. C. Shockley, Optical Pulse Chirp Modification in Degenerate Four-Wave Mixing at Finite Optical Pump Powers, *Appl. Phys. Lett.* **40,** 930 (1982).

E. I. Shtyrkov, Scattering of Light by a Periodic Structure of Excited and Unexcited Atoms, *Pis'ma Zh. Eksp. Teor. Fiz.* **12,** 134 (1970); *JETP Lett.* (*Engl. Transl.*) **12,** 92 (1970).

E. I. Shtyrkov and V. V. Samartsev, Imaging Properties of Dynamic Echo Holograms in Resonant Media, *Opt. Spektrosk.* **40,** 392 (1976); *Opt. Spectrosc.* (*Engl. Transl.*) **40,** 224 (1976).

E. I. Shtyrkov and V. V. Samartsev, Dynamic Holograms on the Superposition States of Atoms, *Phys. Status Solidi A* **45,** 647 (1978).

E. I. Shtyrkov, V. S. Lobkov, and N. G. Yarmukhametov, Grating Induced in Ruby by Interference of Atomic States, *Pis'ma Zh. Eksp. Teor. Fiz.* **27,** 685 (1978); *JETP Lett.* (*Engl. Transl.*) **27,** 648 (1978).

E. I. Shtyrkov, N. L. Nevelskaya, V. S. Lobkov, and N. G. Yarmukhametov, Transient Light-Induced Spatial Gratings by Successive Optical Coherent Pulses, *Phys. Status Solidi B* **98,** 472 (1980).

V. G. Sidorovich, Theory of the "Brillouin Mirror," *Zh. Tekn. Fiz.* **46,** 2168 (1976); *Sov. Phys.—Tech. Phys.* (*Engl. Transl.*) **21,** 1270 (1976).

V. G. Sidorovich, Theory of the Transformation of Light Fields by Amplitude Three-Dimensional Holograms Recorded in Amplifying Media, *Opt. Spektrosk.* **42,** 693 (1977); *Opt. Spectrosc.* (*Engl. Transl.*) **42,** 395 (1977).

V. G. Sidorovich, Reproduction of the Pump Spectrum in Stimulated Raman Scattering, *Kvantovaya Elektron.* **5,** 1370 (1978); *Sov. J. Quantum Electron.* (*Engl. Transl.*) **8,** 784 (1978).

V. G. Sidorovich and V. V. Shkunov, Spectral Selectivity of 3-D Holograms, *Opt. Spektrosk.* **44,** 1001 (1978); *Opt. Spectrosc.* (*Engl. Transl.*) **44,** 586 (1978).

V. G. Sidorovich and V. V. Shkunov, "Capture" of a Stokes Pump Wave in a Stimulated Raman Scattering Amplifier, *Zh. Tekh. Fiz.* **49,** 816 (1979); *Sov. Phys.—Tech. Phys.* (*Engl. Transl.*) **24,** 472 (1979).

A. E. Siegman, Dynamic Interferometery and Differential Holography of Irregular Phase Objects Using Phase Conjugate Reflection, *Opt. Commun.* **31,** 257 (1979).

Y. Silberberg and I. Bar-Joseph, Transient Effects in Degenerate Four-Wave Mixing in Saturable Absorber, *IEEE J. Quantum Electron.* **QE-17,** 1967 (1981).

Y. Silberberg and I. Bar-Joseph, Low-Power Phase Conjugation in Thin Films of Saturable absorbers, *Opt. Commun.* **39,** 265 (1981).

M. Slatkine, I. J. Bigio, B. J. Feldman, and R. A. Fisher, Efficient Phase Conjugation of an Ultraviolet XeF Laser Beam by Stimulated Brillouin Scattering, *Opt. Lett.* **7,** 108 (1982).

A. L. Smirl, T. F. Boggess, and F. A. Hopf, Generation of Forward-Traveling Phase-Conjugate Wave in Germanium, *Opt. Commun.* **34,** 463 (1980).

P. W. Smith, A. Ashkin, and W. J. Tomlinson, Four-Wave Mixing in an Artificial Kerr Medium, *Opt. Lett.* **6,** 284 (1981).

P. W. Smith, P. J. Maloney, and A. Ashkin, Use of a Liquid Suspension of Dielectric Spheres as an Artificial Kerr Medium, *Opt. Lett.* **7,** 347 (1982).

P. W. Smith, W. J. Tomlinson, D. J. Eilenberger, and P. J. Maloney, Measurement of Electronic Optical Kerr Coefficients, *Opt. Lett.* **6,** 581 (1981).

A. I. Sokolovskaya, G. L. Brekhovskikh, and A. D. Kudryavtseva, Restoration of Wave Front of Light Beams During Stimulated Raman Scattering, *Dokl. Akad. Nauk SSSR* **233,** 356 (1977); *Sov. Phys. Dokl.* (*Engl. Transl.*) **22,** 156 (1977).

A. I. Sokolovskaya, G. L. Brekhovskikh, and A. D. Kudryavtseva, Light Beam Wavefront Reconstruction and Real Volume Image Reconstruction of the Object at the Stimulated Raman Scattering, *Opt. Commun.* **24,** 74 (1978).

M. S. Soskin and A. I. Khizhnyak, Interaction of Four Counterpropagating Plane Waves in a Medium with an Instantaneous Cubic Nonlinearity, *Kvantovaya Elektron.* **7,** 42 (1980); *Sov. J. Quantum Electron.* (*Engl. Transl.*) **10,** 21 (1980).

D. G. Steel and J. F. Lam, Degenerate Four-Wave Mixing in Plasmas, *Opt. Lett.* **4,** 363 (1979).

D. G. Steel and J. F. Lam, Two-Photon Coherent Transient Measurements of the Nonradiative Collisionless Dephasing Rate in SF_6 via Doppler-Free Degenerate Four-Wave Mixing, *Phys. Fev. Lett.* **43,** 1588 (1979).

D. G. Steel and J. F. Lam, Multiline Phase Conjugation in Resonant Materials, *Opt. Lett.* **5,** 297 (1980).

D. G. Steel and J. F. Lam, Saturation Effects and Inhomogeneous Broadening in Doppler-Free Degenerate Four-Wave Mixing, *Opt. Commun.* **40,** 77 (1981).

D. G. Steel and R. C. Lind, Multiresonant Behavior in Nearly Degenerate Four-Wave Mixing, *Opt. Lett.* **6,** 587 (1981).

D. G. Steel, R. C. Lind, J. F. Lam, and C. R. Giuliano, Polarization-Rotation and Thermal-Motion Studies via Resonant Degenerate Four-Wave Mixing, *Appl. Phys. Lett.* **35,** 376 (1979).

D. G. Steel, R. C. Lind, and J. F. Lam, Degenerate Four-Wave Mixing in a Resonant Homogeneously Broadened System, *Phys. Rev. A* **23,** 2513 (1981).

B. I. Stepanov, E. V. Ivakin, and A. S. Rubanov, Recording Two-Dimensional and Three-Dimensional Dynamic Holograms in Bleachable Substances, *Dokl. Akad. Nauk SSSR* **196,** 567 (1971); *Sov. Phys.—Dokl.* (*Engl. Transl.*) **16,** 46 (1971).

B. R. Suydam and R. A. Fisher, Transient Response of Kerr-Like Phase Conjugators: A Review, *Opt. Eng.* **21,** 184 (1982).

N. Tan-No and H. Inaba, Generation and Application of Phase Conjugation Wavefronts in the Optical Range, *Oyo Butsuri* **49,** 831 (1980).

N. Tan-No, T. Hishimiya, and H. Inaba, Dispersion-Free Amplification and Oscillation in Phase-Conjugate Four-Wave Mixing in an Atomic Vapor Doublet, *IEEE J. Quantum Electron.* **QE-16,** 147 (1980).

J. O. Tocho, W. Sibbett, and D. J. Bradley, Picosecond Phase-Conjugate Reflection from Organic Dye Saturable Absorbers, *Opt. Commun.* **34,** 122 (1980).

J. O. Tocho, W. Sibbett, and D. J. Bradley, Thermal Effects in Phase-Conjugation in Saturable Absorbers with Picosecond Pulses, *Opt. Commun.* **37,** 67 (1981).

A. Tomita, Phase Conjugation Using Gain Saturation of a Nd:YAG Laser, *Appl. Phys. Lett.* **34,** 463 (1979).

R. Trebino and A. E. Siegman, Phase-Conjugate Reflection at Arbitrary Angles Using TEM_{00} Pump Beams, *Opt. Commun.* **32,** 1 (1980).

K. Ujihara, Four-Wave Mixing and Two-Dimensional Phase Conjugation of Surface Plasmons, Opt. Commun. **42,** 1 (1982).

N. D. Ustinov and I. N. Matveev, Nonlinear Optical Methods in Conversion of Spatially Modulated Signals, *Kvantovaya Elektron.* **4,** 2595 (1977); *Sov. J. Quantum Electron.* (*Engl. Transl.*) **7,** 1483 (1977).

H. Vanherzeele and J. L. Van Eck, Pulse Ompression by Intracavity Degenerate Four-Wave Mixing, *Appl. Opt.* **20,** 524 (1981).

H. Vanherzeele, J. L. Van Eck, and A. E. Siegman, Mode-Locked Laser Oscillation Using Self-Pumped Phase-Conjugate Reflection, *Opt. Lett.* **6,** 467 (1981).

M. V. Vasil'ev, A. L. Gyulameryan, A. V. Mamaev, V. V. Ragul'skii, M. Semenov, and V. G. Sidorovich, Recording of Phase Fluctuations of Stimulated Scattered Light, *Pis'ma Zh. Eksp. Teor. Fiz.* **31,** 673 (1980); *JETP Lett.* (*Engl. Transl.*) **31,** 634 (1980).

A. P. Veduta and B. P. Kirsanov, Four-Photon Parametric Frequency Selection in Wide Stimulated Emission Lines, *Kvantovaya Elektron.* **1,** 73 (1971); *Sov. J. Quantum Electron.* (*Engl. Transl.*) **1,** 256 (1971).

K. I. Volyak and G. A. Lyakhov, Fluctuations in Transient Nonlinear Interactions of Opposed Waves, *Kvantovaya Elektron.* **3,** 2470 (1976); *Sov. J. Quantum Electron.* (*Engl. Transl.*) **6,** 1351 (1977).

N. S. Vorobiev, I. S. Ruddock, and R. Illingworth, Generation of Picosecond Optical Phase Conjugate Pulses Using cw Mode-Locked Lasers, *Opt. Commun.* **41,** 216 (1982).

E. S. Voronin, V. V. Ivakhnik, V. M. Petnikova, V. S. Solomatin, and V. V. Shuvalov, Compensation for Phase Distortions by Three-Frequency Parametric Interaction, *Kvantovaya Elektron.* **6,** 1304 (1979); *Sov. J. Quantum Electron.* (*Engl. Transl.*) **9,** 765 (1979).

E. S. Voronin, V. V. Ivakhnik, V. M. Petnikova, V. S. Solomatin, and V. V. Shuvalov, Optimization of a System for Compensation of Phase Distortions Caused by Extended Inhomogeneities, *Kvantovaya Elektron.* **7,** 1543 (1980); *Sov. J. Quantum Electron.* (*Engl. Transl.*) **10,** 887 (1980).

E. S. Voronin, V. V. Ivakhnik, V. M. Petnikova, V. S. Solomatin, and V. V. Shuvalov, Possibility of Compensation of Nonlinear Phase Distortions by Parametric Radiation Converters, *Kvantovaya Elektron.* **8,** 443 (1981); *Sov. J. Quantum Electron.* (*Engl. Transl.*) **11,** 273 (1981).

E. S. Voronin, V. M. Petnikova, and V. V. Shuvalov, Use of Degenerate Parametric Processes for Wavefront Correction (Review), *Kvantovaya Elektron.* **8,** 917 (1981); *Sov. J. Quantum Electron.* (*Engl. Transl.*) **11,** 551 (1981).

M. A. Vorontsov, Phase Conjugation Method for the Compensation of Thermal Defocusing of the Light Beams, *Kvantovaya Elektron.* **6,** 2078 (1979); *Sov. J. Quantum Electron.* (*Engl. Transl.*) **9,** 1221 (1979).

S. M. Wandzura, Effects of Atomic Motion on Wavefront Conjugation by Resonantly Enhanced Degenerate Four-Wave Mixing, *Opt. Lett.* **4,** 208 (1979).

V. Wang, Nonlinear Optical Phase Conjugation for Laser Systems, *Opt. Eng.* **17,** 267 (1978).

V. Wang, and C. R. Giuliano, Correction of Phase Aberrations via Stimulated Brillouin Scattering, *Opt. Lett.* **2,** 4 (1978).

D. E. Watkins, J. F. Figueira, and S. J. Thomas, Observation of Resonantly Enhanced Degenerate Four-Wave Mixing in Doped Alkali-Halides, *Opt. Lett.* **5,** 169 (1980).

D. E. Watkins, C. R. Phipps, Jr., and S. J. Thomas, Observation of Amplified Reflection via Degenerate Four-Wave Mixing at CO Laser Wavelengths in Germanium, *Opt. Lett.* **6,** 76 (1981).

J. O. White and A. Yariv, Real-Time Image Processing via Four-Wave Mixing in a Photorefractive Medium, *Appl. Phys. Lett.* **37,** 5 (1980).

J. O. White and A. Yariv, Spatial Information Processing and Distortion Correction via Four-Wave Mixing, *Opt. Eng.* **21,** 224 (1982).

J. O. White, M. Cronin-Golomb, B. Fischer, and A. Yariv, Coherent Oscillation by Self-Induced Gratings in the Photorefractive Crystal $BaTiO_3$, *Appl. Phys. Lett.* **40,** 450 (1982).

T. Wilson and L. Solymar, Two-Dimensional Coupled Differential Equations for Degenerate Four-Wave Mixing, *Appl. Phys. Lett.* **38,** 669 (1981).

H. G. Winful and J. H. Marburger, Hysteresis and Optical Bistability in Degenerate Four-Wave Mixing, *Appl. Phys. Lett.* **36,** 613 (1980).

J. P. Woerdman, Formation of a Transient Free Carrier Hologram in Si, *Opt. Commun.* **2,** 212 (1971).

J. P. Woerdman and M. F. H. Schuurmans, Effect of Saturation of the Spectrum of Degenerate Four-Wave Mixing in Atomic Sodium , *Opt. Lett.* **6,** 239 (1981).

E. Wolf, Phase-Conjugacy and Symmetries in Spatially Bandlimited Wavefields Containing No Evanescent Components, *J. Opt. Soc. Am.* **70,** 1311 (1980).

E. Wolf and W. H. Carter, Comments on the Theory of Phase-Conjugated Waves, *Opt. Commun.* **40,** 397 (1982).

C. K. Wu, Y. Z. Cui, and Z. Y. Wang, Degenerate Four-Wave Mixing and Phase Conjugation in Organic Dye Solutions, *Wu Li Hsueh Pao* **29,** 937 (1980).

C. K. Wu, J. Y. Fan, and Z. Y. Wang, Generation of Phase Conjugated Backward Wave by Degenerate Four-Wave Mixing in Chlorophyll Solution, *Wu Li Hsueh Pao* **29,** 305 (1980); *Chin. J. Phys.* (*Peking*) (*Engl. Trans.*) **1,** 32 (1981).

C. K. Wu, J. Y. Fan, and Z. Y. Wang, A High Efficiency Stimulated Raman Scattering Source, *Wu Li Hsueh Pao* **29,** 588 (1980); *Chin. J. Phys.* (*Peking*) (*Engl. Trans.*) **1,** 369 (1981).

C. K. Wu, J. Y. Fan, and Z. Y. Wang, Investigation of Degenerate Four-Wave Mixing and Phase Conjugation in Organic Dye Solutions, *Wu Li Hsueh Pao* **30,** 189 (1981); *Chin. J. Phys.* (*Peking*) (*Engl. Trans.*) **1,** 1040 (1981).

C. K. Wu, Y. T. Long, S. H. Zhen, and W. Z. Ying, Degenerate Four-Wave Mixing in Organic Dye Solutions, *Laser J.* (*China*) **6,** 12 (1979).

C. K. Wu, Y. T. Long, S. H. Zhen, and W. Z. Ying, Principles of Generating the Four-Wave Mixing Conjugate Wave, *Laser J.* (*China*) **6,** 10 (1979).

C. K. Wu, Z. Y. Wang, and J. Y. Fan, Masurement of Third-Order Susceptibility by Degenerate Four-Wave Mixing, *Wu Li Hsueh Pao* **29,** 508 (1980); *Chin. J. Phys.* (*Peking*) (*Engl. Trans.*) **1,** 366 (1981).

A. Yariv, Three-Dimensional Pictorial Transmission in Optical Fibers, *Appl. Phys. Lett.* **28,** 88 (1976).

A. Yariv, On Transmission and Recovery of Three-Dimensional Image Information in Optical Waveguides, *J. Opt. Soc. Am.* **66,** 301 (1976).

A. Yariv, compensation for Atmospheric Degradation of Optical Beam Transmission by Nonlinear Optical Mixing, *Opt. Commun.* **21,** 49 (1977).

A. Yariv, Compensation for Optical Propagation Distortion Through Phase Adaptation by Nonlinear techniques (PANT), *Phys. Quantum Electron.* **6,** 175 (1978).

A. Yariv, Phase Conjugate Optics and Real-Time Holography, *IEEE J. Quantum Electron.* **QE-14,** 650 (1978).

A. Yariv, Comments on "Phase-Conjugate Optics and Real-Time Holography;" Author's Reply, *IEEE J. Quantum Electron.* **QE-15,** 524 (1979).

A. Yariv, Reply to the Paper "Comments on the Theory of Phase-Conjugated Waves," by E. Wolf and W. H. Carter, *Opt. Commun.* **40,** 401 (1982).

A. Yariv and J. AuYeung, Transient Four-Wave Mixing and Real Time Holography in Atomic Systems, *IEEE J. Quantum Electron.* **QE-15,** 224 (1979).

A. Yariv and D. M. Pepper, Amplified Reflection, Phase Conjugation, and Oscillation in Degenerate Four-Wave Mixing, *Opt. Lett.* **1,** 16 (1977).

A. Yariv, J. AuYeung, D. Fekete, and D. M. Pepper, Image Phase Compensation and Real-Time Holography by Four-Wave Mixing in Optical Fibers, *Appl. Phys. Lett.* **32,** 635 (1978).

A. Yariv, D. Fekete, and D. M. Pepper, Compensation for Channel Dispersion by Nonlinear Optical Phase Conjugation, *Opt. Lett.* **4,** 52 (1979).

H. P. Yuen and J. H. Shapiro, Generation and Detection of Two-Photon Coherent States in Degenerate Four-Wave Mixing, *Opt. Lett.* **4,** 334 (1979).

B. Ya. Zel'dovich and V. V. Shkunov, Wavefront Reproduction in Stimulated Raman Scattering, *Kvantovaya Elektron.* **4,** 1090 (1977); *Sov. J. Quantum Electron.* (*Engl. Transl.*) **7,** 610 (1977).

B. Ya. Zel'dovich and V. V. Shkunov, Reversal of the Wave Front of Light in the Case of Depolarized Pumping, *Zh. Eksp. Teor. Fiz.* **75,** 428 (1978); *Sov. Phys.—JETP* (*Engl. Transl.*) **48,** 214 (1978).

B. Ya. Zel'dovich and V. V. Shkunov, Limits of Existence of Wavefront Reversal in Stimulated Light Scattering, *Kvantovaya Elektron.* **5,** 36 (1978); *Sov. J. Quantum Electron.* (*Engl. Transl.*) **8,** 15 (1978).

B. Ya. Zel'dovich and V. V. Shkunov, Spatial-Polarization Wavefront Reversal in Four-Photon Interaction, *Kvantovaya Elektron.* **6,** 629 (1979); *Sov. J. Quantum Electron.* (*Engl. Transl.*) **9,** 379 (1979).

B. Ya. Zel'dovich and V. V. Shkunov, Appearance of Conjugate Images in a Volume Reflection Hologram, *Kvantovaya Elektron.* **6,** 1533 (1979); *Sov. J. Quantum Electron.* (*Engl. Transl.*) **9,** 897 (1979).

B. Ya. Zel'dovich and N. V. Tabiryan, Possibility of Optical Wavefront Reversal with the Aid of Liquid–Crystal Transparencies, *Kvantovaya Elektron.* **8,** 421 (1981); *Sov. J. Quantum Electron.* (*Engl. Transl.*) **11,** 257 (1981).

B. Ya. Zel'dovich and T. V. Yakovleva, Small-Scale Distortions in Wavefront Reversal of a Beam with Incomplete Spatial Modulation (Stimulated Brillouin Backscattering, Theory), *Kvantovaya Elektron.* **7,** 316 (1980); *Sov. J. Quantum Electron.* (*Engl. Transl.*) **10,** 181 (1980).

B. Ya. Zel'dovich and T. V. Yakovleva, Mode Theory of Volume Holograms Allowing for Nonlinearity of the Photographic Process, *Kvantovaya Elektron.* **7,** 519 (1980); *Sov. J. Quantum Electron.* (*Engl. Transl.*) **10,** 295 (1980).

B. Ya. Zel'dovich and T. V. Yakovleva, Spatial-Polarization Wavefront Reversal in Stimulated Scattering of the Rayleigh Wing, *Kvantovaya Elektron.* **7,** 880 (1980); *Sov. J. Quantum Electron.* (*Engl. Transl.*) **10,** 501 (1980).

B. Ya. Zel'dovich & T. V. Yakovleva, Calculations of the Accuracy of Wavefront Reversal Utilizing Pump Modulation with One-Dimensional Modulation, *Kvantovaya Elektron.* **8,** 314 (1981); *Sov. J. Quantum Electron.* (*Engl. Transl.*) **11,** 186 (1981).

B. Ya. Zel'dovich and T. V. Yakovleva, Influence of Linear Absorption and Reflection on the Characteristics of Four-Wavefront Reversal, *Kvantovaya Elektron.* **8,** 1891 (1981); *Sov. J. Quantum Electron.* (*Engl. Transl.*) **11,** 1144 (1981).

B. Ya. Zel'dovich, N. A. Mel'nikov, N. F. Pilipetskii, and V. V. Ragul'skii, Observation of Wave-Front Inversion in Stimulated Raman Scattering Light, *Pis'ma Zh. Eksp. Teor. Fiz.* **25,** 41 (1977); *JETP Lett.* (*Engl. Transl.*) **25,** 36 (1977).

B. Ya. Zel'dovich, M. A. Orlova, and V. V. Shkunov, Nonstationary Theory and Calculation of the Time of Establishment of Four-Wave Wave-Front Reversal, *Dokl. Akad. Nauk SSSR* **252,** 592 (1980); *Sov. Phys.—Dokl.* (*Engl. Transl.*) **25,** 390 (1980).

B. Ya. Zel'dovich, N. F. Pilipetskii, V. V. Ragul'skii, and V. V. Shkunov, Wavefront Reversal by Nonlinear Optics Methods, *Kvantovaya Elektron.* **5,** 1800 (1978); *Sov. J. Quantum Electron.* (*Engl. Transl.*) **8,** 1021 (1978).

B. Ya. Zel'dovich, N. F. Pilipetskii, A. N. Sudarkin, and V. V. Shkunov, Wave Front Reversal by an Interface, *Dokl. Akad. Nauk SSSR* **252,** 92 (1980); *Sov. Phys.—Dokl.* (*Engl. Transl.*) **25,** 377 (1980).

B. Ya. Zel'dovich, V. I. Popovichev, V. V. Ragul'skii, and F. S. Faizullov, Connection Between the Wave Fronts of the Reflected and Exciting Light in Stimulated Mandel'shtam–Brillouin Scattering, *Pis'ma Zh. Eksp. Teor. Fiz.* **15,** 160 (1972); *JETP Lett.* (*Engl. Transl.*) **15,** 109 (1972).

B. Ya. Zel'dovich, I. G. Zubarev, G. A. Pasmanik, and V. G. Sidorovich, Comments on Papers by V. B. Gerasimov *et al., Kvantovaya Elektron.* **6,** 592 (1979). *Sov. J. Quantum Electron.* (*Engl. Transl.*) **9,** 351 (1979).

Q. D. Zhong, Double Degenerate Four-Wave Mixing in Plasma, *Actc Phys. Sin.* **30,** 133 (1981); *Chin. Phys.* (*Engl. Transl.*) **2,** 141 (1982).

G.-S. Zhou and L. W. Casperson, Modes of a Laser Resonator with a Retroreflective Mirror, *J. Appl. Opt.* **20,** 1621 (1981).

I. G. Zubarev, A. B. Mironov, and S. I. Mikhailov, Single-Mode Pulse-Periodic Oscillator-Amplifier System with Wavefront Reversal, *Kvantovaya Elektron.* (*Moscow*) **7,** 2035 (1980); *Sov. J. Quantum Electron.* **10,** 1179 (1980).

G. M. Zverev and A. D. Martynov, Investigation of Stimulated Mandel'shtam–Brillouin Scattering Thresholds for Different Media at Wavelengths 0.35, 0.69, and 1.06 μ, *Pis'ma Zh. Eksp. Teor. Fiz.* **6,** 931 (1967); *JETP Lett.* (*Engl. Transl.*) **6,** 351 (1967).

Index